KU-736-339

Springer Series in Statistics

Advisors:
J. Berger, S. Fienberg, J. Gani
K. Krickeberg, I. Olkin, B. Singer

Springer
New York
Berlin
Heidelberg
Barcelona
Budapest
Hong Kong
London
Milan
Paris
Santa Clara
Singapore
Tokyo

Springer Series in Statistics

Andersen/Borgan/Gill/Keiding: Statistical Models Based on Counting Processes.
Andrews/Herzberg: Data: A Collection of Problems from Many Fields for the Student and Research Worker.
Anscombe: Computing in Statistical Science through APL.
Berger: Statistical Decision Theory and Bayesian Analysis, 2nd edition.
Bolfarine/Zacks: Prediction Theory for Finite Populations.
Brémaud: Point Processes and Queues: Martingale Dynamics.
Brockwell/Davis: Time Series: Theory and Methods, 2nd edition.
Daley/Vere-Jones: An Introduction to the Theory of Point Processes.
Dzhaparidze: Parameter Estimation and Hypothesis Testing in Spectral Analysis of Stationary Time Series.
Fahrmeir/Tutz: Multivariate Statistical Modelling Based on Generalized Linear Models.
Farrell: Multivariate Calculation.
Federer: Statistical Design and Analysis for Intercropping Experiments.
Fienberg/Hoaglin/Kruskal/Tanur (Eds.): A Statistical Model: Frederick Mosteller's Contributions to Statistics, Science and Public Policy.
Fisher/Sen: The Collected Works of Wassily Hoeffding.
Good: Permutation Tests: A Practical Guide to Resampling Methods for Testing Hypotheses.
Goodman/Kruskal: Measures of Association for Cross Classifications.
Grandell: Aspects of Risk Theory.
Hall: The Bootstrap and Edgeworth Expansion.
Härdle: Smoothing Techniques: With Implementation in S.
Hartigan: Bayes Theory.
Heyer: Theory of Statistical Experiments.
Jolliffe: Principal Component Analysis.
Kolen/Brennan: Test Equating: Methods and Practices.
Kotz/Johnson (Eds.): Breakthroughs in Statistics Volume I.
Kotz/Johnson (Eds.): Breakthroughs in Statistics Volume II.
Kres: Statistical Tables for Multivariate Analysis.
Le Cam: Asymptotic Methods in Statistical Decision Theory.
Le Cam/Yang: Asymptotics in Statistics: Some Basic Concepts.
Longford: Models for Uncertainty in Educational Testing.
Manoukian: Modern Concepts and Theorems of Mathematical Statistics.
Miller, Jr.: Simultaneous Statistical Inference, 2nd edition.
Mosteller/Wallace: Applied Bayesian and Classical Inference: The Case of *The Federalist Papers.*
Mueller: Basic Principles of Structural Equation Modelling.
Pollard: Convergence of Stochastic Processes.

(continued after index)

Peter J. Brockwell
Richard A. Davis

Time Series: Theory and Methods

Second Edition

With 124 Illustrations

UNIVERSITY OF WOLVERHAMPTON LIBRARY
Acc No. 2027813
CLASS 521
519.55 BRO
CONTROL 0387974296
DATE 15 AUG 1996
SITE RS

Springer

Peter J. Brockwell
Richard A. Davis
Department of Statistics
Colorado State University
Fort Collins, CO 80523
USA

Mathematical Subject Classification: 62-01, 62M10

Library of Congress Cataloging-in-Publication Data
Brockwell, Peter J.
Time series : theory and methods / Peter J. Brockwell, Richard A. Davis.
p. cm. — (Springer series in statistics)
"Second edition"—Pref.
Includes bibliographical references and index.
ISBN 0-387-97429-6 (USA). — ISBN 3-540-97429-6 (EUR)
1. Time-series analysis. I. Davis, Richard A. II. Title. III. Series.
QA280.B76 1991 90-25821
519.5′5—dc20

Printed on acid-free paper.

© 1987, 1991 by Springer-Verlag New York, Inc.
All rights reserved. This work may not be translated or copied in whole or in part without the written permission of the publisher (Springer-Verlag New York, Inc., 175 Fifth Avenue, New York, NY 10010, USA), except for brief excerpts in connection with reviews or scholarly analysis. Use in connection with any form of information storage and retrieval, electronic adaptation, computer software, or by similar or dissimilar methodology now known or hereafter developed is forbidden.
The use of general descriptive names, trade names, tradmarks, etc., in this publication, even if the former are not especially identified, is not to be taken as a sign that such names, as understood by the Trade Marks and Merchandise Marks Act, may accordingly be used freely by anyone.

Typeset by Asco Trade Typesetting Ltd., Hong Kong.
Printed and bound by R.R. Donnelley & Sons, Harrisonburg, Virginia.
Printed in the United States of America.

9 8 7 6 5 4 (Fourth corrected printing, 1995)

ISBN 0-387-97429-6 Springer-Verlag New York Berlin Heidelberg
ISBN 3-540-97429-6 Springer-Verlag Berlin Heidelberg New York

To our families

Preface to the Second Edition

This edition contains a large number of additions and corrections scattered throughout the text, including the incorporation of a new chapter on state-space models. The companion diskette for the IBM PC has expanded into the software package *ITSM: An Interactive Time Series Modelling Package for the PC*, which includes a manual and can be ordered from Springer-Verlag.*

We are indebted to many readers who have used the book and programs and made suggestions for improvements. Unfortunately there is not enough space to acknowledge all who have contributed in this way; however, special mention must be made of our prize-winning fault-finders, Sid Resnick and F. Pukelsheim. Special mention should also be made of Anthony Brockwell, whose advice and support on computing matters was invaluable in the preparation of the new diskettes. We have been fortunate to work on the new edition in the excellent environments provided by the University of Melbourne and Colorado State University. We thank Duane Boes particularly for his support and encouragement throughout, and the Australian Research Council and National Science Foundation for their support of research related to the new material. We are also indebted to Springer-Verlag for their constant support and assistance in preparing the second edition.

Fort Collins, Colorado
November, 1990

P.J. Brockwell
R.A. Davis

* *ITSM: An Interactive Time Series Modelling Package for the PC* by P.J. Brockwell and R.A. Davis. ISBN: 0-387-97482-2; 1991.

Preface to the First Edition

We have attempted in this book to give a systematic account of linear time series models and their application to the modelling and prediction of data collected sequentially in time. The aim is to provide specific techniques for handling data and at the same time to provide a thorough understanding of the mathematical basis for the techniques. Both time and frequency domain methods are discussed but the book is written in such a way that either approach could be emphasized. The book is intended to be a text for graduate students in statistics, mathematics, engineering, and the natural or social sciences. It has been used both at the M.S. level, emphasizing the more practical aspects of modelling, and at the Ph.D. level, where the detailed mathematical derivations of the deeper results can be included.

Distinctive features of the book are the extensive use of elementary Hilbert space methods and recursive prediction techniques based on innovations, use of the exact Gaussian likelihood and AIC for inference, a thorough treatment of the asymptotic behavior of the maximum likelihood estimators of the coefficients of univariate ARMA models, extensive illustrations of the techniques by means of numerical examples, and a large number of problems for the reader. The companion diskette contains programs written for the IBM PC, which can be used to apply the methods described in the text. Data sets can be found in the Appendix, and a more extensive collection (including most of those used for the examples in Chapters 1, 9, 10, 11 and 12) is on the diskette. Simulated ARMA series can easily be generated and filed using the program PEST. Valuable sources of additional time-series data are the collections of Makridakis et al. (1984) and Working Paper 109 (1984) of Scientific Computing Associates, DeKalb, Illinois.

Most of the material in the book is by now well-established in the time series literature and we have therefore not attempted to give credit for all the results discussed. Our indebtedness to the authors of some of the well-known existing books on time series, in particular Anderson, Box and Jenkins, Fuller, Grenander and Rosenblatt, Hannan, Koopmans and Priestley will however be apparent. We were also fortunate to have access to notes on time series by W. Dunsmuir. To these and to the many other sources that have influenced our presentation of the subject we express our thanks.

Recursive techniques based on the Kalman filter and state-space representations of ARMA processes have played an important role in many recent developments in time series analysis. In particular the Gaussian likelihood of a time series can be expressed very simply in terms of the one-step linear predictors and their mean squared errors, both of which can be computed recursively using a Kalman filter. Instead of using a state-space representation for recursive prediction we utilize the innovations representation of an arbitrary Gaussian time series in order to compute best linear predictors and exact Gaussian likelihoods. This approach, developed by Rissanen and Barbosa, Kailath, Ansley and others, expresses the value of the series at time t in terms of the one-step prediction errors up to that time. This representation provides insight into the structure of the time series itself as well as leading to simple algorithms for simulation, prediction and likelihood calculation.

These algorithms are used in the parameter estimation program (PEST) found on the companion diskette. Given a data set of up to 2300 observations, the program can be used to find preliminary, least squares and maximum Gaussian likelihood estimators of the parameters of any prescribed ARIMA model for the data, and to predict future values. It can also be used to simulate values of an ARMA process and to compute and plot its theoretical autocovariance and spectral density functions. Data can be plotted, differenced, deseasonalized and detrended. The program will also plot the sample autocorrelation and partial autocorrelation functions of both the data itself and the residuals after model-fitting. The other time-series programs are SPEC, which computes spectral estimates for univariate or bivariate series based on the periodogram, and TRANS, which can be used either to compute and plot the sample cross-correlation function of two series, or to perform least squares estimation of the coefficients in a transfer function model relating the second series to the first (see Section 12.2). Also included on the diskette is a screen editing program (WORD6), which can be used to create arbitrary data files, and a collection of data files, some of which are analyzed in the book. Instructions for the use of these programs are contained in the file HELP on the diskette.

For a one-semester course on time-domain analysis and modelling at the M.S. level, we have used the following sections of the book:

1.1–1.6; 2.1–2.7; 3.1–3.5; 5.1–5.5; 7.1, 7.2; 8.1–8.9; 9.1–9.6

(with brief reference to Sections 4.2 and 4.4). The prerequisite for this course is a knowledge of probability and statistics at the level of the book *Introduction to the Theory of Statistics* by Mood, Graybill and Boes.

For a second semester, emphasizing frequency-domain analysis and multivariate series, we have used

4.1–4.4, 4.6–4.10; 10.1–10.7; 11.1–11.7; selections from Chap. 12.

At the M.S. level it has not been possible (or desirable) to go into the mathematical derivation of all the results used, particularly those in the starred sections, which require a stronger background in mathematical analysis and measure theory. Such a background is assumed in all of the starred sections and problems.

For Ph.D. students the book has been used as the basis for a more theoretical one-semester course covering the starred sections from Chapters 4 through 11 and parts of Chapter 12. The prerequisite for this course is a knowledge of measure-theoretic probability.

We are greatly indebted to E.J. Hannan, R.H. Jones, S.I. Resnick, S.Tavaré and D. Tjøstheim, whose comments on drafts of Chapters 1–8 led to substantial improvements. The book arose out of courses taught in the statistics department at Colorado State University and benefitted from the comments of many students. The development of the computer programs would not have been possible without the outstanding work of Joe Mandarino, the architect of the computer program PEST, and Anthony Brockwell, who contributed WORD6, graphics subroutines and general computing expertise. We are indebted also to the National Science Foundation for support for the research related to the book, and one of us (P.J.B.) to Kuwait University for providing an excellent environment in which to work on the early chapters. For permission to use the optimization program UNC22MIN we thank R. Schnabel of the University of Colorado computer science department. Finally we thank Pam Brockwell, whose contributions to the manuscript went far beyond those of typist, and the editors of Springer-Verlag, who showed great patience and cooperation in the final production of the book.

Fort Collins, Colorado
October 1986

P.J. Brockwell
R.A. Davis

Contents

CHAPTER 1

Stationary Time Series

In this chapter we introduce some basic ideas of time series analysis and stochastic processes. Of particular importance are the concepts of stationarity and the autocovariance and sample autocovariance functions. Some standard techniques are described for the estimation and removal of trend and seasonality (of known period) from an observed series. These are illustrated with reference to the data sets in Section 1.1. Most of the topics covered in this chapter will be developed more fully in later sections of the book. The reader who is not already familiar with random vectors and multivariate analysis should first read Section 1.6 where a concise account of the required background is given. Notice our convention that an n-dimensional random vector is assumed (unless specified otherwise) to be a column vector $\mathbf{X} = (X_1, X_2, \ldots, X_n)'$ of random variables. If S is an arbitrary set then we shall use the notation S^n to denote both the set of n-component column vectors with components in S and the set of n-component row vectors with components in S.

§1.1 Examples of Time Series

A time series is a set of observations x_t, each one being recorded at a specified time t. A discrete-time series (the type to which this book is primarily devoted) is one in which the set T_0 of times at which observations are made is a discrete set, as is the case for example when observations are made at fixed time intervals. Continuous-time series are obtained when observations are recorded continuously over some time interval, e.g. when $T_0 = [0, 1]$. We shall use the notation $x(t)$ rather than x_t if we wish to indicate specifically that observations are recorded continuously.

EXAMPLE 1.1.1 (Current Through a Resistor). If a sinusoidal voltage $v(t) = a\cos(vt + \theta)$ is applied to a resistor of resistance r and the current recorded continuously we obtain a continuous time series

$$x(t) = r^{-1}a\cos(vt + \theta).$$

If observations are made only at times 1, 2, ..., the resulting time series will be discrete. Time series of this particularly simple type will play a fundamental role in our later study of stationary time series.

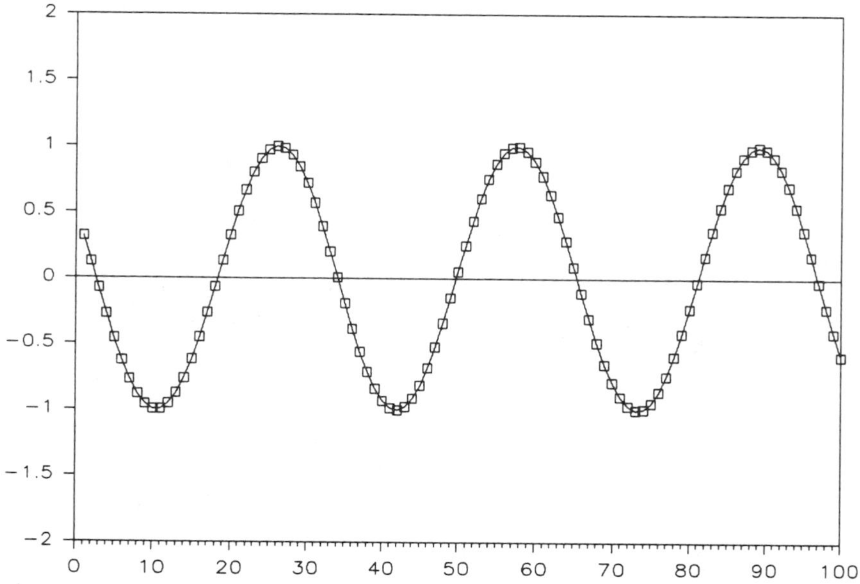

Figure 1.1. 100 observations of the series $x(t) = \cos(.2t + \pi/3)$.

EXAMPLE 1.1.2 (Population x_t of the U.S.A., 1790–1980).

t	x_t	t	x_t
1790	3,929,214	1890	62,979,766
1800	5,308,483	1900	76,212,168
1810	7,239,881	1910	92,228,496
1820	9,638,453	1920	106,021,537
1830	12,860,702	1930	123,202,624
1840	17,063,353	1940	132,164,569
1850	23,191,876	1950	151,325,798
1860	31,443,321	1960	179,323,175
1870	38,558,371	1970	203,302,031
1880	50,189,209	1980	226,545,805

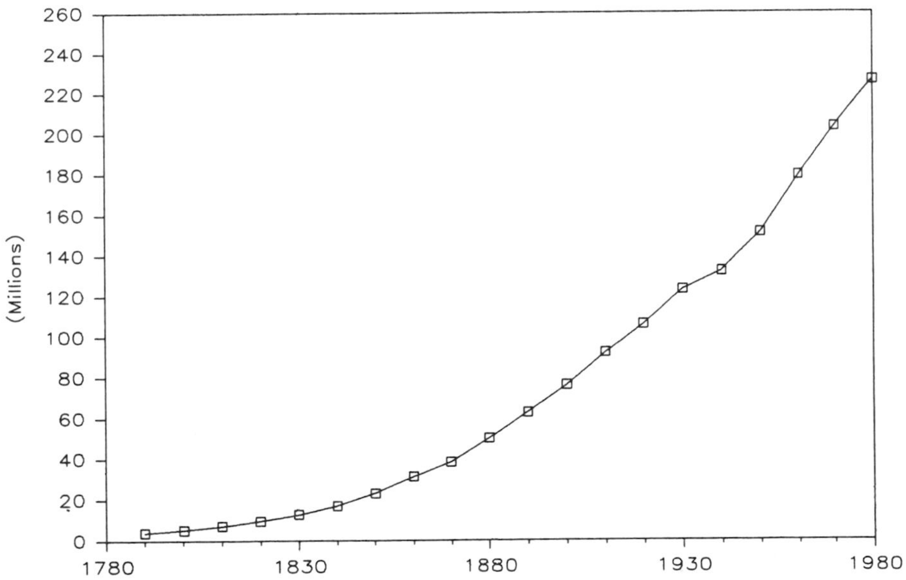

Figure 1.2. Population of the U.S.A. at ten-year intervals, 1790–1980 (U.S. Bureau of the Census).

EXAMPLE 1.1.3 (Strikes in the U.S.A., 1951–1980).

t	x_t	t	x_t
1951	4737	1966	4405
1952	5117	1967	4595
1953	5091	1968	5045
1954	3468	1969	5700
1955	4320	1970	5716
1956	3825	1971	5138
1957	3673	1972	5010
1958	3694	1973	5353
1959	3708	1974	6074
1960	3333	1975	5031
1961	3367	1976	5648
1962	3614	1977	5506
1963	3362	1978	4230
1964	3655	1979	4827
1965	3963	1980	3885

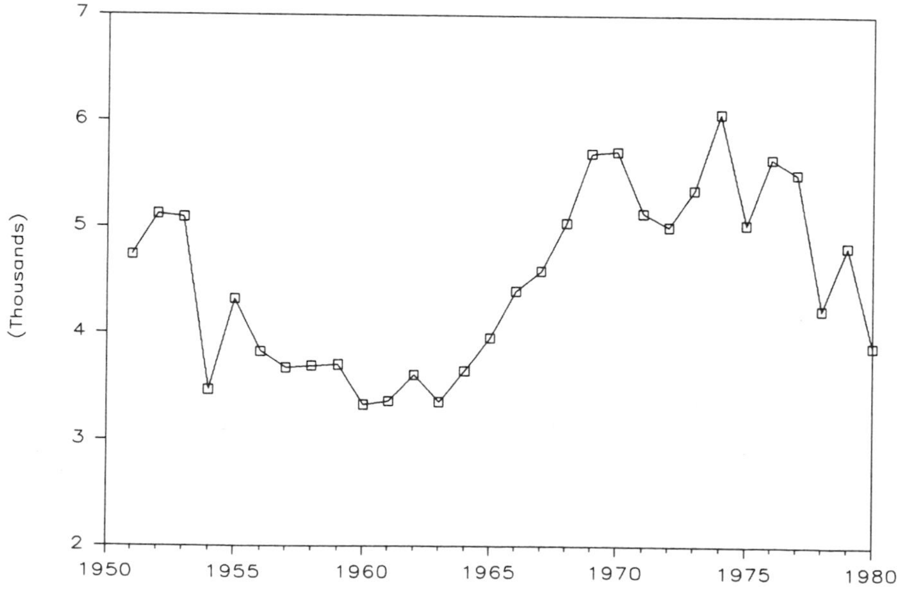

Figure 1.3. Strikes in the U.S.A., 1951–1980 (Bureau of Labor Statistics, U.S. Labor Department).

EXAMPLE 1.1.4 (All Star Baseball Games, 1933–1980).

$$x_t = \begin{cases} 1 & \text{if the National League won in year } t, \\ -1 & \text{if the American League won in year } t. \end{cases}$$

$t - 1900$	33	34	35	36	37	38	39	40	41	42	43	44	45	46	47	48
x_t	−1	−1	−1	1	−1	1	−1	1	−1	−1	−1	1	†	−1	−1	−1
$t - 1900$	49	50	51	52	53	54	55	56	57	58	59	60	61	62	63	64
x_t	−1	1	1	1	1	−1	1	1	−1	−1	*	*	*	*	1	1
$t - 1900$	65	66	67	68	69	70	71	72	73	74	75	76	77	78	79	80
x_t	1	1	1	1	1	1	−1	1	1	1	1	1	1	1	1	1

† = no game.

* = two games scheduled.

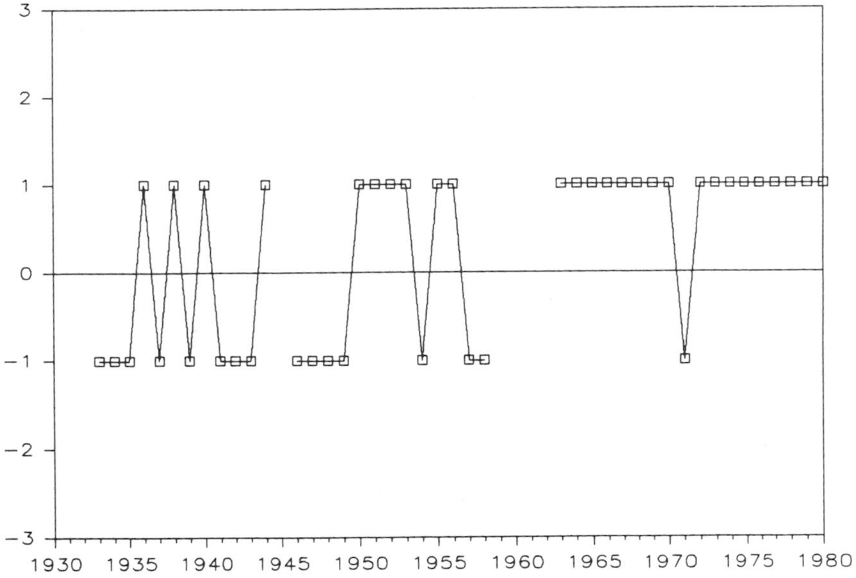

Figure 1.4. Results x_t, Example 1.1.4, of All-star baseball games, 1933–1980.

EXAMPLE 1.1.5 (Wölfer Sunspot Numbers, 1770–1869).

1770	101	1790	90	1810	0	1830	71	1850	66
1771	82	1791	67	1811	1	1831	48	1851	64
1772	66	1792	60	1812	5	1832	28	1852	54
1773	35	1793	47	1813	12	1833	8	1853	39
1774	31	1794	41	1814	14	1834	13	1854	21
1775	7	1795	21	1815	35	1835	57	1855	7
1776	20	1796	16	1816	46	1836	122	1856	4
1777	92	1797	6	1817	41	1837	138	1857	23
1778	154	1798	4	1818	30	1838	103	1858	55
1779	125	1799	7	1819	24	1839	86	1859	94
1780	85	1800	14	1820	16	1840	63	1860	96
1781	68	1801	34	1821	7	1841	37	1861	77
1782	38	1802	45	1822	4	1842	24	1862	59
1783	23	1803	43	1823	2	1843	11	1863	44
1784	10	1804	48	1824	8	1844	15	1864	47
1785	24	1805	42	1825	17	1845	40	1865	30
1786	83	1806	28	1826	36	1846	62	1866	16
1787	132	1807	10	1827	50	1847	98	1867	7
1788	131	1808	8	1828	62	1848	124	1868	37
1789	118	1809	2	1829	67	1849	96	1869	74

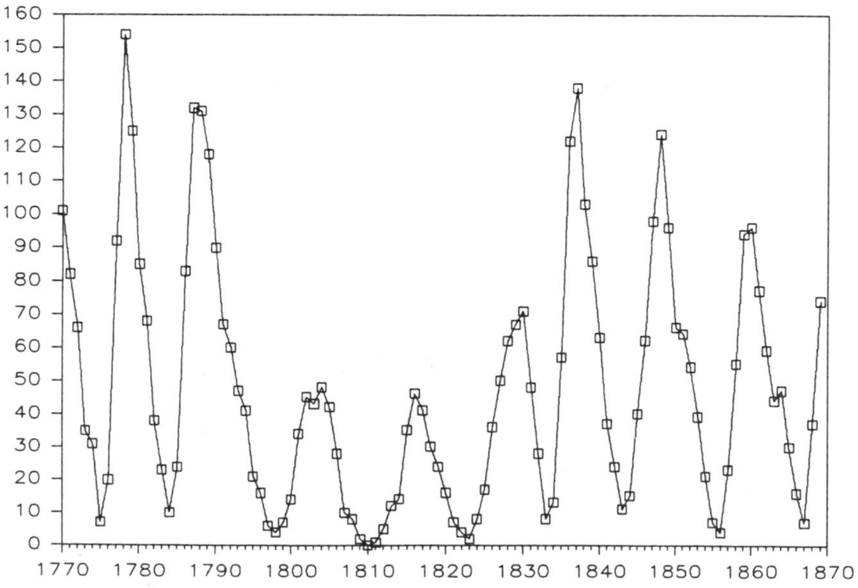

Figure 1.5. The Wölfer sunspot numbers, 1770–1869.

EXAMPLE 1.1.6 (Monthly Accidental Deaths in the U.S.A., 1973–1978).

	1973	1974	1975	1976	1977	1978
Jan.	9007	7750	8162	7717	7792	7836
Feb.	8106	6981	7306	7461	6957	6892
Mar.	8928	8038	8124	7776	7726	7791
Apr.	9137	8422	7870	7925	8106	8129
May	10017	8714	9387	8634	8890	9115
Jun.	10826	9512	9556	8945	9299	9434
Jul.	11317	10120	10093	10078	10625	10484
Aug.	10744	9823	9620	9179	9302	9827
Sep.	9713	8743	8285	8037	8314	9110
Oct.	9938	9129	8433	8488	8850	9070
Nov.	9161	8710	8160	7874	8265	8633
Dec.	8927	8680	8034	8647	8796	9240

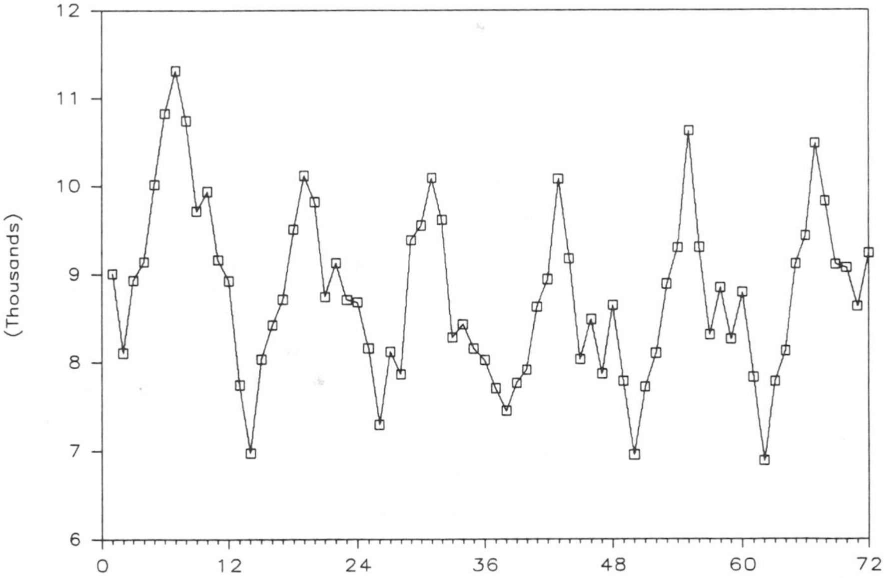

Figure 1.6. Monthly accidental deaths in the U.S.A., 1973–1978 (National Safety Council).

These examples are of course but a few of the multitude of time series to be found in the fields of engineering, science, sociology and economics. Our purpose in this book is to study the techniques which have been developed for drawing inferences from such series. Before we can do this however, it is necessary to set up a hypothetical mathematical model to represent the data. Having chosen a model (or family of models) it then becomes possible to estimate parameters, check for goodness of fit to the data and possibly to use the fitted model to enhance our understanding of the mechanism generating the series. Once a satisfactory model has been developed, it may be used in a variety of ways depending on the particular field of application. The applications include separation (filtering) of noise from signals, prediction of future values of a series and the control of future values.

The six examples given show some rather striking differences which are apparent if one examines the graphs in Figures 1.1–1.6. The first gives rise to a smooth sinusoidal graph oscillating about a constant level, the second to a roughly exponentially increasing graph, the third to a graph which fluctuates erratically about a nearly constant or slowly rising level, and the fourth to an erratic series of minus ones and ones. The fifth graph appears to have a strong cyclic component with period about 11 years and the last has a pronounced seasonal component with period 12.

In the next section we shall discuss the general problem of constructing mathematical models for such data.

§1.2 Stochastic Processes

The first step in the analysis of a time series is the selection of a suitable mathematical model (or class of models) for the data. To allow for the possibly unpredictable nature of future observations it is natural to suppose that each observation x_t is a realized value of a certain random variable X_t. The time series $\{x_t, t \in T_0\}$ is then a realization of the family of random variables $\{X_t, t \in T_0\}$. These considerations suggest modelling the data as a realization (or part of a realization) of a stochastic process $\{X_t, t \in T\}$ where $T \supseteq T_0$. To clarify these ideas we need to define precisely what is meant by a stochastic process and its realizations. In later sections we shall restrict attention to special classes of processes which are particularly useful for modelling many of the time series which are encountered in practice.

Definition 1.2.1 (Stochastic Process). A stochastic process is a family of random variables $\{X_t, t \in T\}$ defined on a probability space $(\Omega, \mathscr{F}, P)$.

Remark 1. In time series analysis the index (or parameter) set T is a set of time points, very often $\{0, \pm 1, \pm 2, \ldots\}$, $\{1, 2, 3, \ldots\}$, $[0, \infty)$ or $(-\infty, \infty)$. Stochastic processes in which T is not a subset of $\mathbb{R}$ are also of importance. For example in geophysics stochastic processes with T the surface of a sphere are used to represent variables indexed by their location on the earth's surface. In this book however the index set T will always be a subset of $\mathbb{R}$.

Recalling the definition of a random variable we note that for each fixed $t \in T$, X_t is in fact a function $X_t(\cdot)$ on the set Ω. On the other hand, for each fixed $\omega \in \Omega$, $X_{\cdot}(\omega)$ is a function on T.

Definition 1.2.2 (Realizations of a Stochastic Process). The functions $\{X_{\cdot}(\omega), \omega \in \Omega\}$ on T are known as the realizations or sample-paths of the process $\{X_t, t \in T\}$.

Remark 2. We shall frequently use the term time series to mean both the data and the process of which it is a realization.

The following examples illustrate the realizations of some specific stochastic processes. The first two could be considered as possible models for the time series of Examples 1.1.1 and 1.1.4 respectively.

EXAMPLE 1.2.1 (Sinusoid with Random Phase and Amplitude). Let A and Θ be independent random variables with $A \geq 0$ and Θ distributed uniformly on $[0, 2\pi)$. A stochastic process $\{X(t), t \in \mathbb{R}\}$ can then be defined in terms of A and Θ for any given $v \geq 0$ and $r > 0$ by

$$X_t = r^{-1} A \cos(vt + \Theta), \tag{1.2.1}$$

or more explicitly,

$$X_t(\omega) = r^{-1} A(\omega) \cos(vt + \Theta(\omega)), \tag{1.2.2}$$

where ω is an element of the probability space Ω on which A and Θ are defined.

The realizations of the process defined by 1.2.2 are the functions of t obtained by fixing ω, i.e. functions of the form

$$x(t) = r^{-1} a \cos(vt + \theta).$$

The time series plotted in Figure 1.1 is one such realization.

EXAMPLE 1.2.2 (A Binary Process). Let $\{X_t, t = 1, 2, \ldots\}$ be a sequence of independent random variables for each of which

$$P(X_t = 1) = P(X_t = -1) = \tfrac{1}{2}. \tag{1.2.3}$$

In this case it is not so obvious as in Example 1.2.1 that there exists a probability space $(\Omega, \mathscr{F}, P)$ with random variables $X_1, X_2, \ldots$ defined on Ω having the required joint distributions, i.e. such that

$$P(X_1 = i_1, X_2 = i_2, \ldots, X_n = i_n) = 2^{-n}, \tag{1.2.4}$$

for every n-tuple $(i_1, \ldots, i_n)$ of 1's and -1's. The existence of such a process is however guaranteed by Kolmogorov's theorem which is stated below and discussed further in Section 1.7.

The time series obtained by tossing a penny repeatedly and scoring $+1$ for each head, -1 for each tail is usually modelled as a realization of the process defined by (1.2.4). Each realization of this process is a sequence of 1's and -1's.

A priori we might well consider this process as a model for the All Star baseball games, Example 1.1.4. However even a cursory inspection of the results from 1963 onwards casts serious doubt on the hypothesis $P(X_t = 1) = \frac{1}{2}$.

Example 1.2.3 (Random Walk). The simple symmetric random walk $\{S_t, t = 0, 1, 2, \ldots\}$ is defined in terms of Example 1.2.2 by $S_0 = 0$ and

$$S_t = \sum_{i=1}^{t} X_i, \qquad t \geq 1. \tag{1.2.5}$$

The general random walk is defined in the same way on replacing $X_1, X_2, \ldots$ by a sequence of independently and identically distributed random variables whose distribution is not constrained to satisfy (1.2.3). The existence of such an independent sequence is again guaranteed by Kolmogorov's theorem (see Problem 1.18).

Example 1.2.4 (Branching Processes). There is a large class of processes, known as branching processes, which in their most general form have been applied with considerable success to the modelling of population growth (see for example Jagers (1976)). The simplest such process is the Bienaymé–Galton–Watson process defined by the equations $X_0 = x$ (the population size in generation zero) and

$$X_{t+1} = \sum_{j=1}^{X_t} Z_{t,j}, \qquad t = 0, 1, 2, \ldots, \tag{1.2.6}$$

where $Z_{t,j}$, $t = 0, 1, \ldots$, $j = 1, 2, \ldots$ are independently and identically distributed non-negative integer-valued random variables, $Z_{t,j}$, representing the number of offspring of the jth individual born in generation t.

In the first example we were able to define $X_t(\omega)$ quite explicitly for each t and ω. Very frequently however we may wish (or be forced) to specify instead the collection of all joint distributions of all finite-dimensional vectors $(X_{t_1}, X_{t_2}, \ldots, X_{t_n})$, $\mathbf{t} = (t_1, \ldots, t_n) \in T^n$, $n \in \{1, 2, \ldots\}$. In such a case we need to be sure that a stochastic process (see Definition 1.2.1) with the specified distributions really does exist. Kolmogorov's theorem, which we state here and discuss further in Section 1.7, guarantees that this is true under minimal conditions on the specified distribution functions. Our statement of Kolmogorov's theorem is simplified slightly by the assumption (Remark 1) that T is a subset of $\mathbb{R}$ and hence a linearly ordered set. If T were not so ordered an additional "permutation" condition would be required (a statement and proof of the theorem for arbitrary T can be found in numerous books on probability theory, for example Lamperti, 1966).

Definition 1.2.3 (The Distribution Functions of a Stochastic Process $\{X_t, t \in T \subset \mathbb{R}\}$). Let $\mathscr{T}$ be the set of all vectors $\{\mathbf{t} = (t_1, \ldots, t_n)' \in T^n : t_1 < t_2 < \cdots < t_n, n = 1, 2, \ldots\}$. Then the (finite-dimensional) distribution functions of $\{X_t, t \in T\}$ are the functions $\{F_{\mathbf{t}}(\cdot), \mathbf{t} \in \mathscr{T}\}$ defined for $\mathbf{t} = (t_1, \ldots, t_n)'$ by

$$F_{\mathbf{t}}(\mathbf{x}) = P(X_{t_1} \le x_1, \ldots, X_{t_n} \le x_n), \qquad \mathbf{x} = (x_1, \ldots, x_n)' \in \mathbb{R}^n. \qquad (1.2.7)$$

Theorem 1.2.1 (Kolmogorov's Theorem). *The probability distribution functions $\{F_{\mathbf{t}}(\cdot), \mathbf{t} \in \mathscr{T}\}$ are the distribution functions of some stochastic process if and only if for any $n \in \{1, 2, \ldots\}$, $\mathbf{t} = (t_1, \ldots, t_n)' \in \mathscr{T}$ and $1 \le i \le n$,*

$$\lim_{x_i \to \infty} F_{\mathbf{t}}(\mathbf{x}) = F_{\mathbf{t}(i)}(\mathbf{x}(i)) \qquad (1.2.8)$$

where $\mathbf{t}(i)$ and $\mathbf{x}(i)$ are the $(n-1)$-component vectors obtained by deleting the ith components of $\mathbf{t}$ and $\mathbf{x}$ respectively.

If $\phi_{\mathbf{t}}(\cdot)$ is the characteristic function corresponding to $F_{\mathbf{t}}(\cdot)$, i.e.

$$\phi_{\mathbf{t}}(\mathbf{u}) = \int_{\mathbb{R}^n} e^{i\mathbf{u}'\mathbf{x}} F_{\mathbf{t}}(dx_1, \ldots, dx_n), \qquad \mathbf{u} = (u_1, \ldots, u_n)' \in \mathbb{R}^n,$$

then (1.2.8) can be restated in the equivalent form,

$$\lim_{u_i \to 0} \phi_{\mathbf{t}}(\mathbf{u}) = \phi_{\mathbf{t}(i)}(\mathbf{u}(i)), \qquad (1.2.9)$$

where $\mathbf{u}(i)$ is the $(n-1)$-component vector obtained by deleting the ith component of $\mathbf{u}$.

Condition (1.2.8) is simply the "consistency" requirement that each function $F_{\mathbf{t}}(\cdot)$ should have marginal distributions which coincide with the specified lower dimensional distribution functions.

§1.3 Stationarity and Strict Stationarity

When dealing with a finite number of random variables, it is often useful to compute the covariance matrix (see Section 1.6) in order to gain insight into the dependence between them. For a time series $\{X_t, t \in T\}$ we need to extend the concept of covariance matrix to deal with infinite collections of random variables. The autocovariance function provides us with the required extension.

Definition 1.3.1 (The Autocovariance Function). If $\{X_t, t \in T\}$ is a process such that $\operatorname{Var}(X_t) < \infty$ for each $t \in T$, then the autocovariance function $\gamma_X(\cdot, \cdot)$ of $\{X_t\}$ is defined by

$$\gamma_X(r, s) = \operatorname{Cov}(X_r, X_s) = E[(X_r - EX_r)(X_s - EX_s)], \qquad r, s \in T. \qquad (1.3.1)$$

Definition 1.3.2 (Stationarity). The time series $\{X_t, t \in \mathbb{Z}\}$, with index set $\mathbb{Z} = \{0, \pm 1, \pm 2, \dots\}$, is said to be stationary if

$$\text{(i)}\ E|X_t|^2 < \infty \quad \text{for all } t \in \mathbb{Z},$$

$$\text{(ii)}\ EX_t = m \quad \text{for all } t \in \mathbb{Z},$$

and

$$\text{(iii)}\ \gamma_X(r, s) = \gamma_X(r + t, s + t) \quad \text{for all } r, s, t \in \mathbb{Z}.$$

Remark 1. Stationarity as just defined is frequently referred to in the literature as weak stationarity, covariance stationarity, stationarity in the wide sense or second-order stationarity. For us however the term stationarity, without further qualification, will always refer to the properties specified by Definition 1.3.2.

Remark 2. If $\{X_t, t \in \mathbb{Z}\}$ is stationary then $\gamma_X(r, s) = \gamma_X(r - s, 0)$ for all $r, s \in \mathbb{Z}$. It is therefore convenient to redefine the autocovariance function of a stationary process as the function of just one variable,

$$\gamma_X(h) \equiv \gamma_X(h, 0) = \operatorname{Cov}(X_{t+h}, X_t) \quad \text{for all } t, h \in \mathbb{Z}.$$

The function $\gamma_X(\cdot)$ will be referred to as the autocovariance function of $\{X_t\}$ and $\gamma_X(h)$ as its value at "lag" h. The autocorrelation function (acf) of $\{X_t\}$ is defined analogously as the function whose value at lag h is

$$\rho_X(h) \equiv \gamma_X(h)/\gamma_X(0) = \operatorname{Corr}(X_{t+h}, X_t) \quad \text{for all } t, h \in \mathbb{Z}.$$

Remark 3. It will be noticed that we have defined stationarity only in the case when $T = \mathbb{Z}$. It is not difficult to define stationarity using a more general index set, but for our purposes this will not be necessary. If we wish to model a set of data $\{x_t, t \in T \subset \mathbb{Z}\}$ as a realization of a stationary process, we can always consider it to be part of a realization of a stationary process $\{X_t, t \in \mathbb{Z}\}$.

Another important and frequently used notion of stationarity is introduced in the following definition.

Definition 1.3.3 (Strict Stationarity). The time series $\{X_t, t \in \mathbb{Z}\}$ is said to be strictly stationary if the joint distributions of $(X_{t_1}, \dots, X_{t_k})'$ and $(X_{t_1+h}, \dots, X_{t_k+h})'$ are the same for all positive integers k and for all $t_1, \dots, t_k, h \in \mathbb{Z}$.

Strict stationarity means intuitively that the graphs over two equal-length time intervals of a realization of the time series should exhibit similar statistical characteristics. For example, the proportion of ordinates not exceeding a given level x should be roughly the same for both intervals.

Remark 4. Definition 1.3.3 is equivalent to the statement that $(X_1, \dots, X_k)'$ and $(X_{1+h}, \dots, X_{k+h})'$ have the same joint distribution for all positive integers k and integers h.

The Relation Between Stationarity and Strict Stationarity

If $\{X_t\}$ is strictly stationary it immediately follows, on taking $k = 1$ in Definition 1.3.3, that X_t has the same distribution for each $t \in \mathbb{Z}$. If $E|X_t|^2 < \infty$ this implies in particular that EX_t and $\text{Var}(X_t)$ arc both constant. Moreover, taking $k = 2$ in Definition 1.3.3, we find that X_{t+h} and X_t have the same joint distribution and hence the same covariance for all $h \in \mathbb{Z}$. Thus a strictly stationary process with finite second moments is stationary.

The converse of the previous statement is not true. For example if $\{X_t\}$ is a sequence of independent random variables such that X_t is exponentially distributed with mean one when t is odd and normally distributed with mean one and variance one when t is even, then $\{X_t\}$ is stationary with $\gamma_X(0) = 1$ and $\gamma_X(h) = 0$ for $h \neq 0$. However since X_1 and X_2 have different distributions, $\{X_t\}$ cannot be strictly stationary.

There is one important case however in which stationarity does imply strict stationarity.

Definition 1.3.4 (Gaussian Time Series). The process $\{X_t\}$ is a Gaussian time series if and only if the distribution functions of $\{X_t\}$ are all multivariate normal.

If $\{X_t, t \in \mathbb{Z}\}$ is a stationary Gaussian process then $\{X_t\}$ is strictly stationary, since for all $n \in \{1, 2, \dots\}$ and for all $h, t_1, t_2, \dots \in \mathbb{Z}$, the random vectors $(X_{t_1}, \dots, X_{t_n})'$ and $(X_{t_1+h}, \dots, X_{t_n+h})'$ have the same mean and covariance matrix, and hence the same distribution.

EXAMPLE 1.3.1. Let $X_t = A\cos(\theta t) + B\sin(\theta t)$ where A and B are two uncorrelated random variables with zero means and unit variances with $\theta \in [-\pi, \pi]$. This time series is stationary since

$$\begin{aligned}\text{Cov}(X_{t+h}, X_t) &= \text{Cov}(A\cos(\theta(t+h)) + B\sin(\theta(t+h)), A\cos(\theta t) + B\sin(\theta t)) \\ &= \cos(\theta t)\cos(\theta(t+h)) + \sin(\theta t)\sin(\theta(t+h)) \\ &= \cos(\theta h),\end{aligned}$$

which is independent of t.

EXAMPLE 1.3.2. Starting with an independent and identically distributed sequence of zero-mean random variables Z_t with finite variance σ_Z^2, define $X_t = Z_t + \theta Z_{t-1}$. Then the autocovariance function of X_t is given by

$$\begin{aligned}\text{Cov}(X_{t+h}, X_t) &= \text{Cov}(Z_{t+h} + \theta Z_{t+h-1}, Z_t + \theta Z_{t-1}) \\ &= \begin{cases} (1+\theta^2)\sigma_Z^2 & \text{if } h = 0, \\ \theta\sigma_Z^2 & \text{if } h = \pm 1, \\ 0 & \text{if } |h| > 1, \end{cases}\end{aligned}$$

and hence $\{X_t\}$ is stationary. In fact it can be shown that $\{X_t\}$ is strictly stationary (see Problem 1.1).

EXAMPLE 1.3.3. Let

$$X_t = \begin{cases} Y_t & \text{if } t \text{ is even,} \\ Y_t + 1 & \text{if } t \text{ is odd.} \end{cases}$$

where $\{Y_t\}$ is a stationary time series. Although $\text{Cov}(X_{t+h}, X_t) = \gamma_Y(h)$, $\{X_t\}$ is not stationary for it does not have a constant mean.

EXAMPLE 1.3.4. Referring to Example 1.2.3, let S_t be the random walk $S_t = X_1 + X_2 + \cdots + X_t$, where $X_1, X_2, \ldots,$ are independent and identically distributed with mean zero and variance σ^2. For $h > 0$,

$$\begin{aligned} \text{Cov}(S_{t+h}, S_t) &= \text{Cov}\left(\sum_{i=1}^{t+h} X_i, \sum_{j=1}^{t} X_j\right) \\ &= \text{Cov}\left(\sum_{i=1}^{t} X_i, \sum_{j=1}^{t} X_j\right) \\ &= \sigma^2 t \end{aligned}$$

and thus S_t is not stationary.

Stationary processes play a crucial role in the analysis of time series. Of course many observed time series (see Section 1.1) are decidedly non-stationary in appearance. Frequently such data sets can be transformed by the techniques described in Section 1.4 into series which can reasonably be modelled as realizations of some stationary process. The theory of stationary processes (developed in later chapters) is then used for the analysis, fitting and prediction of the resulting series. In all of this the autocovariance function is a primary tool. Its properties will be discussed in Section 1.5.

§1.4 The Estimation and Elimination of Trend and Seasonal Components

The first step in the analysis of any time series is to plot the data. If there are apparent discontinuities in the series, such as a sudden change of level, it may be advisable to analyze the series by first breaking it into homogeneous segments. If there are outlying observations, they should be studied carefully to check whether there is any justification for discarding them (as for example if an observation has been recorded of some other process by mistake). Inspection of a graph may also suggest the possibility of representing the data as a realization of the process (the "classical decomposition" model),

$$X_t = m_t + s_t + Y_t, \tag{1.4.1}$$

where m_t is a slowly changing function known as a "trend component", s_t is a function with known period d referred to as a "seasonal component", and Y_t is a "random noise component" which is stationary in the sense of Definition 1.3.2. If the seasonal and noise fluctuations appear to increase with the level of the process then a preliminary transformation of the data is often used to make the transformed data compatible with the model (1.4.1). See for example the airline passenger data, Figure 9.7, and the transformed data, Figure 9.8, obtained by applying a logarithmic transformation. In this section we shall discuss some useful techniques for identifying the components in (1.4.1).

Our aim is to estimate and extract the deterministic components m_t and s_t in the hope that the residual or noise component Y_t will turn out to be a stationary random process. We can then use the theory of such processes to find a satisfactory probabilistic model for the process $\{Y_t\}$, to analyze its properties, and to use it in conjunction with m_t and s_t for purposes of prediction and control of $\{X_t\}$.

An alternative approach, developed extensively by Box and Jenkins (1970), is to apply difference operators repeatedly to the data $\{x_t\}$ until the differenced observations resemble a realization of some stationary process $\{W_t\}$. We can then use the theory of stationary processes for the modelling, analysis and prediction of $\{W_t\}$ and hence of the original process. The various stages of this procedure will be discussed in detail in Chapters 8 and 9.

The two approaches to trend and seasonality removal, (a) by estimation of m_t and s_t in (1.4.1) and (b) by differencing the data $\{x_t\}$, will now be illustrated with reference to the data presented in Section 1.1.

Elimination of a Trend in the Absence of Seasonality

In the absence of a seasonal component the model (1.4.1) becomes

$$X_t = m_t + Y_t, \qquad t = 1, \dots, n \tag{1.4.2}$$

where, without loss of generality, we can assume that $EY_t = 0$.

Method 1 (Least Squares Estimation of m_t). In this procedure we attempt to fit a parametric family of functions, e.g.

$$m_t = a_0 + a_1 t + a_2 t^2, \tag{1.4.3}$$

to the data by choosing the parameters, in this illustration a_0, a_1 and a_2, to minimize $\sum_t (x_t - m_t)^2$.

Fitting a function of the form (1.4.3) to the population data of Figure 1.2, $1790 \le t \le 1980$ gives the estimated parameter values,

$$\hat{a}_0 = 2.097911 \times 10^{10},$$

$$\hat{a}_1 = -2.334962 \times 10^{7},$$

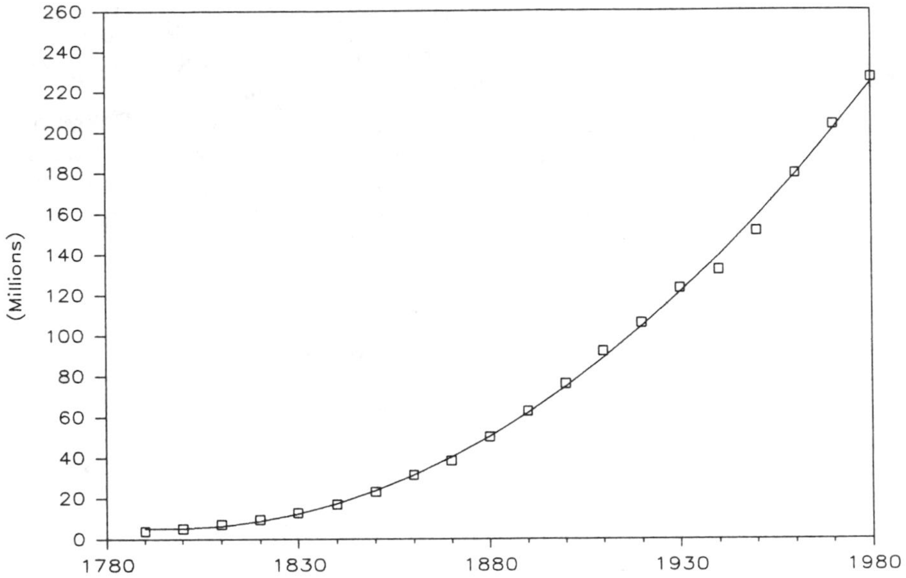

Figure 1.7. Population of the U.S.A., 1790–1980, showing the parabola fitted by least squares.

and

$$\hat{a}_2 = 6.498591 \times 10^3.$$

A graph of the fitted function is shown with the original data in Figure 1.7. The estimated values of the noise process Y_t, $1790 \le t \le 1980$, are the residuals obtained by subtraction of $\hat{m}_t = \hat{a}_0 + \hat{a}_1 t + \hat{a}_2 t^2$ from x_t.

The trend component $\hat{m}_t$ furnishes us with a natural predictor of future values of X_t. For example if we estimate Y_{1990} by its mean value (i.e. zero) we obtain the estimate,

$$\hat{m}_{1990} = 2.484 \times 10^8,$$

for the population of the U.S.A. in 1990. However if the residuals $\{\hat{Y}_t\}$ are highly correlated we may be able to use their values to give a better estimate of Y_{1990} and hence of X_{1990}.

Method 2 (Smoothing by Means of a Moving Average). Let q be a non-negative integer and consider the two-sided moving average,

$$W_t = (2q + 1)^{-1} \sum_{j=-q}^{q} X_{t+j}, \tag{1.4.4}$$

of the process $\{X_t\}$ defined by (1.4.2). Then for $q + 1 \le t \le n - q$,

$$\begin{aligned} W_t &= (2q + 1)^{-1} \sum_{j=-q}^{q} m_{t+j} + (2q + 1)^{-1} \sum_{j=-q}^{q} Y_{t+j} \\ &\simeq m_t, \end{aligned} \tag{1.4.5}$$

assuming that m_t is approximately linear over the interval $[t - q, t + q]$ and that the average of the error terms over this interval is close to zero.

The moving average thus provides us with the estimates

$$\hat{m}_t = (2q + 1)^{-1} \sum_{j=-q}^{q} X_{t+j}, \qquad q + 1 \le t \le n - q. \tag{1.4.6}$$

Since X_t is not observed for $t \le 0$ or $t > n$ we cannot use (1.4.6) for $t \le q$ or $t > n - q$. The program SMOOTH deals with this problem by defining $X_t := X_1$ for $t < 1$ and $X_t := X_n$ for $t > n$. The results of applying this program to the strike data of Figure 1.3 are shown in Figure 1.8. The estimated noise terms, $\hat{Y}_t = X_t - \hat{m}_t$, are shown in Figure 1.9. As expected, they show no apparent trend.

For any fixed $a \in [0, 1]$, the one-sided moving averages $\hat{m}_t$, $t = 1, \ldots, n$, defined by the recursions,

$$\hat{m}_t = aX_t + (1 - a)\hat{m}_{t-1}, \qquad t = 2, \ldots, n, \tag{1.4.7}$$

and

$$\hat{m}_1 = X_1, \tag{1.4.8}$$

can also be computed using the program SMOOTH. Application of (1.4.7) and (1.4.8) is often referred to as exponential smoothing, since it follows from these recursions that, for $t \ge 2$, $\hat{m}_t = \sum_{j=0}^{t-2} a(1 - a)^j X_{t-j} + (1 - a)^{t-1} X_1$, a weighted moving average of $X_t, X_{t-1}, \ldots$, with weights decreasing exponentially (except for the last one).

It is useful to think of $\{\hat{m}_t\}$ in (1.4.6) as a process obtained from $\{X_t\}$ by application of a linear operator or linear filter, $\hat{m}_t = \sum_{j=-\infty}^{\infty} a_j X_{t+j}$ with

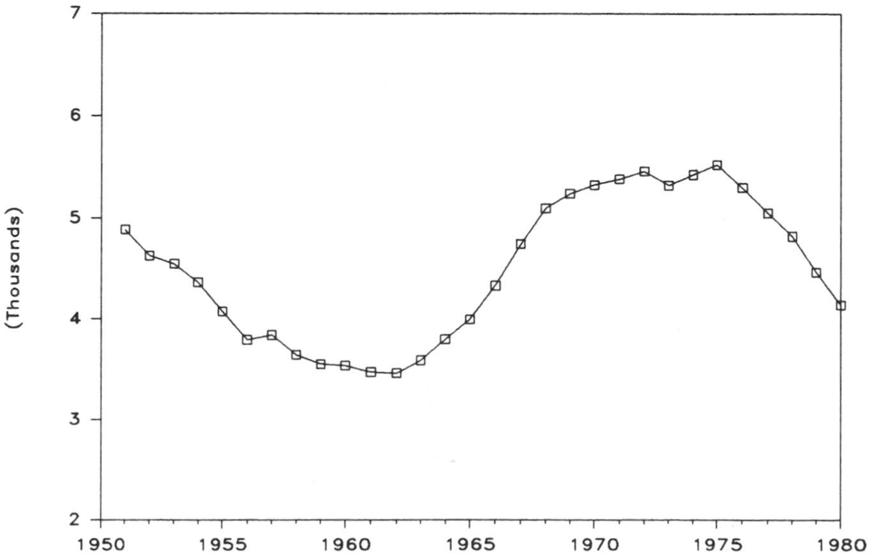

Figure 1.8. Simple 5-term moving average $\hat{m}_t$ of the strike data from Figure 1.3.

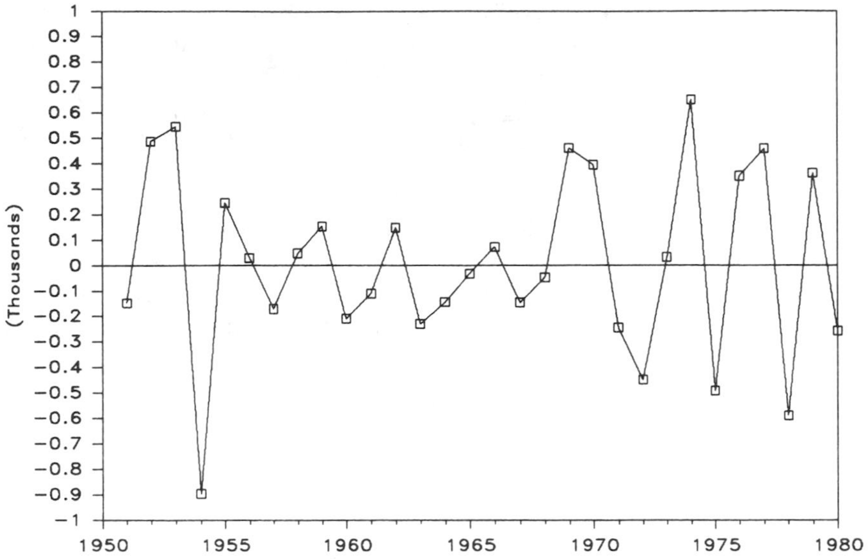

Figure 1.9. Residuals, $Y_t = x_t - \hat{m}_t$, after subtracting the 5-term moving average from the strike data.

weights $a_j = (2q + 1)^{-1}$, $-q \le j \le q$, and $a_j = 0$, $|j| > q$. This particular filter is a "low-pass" filter since it takes the data $\{x_t\}$ and removes from it the rapidly fluctuating (or high frequency) component $\{\hat{Y}_t\}$, to leave the slowly varying estimated trend term $\{\hat{m}_t\}$ (see Figure 1.10).

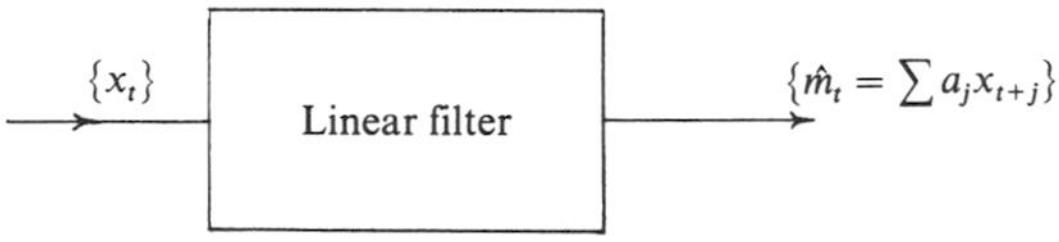

Figure 1.10. Smoothing with a low-pass linear filter.

The particular filter (1.4.6) is only one of many which could be used for smoothing. For large q, provided $(2q + 1)^{-1} \sum_{i=-q}^{q} Y_{t+i} \simeq 0$, it will not only attenuate noise but at the same time will allow linear trend functions $m_t = at + b$, to pass without distortion. However we must beware of choosing q to be too large since if m_t is not linear, the filtered process, although smooth, will not be a good estimate of m_t. By clever choice of the weights $\{a_j\}$ it is possible to design a filter which will not only be effective in attenuating noise from the data, but which will also allow a larger class of trend functions (for example all polynomials of degree less than or equal to 3) to pass undistorted through the filter. The Spencer 15-point moving average for example has weights

$$a_i = 0, \qquad |i| > 7,$$

with

$$a_i = a_{-i}, \qquad |i| \leq 7,$$

and

$$[a_0, a_1, \ldots, a_7] = \tfrac{1}{320}[74, 67, 46, 21, 3, -5, -6, -3]. \tag{1.4.9}$$

Applied to the process (1.4.3) with $m_t = at^3 + bt^2 + ct + d$, it gives

$$\begin{aligned}\sum_{i=-7}^{7} a_i X_{t+i} &= \sum_{i=-7}^{7} a_i m_{t+i} + \sum_{i=-7}^{7} a_i Y_{t+i} \\ &\simeq \sum_{i=-7}^{7} a_i m_{t+i}, \\ &= m_t,\end{aligned}$$

where the last step depends on the assumed form of m_t (Problem 1.2). Further details regarding this and other smoothing filters can be found in Kendall and Stuart, Volume 3, Chapter 46.

Method 3 (Differencing to Generate Stationary Data). Instead of attempting to remove the noise by smoothing as in Method 2, we now attempt to eliminate the trend term by differencing. We define the first difference operator ∇ by

$$\nabla X_t = X_t - X_{t-1} = (1 - B)X_t, \tag{1.4.10}$$

where B is the backward shift operator,

$$BX_t = X_{t-1}. \tag{1.4.11}$$

Powers of the operators B and ∇ are defined in the obvious way, i.e. $B^j(X_t) = X_{t-j}$ and $\nabla^j(X_t) = \nabla(\nabla^{j-1}(X_t)), j \geq 1$ with $\nabla^0(X_t) = X_t$. Polynomials in B and ∇ are manipulated in precisely the same way as polynomial functions of real variables. For example

$$\begin{aligned}\nabla^2 X_t = \nabla(\nabla X_t) &= (1 - B)(1 - B)X_t = (1 - 2B + B^2)X_t \\ &= X_t - 2X_{t-1} + X_{t-2}.\end{aligned}$$

If the operator ∇ is applied to a linear trend function $m_t = at + b$, then we obtain the constant function $\nabla m_t = a$. In the same way any polynomial trend of degree k can be reduced to a constant by application of the operator ∇^k (Problem 1.4).

Starting therefore with the model $X_t = m_t + Y_t$ where $m_t = \sum_{j=0}^{k} a_j t^j$ and Y_t is stationary with mean zero, we obtain

$$\nabla^k X_t = k!\, a_k + \nabla^k Y_t,$$

a stationary process with mean $k!a_k$. These considerations suggest the possibility, given any sequence $\{x_t\}$ of data, of applying the operator ∇ repeatedly until we find a sequence $\{\nabla^k x_t\}$ which can plausibly be modelled as a realization of a stationary process. It is often found in practice that the

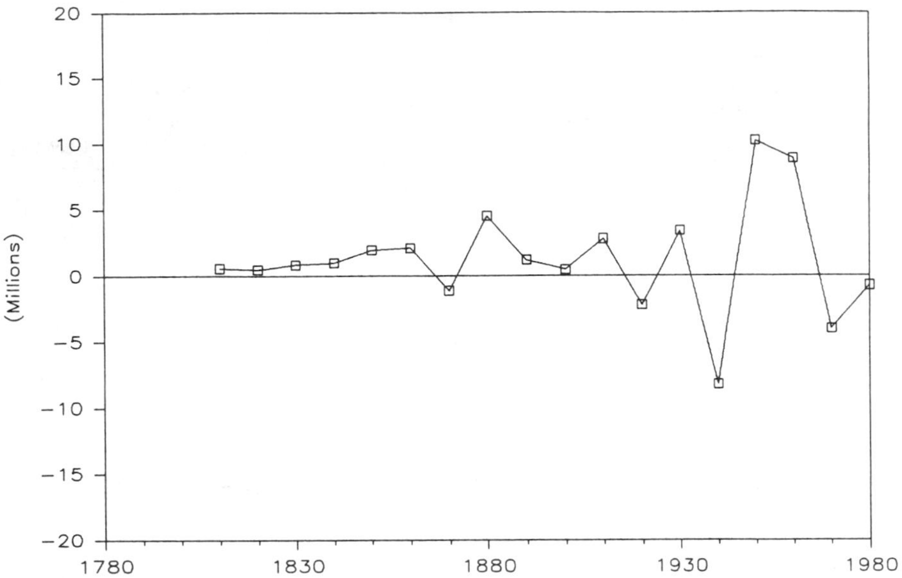

Figure 1.11. The twice-differenced series derived from the population data of Figure 1.2.

order k of differencing required is quite small, frequently one or two. (This depends on the fact that many functions can be well approximated, on an interval of finite length, by a polynomial of reasonably low degree.)

Applying this technique to the twenty population values $\{x_n, n = 1, \ldots, 20\}$ of Figure 1.2 we find that two differencing operations are sufficient to produce a series with no apparent trend. The differenced data, $\nabla^2 x_n = x_n - 2x_{n-1} + x_{n-2}$, are plotted in Figure 1.11. Notice that the magnitude of the fluctuations in $\nabla^2 x_n$ increase with the value of x_n. This effect can be suppressed by first taking natural logarithms, $y_n = \ln x_n$, and then applying the operator ∇^2 to the series $\{y_n\}$. (See also Section 9.2(a).)

Elimination of both Trend and Seasonality

The methods described for the removal of trend can be adapted in a natural way to eliminate both trend and seasonality in the general model

$$X_t = m_t + s_t + Y_t, \tag{1.4.12}$$

where $EY_t = 0$, $s_{t+d} = s_t$ and $\sum_{j=1}^{d} s_j = 0$. We illustrate these methods, with reference to the accident data of Example 1.1.6 (Figure 1.6) for which the period d of the seasonal component is clearly 12.

It will be convenient in Method 1 to index the data by year and month. Thus $x_{j,k}, j = 1, \ldots, 6, k = 1, \ldots, 12$ will denote the number of accidental deaths reported for the kth month of the jth year, $(1972 + j)$. In other words we define

$$x_{j,k} = x_{k+12(j-1)}, \qquad j = 1, \ldots, 6, \quad k = 1, \ldots, 12.$$

Method S1 (The Small Trend Method). If the trend is small (as in the accident data) it is not unreasonable to suppose that the trend term is constant, say m_j, for the jth year. Since $\sum_{k=1}^{12} s_k = 0$, we are led to the natural unbiased estimate

$$\hat{m}_j = \frac{1}{12} \sum_{k=1}^{12} x_{j,k}, \tag{1.4.13}$$

while for s_k, $k = 1, \ldots, 12$ we have the estimates,

$$\hat{s}_k = \frac{1}{6} \sum_{j=1}^{6} (x_{j,k} - \hat{m}_j), \tag{1.4.14}$$

which automatically satisfy the requirement that $\sum_{k=1}^{12} \hat{s}_k = 0$. The estimated error term for month k of the jth year is of course

$$\hat{Y}_{j,k} = x_{j,k} - \hat{m}_j - \hat{s}_k, \qquad j = 1, \ldots, 6, \quad k = 1, \ldots, 12. \tag{1.4.15}$$

The generalization of (1.4.13)–(1.4.15) to data with seasonality having a period other than 12 should be apparent.

In Figures 1.12, 1.13 and 1.14 we have plotted respectively the detrended observations $x_{j,k} - \hat{m}_j$, the estimated seasonal components $\hat{s}_k$, and the de-

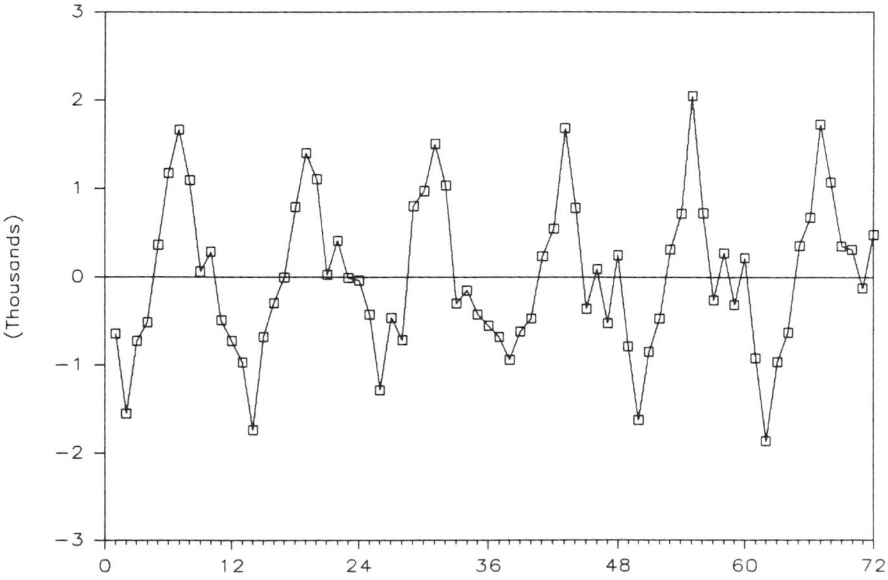

Figure 1.12. Monthly accidental deaths from Figure 1.6 after subtracting the trend estimated by Method S1.

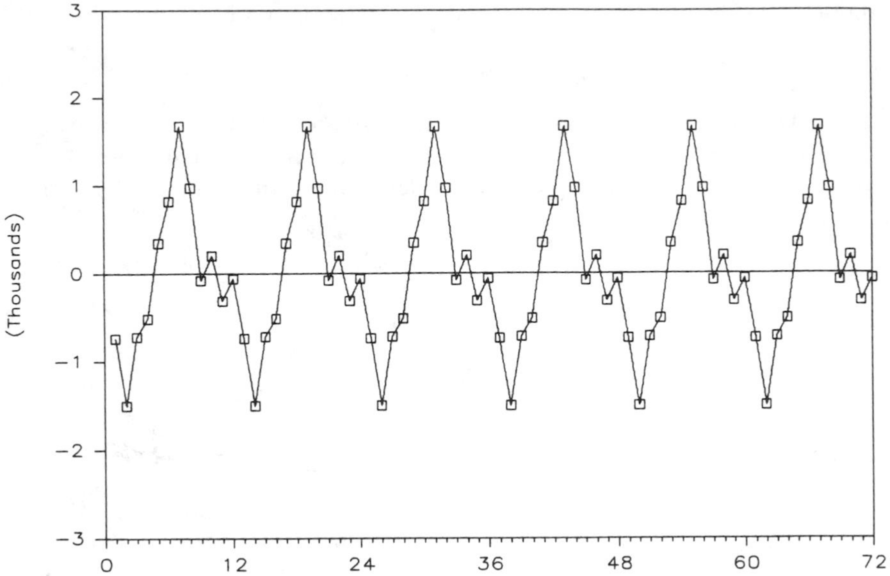

Figure 1.13. The seasonal component of the monthly accidental deaths, estimated by Method S1.

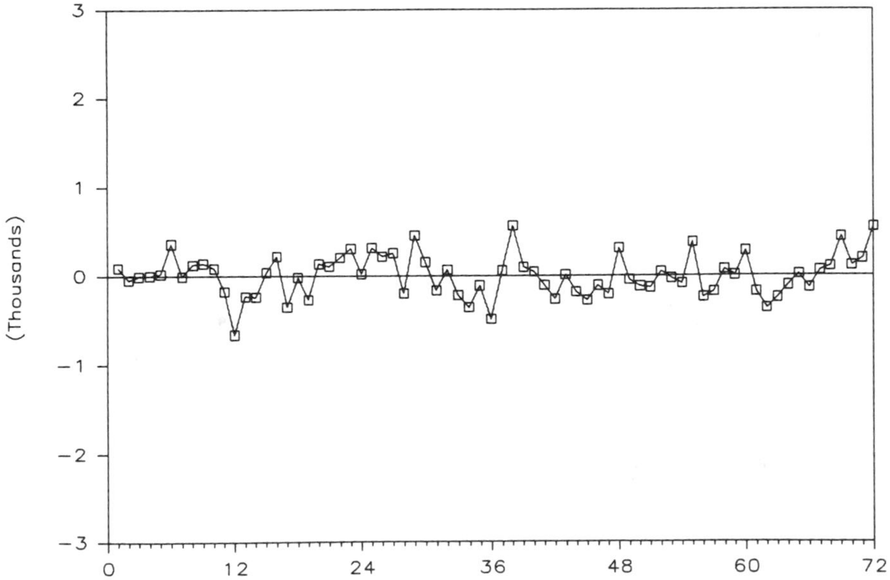

Figure 1.14. The detrended and deseasonalized monthly accidental deaths (Method S1).

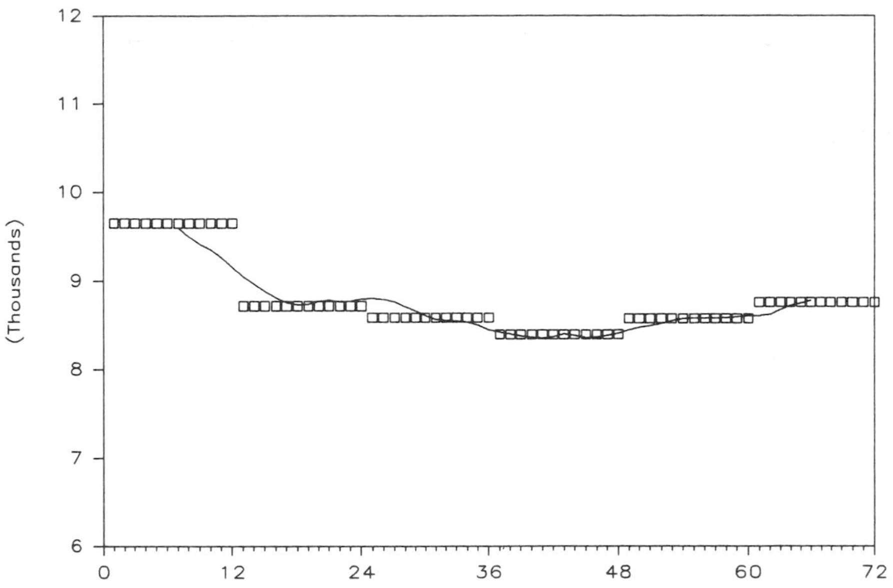

Figure 1.15. Comparison of the moving average and piecewise constant estimates of trend for the monthly accidental deaths.

trended, deseasonalized observations $\hat{Y}_{j,k} = x_{j,k} - \hat{m}_j - \hat{s}_k$. The latter have no apparent trend or seasonality.

Method S2 (Moving Average Estimation). The following technique is preferable to Method S1 since it does not rely on the assumption that m_t is nearly constant over each cycle. It is the basis for the "classical decomposition" option in the time series identification section of the program PEST.

Suppose we have observations $\{x_1, \ldots, x_n\}$. The trend is first estimated by applying a moving average filter specially chosen to eliminate the seasonal component and to dampen the noise. If the period d is even, say $d = 2q$, then we use

$$\hat{m}_t = (0.5x_{t-q} + x_{t-q+1} + \cdots + x_{t+q-1} + 0.5x_{t+q})/d, \qquad q < t \le n - q. \tag{1.4.16}$$

If the period is odd, say $d = 2q + 1$, then we use the simple moving average (1.4.6).

In Figure 1.15 we show the trend estimate $\hat{m}_t$, $6 < t \le 66$, for the accidental deaths data obtained from (1.4.16). Also shown is the piecewise constant estimate obtained from Method S1.

The second step is to estimate the seasonal component. For each $k = 1, \ldots, d$ we compute the average w_k of the deviations $\{(x_{k+jd} - \hat{m}_{k+jd}) : q < k + jd \le n - q\}$. Since these average deviations do not necessarily sum to zero, we

Table 1.1. Estimated Seasonal Components for the Accidental Deaths Data

k	1	2	3	4	5	6	7	8	9	10	11	12
$\hat{s}_k$ (Method S1)	−744	−1504	−724	−523	338	808	1665	961	−87	197	−321	−67
$\hat{s}_k$ (Method S2)	−804	−1522	−737	−526	343	746	1680	987	−109	258	−259	−57

estimate the seasonal component s_k as

$$\hat{s}_k = w_k - d^{-1} \sum_{i=1}^{d} w_i, \qquad k = 1, \ldots, d, \tag{1.4.17}$$

and $\hat{s}_k = \hat{s}_{k-d}$, $k > d$.

The deseasonalized data is then defined to be the original series with the estimated seasonal component removed, i.e.

$$d_t = x_t - \hat{s}_t, \qquad t = 1, \ldots, n. \tag{1.4.18}$$

Finally we reestimate the trend from $\{d_t\}$ either by applying a moving average filter as described earlier for non-seasonal data, or by fitting a polynomial to the series $\{d_t\}$. The program PEST allows the options of fitting a linear or quadratic trend $\hat{m}_t$. The estimated noise terms are then

$$\hat{Y}_t = x_t - \hat{m}_t - \hat{s}_t, \qquad t = 1, \ldots, n.$$

The results of applying Methods S1 and S2 to the accidental deaths data are quite similar, since in this case the piecewise constant and moving average estimates of m_t are reasonably close (see Figure 1.15).

A comparison of the estimates of s_k, $k = 1, \ldots, 12$, obtained by Methods S1 and S2 is made in Table 1.1.

Method S3 (Differencing at Lag d). The technique of differencing which we applied earlier to non-seasonal data can be adapted to deal with seasonality of period d by introducing the lag-d difference operator ∇_d defined by

$$\nabla_d X_t = X_t - X_{t-d} = (1 - B^d) X_t. \tag{1.4.19}$$

(This operator should not be confused with the operator $\nabla^d = (1 - B)^d$ defined earlier.)

Applying the operator ∇_d to the model,

$$X_t = m_t + s_t + Y_t,$$

where $\{s_t\}$ has period d, we obtain

$$\nabla_d X_t = m_t - m_{t-d} + Y_t - Y_{t-d},$$

which gives a decomposition of the difference $\nabla_d X_t$ into a trend component $(m_t - m_{t-d})$ and a noise term $(Y_t - Y_{t-d})$. The trend, $m_t - m_{t-d}$, can then be eliminated using the methods already described, for example by application of some power of the operator ∇.

Figure 1.16 shows the result of applying the operator ∇_{12} to the accidental

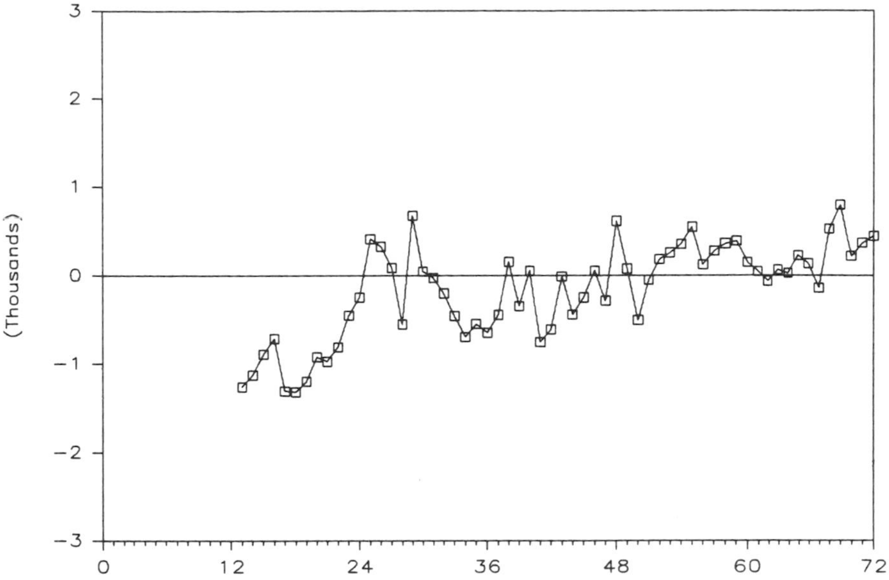

Figure 1.16. The differenced series $\{\nabla_{12}x_t, t = 13,\ldots,72\}$ derived from the monthly accidental deaths $\{x_t, t = 1,\ldots,72\}$.

deaths data. The seasonal component evident in Figure 1.6 is absent from the graph of $\nabla_{12}x_t$, $13 \leq t \leq 72$. There still appears to be a non-decreasing trend however. If we now apply the operator ∇ to $\nabla_{12}x_t$ and plot the resulting differences $\nabla\nabla_{12}x_t$, $t = 14, \ldots, 72$, we obtain the graph shown in Figure 1.17, which has no apparent trend or seasonal component. In Chapter 9 we shall show that the differenced series can in fact be well represented by a stationary time series model.

In this section we have discussed a variety of methods for estimating and/or removing trend and seasonality. The particular method chosen for any given data set will depend on a number of factors including whether or not estimates of the components of the series are required and whether or not it appears that the data contains a seasonal component which does not vary with time. The program PEST allows two options, one which decomposes the series as described in Method S2, and the other which proceeds by successive differencing of the data as in Methods 3 and S3.

§1.5 The Autocovariance Function of a Stationary Process

In this section we study the properties of the autocovariance function introduced in Section 1.3.

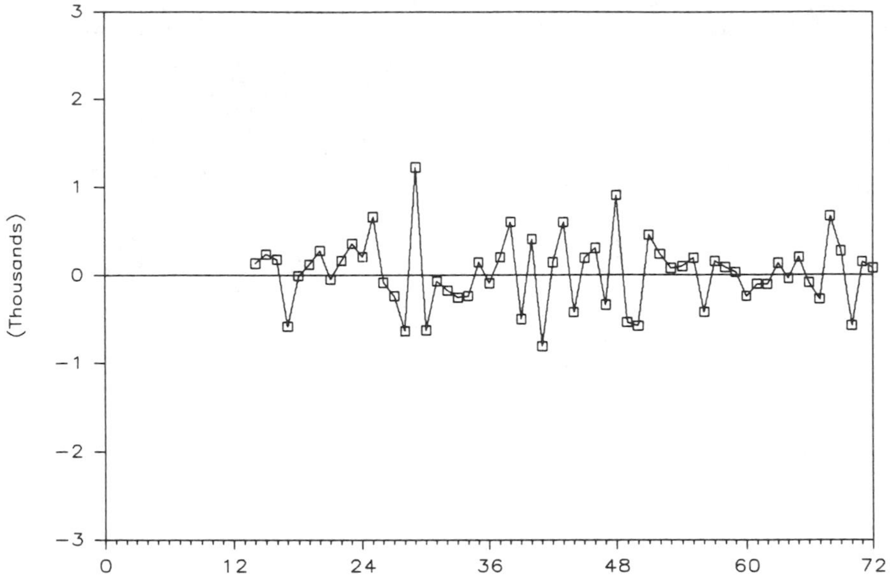

Figure 1.17. The differenced series $\{\nabla\nabla_{12}x_t, t = 14, \ldots, 72\}$ derived from the monthly accidental deaths $\{x_t, t = 1, \ldots, 72\}$.

Proposition 1.5.1 (Elementary Properties). *If $\gamma(\cdot)$ is the autocovariance function of a stationary process $\{X_t, t \in \mathbb{Z}\}$, then*

$$\gamma(0) \geq 0, \tag{1.5.1}$$

$$|\gamma(h)| \leq \gamma(0) \quad \text{for all } h \in \mathbb{Z}, \tag{1.5.2}$$

and $\gamma(\cdot)$ is even, i.e.

$$\gamma(h) = \gamma(-h) \quad \text{for all } h \in \mathbb{Z}. \tag{1.5.3}$$

PROOF. The first property is a statement of the obvious fact that $\text{Var}(X_t) \geq 0$, the second is an immediate consequence of the Cauchy–Schwarz inequality,

$$|\text{Cov}(X_{t+h}, X_t)| \leq (\text{Var}(X_{t+h}))^{1/2}(\text{Var}(X_t))^{1/2}$$

and the third is established by observing that

$$\gamma(-h) = \text{Cov}(X_{t-h}, X_t) = \text{Cov}(X_t, X_{t+h}) = \gamma(h). \qquad \square$$

Autocovariance functions also have the more subtle property of non-negative definiteness.

Definition 1.5.1 (Non-Negative Definiteness). A real-valued function on the integers, $\kappa : \mathbb{Z} \to \mathbb{R}$, is said to be non-negative definite if and only if

$$\sum_{i,j=1}^{n} a_i \kappa(t_i - t_j) a_j \geq 0 \tag{1.5.4}$$

for all positive integers n and for all vectors $\mathbf{a} = (a_1, \ldots, a_n)' \in \mathbb{R}^n$ and $\mathbf{t} = (t_1, \ldots, t_n)' \in \mathbb{Z}^n$ or if and only if $\sum_{i,j=1}^{n} a_i \kappa(i - j) a_j \geq 0$ for all such n and $\mathbf{a}$.

Theorem 1.5.1 (Characterization of Autocovariance Functions). *A real-valued function defined on the integers is the autocovariance function of a stationary time series if and only if it is even and non-negative definite.*

PROOF. To show that the autocovariance function $\gamma(\cdot)$ of any stationary time series $\{X_t\}$ is non-negative definite, we simply observe that if $\mathbf{a} = (a_1, \ldots, a_n)' \in \mathbb{R}^n$, $\mathbf{t} = (t_1, \ldots, t_n)' \in \mathbb{Z}^n$, and $\mathbf{Z_t} = (X_{t_1} - EX_{t_1}, \ldots, X_{t_n} - EX_{t_n})'$, then

$$\begin{aligned} 0 &\leq \operatorname{Var}(\mathbf{a}'\mathbf{Z_t}) \\ &= \mathbf{a}'E\mathbf{Z_t}\mathbf{Z_t}'\mathbf{a} \\ &= \mathbf{a}'\Gamma_n\mathbf{a} \\ &= \sum_{i,j=1}^{n} a_i \gamma(t_i - t_j) a_j, \end{aligned}$$

where $\Gamma_n = [\gamma(t_i - t_j)]_{i,j=1}^{n}$ is the covariance matrix of $(X_{t_1}, \ldots, X_{t_n})'$.

To establish the converse, let $\kappa: \mathbb{Z} \to \mathbb{R}$ be an even non-negative definite function. We need to show that there exists a stationary process with $\kappa(\cdot)$ as its autocovariance function, and for this we shall use Kolmogorov's theorem. For each positive integer n and each $\mathbf{t} = (t_1, \ldots, t_n)' \in \mathbb{Z}^n$ such that $t_1 < t_2 < \cdots < t_n$, let $F_\mathbf{t}$ be the distribution function on $\mathbb{R}^n$ with characteristic function

$$\phi_\mathbf{t}(\mathbf{u}) = \exp(-\mathbf{u}'K\mathbf{u}/2),$$

where $\mathbf{u} = (u_1, \ldots, u_n)' \in \mathbb{R}^n$ and $K = [\kappa(t_i - t_j)]_{i,j=1}^{n}$. Since κ is non-negative definite, the matrix K is also non-negative definite and consequently $\phi_\mathbf{t}$ is the characteristic function of an n-variate normal distribution with mean zero and covariance matrix K (see Section 1.6). Clearly, in the notation of Theorem 1.2.1,

$$\phi_{\mathbf{t}(i)}(\mathbf{u}(i)) = \lim_{u_i \to 0} \phi_\mathbf{t}(\mathbf{u}) \quad \text{for each } \mathbf{t} \in \mathcal{T},$$

i.e. the distribution functions $F_\mathbf{t}$ are consistent, and so by Kolmogorov's theorem there exists a time series $\{X_t\}$ with distribution functions $F_\mathbf{t}$ and characteristic functions $\phi_\mathbf{t}$, $t \in \mathcal{T}$. In particular the joint distribution of X_i and X_j is bivariate normal with mean $\mathbf{0}$ and covariance matrix

$$\begin{bmatrix} \kappa(0) & \kappa(i-j) \\ \kappa(i-j) & \kappa(0) \end{bmatrix},$$

which shows that $\operatorname{Cov}(X_i, X_j) = \kappa(i - j)$ as required. $\square$

Remark 1. As shown in the proof of Theorem 1.5.1, for every autocovariance function $\gamma(\cdot)$, there exists a stationary Gaussian time series with $\gamma(\cdot)$ as its autocovariance function.

Remark 2. To verify that a given function is non-negative definite it is sometimes simpler to specify a stationary process with the given autocovariance function than to check Definition 1.5.1. For example the function $\kappa(h) = \cos(\theta h)$, $h \in \mathbb{Z}$, is the autocovariance function of the process in Example 1.3.1 and is therefore non-negative definite. Direct verification by means of Definition 1.5.1 however is more difficult. Another simple criterion for checking non-negative definiteness is Herglotz's theorem, which will be proved in Section 4.3.

Remark 3. An autocorrelation function $\rho(\cdot)$ has all the properties of an autocovariance function and satisfies the additional condition $\rho(0) = 1$.

EXAMPLE 1.5.1. Let us show that the real-valued function on $\mathbb{Z}$,

$$\kappa(h) = \begin{cases} 1 & \text{if } h = 0, \\ \rho & \text{if } h = \pm 1, \\ 0 & \text{otherwise,} \end{cases}$$

is an autocovariance function if and only if $|\rho| \le \frac{1}{2}$.

If $|\rho| \le \frac{1}{2}$ then $\kappa(\cdot)$ is the autocovariance function of the process defined in Example 1.3.2 with $\sigma^2 = (1 + \theta^2)^{-1}$ and $\theta = (2\rho)^{-1}(1 \pm \sqrt{1 - 4\rho^2})$.

If $\rho > \frac{1}{2}$, $K = [\kappa(i - j)]_{i,j=1}^n$ and $\mathbf{a}$ is the n-component vector $\mathbf{a} = (1, -1, 1, -1, \ldots)'$, then

$$\mathbf{a}'K\mathbf{a} = n - 2(n - 1)\rho < 0 \quad \text{for } n > 2\rho/(2\rho - 1),$$

which shows that $\kappa(\cdot)$ is not non-negative definite and therefore, by Theorem 1.5.1 is not an autocovariance function.

If $\rho < -\frac{1}{2}$, the same argument using the n-component vector $\mathbf{a} = (1, 1, 1, \ldots)'$ again shows that $\kappa(\cdot)$ is not non-negative definite.

The Sample Autocovariance Function of an Observed Series

From the observations $\{x_1, x_2, \ldots, x_n\}$ of a stationary time series $\{X_t\}$ we frequently wish to estimate the autocovariance function $\gamma(\cdot)$ of the underlying process $\{X_t\}$ in order to gain information concerning its dependence structure. This is an important step towards constructing an appropriate mathematical model for the data. The estimate of $\gamma(\cdot)$ which we shall use is the sample autocovariance function.

Definition 1.5.2. The sample autocovariance function of $\{x_1, \ldots, x_n\}$ is defined by

$$\hat{\gamma}(h) := n^{-1} \sum_{j=1}^{n-h} (x_{j+h} - \bar{x})(x_j - \bar{x}), \qquad 0 \le h < n,$$

and $\hat{\gamma}(h) = \hat{\gamma}(-h)$, $-n < h \le 0$, where $\bar{x}$ is the sample mean $\bar{x} = n^{-1} \sum_{j=1}^{n} x_j$.

Remark 4. The divisor n is used rather than $(n - h)$ since this ensures that the matrix $\hat{\Gamma}_n := [\hat{\gamma}(i-j)]_{i,j=1}^{n}$ is non-negative definite (see Section 7.2).

Remark 5. The sample autocorrelation function is defined in terms of the sample autocovariance function as

$$\hat{\rho}(h) := \hat{\gamma}(h)/\hat{\gamma}(0), \qquad |h| < n.$$

The corresponding matrix $\hat{R}_n := [\hat{\rho}(i-j)]_{i,j=1}^{n}$ is then also non-negative definite.

Remark 6. The large-sample properties of the estimators $\hat{\gamma}(h)$ and $\hat{\rho}(h)$ are discussed in Chapter 7.

Example 1.5.2. Figure 1.18(a) shows 300 simulated observations of the series $X_t = Z_t + \theta Z_{t-1}$ of Example 1.3.2 with $\theta = 0.95$ and $Z_t \sim N(0, 1)$. Figure 1.18(b) shows the corresponding sample autocorrelation function at lags 0, ..., 40. Notice the similarity between $\hat{\rho}(\cdot)$ and the function $\rho(\cdot)$ computed as described in Example 1.3.2 ($\rho(h) = 1$ for $h = 0$, .4993 for $h = \pm 1$, 0 otherwise).

Example 1.5.3. Figures 1.19(a) and 1.19(b) show simulated observations and the corresponding sample autocorrelation function for the process $X_t = Z_t + \theta Z_{t-1}$, this time with $\theta = -0.95$ and $Z_t \sim N(0, 1)$. The similarity between $\hat{\rho}(\cdot)$ and $\rho(\cdot)$ is again apparent.

Remark 7. Notice that the realization of Example 1.5.2 is less rapidly fluctuating than that of Example 1.5.3. This is to be expected from the two autocorrelation functions. Positive autocorrelation at lag 1 reflects a tendency for successive observations to lie on the same side of the mean, while negative autocorrelation at lag 1 reflects a tendency for successive observations to lie on opposite sides of the mean. Other properties of the sample-paths are also reflected in the autocorrelation (and sample autocorrelation) functions. For example the sample autocorrelation function of the Wölfer sunspot series (Figure 1.20) reflects the roughly periodic behaviour of the data (Figure 1.5).

Remark 8. The sample autocovariance and autocorrelation functions can be computed for *any* data set $\{x_1, \ldots, x_n\}$ and are not restricted to realizations of a stationary process. For data containing a trend, $|\hat{\rho}(h)|$ will exhibit slow decay as h increases, and for data with a substantial deterministic periodic component, $\hat{\rho}(h)$ will exhibit similar behaviour with the same periodicity. Thus $\hat{\rho}(\cdot)$ can be useful as an indicator of non-stationarity (see also Section 9.1).

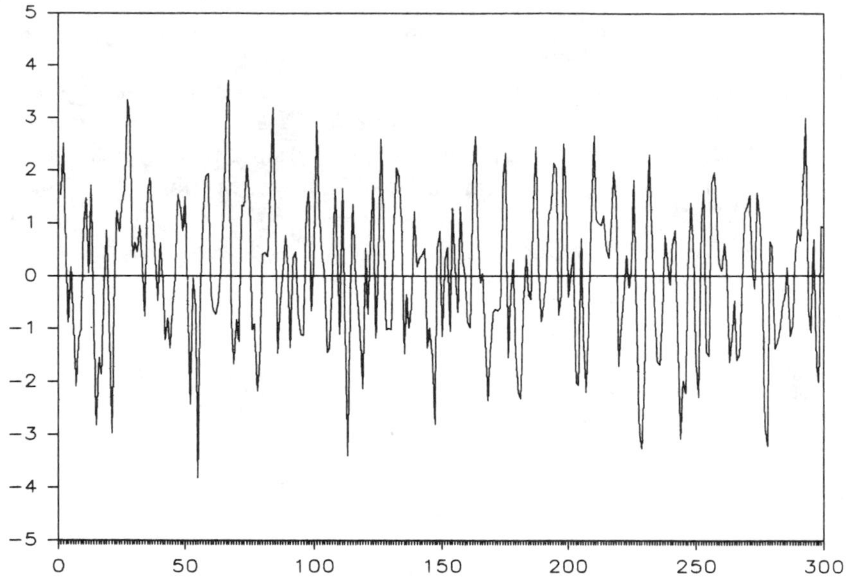

(a)

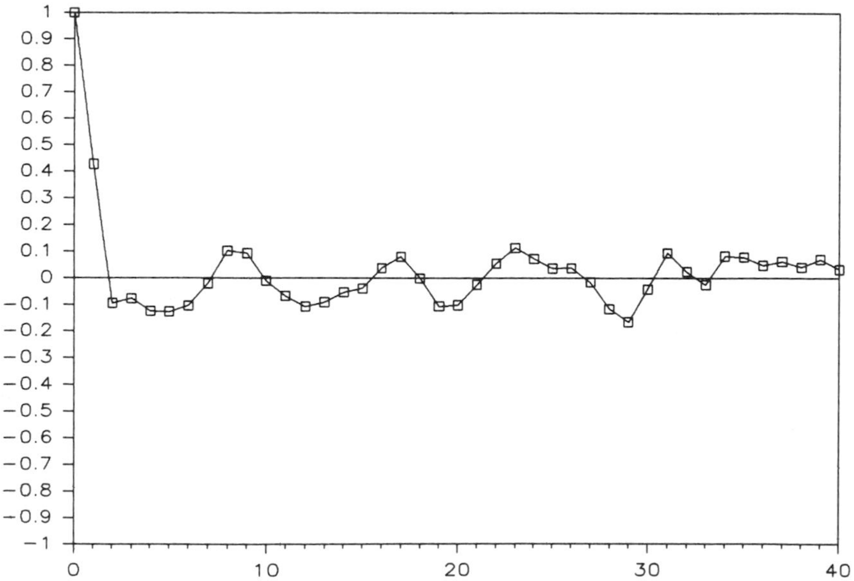

(b)

Figure 1.18. (a) 300 observations of the series $X_t = Z_t + .95Z_{t-1}$, Example 1.5.2. (b) The sample autocorrelation function $\hat{\rho}(h)$, $0 \le h \le 40$.

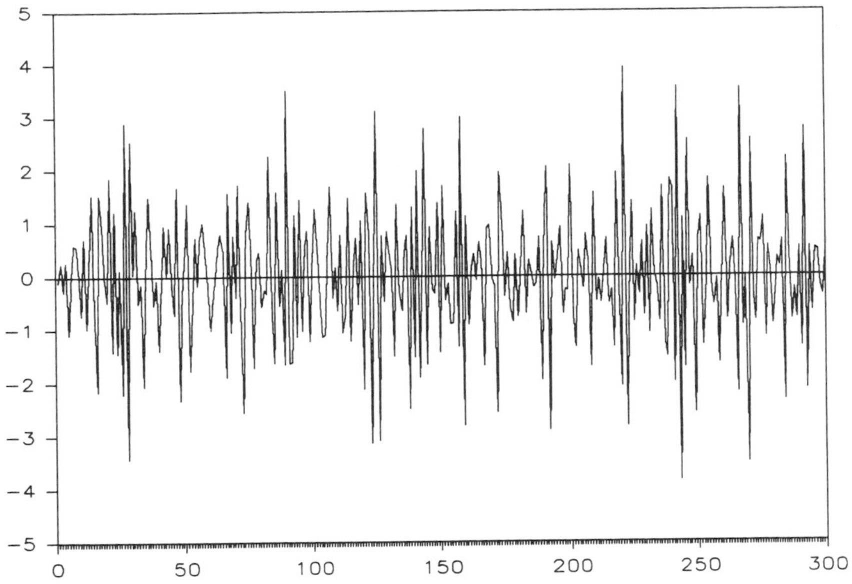

(a)

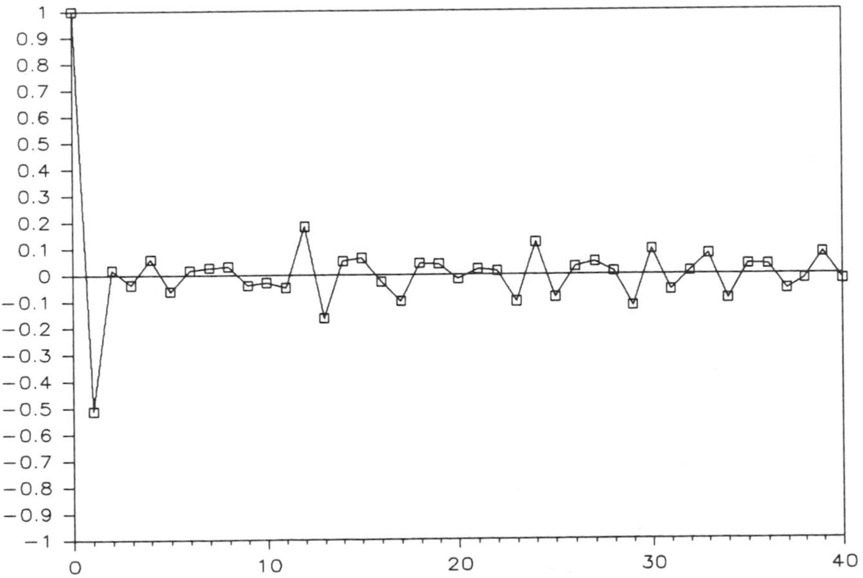

(b)

Figure 1.19. (a) 300 observations of the series $X_t = Z_t - .95Z_{t-1}$, Example 1.5.3. (b) The sample autocorrelation function $\hat{\rho}(h)$, $0 \leq h \leq 40$.

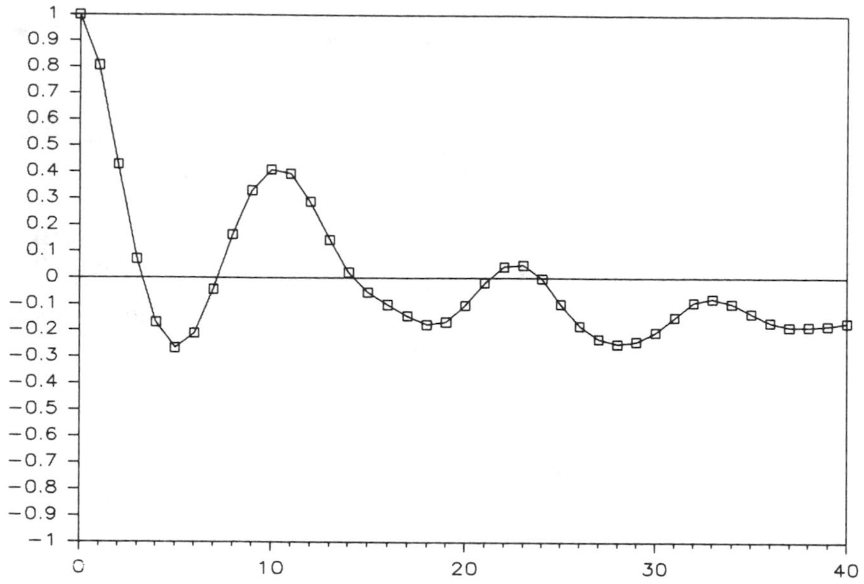

Figure 1.20. The sample autocorrelation function of the Wölfer sunspot numbers (see Figure 1.5).

§1.6 The Multivariate Normal Distribution

An n-dimensional random vector is a column vector, $\mathbf{X} = (X_1, \ldots, X_n)'$, each of whose components is a random variable. If $E|X_i| < \infty$ for each i, then we define the mean or expected value of $\mathbf{X}$ to be the column vector,

$$\boldsymbol{\mu}_{\mathbf{X}} = E\mathbf{X} = (EX_1, \ldots, EX_n)'. \tag{1.6.1}$$

In the same way we define the expected value of any array whose elements are random variables (e.g. a matrix of random variables) to be the same array with each random variable replaced by its expected value (assuming each expectation exists).

If $\mathbf{X} = (X_1, \ldots, X_n)'$ and $\mathbf{Y} = (Y_1, \ldots, Y_m)'$ are random vectors such that $E|X_i|^2 < \infty$, $i = 1, \ldots, n$, and $E|Y_i|^2 < \infty$, $i = 1, \ldots, m$, we define the covariance matrix of $\mathbf{X}$ and $\mathbf{Y}$ to be the matrix,

$$\begin{aligned}\Sigma_{\mathbf{XY}} = \operatorname{Cov}(\mathbf{X}, \mathbf{Y}) &= E[(\mathbf{X} - E\mathbf{X})(\mathbf{Y} - E\mathbf{Y})'] \\ &= E(\mathbf{XY}') - (E\mathbf{X})(E\mathbf{Y})'.\end{aligned} \tag{1.6.2}$$

The (i,j)-element of $\Sigma_{\mathbf{XY}}$ is the covariance, $\operatorname{Cov}(X_i, Y_j) = E(X_i Y_j) - E(X_i)E(Y_j)$. In the special case when $\mathbf{Y} = \mathbf{X}$, $\operatorname{Cov}(\mathbf{X}, \mathbf{Y})$ reduces to the covariance matrix of $\mathbf{X}$.

Proposition 1.6.1. *If* $\mathbf{a}$ *is an m-component column vector, B is an* $m \times n$ *matrix and* $\mathbf{X} = (X_1, \ldots, X_n)'$ *where* $E|X_i|^2 < \infty$, $i = 1, \ldots, n$, *then the random vector,*

$$\mathbf{Y} = \mathbf{a} + B\mathbf{X}, \tag{1.6.3}$$

has mean

$$E\mathbf{Y} = \mathbf{a} + BE\mathbf{X}, \tag{1.6.4}$$

and covariance matrix,

$$\Sigma_{\mathbf{YY}} = B\Sigma_{\mathbf{XX}}B'. \tag{1.6.5}$$

PROOF. Problem 1.15.

Proposition 1.6.2. *The covariance matrix* $\Sigma_{\mathbf{XX}}$ *is symmetric and non-negative definite, i.e.* $\mathbf{b}'\Sigma_{\mathbf{XX}}\mathbf{b} \geq 0$ *for all* $\mathbf{b} = (b_1, \ldots, b_n)' \in \mathbb{R}^n$.

PROOF. The symmetry of $\Sigma_{\mathbf{XX}}$ is apparent from the definition. To prove non-negative definiteness let $\mathbf{b} = (b_1, \ldots, b_n)'$ be an arbitrary vector in $\mathbb{R}^n$. Then by Proposition 1.6.1

$$\mathbf{b}'\Sigma_{\mathbf{XX}}\mathbf{b} = \text{Var}(\mathbf{b}'\mathbf{X}) \geq 0. \tag{1.6.6}$$

□

Proposition 1.6.3. *Any symmetric, non-negative definite* $n \times n$ *matrix* Σ *can be written in the form*

$$\Sigma = P\Lambda P', \tag{1.6.7}$$

where P is an orthogonal matrix (i.e. $P' = P^{-1}$*) and* Λ *is a diagonal matrix* $\Lambda = \text{diag}(\lambda_1, \ldots, \lambda_n)$ *in which* $\lambda_1, \ldots, \lambda_n$ *are the eigenvalues (all non-negative) of* Σ.

PROOF. This proposition is a standard result from matrix theory and for a proof we refer the reader to Graybill (1983). We observe here only that if $\mathbf{p}_i$, $i = 1, \ldots, n$, is a set of orthonormal right eigenvectors of Σ corresponding to the eigenvalues $\lambda_1, \ldots, \lambda_n$ respectively, then P may be chosen as the $n \times n$ matrix whose i^{th} column is $\mathbf{p}_i$, $i = 1, \ldots, n$. □

Remark 1. Using the factorization (1.6.7) and the fact that $\det P = \det P' = 1$, we immediately obtain the result,

$$\det \Sigma = \lambda_1 \lambda_2 \ldots \lambda_n.$$

Definition 1.6.1 (The Multivariate Normal Distribution). The random vector $\mathbf{Y} = (Y_1, \ldots, Y_n)'$ is said to be multivariate normal, or to have a multivariate normal distribution, if and only if there exist a column vector $\mathbf{a}$, a matrix B and a random vector $\mathbf{X} = (X_1, \ldots, X_m)'$ with independent standard normal components, such that

$$\mathbf{Y} = \mathbf{a} + B\mathbf{X}. \tag{1.6.8}$$

Remark 2. The components $X_1, \ldots, X_m$ of $\mathbf{X}$ in (1.6.8) must have the joint density

$$f_{\mathbf{X}}(\mathbf{x}) = (2\pi)^{-m/2} \exp\left(-\sum_{j=1}^{m} x_j^2/2 \right), \qquad \mathbf{x} = (x_1, \ldots, x_m)' \in \mathbb{R}^m, \tag{1.6.9}$$

and corresponding characteristic function,

$$\phi_{\mathbf{X}}(\mathbf{u}) \equiv Ee^{i\mathbf{u}'\mathbf{X}} = \exp\left(-\sum_{j=1}^{m} u_j^2/2 \right), \qquad \mathbf{u} = (u_1, \ldots, u_m)' \in \mathbb{R}^m. \tag{1.6.10}$$

Remark 3. It is clear from the definition that if $\mathbf{Y}$ has a multivariate normal distribution and if D is any $k \times n$ matrix and $\mathbf{c}$ any $k \times 1$ vector, then $\mathbf{Z} = \mathbf{c} + D\mathbf{Y}$ is a k-component multivariate normal random vector.

Remark 4. If $\mathbf{Y}$ is multivariate normal with representation (1.6.8), then by Proposition 1.6.1, $E\mathbf{Y} = \mathbf{a}$ and $\Sigma_{\mathbf{YY}} = BB'$.

Proposition 1.6.4. *If $\mathbf{Y} = (Y_1, \ldots, Y_n)'$ is a multivariate normal random vector such that $E\mathbf{Y} = \boldsymbol{\mu}$ and $\Sigma_{\mathbf{YY}} = \Sigma$, then the characteristic function of $\mathbf{Y}$ is*

$$\phi_{\mathbf{Y}}(\mathbf{u}) = \exp(i\mathbf{u}'\boldsymbol{\mu} - \tfrac{1}{2}\mathbf{u}'\Sigma\mathbf{u}), \qquad \mathbf{u} = (u_1, \ldots, u_n)' \in \mathbb{R}^n. \tag{1.6.11}$$

If $\det \Sigma > 0$ *then* $\mathbf{Y}$ *has the density,*

$$f_{\mathbf{Y}}(\mathbf{y}) = (2\pi)^{-n/2} (\det \Sigma)^{-1/2} \exp[-\tfrac{1}{2}(\mathbf{y} - \boldsymbol{\mu})'\Sigma^{-1}(\mathbf{y} - \boldsymbol{\mu})]. \tag{1.6.12}$$

PROOF. If $\mathbf{Y}$ is multivariate normal with representation (1.6.8) then

$$\phi_{\mathbf{Y}}(\mathbf{u}) = E\exp[i\mathbf{u}'(\mathbf{a} + B\mathbf{X})] = \exp(i\mathbf{u}'\mathbf{a})E\exp(i\mathbf{u}'B\mathbf{X}).$$

Using (1.6.10) with $\mathbf{u}$ ($\in \mathbb{R}^m$) replaced by $B'\mathbf{u}$ ($\mathbf{u} \in \mathbb{R}^n$) in order to evaluate the last term, we obtain

$$\phi_{\mathbf{Y}}(\mathbf{u}) = \exp(i\mathbf{u}'\mathbf{a})\exp(-\tfrac{1}{2}\mathbf{u}'BB'\mathbf{u}),$$

which reduces to (1.6.11) by Remark 4.

If $\det \Sigma > 0$, then by Proposition 1.6.3 we have the factorization,

$$\Sigma = P\Lambda P',$$

where $PP' = I_n$, the $n \times n$ identity matrix, $\Lambda = \operatorname{diag}(\lambda_1, \ldots, \lambda_n)$ and each $\lambda_i > 0$. If we define $\Lambda^{-1/2} = \operatorname{diag}(\lambda_1^{-1/2}, \ldots, \lambda_n^{-1/2})$ and

$$\Sigma^{-1/2} = P\Lambda^{-1/2}P',$$

then it is easy to check that $\Sigma^{-1/2}\Sigma\Sigma^{-1/2} = I_n$. From Proposition 1.6.1 and Remark 3 we conclude that the random vector

$$\mathbf{Z} = \Sigma^{-1/2}(\mathbf{Y} - \boldsymbol{\mu}) \tag{1.6.13}$$

is multivariate normal with $E\mathbf{Z} = \mathbf{0}$ and $\Sigma_{\mathbf{ZZ}} = I_n$. Application of the result (1.6.11) now shows that $\mathbf{Z}$ has the characteristic function $\phi_{\mathbf{Z}}(\mathbf{u}) = \exp(-\mathbf{u}'\mathbf{u}/2)$, whence it follows that $\mathbf{Z}$ has the probability density (1.6.9) with $m = n$. In view of the relation (1.6.13), the density of $\mathbf{Y}$ is given by

$$\begin{aligned} f_{\mathbf{Y}}(\mathbf{y}) &= |\det \Sigma^{-1/2}| f_{\mathbf{Z}}(\Sigma^{-1/2}(\mathbf{y} - \boldsymbol{\mu})) \\ &= (\det \Sigma)^{-1/2}(2\pi)^{-n/2} \exp[-\tfrac{1}{2}(\mathbf{y} - \boldsymbol{\mu})'\Sigma^{-1}(\mathbf{y} - \boldsymbol{\mu})] \end{aligned}$$

as required. □

Remark 5. The transformation (1.6.13) which maps $\mathbf{Y}$ into a vector of independent standard normal random variables is clearly a generalization of the transformation $Z = \sigma^{-1}(Y - \mu)$ which standardizes a single normal random variable with mean μ and variance σ^2.

Remark 6. Given any vector $\boldsymbol{\mu} \in \mathbb{R}^n$ and any symmetric non-negative definite $n \times n$ matrix Σ, there exists a multivariate normal random vector with mean $\boldsymbol{\mu}$ and covariance matrix Σ. To construct such a random vector from a vector $\mathbf{X} = (X_1, \ldots, X_n)'$ with independent standard normal components we simply choose $\mathbf{a} = \boldsymbol{\mu}$ and $B = \Sigma^{1/2}$ in (1.6.8), where $\Sigma^{1/2}$, in the terminology of Proposition 1.6.3, is the matrix $P\Lambda^{1/2}P'$ with $\Lambda^{1/2} = \operatorname{diag}(\lambda_1^{1/2}, \ldots, \lambda_n^{1/2})$.

Remark 7. Proposition 1.6.4 shows that a multivariate normal distribution is uniquely determined by its mean and covariance matrix. If $\mathbf{Y}$ is multivariate normal, $E\mathbf{Y} = \boldsymbol{\mu}$ and $\Sigma_{\mathbf{YY}} = \Sigma$, we shall therefore say that $\mathbf{Y}$ has the multivariate normal distribution with mean $\boldsymbol{\mu}$ and covariance matrix Σ, or more succinctly,

$$\mathbf{Y} \sim N(\boldsymbol{\mu}, \Sigma).$$

EXAMPLE 1.6.1 (The Bivariate Normal Distribution). The random vector $\mathbf{Y} = (Y_1, Y_2)'$ is bivariate normal with mean $\boldsymbol{\mu} = (\mu_1, \mu_2)'$ and covariance matrix

$$\Sigma = \begin{bmatrix} \sigma_1^2 & \rho\sigma_1\sigma_2 \\ \rho\sigma_1\sigma_2 & \sigma_2^2 \end{bmatrix}, \qquad \sigma_1, \sigma_2 \geq 0, \quad -1 \leq \rho \leq 1, \tag{1.6.14}$$

if and only if $\mathbf{Y}$ has the characteristic function (from (1.6.11))

$$\phi_{\mathbf{Y}}(\mathbf{u}) = \exp[i(u_1\mu_1 + u_2\mu_2) - \tfrac{1}{2}(u_1^2\sigma_1^2 + 2u_1u_2\rho\sigma_1\sigma_2 + u_2^2\sigma_2^2)]. \tag{1.6.15}$$

The parameters σ_1, σ_2 and ρ are the standard deviations and correlation of the components Y_1 and Y_2. Since every symmetric non-negative definite 2×2 matrix can be written in the form (1.6.14), it follows that every bivariate normal random vector has a characteristic function of the form (1.6.15). If $\sigma_1 \neq 0$, $\sigma_2 \neq 0$ and $-1 < \rho < 1$ then Σ has an inverse,

$$\Sigma^{-1} = (1-\rho^2)^{-1}\begin{bmatrix} \sigma_1^{-2} & -\rho\sigma_1^{-1}\sigma_2^{-1} \\ -\rho\sigma_1^{-1}\sigma_2^{-1} & \sigma_2^{-2} \end{bmatrix}, \tag{1.6.16}$$

and so by (1.6.12), $\mathbf{Y}$ has the probability density,

$$f_{\mathbf{Y}}(\mathbf{y}) = \frac{(1-\rho^2)^{-1/2}}{2\pi\sigma_1\sigma_2}\exp\left\{-\frac{1}{2(1-\rho^2)}\left[\left(\frac{y_1-\mu_1}{\sigma_1}\right)^2 - 2\rho\left(\frac{y_1-\mu_1}{\sigma_1}\right)\left(\frac{y_2-\mu_2}{\sigma_2}\right) + \left(\frac{y_2-\mu_2}{\sigma_2}\right)^2\right]\right\}. \tag{1.6.17}$$

Proposition 1.6.5. *The random vector* $\mathbf{Y} = (Y_1, \ldots, Y_n)'$ *is multivariate normal with mean* $\boldsymbol{\mu}$ *and covariance matrix* Σ *if and only if for each* $\mathbf{a} = (a_1, \ldots, a_n)' \in \mathbb{R}^n$, $\mathbf{a}'\mathbf{Y}$ *has a univariate normal distribution with mean* $\mathbf{a}'\boldsymbol{\mu}$ *and variance* $\mathbf{a}'\Sigma\mathbf{a}$.

PROOF. The necessity of the condition has already been established. To prove the sufficiency we shall show that $\mathbf{Y}$ has the appropriate characteristic function. For any $\mathbf{a} \in \mathbb{R}^n$ we are assuming that $\mathbf{a}'Y \sim N(\mathbf{a}'\boldsymbol{\mu}, \mathbf{a}'\Sigma\mathbf{a})$, or equivalently that

$$E\exp(it\mathbf{a}'\mathbf{Y}) = \exp(it\mathbf{a}'\boldsymbol{\mu} - \tfrac{1}{2}t^2\mathbf{a}'\Sigma\mathbf{a}). \tag{1.6.18}$$

Setting $t = 1$ in (1.6.18) we obtain the required characteristic function of $\mathbf{Y}$, viz.

$$E\exp(i\mathbf{a}'\mathbf{Y}) = \exp(i\mathbf{a}'\boldsymbol{\mu} - \tfrac{1}{2}\mathbf{a}'\Sigma\mathbf{a}).$$ □

Another important property of multivariate normal distributions (one which we shall use heavily) is that all conditional distributions are again multivariate normal. In the following proposition we shall suppose that $\mathbf{Y}$ is partitioned into two subvectors,

$$\mathbf{Y} = \begin{bmatrix} \mathbf{Y}^{(1)} \\ \mathbf{Y}^{(2)} \end{bmatrix}.$$

Correspondingly we can write the mean and covariance matrix of $\mathbf{Y}$ as

$$\boldsymbol{\mu} = \begin{bmatrix} \boldsymbol{\mu}^{(1)} \\ \boldsymbol{\mu}^{(2)} \end{bmatrix} \quad \text{and} \quad \Sigma = \begin{bmatrix} \Sigma_{11} & \Sigma_{12} \\ \Sigma_{21} & \Sigma_{22} \end{bmatrix}$$

where $\boldsymbol{\mu}^{(i)} = E\mathbf{Y}^{(i)}$ and $\Sigma_{ij} = E(\mathbf{Y}^{(i)} - \boldsymbol{\mu}^{(i)})(\mathbf{Y}^{(j)} - \boldsymbol{\mu}^{(j)})'$.

Proposition 1.6.6.

(i) $\mathbf{Y}^{(1)}$ *and* $\mathbf{Y}^{(2)}$ *are independent if and only if* $\Sigma_{12} = 0$.
(ii) *If* $\det \Sigma_{22} > 0$ *then the conditional distribution of* $\mathbf{Y}^{(1)}$ *given* $\mathbf{Y}^{(2)}$ *is* $N(\boldsymbol{\mu}^{(1)} + \Sigma_{12}\Sigma_{22}^{-1}(\mathbf{Y}^{(2)} - \boldsymbol{\mu}^{(2)}), \Sigma_{11} - \Sigma_{12}\Sigma_{22}^{-1}\Sigma_{21})$.

PROOF. (i) If $\mathbf{Y}^{(1)}$ and $\mathbf{Y}^{(2)}$ are independent, then

$$\Sigma_{12} = E(\mathbf{Y}^{(1)} - \boldsymbol{\mu}^{(1)})E(\mathbf{Y}^{(2)} - \boldsymbol{\mu}^{(2)})' = 0.$$

Conversely if $\Sigma_{12} = 0$ then the characteristic function $\phi_{\mathbf{Y}}(\mathbf{u})$, as specified by

Proposition 1.6.4, factorizes into

$$\phi_{\mathbf{Y}}(\mathbf{u}) = \phi_{\mathbf{Y}^{(1)}}(\mathbf{u}^{(1)})\phi_{\mathbf{Y}^{(2)}}(\mathbf{u}^{(2)}),$$

establishing the independence of $\mathbf{Y}^{(1)}$ and $\mathbf{Y}^{(2)}$.

(ii) If we define

$$\mathbf{X} = \mathbf{Y}^{(1)} - \boldsymbol{\mu}^{(1)} - \Sigma_{12}\Sigma_{22}^{-1}(\mathbf{Y}^{(2)} - \boldsymbol{\mu}^{(2)}), \tag{1.6.19}$$

then clearly

$$\begin{bmatrix} \mathbf{X} \\ \mathbf{Y}^{(2)} \end{bmatrix} \sim N\left(\begin{bmatrix} \mathbf{0} \\ \boldsymbol{\mu}^{(2)} \end{bmatrix}, \begin{bmatrix} \Sigma_{11} - \Sigma_{12}\Sigma_{22}^{-1}\Sigma_{21} & 0 \\ 0 & \Sigma_{22} \end{bmatrix}\right),$$

so that $\mathbf{X}$ and $\mathbf{Y}^{(2)}$ are independent by (i). Using the relation (1.6.19) we can express the conditional characteristic function of $\mathbf{Y}^{(1)}$ given $\mathbf{Y}^{(2)}$ as

$$\begin{aligned} E(\exp(i\mathbf{u}'\mathbf{Y}^{(1)})|\mathbf{Y}^{(2)}) &= E(\exp[i\mathbf{u}'\mathbf{X} + i\mathbf{u}'(\boldsymbol{\mu}^{(1)} + \Sigma_{12}\Sigma_{22}^{-1}(\mathbf{Y}^{(2)} - \boldsymbol{\mu}^{(2)}))]|\mathbf{Y}^{(2)}) \\ &= \exp[i\mathbf{u}'(\boldsymbol{\mu}^{(1)} + \Sigma_{12}\Sigma_{22}^{-1}(\mathbf{Y}^{(2)} - \boldsymbol{\mu}^{(2)}))]E\exp(i\mathbf{u}'\mathbf{X}|\mathbf{Y}^{(2)}), \end{aligned}$$

where the last line is obtained by taking a factor dependent only on $\mathbf{Y}^{(2)}$ outside the conditional expectation. Now since $\mathbf{X}$ and $\mathbf{Y}^{(2)}$ are independent,

$$E(\exp(i\mathbf{u}'\mathbf{X})|\mathbf{Y}^{(2)}) = E\exp(i\mathbf{u}'\mathbf{X}) = \exp[-\tfrac{1}{2}\mathbf{u}'(\Sigma_{11} - \Sigma_{12}\Sigma_{22}^{-1}\Sigma_{21})\mathbf{u}],$$

so

$$\begin{aligned} &E(\exp(i\mathbf{u}'\mathbf{Y}^{(1)})|\mathbf{Y}^{(2)}) \\ &\quad = \exp[i\mathbf{u}'(\boldsymbol{\mu}^{(1)} + \Sigma_{12}\Sigma_{22}^{-1}(\mathbf{Y}^{(2)} - \boldsymbol{\mu}^{(2)})) - \tfrac{1}{2}\mathbf{u}'(\Sigma_{11} - \Sigma_{12}\Sigma_{22}^{-1}\Sigma_{21})\mathbf{u}], \end{aligned}$$

completing the proof. □

EXAMPLE 1.6.2. For the bivariate normal random vector $\mathbf{Y}$ discussed in Example 1.6.1 we immediately deduce from Proposition 1.6.6 that Y_1 and Y_2 are independent if and only if $\rho\sigma_1\sigma_2 = 0$. If $\sigma_1 > 0$, $\sigma_2 > 0$ and $\rho > 0$ then conditional on Y_2, Y_1 is normal with mean

$$E(Y_1|Y_2) = \mu_1 + \rho\sigma_1\sigma_2^{-1}(Y_2 - \mu_2),$$

and variance

$$\mathrm{Var}(Y_1|Y_2) = \sigma_1^2(1 - \rho^2).$$

§1.7* Applications of Kolmogorov's Theorem

In this section we illustrate the use of Theorem 1.2.1 to establish the existence of two important processes, Brownian motion and the Poisson process.

Definition 1.7.1 (Standard Brownian Motion). Standard Brownian motion starting at level zero is a process $\{B(t), t \geq 0\}$ satisfying the conditions

(a) $B(0) = 0$,
(b) $B(t_2) - B(t_1), B(t_3) - B(t_2), \ldots, B(t_n) - B(t_{n-1})$, are independent for every $n \in \{3, 4, \ldots\}$ and every $\mathbf{t} = (t_1, \ldots, t_n)'$ such that $0 \le t_1 < t_2 < \cdots < t_n$,
(c) $B(t) - B(s) \sim N(0, t - s)$ for $t \ge s$.

To establish the existence of such a process we observe that conditions (a), (b) and (c) are satisfied if and only if, for every $\mathbf{t} = (t_1, \ldots, t_n)'$ such that $0 \le t_1 < \cdots < t_n$, the characteristic function of $(B(t_1), \ldots, B(t_n))$ is

$$\begin{aligned} \phi_{\mathbf{t}}(\mathbf{u}) &= E\exp[iu_1 B(t_1) + \cdots + iu_n B(t_n)] \\ &= E\exp[iu_1\Delta_1 + iu_2(\Delta_1 + \Delta_2) + \cdots + iu_n(\Delta_1 + \cdots + \Delta_n)] \\ &\qquad \text{(where } \Delta_j = B(t_j) - B(t_{j-1}), j \ge 1, \text{ and } t_0 = 0) \qquad (1.7.1) \\ &= E\exp[i\Delta_1(u_1 + \cdots + u_n) + i\Delta_2(u_2 + \cdots + u_n) + \cdots + i\Delta_n u_n] \\ &= \exp\left[-\frac{1}{2}\sum_{j=1}^{n}(u_j + \cdots + u_n)^2(t_j - t_{j-1})\right]. \end{aligned}$$

It is trivial to check that the characteristic functions $\phi_{\mathbf{t}}(\cdot)$ satisfy the consistency condition (1.2.9) and so by Kolmogorov's theorem there exists a process with characteristic functions $\phi_{\mathbf{t}}(\cdot)$, or equivalently with the properties (a), (b) and (c).

Definition 1.7.2 (Brownian Motion with Drift). Brownian motion with drift μ, variance parameter σ^2 and initial level x is process $\{Y(t), t \ge 0\}$ where

$$Y(t) = x + \mu t + \sigma B(t),$$

and $B(t)$ is standard Brownian motion.

The existence of Brownian motion with drift follows at once from that of standard Brownian motion.

Definition 1.7.3 (Poisson Process). A Poisson process with mean rate λ (>0) is a process $\{N(t), t \ge 0\}$ satisfying the conditions

(a) $N(0) = 0$,
(b) $N(t_2) - N(t_1)$, $N(t_3) - N(t_2)$, $\ldots$, $N(t_n) - N(t_{n-1})$, are independent for every $n \in \{3,4,\ldots\}$ and every $\mathbf{t} = (t_1,\ldots,t_n)'$ such that $0 \le t_1 < t_2 < \cdots < t_n$,
(c) $N(t) - N(s)$ has the Poisson distribution with mean $\lambda(t - s)$ for $t \ge s$.

The proof of the existence of a Poisson process follows precisely the same steps as the proof of the existence of standard Brownian motion. For the Poisson process however the characteristic function of the increment $\Delta_j = N(t_j) - N(t_{j-1})$ is

$$E\exp(iu\Delta_j) = \exp\{-\lambda(t_j - t_{j-1})(1 - e^{iu})\}.$$

In fact the same proof establishes the existence of a process $\{Z(t), t \ge 0\}$

satisfying conditions (a) and (b) of Definition 1.7.1 provided the increments $\Delta_j = Z(t_j) - Z(t_{j-1})$ have characteristic function of the form

$$E\exp(iu\Delta_j) = \exp\{(t_j - t_{j-1})\Psi(u)\}.$$

Problems

1.1. Suppose that $X_t = Z_t + \theta Z_{t-1}, t = 1, 2, \ldots$, where $Z_0, Z_1, Z_2, \ldots$, are independent random variables, each with moment generating function $E\exp(\lambda Z_i) = m(\lambda)$.
(a) Express the joint moment generating function $E\exp(\sum_{i=1}^n \lambda_i X_i)$ in terms of the function $m(\cdot)$.
(b) Deduce from (a) that $\{X_t\}$ is strictly stationary.

1.2. (a) Show that a linear filter $\{a_j\}$ passes an arbitrary polynomial of degree k without distortion, i.e.

$$m_t = \sum_j a_j m_{t-j}$$

for all k^{th} degree polynomials $m_t = c_0 + c_1 t + \cdots + c_k t^k$, if and only if

$$\begin{cases} \sum_j a_j = 1, \\ \sum_j j^r a_j = 0, \quad \text{for } r = 1, \ldots, k. \end{cases}$$

(b) Show that the Spencer 15-point moving average filter $\{a_j\}$ does not distort a cubic trend.

1.3. Suppose that $m_t = c_0 + c_1 t + c_2 t^2, t = 0, \pm 1, \ldots$.
(a) Show that

$$m_t = \sum_{i=-2}^{2} a_i m_{t+i} = \sum_{i=-3}^{3} b_i m_{t+i}, \qquad t = 0, \pm 1, \ldots,$$

where $a_2 = a_{-2} = -\frac{3}{35}$, $a_1 = a_{-1} = \frac{12}{35}$, $a_0 = \frac{17}{35}$, and $b_3 = b_{-3} = -\frac{2}{21}$, $b_2 = b_{-2} = \frac{3}{21}$, $b_1 = b_{-1} = \frac{6}{21}$, $b_0 = \frac{7}{21}$.
(b) Suppose that $X_t = m_t + Z_t$ where $\{Z_t, t = 0, \pm 1, \ldots\}$ is an independent sequence of normal random variables, each with mean 0 and variance σ^2. Let $U_t = \sum_{i=-2}^{2} a_i X_{t+i}$ and $V_t = \sum_{i=-3}^{3} b_i X_{t+i}$.
(i) Find the means and variances of U_t and V_t.
(ii) Find the correlations between U_t and U_{t+1} and between V_t and V_{t+1}.
(iii) Which of the two filtered series $\{U_t\}$ and $\{V_t\}$ would you expect to be smoother in appearance?

1.4. If $m_t = \sum_{k=0}^{p} c_k t^k, t = 0, \pm 1, \ldots$, show that ∇m_t is a polynomial of degree $(p - 1)$ in t and hence that $\nabla^{p+1} m_t = 0$.

1.5. Design a symmetric moving average filter which eliminates seasonal components with period 3 and which at the same time passes quadratic trend functions without distortion.

1.6. (a) Use the programs WORD6 and PEST to plot the series with values $\{x_1, \ldots, x_{30}\}$ given by

1–10	486	474	434	441	435	401	414	414	386	405
11–20	411	389	414	426	410	441	459	449	486	510
21–30	506	549	579	581	630	666	674	729	771	785

This series is the sum of a quadratic trend and a period-three seasonal component.

(b) Apply the filter found in Problem 1.5 to the preceding series and plot the result. Comment on the result.

1.7. Let Z_t, $t = 0, \pm 1, \ldots$, be independent normal random variables each with mean 0 and variance σ^2 and let a, b and c be constants. Which, if any, of the following processes are stationary? For each stationary process specify the mean and autocovariance function.

(a) $X_t = a + bZ_t + cZ_{t-1}$, (b) $X_t = a + bZ_0$,
(c) $X_t = Z_1 \cos(ct) + Z_2 \sin(ct)$, (d) $X_t = Z_0 \cos(ct)$,
(e) $X_t = Z_t \cos(ct) + Z_{t-1} \sin(ct)$, (f) $X_t = Z_t Z_{t-1}$.

1.8. Let $\{Y_t\}$ be a stationary process with mean zero and let a and b be constants.

(a) If $X_t = a + bt + s_t + Y_t$ where s_t is a seasonal component with period 12, show that $\nabla\nabla_{12}X_t = (1 - B)(1 - B^{12})X_t$ is stationary.

(b) If $X_t = (a + bt)s_t + Y_t$ where s_t is again a seasonal component with period 12, show that $\nabla_{12}^2 X_t = (1 - B^{12})(1 - B^{12})X_t$ is stationary.

1.9. Use the program PEST to analyze the accidental deaths data by "classical decomposition".

(a) Plot the data.

(b) Find estimates $\hat{s}_t$, $t = 1, \ldots, 12$, for the classical decomposition model, $X_t = m_t + s_t + Y_t$, where $s_t = s_{t+12}$, $\sum_{t=1}^{12} s_t = 0$ and $EY_t = 0$.

(c) Plot the deseasonalized data, $X_t - \hat{s}_t$, $t = 1, \ldots, 72$.

(d) Fit a parabola by least squares to the deseasonalized data and use it as your estimate $\hat{m}_t$ of m_t.

(e) Plot the residuals $\hat{Y}_t = X_t - \hat{m}_t - \hat{s}_t$, $t = 1, \ldots, 72$.

(f) Compute the sample autocorrelation function of the residuals $\hat{\rho}(h)$, $h = 0, \ldots, 20$.

(g) Use your fitted model to predict X_t, $t = 73, \ldots, 84$ (using predicted noise values of zero).

1.10. Let $X_t = a + bt + Y_t$, where $\{Y_t, t = 0, \pm 1, \ldots\}$ is an independent and identically distributed sequence of random variables with mean 0 and variance σ^2, and a and b are constants. Define

$$W_t = (2q + 1)^{-1} \sum_{j=-q}^{q} X_{t+j}.$$

Compute the mean and autocovariance function of $\{W_t\}$. Notice that although $\{W_t\}$ is not stationary, its autocovariance function $\gamma(t + h, t) = \text{Cov}(W_{t+h}, W_t)$ does not depend on t. Plot the autocorrelation function $\rho(h) = \text{Corr}(W_{t+h}, W_t)$. Discuss your results in relation to the smoothing of a time series.

1.11. If $\{X_t\}$ and $\{Y_t\}$ are uncorrelated stationary sequences, i.e. if X_s and Y_t are uncorrelated for every s and t, show that $\{X_t + Y_t\}$ is stationary with autocovariance function equal to the sum of the autocovariance functions of $\{X_t\}$ and $\{Y_t\}$.

1.12. Which, if any, of the following functions defined on the integers is the autocovariance function of a stationary time series?

(a) $f(h) = \begin{cases} 1 & \text{if } h = 0, \\ 1/h & \text{if } h \neq 0. \end{cases}$ (b) $f(h) = (-1)^{|h|}$

(c) $f(h) = 1 + \cos\frac{\pi h}{2} + \cos\frac{\pi h}{4}$ (d) $f(h) = 1 + \cos\frac{\pi h}{2} - \cos\frac{\pi h}{4}$

(e) $f(h) = \begin{cases} 1 & \text{if } h = 0, \\ .4 & \text{if } h = \pm 1, \\ 0 & \text{otherwise.} \end{cases}$ (f) $f(h) = \begin{cases} 1 & \text{if } h = 0, \\ .6 & \text{if } h = \pm 1, \\ 0 & \text{otherwise.} \end{cases}$

1.13. Let $\{S_t, t = 0, 1, 2, \ldots\}$ be the random walk with constant drift μ, defined by $S_0 = 0$ and

$$S_t = \mu + S_{t-1} + X_t, \qquad t = 1, 2, \ldots,$$

where $X_1, X_2, \ldots$ are independent and identically distributed random variables with mean 0 and variance σ^2. Compute the mean of S_t and the autocovariance function of the process $\{S_t\}$. Show that $\{\nabla S_t\}$ is stationary and compute its mean and autocovariance function.

1.14. If $X_t = a + bt$, $t = 1, 2, \ldots, n$, where a and b are constants, show that the sample autocorrelations have the property $\hat{\rho}(k) \to 1$ as $n \to \infty$ for each fixed k.

1.15. Prove Proposition 1.6.1.

1.16. (a) If $Z \sim N(0, 1)$ show that Z^2 has moment generating function $Ee^{tZ^2} = (1 - 2t)^{-1/2}$ for $t < \frac{1}{2}$, thus showing that Z^2 has the chi-squared distribution with 1 degree of freedom.
(b) If $Z_1, \ldots, Z_n$ are independent $N(0, 1)$ random variables, prove that $Z_1^2 + \cdots + Z_n^2$ has the chi-squared distribution with n degrees of freedom by showing that its moment generating function is equal to $(1 - 2t)^{-n/2}$ for $t < \frac{1}{2}$.
(c) Suppose that $\mathbf{X} = (X_1, \ldots, X_n)' \sim N(\boldsymbol{\mu}, \Sigma)$ with Σ non-singular. Using (1.6.13), show that $(\mathbf{X} - \boldsymbol{\mu})'\Sigma^{-1}(\mathbf{X} - \boldsymbol{\mu})$ has the chi-squared distribution with n degrees of freedom.

1.17. If $\mathbf{X} = (X_1, \ldots, X_n)'$ is a random vector with covariance matrix Σ, show that Σ is singular if and only if there exists a non-zero vector $\mathbf{b} = (b_1, \ldots, b_n)' \in \mathbb{R}^n$ such that $\text{Var}(\mathbf{b}'\mathbf{X}) = 0$.

1.18.* Let F be any distribution function, let T be the index set $T = \{1, 2, 3, \ldots\}$ and let $\mathscr{T}$ be as in Definition 1.2.3. Show that the functions $F_{\mathbf{t}}$, $\mathbf{t} \in \mathscr{T}$, defined by

$$F_{t_1 \ldots t_n}(x_1, \ldots, x_n) := F(x_1) \cdots F(x_n), \qquad x_1, \ldots, x_n \in \mathbb{R},$$

constitute a family of distribution functions, consistent in the sense of (1.2.8). By Kolmogorov's theorem this establishes that there exists a sequence of independent random variables $\{X_1, X_2, \ldots\}$ defined on some probability space and such that $P(X_i \leq x) = F(x)$ for all i and for all $x \in \mathbb{R}$.

CHAPTER 2

Hilbert Spaces

Although it is possible to study time series analysis without explicit use of Hilbert space terminology and techniques, there are great advantages to be gained from a Hilbert space formulation. These are largely derived from our familiarity with two- and three-dimensional Euclidean geometry and in particular with the concepts of orthogonality and orthogonal projections in these spaces. These concepts, appropriately extended to infinite-dimensional Hilbert spaces, play a central role in the study of random variables with finite second moments and especially in the theory of prediction of stationary processes.

Intuition gained from Euclidean geometry can often be used to make apparently complicated algebraic results in time series analysis geometrically obvious. It frequently serves also as a valuable guide in the development and construction of algorithms.

This chapter is therefore devoted to a study of those aspects of Hilbert space theory which are needed for a geometric understanding of the later chapters in this book. The results developed here will also provide an adequate background for a geometric approach to many other areas of statistics, for example the general linear model (see Section 2.6). For the reader who wishes to go deeper into the theory of Hilbert space we recommend the book by Simmons (1963).

§2.1 Inner-Product Spaces and Their Properties

Definition 2.1.1 (Inner-Product Space). A complex vector space $\mathscr{H}$ is said to be an inner-product space if for each pair of elements x and y in $\mathscr{H}$, there is a complex number $\langle x, y\rangle$, called the inner product of x and y, such that

(a) $\langle x, y\rangle = \overline{\langle y, x\rangle}$, the bar denoting complex conjugation,
(b) $\langle x + y, z\rangle = \langle x, z\rangle + \langle y, z\rangle$ for all $x, y, z \in \mathscr{H}$,
(c) $\langle \alpha x, y\rangle = \alpha\langle x, y\rangle$ for all $x, y \in \mathscr{H}$ and $\alpha \in \mathbb{C}$,
(d) $\langle x, x\rangle \geq 0$ for all $x \in \mathscr{H}$,
(e) $\langle x, x\rangle = 0$ if and only if $x = 0$.

Remark 1. A real vector space $\mathscr{H}$ is an inner-product space if for each $x, y \in \mathscr{H}$ there exists a *real* number $\langle x, y\rangle$ satisfying conditions (a)–(e). Of course condition (a) reduces in this case to $\langle x, y\rangle = \langle y, x\rangle$.

Remark 2. The inner product is a natural generalization of the inner or scalar product of two vectors in n-dimensional Euclidean space. Since many of the properties of Euclidean space carry over to inner-product spaces, it will be helpful to keep Euclidean space in mind in all that follows.

EXAMPLE 2.1.1 (Euclidean Space). The set of all column vectors

$$\mathbf{x} = (x_1, \ldots, x_k)' \in \mathbb{R}^k,$$

is a real inner-product space if we define

$$\langle \mathbf{x}, \mathbf{y}\rangle = \sum_{i=1}^{k} x_i y_i. \tag{2.1.1}$$

Equation (2.1.1) defines the usual scalar product of elements of $\mathbb{R}^k$. It is a simple matter to check that the conditions (a)–(e) are all satisfied.

In the same way it is easy to see that the set of all complex k-dimensional column vectors

$$\mathbf{z} = (z_1, \ldots, z_k)' \in \mathbb{C}^k$$

is a complex inner-product space if we define

$$\langle \mathbf{w}, \mathbf{z}\rangle = \sum_{i=1}^{k} w_i \bar{z}_i. \tag{2.1.2}$$

Definition 2.1.2 (Norm). The norm of an element x of an inner-product space is defined to be

$$\|x\| = \sqrt{\langle x, x\rangle}. \tag{2.1.3}$$

In the Euclidean space $\mathbb{R}^k$ the norm of the vector is simply its length, $\|\mathbf{x}\| = (\sum_{i=1}^{k} x_i^2)^{1/2}$.

The Cauchy–Schwarz Inequality. If $\mathscr{H}$ is an inner-product space, then

$$|\langle x, y\rangle| \leq \|x\|\,\|y\| \quad \text{for all } x, y \in \mathscr{H}, \tag{2.1.4}$$

and

$$|\langle x, y\rangle| = \|x\|\,\|y\| \quad \text{if and only if } x = y\langle x, y\rangle/\langle y, y\rangle. \tag{2.1.5}$$

PROOF. The following proof for complex $\mathscr{H}$ remains valid (although it could be slightly simplified) in the case when $\mathscr{H}$ is real.

Let $a = \|y\|^2$, $b = |\langle x, y\rangle|$ and $c = \|x\|^2$. The polar representation of $\langle x, y\rangle$ is then

$$\langle x, y\rangle = be^{i\theta} \quad \text{for some } \theta \in (-\pi, \pi].$$

Now for all $r \in \mathbb{R}$,

$$\begin{aligned}\langle x - re^{i\theta}y, x - re^{i\theta}y\rangle &= \langle x, x\rangle - re^{i\theta}\langle y, x\rangle - re^{-i\theta}\langle x, y\rangle + r^2\langle y, y\rangle \\ &= c - 2rb + r^2a,\end{aligned} \tag{2.1.6}$$

and using elementary calculus, we deduce from this that

$$0 \le \min_{r \in \mathbb{R}}(c - 2rb + r^2a) = c - b^2/a,$$

thus establishing (2.1.4).

The minimum value, $c - b^2/a$, of $c - 2rb + r^2a$ is achieved when $r = b/a$. If equality is achieved in (2.1.4) then $c - b^2/a = 0$. Setting $r = b/a$ in (2.1.6) we then obtain

$$\langle x - ye^{i\theta}b/a, x - ye^{i\theta}b/a\rangle = 0,$$

which, by property (e) of inner products, implies that

$$x = ye^{i\theta}b/a = y\langle x, y\rangle/\langle y, y\rangle.$$

Conversely if $x = y\langle x, y\rangle/\langle y, y\rangle$ (or equivalently if x is *any* scalar multiple of y), it is obvious that there is equality in (2.1.4). □

EXAMPLE 2.1.2 (The Angle between Elements of a Real Inner-Product Space). In the inner-product space $\mathbb{R}^3$ of Example 2.1.1, the angle between two non-zero vectors $\mathbf{x}$ and $\mathbf{y}$ is the angle in $[0, \pi]$ whose cosine is $\sum_{i=1}^3 x_i y_i / (\|\mathbf{x}\|\,\|\mathbf{y}\|)$. Analogously we define the angle between non-zero elements x and y of any real inner-product space to be

$$\theta = \cos^{-1}[\langle x, y\rangle/(\|x\|\,\|y\|)]. \tag{2.1.7}$$

In particular x and y are said to be *orthogonal* if and only if $\langle x, y\rangle = 0$. For non-zero vectors x and y this is equivalent to the statement that $\theta = \pi/2$.

The Triangle Inequality. If $\mathscr{H}$ is an inner-product space, then

$$\|x + y\| \le \|x\| + \|y\| \quad \text{for all } x, y \in \mathscr{H}. \tag{2.1.8}$$

PROOF.

$$\begin{aligned}\|x + y\|^2 &= \langle x + y, x + y\rangle \\ &= \langle x, x\rangle + \langle x, y\rangle + \langle y, x\rangle + \langle y, y\rangle \\ &\le \|x\|^2 + 2\|x\|\,\|y\| + \|y\|^2\end{aligned}$$

by the Cauchy–Schwarz inequality. □

Proposition 2.1.1 (Properties of the Norm). *If $\mathscr{H}$ is a complex (respectively real) inner-product space and $\|x\|$ is defined as in* (2.1.3), *then*

(a) $\|x + y\| \le \|x\| + \|y\|$ *for all* $x, y \in \mathscr{H}$,
(b) $\|\alpha x\| = |\alpha| \, \|x\|$ *for all* $x \in \mathscr{H}$ *and all* $\alpha \in \mathbb{C}$ ($\alpha \in \mathbb{R}$),
(c) $\|x\| \ge 0$ *for all* $x \in \mathscr{H}$,
(d) $\|x\| = 0$ *if and only if* $x = 0$.

(*These properties justify the use of the terminology "norm" for* $\|x\|$.)

Proof. The first property is a restatement of the triangle inequality and the others follow at once from Definition (2.1.3). □

The Parallelogram Law. If $\mathscr{H}$ is an inner-product space, then

$$\|x + y\|^2 + \|x - y\|^2 = 2\|x\|^2 + 2\|y\|^2 \quad \text{for all } x, y \in \mathscr{H}. \tag{2.1.9}$$

Proof. Problem 2.1. Note that (2.1.9) is not a consequence of the properties (a), (b), (c) and (d) of the norm. It depends on the particular form (2.1.3) of the norm as defined for elements of an inner-product space. □

Definition 2.1.3 (Convergence in Norm). A sequence $\{x_n, n = 1, 2, \ldots\}$ of elements of an inner-product space $\mathscr{H}$ is said to converge in norm to $x \in \mathscr{H}$ if $\|x_n - x\| \to 0$ as $n \to \infty$.

Proposition 2.1.2 (Continuity of the Inner Product). *If $\{x_n\}$ and $\{y_n\}$ are sequences of elements of the inner-product space $\mathscr{H}$ such that $\|x_n - x\| \to 0$ and $\|y_n - y\| \to 0$, then*

$$\text{(a)} \quad \|x_n\| \to \|x\|$$

and

$$\text{(b)} \quad \langle x_n, y_n \rangle \to \langle x, y \rangle.$$

Proof. From the triangle inequality it follows that $\|x\| \le \|x - y\| + \|y\|$ and $\|y\| \le \|y - x\| + \|x\|$. These statements imply that

$$\|x - y\| \ge \big| \|x\| - \|y\| \big|, \tag{2.1.10}$$

from which (a) follows immediately. Now

$$\begin{aligned} |\langle x_n, y_n \rangle - \langle x, y \rangle| &= |\langle x_n, y_n - y \rangle + \langle x_n - x, y \rangle| \\ &\le |\langle x_n, y_n - y \rangle| + |\langle x_n - x, y \rangle| \\ &\le \|x_n\| \, \|y_n - y\| + \|x_n - x\| \, \|y\|, \end{aligned}$$

where the last line follows from the Cauchy–Schwarz inequality. Observing from (a) that $\|x_n\| \to \|x\|$, we conclude that

$$|\langle x_n, y_n \rangle - \langle x, y \rangle| \to 0 \quad \text{as } n \to \infty. \qquad \square$$

§2.2 Hilbert Spaces

An inner-product space with the additional property of completeness is called a Hilbert space. To define completeness we first need the concept of a Cauchy sequence.

Definition 2.2.1 (Cauchy Sequence). A sequence $\{x_n, n = 1, 2, \dots\}$ of elements of an inner-product space is said to be a Cauchy sequence if

$$\|x_n - x_m\| \to 0 \quad \text{as } m, n \to \infty,$$

i.e. if for every $\varepsilon > 0$ there exists a positive integer $N(\varepsilon)$ such that

$$\|x_n - x_m\| < \varepsilon \quad \text{for all } m, n > N(\varepsilon).$$

Definition 2.2.2 (Hilbert Space). A Hilbert space $\mathscr{H}$ is an inner-product space which is complete, i.e. an inner-product space in which every Cauchy sequence $\{x_n\}$ converges in norm to some element $x \in \mathscr{H}$.

EXAMPLE 2.2.1 (Euclidean Space). The completeness of the inner-product space $\mathbb{R}^k$ defined in Example 2.1.1 can be verified as follows. If $\mathbf{x}_n = (x_{n1}, x_{n2}, \dots, x_{nk})' \in \mathbb{R}^k$ satisfies

$$\|\mathbf{x}_n - \mathbf{x}_m\|^2 = \sum_{i=1}^{k} |x_{ni} - x_{mi}|^2 \to 0 \quad \text{as } m, n \to \infty,$$

then each of the components must satisfy

$$|x_{ni} - x_{mi}| \to 0 \quad \text{as } m, n \to \infty.$$

By the completeness of $\mathbb{R}$, there exists $x_i \in \mathbb{R}$ such that

$$|x_{ni} - x_i| \to 0 \quad \text{as } n \to \infty,$$

and hence if $\mathbf{x} = (x_1, \dots, x_k)$, then

$$\|\mathbf{x}_n - \mathbf{x}\| \to 0 \quad \text{as } n \to \infty.$$

Completeness of the complex inner-product space $\mathbb{C}^k$ can be checked in the same way. Thus $\mathbb{R}^k$ and $\mathbb{C}^k$ are both Hilbert spaces.

EXAMPLE 2.2.2 (The Space $L^2(\Omega, \mathscr{F}, P)$). Consider a probability space $(\Omega, \mathscr{F}, P)$ and the collection C of all random variables X defined on Ω and satisfying the condition,

$$EX^2 = \int_{\Omega} X(\omega)^2 P(d\omega) < \infty.$$

With the usual notion of multiplication by a real scalar and addition of random variables, it is clear that C is a vector space since

$$E(aX)^2 = a^2 EX^2 < \infty \quad \text{for all } a \in \mathbb{R} \text{ and } X \in C,$$

and, from the inequality $(X + Y)^2 \le 2X^2 + 2Y^2$,

$$E(X + Y)^2 \le 2EX^2 + 2EY^2 < \infty \quad \text{for all } X, Y \in C.$$

The other properties required of a vector space are easily checked. In particular C has a zero element, the random variable which is identically zero on Ω.

For any two elements X, $Y \in C$ we now dcfine

$$\langle X, Y \rangle = E(XY). \tag{2.2.1}$$

It is easy to check that $\langle X, Y \rangle$ satisfies all the properties of an inner product except for the last. If $\langle X, X \rangle = 0$ then it does not follow that $X(\omega) = 0$ for all ω, but only that $P(X = 0) = 1$. This difficulty is circumvented by saying that the random variables X and Y are equivalent if $P(X = Y) = 1$. This equivalence relation partitions C into classes of random variables such that any two random variables in the same class are equal with probability one. The space L^2 (or more specifically $L^2(\Omega, \mathscr{F}, P)$) is the collection of these equivalence classes with inner product defined by (2.2.1). Since each class is uniquely determined by specifying any one of the random variables in it, we shall continue to use the notation X, Y for elements of L^2 and to call them random variables (or functions) although it is sometimes important to remember that X stands for the collection of all random variables which are equivalent to X.

Norm convergence of a sequence $\{X_n\}$ of elements of L^2 to the limit X means

$$\|X_n - X\|^2 = E|X_n - X|^2 \to 0 \quad \text{as } n \to \infty.$$

Norm convergence of X_n to X in an L^2 space is called mean-square convergence and is written as $X_n \xrightarrow{\text{m.s.}} X$.

To complete the proof that L^2 is a Hilbert space we need to establish completeness, i.e. that if $\|X_m - X_n\|^2 \to 0$ as $m, n \to \infty$, then there exists $X \in L^2$ such that $X_n \xrightarrow{\text{m.s.}} X$. This is indeed true but not so easy to prove as the completeness of $\mathbb{R}^k$. We therefore defer the proof to Section 2.10.

EXAMPLE 2.2.3 (Complex L^2 Spaces). The space of complex-valued random variables X on $(\Omega, \mathscr{F}, P)$ satisfying $E|X|^2 < \infty$ is a complex Hilbert space if we define an inner product by

$$\langle X, Y \rangle = E(X\bar{Y}). \tag{2.2.2}$$

In fact if μ is any finite non-zero measure on the measurable space $(\Omega, \mathscr{F})$, and if D is the class of complex-valued functions on Ω such that

$$\int_\Omega |f|^2 \, d\mu < \infty \tag{2.2.3}$$

(with identification of functions f and g such that $\int_\Omega |f - g|^2 \, d\mu = 0$), then D becomes a Hilbert space if we define the inner product to be

$$\langle f, g \rangle = \int_\Omega f\bar{g}\, d\mu. \tag{2.2.4}$$

This space will be referred to as the complex Hilbert space $L^2(\Omega, \mathscr{F}, \mu)$. (The real Hilbert space $L^2(\Omega, \mathscr{F}, \mu)$ is obtained if D is replaced by the real-valued functions satisfying (2.2.3). The definition of $\langle f, g \rangle$ then reduces to $\int_\Omega fg\, d\mu$.)

Remark 1. The terms $L^2(\Omega, \mathscr{F}, P)$ and $L^2(\Omega, \mathscr{F}, \mu)$ will be reserved for the respective real Hilbert spaces unless we state specifically that reference is being made to the corresponding complex spaces.

Proposition 2.2.1 (Norm Convergence and the Cauchy Criterion). *If $\{x_n\}$ is a sequence of elements belonging to a Hilbert space $\mathscr{H}$, then $\{x_n\}$ converges in norm if and only if $\|x_n - x_m\| \to 0$ as $m, n \to \infty$.*

PROOF. The sufficiency of the Cauchy criterion is simply a restatement of the completeness of $\mathscr{H}$. The necessity is an elementary consequence of the triangle inequality. Thus if $\|x_n - x\| \to 0$,

$$\|x_n - x_m\| \le \|x_n - x\| + \|x - x_m\| \to 0 \quad \text{as } m, n \to 0. \qquad \square$$

EXAMPLE 2.2.4. The Cauchy criterion is used primarily in checking for the norm convergence of a sequence whose limit is not specified. Consider for example the sequence

$$S_n = \sum_{i=1}^{n} a_i X_i \tag{2.2.5}$$

where $\{X_i\}$ is a sequence of independent $N(0, 1)$ random variables. It is easy to see that with the usual definition of the L^2-norm,

$$\|S_m - S_n\|^2 = \sum_{i=n+1}^{m} a_i^2, \qquad m > n,$$

and so by the Cauchy criterion $\{S_n\}$ has a mean-square limit if and only if for every $\varepsilon > 0$, there exists $N(\varepsilon) > 0$ such that $\sum_{i=n+1}^{m} a_i^2 < \varepsilon$ for $m > n > N(\varepsilon)$. Thus $\{S_n\}$ converges in mean square if and only if $\sum_{i=1}^{\infty} a_i^2 < \infty$.

§2.3 The Projection Theorem

We begin this section with two examples which illustrate the use of the projection theorem in particular Hilbert spaces. The general result is then established as Theorem 2.3.1.

EXAMPLE 2.3.1 (Linear Approximation in $\mathbb{R}^3$). Suppose we are given three vectors in $\mathbb{R}^3$,

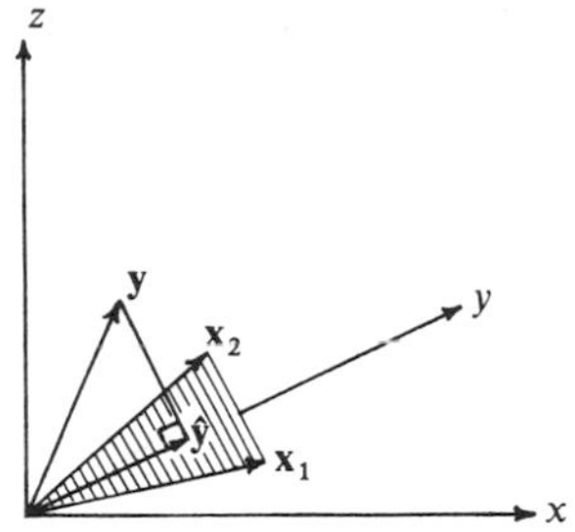

Figure 2.1. The best linear approximation $\hat{\mathbf{y}} = \alpha_1 \mathbf{x}_1 + \alpha_2 \mathbf{x}_2$, to $\mathbf{y}$.

$$\mathbf{y} = (\tfrac{1}{4}, \tfrac{1}{4}, 1)',$$
$$\mathbf{x}_1 = (1, 0, \tfrac{1}{4})',$$
$$\mathbf{x}_2 = (0, 1, \tfrac{1}{4})',$$

Our problem is to find the linear combination $\hat{\mathbf{y}} = \alpha_1 \mathbf{x}_1 + \alpha_2 \mathbf{x}_2$ which is closest to $\mathbf{y}$ in the sense that $S = \|\mathbf{y} - \alpha_1 \mathbf{x}_1 - \alpha_2 \mathbf{x}_2\|^2$ is minimized.

One approach to this problem is to write S in the form $S = (\frac{1}{4} - \alpha_1)^2 + (\frac{1}{4} - \alpha_2)^2 + (1 - \frac{1}{4}\alpha_1 - \frac{1}{4}\alpha_2)^2$ and then to use calculus to minimize with respect to α_1 and α_2. In the alternative geometric approach to the problem we observe that the required vector $\hat{\mathbf{y}} = \alpha_1 \mathbf{x}_1 + \alpha_2 \mathbf{x}_2$ is the vector in the plane determined by $\mathbf{x}_1$ and $\mathbf{x}_2$ such that $\mathbf{y} - \alpha_1 \mathbf{x}_1 - \alpha_2 \mathbf{x}_2$ is orthogonal to the plane of $\mathbf{x}_1$ and $\mathbf{x}_2$ (see Figure 2.1). The orthogonality condition may be stated as

$$\langle \mathbf{y} - \alpha_1 \mathbf{x}_1 - \alpha_2 \mathbf{x}_2, \mathbf{x}_i \rangle = 0, \qquad i = 1, 2, \tag{2.3.1}$$

or equivalently

$$\alpha_1 \langle \mathbf{x}_1, \mathbf{x}_1 \rangle + \alpha_2 \langle \mathbf{x}_2, \mathbf{x}_1 \rangle = \langle \mathbf{y}, \mathbf{x}_1 \rangle,$$
$$\alpha_1 \langle \mathbf{x}_1, \mathbf{x}_2 \rangle + \alpha_2 \langle \mathbf{x}_2, \mathbf{x}_2 \rangle = \langle \mathbf{y}, \mathbf{x}_2 \rangle.$$

For the particular vectors $\mathbf{x}_1$, $\mathbf{x}_2$ and $\mathbf{y}$ specified, these equations become

$$\tfrac{17}{16}\alpha_1 + \tfrac{1}{16}\alpha_2 = \tfrac{1}{2},$$
$$\tfrac{1}{16}\alpha_1 + \tfrac{17}{16}\alpha_2 = \tfrac{1}{2},$$

from which we deduce that $\alpha_1 = \alpha_2 = \frac{4}{9}$, and $\hat{\mathbf{y}} = (\frac{4}{9}, \frac{4}{9}, \frac{2}{9})'$.

EXAMPLE 2.3.2 (Linear Approximation in $L^2(\Omega, \mathscr{F}, P)$). Now suppose that X_1, X_2 and Y are random variables in $L^2(\Omega, \mathscr{F}, P)$. If only X_1 and X_2 are observed we may wish to estimate the value of Y by using the linear combination $\hat{Y} = \alpha_1 X_1 + \alpha_2 X_2$ which minimizes the mean squared error,

$$S = E|Y - \alpha_1 X_1 - \alpha_2 X_2|^2 = \|Y - \alpha_1 X_1 - \alpha_2 X_2\|^2.$$

As in Example 2.3.1 there are at least two possible approaches to this problem. The first is to write

$$S = EY^2 + \alpha_1^2 EX_1^2 + \alpha_2^2 EX_2^2 - 2\alpha_1 E(YX_1) - 2\alpha_2 E(YX_2) + \alpha_1\alpha_2 E(X_1X_2),$$

and then to minimize with respect to α_1 and α_2 by setting the appropriate derivatives equal to zero. However it is also possible to use the same geometric approach as in Example 2.3.1. Our aim is to find an element $\hat{Y}$ in the set

$$\mathcal{M} = \{X \in L^2(\Omega, \mathcal{F}, P) : X = a_1X_1 + a_2X_2 \text{ for some } a_1, a_2 \in \mathbb{R}\},$$

whose "squared distance" from Y, $\|Y - \hat{Y}\|^2$, is as small as possible. By analogy with Example 2.3.1 we might expect $\hat{Y}$ to have the property that $Y - \hat{Y}$ is orthogonal to all elements of $\mathcal{M}$. The validity of this analogy, and the extent to which it may be applied in more general situations, is established in Theorem 2.3.1 (the projection theorem). Applying it to our present problem, we can write

$$\langle Y - \alpha_1X_1 - \alpha_2X_2, X\rangle = 0 \quad \text{for all } X \in \mathcal{M}, \tag{2.3.2}$$

or equivalently, by the linearity of the inner product,

$$\langle Y - \alpha_1X_1 - \alpha_2X_2, X_i\rangle = 0, \qquad i = 1, 2. \tag{2.3.3}$$

These are the same equations for α_1 and α_2 as (2.3.1), although the inner product is of course defined differently in (2.3.3). In terms of expectations we can rewrite (2.3.3) in the form

$$\alpha_1 E(X_1^2) + \alpha_2 E(X_2X_1) = E(YX_1),$$

$$\alpha_1 E(X_1X_2) + \alpha_2 E(X_2^2) = E(YX_2),$$

from which α_1 and α_2 are easily found.

Before establishing the projection theorem for a general Hilbert space we need to introduce a certain amount of new terminology.

Definition 2.3.1 (Closed Subspace). A linear subspace $\mathcal{M}$ of a Hilbert space $\mathcal{H}$ is said to be a closed subspace of $\mathcal{H}$ if $\mathcal{M}$ contains all of its limit points (i.e. if $x_n \in \mathcal{M}$ and $\|x_n - x\| \to 0$ imply that $x \in \mathcal{M}$).

Definition 2.3.2 (Orthogonal Complement). The orthogonal complement of a subset $\mathcal{M}$ of $\mathcal{H}$ is defined to be the set $\mathcal{M}^\perp$ of all elements of $\mathcal{H}$ which are orthogonal to every element of $\mathcal{M}$. Thus

$$x \in \mathcal{M}^\perp \text{ if and only if } \langle x, y\rangle = 0 \text{ (written } x \perp y) \text{ for all } y \in \mathcal{M}. \tag{2.3.4}$$

Proposition 2.3.1. *If $\mathcal{M}$ is any subset of a Hilbert space $\mathcal{H}$ then $\mathcal{M}^\perp$ is a closed subspace of $\mathcal{H}$.*

PROOF. It is easy to check from (2.3.4) that $0 \in \mathcal{M}^\perp$ and that if $x_1, x_2 \in \mathcal{M}^\perp$ then all linear combinations of x_1 and x_2 belong to $\mathcal{M}^\perp$. Hence $\mathcal{M}^\perp$ is a subspace of $\mathcal{H}$. If $x_n \in \mathcal{M}^\perp$ and $\|x_n - x\| \to 0$, then by continuity of the inner product (Proposition 2.1.2), $\langle x, y\rangle = 0$ for all $y \in \mathcal{M}$, so $x \in \mathcal{M}^\perp$ and hence $\mathcal{M}^\perp$ is closed. □

Theorem 2.3.1 (The Projection Theorem). *If $\mathcal{M}$ is a closed subspace of the Hilbert space $\mathcal{H}$ and $x \in \mathcal{H}$, then*

(i) *there is a unique element $\hat{x} \in \mathcal{M}$ such that*

$$\|x - \hat{x}\| = \inf_{y \in \mathcal{M}} \|x - y\|, \tag{2.3.5}$$

and

(ii) $\hat{x} \in \mathcal{M}$ *and* $\|x - \hat{x}\| = \inf_{y \in \mathcal{M}} \|x - y\|$ *if and only if* $\hat{x} \in \mathcal{M}$ *and* $(x - \hat{x}) \in \mathcal{M}^{\perp}$.

[*The element $\hat{x}$ is called the (orthogonal) projection of x onto $\mathcal{M}$.*]

PROOF. (i) If $d = \inf_{y \in \mathcal{M}} \|x - y\|^2$ then there is a sequence $\{y_n\}$ of elements of $\mathcal{M}$ such that $\|y_n - x\|^2 \to d$. Apply the parallelogram law (2.1.9), and using the fact that $(y_m + y_n)/2 \in \mathcal{M}$, we can write

$$\begin{aligned} 0 \le \|y_m - y_n\|^2 &= -4\|(y_m + y_n)/2 - x\|^2 + 2(\|y_n - x\|^2 + \|y_m - x\|^2) \\ &\le -4d + 2(\|y_n - x\|^2 + \|y_m - x\|^2) \\ &\to 0 \text{ as } m, n \to \infty. \end{aligned}$$

Consequently, by the Cauchy criterion, there exists $\hat{x} \in \mathcal{H}$ such that $\|y_n - \hat{x}\| \to 0$. Since $\mathcal{M}$ is closed we know that $\hat{x} \in \mathcal{M}$, and by continuity of the inner product

$$\|x - \hat{x}\|^2 = \lim_{n \to \infty} \|x - y_n\|^2 = d.$$

To establish uniqueness, suppose that $\hat{y} \in \mathcal{M}$ and that $\|x - \hat{y}\|^2 = \|x - \hat{x}\|^2 = d$. Then, applying the parallelogram law again,

$$\begin{aligned} 0 \le \|\hat{x} - \hat{y}\|^2 &= -4\|(\hat{x} + \hat{y})/2 - x\|^2 + 2(\|\hat{x} - x\|^2 + \|\hat{y} - x\|^2) \\ &\le -4d + 4d = 0. \end{aligned}$$

Hence $\hat{y} = \hat{x}$.

(ii) If $\hat{x} \in \mathcal{M}$ and $(x - \hat{x}) \in \mathcal{M}^{\perp}$ then $\hat{x}$ is the unique element of $\mathcal{M}$ defined in (i) since for any $y \in \mathcal{M}$,

$$\begin{aligned} \|x - y\|^2 &= \langle x - \hat{x} + \hat{x} - y, x - \hat{x} + \hat{x} - y \rangle \\ &= \|x - \hat{x}\|^2 + \|\hat{x} - y\|^2 \\ &\ge \|x - \hat{x}\|^2, \end{aligned}$$

with equality if and only if $y = \hat{x}$.

Conversely if $\hat{x} \in \mathcal{M}$ and $(x - \hat{x}) \notin \mathcal{M}^{\perp}$ then $\hat{x}$ is *not* the element of $\mathcal{M}$ closest to x since

$$\tilde{x} = \hat{x} + ay/\|y\|^2$$

is closer, where y is any element of $\mathcal{M}$ such that $\langle x - \hat{x}, y \rangle \neq 0$ and

$a = \langle x - \hat{x}, y\rangle$. To see this we write

$$\begin{aligned}\|x - \tilde{x}\|^2 &= \langle x - \hat{x} + \hat{x} - \tilde{x}, x - \hat{x} + \hat{x} - \tilde{x}\rangle \\ &= \|x - \hat{x}\|^2 + |a|^2/\|y\|^2 + 2\,\mathrm{Re}\langle x - \hat{x}, \hat{x} - \tilde{x}\rangle \\ &= \|x - \hat{x}\|^2 - |a|^2/\|y\|^2 \\ &< \|x - \hat{x}\|^2.\end{aligned}$$

□

Corollary 2.3.1 (The Projection Mapping of $\mathscr{H}$ onto $\mathscr{M}$). *If $\mathscr{M}$ is a closed subspace of the Hilbert space $\mathscr{H}$ and I is the identity mapping on $\mathscr{H}$, then there is a unique mapping $P_{\mathscr{M}}$ of $\mathscr{H}$ onto $\mathscr{M}$ such that $I - P_{\mathscr{M}}$ maps $\mathscr{H}$ onto $\mathscr{M}^\perp$. $P_{\mathscr{M}}$ is called the projection mapping of $\mathscr{H}$ onto $\mathscr{M}$.*

PROOF. By Theorem 2.3.1, for each $x \in \mathscr{H}$ there is a unique $\hat{x} \in \mathscr{M}$ such that $x - \hat{x} \in \mathscr{M}^\perp$. The required mapping is therefore

$$P_{\mathscr{M}}x = \hat{x}, \qquad x \in \mathscr{H}. \tag{2.3.6}$$

□

Proposition 2.3.2 (Properties of Projection Mappings). *Let $\mathscr{H}$ be a Hilbert space and let $P_{\mathscr{M}}$ denote the projection mapping onto a closed subspace $\mathscr{M}$. Then*

(i) $P_{\mathscr{M}}(\alpha x + \beta y) = \alpha P_{\mathscr{M}}x + \beta P_{\mathscr{M}}y, \quad x, y \in \mathscr{H}, \quad \alpha, \beta \in \mathbb{C}$,
(ii) $\|x\|^2 = \|P_{\mathscr{M}}x\|^2 + \|(I - P_{\mathscr{M}})x\|^2$,
(iii) *each $x \in \mathscr{H}$ has a unique representation as a sum of an element of $\mathscr{M}$ and an element of $\mathscr{M}^\perp$*, i.e.

$$x = P_{\mathscr{M}}x + (I - P_{\mathscr{M}})x, \tag{2.3.7}$$

(iv) $P_{\mathscr{M}}x_n \to P_{\mathscr{M}}x$ *if* $\|x_n - x\| \to 0$,
(v) $x \in \mathscr{M}$ *if and only if* $P_{\mathscr{M}}x = x$,
(vi) $x \in \mathscr{M}^\perp$ *if and only if* $P_{\mathscr{M}}x = 0$,
and
(vii) $\mathscr{M}_1 \subseteq \mathscr{M}_2$ *if and only if* $P_{\mathscr{M}_1}P_{\mathscr{M}_2}x = P_{\mathscr{M}_1}x$ *for all* $x \in \mathscr{H}$.

PROOF. (i) $\alpha P_{\mathscr{M}}x + \beta P_{\mathscr{M}}y \in \mathscr{M}$ since $\mathscr{M}$ is a linear subspace of $\mathscr{H}$. Also

$$\begin{aligned}\alpha x + \beta y - (\alpha P_{\mathscr{M}}x + \beta P_{\mathscr{M}}y) &= \alpha(x - P_{\mathscr{M}}x) + \beta(y - P_{\mathscr{M}}y) \\ &\in \mathscr{M}^\perp,\end{aligned}$$

since $\mathscr{M}^\perp$ is a linear subspace of $\mathscr{H}$ by Proposition 2.3.1. These two properties identify $\alpha P_{\mathscr{M}}x + \beta P_{\mathscr{M}}y$ as the projection $P_{\mathscr{M}}(\alpha x + \beta y)$.

(ii) This is an immediate consequence of the orthogonality of $P_{\mathscr{M}}x$ and $(I - P_{\mathscr{M}})x$.

(iii) One such representation is clearly $x = P_{\mathscr{M}}x + (I - P_{\mathscr{M}})x$. If $x = y + z$, $y \in \mathscr{M}$, $z \in \mathscr{M}^\perp$ is another, then

$$y - P_{\mathscr{M}}x + z - (I - P_{\mathscr{M}})x = 0.$$

Taking inner products of each side with $y - P_{\mathscr{M}}x$ gives $\|y - P_{\mathscr{M}}x\|^2 = 0$, since $z - (I - P_{\mathscr{M}})x \in \mathscr{M}^{\perp}$. Hence $y = P_{\mathscr{M}}x$ and $z = (I - P_{\mathscr{M}})x$.

(iv) By (ii), $\|P_{\mathscr{M}}(x_n - x)\|^2 \le \|x_n - x\|^2 \to 0$ if $\|x_n - x\| \to 0$.

(v) $x \in \mathscr{M}$ if and only if the unique representation $x = y + z$, $y \in \mathscr{M}$, $z \in \mathscr{M}^{\perp}$, is such that $y = x$ and $z = 0$, i.e. if and only if $P_{\mathscr{M}}x = x$.

(vi) Repeat the argument in (v) with $y = 0$ and $z = x$.

(vii) $x = P_{\mathscr{M}_2}x + (I - P_{\mathscr{M}_2})x$. Projecting each side onto $\mathscr{M}_1$ we obtain

$$P_{\mathscr{M}_1}x = P_{\mathscr{M}_1}P_{\mathscr{M}_2}x + P_{\mathscr{M}_1}(I - P_{\mathscr{M}_2})x.$$

Hence $P_{\mathscr{M}_1}x = P_{\mathscr{M}_1}P_{\mathscr{M}_2}x$ for all $x \in \mathscr{H}$ if and only if $P_{\mathscr{M}_1}y = 0$ for all $y \in \mathscr{M}_2^{\perp}$, i.e. if and only if $\mathscr{M}_2^{\perp} \subseteq \mathscr{M}_1^{\perp}$, i.e. if and only if $\mathscr{M}_1 \subseteq \mathscr{M}_2$. □

The Prediction Equations. Given a Hilbert space $\mathscr{H}$, a closed subspace $\mathscr{M}$, and an element $x \in \mathscr{H}$, Theorem 2.3.1 shows that the element of $\mathscr{M}$ closest to x is the unique element $\hat{x} \in \mathscr{M}$ such that

$$\langle x - \hat{x}, y \rangle = 0 \quad \text{for all } y \in \mathscr{M}. \tag{2.3.8}$$

The equations (2.3.1) and (2.3.2) which arose in Examples 2.3.1 and 2.3.2 are special cases of (2.3.8). In later chapters we shall constantly be making use of the equations (2.3.8), interpreting $\hat{x} = P_{\mathscr{M}}x$ as the best predictor of x in the subspace $\mathscr{M}$.

Remark 1. It is helpful to visualize the projection theorem in terms of Figure 2.1, which depicts the special case in which $\mathscr{H} = \mathbb{R}^3$, $\mathscr{M}$ is the plane containing $\mathbf{x}_1$ and $\mathbf{x}_2$, and $\hat{\mathbf{y}} = P_{\mathscr{M}}\mathbf{y}$. The prediction equation (2.3.8) is simply the statement (obvious in this particular example) that $\mathbf{y} - \hat{\mathbf{y}}$ must be orthogonal to $\mathscr{M}$. The projection theorem tells us that $\hat{y} = P_{\mathscr{M}}y$ is uniquely determined by this condition for *any* Hilbert space $\mathscr{H}$ and closed subspace $\mathscr{M}$. This justifies in particular our use of equations (2.3.2) in Example 2.3.2. As we shall see later (especially in Chapter 5), the projection theorem plays a fundamental role in all problems involving the approximation or prediction of random variables with finite variance.

Example 2.3.3 (Minimum Mean Squared Error Linear Prediction of a Stationary Process). Let $\{X_t, t = 0, \pm 1, \dots\}$ be a stationary process on $(\Omega, \mathscr{F}, P)$ with mean zero and autocovariance function $\gamma(\cdot)$, and consider the problem of finding the linear combination $\hat{X}_{n+1} = \sum_{j=1}^{n} \phi_{nj} X_{n+1-j}$ which best approximates X_{n+1} in the sense that $E|X_{n+1} - \sum_{j=1}^{n} \phi_{nj} X_{n+1-j}|^2$ is minimum. This problem is easily solved with the aid of the projection theorem by taking $\mathscr{H} = L^2(\Omega, \mathscr{F}, P)$ and $\mathscr{M} = \{\sum_{j=1}^{n} \alpha_j X_{n+1-j} : \alpha_1, \dots, \alpha_n \in \mathbb{R}\}$. Since minimization of $E|X_{n+1} - \hat{X}_{n+1}|^2$ is identical to minimization of the squared norm $\|X_{n+1} - \hat{X}_{n+1}\|^2$, we see at once that $\hat{X}_{n+1} = P_{\mathscr{M}} X_{n+1}$. The prediction equations (2.3.8) are

$$\left\langle X_{n+1} - \sum_{j=1}^{n} \phi_{nj} X_{n+1-j}, Y \right\rangle = 0 \quad \text{for all } Y \in \mathscr{M},$$

which, by the linearity of the inner product, are equivalent to the n equations

$$\left\langle X_{n+1} - \sum_{j=1}^{n} \phi_{nj} X_{n+1-j}, X_k \right\rangle = 0, \qquad k = n, n-1, \ldots, 1.$$

Recalling the definition $\langle X, Y \rangle = E(XY)$ of the inner product in $L^2(\Omega, \mathscr{F}, P)$, we see that the prediction equations can be written in the form

$$\Gamma_n \phi_n = \gamma_n \tag{2.3.9}$$

where $\boldsymbol{\phi}_n = (\phi_{n1}, \ldots, \phi_{nn})'$, $\gamma_n = (\gamma(1), \ldots, \gamma(n))'$ and $\Gamma_n = [\gamma(i-j)]_{i,j=1}^{n}$. The projection theorem guarantees that there is *at least* one solution ϕ_n of (2.3.9). If Γ_n is singular then (2.3.9) will have infinitely many solutions, but the projection theorem guarantees that every solution will give the same (uniquely defined) predictor $\hat{X}_{n+1}$.

EXAMPLE 2.3.4. To illustrate the last assertion of Example 2.3.3, consider the stationary process

$$X_t = A\cos(\omega t) + B\sin(\omega t),$$

where $\omega \in (0, \pi)$ is constant and A, B are uncorrelated random variables with mean 0 and variance σ^2. We showed in Example 1.3.1 that for this process, $\gamma(h) = \sigma^2 \cos(\omega h)$. It is easy to check from (2.3.9) (see Problem 2.6) that

$$\phi_1 = \cos\omega \quad \text{and} \quad \phi_2 = (2\cos\omega, -1)'.$$

Thus

$$\hat{X}_3 = (2\cos\omega)X_2 - X_1.$$

The mean squared error of $\hat{X}_3$ is

$$E(X_3 - (2\cos\omega)X_2 + X_1)^2 = 0,$$

showing that for this process we have the identity,

$$X_3 = (2\cos\omega)X_2 - X_1. \tag{2.3.10}$$

The same argument and the stationarity of $\{X_t\}$ show that

$$\hat{X}_4 = (2\cos\omega)X_3 - X_2, \tag{2.3.11}$$

again with mean squared error zero. Because of the relation (2.3.10) there are infinitely many ways to reexpress $\hat{X}_4$ in terms of X_1, X_2 and X_3. This is reflected by the fact that Γ_3 is singular for this process and (2.3.9) has infinitely many solutions for ϕ_3.

§2.4 Orthonormal Sets

Definition 2.4.1 (Closed Span). The closed span $\overline{\text{sp}}\{x_t, t \in T\}$ of any subset $\{x_t, t \in T\}$ of a Hilbert space $\mathscr{H}$ is defined to be the smallest closed subspace of $\mathscr{H}$ which contains each element x_t, $t \in T$.

Remark 1. The closed span of a finite set $\{x_1, \ldots, x_n\}$ is the set of all linear combinations, $y = \alpha_1 x_1 + \cdots + \alpha_n x_n$, $\alpha_1, \ldots, \alpha_n \in \mathbb{C}$ (or $\mathbb{R}$ if $\mathscr{H}$ is real). See Problem 2.7. If for example $\mathbf{x}_1, \mathbf{x}_2 \in \mathbb{R}^3$ and $\mathbf{x}_1$ is not a scalar multiple of $\mathbf{x}_2$ then $\overline{\mathrm{sp}}\{\mathbf{x}_1, \mathbf{x}_2\}$ is the plane containing $\mathbf{x}_1$ and $\mathbf{x}_2$.

Remark 2. If $\mathscr{M} = \overline{\mathrm{sp}}\{x_1, \ldots, x_n\}$, then for any given $x \in \mathscr{H}$, $P_{\mathscr{M}}x$ is the unique element of the form

$$P_{\mathscr{M}}x = \alpha_1 x_1 + \cdots + \alpha_n x_n,$$

such that

$$\langle x - P_{\mathscr{M}}x, y \rangle = 0, \qquad y \in \mathscr{M},$$

or equivalently such that

$$\langle P_{\mathscr{M}}x, x_j \rangle = \langle x, x_j \rangle, \qquad j = 1, \ldots, n. \tag{2.4.1}$$

The equations (2.4.1) can be rewritten as a set of linear equations for $\alpha_1, \ldots, \alpha_n$, viz.

$$\sum_{i=1}^{n} \alpha_i \langle x_i, x_j \rangle = \langle x, x_j \rangle, \qquad j = 1, \ldots, n. \tag{2.4.2}$$

By the projection theorem the system (2.4.2) has at least one solution for $\alpha_1, \ldots, \alpha_n$. The uniqueness of $P_{\mathscr{M}}x$ implies that all solutions of (2.4.2) must yield the same element $\alpha_1 x_1 + \cdots + \alpha_n x_n$.

Definition 2.4.2 (Orthonormal Set). A set $\{e_t, t \in T\}$ of elements of an inner-product space is said to be orthonormal if for every $s, t \in T$,

$$\langle e_s, e_t \rangle = \begin{cases} 1 & \text{if } s = t, \\ 0 & \text{if } s \neq t. \end{cases} \tag{2.4.3}$$

EXAMPLE 2.4.1. The set of vectors $\{(1,0,0)', (0,1,0)', (0,0,1)'\}$ is an orthonormal set in $\mathbb{R}^3$.

EXAMPLE 2.4.2. Any sequence $\{Z_t, t \in \mathbb{Z}\}$ of independent standard normal random variables is an orthonormal set in $L^2(\Omega, \mathscr{F}, P)$.

Theorem 2.4.1. *If $\{e_1, \ldots, e_k\}$ is an orthonormal subset of the Hilbert space $\mathscr{H}$ and $\mathscr{M} = \overline{\mathrm{sp}}\{e_1, \ldots, e_k\}$, then*

$$P_{\mathscr{M}}x = \sum_{i=1}^{k} \langle x, e_i \rangle e_i \quad \textit{for all } x \in \mathscr{H}, \tag{2.4.4}$$

$$\|P_{\mathscr{M}}x\|^2 = \sum_{i=1}^{k} |\langle x, e_i \rangle|^2 \quad \textit{for all } x \in \mathscr{H}, \tag{2.4.5}$$

$$\left\| x - \sum_{i=1}^{k} \langle x, e_i \rangle e_i \right\| \leq \left\| x - \sum_{i=1}^{k} c_i e_i \right\| \quad \textit{for all } x \in \mathscr{H}, \tag{2.4.6}$$

and for all $c_1, \dots, c_k \in \mathbb{C}$ *(or* $\mathbb{R}$ *if* $\mathscr{H}$ *is real). Equality holds in* (2.4.6) *if and only if* $c_i = \langle x, e_i \rangle$, $i = 1, \dots, k$.

The numbers $\langle x, e_i \rangle$ *are sometimes called the Fourier coefficients of x relative to the set* $\{e_1, \dots, e_k\}$.

PROOF. To establish (2.4.4) it suffices by Remark 2 to check that $P_{\mathscr{M}}x$ as defined by (2.4.4) satisfies the prediction equations (2.4.1), i.e. that

$$\left\langle \sum_{i=1}^{k} \langle x, e_i \rangle e_i, e_j \right\rangle = \langle x, e_j \rangle, \qquad j = 1, \dots, k.$$

But this is an immediate consequence of the orthonormality condition (2.4.3).

The proof of (2.4.5) is a routine computation using properties of the inner product and the assumed orthonormality of $\{e_1, \dots, e_k\}$.

By Theorem 2.3.1 (ii), $\|x - P_{\mathscr{M}}x\| \le \|x - y\|$ for all $y \in \mathscr{M}$, and this is precisely the inequality (2.4.6). By Theorem 2.3.1 (ii) again, there is equality in (2.4.6) if and only if

$$\sum_{i=1}^{k} c_i e_i = P_{\mathscr{M}}x = \sum_{i=1}^{k} \langle x, e_i \rangle e_i. \tag{2.4.7}$$

Taking inner products of each side with e_j and recalling the orthonormality assumption, we immediately find that (2.4.7) is equivalent to the condition $c_j = \langle x, e_j \rangle, j = 1, \dots, k$. □

Corollary 2.4.1 (Bessel's Inequality). *If x is any element of a Hilbert space* $\mathscr{H}$ *and* $\{e_1, \dots, e_k\}$ *is an orthonormal subset of* $\mathscr{H}$ *then*

$$\sum_{i=1}^{k} |\langle x, e_i \rangle|^2 \le \|x\|^2. \tag{2.4.8}$$

PROOF. This follows at once from (2.4.5) and Proposition 2.3.2 (ii). □

Definition 2.4.3. (Complete Orthonormal Set). If $\{e_t, t \in T\}$ is an orthonormal subset of the Hilbert space $\mathscr{H}$ and if $\mathscr{H} = \overline{\text{sp}}\{e_t, t \in T\}$ then we say that $\{e_t, t \in T\}$ is a complete orthonormal set or an orthonormal basis for $\mathscr{H}$.

Definition 2.4.4 (Separability). The Hilbert space $\mathscr{H}$ is separable if $\mathscr{H} = \overline{\text{sp}}\{e_t, t \in T\}$ with $\{e_t, t \in T\}$ a finite or countably infinite orthonormal set.

Theorem 2.4.2. *If* $\mathscr{H}$ *is the separable Hilbert space* $\mathscr{H} = \overline{\text{sp}}\{e_1, e_2, \dots\}$ *where* $\{e_i, i = 1, 2, \dots\}$ *is an orthonormal set, then*

(i) *the set of all finite linear combinations of* $\{e_1, e_2, \dots\}$ *is dense in* $\mathscr{H}$, *i.e. for each* $x \in \mathscr{H}$ *and* $\varepsilon > 0$, *there exists a positive integer k and constants* $c_1, \dots, c_k$ *such that*

$$\left\| x - \sum_{i=1}^{k} c_i e_i \right\| < \varepsilon, \tag{2.4.9}$$

(ii) $x = \sum_{i=1}^{\infty} \langle x, e_i \rangle e_i$ *for each* $x \in \mathscr{H}$, *i.e.* $\|x - \sum_{i=1}^{n} \langle x, e_i \rangle e_i\| \to 0$ *as* $n \to \infty$,
(iii) $\|x\|^2 = \sum_{i=1}^{\infty} |\langle x, e_i \rangle|^2$ *for each* $x \in \mathscr{H}$,
(iv) $\langle x, y \rangle = \sum_{i=1}^{\infty} \langle x, e_i \rangle \langle e_i, y \rangle$ *for each* $x, y \in \mathscr{H}$, *and*
(v) $x = 0$ *if and only if* $\langle x, e_i \rangle = 0$ *for all* $i = 1, 2, \ldots$.

The result (iv) *is known as* ***Parseval's identity***.

PROOF. (i) If $S = \bigcup_{j=1}^{\infty} \overline{\text{sp}}\{e_1, \ldots, e_j\}$, the set of all finite linear combinations of $\{e_1, e_2, \ldots\}$, then the closure $\bar{S}$ of S is a closed subspace of $\mathscr{H}$ (Problem 2.17) containing $\{e_i, i = 1, 2, \ldots\}$. Since $\mathscr{H}$ is by assumption the smallest such closed subspace, we conclude that $\bar{S} = \mathscr{H}$.

(ii) By Bessel's inequality (2.4.8), $\sum_{i=1}^{k} |\langle x, e_i \rangle|^2 \le \|x\|^2$ for all positive integers k. Hence $\sum_{i=1}^{\infty} |\langle x, e_i \rangle|^2 \le \|x\|^2$. From (2.4.6) and (2.4.9) we conclude that for each $\varepsilon > 0$ there exists a positive integer k such that

$$\left\| x - \sum_{i=1}^{k} \langle x, e_i \rangle e_i \right\| < \varepsilon.$$

Now by Theorem 2.4.1, $\sum_{i=1}^{n} \langle x, e_i \rangle e_i = P_{\mathscr{M}} x$ where $\mathscr{M} = \overline{\text{sp}}\{e_1, \ldots, e_n\}$, and since for $k \le n$, $\sum_{i=1}^{k} \langle x, e_i \rangle e_i \in \mathscr{M}$, we also have

$$\left\| x - \sum_{i=1}^{n} \langle x, e_i \rangle e_i \right\| < \varepsilon \quad \text{for all } n \ge k. \tag{2.4.10}$$

(iii) From (2.4.10) we can write, for $n \ge k$,

$$\begin{aligned} \|x\|^2 &= \left\| x - \sum_{i=1}^{n} \langle x, e_i \rangle e_i \right\|^2 + \left\| \sum_{i=1}^{n} \langle x, e_i \rangle e_i \right\|^2 \\ &< \varepsilon^2 + \sum_{i=1}^{\infty} |\langle x, e_i \rangle|^2. \end{aligned}$$

Since $\varepsilon > 0$ was arbitrary, we deduce that

$$\|x\|^2 \le \sum_{i=1}^{\infty} |\langle x, e_i \rangle|^2,$$

which together with the reversed inequality proved in (ii), establishes (iii).

(iv) The result (2.4.10) established in (iii) states that $\|\sum_{i=1}^{n} \langle x, e_i \rangle e_i - x\| \to 0$ as $n \to \infty$ for each $x \in \mathscr{H}$. By continuity of the inner product we therefore have, for each $x, y \in \mathscr{H}$,

$$\begin{aligned} \langle x, y \rangle &= \lim_{n \to \infty} \left\langle \sum_{i=1}^{n} \langle x, e_i \rangle e_i, \sum_{j=1}^{n} \langle y, e_j \rangle e_j \right\rangle \\ &= \lim_{n \to \infty} \sum_{i=1}^{n} \langle x, e_i \rangle \langle e_i, y \rangle \\ &= \sum_{i=1}^{\infty} \langle x, e_i \rangle \langle e_i, y \rangle. \end{aligned}$$

(v) This result is an immediate consequence of (ii). □

Remark 3. Separable Hilbert spaces are frequently encountered as the closed spans of countable subsets of possibly non-separable Hilbert spaces.

§2.5 Projection in $\mathbb{R}^n$

In Examples 2.1.1, 2.1.2 and 2.2.1 we showed that $\mathbb{R}^n$ is a Hilbert space with the inner product

$$\langle \mathbf{x}, \mathbf{y} \rangle = \sum_{i=1}^{n} x_i y_i, \tag{2.5.1}$$

the corresponding squared norm

$$\|\mathbf{x}\|^2 = \sum_{i=1}^{n} x_i^2, \tag{2.5.2}$$

and the angle between $\mathbf{x}$ and $\mathbf{y}$,

$$\theta = \cos^{-1}\left(\frac{\langle \mathbf{x}, \mathbf{y} \rangle}{\|\mathbf{x}\| \, \|\mathbf{y}\|}\right). \tag{2.5.3}$$

Every closed subspace $\mathcal{M}$ of the Hilbert space $\mathbb{R}^n$ can be expressed by means of Gram–Schmidt orthogonalization (see for example Simmons (1963)) as $\mathcal{M} = \overline{\text{sp}}\{\mathbf{e}_1, \ldots, \mathbf{e}_m\}$ where $\{\mathbf{e}_1, \ldots, \mathbf{e}_m\}$ is an orthonormal subset of $\mathcal{M}$ and m $(\leq n)$ is called the dimension of $\mathcal{M}$ (see also Problem 2.14). If $m < n$ then there is an orthonormal subset $\{\mathbf{e}_{m+1}, \ldots, \mathbf{e}_n\}$ of $\mathcal{M}^\perp$ such that $\mathcal{M}^\perp = \overline{\text{sp}}\{\mathbf{e}_{m+1}, \ldots, \mathbf{e}_n\}$. By Proposition 2.3.2 (iii) every $\mathbf{x} \in \mathbb{R}^n$ can be expressed uniquely as a sum of two elements of $\mathcal{M}$ and $\mathcal{M}^\perp$ respectively, namely

$$\mathbf{x} = P_{\mathcal{M}}\mathbf{x} + (I - P_{\mathcal{M}})\mathbf{x}, \tag{2.5.4}$$

where, by Theorem 2.4.1,

$$P_{\mathcal{M}}\mathbf{x} = \sum_{i=1}^{m} \langle \mathbf{x}, \mathbf{e}_i \rangle \mathbf{e}_i \tag{2.5.5}$$

and

$$(I - P_{\mathcal{M}})\mathbf{x} = \sum_{i=m+1}^{n} \langle \mathbf{x}, \mathbf{e}_i \rangle \mathbf{e}_i. \tag{2.5.6}$$

The following theorem enables us to compute $P_{\mathcal{M}}\mathbf{x}$ directly from any specified set of vectors $\{\mathbf{x}_1, \ldots, \mathbf{x}_m\}$ spanning $\mathcal{M}$.

Theorem 2.5.1. *If* $\mathbf{x}_i \in \mathbb{R}^n$, $i = 1, \ldots, m$, *and* $\mathcal{M} = \overline{\text{sp}}\{\mathbf{x}_1, \ldots, \mathbf{x}_m\}$ *then*

$$P_{\mathcal{M}}\mathbf{x} = X\boldsymbol{\beta}, \tag{2.5.7}$$

where X *is the* $n \times m$ *matrix whose* j^{th} *column is* $\mathbf{x}_j$ *and*

$$X'X\boldsymbol{\beta} = X'\mathbf{x}. \tag{2.5.8}$$

Equation (2.5.8) *has at least one solution for* $\boldsymbol{\beta}$ *but* $X\boldsymbol{\beta}$ *is the same for all solutions. There is exactly one solution of* (2.5.8) *if and only if* $X'X$ *is non-singular and in this case*

$$P_{\mathscr{M}}\mathbf{x} = X(X'X)^{-1}X'\mathbf{x}. \tag{2.5.9}$$

PROOF. Since $P_{\mathscr{M}}\mathbf{x} \in \mathscr{M}$, we can write

$$P_{\mathscr{M}}\mathbf{x} = \sum_{i=1}^{m} \beta_i \mathbf{x}_i = X\boldsymbol{\beta}, \qquad \text{for some } \boldsymbol{\beta} = (\beta_1, \ldots, \beta_m)' \in \mathbb{R}^m. \tag{2.5.10}$$

The prediction equations (2.3.8) are equivalent in this case to

$$\langle X\boldsymbol{\beta}, \mathbf{x}_j \rangle = \langle \mathbf{x}, \mathbf{x}_j \rangle, \qquad j = 1, \ldots, m, \tag{2.5.11}$$

and in matrix form these equations can be written

$$X'X\boldsymbol{\beta} = X'\mathbf{x}. \tag{2.5.12}$$

The existence of at least one solution for $\boldsymbol{\beta}$ is guaranteed by the existence of the projection $P_{\mathscr{M}}\mathbf{x}$. The fact that $X\boldsymbol{\beta}$ is the same for all solutions is guaranteed by the uniqueness of $P_{\mathscr{M}}\mathbf{x}$. The last statement of the theorem follows at once from (2.5.7) and (2.5.8). □

Remark 1. If $\{\mathbf{x}_1, \ldots, \mathbf{x}_m\}$ is an orthonormal set then $X'X$ is the identity matrix and so we find that

$$P_{\mathscr{M}}\mathbf{x} = XX'\mathbf{x} = \sum_{i=1}^{m} \langle \mathbf{x}, \mathbf{x}_i \rangle \mathbf{x}_i,$$

in accordance with (2.5.5)

Remark 2. If $\{\mathbf{x}_1, \ldots, \mathbf{x}_m\}$ is a linearly independent set then there must be a unique vector $\boldsymbol{\beta}$ such that $P_{\mathscr{M}}\mathbf{x} = X\boldsymbol{\beta}$. This means that (2.5.8) must have a unique solution, which in turn implies that $X'X$ is non-singular and

$$P_{\mathscr{M}}\mathbf{x} = X(X'X)^{-1}X'\mathbf{x} \quad \text{for all } \mathbf{x} \in \mathbb{R}^n.$$

The matrix $X(X'X)^{-1}X'$ must be the same for all linearly independent sets $\{x_1, \ldots, x_m\}$ spanning $\mathscr{M}$ since $P_{\mathscr{M}}$ is a uniquely defined mapping on $\mathbb{R}^n$.

Remark 3. Given a real $n \times n$ matrix M, how can we tell whether or not there is a subspace $\mathscr{M}$ of $\mathbb{R}^n$ such that $M\mathbf{x} = P_{\mathscr{M}}\mathbf{x}$ for all $\mathbf{x} \in \mathbb{R}^n$? If there is such a subspace we say that M is a projection matrix. Such matrices are characterized in the next theorem.

Theorem 2.5.2. *The* $n \times n$ *matrix* M *is a projection matrix if and only if*

$$\text{(a)} \quad M' = M$$

and

$$\text{(b)} \quad M^2 = M.$$

Proof. If M is the projection matrix corresponding to some subspace $\mathcal{M}$ then by Remark 2 it can be written in the form $X(X'X)^{-1}X'$ where X is any matrix having linearly independent columns which span $\mathcal{M}$. It is easily verified that (a) and (b) are then satisfied.

Suppose now that (a) and (b) are satisfied. We shall show that $M\mathbf{x} = P_{\mathcal{M}}\mathbf{x}$ for all $\mathbf{x} \in \mathbb{R}^n$ where $\mathcal{M}$ is the range of M defined by

$$R(M) = \{M\mathbf{x} : \mathbf{x} \in \mathbb{R}^n\}.$$

First observe that $M\mathbf{x} \in R(M)$ by definition. Secondly we know that for any $\mathbf{y} \in R(M)$ there exists $\mathbf{w} \in \mathbb{R}^n$ such that $\mathbf{y} = M\mathbf{w}$. Hence

$$\langle \mathbf{x} - M\mathbf{x}, \mathbf{y} \rangle = \langle \mathbf{x} - M\mathbf{x}, M\mathbf{w} \rangle = \mathbf{x}'(I - M)'M\mathbf{w} = 0 \quad \text{for all } \mathbf{y} \in R(M),$$

showing that $M\mathbf{x}$ is indeed the required projection. □

§2.6 Linear Regression and the General Linear Model

Consider the problem of finding the "best" straight line

$$y = \theta_1 x + \theta_2, \tag{2.6.1}$$

or equivalently the best values $\hat{\theta}_1$, $\hat{\theta}_2$ of $\theta_1, \theta_2 \in \mathbb{R}$, to fit a given set of data points (x_i, y_i), $i = 1, \ldots, n$. In least squares regression the best estimates $\hat{\theta}_1$, $\hat{\theta}_2$ are defined to be values of θ_1, θ_2 which minimize the sum,

$$S(\theta_1, \theta_2) = \sum_{i=1}^{n} (y_i - \theta_1 x_i - \theta_2)^2,$$

of squared deviations of the observations y_i from the fitted values $\theta_1 x_i + \theta_2$.

This problem reduces to that of computing a projection in $\mathbb{R}^n$ as is easily seen by writing $S(\theta_1, \theta_2)$ in the equivalent form

$$S(\theta_1, \theta_2) = \|\mathbf{y} - \theta_1 \mathbf{x} - \theta_2 \mathbf{1}\|^2, \tag{2.6.2}$$

where $\mathbf{x} = (x_1, \ldots, x_n)'$, $\mathbf{1} = (1, \ldots, 1)'$ and $\mathbf{y} = (y_1, \ldots, y_n)'$. By the projection theorem there is a unique vector of the form $(\hat{\theta}_1 \mathbf{x} + \hat{\theta}_2 \mathbf{1})$ which minimizes $S(\theta_1, \theta_2)$, namely $P_{\mathcal{M}}\mathbf{y}$ where $\mathcal{M} = \overline{\text{sp}}\{\mathbf{x}, \mathbf{1}\}$.

Defining X to be the $n \times 2$ matrix $X = [\mathbf{x}, \mathbf{1}]$ and $\hat{\boldsymbol{\theta}}$ to be the column vector $\hat{\boldsymbol{\theta}} = (\hat{\theta}_1, \hat{\theta}_2)'$, we deduce from Theorem 2.5.1 that

$$P_{\mathcal{M}}\mathbf{y} = X\hat{\boldsymbol{\theta}}$$

where

$$X'X\hat{\boldsymbol{\theta}} = X'\mathbf{y}. \tag{2.6.3}$$

There is a unique solution $\hat{\boldsymbol{\theta}}$ if and only if $X'X$ is non-singular. In this case

$$\hat{\boldsymbol{\theta}} = (X'X)^{-1}X'\mathbf{y}. \tag{2.6.4}$$

If $X'X$ is singular there are infinitely many solutions of (2.6.4), however by the uniqueness of $P_{\mathcal{M}}\mathbf{y}$, $X\hat{\boldsymbol{\theta}}$ is the same for all of them.

The argument just given applies equally well to least squares estimation for the general linear model. The general problem is as follows. Given a set of data points

$$(x_i^{(1)}, x_i^{(2)}, \ldots, x_i^{(m)}, y_i), \qquad i = 1, \ldots, n; m \le n,$$

we are required to find a value $\hat{\boldsymbol{\theta}} = (\hat{\theta}_1, \ldots, \hat{\theta}_m)'$ of $\boldsymbol{\theta} = (\theta_1, \ldots, \theta_m)'$ which minimizes

$$\begin{aligned} S(\boldsymbol{\theta}) &= \sum_{i=1}^{n} (y_i - \theta_1 x_i^{(1)} - \cdots - \theta_m x_i^{(m)})^2 \\ &= \|\mathbf{y} - \theta_1 \mathbf{x}^{(1)} - \cdots - \theta_m \mathbf{x}^{(m)}\|^2, \end{aligned}$$

where $\mathbf{y} = (y_1, \ldots, y_n)'$ and $\mathbf{x}^{(j)} = (x_1^{(j)}, \ldots, x_n^{(j)})', j = 1, \ldots, m$. By the projection theorem there is a unique vector of the form $(\hat{\theta}_1 \mathbf{x}^{(1)} + \cdots + \hat{\theta}_m \mathbf{x}^{(m)})$ which minimizes $S(\boldsymbol{\theta})$, namely $P_{\mathcal{M}}\mathbf{y}$ where $\mathcal{M} = \overline{\operatorname{sp}}\{\mathbf{x}^{(1)}, \ldots, \mathbf{x}^{(m)}\}$.

Defining X to be the $n \times m$ matrix $X = [\mathbf{x}^{(1)}, \ldots, \mathbf{x}^{(m)}]$ and $\hat{\boldsymbol{\theta}}$ to be the column vector $\hat{\boldsymbol{\theta}} = (\theta_1, \ldots, \theta_m)'$, we deduce from Theorem 2.5.1 that

$$P_{\mathcal{M}}\mathbf{y} = X\hat{\boldsymbol{\theta}}$$

where

$$X'X\hat{\boldsymbol{\theta}} = X'\mathbf{y} \tag{2.6.5}$$

As in the special case of fitting a straight line, $\hat{\boldsymbol{\theta}}$ is uniquely defined if and only if $X'X$ is non-singular, in which case

$$\hat{\boldsymbol{\theta}} = (X'X)^{-1}X'\mathbf{y}. \tag{2.6.6}$$

If $X'X$ is singular then there are infinitely many solutions of (2.6.5) but $X\hat{\boldsymbol{\theta}}$ is the same for all of them.

In spite of the assumed linearity in the parameters $\theta_1, \ldots, \theta_m$, the applications of the general linear model are very extensive. As a simple illustration, let us fit a quadratic function,

$$y = \theta_1 x^2 + \theta_2 x + \theta_3,$$

to the data

x	0	1	2	3	4
y	1	0	3	5	8

The matrix X for this problem is

$$X = \begin{bmatrix} 0 & 0 & 1 \\ 1 & 1 & 1 \\ 4 & 2 & 1 \\ 9 & 3 & 1 \\ 16 & 4 & 1 \end{bmatrix} \quad \text{giving } (X'X)^{-1} = \frac{1}{140}\begin{bmatrix} 10 & -40 & 20 \\ -40 & 174 & -108 \\ 20 & -108 & 124 \end{bmatrix}.$$

The least squares estimate $\hat{\boldsymbol{\theta}} = (\hat{\theta}_1, \hat{\theta}_2, \hat{\theta}_3)'$ is therefore unique and is found from (2.6.6) to be

$$\hat{\boldsymbol{\theta}} = (0.5, -0.1, 0.6)'.$$

The vector of fitted values $X\hat{\boldsymbol{\theta}} = P_{\mathscr{M}}\mathbf{y}$ is given by

$$X\hat{\boldsymbol{\theta}} = (0.6, 1, 2.4, 4.8, 8.2)',$$

as compared with the vector of observations,

$$\mathbf{y} = (1, 0, 3, 5, 8)'.$$

§2.7 Mean Square Convergence, Conditional Expectation and Best Linear Prediction in $L^2(\Omega, \mathscr{F}, P)$

All results in this section will be stated for the real Hilbert space $L^2 = L^2(\Omega, \mathscr{F}, P)$ with inner product $\langle X, Y \rangle = E(XY)$. The reader should have no difficulty however in writing down analogous results for the complex space $L^2(\Omega, \mathscr{F}, P)$ with inner product $\langle X, Y \rangle = E(X\bar{Y})$. As indicated in Example 2.2.2, mean square convergence is just another name for norm convergence in L^2, i.e. if $X_n, X \in L^2$, then

$$X_n \xrightarrow{\text{m.s.}} X \quad \text{if and only if } \|X_n - X\|^2 = E|X_n - X|^2 \to 0 \text{ as } n \to \infty. \tag{2.7.1}$$

By simply restating properties already established for norm convergence we obtain the following proposition.

Proposition 2.7.1 (Properties of Mean Square Convergence).
(a) *X_n converges in mean square if and only if* $E|X_m - X_n|^2 \to 0$ as $m, n \to \infty$.
(b) *If $X_n \xrightarrow{\text{m.s.}} X$ and $Y_n \xrightarrow{\text{m.s.}} Y$ then as $n \to \infty$,*

$$\text{(i) } EX_n = \langle X_n, 1 \rangle \to \langle X, 1 \rangle = EX,$$

$$\text{(ii) } E|X_n|^2 = \langle X_n, X_n \rangle \to \langle X, X \rangle = E|X|^2,$$

and

$$\text{(iii) } E(X_n Y_n) = \langle X_n, Y_n \rangle \to \langle X, Y \rangle = E(XY).$$

Definition 2.7.1 (Best Mean Square Predictor of Y). If $\mathscr{M}$ is a closed subspace of L^2 and $Y \in L^2$, then the best mean square predictor of Y in $\mathscr{M}$ is the element $\hat{Y} \in \mathscr{M}$ such that

$$\|Y - \hat{Y}\|^2 = \inf_{Z \in \mathscr{M}} \|Y - Z\|^2 = \inf_{Z \in \mathscr{M}} E|Y - Z|^2. \tag{2.7.2}$$

The projection theorem immediately identifies the unique best predictor of Y in $\mathscr{M}$ as $P_{\mathscr{M}} Y$. By imposing a little more structure on the closed subspace

$\mathcal{M}$, we are led from Definition 2.7.1 to the notions of conditional expectation and best linear predictor.

Definition 2.7.2 (The Conditional Expectation, $E_{\mathcal{M}}X$). If $\mathcal{M}$ is a closed subspace of L^2 containing the constant functions, and if $X \in L^2$, then we define the conditional expectation of X given $\mathcal{M}$ to be the projection,

$$E_{\mathcal{M}}X = P_{\mathcal{M}}X. \tag{2.7.3}$$

Using the definition of the inner product in L^2 and the prediction equations (2.3.8) we can state equivalently that $E_{\mathcal{M}}X$ is the unique element of $\mathcal{M}$ such that

$$E(WE_{\mathcal{M}}X) = E(WX) \quad \text{for all } W \in \mathcal{M}. \tag{2.7.4}$$

Obviously the operator $E_{\mathcal{M}}$ on L^2 has all the properties of a projection operator, in particular (see Proposition 2.3.2)

$$E_{\mathcal{M}}(aX + bY) = aE_{\mathcal{M}}X + bE_{\mathcal{M}}Y, \qquad a, b \in \mathbb{R}, \tag{2.7.5}$$

$$E_{\mathcal{M}}X_n \xrightarrow{\text{m.s.}} E_{\mathcal{M}}X \quad \text{if } X_n \xrightarrow{\text{m.s.}} X \tag{2.7.6}$$

and

$$E_{\mathcal{M}_1}(E_{\mathcal{M}_2}X) = E_{\mathcal{M}_1}X \quad \text{if } \mathcal{M}_1 \subseteq \mathcal{M}_2. \tag{2.7.7}$$

Notice also that

$$E_{\mathcal{M}}1 = 1 \tag{2.7.8}$$

and if $\mathcal{M}_0$ is the closed subspace of L^2 consisting of all the constant functions, then an application of the prediction equations (2.3.8) gives

$$E_{\mathcal{M}_0}X = EX. \tag{2.7.9}$$

Definition 2.7.3 (The Conditional Expectation $E(X|Z)$). If Z is a random variable on $(\Omega, \mathcal{F}, P)$ and $X \in L^2(\Omega, \mathcal{F}, P)$ then the conditional expectation of X given Z is defined to be

$$E(X|Z) = E_{\mathcal{M}(Z)}X, \tag{2.7.10}$$

where $\mathcal{M}(Z)$ is the closed subspace of L^2 consisting of all random variables in L^2 which can be written in the form $\phi(Z)$ for some Borel function $\phi: \mathbb{R} \to \mathbb{R}$. (For the proof that $\mathcal{M}(Z)$ is a closed subspace see Problem 2.25.)

The operator $E_{\mathcal{M}(Z)}$ has all the properties (2.7.5)–(2.7.8), and in addition

$$E_{\mathcal{M}(Z)}X \geq 0 \quad \text{if } X \geq 0. \tag{2.7.11}$$

Definition 2.7.3 can be extended in a fairly obvious way as follows: if $Z_1, \dots, Z_n$ are random variables on $(\Omega, \mathcal{F}, P)$ and $X \in L^2$, then we define

$$E(X|Z_1, \dots, Z_n) = E_{\mathcal{M}(Z_1, \dots, Z_n)}(X), \tag{2.7.12}$$

where $\mathcal{M}(Z_1, \dots, Z_n)$ is the closed subspace of L^2 consisting of all random variables in L^2 of the form $\phi(Z_1, \dots, Z_n)$ for some Borel function $\phi: \mathbb{R}^n \to \mathbb{R}$. The properties of $E_{\mathcal{M}(Z)}$ listed above all carry over to $E_{\mathcal{M}(Z_1,\dots,Z_n)}$.

Conditional Expectation and Best Linear Prediction. By the projection theorem, the conditional expectation $E_{\mathcal{M}(Z_1,\dots,Z_n)}(X)$ is the best mean square predictor of X in $\mathcal{M}(Z_1, \dots, Z_n)$, i.e. it is the best function of $Z_1, \dots, Z_n$ (in the m.s. sense) for predicting X. However the determination of projections on $\mathcal{M}(Z_1, \dots, Z_n)$ is usually very difficult because of the complex nature of the equations (2.7.4). On the other hand if $Z_1, \dots, Z_n \in L^2$, it is relatively easy to compute instead the projection of X on $\overline{\text{sp}}\{1, Z_1, \dots, Z_n\} \subseteq \mathcal{M}(Z_1, \dots, Z_n)$ since we can write

$$P_{\overline{\text{sp}}\{1,Z_1,\dots,Z_n\}}(X) = \sum_{i=0}^{n} \alpha_i Z_i, \quad Z_0 = 1, \tag{2.7.13}$$

where $\alpha_0, \dots, \alpha_n$ satisfy

$$\left\langle \sum_{i=0}^{n} \alpha_i Z_i, Z_j \right\rangle = \langle X, Z_j \rangle, \quad j = 0, 1, \dots, n, \tag{2.7.14}$$

or equivalently,

$$\sum_{i=0}^{n} \alpha_i E(Z_i Z_j) = E(XZ_j), \quad j = 0, 1, \dots, n. \tag{2.7.15}$$

The projection theorem guarantees that a solution $(\alpha_0, \dots, \alpha_n)$ exists. Any solution, when substituted into (2.7.13) gives the required projection, known as the best *linear* predictor of X in terms of 1, $Z_1, \dots, Z_n$. As a projection of X onto a subspace of $\mathcal{M}(Z_1, \dots, Z_n)$ it can never have smaller mean squared error than $E_{\mathcal{M}(Z_1,\dots,Z_n)}X$. Nevertheless it is of great importance for the following reasons:

(a) it is easier to calculate than $E_{\mathcal{M}(Z_1,\dots,Z_n)}(X)$,
(b) it depends only on the first and second order moments, EX, EZ_i, $E(Z_iZ_j)$ and $E(XZ_j)$ of the joint distribution of $(X, Z_1, \dots, Z_n)$,
(c) if $(X, Z_1, \dots, Z_n)'$ has a multivariate normal distribution then (see Problem 2.20),

$$P_{\overline{\text{sp}}\{1,Z_1,\dots,Z_n\}}(X) = E_{\mathcal{M}(Z_1,\dots,Z_n)}(X).$$

Best linear predictors are defined more generally as follows:

Definition 2.7.4 (Best Linear Predictor of X in Terms of $\{Z_\lambda, \lambda \in \Lambda\}$). If $X \in L^2$ and $Z_\lambda \in L^2$ for all $\lambda \in \Lambda$, then the best linear predictor of X in terms of $\{Z_\lambda, \lambda \in \Lambda\}$ is defined to be the element of $\overline{\text{sp}}\{Z_\lambda, \lambda \in \Lambda\}$ with smallest mean square distance from X. By the projection theorem this is just $P_{\overline{\text{sp}}\{Z_\lambda, \lambda \in \Lambda\}}X$.

EXAMPLE 2.7.1. Suppose $Y = X^2 + Z$ where X and Z are independent standard normal random variables. The best predictor of Y in terms of X

is $E(Y|X) = X^2$. (The reader should check that the defining properties of $E(Y|X) = E_{\mathcal{M}(X)}Y$ are satisfied by X^2, i.e. that $X^2 \in \mathcal{M}(X)$ and that (2.7.4) is satisfied with $\mathcal{M} = \mathcal{M}(X)$.) On the other hand the best *linear* predictor of Y in terms of $\{1, X\}$ is

$$P_{\overline{\text{sp}}\{1,X\}} Y = aX + b,$$

where, by the prediction equations (2.7.15),

$$\langle aX + b, X \rangle = \langle Y, X \rangle = E(YX) = 0$$

and

$$\langle aX + b, 1 \rangle = \langle Y, 1 \rangle = E(Y) = 1.$$

Hence $a = 0$ and $b = 1$ so that

$$P_{\overline{\text{sp}}\{1,X\}} Y = 1.$$

The mean squared errors of the two predictors are

$$\|E(Y|X) - Y\|^2 = E(Z^2) = 1,$$

and

$$\|Y - P_{\overline{\text{sp}}\{1,X\}} Y\|^2 = \|Y\|^2 - 1 = E(X^4) + E(Z^2) - 1 = 3,$$

showing the substantial superiority of the best predictor over the best linear predictor in this case.

Remark 1. The conditional expectation operators $E_{\mathcal{M}(Z)}$ and $E_{\mathcal{M}(Z_1,\ldots,Z_n)}$ are usually defined on the space $L^1(\Omega, \mathcal{F}, P)$ of random variables X such that $E|X| < \infty$ (see e.g. Breiman (1968), Chapter 4). The restrictions of these operators to $L^2(\Omega, \mathcal{F}, P)$ coincide with $E_{\mathcal{M}(Z)}$ and $E_{\mathcal{M}(Z_1,\ldots,Z_n)}$ as we have defined them.

§2.8 Fourier Series

Consider the complex Hilbert space $L^2[-\pi, \pi] = L^2([-\pi, \pi], \mathcal{B}, U)$ where $\mathcal{B}$ consists of the Borel subsets of $[-\pi, \pi]$, U is the uniform probability measure $U(dx) = (2\pi)^{-1}\,dx$, and the inner product of f, $g \in L^2[-\pi, \pi]$ is defined as usual by

$$\langle f, g \rangle = Ef\bar{g} = \frac{1}{2\pi}\int_{-\pi}^{\pi} f(x)\bar{g}(x)\,dx. \tag{2.8.1}$$

The functions $\{e_n, n \in \mathbb{Z}\}$ defined by

$$e_n(x) = e^{inx}, \tag{2.8.2}$$

are orthonormal in $L^2[-\pi,\pi]$ since

$$\langle e_m, e_n\rangle = \frac{1}{2\pi}\int_{-\pi}^{\pi} e^{i(m-n)x}\,dx$$
$$= \frac{1}{2\pi}\int_{-\pi}^{\pi} [\cos(m-n)x + i\sin(m-n)x]\,dx$$
$$= \begin{cases} 1 & \text{if } m = n \\ 0 & \text{if } m \neq n. \end{cases}$$

Definition 2.8.1 (Fourier Approximations and Coefficients). The nth order Fourier approximation to any function $f \in L^2[-\pi,\pi]$ is defined to be the projection of f onto $\overline{\text{sp}}\{e_j, |j| \leq n\}$, which by Theorem 2.4.1 is

$$S_n f = \sum_{j=-n}^{n} \langle f, e_j\rangle e_j. \tag{2.8.3}$$

The coefficients

$$\langle f, e_j\rangle = \frac{1}{2\pi}\int_{-\pi}^{\pi} f(x)e^{-ijx}\,dx \tag{2.8.4}$$

are called the Fourier coefficients of the function f.

We can write (2.8.3) a little more explicitly in the form

$$S_n f(x) = \sum_{j=-n}^{n} \langle f, e_j\rangle e^{ijx}, \qquad x \in [-\pi,\pi], \tag{2.8.5}$$

and one is naturally led to investigate the senses (if any) in which the sequence of functions $\{S_n f\}$ converges to f as $n \to \infty$. In this section we shall restrict attention to mean square convergence, deferring questions of pointwise and uniform convergence to Section 2.11.

Theorem 2.8.1. (a) *The sequence $\{S_n f\}$ has a mean square limit as $n \to \infty$ which we shall denote by $\sum_{j=-\infty}^{\infty} \langle f, e_j\rangle e_j$ or Sf.*

(b) $Sf = f$.

PROOF. (a) From Bessel's inequality (2.4.8) we have $\sum_{|j|\leq n} |\langle f, e_j\rangle|^2 \leq \|f\|^2$ for all n which implies that $\sum_{j=-\infty}^{\infty} |\langle f, e_j\rangle|^2 < \infty$. Hence for $n > m \geq 1$,

$$\|S_n f - S_m f\|^2 \leq \sum_{|j|>m} |\langle f, e_j\rangle|^2 \to 0 \quad \text{as } m \to \infty,$$

showing that $\{S_n f\}$ is a Cauchy sequence and therefore has a mean square limit.

(b) For $|j| \leq n$, $\langle S_n f, e_j\rangle = \langle f, e_j\rangle$, so by continuity of the inner product

$$\langle Sf, e_j\rangle = \lim_{n\to\infty} \langle S_n f, e_j\rangle = \langle f, e_j\rangle \quad \text{for all } j \in \mathbb{Z}.$$

In Theorem 2.11.2 we shall show that $\langle g, e_j \rangle = 0$ for all $j \in \mathbb{Z}$ implies that $g = 0$. Hence $Sf - f = 0$. □

Corollary 2.8.1. $L^2[-\pi, \pi] = \overline{\text{sp}}\{e_j, j \in \mathbb{Z}\}$.

Proof. Any $f \in L^2[-\pi, \pi]$ can be expressed as the mean square limit of $S_n f$ where $S_n f \in \overline{\text{sp}}\{e_j, j \in \mathbb{Z}\}$. Since $\overline{\text{sp}}\{e_j, j \in \mathbb{Z}\}$ is by definition closed it must contain f. Hence $\overline{\text{sp}}\{e_j, j \in \mathbb{Z}\} \supseteq L^2[-\pi, \pi]$. □

Corollary 2.8.2. (a) $\|f\|^2 = \sum_{j=-\infty}^{\infty} |\langle f, e_j \rangle|^2$.
(b) $\langle f, g \rangle = \sum_{j=-\infty}^{\infty} \langle f, e_j \rangle \overline{\langle g, e_j \rangle}$.

Proof. Corollary 2.8.1 implies that the conditions of Theorem 2.4.2 are satisfied. □

§2.9 Hilbert Space Isomorphisms

Definition 2.9.1 (Isomorphism). An isomorphism of the Hilbert space $\mathscr{H}_1$ onto the Hilbert space $\mathscr{H}_2$ is a one to one mapping T of $\mathscr{H}_1$ onto $\mathscr{H}_2$ such that for all $f_1, f_2 \in \mathscr{H}_1$,

$$\text{(a)} \quad T(af_1 + bf_2) = aTf_1 + bTf_2 \text{ for all scalars } a \text{ and } b$$

and

$$\text{(b)} \quad \langle Tf_1, Tf_2 \rangle = \langle f_1, f_2 \rangle.$$

We say that $\mathscr{H}_1$ and $\mathscr{H}_2$ are isomorphic if there is an isomorphism T of $\mathscr{H}_1$ onto $\mathscr{H}_2$. The inverse mapping T^{-1} is then an isomorphism of $\mathscr{H}_2$ onto $\mathscr{H}_1$.

Remark 1. In this book we shall always use the term isomorphism to indicate that both (a) and (b) are satisfied. Elsewhere the term is frequently used to denote a mapping satisfying (a) only.

Example 2.9.1 (The Space l^2). Let l^2 denote the complex Hilbert space of sequences $\{z_n, n = 1, 2, \ldots\}$, $z_n \in \mathbb{C}$, $\sum_{n=1}^{\infty} |z_n^2| < \infty$, with inner product

$$\langle \{y_n\}, \{z_n\} \rangle = \sum_{i=1}^{\infty} y_i \bar{z}_i.$$

(For the proof that l^2 is a separable Hilbert space see Problem 2.23.) If now $\mathscr{H}$ is any Hilbert space with an orthonormal basis $\{e_n, n = 1, 2, \ldots\}$ then the mapping $T: \mathscr{H} \to l^2$ defined by

$$Th = \{\langle h, e_n \rangle\} \tag{2.9.1}$$

is an isomorphism of $\mathscr{H}$ onto l^2 (see Problem 2.24). Thus every separable Hilbert space is isomorphic to l^2.

Properties of Isomorphisms. Suppose T is an isomorphism of $\mathscr{H}_1$ onto $\mathscr{H}_2$. We then have the following properties, all of which follow at once from the definitions:

(i) If $\{e_n\}$ is a complete orthonormal set in $\mathscr{H}_1$ then $\{Te_n\}$ is a complete orthonormal set in $\mathscr{H}_2$.
(ii) $\|Tx\| = \|x\|$ for all $x \in \mathscr{H}_1$.
(iii) $\|x_n - x\| \to 0$ if and only if $\|Tx_n - Tx\| \to 0$.
(iv) $\{x_n\}$ is a Cauchy sequence if and only if $\{Tx_n\}$ is a Cauchy sequence.
(v) $TP_{\overline{\text{sp}}\{x_\lambda, \lambda \in \Lambda\}}(x) = P_{\overline{\text{sp}}\{Tx_\lambda, \lambda \in \Lambda\}}(Tx)$.

The last property is the basis for the spectral theory of prediction of a stationary process $\{X_t, t \in \mathbb{Z}\}$ (Section 5.6), in which we use the fact that the mapping

$$TX_t = e^{it\cdot}$$

defines an isomorphism of a certain Hilbert space of random variables onto a Hilbert space $L^2([-\pi, \pi], \mathscr{B}, \mu)$ with μ a finite measure. The problem of computing projections in the former space can then be tranformed by means of (v) into the problem of computing projections in the latter.

§2.10* The Completeness of $L^2(\Omega, \mathscr{F}, P)$

We need to show that if $X_n \in L^2, n = 1, 2, \ldots,$ and $\|X_n - X_m\| \to 0$ as $m, n \to \infty$, then there exists $X \in L^2$ such that $X_n \xrightarrow{\text{m.s.}} X$. This will be shown by identifying X as the limit of a sufficiently rapidly converging subsequence of $\{X_n\}$. We first need a proposition.

Proposition 2.10.1. *If $X_n \in L^2$ and $\|X_{n+1} - X_n\| \le 2^{-n}$, $n = 1, 2, \ldots,$ then there is a random variable X on $(\Omega, \mathscr{F}, P)$ such that $X_n \to X$ with probability one.*

PROOF. Let $X_0 = 0$. Then $X_n = \sum_{j=1}^{n} (X_j - X_{j-1})$. Now $\sum_{j=1}^{\infty} |X_j - X_{j-1}|$ is finite with probability one since, by the monotone convergence theorem and the Cauchy–Schwarz inequality,

$$E \sum_{j=1}^{\infty} |X_j - X_{j-1}| = \sum_{j=1}^{\infty} E|X_j - X_{j-1}| \le \sum_{j=1}^{\infty} \|X_j - X_{j-1}\| \le \|X_1\| + \sum_{j=1}^{\infty} 2^{-j} < \infty.$$

It follows that $\lim_{n\to\infty} \sum_{j=1}^{n} |X_j - X_{j-1}|$ (and hence $\lim_{n\to\infty} \sum_{j=1}^{n} (X_j - X_{j-1}) = \lim_{n\to\infty} X_n$) exists and is finite with probability one. □

Theorem 2.10.1. *$L^2(\Omega, \mathscr{F}, P)$ is complete.*

PROOF. If $\{X_n\}$ is a Cauchy sequence in L^2 then we can find integers $n_1, n_2, \ldots,$ such that $n_1 < n_2 < \cdots$ and

$$\|X_n - X_m\| \le 2^{-k} \quad \text{for } n, m > n_k. \tag{2.10.1}$$

(First choose n_1 to satisfy (2.10.1) with $k = 1$, then successively choose $n_2, n_3, \ldots$, to satisfy the appropriate conditions.)

By Proposition 2.10.1 there is a random variable X such that $X_{n_k} \to X$ with probability one as $k \to \infty$. Now

$$\|X_n - X\|^2 = \int |X_n - X|^2 \, dP = \int \liminf_{k\to\infty} |X_n - X_{n_k}|^2 \, dP,$$

and so by Fatou's lemma,

$$\|X_n - X\|^2 \le \liminf_{k\to\infty} \|X_n - X_{n_k}\|^2. \tag{2.10.2}$$

The right-hand side of (2.10.2) can be made arbitrarily small by choosing n large enough since $\{X_n\}$ is a Cauchy sequence. Consequently $\|X_n - X\|^2 \to 0$. The fact that $E|X|^2 < \infty$ follows from the triangle inequality

$$\|X\| \le \|X_n - X\| + \|X_n\|,$$

the right-hand side of which is certainly finite for large enough n. □

§2.11* Complementary Results for Fourier Series

The terminology and notation of Section 2.8 will be retained throughout this section. We begin with the classical result that trigonometric polynomials are uniformly dense in the space of continuous functions f which are defined on $[-\pi, \pi]$ and which satisfy the condition $f(\pi) = f(-\pi)$.

Theorem 2.11.1. *Let f be a continuous function on $[-\pi, \pi]$ such that $f(\pi) = f(-\pi)$. Then*

$$n^{-1}(S_0 f + S_1 f + \cdots + S_{n-1} f) \to f \tag{2.11.1}$$

uniformly on $[-\pi, \pi]$ as $n \to \infty$.

PROOF. By definition of the nth order Fourier approximation,

$$\begin{aligned} S_n f(x) &= \sum_{|j|\le n} \langle f, e_j \rangle e_j \\ &= (2\pi)^{-1} \int_{-\pi}^{\pi} f(y) \sum_{|j|\le n} e^{ij(x-y)} \, dy, \end{aligned}$$

which by defining $f(x) = f(x + 2\pi)$, $x \in \mathbb{R}$, can be rewritten as

$$S_n f(x) = (2\pi)^{-1} \int_{-\pi}^{\pi} f(x - y) D_n(y) \, dy, \tag{2.11.2}$$

where $D_n(y)$ is the Dirichlet kernel,

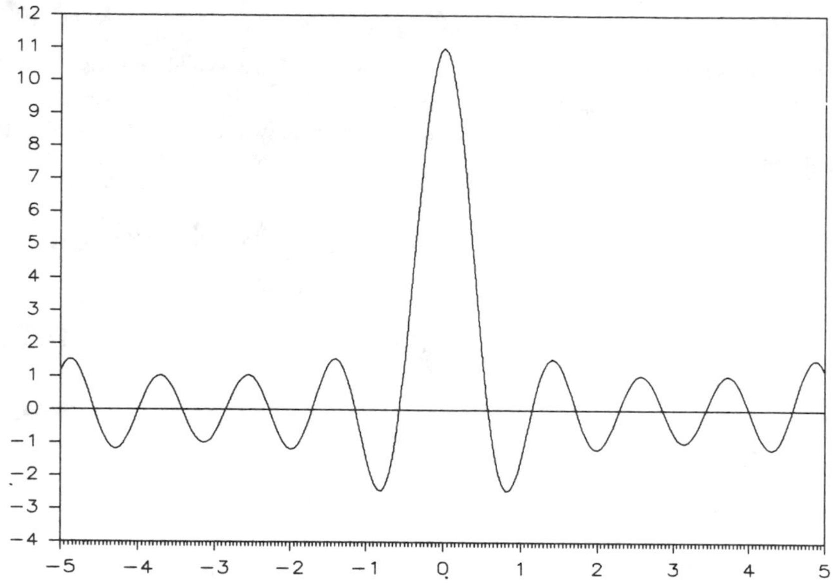

Figure 2.2. The Dirichlet kernel $D_5(x)$, $-5 \le x \le 5$ ($D_n(\cdot)$ has period 2π).

$$D_n(y) = \sum_{|j| \le n} e^{ijy} = \frac{e^{i(n+1/2)y} - e^{-i(n+1/2)y}}{e^{iy/2} - e^{-iy/2}} = \begin{cases} \dfrac{\sin[(n+\frac{1}{2})y]}{\sin(\frac{1}{2}y)} & \text{if } y \ne 0, \\ 2n+1 & \text{if } y = 0. \end{cases} \tag{2.11.3}$$

A graph of the function D_n is shown in Figure 2.2. For the function $f(x) \equiv 1$, $\langle f, e_0 \rangle = 1$ and $\langle f, e_j \rangle = 0$, $j \ne 0$. Hence $S_n 1(x) \equiv 1$, and substituting this in (2.11.2) we find that

$$(2\pi)^{-1} \int_{-\pi}^{\pi} D_n(y)\, dy = 1. \tag{2.11.4}$$

Making use of (2.11.2) we can now write

$$n^{-1}(S_0 f(x) + \cdots + S_{n-1} f(x)) = \int_{-\pi}^{\pi} f(x-y) K_n(y)\, dy, \tag{2.11.5}$$

where $K_n(y)$ is the Fejer kernel,

$$K_n(y) = \frac{1}{2\pi n} \sum_{j=0}^{n-1} D_j(y) = \frac{\sum_{j=0}^{n-1} \sin[(j+\frac{1}{2})y]}{2\pi n \sin(\frac{1}{2}y)}.$$

Evaluating the sum with the aid of the identity,

$$2\sin(\tfrac{1}{2}y)\sin[(j+\tfrac{1}{2})y] = \cos(jy) - \cos[(j+1)y],$$

we find that

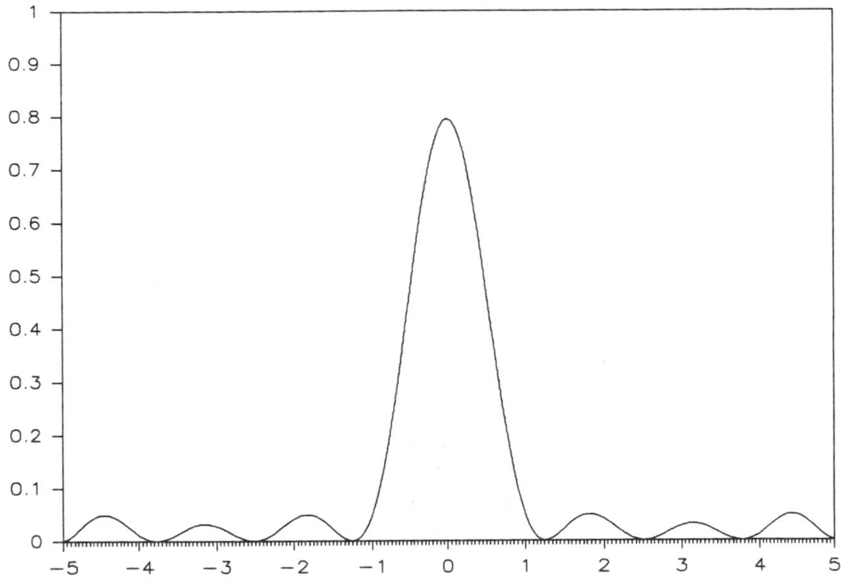

Figure 2.3. The Fejer kernel $K_5(x)$, $-5 \le x \le 5$ ($K_n(\cdot)$ has period 2π).

$$K_n(y) = \begin{cases} \dfrac{1}{2\pi n} \dfrac{\sin^2(ny/2)}{\sin^2(y/2)} & \text{if } y \neq 0, \\[2ex] \dfrac{n}{2\pi} & \text{if } y = 0. \end{cases} \tag{2.11.6}$$

The Fejer kernel is shown in Figure 2.3. It has the properties,

(a) $K_n(y) \ge 0$ (unlike $D_n(y)$),
(b) $K_n(\cdot)$ has period 2π,
(c) $K_n(\cdot)$ is an even function
(d) $\int_{-\pi}^{\pi} K_n(y)\,dy = 1$,
(e) for each $\delta > 0$, $\int_{-\delta}^{\delta} K_n(y)\,dy \to 1$ as $n \to \infty$.

The first three properties are evident from (2.11.6). Property (d) is obtained by setting $f(x) \equiv 1$ in (2.11.5). To establish (e), observe that

$$K_n(y) \le \frac{1}{2\pi n \sin^2(\delta/2)} \quad \text{for } 0 < \delta < |y| \le \pi.$$

For each $\delta > 0$ this inequality implies that

$$\int_{-\pi}^{-\delta} K_n(y)\,dy + \int_{\delta}^{\pi} K_n(y)\,dy \to 0 \quad \text{as } n \to \infty,$$

which, together with property (d), proves (e).

Now for any continuous function f with period 2π, we have from (2.11.5) and property (d) of $K_n(.)$,

$$\begin{aligned}\Delta_n(x) &\equiv |n^{-1}(S_0 f(x) + \cdots + S_{n-1} f(x)) - f(x)| \\ &= \left|\int_{-\pi}^{\pi} f(x-y) K_n(y)\,dy - f(x)\right| \\ &= \left|\int_{-\pi}^{\pi} [f(x-y) - f(x)] K_n(y)\,dy\right|.\end{aligned}$$

Hence for each $\delta > 0$,

$$\begin{aligned}\Delta_n(x) \le &\left|\int_{-\delta}^{\delta} [f(x-y) - f(x)] K_n(y)\,dy\right| \\ &+ \left|\int_{[-\pi,\pi]\backslash(-\delta,\delta)} [f(x-y) - f(x)] K_n(y)\,dy\right|.\end{aligned} \tag{2.11.7}$$

Since a continuous function with period 2π is uniformly continuous, we can choose for any $\varepsilon > 0$, a value of δ such that $\sup_{-\pi \le x \le \pi} |f(x-y) - f(x)| < \varepsilon$ whenever $|y| < \delta$. The first term on the right of (2.11.7) is then bounded by $\varepsilon \int_{-\pi}^{\pi} K_n(y)\,dy$ and the second by $2M(1 - \int_{-\delta}^{\delta} K_n(y)\,dy)$ where $M = \sup_{-\pi \le x \le \pi} |f(x)|$. Hence

$$\begin{aligned}\sup_{-\pi \le x \le \pi} \Delta_n(x) &\le \varepsilon \int_{-\delta}^{\delta} K_n(dy)\,dy + 2M\left(1 - \int_{-\delta}^{\delta} K_n(y)\,dy\right) \\ &\to \varepsilon \quad \text{as } n \to \infty.\end{aligned}$$

But since ε was arbitrary and $\Delta_n(x) \ge 0$, we conclude that $\Delta_n(x) \to 0$ uniformly on $[-\pi, \pi]$ as required. □

Remark 1. Under additional smoothness conditions on f, $S_n f$ may converge to f in a much stronger sense. For example if the derivative f' exists and $f' \in L^2[-\pi, \pi]$, then $S_n f$ converges absolutely and uniformly to f (see Churchill (1969) and Problem 2.22).

Theorem 2.11.2. *If $f \in L^2[-\pi, \pi]$ and $\langle f, e_j \rangle = 0$ for all $j \in \mathbb{Z}$, then $f = 0$ almost everywhere.*

Proof. It sufficies to show that $\int_A f(x)\,dx = 0$ for all Borel subsets A of $[-\pi, \pi]$ or, equivalently, by a monotone class argument (see Billingsley (1986)),

$$(2\pi)^{-1} \int_a^b f(x)\,dx = \langle f, I_{[a,b]} \rangle = 0 \tag{2.11.8}$$

for all subintervals $[a, b]$ of $[-\pi, \pi]$. Here $I_{[a,b]}$ denotes the indicator function of $[a, b]$.

To establish (2.11.8) we first show that $\langle f, g \rangle = 0$ for any continuous function g on $[-\pi, \pi]$ with $g(-\pi) = g(\pi)$. By Theorem 2.11.1 we know that

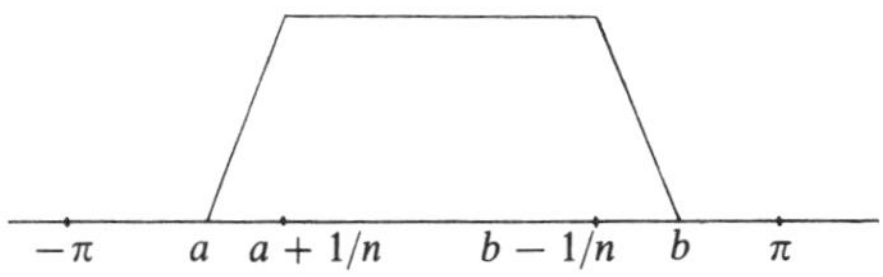

Figure 2.4. The continuous function h_n approximating $I_{[a,b]}$.

for g continuous, $g_n = n^{-1}(S_0 g + \cdots + S_{n-1} g) \to g$ uniformly on $[-\pi, \pi]$, implying in particular that

$$g_n \xrightarrow{\text{m.s.}} g.$$

By assumption $\langle f, g_n \rangle = 0$, so by continuity of the inner product,

$$\langle f, g \rangle = \lim_{n \to \infty} \langle f, g_n \rangle = 0.$$

The next step is to find a sequence $\{h_n\}$ of continuous functions such that $h_n \xrightarrow{\text{m.s.}} I_{[a,b]}$. One such sequence is defined by

$$h_n(x) = \begin{cases} 0 & \text{if } -\pi \le x \le a, \\ n(x-a) & \text{if } a \le x \le a + 1/n, \\ 1 & \text{if } a + 1/n \le x \le b - 1/n, \\ -n(x-b) & \text{if } b - 1/n \le x \le b, \\ 0 & \text{if } b \le x \le \pi, \end{cases}$$

since $\|I_{[a,b]} - h_n\|^2 \le (1/2\pi)(2/n) \to 0$ as $n \to \infty$. (See Figure 2.4.) Using the continuity of the inner product again,

$$\langle f, I_{[a,b]} \rangle = \lim_{n \to \infty} \langle f, h_n \rangle = 0. \qquad \square$$

Problems

2.1. Prove the parallelogram law (2.1.9).

2.2. If $\{X_t, t = 0, \pm 1, \ldots\}$ is a stationary process with mean zero and autocovariance function $\gamma(\cdot)$, show that $Y_n = \sum_{k=1}^{n} a_k X_k$ converges in mean square if $\sum_{i=0}^{\infty} \sum_{j=0}^{\infty} a_i a_j \gamma(i-j)$ is finite.

2.3. Show that if $\{X_t, t = 0, \pm 1, \ldots\}$ is stationary and $|\theta| < 1$ then for each n, $\sum_{j=1}^{m} \theta^j X_{n+1-j}$ converges in mean square as $m \to \infty$.

2.4. If $\mathcal{M}$ is a closed subspace of the Hilbert space $\mathcal{H}$, show that $(\mathcal{M}^\perp)^\perp = \mathcal{M}$.

2.5. If $\mathcal{M}$ is a closed subspace of the Hilbert space $\mathcal{H}$ and $x \in \mathcal{H}$, prove that

$$\min_{y \in \mathcal{M}} \|x - y\| = \max\{|\langle x, z \rangle| : z \in \mathcal{M}^\perp, \|z\| = 1\}.$$

2.6. Verify the calculations of ϕ_1 and ϕ_2 in Example 2.3.4. Also check that $X_3 = (2\cos\omega)X_2 - X_1$.

2.7. If $\mathscr{H}$ is a complex Hilbert space and $x_i \in \mathscr{H}, i = 1, \ldots, n$, show that $\overline{\text{sp}}\{x_1, \ldots, x_n\} = \{\sum_{j=1}^n \alpha_j x_j : \alpha_j \in \mathbb{C}, j = 1, \ldots, n\}$.

2.8. Suppose that $\{X_t, t = 1, 2, \ldots\}$ is a stationary process with mean zero. Show that $P_{\overline{\text{sp}}\{1, X_1, \ldots, X_n\}} X_{n+1} = P_{\overline{\text{sp}}\{X_1, \ldots, X_n\}} X_{n+1}$.

2.9. (a) Let $\mathscr{H} = L^2([-1,1], \mathscr{B}[-1,1], \mu)$ where $d\mu = dx$ is Lebesgue measure on $[-1,1]$. Use the prediction equations to find constants α_0, α_1 and α_2 which minimize

$$\int_{-1}^{1} |e^x - \alpha_0 - \alpha_1 x - \alpha_2 x^2|^2 \, dx.$$

(b) Find $\max_{\{g \in \mathscr{M}^\perp, \|g\| = 1\}} \int_{-1}^{1} e^x g(x)\, dx$ where $\mathscr{M} = \overline{\text{sp}}\{1, x, x^2\}$.

2.10. If $X_t = Z_t - \theta Z_{t-1}$, where $|\theta| < 1$ and $\{Z_t, t = 0, \pm 1, \ldots\}$ is a sequence of uncorrelated random variables, each with mean 0 and variance σ^2, show by checking the prediction equations that the best mean square predictor of X_{n+1} in $\overline{\text{sp}}\{X_j, -\infty < j \le n\}$ is

$$\hat{X}_{n+1} = -\sum_{j=1}^{\infty} \theta^j X_{n+1-j}.$$

What is the mean squared error of $\hat{X}_{n+1}$?

2.11. If X_t is defined as in Problem 2.10 with $\theta = 1$, find the best mean square predictor of X_{n+1} in $\overline{\text{sp}}\{X_j, 1 \le j \le n\}$ and the corresponding mean squared error.

2.12. If $X_t = \phi_1 X_{t-1} + \phi_2 X_{t-2} + \cdots + \phi_p X_{t-p} + Z_t$, $t = 0, \pm 1, \ldots$ where $\{Z_t\}$ is a sequence of uncorrelated random variables, each with mean zero and variance σ^2 and such that Z_t is uncorrelated with $\{X_j, j < t\}$ for each t, use the prediction equations to show that the best mean square predictor of X_{n+1} in $\overline{\text{sp}}\{X_j, -\infty < j \le n\}$ is

$$\hat{X}_{n+1} = \phi_1 X_n + \phi_2 X_{n-1} + \cdots + \phi_p X_{n+1-p}.$$

2.13. (Gram–Schmidt orthogonalization). Let $x_1, x_2, \ldots, x_n$ be linearly independent elements of a Hilbert space $\mathscr{H}$ (i.e. elements for which $\|\alpha_1 x_1 + \cdots + \alpha_n x_n\| = 0$ implies that $\alpha_1 = \alpha_2 = \cdots = \alpha_n = 0$). Define

$$w_1 = x_1$$

and

$$w_k = x_k - P_{\overline{\text{sp}}\{w_1, \ldots, w_{k-1}\}} x_k, \quad k > 1.$$

Show that $\{e_k = w_k/\|w_k\|, k = 1, \ldots, n\}$ is an orthonormal set and that $\overline{\text{sp}}\{e_1, \ldots, e_k\} = \overline{\text{sp}}\{x_1, \ldots, x_k\}$ for $1 \le k \le n$.

2.14. Show that every closed subspace $\mathscr{M}$ of $\mathbb{R}^n$ which contains a non-zero vector can be written as $\mathscr{M} = \overline{\text{sp}}\{\mathbf{e}_1, \ldots, \mathbf{e}_m\}$ where $\{\mathbf{e}_1, \ldots, \mathbf{e}_m\}$ is an orthonormal subset of $\mathscr{M}$ and $m\ (\le n)$ is the same for all such representations.

2.15. Let X_1, X_2 and X_3 be three random variables with mean zero and covariance matrix,

$$V = \begin{bmatrix} 14 & -1 & 3 \\ -1 & 5 & -1 \\ 3 & -1 & 1 \end{bmatrix}.$$

Use the Gram–Schmidt orthogonalization process of Problem 2.13 to find three uncorrelated random variables Z_1, Z_2 and Z_3 such that $\overline{\text{sp}}\{X_1\} = \overline{\text{sp}}\{Z_1\}$, $\overline{\text{sp}}\{X_1, X_2\} = \overline{\text{sp}}\{Z_1, Z_2\}$ and $\overline{\text{sp}}\{X_1, X_2, X_3\} = \overline{\text{sp}}\{Z_1, Z_2, Z_3\}$.

2.16. (Hermite polynomials). Let $\mathscr{H} = L^2(\mathbb{R}, \mathscr{B}, \mu)$ where $d\mu = (2\pi)^{-1/2} e^{-x^2/2}\, dx$. Set $f_0(x) \equiv 1$, $f_1(x) = x$, $f_2(x) = x^2$, $f_3(x) = x^3$. Using the Gram–Schmidt orthogonalization process, find polynomials $H_k(x)$ of degree k, $k = 0, 1, 2, 3$ which are orthogonal in $\mathscr{H}$. (Do not however normalize $H_k(x)$ to have unit length.) Verify that $H_k(x) = (-1)^k e^{x^2/2} \dfrac{d^k}{dx^k} e^{-x^2/2}$, $k = 0, 1, 2, 3$.

2.17. Prove the first statement in the proof of Theorem 2.4.2.

2.18. (a) Let x be an element of the Hilbert space $\mathscr{H} = \overline{\text{sp}}\{x_1, x_2, \ldots\}$. Show that $\mathscr{H}$ is separable and that

$$P_{\overline{\text{sp}}\{x_1, \ldots, x_n\}} x \to x \quad \text{as } n \to \infty.$$

(b) If $\{X_t, t = 0, \pm 1, \ldots\}$ is a stationary process show that

$$P_{\overline{\text{sp}}\{X_j, -\infty < j \le n\}} X_{n+1} = \lim_{r \to \infty} P_{\overline{\text{sp}}\{X_j, n-r < j \le n\}} X_{n+1}.$$

2.19. (General linear model). Consider the general linear model

$$\mathbf{Y} = \mathbf{X\theta} + \mathbf{Z},$$

where $\mathbf{Y} = (Y_1, \ldots, Y_n)'$ is the vector of observations, X is a known $n \times m$ matrix of rank $m < n$, $\boldsymbol{\theta} = (\theta_1, \ldots, \theta_m)'$ is an m-vector of parameter values, and $\mathbf{Z} = (Z_1, \ldots, Z_n)'$ is the vector of noise variables. The least squares estimator of $\boldsymbol{\theta}$ is given by equation (2.6.4), i.e.

$$\hat{\boldsymbol{\theta}} = (X'X)^{-1} X'\mathbf{Y}.$$

Assume that $\mathbf{Z} \sim N(\mathbf{0}, \sigma^2 I_n)$ where I_n is the n-dimensional identity matrix.

(a) Show that $\mathbf{Y} \sim N(X\boldsymbol{\theta}, \sigma^2 I_n)$.

(b) Show that $\hat{\boldsymbol{\theta}} \sim N(\boldsymbol{\theta}, \sigma^2 (X'X)^{-1})$.

(c) Show that the projection matrix $P_{\mathscr{M}} = X(X'X)^{-1}X'$ is non-negative definite and has m non-zero eigenvalues all of which are equal to one. Similarly, $I_n - P_{\mathscr{M}}$ is also non-negative definite with $(n - m)$ non-zero eigenvalues all of which are equal to one.

(d) Show that the two vectors of random variables, $P_{\mathscr{M}}(\mathbf{Y} - X\boldsymbol{\theta})$ and $(I_n - P_{\mathscr{M}})\mathbf{Y}$ are independent and that $\sigma^{-2} \| P_{\mathscr{M}}(\mathbf{Y} - X\boldsymbol{\theta}) \|^2$ and $\sigma^{-2} \|(I_n - P_{\mathscr{M}})\mathbf{Y}\|^2$ are independent chi-squared random variables with m and $(n - m)$ degrees of freedom respectively. ($\|\mathbf{Y}\|$ here denotes the Euclidean norm of $\mathbf{Y}$, i.e. $(\sum_{i=1}^n Y_i^2)^{1/2}$.)

(e) Conclude that

$$\frac{(n-m)\|P_{\mathscr{M}}(Y - X\boldsymbol{\theta})\|^2}{m\|\mathbf{Y} - P_{\mathscr{M}}\mathbf{Y}\|^2}$$

has the F distribution with m and $(n-m)$ degree of freedom.

2.20. Suppose $(X, Z_1, \ldots, Z_n)'$ has a multivariate normal distribution. Show that

$$P_{\overline{\text{sp}}\{1, Z_1, \ldots, Z_n\}}(X) = E_{\mathscr{M}(Z_1, \ldots, Z_n)}(X),$$

where the conditional expectation operator $E_{\mathscr{M}(Z_1, \ldots, Z_n)}$ is defined as in Section 2.7.

2.21. Suppose $\{X_t, t = 0, \pm 1, \ldots\}$ is a stationary process with mean zero and auto-covariance function $\gamma(\cdot)$ which is absolutely summable (i.e. $\sum_{h=-\infty}^{\infty} |\gamma(h)| < \infty$). Define f to be the function,

$$f(\lambda) = \frac{1}{2\pi} \sum_{h=-\infty}^{\infty} \gamma(h) e^{-ih\lambda}, \qquad -\pi \le \lambda \le \pi,$$

and show that $\gamma(h) = \int_{-\pi}^{\pi} e^{ih\lambda} f(\lambda)\, d\lambda$.

2.22. (a) If $f \in L^2([-\pi, \pi])$, prove the Riemann–Lebesgue lemma: $\langle f, e_h \rangle \to 0$ as $h \to \infty$, where e_n was defined by (2.8.2).

(b) If $f \in L^2([-\pi, \pi])$ has a continuous derivative $f'(x)$ and $f(\pi) = f(-\pi)$, show that $\langle f, e_h \rangle = (ih)^{-1} \langle f', e_h \rangle$ and hence that $h\langle f, e_h \rangle \to 0$ as $h \to \infty$. Show also that $\sum_{h=-\infty}^{\infty} |\langle f, e_h \rangle| < \infty$ and conclude that $S_n f$ (see Section 2.8) converges uniformly to f.

2.23. Show that the space l^2 (Example 2.9.1) is a separable Hilbert space.

2.24. If $\mathscr{H}$ is any Hilbert space with orthonormal basis $\{e_n, n = 1, 2, \ldots\}$, show that the mapping defined by $Th = \{\langle h, e_n \rangle\}$, $h \in \mathscr{H}$, is an isomorphism of $\mathscr{H}$ onto l^2.

2.25.* Prove that $\mathscr{M}(Z)$ (see Definition 2.7.3) is closed.

CHAPTER 3

Stationary ARMA Processes

In this chapter we introduce an extremely important class of time series $\{X_t, t = 0, \pm 1, \pm 2, \ldots\}$ defined in terms of linear difference equations with constant coefficients. The imposition of this additional structure defines a parametric family of stationary processes, the autoregressive moving average or ARMA processes. For any autocovariance function $\gamma(\cdot)$ such that $\lim_{h\to\infty} \gamma(h) = 0$, and for any integer $k > 0$, it is possible to find an ARMA process with autocovariance function $\gamma_X(\cdot)$ such that $\gamma_X(h) = \gamma(h)$, $h = 0, 1, \ldots, k$. For this (and other) reasons the family of ARMA processes plays a key role in the modelling of time-series data. The linear structure of ARMA processes leads also to a very simple theory of linear prediction which is discussed in detail in Chapter 5.

§3.1 Causal and Invertible ARMA Processes

In many respects the simplest kind of time series $\{X_t\}$ is one in which the random variables X_t, $t = 0, \pm 1, \pm 2, \ldots$ are independently and identically distributed with zero mean and variance σ^2. From a second order point of view i.e. ignoring all properties of the joint distributions of $\{X_t\}$ except those which can be deduced from the moments $E(X_t)$ and $E(X_s X_t)$, such processes are identified with the class of all stationary processes having mean zero and autocovariance function

$$\gamma(h) = \begin{cases} \sigma^2 & \text{if } h = 0, \\ 0 & \text{if } h \neq 0. \end{cases} \tag{3.1.1}$$

Definition 3.1.1. The process $\{Z_t\}$ is said to be white noise with mean 0 and variance σ^2, written

$$\{Z_t\} \sim \mathrm{WN}(0, \sigma^2), \tag{3.1.2}$$

if and only if $\{Z_t\}$ has zero mean and covariance function (3.1.1).

If the random variables Z_t are independently and identically distributed with mean 0 and variance σ^2 then we shall write

$$\{Z_t\} \sim \mathrm{IID}(0, \sigma^2). \tag{3.1.3}$$

A very wide class of stationary processes can be generated by using white noise as the forcing terms in a set of linear difference equations. This leads to the notion of an autoregressive-moving average (ARMA) process.

Definition 3.1.2 (The ARMA (p, q) Process). The process $\{X_t, t = 0, \pm 1, \pm 2, \ldots\}$ is said to be an ARMA(p, q) process if $\{X_t\}$ is stationary and if for every t,

$$X_t - \phi_1 X_{t-1} - \cdots - \phi_p X_{t-p} = Z_t + \theta_1 Z_{t-1} + \cdots + \theta_q Z_{t-q}, \tag{3.1.4}$$

where $\{Z_t\} \sim \mathrm{WN}(0, \sigma^2)$. We say that $\{X_t\}$ is an ARMA(p, q) process with mean μ if $\{X_t - \mu\}$ is an ARMA(p, q) process.

The equations (3.1.4) can be written symbolically in the more compact form

$$\phi(B)X_t = \theta(B)Z_t, \qquad t = 0, \pm 1, \pm 2, \ldots, \tag{3.1.5}$$

where ϕ and θ are the pth and q^{th} degree polynomials

$$\phi(z) = 1 - \phi_1 z - \cdots - \phi_p z^p \tag{3.1.6}$$

and

$$\theta(z) = 1 + \theta_1 z + \cdots + \theta_q z^q \tag{3.1.7}$$

and B is the backward shift operator defined by

$$B^j X_t = X_{t-j}, \qquad j = 0, \pm 1, \pm 2, \ldots. \tag{3.1.8}$$

The polynomials ϕ and θ will be referred to as the autoregressive and moving average polynomials respectively of the difference equations (3.1.5).

EXAMPLE 3.1.1 (The MA(q) Process). If $\phi(z) \equiv 1$ then

$$X_t = \theta(B)Z_t \tag{3.1.9}$$

and the process is said to be a moving-average process of order q (or MA(q)). It is quite clear in this case that the difference equations have the unique solution (3.1.9). Moreover the solution $\{X_t\}$ is a stationary process since (defining $\theta_0 = 1$ and $\theta_j = 0$ for $j > q$), we see that

$$EX_t = \sum_{j=0}^{q} \theta_j E Z_{t-j} = 0$$

and

$$\operatorname{Cov}(X_{t+h}, X_t) = \begin{cases} \sigma^2 \sum_{j=0}^{q-|h|} \theta_j \theta_{j+|h|} & \text{if } |h| \le q, \\ 0 & \text{if } |h| > q. \end{cases}$$

A realization of $\{X_1, \dots, X_{100}\}$ with $q = 1, \theta_1 = -.8$ and $Z_t \sim N(0, 1)$ is shown in Figure 3.1(a). The autocorrelation function of the process is shown in Figure 3.1(b).

Example 3.1.2 (The AR(p) Process). If $\theta(z) \equiv 1$ then

$$\phi(B)X_t = Z_t \tag{3.1.10}$$

and the process is said to be an autoregressive process of order p (or AR(p)). In this case (as in the general case to be considered in Theorems 3.1.1–3.1.3) the existence and uniqueness of a stationary solution of (3.1.10) needs closer investigation. We illustrate by examining the case $\phi(z) = 1 - \phi_1 z$, i.e.

$$X_t = Z_t + \phi_1 X_{t-1}. \tag{3.1.11}$$

Iterating (3.1.11) we obtain

$$\begin{aligned} X_t &= Z_t + \phi_1 Z_{t-1} + \phi_1^2 X_{t-2} \\ &= \cdots \\ &= Z_t + \phi_1 Z_{t-1} + \cdots + \phi_1^k Z_{t-k} + \phi_1^{k+1} X_{t-k-1}. \end{aligned}$$

If $|\phi_1| < 1$ and $\{X_t\}$ is stationary then $\|X_t\|^2 = E(X_t^2)$ is constant so that

$$\left\| X_t - \sum_{j=0}^{k} \phi_1^j Z_{t-j} \right\|^2 = \phi_1^{2k+2} \|X_{t-k-1}\|^2 \to 0 \quad \text{as } k \to \infty.$$

Since $\sum_{j=0}^{\infty} \phi_1^j Z_{t-j}$ is mean-square convergent (by the Cauchy criterion), we conclude that

$$X_t = \sum_{j=0}^{\infty} \phi_1^j Z_{t-j}. \tag{3.1.12}$$

Equation (3.1.12) is valid not only in the mean square sense but also (by Proposition 3.1.1 below) with probability one, i.e.

$$X_t(\omega) = \sum_{j=0}^{\infty} \phi_1^j Z_{t-j}(\omega) \quad \text{for all } \omega \notin E,$$

where E is a subset of the underlying probability space with probability zero. All the convergent series of random variables encountered in this chapter will (by Proposition 3.1.1) be both mean square convergent and absolutely convergent with probability one. Now $\{X_t\}$ defined by (3.1.12) is stationary since

$$EX_t = \sum_{j=0}^{\infty} \phi_1^j EZ_{t-j} = 0$$

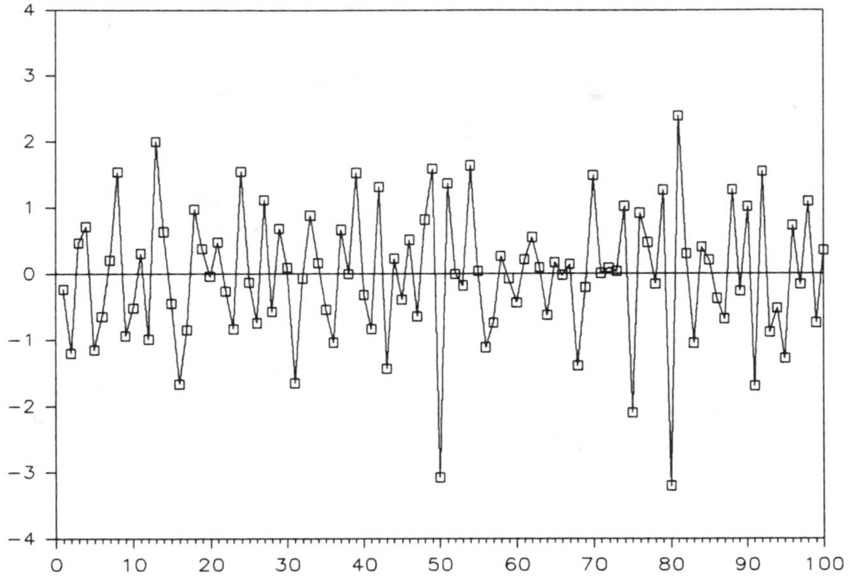

(a)

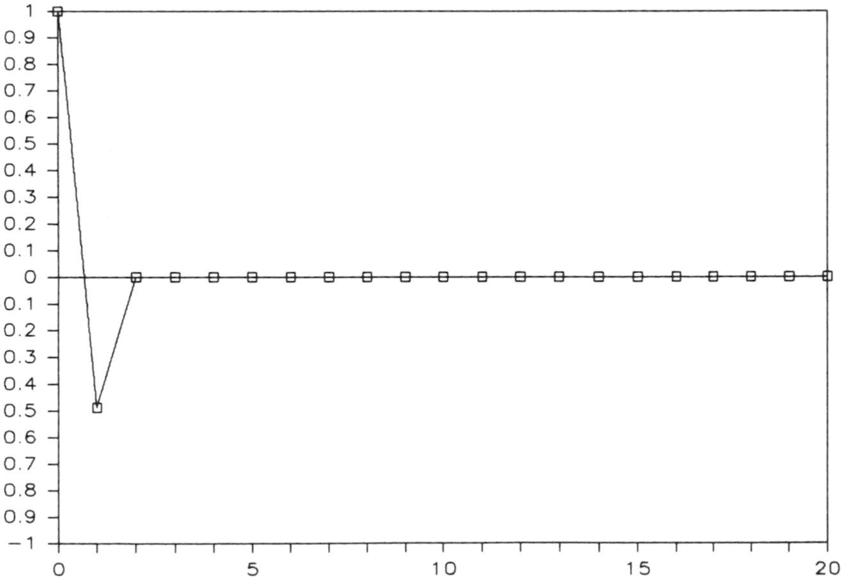

(b)

Figure 3.1. (a) 100 observations of the series $X_t = Z_t - .8Z_{t-1}$, Example 3.1.1. (b) The autocorrelation function of $\{X_t\}$.

and

$$\begin{aligned}\operatorname{Cov}(X_{t+h}, X_t) &= \lim_{n\to\infty} E\left[\left(\sum_{j=0}^{n} \phi_1^j Z_{t+h-j}\right)\left(\sum_{k=0}^{n} \phi_1^k Z_{t-k}\right)\right]\\ &= \sigma^2 \phi_1^{|h|} \sum_{j=0}^{\infty} \phi_1^{2j}\\ &= \sigma^2 \phi_1^{|h|}/(1-\phi_1^2).\end{aligned}$$

Moreover $\{X_t\}$ as defined by (3.1.12) satisfies the difference equations (3.1.11) and is therefore the unique stationary solution. A realization of the process with $\phi_1 = .9$ and $Z_t \sim N(0, 1)$ is shown in Figure 3.2(a). The autocorrelation function of the same process is shown in Figure 3.2(b).

In the case when $|\phi_1| > 1$ the series (3.1.12) does not converge in L^2. However we can rewrite (3.1.11) in the form

$$X_t = -\phi_1^{-1} Z_{t+1} + \phi_1^{-1} X_{t+1}. \tag{3.1.13}$$

Iterating (3.1.13) gives

$$\begin{aligned}X_t &= -\phi_1^{-1} Z_{t+1} - \phi_1^{-2} Z_{t+2} + \phi_1^{-2} X_{t+2}\\ &= \cdots\\ &= -\phi_1^{-1} Z_{t+1} - \cdots - \phi_1^{-k-1} Z_{t+k+1} + \phi_1^{-k-1} X_{t+k+1},\end{aligned}$$

which shows, by the same arguments as in the preceding paragraph, that

$$X_t = -\sum_{j=1}^{\infty} \phi_1^{-j} Z_{t+j} \tag{3.1.14}$$

is the unique stationary solution of (3.1.11). This solution should not be confused with the non-stationary solution $\{X_t, t = 0, \pm 1, \ldots\}$ of (3.1.11) obtained when X_0 is any specified random variable which is uncorrelated with $\{Z_t\}$.

The stationary solution (3.1.14) is frequently regarded as unnatural since X_t as defined by (3.1.14) is correlated with $\{Z_s, s > t\}$, a property not shared by the solution (3.1.12) obtained when $|\phi| < 1$. It is customary therefore when modelling stationary time series to restrict attention to AR(1) processes with $|\phi_1| < 1$ for which X_t has the representation (3.1.12) in terms of $\{Z_s, s \le t\}$. Such processes are called causal or future-independent autoregressive processes. It should be noted that every AR(1) process with $|\phi_1| > 1$ can be reexpressed as an AR(1) process with $|\phi_1| < 1$ and a new white noise sequence (Problem 3.3). From a second-order point of view therefore, nothing is lost by eliminating AR(1) processes with $|\phi_1| > 1$ from consideration.

If $|\phi_1| = 1$ there is no stationary solution of (3.1.11) (Problem 3.4). Consequently there is no such thing as an AR(1) with $|\phi_1| = 1$ according to our Definition 3.1.2.

The concept of causality will now be defined for a general ARMA(p, q) process.

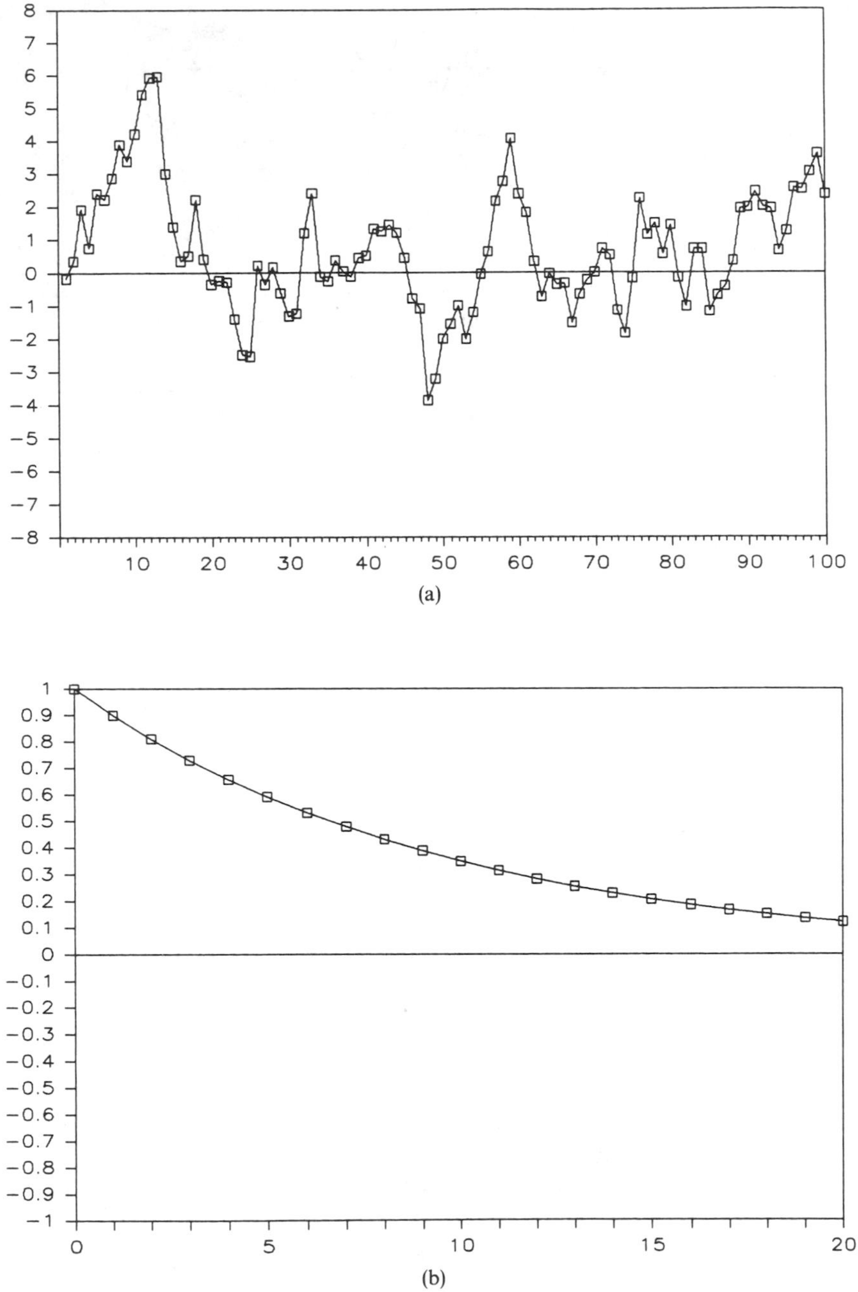

Figure 3.2. (a) 100 observations of the series $X_t - .9X_{t-1} = Z_t$, Example 3.1.2. (b) The autocorrelation function of $\{X_t\}$.

Definition 3.1.3. An ARMA(p, q) process defined by the equations $\phi(B)X_t = \theta(B)Z_t$ is said to be *causal* (or more specifically to be a causal function of $\{Z_t\}$) if there exists a sequence of constants $\{\psi_j\}$ such that $\sum_{j=0}^{\infty} |\psi_j| < \infty$ and

$$X_t = \sum_{j=0}^{\infty} \psi_j Z_{t-j}, \qquad t = 0, \pm 1, \dots. \tag{3.1.15}$$

It should be noted that causality is a property not of the process $\{X_t\}$ alone but rather of the relationship between the two processes $\{X_t\}$ and $\{Z_t\}$ appearing in the defining ARMA equations. In the terminology of Section 4.10 we can say that $\{X_t\}$ is causal if it is obtained from $\{Z_t\}$ by application of a causal linear filter. The following proposition clarifies the meaning of the sum appearing in (3.1.15).

Proposition 3.1.1. *If $\{X_t\}$ is any sequence of random variables such that $\sup_t E|X_t| < \infty$, and if $\sum_{j=-\infty}^{\infty} |\psi_j| < \infty$, then the series*

$$\psi(B)X_t = \sum_{j=-\infty}^{\infty} \psi_j B^j X_t = \sum_{j=-\infty}^{\infty} \psi_j X_{t-j}, \tag{3.1.16}$$

converges absolutely with probability one. If in addition $\sup_t E|X_t|^2 < \infty$ *then the series converges in mean square to the same limit.*

Proof. The monotone convergence theorem and finiteness of $\sup_t E|X_t|$ give

$$\begin{aligned} E\left(\sum_{j=-\infty}^{\infty} |\psi_j||X_{t-j}|\right) &= \lim_{n\to\infty} E\left(\sum_{j=-n}^{n} |\psi_j||X_{t-j}|\right) \\ &\le \lim_{n\to\infty}\left(\sum_{j=-n}^{n} |\psi_j|\right)\sup_t E|X_t| \\ &< \infty, \end{aligned}$$

from which it follows that $\sum_{j=-\infty}^{\infty} |\psi_j||X_{t-j}|$ and $\psi(B)X_t = \sum_{j=-\infty}^{\infty} \psi_j X_{t-j}$ are both finite with probability one.

If $\sup_t E|X_t|^2 < \infty$ and $n > m > 0$, then

$$\begin{aligned} E\left|\sum_{m<|j|\le n} \psi_j X_{t-j}\right|^2 &= \sum_{m<|j|\le n}\ \sum_{m<|k|\le n} \psi_j \bar{\psi}_k E(X_{t-j}\bar{X}_{t-k}) \\ &\le \sup_t E|X_t|^2 \left(\sum_{m<j\le n} |\psi_j|\right)^2 \\ &\to 0 \quad \text{as } m, n \to \infty, \end{aligned}$$

and so by the Cauchy criterion the series (3.1.16) converges in mean square. If S denotes the mean square limit, then by Fatou's lemma,

$$E|S - \psi(B)X_t|^2 = E \liminf_{n\to\infty} \left|S - \sum_{j=-n}^{n} \psi_j X_{t-j}\right|^2$$

$$\leq \liminf_{n\to\infty} E\left|S - \sum_{j=-n}^{n} \psi_j X_{t-j}\right|^2$$

$$= 0,$$

showing that the limits S and $\psi(B)X_t$ are equal with probability one. □

Proposition 3.1.2. *If $\{X_t\}$ is a stationary process with autocovariance function $\gamma(\cdot)$ and if $\sum_{j=-\infty}^{\infty} |\psi_j| < \infty$, then for each $t \in \mathbb{Z}$ the series (3.1.16) converges absolutely with probability one and in mean square to the same limit. If*

$$Y_t = \psi(B)X_t$$

then the process $\{Y_t\}$ is stationary with autocovariance function

$$\gamma_Y(h) = \sum_{j,k=-\infty}^{\infty} \psi_j \psi_k \gamma(h - j + k).$$

Proof. The convergence assertions follow at once from Proposition 3.1.1 and the observation that if $\{X_t\}$ is stationary then

$$E|X_t| \leq (E|X_t|^2)^{1/2} = c,$$

where c is finite and independent of t.

To check the stationarity of $\{Y_t\}$ we observe, using the mean square convergence of (3.1.16) and continuity of the inner product, that

$$EY_t = \lim_{n\to\infty} \sum_{j=-n}^{n} \psi_j EX_{t-j} = \left(\sum_{j=-\infty}^{\infty} \psi_j\right) EX_t,$$

and

$$E(Y_{t+h}Y_t) = \lim_{n\to\infty} E\left[\left(\sum_{j=-n}^{n} \psi_j X_{t+h-j}\right)\left(\sum_{k=-n}^{n} \psi_k X_{t-k}\right)\right]$$

$$= \sum_{j,k=-\infty}^{\infty} \psi_j \psi_k (\gamma(h - j + k) + (EX_t)^2).$$

Thus EY_t and $E(Y_{t+h}Y_t)$ are both finite and independent of t. The autocovariance function $\gamma_Y(\cdot)$ of $\{Y_t\}$ is given by

$$\gamma_Y(h) = E(Y_{t+h}Y_t) - EY_{t+h} \cdot EY_t = \sum_{j,k=-\infty}^{\infty} \psi_j \psi_k \gamma(h - j + k). \qquad \square$$

It is an immediate corollary of Proposition 3.1.2 that operators such as $\psi(B) = \sum_{j=-\infty}^{\infty} \psi_j B^j$ with $\sum_{j=-\infty}^{\infty} |\psi_j| < \infty$, when applied to stationary processes, are not only meaningful but also inherit the algebraic properties of power series. In particular if $\sum_{j=-\infty}^{\infty} |\alpha_j| < \infty$, $\sum_{j=-\infty}^{\infty} |\beta_j| < \infty$, $\alpha(z) = \sum_{j=-\infty}^{\infty} \alpha_j z^j$, $\beta(z) = \sum_{j=-\infty}^{\infty} \beta_j z^j$ and

$$\alpha(z)\beta(z) = \psi(z), \qquad |z| \leq 1,$$

then $\alpha(B)\beta(B)X_t$ is well-defined and

$$\alpha(B)\beta(B)X_t = \beta(B)\alpha(B)X_t = \psi(B)X_t.$$

The following theorem gives necessary and sufficient conditions for an ARMA process to be causal. It also gives an explicit representation of X_t in terms of $\{Z_s, s \le t\}$.

Theorem 3.1.1. *Let $\{X_t\}$ be an* ARMA(p, q) *process for which the polynomials $\phi(\cdot)$ and $\theta(\cdot)$ have no common zeroes. Then $\{X_t\}$ is causal if and only if $\phi(z) \neq 0$ for all $z \in \mathbb{C}$ such that $|z| \le 1$. The coefficients $\{\psi_j\}$ in* (3.1.15) *are determined by the relation*

$$\psi(z) = \sum_{j=0}^{\infty} \psi_j z^j = \theta(z)/\phi(z), \qquad |z| \le 1. \tag{3.1.17}$$

(The numerical calculation of the coefficients ψ_j is discussed in Section 3.3.)

PROOF. First assume that $\phi(z) \neq 0$ if $|z| \le 1$. This implies that there exists $\varepsilon > 0$ such that $1/\phi(z)$ has a power series expansion,

$$1/\phi(z) = \sum_{j=0}^{\infty} \xi_j z^j = \xi(z), \qquad |z| < 1 + \varepsilon.$$

Consequently $\xi_j(1 + \varepsilon/2)^j \to 0$ as $j \to \infty$ so that there exists $K \in (0, \infty)$ for which

$$|\xi_j| < K(1 + \varepsilon/2)^{-j} \quad \text{for all } j = 0, 1, 2, \ldots.$$

In particular we have $\sum_{j=0}^{\infty} |\xi_j| < \infty$ and $\xi(z)\phi(z) \equiv 1$ for $|z| \le 1$. By Proposition 3.1.2 we can therefore apply the operator $\xi(B)$ to both sides of the equation $\phi(B)X_t = \theta(B)Z_t$ to obtain

$$X_t = \xi(B)\theta(B)Z_t.$$

Thus we have the desired representation,

$$X_t = \sum_{j=0}^{\infty} \psi_j Z_{t-j}$$

where the sequence $\{\psi_j\}$ is determined by (3.1.17).

Now assume that $\{X_t\}$ is causal, i.e. $X_t = \sum_{j=0}^{\infty} \psi_j Z_{t-j}$ for some sequence $\{\psi_j\}$ such that $\sum_{j=0}^{\infty} |\psi_j| < \infty$. Then

$$\theta(B)Z_t = \phi(B)X_t = \phi(B)\psi(B)Z_t.$$

If we let $\eta(z) = \phi(z)\psi(z) = \sum_{j=0}^{\infty} \eta_j z^j$, $|z| \le 1$, we can rewrite this equation as

$$\sum_{j=0}^{q} \theta_j Z_{t-j} = \sum_{j=0}^{\infty} \eta_j Z_{t-j},$$

and taking inner products of each side with Z_{t-k} (recalling that $\{Z_t\} \sim \text{WN}(0, \sigma^2)$) we obtain $\eta_k = \theta_k$, $k = 0, \ldots, q$ and $\eta_k = 0$, $k > q$. Hence

$$\theta(z) = \eta(z) = \phi(z)\psi(z), \qquad |z| \le 1.$$

Since $\theta(z)$ and $\phi(z)$ have no common zeroes and since $|\psi(z)| < \infty$ for $|z| \le 1$, we conclude that $\phi(z)$ cannot be zero for $|z| \le 1$. □

Remark 1. If $\{X_t\}$ is an ARMA process for which the polynomials $\phi(\cdot)$ and $\theta(\cdot)$ have common zeroes, then there are two possibilities:

(a) none of the common zeroes lie on the unit circle, in which case (Problem 3.6) $\{X_t\}$ is the unique stationary solution of the ARMA equations with no common zeroes, obtained by cancelling the common factors of $\phi(\cdot)$ and $\theta(\cdot)$,

(b) at least one of the common zeroes lies on the unit circle, in which case the ARMA equations may have more than one stationary solution (see Problem 3.24).

Consequently ARMA processes for which $\phi(\cdot)$ and $\theta(\cdot)$ have common zeroes are rarely considered.

Remark 2. The first part of the proof of Theorem 3.1.1 shows that if $\{X_t\}$ is a stationary solution of the ARMA equations with $\phi(z) \ne 0$ for $|z| \le 1$, then we must have $X_t = \sum_{j=0}^{\infty} \psi_j Z_{t-j}$ where $\{\psi_j\}$ is defined by (3.1.17). Conversely if $X_t = \sum_{j=0}^{\infty} \psi_j Z_{t-j}$ then $\phi(B)X_t = \phi(B)\psi(B)Z_t = \theta(B)Z_t$. Thus the process $\{\psi(B)Z_t\}$ is the *unique* stationary solution of the ARMA equations if $\phi(z) \ne 0$ for $|z| \le 1$.

Remark 3. We shall see later (Problem 4.28) that if $\phi(\cdot)$ and $\theta(\cdot)$ have no common zeroes and if $\phi(z) = 0$ for some $z \in \mathbb{C}$ with $|z| = 1$, then there is no stationary solution of $\phi(B)X_t = \theta(B)Z_t$.

We now introduce another concept which is closely related to that of causality.

Definition 3.1.4. An ARMA(p, q) process defined by the equations $\phi(B)X_t = \theta(B)Z_t$ is said to be *invertible* if there exists a sequence of constants $\{\pi_j\}$ such that $\sum_{j=0}^{\infty} |\pi_j| < \infty$ and

$$Z_t = \sum_{j=0}^{\infty} \pi_j X_{t-j}, \qquad t = 0, \pm 1, \ldots. \tag{3.1.18}$$

Like causality, the property of invertibility is not a property of the process $\{X_t\}$ alone, but of the relationship between the two processes $\{X_t\}$ and $\{Z_t\}$ appearing in the defining ARMA equations. The following theorem gives necessary and sufficient conditions for invertibility and specifies the coefficients π_j in the representation (3.1.18).

Theorem 3.1.2. *Let $\{X_t\}$ be an* ARMA(p, q) *process for which the polynomials $\phi(\cdot)$ and $\theta(\cdot)$ have no common zeroes. Then $\{X_t\}$ is invertible if and only if*

$\theta(z) \neq 0$ for all $z \in \mathbb{C}$ such that $|z| \leq 1$. The coefficients $\{\pi_j\}$ in (3.1.18) *are determined by the relation*

$$\pi(z) = \sum_{j=0}^{\infty} \pi_j z^j = \phi(z)/\theta(z), \qquad |z| \leq 1. \tag{3.1.19}$$

(The coefficients $\{\pi_j\}$ can be calculated from recursion relations analogous to those for $\{\psi_j\}$ (see Problem 3.7).)

PROOF. First assume that $\theta(z) \neq 0$ if $|z| \leq 1$. By the same argument as in the proof of Theorem 3.1.1, $1/\theta(z)$ has a power series expansion

$$1/\theta(z) = \sum_{j=0}^{\infty} \eta_j z^j = \eta(z), \qquad |z| < 1 + \varepsilon,$$

for some $\varepsilon > 0$. Since $\sum_{j=0}^{\infty} |\eta_j| < \infty$, Proposition 3.1.2 allows us to apply $\eta(B)$ to both sides of the equation $\phi(B)X_t = \theta(B)Z_t$ to obtain

$$\eta(B)\phi(B)X_t = \eta(B)\theta(B)Z_t = Z_t.$$

Thus we have the desired representation

$$Z_t = \sum_{j=0}^{\infty} \pi_j X_{t-j},$$

where the sequence $\{\pi_j\}$ is determined by (3.1.19).

Conversely if $\{X_t\}$ is invertible then $Z_t = \sum_{j=0}^{\infty} \pi_j X_{t-j}$ for some sequence $\{\pi_j\}$ such that $\sum_{j=0}^{\infty} |\pi_j| < \infty$. Then

$$\phi(B)Z_t = \pi(B)\phi(B)X_t = \pi(B)\theta(B)Z_t.$$

Setting $\xi(z) = \pi(z)\theta(z) = \sum_{j=0}^{\infty} \xi_j z^j$, $|z| \leq 1$, we can rewrite this equation as

$$\sum_{j=0}^{p} \phi_j Z_{t-j} = \sum_{j=0}^{\infty} \xi_j Z_{t-j},$$

and taking inner products of each side with Z_{t-k} we obtain $\xi_k = \phi_k$, $k = 0, \ldots, p$ and $\xi_k = 0$, $k > p$. Hence

$$\phi(z) = \xi(z) = \pi(z)\theta(z), \qquad |z| \leq 1.$$

Since $\phi(z)$ and $\theta(z)$ have no common zeroes and since $|\pi(z)| < \infty$ for $|z| \leq 1$, we conclude that $\theta(z)$ cannot be zero for $|z| \leq 1$. □

Remark 4. If $\{X_t\}$ is a stationary solution of the equations

$$\phi(B)X_t = \theta(B)Z_t, \qquad \{Z_t\} \sim \text{WN}(0, \sigma^2), \tag{3.1.20}$$

and if $\phi(z)\theta(z) \neq 0$ for $|z| \leq 1$, then

$$X_t = \sum_{j=0}^{\infty} \psi_j Z_{t-j}$$

and

$$Z_t = \sum_{j=0}^{\infty} \pi_j X_{t-j},$$

where $\sum_{j=0}^{\infty} \psi_j z^j = \theta(z)/\phi(z)$ and $\sum_{j=0}^{\infty} \pi_j z^j = \phi(z)/\theta(z)$, $|z| \le 1$.

Remark 5. If $\{X_t\}$ is any ARMA process, $\phi(B)X_t = \theta(B)Z_t$, with $\phi(z)$ non-zero for all z such that $|z| = 1$, then it is possible to find polynomials $\tilde{\phi}(\cdot)$, $\tilde{\theta}(\cdot)$ and a white noise process $\{Z_t^*\}$ such that $\tilde{\phi}(B)X_t = \tilde{\theta}(B)Z_t^*$ and such that $\{X_t\}$ is a causal function of $\{Z_t^*\}$. If in addition $\theta(z)$ is non-zero when $|z| = 1$ then $\tilde{\theta}(\cdot)$ can be chosen in such a way that $\{X_t\}$ is also an invertible function of $\{Z_t^*\}$, i.e. such that $\tilde{\theta}(z)$ is non-zero for $|z| \le 1$ (see Proposition 3.5.1). If $\{Z_t\} \sim \text{IID}(0, \sigma^2)$ it is not true in general that $\{Z_t^*\}$ is independent (Breidt and Davis (1990)). It is true, however, if $\{Z_t\}$ is Gaussian (see Problem 3.18).

Remark 6. Theorem 3.1.2 can be extended to include the case when the moving average polynomial has zeroes on the unit circle if we extend the definition of invertibility to require only that $Z_t \in \overline{\text{sp}}\{X_s, -\infty < s \le t\}$. Under this definition, an ARMA process is invertible if and only if $\theta(z) \ne 0$ for all $|z| < 1$ (see Problem 3.8 and Propositions 4.4.1 and 4.4.3).

In view of Remarks 4 and 5 we shall focus attention on causal invertible ARMA processes except when the contrary is explicitly indicated. We conclude this section however with a discussion of the more general case when causality and invertibility are not assumed. Recall from Remark 3 that if $\phi(\cdot)$ and $\theta(\cdot)$ have no common zeroes and if $\phi(z) = 0$ for some $z \in \mathbb{C}$ with $|z| = 1$, then there is no stationary solution of $\phi(B)X_t = \theta(B)Z_t$. If on the other hand $\phi(z) \ne 0$ for all $z \in \mathbb{C}$ such that $|z| = 1$, then a well-known result from complex analysis guarantees the existence of $r > 1$ such that

$$\theta(z)\phi(z)^{-1} = \sum_{j=-\infty}^{\infty} \psi_j z^j = \psi(z), \qquad r^{-1} < |z| < r, \tag{3.1.21}$$

the Laurent series being absolutely convergent in the specified annulus (see e.g. Ahlfors (1953)). The existence of this Laurent expansion plays a key role in the proof of the following theorem.

Theorem 3.1.3. *If $\phi(z) \ne 0$ for all $z \in \mathbb{C}$ such that $|z| = 1$, then the* ARMA *equations $\phi(B)X_t = \theta(B)Z_t$ have the unique stationary solution,*

$$X_t = \sum_{j=-\infty}^{\infty} \psi_j Z_{t-j}, \tag{3.1.22}$$

where the coefficients ψ_j are determined by (3.1.21).

PROOF. By Proposition 3.1.2, $\{X_t\}$ as defined by (3.1.22) is a stationary process. Applying the operator $\phi(B)$ to each side of (3.1.22) and noting, again by

Proposition 3.1.2, that $\phi(B)\psi(B)Z_t = \theta(B)Z_t$, we obtain

$$\phi(B)X_t = \theta(B)Z_t. \tag{3.1.23}$$

Hence $\{X_t\}$ is a stationary solution of the ARMA equations.

To prove the converse let $\{X_t\}$ be any stationary solution of (3.1.23). Since $\phi(z) \neq 0$ for all $z \in \mathbb{C}$ such that $|z| = 1$, there exists $\delta > 1$ such that the series $\sum_{j=-\infty}^{\infty} \xi_j z^j = \phi(z)^{-1} = \xi(z)$ is absolutely convergent for $\delta^{-1} < |z| < \delta$. We can therefore apply the operator $\xi(B)$ to each side of (3.1.23) to get

$$\xi(B)\phi(B)X_t = \xi(B)\theta(B)Z_t,$$

or equivalently

$$X_t = \psi(B)Z_t. \qquad \square$$

§3.2 Moving Average Processes of Infinite Order

In this section we extend the notion of MA(q) process introduced in Section 3.1 by allowing q to be infinite.

Definition 3.2.1. If $\{Z_t\} \sim \mathrm{WN}(0, \sigma^2)$ then we say that $\{X_t\}$ is a moving average (MA(∞)) of $\{Z_t\}$ if there exists a sequence $\{\psi_j\}$ with $\sum_{j=0}^{\infty} |\psi_j| < \infty$ such that

$$X_t = \sum_{j=0}^{\infty} \psi_j Z_{t-j}, \qquad t = 0, \pm 1, \pm 2, \ldots. \tag{3.2.1}$$

Example 3.2.1. The MA(q) process defined by (3.1.9) is a moving average of $\{Z_t\}$ with $\psi_j = \theta_j, j = 0, 1, \ldots, q$ and $\psi_j = 0, j > q$.

Example 3.2.2. The AR(1) process with $|\phi| < 1$ is a moving average of $\{Z_t\}$ with $\psi_j = \phi^j, j = 0, 1, 2, \ldots$.

Example 3.2.3. By Theorem 3.1.1 the causal ARMA(p, q) process $\phi(B)X_t = \theta(B)Z_t$ is a moving average of $\{Z_t\}$ with $\sum_{j=0}^{\infty} \psi_j z^j = \theta(z)/\phi(z)$, $|z| \leq 1$.

It should be emphasized that in the definition of MA(∞) of $\{Z_t\}$ it is required that X_t should be expressible in terms of Z_s, $s \leq t$, only. It is for this reason that we need the assumption of causality in Example 3.2.3. However, even for non-causal ARMA processes, it is possible to find a white noise sequence $\{Z_t^*\}$ such that X_t is a moving average of $\{Z_t^*\}$ (Proposition 3.5.1). Moreover, as we shall see in Section 5.7, a large class of stationary processes have MA(∞) representations. We consider a special case in the following proposition.

Proposition 3.2.1. *If $\{X_t\}$ is a zero-mean stationary process with autocovariance function $\gamma(\cdot)$ such that $\gamma(h) = 0$ for $|h| > q$ and $\gamma(q) \neq 0$, then $\{X_t\}$ is an* MA(q)

process, i.e. there exists a white noise process $\{Z_t\}$ *such that*

$$X_t = Z_t + \theta_1 Z_{t-1} + \cdots + \theta_q Z_{t-q}. \tag{3.2.2}$$

PROOF. For each t, define the subspace $\mathscr{M}_t = \overline{\text{sp}}\{X_s, -\infty < s \le t\}$ of L^2 and set

$$Z_t = X_t - P_{\mathscr{M}_{t-1}} X_t. \tag{3.2.3}$$

Clearly $Z_t \in \mathscr{M}_t$, and by definition of $P_{\mathscr{M}_{t-1}}$, $Z_t \in \mathscr{M}_{t-1}^{\perp}$. Thus if $s < t$, $Z_s \in \mathscr{M}_s \subset \mathscr{M}_{t-1}$ and hence $EZ_sZ_t = 0$. Moreover, by Problem 2.18

$$P_{\overline{\text{sp}}\{X_s, s=t-n,\dots,t-1\}} X_t \xrightarrow{\text{m.s.}} P_{\mathscr{M}_{t-1}} X_t \quad \text{as } n \to \infty,$$

so that by stationarity and the continuity of the L^2 norm,

$$\begin{aligned}
\|Z_{t+1}\| &= \|X_{t+1} - P_{\mathscr{M}_t} X_{t+1}\| \\
&= \lim_{n\to\infty} \|X_{t+1} - P_{\overline{\text{sp}}\{X_s, s=t+1-n,\dots,t\}} X_{t+1}\| \\
&= \lim_{n\to\infty} \|X_t - P_{\overline{\text{sp}}\{X_s, s=t-n,\dots,t-1\}} X_t\| \\
&= \|X_t - P_{\mathscr{M}_{t-1}} X_t\| = \|Z_t\|.
\end{aligned}$$

Defining $\sigma^2 = \|Z_t\|^2$, we conclude that $\{Z_t\} \sim \text{WN}(0, \sigma^2)$.

Now by (3.2.3), it follows that

$$\begin{aligned}
\mathscr{M}_{t-1} &= \overline{\text{sp}}\{X_s, s < t-1, Z_{t-1}\} \\
&= \overline{\text{sp}}\{X_s, s < t-q, Z_{t-q}, \dots, Z_{t-1}\}
\end{aligned}$$

and consequently $\mathscr{M}_{t-1}$ can be decomposed into the two orthogonal subspaces, $\mathscr{M}_{t-q-1}$ and $\overline{\text{sp}}\{Z_{t-q}, \dots, Z_{t-1}\}$. Since $\gamma(h) = 0$ for $|h| > q$, it follows that $X_t \perp \mathscr{M}_{t-q-1}$ and so by Proposition 2.3.2 and Theorem 2.4.1,

$$\begin{aligned}
P_{\mathscr{M}_{t-1}} X_t &= P_{\mathscr{M}_{t-q-1}} X_t + P_{\overline{\text{sp}}\{Z_{t-q},\dots,Z_{t-1}\}} X_t \\
&= 0 + \sigma^{-2} E(X_t Z_{t-1}) Z_{t-1} + \cdots + \sigma^{-2} E(X_t Z_{t-q}) Z_{t-q} \\
&= \theta_1 Z_{t-1} + \cdots + \theta_q Z_{t-q}
\end{aligned}$$

where $\theta_j := \sigma^{-2} E(X_t Z_{t-j})$, which by stationarity is independent of t for $j = 1, \dots, q$. Substituting for $P_{\mathscr{M}_{t-1}} X_t$ in (3.2.3) gives (3.2.2). □

Remark. If $\{X_t\}$ has the same autocovariance function as that of an ARMA(p, q) process, then $\{X_t\}$ is also an ARMA(p, q) process. In other words, there exists a white noise sequence $\{Z_t\}$ and coefficients $\phi_1, \dots, \phi_p$, $\theta_1, \dots, \theta_q$ such that

$$X_t - \phi_1 X_{t-1} - \cdots - \phi_p X_{t-p} = Z_t + \theta_1 Z_{t-1} + \cdots + \theta_q Z_{t-q}$$

(see Problem 3.19).

The following theorem is an immediate consequence of Proposition 3.1.2.

Theorem 3.2.1. *The* MA(∞) *process defined by* (3.2.1) *is stationary with mean zero and autocovariance function*

$$\gamma(k) = \sigma^2 \sum_{j=0}^{\infty} \psi_j \psi_{j+|k|}. \tag{3.2.4}$$

Notice that Theorem 3.2.1 together with Example 3.2.3 completely determines the autocovariance function γ of any causal ARMA(p, q) process. We shall discuss the calculation of γ in more detail in Section 3.3.

The notion of AR(p) process introduced in Section 3.1 can also be extended to allow p to be infinite. In particular we note from Theorem 3.1.2 that any invertible ARMA(p, q) process satisfies the equations

$$X_t + \sum_{j=1}^{\infty} \pi_j X_{t-j} = Z_t, \qquad t = 0, \pm 1, \pm 2, \ldots$$

which have the same form as the AR(p) equations (3.1.10) with $p = \infty$.

§3.3 Computing the Autocovariance Function of an ARMA(p, q) Process

We now give three methods for computing the autocovariance function of an ARMA process. In practice, the third method is the most convenient for obtaining numerical values and the second is the most convenient for obtaining a solution in closed form.

First Method. The autocovariance function γ of the causal ARMA(p, q) process $\phi(B)X_t = \theta(B)Z_t$ was shown in Section 3.2 to satisfy

$$\gamma(k) = \sigma^2 \sum_{j=0}^{\infty} \psi_j \psi_{j+|k|}, \tag{3.3.1}$$

where

$$\psi(z) = \sum_{j=0}^{\infty} \psi_j z^j = \theta(z)/\phi(z) \qquad \text{for } |z| \le 1, \tag{3.3.2}$$

$\theta(z) = 1 + \theta_1 z + \cdots + \theta_q z^q$ and $\phi(z) = 1 - \phi_1 z - \cdots - \phi_p z^p$. In order to determine the coefficients ψ_j we can rewrite (3.3.2) in the form $\psi(z)\phi(z) = \theta(z)$ and equate coefficients of z^j to obtain (defining $\theta_0 = 1$, $\theta_j = 0$ for $j > q$ and $\phi_j = 0$ for $j > p$),

$$\psi_j - \sum_{0<k\le j} \phi_k \psi_{j-k} = \theta_j, \qquad 0 \le j < \max(p, q+1) \tag{3.3.3}$$

and

$$\psi_j - \sum_{0<k\le p} \phi_k \psi_{j-k} = 0, \qquad j \ge \max(p, q+1). \tag{3.3.4}$$

These equations can easily be solved successively for $\psi_0, \psi_1, \psi_2, \ldots$. Thus

$$\begin{aligned}
\psi_0 &= \theta_0 = 1,\\
\psi_1 &= \theta_1 + \psi_0\phi_1 = \theta_1 + \phi_1,\\
\psi_2 &= \theta_2 + \psi_0\phi_2 + \psi_1\phi_1 = \theta_2 + \phi_2 + \theta_1\phi_1 + \phi_1^2,\\
&\ldots.
\end{aligned} \tag{3.3.5}$$

Alternatively the general solution (3.3.4) can be written down, with the aid of Section 3.6 as

$$\psi_n = \sum_{i=1}^{k} \sum_{j=0}^{r_i-1} \alpha_{ij} n^j \xi_i^{-n}, \qquad n \ge \max(p, q+1) - p, \tag{3.3.6}$$

where $\xi_i, i = 1, \ldots, k$ are the distinct zeroes of $\phi(z)$ and r_i is the multiplicity of ξ_i (so that in particular we must have $\sum_{i=1}^{k} r_i = p$). The p constants α_{ij} and the coefficients $\psi_j, 0 \le j < \max(p, q+1) - p$, are then determined uniquely by the $\max(p, q+1)$ boundary conditions (3.3.3). This completes the determination of the sequence $\{\psi_j\}$ and hence, by (3.3.1), of the autocovariance function γ.

EXAMPLE 3.3.1. $(1 - B + \frac{1}{4}B^2)X_t = (1 + B)Z_t$. The equations (3.3.3) take the form

$$\psi_0 = \theta_0 = 1,$$

$$\psi_1 = \theta_1 + \psi_0\phi_1 = \theta_1 + \phi_1 = 2,$$

and (3.3.4) becomes

$$\psi_j - \psi_{j-1} + \tfrac{1}{4}\psi_{j-2} = 0, \qquad j \ge 2.$$

The general solution of (3.3.4) is (see Section 3.6)

$$\psi_n = (\alpha_{10} + n\alpha_{11})2^{-n}, \qquad n \ge 0.$$

The constants α_{10} and α_{11} are found from the boundary conditions $\psi_0 = 1$ and $\psi_1 = 2$ to be

$$\alpha_{10} = 1 \quad \text{and} \quad \alpha_{11} = 3.$$

Hence

$$\psi_n = (1 + 3n)2^{-n}, \qquad n = 0, 1, 2, \ldots.$$

Finally, substituting in (3.3.1), we obtain for $k \ge 0$

$$\begin{aligned}
\gamma(k) &= \sigma^2 \sum_{j=0}^{\infty} (1 + 3j)(1 + 3j + 3k)2^{-2j-k}\\
&= \sigma^2 2^{-k} \sum_{j=0}^{\infty} [(3k+1)4^{-j} + 3(3k+2)j4^{-j} + 9j^2 4^{-j}]\\
&= \sigma^2 2^{-k}[\tfrac{4}{3}(3k+1) + \tfrac{12}{9}(3k+2) + \tfrac{180}{27}]\\
&= \sigma^2 2^{-k}[\tfrac{32}{3} + 8k].
\end{aligned}$$

Second Method. An alternative method for computing the autocovariance function $\gamma(\cdot)$ of the causal ARMA(p,q)

$$\phi(B)X_t = \theta(B)Z_t, \tag{3.3.7}$$

is based on the difference equations for $\gamma(k)$, $k = 0, 1, 2, \ldots$, which are obtained by multiplying each side of (3.3.7) by X_{t-k} and taking expectations, namely

$$\gamma(k) - \phi_1\gamma(k-1) - \cdots - \phi_p\gamma(k-p) = \sigma^2 \sum_{k\le j\le q} \theta_j\psi_{j-k}, \quad 0 \le k < \max(p, q+1), \tag{3.3.8}$$

and

$$\gamma(k) - \phi_1\gamma(k-1) - \cdots - \phi_p\gamma(k-p) = 0, \qquad k \ge \max(p, q+1). \tag{3.3.9}$$

(In evaluating the right-hand sides of these equations we have used the representation $X_t = \sum_{j=0}^{\infty} \psi_j Z_{t-j}$.)

The general solution of (3.3.9) has the same form as (3.3.6), viz.

$$\gamma(h) = \sum_{i=1}^{k} \sum_{j=0}^{r_i-1} \beta_{ij} h^j \xi_i^{-h}, \qquad h \ge \max(p, q+1) - p, \tag{3.3.10}$$

where the p constants β_{ij} and the covariances $\gamma(j)$, $0 \le j < \max(p, q+1) - p$, are uniquely determined from the boundary conditions (3.3.8) after first computing $\psi_0, \psi_1, \ldots, \psi_q$ from (3.3.5).

EXAMPLE 3.3.2. $(1 - B + \frac{1}{4}B^2)X_t = (1 + B)Z_t$. The equations (3.3.9) become

$$\gamma(k) - \gamma(k-1) + \tfrac{1}{4}\gamma(k-2) = 0, \qquad k \ge 2,$$

with general solution

$$\gamma(n) = (\beta_{10} + \beta_{11}n)2^{-n}, \qquad n \ge 0. \tag{3.3.11}$$

The boundary conditions (3.3.8) are

$$\gamma(0) - \gamma(1) + \tfrac{1}{4}\gamma(2) = \sigma^2(\psi_0 + \psi_1),$$

$$\gamma(1) - \gamma(0) + \tfrac{1}{4}\gamma(1) = \sigma^2\psi_0,$$

where from (3.3.5), $\psi_0 = 1$ and $\psi_1 = \theta_1 + \phi_1 = 2$. Replacing $\gamma(0)$, $\gamma(1)$ and $\gamma(2)$ in accordance with the general solution (3.3.11) we obtain

$$3\beta_{10} - 2\beta_{11} = 16\sigma^2,$$

$$-3\beta_{10} + 5\beta_{11} = 8\sigma^2,$$

whence $\beta_{11} = 8\sigma^2$ and $\beta_{10} = 32\sigma^2/3$. Finally therefore we obtain the solution

$$\gamma(k) = \sigma^2 2^{-k}[\tfrac{32}{3} + 8k],$$

as found in Example 3.3.1 using the first method.

EXAMPLE 3.3.3 (The Autocovariance Function of an MA(q) Process). By Theorem 3.2.1 the autocovariance function of the process

$$X_t = \sum_{j=0}^{q} \theta_j Z_{t-j}, \qquad \{Z_t\} \sim \mathrm{WN}(0, \sigma^2),$$

has the extremely simple form

$$\gamma(k) = \begin{cases} \sigma^2 \sum_{j=0}^{q} \theta_j \theta_{j+|k|}, & |k| \le q, \\ 0, & |k| > q. \end{cases} \tag{3.3.12}$$

where θ_0 is defined to be 1 and $\theta_j, j > q$, is defined to be zero.

EXAMPLE 3.3.4 (The Autocovariance Function of an AR(p) Process). From (3.3.10) we know that the causal AR(p) process

$$\phi(B)X_t = Z_t,$$

has an autocovariance function of the form

$$\gamma(h) = \sum_{i=1}^{k} \sum_{j=0}^{r_i - 1} \beta_{ij} h^j \xi_i^{-h}, \qquad h \ge 0, \tag{3.3.13}$$

where ξ_i, $i = 1, \ldots, k$, are the zeroes (possibly complex) of $\phi(z)$, and r_i is the multiplicity of ξ_i. The constants β_{ij} are found from (3.3.8).

By changing the autoregressive polynomial $\phi(\cdot)$ and allowing p to be arbitrarily large it is possible to generate a remarkably large variety of covariance functions $\gamma(\cdot)$. This is extremely important when we attempt to find a process whose autocovariance function "matches" the sample autocovariances of a given data set. The general problem of finding a suitable ARMA process to represent a given set of data is discussed in detail in Chapters 8 and 9. In particular we shall prove in Section 8.1 that if $\gamma(\cdot)$ is *any* covariance function such that $\gamma(h) \to 0$ as $h \to \infty$, then for any k there is a causal AR(k) process whose autocovariance function at lags 0, 1, ..., k, coincides with $\gamma(j)$, $j = 0, 1, \ldots, k$.

We note from (3.3.13) that the rate of convergence of $\gamma(n)$ to zero as $n \to \infty$ depends on the zeroes ξ_i which are closest to the unit circle. (The causality condition guarantees that $|\xi_i| > 1$, $i = 1, \ldots, k$.) If $\phi(\cdot)$ has a zero close to the unit circle then the corresponding term or terms of (3.3.13) will decay in absolute value very slowly. Notice also that simple real zeroes of $\phi(\cdot)$ contribute terms to (3.3.13) which decrease geometrically with h. A pair of complex conjugate zeroes together contribute a geometrically damped sinusoidal term. We shall illustrate these possibilities numerically in Example 3.3.5 with reference to an AR(2) process.

EXAMPLE 3.3.5 (An Autoregressive Process with $p = 2$). For the causal AR(2),

$$(1 - \xi_1^{-1}B)(1 - \xi_2^{-1}B)X_t = Z_t, \qquad |\xi_1|, |\xi_2| > 1, \xi_1 \neq \xi_2,$$

we easily find from (3.3.13) and (3.3.8), using the relations

$$\phi_1 = \xi_1^{-1} + \xi_2^{-1},$$

and

$$\phi_2 = -\xi_1^{-1}\xi_2^{-1},$$

that

$$\gamma(h) = \frac{\sigma^2 \xi_1^2 \xi_2^2}{(\xi_1\xi_2 - 1)(\xi_2 - \xi_1)}[(\xi_1^2 - 1)^{-1}\xi_1^{1-h} - (\xi_2^2 - 1)^{-1}\xi_2^{1-h}]. \quad (3.3.14)$$

Figure 3.3 illustrates some of the possible forms of $\gamma(\cdot)$ for different values of ξ_1 and ξ_2. Notice that if $\xi_1 = re^{i\theta}$ and $\xi_2 = re^{-i\theta}$, $0 < \theta < \pi$, then we can rewrite (3.3.14) in the more illuminating form,

$$\gamma(h) = \frac{\sigma^2 r^4 \cdot r^{-h} \sin(h\theta + \psi)}{(r^2 - 1)(r^4 - 2r^2 \cos 2\theta + 1)^{1/2} \sin \theta}, \quad (3.3.15)$$

where

$$\tan \psi = \frac{r^2 + 1}{r^2 - 1} \tan \theta \quad (3.3.16)$$

and $\cos \psi$ has the same sign as $\cos \theta$.

Third Method. The numerical determination of the autocovariance function $\gamma(\cdot)$ from equations (3.3.8) and (3.3.9) can be carried out readily by first finding $\gamma(0), \ldots, \gamma(p)$ from the equations with $k = 0, 1, \ldots, p$, and then using the subsequent equations to determine $\gamma(p + 1), \gamma(p + 2), \ldots$ recursively.

EXAMPLE 3.3.6. For the process considered in Examples 3.3.1 and 3.3.2 the equations (3.3.8) and (3.3.9) with $k = 0, 1, 2$ are

$$\gamma(0) - \gamma(1) + \tfrac{1}{4}\gamma(2) = 3\sigma^2,$$

$$\gamma(1) - \gamma(0) + \tfrac{1}{4}\gamma(1) = \sigma^2,$$

$$\gamma(2) - \gamma(1) + \tfrac{1}{4}\gamma(0) = 0,$$

with solution $\gamma(0) = 32\sigma^2/3$, $\gamma(1) = 28\sigma^2/3$, $\gamma(2) = 20\sigma^2/3$. The higher lag autocovariances can now easily be found recursively from the equations

$$\gamma(k) = \gamma(k - 1) - \tfrac{1}{4}\gamma(k - 2), \qquad k = 3, 4, \ldots.$$

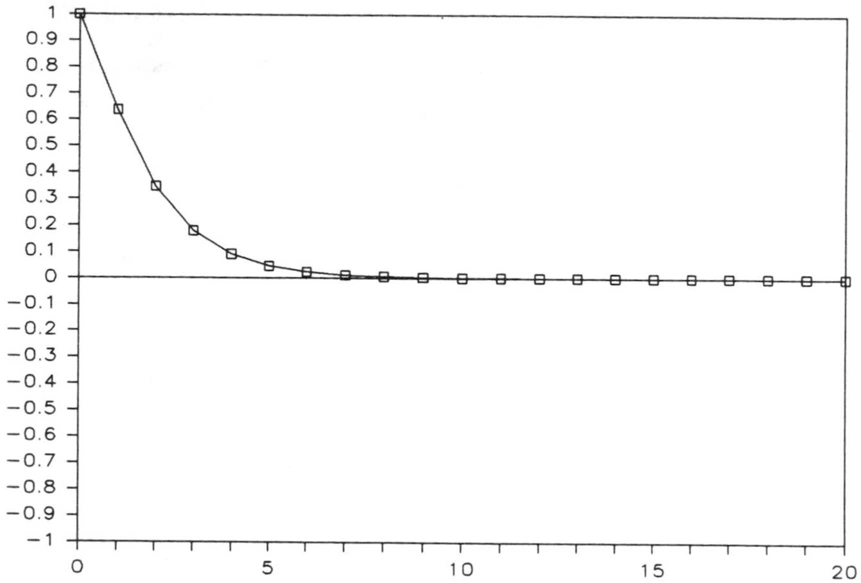

(a)

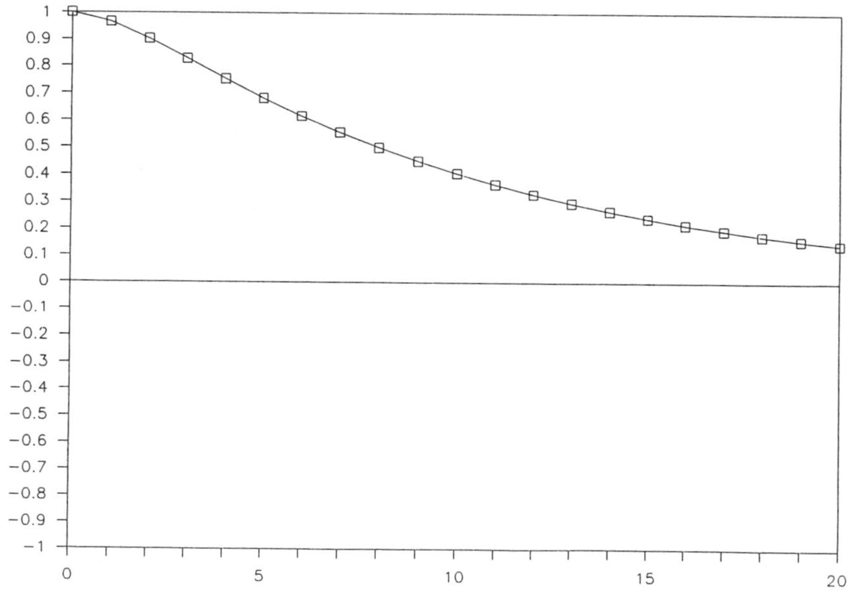

(b)

Figure 3.3. Autocorrelation functions $\gamma(h)/\gamma(0)$, $h = 0, \ldots, 20$, of the AR(2) process $(1 - \xi_1^{-1}B)(1 - \xi_2^{-1}B)X_t = Z_t$ when (a) $\xi_1 = 2$ and $\xi_2 = 5$, (b) $\xi_1 = \frac{10}{9}$ and $\xi_2 = 2$, (c) $\xi_1 = -\frac{10}{9}$ and $\xi_2 = 2$, (d) $\xi_1, \xi_2 = 2(1 \pm i\sqrt{3})/3$.

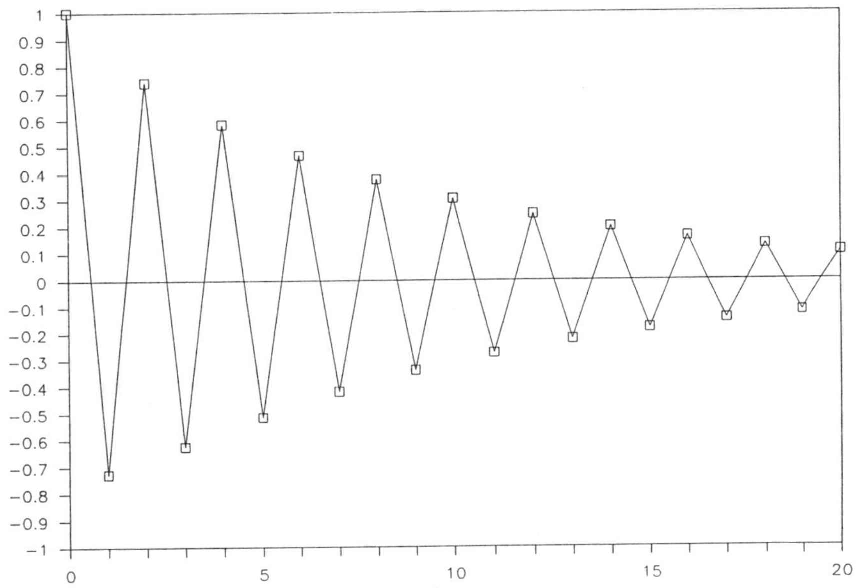

(c)

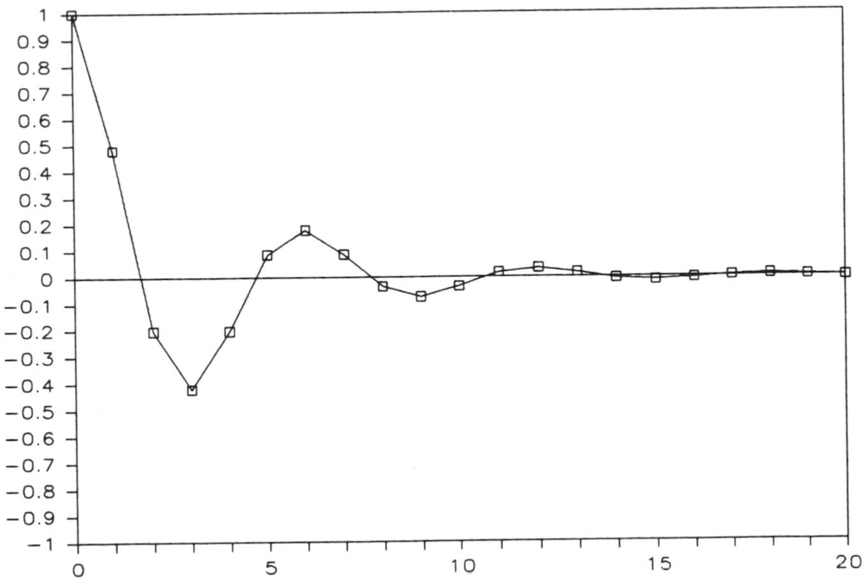

(d)

Figure 3.3 *Continued*

§3.4 The Partial Autocorrelation Function

The partial autocorrelation function, like the autocorrelation function, conveys vital information regarding the dependence structure of a stationary process. Like the autocorrelation function it also depends only on the second order properties of the process. The partial autocorrelation $\alpha(k)$ at lag k may be regarded as the correlation between X_1 and X_{k+1}, adjusted for the intervening observations $X_2, \ldots, X_k$. The idea is made precise in the following definition.

Definition 3.4.1. The partial autocorrelation function (pacf) $\alpha(\cdot)$ of a stationary time series is defined by

$$\alpha(1) = \operatorname{Corr}(X_2, X_1) = \rho(1),$$

and

$$\alpha(k) = \operatorname{Corr}(X_{k+1} - P_{\overline{\text{sp}}\{1, X_2, \ldots, X_k\}} X_{k+1}, X_1 - P_{\overline{\text{sp}}\{1, X_2, \ldots, X_k\}} X_1), \qquad k \geq 2,$$

where the projections $P_{\overline{\text{sp}}\{1, X_2, \ldots, X_k\}} X_{k+1}$ and $P_{\overline{\text{sp}}\{1, X_2, \ldots, X_k\}} X_1$ can be found from (2.7.13) and (2.7.14). The value $\alpha(k)$ is known as the partial autocorrelation at lag k.

The partial autocorrelation $\alpha(k)$, $k \geq 2$, is thus the correlation of the two residuals obtained after regressing X_{k+1} and X_1 on the intermediate observations $X_2, \ldots, X_k$. Recall that if the stationary process has zero mean then $P_{\overline{\text{sp}}\{1, X_2, \ldots, X_k\}}(\cdot) = P_{\overline{\text{sp}}\{X_2, \ldots, X_k\}}(\cdot)$ (see Problem 2.8).

EXAMPLE 3.4.1. Let $\{X_t\}$ be the zero mean AR(1) process

$$X_t = .9X_{t-1} + Z_t.$$

For this example

$$\begin{aligned}\alpha(1) &= \operatorname{Corr}(X_2, X_1) \\ &= \operatorname{Corr}(.9X_1 + Z_2, X_1) \\ &= .9\end{aligned}$$

since $\operatorname{Corr}(Z_2, X_1) = 0$. Moreover $P_{\overline{\text{sp}}\{X_2, \ldots, X_k\}} X_{k+1} = .9X_k$ by Problem 2.12 and $P_{\overline{\text{sp}}\{X_2, \ldots, X_k\}} X_1 = .9X_2$ since $(X_1, X_2, \ldots, X_k)'$ has the same covariance matrix as $(X_{k+1}, X_k, \ldots, X_2)'$. Hence for $k \geq 2$,

$$\begin{aligned}\alpha(k) &= \operatorname{Corr}(X_{k+1} - .9X_k, X_1 - .9X_2) \\ &= \operatorname{Corr}(Z_{k+1}, X_1 - .9X_2) \\ &= 0.\end{aligned}$$

A realization of 100 observations $\{X_t, t = 1, \ldots, 100\}$ was displayed in Figure 3.2. Scatter diagrams of (X_{t-1}, X_t) and (X_{t-2}, X_t) are shown in Figures 3.4 and 3.5 respectively. The sample correlation $\hat{\rho}(1) = \sum_{t=1}^{99} (X_t - \bar{X})(X_{t+1} - \bar{X}) / [\sum_{t=1}^{100} (X_t - \bar{X})^2]$ for Figure 3.4 is .814 (as compared with the corresponding

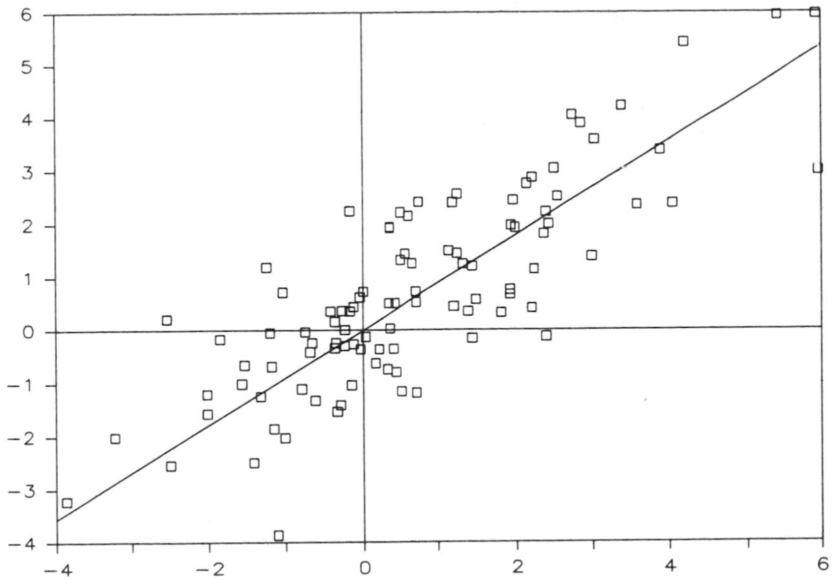

Figure 3.4. Scatter plot of the points (x_{t-1}, x_t) for the data of Figure 3.2, showing the line $x_t = .9x_{t-1}$.

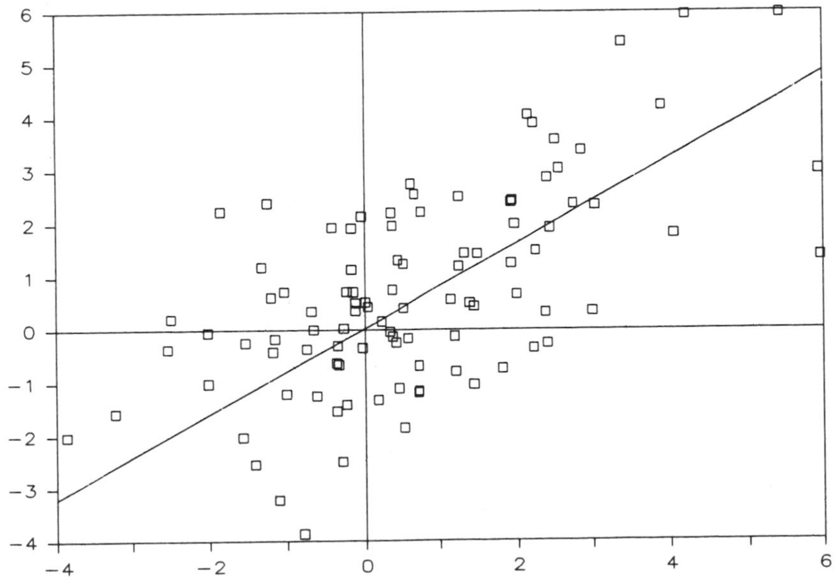

Figure 3.5. Scatter plot of the points (x_{t-2}, x_t) for the data of Figure 3.2, showing the line $x_t = .81x_{t-2}$.

theoretical correlation $\rho(1) = .9$). Likewise the sample correlation $\hat{\rho}(2) = \sum_{t=1}^{98} (X_t - \bar{X})(X_{t+2} - \bar{X})/[\sum_{t=1}^{100} (X_t - \bar{X})^2]$ for Figure 3.5 is .605 as compared with the theoretical correlation $\rho(2) = .81$. In Figure 3.6 we have plotted the points $(X_{t-2} - .9X_{t-1}, X_t - .9X_{t-1})$. It is apparent from the graph that the sample correlation between these variables is very small as expected from the fact that the theoretical partial autocorrelation at lag 2, i.e. $\alpha(2)$, is zero. One could say that the correlation between X_{t-2} and X_t is entirely eliminated when we remove the information in both variables explained by X_{t-1}.

EXAMPLE 3.4.2 (An MA(1) Process). For the moving average process,

$$X_t = Z_t + \theta Z_{t-1}, \qquad |\theta| < 1, \{Z_t\} \sim \text{WN}(0, \sigma^2),$$

we have

$$\alpha(1) = \rho(1) = \theta/(1 + \theta^2).$$

A simple calculation yields $P_{\overline{sp}\{X_2\}} X_3 = [\theta/(1 + \theta^2)] X_2 = P_{\overline{sp}\{X_2\}} X_1$, whence

$$\begin{aligned} \alpha(2) &= \text{Corr}(X_3 - \theta(1 + \theta^2)^{-1} X_2, X_1 - \theta(1 + \theta^2)^{-1} X_2) \\ &= -\theta^2/(1 + \theta^2 + \theta^4). \end{aligned}$$

More lengthy calculations (Problem 3.23) give

$$\alpha(k) = -\frac{(-\theta)^k (1 - \theta^2)}{1 - \theta^{2(k+1)}}.$$

One hundred observations $\{X_t, t = 1, \ldots, 100\}$ of the process with $\theta = -.8$ and $\rho(1) = -.488$ were displayed in Figure 3.1. The scatter diagram of the points $(X_{t-2} + .488X_{t-1}, X_t + .488X_{t-1})$ is plotted in Figure 3.7 and the sample correlation of the two variables is found to be $-.297$, as compared with the theoretical correlation $\alpha(2) = -(.8)^2/(1 + .8^2 + .8^4) = -.312$.

EXAMPLE 3.4.3 (An AR(p) Process). For the causal AR process

$$X_t - \phi_1 X_{t-1} - \cdots - \phi_p X_{t-p} = Z_t, \qquad \{Z_t\} \sim \text{WN}(0, \sigma^2),$$

we have for $k > p$,

$$P_{\overline{sp}\{X_2, \ldots, X_k\}} X_{k+1} = \sum_{j=1}^{p} \phi_j X_{k+1-j}, \tag{3.4.1}$$

since if $Y \in \overline{sp}\{X_2, \ldots, X_k\}$ then by causality $Y \in \overline{sp}\{Z_j, j \leq k\}$ and

$$\left\langle X_{k+1} - \sum_{j=1}^{p} \phi_j X_{k+1-j}, Y \right\rangle = \langle Z_{k+1}, Y \rangle = 0.$$

For $k > p$ we conclude from (3.4.1) that

$$\begin{aligned} \alpha(k) &= \text{Corr}\left(X_{k+1} - \sum_{j=1}^{p} \phi_j X_{k+1-j}, X_1 - P_{\overline{sp}\{X_2, \ldots, X_k\}} X_1 \right) \\ &= \text{Corr}(Z_{k+1}, X_1 - P_{\overline{sp}\{X_2, \ldots, X_k\}} X_1) \\ &= 0. \end{aligned}$$

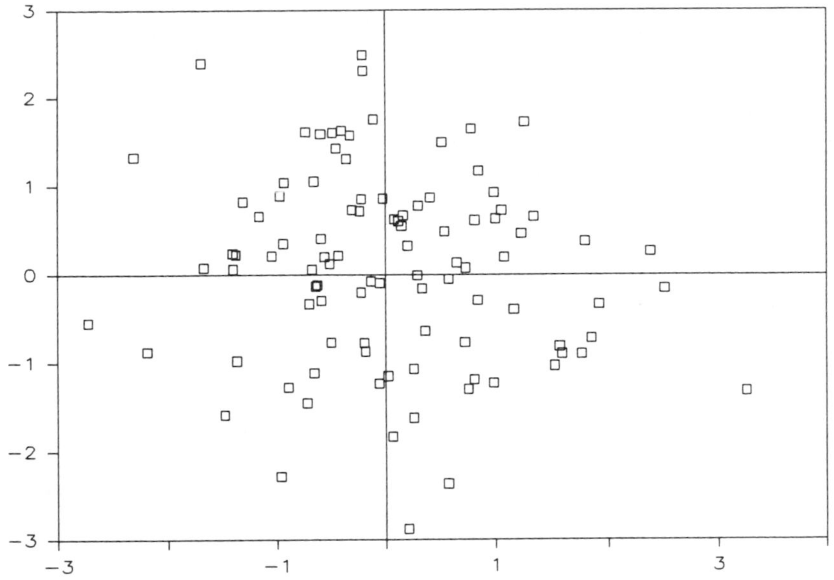

Figure 3.6. Scatter plot of the points $(x_{t-2} - .9x_{t-1}, x_t - .9x_{t-1})$ for the data of Figure 3.2.

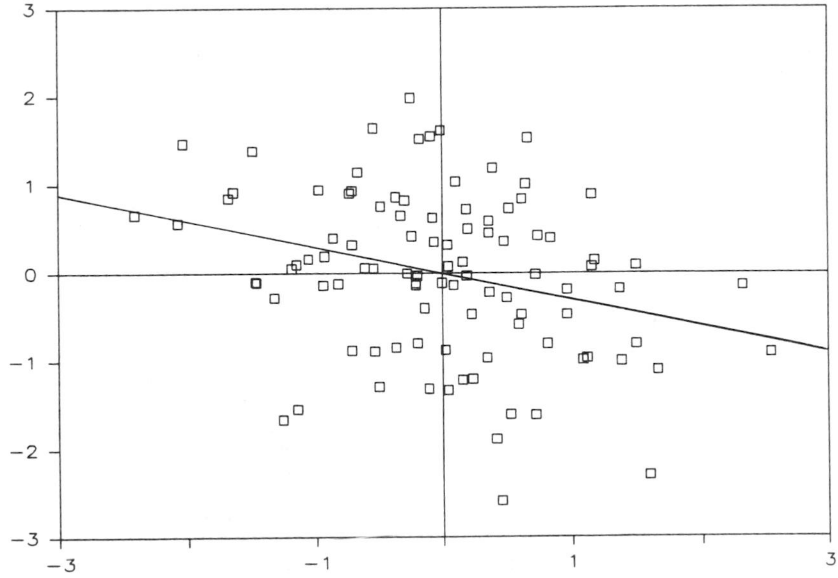

Figure 3.7. Scatter plot of the points $(x_{t-2} + .488x_{t-1}, x_t + .488x_{t-1})$ for the data of Figure 3.1, showing the line $y = -.312x$.

For $k \le p$ the values of $\alpha(k)$ can easily be computed from the equivalent Definition 3.4.2 below, after first determining $\rho(j) = \gamma(j)/\gamma(0)$ as described in Section 3.3.

In contrast with the partial autocorrelation function of an AR(p) process, that of an MA(q) process does not vanish for large lags. It is however bounded in absolute value by a geometrically decreasing function.

An Equivalent Definition of the Partial Autocorrelation Function

Let $\{X_t\}$ be a zero-mean stationary process with autocovariance function $\gamma(\cdot)$ such that $\gamma(h) \to 0$ as $h \to \infty$, and suppose that ϕ_{kj}, $j = 1, \dots, k$; $k = 1, 2, \dots$, are the coefficients in the representation

$$P_{\overline{\text{sp}}\{X_1,\dots,X_k\}} X_{k+1} = \sum_{j=1}^{k} \phi_{kj} X_{k+1-j}.$$

Then from the equations

$$\langle X_{k+1} - P_{\overline{\text{sp}}\{X_1,\dots,X_k\}} X_{k+1}, X_j \rangle = 0, \qquad j = k, \dots, 1,$$

we obtain

$$\begin{bmatrix} \rho(0) & \rho(1) & \rho(2) & \cdots & \rho(k-1) \\ \rho(1) & \rho(0) & \rho(1) & \cdots & \rho(k-2) \\ \vdots & & & & \vdots \\ \rho(k-1) & \rho(k-2) & \rho(k-3) & \cdots & \rho(0) \end{bmatrix} \begin{bmatrix} \phi_{k1} \\ \phi_{k2} \\ \vdots \\ \phi_{kk} \end{bmatrix} = \begin{bmatrix} \rho(1) \\ \rho(2) \\ \vdots \\ \rho(k) \end{bmatrix}, \qquad k \ge 1. \tag{3.4.2}$$

Definition 3.4.2. The partial autocorrelation $\alpha(k)$ of $\{X_t\}$ at lag k is

$$\alpha(k) = \phi_{kk}, \qquad k \ge 1,$$

where ϕ_{kk} is uniquely determined by (3.4.2).

The equivalence of Definitions 3.4.1 and 3.4.2 will be established in Chapter 5, Corollary 5.2.1. The *sample* partial autocorrelation function is defined similarly.

Definition 3.4.3. The sample partial autocorrelation $\hat{\alpha}(k)$ at lag k of $\{x_1, \dots, x_n\}$ is defined, provided $x_i \ne x_j$ for some i and j, by

$$\hat{\alpha}(k) = \hat{\phi}_{kk}, \qquad 1 \le k < n,$$

where $\hat{\phi}_{kk}$ is uniquely determined by (3.4.2) with each $\rho(j)$ replaced by the corresponding sample autocorrelation $\hat{\rho}(j)$.

§3.5 The Autocovariance Generating Function

If $\{X_t\}$ is a stationary process with autocovariance function $\gamma(\cdot)$, then its autocovariance generating function is defined by

$$G(z) = \sum_{k=-\infty}^{\infty} \gamma(k) z^k, \tag{3.5.1}$$

provided the series converges for all z in some annulus $r^{-1} < |z| < r$ with $r > 1$. Frequently the generating function is easy to calculate, in which case the autocovariance at lag k may be determined by identifying the coefficient of either z^k or z^{-k}. Clearly $\{X_t\}$ is white noise if and only if the autocovariance generating function $G(z)$ is constant for all z. If

$$X_t = \sum_{j=-\infty}^{\infty} \psi_j Z_{t-j}, \qquad \{Z_t\} \sim \text{WN}(0, \sigma^2), \tag{3.5.2}$$

and there exists $r > 1$ such that

$$\sum_{j=-\infty}^{\infty} |\psi_j| z^j < \infty, \qquad r^{-1} < |z| < r, \tag{3.5.3}$$

the generating function $G(\cdot)$ takes a very simple form. It is easy to see that

$$\gamma(k) = \text{Cov}(X_{t+k}, X_t) = \sigma^2 \sum_{j=-\infty}^{\infty} \psi_j \psi_{j+|k|},$$

and hence that

$$\begin{aligned} G(z) &= \sigma^2 \sum_{k=-\infty}^{\infty} \sum_{j=-\infty}^{\infty} \psi_j \psi_{j+|k|} z^k \\ &= \sigma^2 \left[\sum_{j=-\infty}^{\infty} \psi_j^2 + \sum_{k=1}^{\infty} \sum_{j=-\infty}^{\infty} \psi_j \psi_{j+k} (z^k + z^{-k}) \right] \\ &= \sigma^2 \left(\sum_{j=-\infty}^{\infty} \psi_j z^j \right) \left(\sum_{k=-\infty}^{\infty} \psi_k z^{-k} \right). \end{aligned}$$

Defining

$$\psi(z) = \sum_{j=-\infty}^{\infty} \psi_j z^j, \qquad r^{-1} < |z| < r,$$

we can write this result more neatly in the form

$$G(z) = \sigma^2 \psi(z) \psi(z^{-1}), \qquad r^{-1} < |z| < r. \tag{3.5.4}$$

Example 3.5.1 (The Autocovariance Generating Function of an ARMA(p, q) Process). By Theorem 3.1.3 and (3.1.21), any ARMA process $\phi(B)X_t = \theta(B)Z_t$ for which $\phi(z) \neq 0$ when $|z| = 1$ can be written in the form (3.5.2) with

$$\psi(z) = \theta(z)/\phi(z), \qquad r^{-1} < |z| < r$$

for some $r > 1$. Hence from (3.5.4)

$$G(z) = \sigma^2 \frac{\theta(z)\theta(z^{-1})}{\phi(z)\phi(z^{-1})}, \qquad r^{-1} < |z| < r. \tag{3.5.5}$$

In particular for the MA(2) process

$$X_t = Z_t + \theta_1 Z_{t-1} + \theta_2 Z_{t-2},$$

we have

$$\begin{aligned} G(z) &= \sigma^2(1 + \theta_1 z + \theta_2 z^2)(1 + \theta_1 z^{-1} + \theta_2 z^{-2}) \\ &= \sigma^2[(1 + \theta_1^2 + \theta_2^2) + (\theta_1 + \theta_1\theta_2)(z + z^{-1}) + \theta_2(z^2 + z^{-2})], \end{aligned}$$

from which we immediately find that

$$\begin{aligned} \gamma(0) &= \sigma^2(1 + \theta_1^2 + \theta_2^2), \\ \gamma(\pm 1) &= \sigma^2\theta_1(1 + \theta_2), \\ \gamma(\pm 2) &= \sigma^2\theta_2 \end{aligned}$$

and

$$\gamma(k) = 0 \quad \text{for } |k| > 2.$$

EXAMPLE 3.5.2. Let $\{X_t\}$ be the non-invertible MA(1) process

$$X_t = Z_t - 2Z_{t-1}, \qquad \{Z_t\} \sim \text{WN}(0, \sigma^2).$$

The process defined by

$$\begin{aligned} Z_t^* &:= (1 - .5B)^{-1}(1 - 2B)Z_t \\ &= (1 - .5B)^{-1}X_t = \sum_{j=0}^{\infty} (.5)^j X_{t-j}, \end{aligned}$$

has autocovariance generating function,

$$\begin{aligned} G(z) &= \frac{(1 - 2z)(1 - 2z^{-1})}{(1 - .5z)(1 - .5z^{-1})}\sigma^2 \\ &= \frac{4(1 - 2z)(1 - 2z^{-1})}{(1 - 2z)(1 - 2z^{-1})}\sigma^2 \\ &= 4\sigma^2. \end{aligned}$$

It follows that $\{Z_t^*\} \sim \text{WN}(0, 4\sigma^2)$ and hence that $\{X_t\}$ has the invertible representation,

$$X_t = Z_t^* - .5Z_{t-1}^*.$$

A corresponding result for ARMA processes is contained in the following proposition.

Proposition 3.5.1. *Let $\{X_t\}$ be the* ARMA(p, q) *process satisfying the equations*

$$\phi(B)X_t = \theta(B)Z_t, \qquad \{Z_t\} \sim \mathrm{WN}(0, \sigma^2),$$

where $\phi(z) \neq 0$ and $\theta(z) \neq 0$ for all $z \in \mathbb{C}$ such that $|z| = 1$. Then there exist polynomials, $\tilde{\phi}(z)$ and $\tilde{\theta}(z)$, nonzero for $|z| \leq 1$, of degree p and q respectively, and a white noise sequence $\{Z_t^*\}$ such that $\{X_t\}$ satisfies the causal invertible equations

$$\tilde{\phi}(B)X_t = \tilde{\theta}(B)Z_t^*.$$

Proof. Define

$$\tilde{\phi}(z) = \phi(z) \prod_{r<j\leq p} \frac{(1 - a_j z)}{(1 - a_j^{-1} z)},$$

$$\tilde{\theta}(z) = \theta(z) \prod_{s<j\leq q} \frac{(1 - b_j z)}{(1 - b_j^{-1} z)},$$

where $a_{r+1}, \ldots, a_p$ and $b_{s+1}, \ldots, b_q$ are the zeroes of $\phi(z)$ and $\theta(z)$ which lie inside the unit circle. Since $\tilde{\phi}(z) \neq 0$ and $\tilde{\theta}(z) \neq 0$ for all $|z| \leq 1$, it suffices to show that the process defined by

$$\begin{aligned} Z_t^* &= \frac{\tilde{\phi}(B)}{\tilde{\theta}(B)} X_t \\ &= \left(\prod_{r<j\leq p} \frac{1 - a_j B}{1 - a_j^{-1} B} \right) \left(\prod_{s<k\leq q} \frac{1 - b_k^{-1} B}{1 - b_k B} \right) Z_t \end{aligned}$$

is white noise. Using the same calculation as in Example 3.5.2, we find that the autocovariance generating function for $\{Z_t^*\}$ is given by

$$G(z) = \sigma^2 \left(\prod_{r<j\leq p} |a_j|^2 \right) \left(\prod_{s<k\leq q} |b_k|^{-2} \right).$$

Since $G(z)$ is constant, we conclude that $\{Z_t^*\}$ is white noise as asserted. □

§3.6* Homogeneous Linear Difference Equations with Constant Coefficients

In this section we consider the solution $\{h_t\}$ of the kth order linear difference equation

$$h_t + \alpha_1 h_{t-1} + \cdots + \alpha_k h_{t-k} = 0, \qquad t \in T, \tag{3.6.1}$$

where $\alpha_1, \ldots, \alpha_k$ are real constants with $\alpha_k \neq 0$ and T is a subinterval of the integers which without loss of generality we can assume to be $[k, \infty)$, $(-\infty, \infty)$ or $[k, k + r]$, $r > 0$. Introducing the backward shift operator B defined by equation (3.1.8), we can write (3.6.1) in the more compact form

$$\alpha(B)h_t = 0, \qquad t \in T, \tag{3.6.2}$$

where $\alpha(B) = 1 + \alpha_1 B + \cdots + \alpha_k B^k$.

Definition 3.6.1. A set of $m \le k$ solutions, $\{h_t^{(1)}, \ldots, h_t^{(m)}\}$, of (3.6.2) will be called linearly independent if from

$$c_1 h_t^{(1)} + c_2 h_t^{(2)} + \cdots + c_m h_t^{(m)} = 0 \quad \text{for all } t = 0, 1, \ldots, k-1,$$

it follows that $c_1 = c_2 = \cdots = c_m = 0$.

We note that if $\{h_t^1\}$ and $\{h_t^2\}$ are any two solutions of (3.6.2) then $\{c_1 h_t^1 + c_2 h_t^2\}$ is also a solution. Moreover for any specified values of $h_0, h_1, \ldots, h_{k-1}$, henceforth referred to as *initial conditions*, all the remaining values h_t, $t \notin [0, k-1]$, are uniquely determined by one or other of the recursion relations

$$h_t = -\alpha_1 h_{t-1} - \cdots - \alpha_k h_{t-k}, \qquad t = k, k+1, \ldots, \tag{3.6.3}$$

and

$$\alpha_k h_t = -h_{t+k} - \alpha_1 h_{t+k-1} - \cdots - \alpha_{k-1} h_{t+1}, \qquad t = -1, -2, \ldots. \tag{3.6.4}$$

Thus if we can find k linearly independent solutions $\{h_t^{(1)}, \ldots, h_t^{(k)}\}$ of (3.6.2) then by linear independence there will be exactly one set of coefficients $c_1, \ldots, c_k$ such that the solution

$$h_t = c_1 h_t^{(1)} + \cdots + c_k h_t^{(k)}, \tag{3.6.5}$$

has prescribed initial values $h_0, h_1, \ldots, h_{k-1}$. Since these values uniquely determine the entire sequence $\{h_t\}$ we conclude that (3.6.5) is the unique solution of (3.6.2) satisfying the initial conditions. The remainder of this section is therefore devoted to finding a set of k linearly independent solutions of (3.6.2).

Theorem 3.6.1. *If $h_t = (a_0 + a_1 t + \cdots + a_j t^j)m^t$ where $a_0, \ldots, a_j, m$ are (possibly complex-valued) constants, then there are constants $b_0, \ldots, b_{j-1}$ such that*

$$(1 - mB)h_t = (b_0 + b_1 t + \cdots + b_{j-1} t^{j-1})m^t.$$

PROOF.

$$\begin{aligned}(1 - mB)h_t &= (a_0 + a_1 t + \cdots + a_k t^j)m^t - m(a_0 + a_1(t-1) + \cdots \\ &\quad + a_k(t-1)^j)m^{t-1} \\ &= m^t\left[\sum_{r=0}^{j} a_r(t^r - (t-1)^r)\right]\end{aligned}$$

and $\sum_{r=0}^{j} a_r(t^r - (t-1)^r)$ is clearly a polynomial of degree $j-1$. □

Corollary 3.6.1. *The functions* $h_t^{(j)} = t^j \xi^{-t}$, $j = 0, 1, \ldots, k - 1$ *are* k *linearly independent solutions of the difference equation*

$$(1 - \xi^{-1}B)^k h_t = 0. \tag{3.6.6}$$

PROOF. Repeated application of the operator $(1 - \xi^{-1}B)$ to $h_t^{(j)}$ in conjunction with Theorem 3.6.1 establishes that $h_t^{(j)}$ satisfies (3.6.6). If

$$(c_0 + c_1 t + \cdots + c_{k-1}t^{k-1})\xi^{-t} = 0 \quad \text{for } t = 0, 1, \ldots, k - 1,$$

then the polynomial $\sum_{j=0}^{k-1} c_j t^j$, which is of degree less than k, has k zeroes. This is only possible if $c_0 = c_1 = \cdots = c_{k-1} = 0$. □

Solution of the General Equation of Order k

For the general equation (3.6.2), the difference operator $\alpha(B)$ can be written as

$$\alpha(B) = \prod_{i=1}^{j} (1 - \xi_i^{-1}B)^{r_i}$$

where ξ_i, $i = 1, \ldots, j$ are the distinct zeroes of $\alpha(z)$ and r_i is the multiplicity of ξ_i. It follows from Corollary 3.6.1 that $t^n \xi_i^{-t}$, $n = 0, 1, \ldots, r_i - 1$; $i = 1, \ldots, j$, are k solutions of the difference equation (3.6.2) since

$$\alpha(B)t^n \xi_i^{-t} = \prod_{s \neq i} (1 - \xi_s^{-1}B)^{r_s}(1 - \xi_i^{-1}B)^{r_i} t^n \xi_i^{-t} = 0.$$

It is shown below in Theorem 3.6.2 and Corollary 3.6.2 that these solutions are indeed linearly independent and hence that the general solution of (3.6.2) is

$$h_t = \sum_{i=1}^{j} \sum_{n=0}^{r_i-1} c_{in} t^n \xi_i^{-t}. \tag{3.6.7}$$

In order for this general solution to be real, the coefficients corresponding to a pair of complex conjugate roots must themselves be complex conjugates. More specifically if $(\xi_j, \bar{\xi}_j)$ is a pair of complex conjugate zeroes of $\alpha(z)$ and $\xi_j = d\exp(i\theta_j)$, then the corresponding terms in (3.6.7) are

$$\sum_{n=0}^{r_i-1} c_{in} t^n \xi_i^{-t} + \sum_{n=0}^{r_i-1} \bar{c}_{in} t^n \bar{\xi}_i^{-t},$$

which can be rewritten as

$$\sum_{n=0}^{r_i-1} 2[\operatorname{Re}(c_{in})\cos(\theta_i t) + \operatorname{Im}(c_{in})\sin(\theta_i t)] t^n d^{-t},$$

or equivalently as

$$\sum_{n=0}^{r_i-1} a_{in} t^n d^{-t} \cos(\theta_i t + b_{in}),$$

with appropriately chosen constants a_{in} and b_{in}.

EXAMPLE 3.6.1. Suppose h_t satisfies the first order linear difference equation $(1 - \xi^{-1}B)h_t = 0$. Then the general solution is given by $h_t = C\xi^{-t} = h_0\xi^{-t}$. Observe that if $|\xi| > 1$, then h_t decays at an exponential rate as $t \to \infty$.

EXAMPLE 3.6.2. Consider the second order difference equation $(1 + \alpha_1 B + \alpha_2 B^2)h_t = 0$. Since $1 + \alpha_1 B + \alpha_2 B^2 = (1 - \xi_1^{-1}B)(1 - \xi_2^{-1}B)$, the character of the general solution will depend on ξ_1 and ξ_2.

Case 1 ξ_1 and ξ_2 are real and distinct. In this case, $h_t = c_1\xi_1^{-t} + c_2\xi_2^{-t}$ where c_1 and c_2 are determined by the two initial conditions $c_1 + c_2 = h_0$ and $c_1\xi_1^{-1} + c_2\xi_2^{-1} = h_1$. These have a unique solution since $\xi_1 \neq \xi_2$.

Case 2 $\xi_1 = \xi_2$. Using (3.6.7) with $j = 1$ and $r_1 = 2$ we have $h_t = (c_0 + c_1 t)\xi_1^{-t}$.

Case 3 $\xi_1 = \bar{\xi}_2 = de^{i\theta}$, $0 < \theta < 2\pi$. The solution can be written either as $c\xi_1^{-t} + \bar{c}\bar{\xi}_1^{-t}$ or as the sinusoid $h_t = ad^{-t}\cos(\theta t + b)$.

Observe that if $|\xi_1| > 1$ and $|\xi_2| > 1$, then in each of the three cases, h_t approaches zero at a geometric rate as $t \to \infty$. In the third case, h_t is a damped sinusoid. More generally, if the roots of $\alpha(z)$ lie outside the unit circle, then the general solution is a sum of exponentially decaying functions and exponentially damped sinusoids.

We now return to the problem of establishing linear independence of the solutions $t^n\xi_i^{-t}$, $n = 0, 1, \ldots, r_i - 1$; $i = 1, \ldots, j$, of (3.6.2).

Theorem 3.6.2. *If*

$$\sum_{l=1}^{q}\sum_{j=0}^{p} c_{lj}t^j m_l^t = 0 \quad \text{for } t = 0, 1, 2, \ldots \tag{3.6.8}$$

where $m_1, m_2, \ldots, m_q$ *are distinct numbers, then* $c_{lj} = 0$ *for* $l = 1, 2, \ldots, q$; $j = 0, 1, \ldots, p$.

PROOF. Without loss of generality we can assume that $|m_1| \geq |m_2| \geq \cdots \geq |m_q| > 0$. It will be sufficient to show that (3.6.8) implies that

$$c_{1j} = 0, \qquad j = 0, \ldots, p \tag{3.6.9}$$

since if this is the case then equations (3.6.8) reduce to

$$\sum_{l=2}^{q}\sum_{j=0}^{p} c_{lj}t^j m_l^j = 0, \qquad t = 0, 1, 2, \ldots,$$

which in turn imply that $c_{2j} = 0$, $j = 0, \ldots, p$. Repetition of this argument shows then that $c_{lj} = 0, j = 0, \ldots, p; l = 1, \ldots, q$.

To prove that (3.6.8) implies (3.6.9) we need to consider two separate cases.

Case 1 $|m_1| > |m_2|$. Dividing each side of (3.6.8) by $t^p m_1^t$ and letting $t \to \infty$, we find that $c_{1p} = 0$. Setting $c_{1p} = 0$ in (3.6.8), dividing each side by $t^{p-1}m_1^t$ and letting $t \to \infty$, we then obtain $c_{2p} = 0$. Repeating the procedure with divisors $t^{p-2}m_1^t, t^{p-3}m_1^t, \ldots, m_1^t$ (in that order) we find that $c_{1j} = 0, j = 0, 1, \ldots, p$ as required.

Case 2 $|m_1| = |m_2| = \cdots = |m_s| > |m_{s+1}| > 0$, where $s \le q$. In this case we can write $m_j = re^{i\theta_j}$ where $-\pi < \theta_j \le \pi$ and $\theta_1, \ldots, \theta_s$ are all different. Dividing each side of (3.6.8) by $t^p r^t$ and letting $t \to \infty$ we find that

$$\sum_{l=1}^{s} c_{lp} e^{i\theta_l t} \to 0 \quad \text{as } t \to \infty. \tag{3.6.10}$$

We shall now show that this is impossible unless $c_{1p} = c_{2p} = \cdots = c_{sp} = 0$. Set $g_t = \sum_{l=1}^{s} c_{lp} e^{i\theta_l t}$ and let A_n, $n = 0, 1, 2, \ldots$, be the matrix

$$A_n = \begin{bmatrix} e^{i\theta_1 n} & e^{i\theta_2 n} & & e^{i\theta_s n} \\ e^{i\theta_1(n+1)} & e^{i\theta_2(n+1)} & & e^{i\theta_s(n+1)} \\ \vdots & \vdots & & \vdots \\ e^{i\theta_1(n+s-1)} & e^{i\theta_2(n+s-1)} & \cdots & e^{i\theta_s(n+s-1)} \end{bmatrix}. \tag{3.6.11}$$

Observe that $\det A_n = e^{i(\theta_1 + \cdots + \theta_s)n}(\det A_0)$. The matrix A_0 is a Vandermonde matrix (Birkhoff and Mac Lane (1965)) and hence has a non-zero determinant. Applying Cramer's rule to the equation

$$A_n \begin{bmatrix} c_{1p} \\ \vdots \\ c_{sp} \end{bmatrix} = \begin{bmatrix} g_n \\ \vdots \\ g_{n+s-1} \end{bmatrix},$$

we have

$$c_{1p} = \frac{\det M}{\det A_n}, \tag{3.6.12}$$

where

$$M = \begin{bmatrix} g_n & e^{i\theta_2 n} & \cdots & e^{i\theta_3 n} \\ \vdots & \vdots & & \vdots \\ g_{n+s-1} & e^{i\theta_2(n+s-1)} & \cdots & e^{i\theta_s(n+s-1)} \end{bmatrix}.$$

Since $g_n \to 0$ as $n \to \infty$, the numerator in (3.6.12) approaches zero while the denominator remains bounded away from zero because $|\det A_n| = |\det A_0| > 0$. Hence c_{1p} must be zero. The same argument applies to the other coefficients $c_{2p}, \ldots, c_{sp}$ showing that they are all necessarily zero as claimed.

We now divide (3.6.8) by $t^{p-1}r^t$ and repeat the preceding argument, letting $t \to \infty$ to deduce that

$$\sum_{l=1}^{s} c_{l,p-1} e^{i\theta_l t} \to 0 \quad \text{as } t \to \infty,$$

and hence that $c_{l,p-1} = 0, l = 1, \ldots, s$. We then divide by $t^{p-2}r^t, \ldots, r^t$ (in that order), repeating the argument at each stage to deduce that

$$c_{lj} = 0, \quad j = 0, 1, \ldots, p \quad \text{and} \quad l = 1, 2, \ldots, s.$$

This shows that (3.6.8) implies (3.6.9) in this case, thereby completing the proof of the theorem. □

Corollary 3.6.2. *The k solutions $t^n \xi_i^{-t}$, $n = 0, 1, \ldots, r_i - 1$; $i = 1, \ldots, j$, of the difference equation* (3.6.2) *are linearly independent.*

PROOF. We must show that each c_{in} is zero if $\sum_{i=1}^{j} \sum_{n=0}^{r_i - 1} c_{in} t^n \xi_i^{-t} = 0$ for $t = 0, 1, \ldots, k-1$. Setting h_t equal to the double sum we have $\alpha(B)h_t = 0$ and $h_0 = h_1 = \cdots = h_{k-1} = 0$. But by the recursions (3.6.3) and (3.6.4), this necessarily implies that $h_t = 0$ for all t. Direct application of Theorem 3.6.2 with $p = \max\{r_1, \ldots, r_j\}$ completes the proof. □

Problems

3.1. Determine which of the following processes are causal and/or invertible:
   (a) $X_t + .2X_{t-1} - .48X_{t-2} = Z_t$,
   (b) $X_t + 1.9X_{t-1} + .88X_{t-2} = Z_t + .2Z_{t-1} + .7Z_{t-2}$,
   (c) $X_t + .6X_{t-2} = Z_t + 1.2Z_{t-1}$,
   (d) $X_t + 1.8X_{t-1} + .81X_{t-2} = Z_t$,
   (e) $X_t + 1.6X_{t-1} = Z_t - .4Z_{t-1} + .04Z_{t-2}$.

3.2. Show that in order for an AR(2) process with autoregressive polynomial $\phi(z) = 1 - \phi_1 z - \phi_2 z^2$ to be causal, the parameters (ϕ_1, ϕ_2) must lie in the triangular region determined by the intersection of the three regions,

$$\phi_2 + \phi_1 < 1,$$

$$\phi_2 - \phi_1 < 1,$$

$$|\phi_2| < 1.$$

3.3. Let $\{X_t, t = 0, \pm 1, \ldots\}$ be the stationary solution of the non-causal AR(1) equations,

$$X_t = \phi X_{t-1} + Z_t, \qquad \{Z_t\} \sim \text{WN}(0, \sigma^2), \qquad |\phi| > 1.$$

Show that $\{X_t\}$ also satisfies the causal AR(1) equations,

$$X_t = \phi^{-1} X_{t-1} + \tilde{Z}_t, \quad \{\tilde{Z}_t\} \sim \text{WN}(0, \tilde{\sigma}^2),$$

for a suitably chosen white noise process $\{\tilde{Z}_t\}$. Determine $\tilde{\sigma}^2$.

3.4. Show that there is no stationary solution of the difference equations

$$X_t = \phi X_{t-1} + Z_t, \qquad \{Z_t\} \sim \text{WN}(0, \sigma^2),$$

if $\phi = \pm 1$.

3.5. Let $\{Y_t, t = 0, \pm 1, \ldots\}$ be a stationary time series. Show that there exists a stationary solution $\{X_t\}$ of the difference equations,

$$X_t - \phi_1 X_{t-1} - \cdots - \phi_p X_{t-p} = Y_t + \theta_1 Y_{t-1} + \cdots + \theta_q Y_{t-q},$$

if $\phi(z) = 1 - \phi_1 z - \cdots - \phi_p z^p \neq 0$ for $|z| = 1$. Furthermore, if $\phi(z) \neq 0$ for $|z| \leq 1$ show that $\{X_t\}$ is a causal function of $\{Y_t\}$.

3.6. Suppose that $\{X_t\}$ is the ARMA process defined by

$$\phi(B)X_t = \theta(B)Z_t, \qquad \{Z_t\} \sim \text{WN}(0, \sigma^2),$$

where $\phi(\cdot)$ and $\theta(\cdot)$ have no common zeroes and $\phi(z) \neq 0$ for $|z| = 1$. If $\xi(\cdot)$ is any polynomial such that $\xi(z) \neq 0$ for $|z| = 1$, show that the difference equations,

$$\xi(B)\phi(B)Y_t = \xi(B)\theta(B)Z_t,$$

have the unique stationary solution, $\{Y_t\} = \{X_t\}$.

3.7. Suppose $\{X_t\}$ is an invertible ARMA(p, q) process satisfying (3.1.4) with

$$Z_t = \sum_{j=0}^{\infty} \pi_j X_{t-j}.$$

Show that the sequence $\{\pi_j\}$ is determined by the equations

$$\pi_j + \sum_{k=1}^{\min(q,j)} \theta_k \pi_{j-k} = -\phi_j, \qquad j = 0, 1, \ldots$$

where we define $\phi_0 = -1$ and $\theta_k = 0$ for $k > q$ and $\phi_j = 0$ for $j > p$.

3.8. The process $X_t = Z_t - Z_{t-1}$, $\{Z_t\} \sim \text{WN}(0, \sigma^2)$, is not invertible according to Definition 3.1.4. Show however that $Z_t \in \overline{\text{sp}}\{X_j, -\infty < j \leq t\}$ by considering the mean square limit of the sequence $\sum_{j=0}^{n} (1 - j/n)X_{t-j}$ as $n \to \infty$.

3.9. Suppose $\{X_t\}$ is the two-sided moving average

$$X_t = \sum_{j=-\infty}^{\infty} \psi_j Z_{t-j}, \qquad \{Z_t\} \sim \text{WN}(0, \sigma^2)$$

where $\Sigma_j |\psi_j| < \infty$. Show that $\sum_{h=-\infty}^{\infty} |\gamma(h)| < \infty$ where $\gamma(\cdot)$ is the autocovariance function of $\{X_t\}$.

3.10. Let $\{Y_t\}$ be a stationary zero-mean time series. Define

$$X_t = (1 - .4B)Y_t = Y_t - .4Y_{t-1}$$

and

$$W_t = (1 - 2.5B)Y_t = Y_t - 2.5Y_{t-1}.$$

(a) Express the autocovariance functions of $\{X_t\}$ and $\{W_t\}$ in terms of the autocovariance function of $\{Y_t\}$.
(b) Show that $\{X_t\}$ and $\{W_t\}$ have the same autocorrelation functions.
(c) Show that the process $U_t = -\sum_{j=1}^{\infty} (.4)^j X_{t+j}$ satisfies the difference equations $U_t - 2.5U_{t-1} = X_t$.

3.11. Let $\{X_t\}$ be an ARMA process with $\phi(z) \neq 0$, $|z| = 1$, and autocovariance function $\gamma(\cdot)$. Show that there exist constants $C > 0$ and $s \in (0, 1)$ such that $|\gamma(h)| \leq Cs^{|h|}$, $h = 0, \pm 1, \ldots$ and hence that $\sum_{h=-\infty}^{\infty} |\gamma(h)| < \infty$.

3.12. For those processes in Problem 3.1 which are causal, compute and graph their autocorrelation and partial autocorrelation functions using PEST.

3.13. Find the coefficients ψ_j, $j = 0, 1, 2, \ldots$, in the representation

$$X_t = \sum_{j=0}^{\infty} \psi_j Z_{t-j}$$

of the ARMA(2, 1) process,

$$(1 - .5B + .04B^2)X_t = (1 + .25B)Z_t, \qquad \{Z_t\} \sim \text{WN}(0, \sigma^2).$$

3.14. Find the autocovariances $\gamma(j), j = 0, 1, 2, \ldots$, of the AR(3) process,

$$(1 - .5B)(1 - .4B)(1 - .1B)X_t = Z_t, \qquad \{Z_t\} \sim \text{WN}(0, 1).$$

Check your answers for $j = 0, \ldots, 4$ with the aid of the program PEST.

3.15. Find the mean and autocovariance function of the ARMA(2, 1) process,

$$X_t = 2 + 1.3X_{t-1} - .4X_{t-2} + Z_t + Z_{t-1}, \qquad \{Z_t\} \sim \text{WN}(0, \sigma^2).$$

Is the process causal and invertible?

3.16. Let $\{X_t\}$ be the ARMA(1, 1) process,

$$X_t - \phi X_{t-1} = Z_t + \theta Z_{t-1}, \qquad \{Z_t\} \sim \text{WN}(0, \sigma^2),$$

where $|\phi| < 1$ and $|\theta| < 1$. Determine the coefficients $\{\psi_j\}$ in Theorem 3.1.1 and show that the autocorrelation function of $\{X_t\}$ is given by $\rho(1) = (1 + \phi\theta)(\phi + \theta)/(1 + \theta^2 + 2\phi\theta)$, $\rho(h) = \phi^{h-1}\rho(1)$ for $h \geq 1$.

3.17. For an MA(2) process find the largest possible values of $|\rho(1)|$ and $|\rho(2)|$.

3.18. Let $\{X_t\}$ be the moving average process

$$X_t = Z_t - 2Z_{t-1}, \qquad \{Z_t\} \sim \text{IID}(0, 1).$$

(a) If $Z_t^* := (1 - .5B)^{-1}X_t$, show that

$$Z_t^* = X_t - P_{\mathcal{M}_{t-1}}X_t,$$

where $\mathcal{M}_{t-1} = \overline{\text{sp}}\{X_s, -\infty < s < t\}$.

(b) Conclude from (a) that

$$X_t = Z_t^* + \theta Z_{t-1}^*, \qquad \{Z_t^*\} \sim \text{WN}(0, \sigma^2).$$

Specify the values of θ and σ^2.

(c) Find the linear filter which relates $\{Z_t\}$ to $\{Z_t^*\}$, i.e. determine the coefficients $\{\alpha_j\}$ in the representation $Z_t^* = \sum_{j=-\infty}^{\infty} \alpha_j Z_{t-j}$.

(d) If $EZ_t^3 = c$, compute $E((Z_1^*)^2 Z_2^*)$. If $c \neq 0$, are Z_1^* and Z_2^* independent? If $Z_t \sim N(0, 1)$, are Z_1^* and Z_2^* independent?

3.19. Suppose that $\{X_t\}$ and $\{Y_t\}$ are two zero-mean stationary processes with the same autovariance function and that $\{Y_t\}$ is an ARMA(p, q) process. Show that $\{X_t\}$ must also be an ARMA(p, q) process. (Hint: If $\phi_1, \ldots, \phi_p$ are the AR coefficients for $\{Y_t\}$, show that $\{W_t := X_t - \phi_1 X_{t-1} - \cdots - \phi_p X_{t-p}\}$ has an autocovariance function which is zero for lags $|h| > q$. Then apply Proposition 3.2.1 to $\{W_t\}$.)

3.20. (a) Calculate the autocovariance function $\gamma(\cdot)$ of the stationary time series

$$Y_t = \mu + Z_t + \theta_1 Z_{t-1} + \theta_{12} Z_{t-12}, \qquad \{Z_t\} \sim \text{WN}(0, \sigma^2).$$

(b) Use program PEST to compute the sample mean and sample autocovariances $\hat{\gamma}(h), 0 \leq h \leq 20$, of $\{\nabla\nabla_{12}X_t\}$ where $\{X_t, t = 1, \ldots, 72\}$ is the accidental deaths series of Example 1.1.6.

(c) By equating $\hat{\gamma}(1)$, $\hat{\gamma}(11)$ and $\hat{\gamma}(12)$ from part(b) to $\gamma(1)$, $\gamma(11)$ and $\gamma(12)$ respectively from part(a), find a model of the form defined in (a) to represent $\{\nabla\nabla_{12}X_t\}$.

3.21. By matching the autocovariances and sample autocovariances at lags 0 and 1, fit a model of the form

$$X_t - \mu = \phi(X_{t-1} - \mu) + Z_t, \qquad \{Z_t\} \sim \text{WN}(0, \sigma^2),$$

to the strikes data of Example 1.1.3. Use the fitted model to compute the best linear predictor of the number of strikes in 1981. Estimate the mean squared error of your predictor.

3.22. If $X_t = Z_t - \theta Z_{t-1}$, $\{Z_t\} \sim \text{WN}(0, \sigma^2)$ and $|\theta| < 1$, show from the prediction equations that the best linear predictor of X_{n+1} in $\overline{\text{sp}}\{X_1, \dots, X_n\}$ is

$$\hat{X}_{n+1} = \sum_{j=1}^{n} \phi_j X_{n+1-j},$$

where $\phi_1, \dots, \phi_n$ satisfy the difference equations,

$$-\theta\phi_{j-1} + (1 + \theta^2)\phi_j - \theta\phi_{j+1} = 0, \qquad 2 \le j \le n - 1,$$

with boundary conditions,

$$(1 + \theta^2)\phi_n - \theta\phi_{n-1} = 0$$

and

$$(1 + \theta^2)\phi_1 - \theta\phi_2 = -\theta.$$

3.23. Use Definition 3.4.2 and the results of Problem 3.22 to determine the partial autocorrelation function of a moving average of order 1.

3.24. Let $\{X_t\}$ be the stationary solution of $\phi(B)X_t = \theta(B)Z_t$, where $\{Z_t\} \sim \text{WN}(0, \sigma^2)$, $\phi(z) \neq 0$ for all $z \in \mathbb{C}$ such that $|z| = 1$, and $\phi(\cdot)$ and $\theta(\cdot)$ have no common zeroes. If A is any zero-mean random variable in L^2 which is uncorrelated with $\{X_t\}$ and if $|z_0| = 1$, show that the process $\{X_t + Az_0^t\}$ is a complex-valued stationary process (see Definition 4.1.1) and that $\{X_t + Az_0^t\}$ and $\{X_t\}$ both satisfy the equations $(1 - z_0B)\phi(B)X_t = (1 - z_0B)\theta(B)Z_t$.

CHAPTER 4

The Spectral Representation of a Stationary Process

The spectral representation of a stationary process $\{X_t, t = 0, \pm 1, \ldots\}$ essentially decomposes $\{X_t\}$ into a sum of sinusoidal components with uncorrelated random coefficients. In conjunction with this decomposition there is a corresponding decomposition into sinusoids of the autocovariance function of $\{X_t\}$. The spectral decomposition is thus an analogue for stationary stochastic processes of the more familiar Fourier representation of deterministic functions. The analysis of stationary processes by means of their spectral representations is often referred to as the "frequency domain" analysis of time series. It is equivalent to "time domain" analysis, based on the autocovariance function, but provides an alternative way of viewing the process which for some applications may be more illuminating. For example in the design of a structure subject to a randomly fluctuating load it is important to be aware of the presence in the loading force of a large harmonic with a particular frequency to ensure that the frequency in question is not a resonant frequency of the structure. The spectral point of view is particularly advantageous in the analysis of multivariate stationary processes (Chapter 11) and in the analysis of very large data sets, for which numerical calculations can be performed rapidly using the fast Fourier transform (Section 10.7).

§4.1 Complex-Valued Stationary Time Series

It will often be convenient for us to make use of complex-valued stationary processes. Although processes encountered in practice are nearly always real-valued, it is mathematically simpler in spectral analysis to treat them as special cases of complex-valued processes.

Definition 4.1.1. The process $\{X_t\}$ is a complex-valued stationary process if $E|X_t|^2 < \infty$, EX_t is independent of t and $E(X_{t+h}\bar{X}_t)$ is independent of t.

As already pointed out in Example 2.2.3, Remark 1, the complex-valued random variables X on $(\Omega, \mathscr{F}, P)$ satisfying $E|X|^2 < \infty$ constitute a Hilbert space with the inner product

$$\langle X, Y \rangle = E(X\bar{Y}). \tag{4.1.1}$$

Definition 4.1.2. The autocovariance function $\gamma(\cdot)$ of a complex-valued stationary process $\{X_t\}$ is

$$\gamma(h) = E(X_{t+h}\bar{X}_t) - EX_{t+h}E\bar{X}_t. \tag{4.1.2}$$

Notice that Definitions 4.1.1 and 4.1.2 reduce to the corresponding definitions for real processes if $\{X_t\}$ is restricted to be real-valued.

Properties of Complex-Valued Autocovariance Functions

The properties of real-valued autocovariance functions which were established in Section 1.5, can be restated for complex-valued autocovariance functions as follows:

$$\gamma(0) \geq 0, \tag{4.1.3}$$

$$|\gamma(h)| \leq \gamma(0) \quad \text{for all integers } h, \tag{4.1.4}$$

$$\gamma(\cdot) \text{ is a Hermitian function (i.e. } \gamma(h) = \overline{\gamma(-h)}). \tag{4.1.5}$$

We also have an analogue of Theorem 1.5.1, namely

Theorem 4.1.1. *A function $K(\cdot)$ defined on the integers is the autocovariance function of a (possibly complex-valued) stationary time series if and only if $K(\cdot)$ is Hermitian and non-negative definite, i.e. if and only if $K(n) = \overline{K(-n)}$ and*

$$\sum_{i,j=1}^{n} a_i K(i-j)\bar{a}_j \geq 0, \tag{4.1.6}$$

for all positive integers n and all vectors $\mathbf{a} = (a_1, \ldots, a_n)' \in \mathbb{C}^n$.

The proofs of these extensions (which reduce to the analogous results in Section 1.5.1 in the real case) are left as exercises (see Problems 4.1 and 4.30). We shall see (Corollary 4.3.1) that "Hermitian" can be dropped from the statement of Theorem 4.1.1 since the Hermitian property follows from the validity of (4.1.6) for all complex $\mathbf{a}$.

§4.2 The Spectral Distribution of a Linear Combination of Sinusoids

In this section we illustrate the essential features of the spectral representation of an arbitrary stationary process by considering the simple complex-valued process,

$$X_t = \sum_{j=1}^{n} A(\lambda_j)e^{it\lambda_j} \tag{4.2.1}$$

in which $-\pi < \lambda_1 < \lambda_2 < \cdots < \lambda_n = \pi$ and $A(\lambda_1), \ldots, A(\lambda_n)$ are uncorrelated complex-valued random coefficients (possibly zero) such that

$$E(A(\lambda_j)) = 0, \qquad j = 1, \ldots, n,$$

and

$$E(A(\lambda_j)\overline{A(\lambda_j)}) = \sigma_j^2, \qquad j = 1, \ldots, n.$$

For $\{X_t\}$ to be real-valued it is necessary that $A(\lambda_n)$ be real and that $\lambda_j = \lambda_{n-j}$ and $A(\lambda_j) = \overline{A(\lambda_{n-j})}$ for $j = 1, \ldots, n-1$. (Note that $A(\lambda_j)$ and $A(\lambda_{n-j})$ are uncorrelated in spite of the last relation.) In this case (see Problem 4.4),

$$X_t = \sum_{j=1}^{n} (C(\lambda_j)\cos t\lambda_j - D(\lambda_j)\sin t\lambda_j), \tag{4.2.2}$$

where $A(\lambda_j) = C(\lambda_j) + iD(\lambda_j)$, $j = 1, \ldots, n$ and $D(\lambda_n) = 0$.

It is easy to see that the process (4.2.1), and in particular the real-valued process (4.2.2), is stationary since

$$EX_t = 0$$

and

$$E(X_{t+h}\bar{X}_t) = \sum_{j=1}^{n} \sigma_j^2 e^{ih\lambda_j},$$

the latter being independent of t. Rewriting the last expression as a Riemann–Stieltjes integral, we see that the process $\{X_t\}$ defined by (4.2.1) is stationary with autocovariance function,

$$\gamma(h) = \int_{(-\pi,\pi]} e^{ih\nu}\, dF(\nu), \tag{4.2.3}$$

where F is the distribution function,

$$F(\lambda) = \sum_{j:\lambda_j \le \lambda} \sigma_j^2. \tag{4.2.4}$$

Notice that the function F, which is known as the *spectral distribution function of* $\{X_t\}$, assigns all of its mass to the frequency interval $(-\pi, \pi]$. The mass assigned to each frequency in the interval is precisely the variance of the random coefficient corresponding to that frequency in the representation (4.2.1).

The equations (4.2.1) and (4.2.3) are fundamental to the spectral analysis of time series. Equation (4.2.1) is the spectral representation of the process $\{X_t\}$ itself and equation (4.2.3) is the corresponding spectral representation of the covariance function. The spectral distribution function appearing in the latter is related to the random coefficients in (4.2.1) through the equation $F(\lambda) = \sum_{\lambda_j \le \lambda} E|A(\lambda_j)|^2$.

The remarkable feature of this example is that *every* zero-mean stationary process has a representation which is a natural generalization of (4.2.1), namely

$$X_t = \int_{(-\pi,\pi]} e^{it\nu}\, dZ(\nu). \tag{4.2.5}$$

The integral is a stochastic integral with respect to an orthogonal-increment process, a precise definition of which will be given in Section 4.7. Correspondingly the autocovariance function $\gamma_X(\cdot)$ can be expressed as

$$\gamma_X(h) = \int_{(-\pi,\pi]} e^{ih\nu}\, dF(\nu), \tag{4.2.6}$$

where F is a distribution function with $F(-\pi) = 0$ and $F(\pi) = \gamma(0) = E|X_t|^2$. The representation (4.2.6) is easier to establish than (4.2.5) since it does not require the notion of stochastic integration. We shall therefore establish (4.2.6) (Herglotz's theorem) in Section 4.3, deferring the spectral representation of $\{X_t\}$ itself until after we have introduced the definition of the stochastic integral in Section 4.7.

In the special case when $\{X_t\}$ is real there are alternative forms in which we can write (4.2.5) and (4.2.6). In particular if $\gamma_X(h)$ is to be real it is necessary that $F(\cdot)$ be symmetric in the sense that $F(\lambda) = F(\pi^-) - F(-\lambda^-)$, $-\pi < \lambda < \pi$, where $F(\lambda^-)$ is the left limit of F at λ (see Problem 4.25). Equation (4.2.6) can then be expressed as

$$\gamma_X(h) = \int_{(-\pi,\pi]} \cos \nu h\, dF(\nu).$$

Equivalent forms of (4.2.5) when $\{X_t\}$ is real are given in Problem 4.25.

§4.3 Herglotz's Theorem

Theorem 4.1.1 characterizes the complex-valued autocovariance functions on the integers as those functions which are Hermitian and non-negative definite. Herglotz's theorem, which we are about to prove, characterizes them as the functions which can be written in the form (4.2.6) for some bounded distribution function F with mass concentrated on $(-\pi, \pi]$.

Theorem 4.3.1 (Herglotz). *A complex-valued function $\gamma(\cdot)$ defined on the integers is non-negative definite if and only if*

$$\gamma(h) = \int_{(-\pi,\pi]} e^{ihv}\, dF(v) \quad \textit{for all } h = 0, \pm 1, \ldots, \tag{4.3.1}$$

where $F(\cdot)$ is a right-continuous, non-decreasing, bounded function on $[-\pi, \pi]$ and $F(-\pi) = 0$. (The function F is called ***the spectral distribution function*** *of γ and if $F(\lambda) = \int_{-\pi}^{\lambda} f(v)\, dv$, $-\pi \le \lambda \le \pi$, then f is called a* ***spectral density*** *of $\gamma(\cdot)$.)*

PROOF. If $\gamma(\cdot)$ has the representation (4.3.1) then it is clear that $\gamma(\cdot)$ is Hermitian, i.e. $\gamma(-h) = \overline{\gamma(h)}$. Moreover if $a_r \in \mathbb{C}$, $r = 1, \ldots, n$, then

$$\begin{aligned}\sum_{r,s=1}^{n} a_r \gamma(r-s) \bar{a}_s &= \int_{-\pi}^{\pi} \sum_{r,s=1}^{n} a_r \bar{a}_s \exp[iv(r-s)]\, dF(v) \\ &= \int_{-\pi}^{\pi} \left| \sum_{r=1}^{n} a_r \exp[ivr] \right|^2 dF(v) \\ &\ge 0,\end{aligned}$$

so that $\gamma(\cdot)$ is also non-negative definite and therefore, by Theorem 4.1.1, an autocovariance function.

Conversely suppose $\gamma(\cdot)$ is a non-negative definite function on the integers. Then, defining

$$\begin{aligned}f_N(v) &= \frac{1}{2\pi N} \sum_{r,s=1}^{N} e^{-irv} \gamma(r-s) e^{isv} \\ &= \frac{1}{2\pi N} \sum_{|m|<N} (N - |m|) e^{-imv} \gamma(m),\end{aligned}$$

we see from the non-negative definiteness of $\gamma(\cdot)$ that

$$f_N(v) \ge 0 \quad \text{for all } v \in (-\pi, \pi].$$

Let $F_N(\cdot)$ be the distribution function corresponding to the density $f_N(\cdot) I_{(-\pi,\pi]}(\cdot)$. Thus $F_N(\lambda) = 0$, $\lambda \le -\pi$, $F_N(\lambda) = F_N(\pi)$, $\lambda \ge \pi$, and

$$F_N(\lambda) = \int_{-\pi}^{\lambda} f_N(v)\, dv, \qquad -\pi \le \lambda \le \pi.$$

Then for any integer h,

$$\int_{(-\pi,\pi]} e^{ihv}\, dF_N(v) = \frac{1}{2\pi} \sum_{|m|<N} \left(1 - \frac{|m|}{N}\right) \gamma(m) \int_{-\pi}^{\pi} e^{i(h-m)v}\, dv,$$

i.e.

$$\int_{(-\pi,\pi]} e^{ihv}\, dF_N(v) = \begin{cases} \left(1 - \dfrac{|h|}{N}\right) \gamma(h), & |h| < N, \\ 0, & \text{otherwise.} \end{cases} \tag{4.3.2}$$

Since $F_N(\pi) = \int_{(-\pi,\pi]} dF_N(\nu) = \gamma(0) < \infty$ for all N, we can apply Helly's theorem (see e.g. Ash (1972), p. 329) to deduce that there is a distribution function F and a subsequence $\{F_{N_k}\}$ of the sequence $\{F_N\}$ such that for any bounded continuous function g with $g(\pi) = g(-\pi)$,

$$\int_{(-\pi,\pi]} g(\nu)\, dF_{N_k}(\nu) \to \int_{(-\pi,\pi]} g(\nu)\, dF(\nu) \quad \text{as } k \to \infty. \tag{4.3.3}$$

Replacing N by N_k in (4.3.2) and letting $k \to \infty$, we obtain

$$\gamma(h) = \int_{(-\pi,\pi]} e^{ih\nu}\, dF(\nu) \tag{4.3.4}$$

which is the required spectral representation of $\gamma(\cdot)$. □

Corollary 4.3.1. *A complex-valued function $\gamma(\cdot)$ defined on the integers is the autocovariance function of a stationary process $\{X_t, t = 0, \pm 1, \ldots\}$ if and only if either*

(i) *$\gamma(h) = \int_{(-\pi,\pi]} e^{ih\nu}\, dF(\nu)$ for all $h = 0, \pm 1, \ldots$, where F is a right-continuous, non-decreasing, bounded function on $[-\pi, \pi]$ with $F(-\pi) = 0$, or*

(ii) *$\sum_{i,j=1}^{n} a_i \gamma(i-j) \bar{a}_j \geq 0$ for all positive integers n and for all $\mathbf{a} = (a_1, \ldots, a_n)' \in \mathbb{C}^n$.*

The spectral distribution function $F(\cdot)$ (and the corresponding spectral density if there is one) will be referred to as the ***spectral distribution function*** *(and the* ***spectral density****) of both $\gamma(\cdot)$ and of $\{X_t\}$.*

Proof. Herglotz's theorem asserts the equivalence of (i) and (ii). From (i) it follows at once that $\gamma(\cdot)$ is Hermitian. Consequently the conditions of Theorem 4.1.1 are satisfied if and only if $\gamma(\cdot)$ satisfies either (i) or (ii). □

It is important to note that the distribution function $F(\cdot)$ (with $F(-\pi) = 0$) is uniquely determined by $\gamma(n)$, $n = 0, \pm 1, \ldots$. For if F and G are two distribution functions vanishing on $(-\infty, -\pi]$, constant on $[\pi, \infty)$ and such that

$$\gamma(h) = \int_{(-\pi,\pi]} e^{ih\nu}\, dF(\nu) = \int_{(-\pi,\pi]} e^{ih\nu}\, dG(\nu), \qquad h = 0, \pm 1, \ldots,$$

then it follows from Theorem 2.11.1 that

$$\int_{(-\pi,\pi]} \phi(\nu)\, dF(\nu) = \int_{(-\pi,\pi]} \phi(\nu)\, dG(\nu) \quad \text{if } \phi \text{ is continuous with } \phi(\pi) = \phi(-\pi),$$

and hence that $F(\lambda) = G(\lambda)$ for all $\lambda \in (-\infty, \infty)$.

The following theorem is useful for finding F from γ in many important cases (and in particular when γ is the autocovariance function of an ARMA(p, q) process).

Theorem 4.3.2. *If $K(\cdot)$ is any complex-valued function on the integers such that*

$$\sum_{n=-\infty}^{\infty} |K(n)| < \infty, \tag{4.3.5}$$

then

$$K(h) = \int_{-\pi}^{\pi} e^{ih\nu} f(\nu)\, d\nu, \qquad h = 0, \pm 1, \ldots \tag{4.3.6}$$

where

$$f(\lambda) = \frac{1}{2\pi} \sum_{n=-\infty}^{\infty} e^{-in\lambda} K(n). \tag{4.3.7}$$

PROOF.

$$\begin{aligned}\int_{-\pi}^{\pi} e^{ih\nu} f(\nu)\, d\nu &= \int_{-\pi}^{\pi} \frac{1}{2\pi} \sum_{n=-\infty}^{\infty} e^{i(h-n)\nu} K(n)\, d\nu \\ &= \frac{1}{2\pi} \sum_{n=-\infty}^{\infty} K(n) \int_{-\pi}^{\pi} e^{i(h-n)\nu}\, d\nu \\ &= K(h),\end{aligned}$$

since the only non-zero summand is the one for which $n = h$. The interchange of summation and integration is justified by Fubini's theorem since $\int_{-\pi}^{\pi} (1/2\pi) \sum_{n=-\infty}^{\infty} |e^{i(h-n)\nu} K(n)|\, d\nu < \infty$ by (4.3.5). □

Corollary 4.3.2. *An absolutely summable complex-valued function $\gamma(\cdot)$ defined on the integers is the autocovariance function of a stationary process if and only if*

$$f(\lambda) := \frac{1}{2\pi} \sum_{n=-\infty}^{\infty} e^{-in\lambda} \gamma(n) \ge 0 \quad \textit{for all } \lambda \in [-\pi, \pi], \tag{4.3.8}$$

in which case $f(\cdot)$ is the spectral density of $\gamma(\cdot)$.

PROOF. First suppose that $\gamma(\cdot)$ is an autocovariance function. Since $\gamma(\cdot)$ is non-negative definite and absolutely summable,

$$\begin{aligned}0 \le f_N(\lambda) &= \frac{1}{2\pi N} \sum_{r,s=1}^{N} e^{-ir\lambda} \gamma(r-s) e^{is\lambda} \\ &= \frac{1}{2\pi} \sum_{|m|<N} \left(1 - \frac{|m|}{N}\right) e^{-im\lambda} \gamma(m) \\ &\to f(\lambda) \quad \text{as } N \to \infty.\end{aligned}$$

Consequently $f(\lambda) \ge 0$, $-\pi \le \lambda \le \pi$. Also from Theorem 4.3.2 we have $\gamma(h) = \int_{-\pi}^{\pi} e^{ih\nu} f(\nu)\, d\nu$, $h = 0, \pm 1, \ldots$. Hence $f(\cdot)$ is the spectral density of $\gamma(\cdot)$.

On the other hand if we assume only that $\gamma(\cdot)$ is absolutely summable,

Theorem 4.3.2 allows us to write $\gamma(h) = \int_{-\pi}^{\pi} e^{ihv} f(v)\,dv$. If $f(\lambda) \geq 0$ then this integral is of the form (4.3.1) with $F(\lambda) = \int_{-\pi}^{\lambda} f(v)\,dv$. This implies, by Corollary 4.3.1, that $\gamma(\cdot)$ is an autocovariance function with spectral density f. □

Example 4.3.1. Let us prove that the real-valued function

$$K(h) = \begin{cases} 1 & \text{if } h = 0, \\ \rho & \text{if } h = \pm 1, \\ 0 & \text{otherwise,} \end{cases}$$

is an autocovariance function if and only if $|\rho| \leq \frac{1}{2}$.

Since $K(\cdot)$ is absolutely summable we can apply Corollary 4.3.2. Thus

$$\begin{aligned} f(\lambda) &= \frac{1}{2\pi} \sum_{n=-\infty}^{\infty} e^{-in\lambda} K(n) \\ &= \frac{1}{2\pi} \left[1 + 2\rho \cos \lambda \right], \end{aligned}$$

is non-negative for all $\lambda \in [-\pi, \pi]$ if and only if $|\rho| \leq \frac{1}{2}$. Consequently $K(\cdot)$ is an autocovariance function if and only if $|\rho| \leq \frac{1}{2}$ in which case $K(\cdot)$ has the spectral density f computed above. In fact $K(\cdot)$ is the autocovariance function of an MA(1) process (see Example 1.5.1).

Notice that Corollary 4.3.2 provides us with a very powerful tool for checking non-negative definiteness, which can be applied to any absolutely summable function on the integers. It is much simpler and much more informative than direct verification using the definition of non-negative definiteness stated in Theorem 4.1.1.

Corollary 4.3.2 shows in particular that every ARMA(p, q) process has a spectral density (see Problem 3.11). This density is found explicitly in Section 4.4. On the other hand the linear combination of sinusoids (4.2.1) studied in Section 4.2 has the purely discrete spectral distribution function (4.2.4) and therefore there is no corresponding spectral density.

If $\{X_t\}$ is a real-valued stationary process then its autocovariance function $\gamma_X(\cdot)$ is real, implying (as pointed out in Section 4.2) that its spectral distribution function is symmetric in the sense that

$$F_X(\lambda) = F_X(\pi^-) - F_X(-\lambda^-), \qquad -\pi < \lambda < \pi.$$

We can then write

$$\gamma_X(h) = \int_{(-\pi,\pi]} \cos(vh)\,dF_X(v). \tag{4.3.9}$$

In particular if $\gamma_X(\cdot)$ has spectral density $f_X(\lambda)$, $-\pi \leq \lambda \leq \pi$, then $f_X(\lambda) = f_X(-\lambda)$, $-\pi \leq \lambda \leq \pi$, and hence

$$\gamma_X(h) = 2 \int_0^{\pi} f_X(v) \cos(vh)\,dv. \tag{4.3.10}$$

The covariance structure of a real-valued stationary process $\{X_t\}$ is thus determined by $F_X(0^-)$ and $F_X(\lambda)$, $0 \le \lambda \le \pi$ (or by $f_X(\lambda)$, $0 \le \lambda \le \pi$, if the spectral density $f_X(\cdot)$ exists).

Remark. From the above discussion, it follows that a function f defined on $[-\pi, \pi]$ is the spectral density of a real-valued stationary process if and only if

(i) $f(\lambda) = f(-\lambda)$,
(ii) $f(\lambda) \ge 0$, and
(iii) $\int_{-\pi}^{\pi} f(\lambda)\, d\lambda < \infty$.

§4.4 Spectral Densities and ARMA Processes

Theorem 4.4.1. *If $\{Y_t\}$ is any zero-mean, possibly complex-valued stationary process with spectral distribution function $F_Y(\cdot)$, and $\{X_t\}$ is the process*

$$X_t = \sum_{j=-\infty}^{\infty} \psi_j Y_{t-j} \quad \text{where} \quad \sum_{j=-\infty}^{\infty} |\psi_j| < \infty, \tag{4.4.1}$$

then $\{X_t\}$ is stationary with spectral distribution function

$$F_X(\lambda) = \int_{(-\pi, \lambda]} \left| \sum_{j=-\infty}^{\infty} \psi_j e^{-ijv} \right|^2 dF_Y(v), \qquad -\pi \le \lambda \le \pi. \tag{4.4.2}$$

Proof. The argument of Proposition 3.1.2 shows that $\{X_t\}$ is stationary with mean zero and autocovariance function,

$$E(X_{t+h}\bar{X}_t) = \sum_{j,k=-\infty}^{\infty} \psi_j \bar{\psi}_k \gamma_Y(h - j + k), \qquad h = 0, \pm 1, \ldots.$$

Using the spectral representation of $\{\gamma_Y(\cdot)\}$ we can write

$$\begin{aligned} \gamma_X(h) &= \sum_{j,k=-\infty}^{\infty} \psi_j \bar{\psi}_k \int_{(-\pi, \pi]} e^{i(h-j+k)v}\, dF_Y(v) \\ &= \int_{(-\pi, \pi]} \left(\sum_{j=-\infty}^{\infty} \psi_j e^{-ijv} \right) \left(\sum_{k=-\infty}^{\infty} \bar{\psi}_k e^{ikv} \right) e^{ihv}\, dF_Y(v) \\ &= \int_{(-\pi, \pi]} e^{ihv} \left| \sum_{j=-\infty}^{\infty} \psi_j e^{-ijv} \right|^2 dF_Y(v), \end{aligned}$$

which immediately identifies $F_X(\cdot)$ defined by (4.4.2) as the spectral distribution function of $\{X_t\}$. □

If $\{Y_t\}$ has a spectral density $f_Y(\cdot)$ and if $\{X_t\}$ is defined by (4.4.1), then $\{X_t\}$ also has a spectral density $f_X(\cdot)$ given by

$$f_X(\lambda) = |\psi(e^{-i\lambda})|^2 f_Y(\lambda), \tag{4.4.3}$$

where $\psi(e^{-i\lambda}) = \sum_{j=-\infty}^{\infty} \psi_j e^{-ij\lambda}$. The operator $\psi(B) = \sum_{j=-\infty}^{\infty} \psi_j B^j$ applied to $\{Y_t\}$ in (4.4.1) is often called a *time-invariant linear filter* with weights $\{\psi_j\}$. The function $\psi(e^{-i\cdot})$ is called the *transfer function* of the filter and the squared modulus $|\psi(e^{-i\cdot})|^2$ is referred to as the *power transfer function* of the filter. Time-invariant linear filters will be discussed in more detail in Section 4.10.

As an application of Theorem 4.4.1 we can now derive the spectral density of an arbitrary ARMA(p, q) process.

Theorem 4.4.2 (Spectral Density of an ARMA(p, q) Process). *Let $\{X_t\}$ be an* ARMA(p, q) *process (not necessarily causal or invertible) satisfying*

$$\phi(B)X_t = \theta(B)Z_t, \qquad \{Z_t\} \sim \text{WN}(0, \sigma^2), \tag{4.4.4}$$

where $\phi(z) = 1 - \phi_1 z - \cdots - \phi_p z^p$ and $\theta(z) = 1 + \theta_1 z + \cdots + \theta_q z^q$ have no common zeroes and $\phi(z)$ has no zeroes on the unit circle. Then $\{X_t\}$ has spectral density

$$f_X(\lambda) = \frac{\sigma^2}{2\pi} \frac{|\theta(e^{-i\lambda})|^2}{|\phi(e^{-i\lambda})|^2}, \qquad -\pi \le \lambda \le \pi. \tag{4.4.5}$$

(Because the spectral density of an ARMA *process is a ratio of trigonometric polynomials it is often called a* ***rational spectral density***.)

PROOF. First recall from Section 3.1 that the stationary solution of (4.4.4) can be written as $X_t = \sum_{j=-\infty}^{\infty} \psi_j Z_{t-j}$ where $\sum_{j=-\infty}^{\infty} |\psi_j| < \infty$. Since $\{Z_t\}$ has spectral density $\sigma^2/(2\pi)$ (Problem 4.6), Theorem 4.4.1 implies that $\{X_t\}$ has a spectral density. This also follows from Corollary 4.3.2 and Problem 3.11. Setting $U_t = \phi(B)X_t = \theta(B)Z_t$ and applying Theorem 4.4.1, we obtain

$$f_U(\lambda) = |\phi(e^{-i\lambda})|^2 f_X(\lambda) = |\theta(e^{-i\lambda})|^2 f_Z(\lambda). \tag{4.4.6}$$

Since $\phi(e^{-i\lambda}) \neq 0$ for all $\lambda \in [-\pi, \pi]$ we can divide (4.4.6) by $|\phi(e^{-i\lambda})|^2$ to obtain (4.4.5). □

EXAMPLE 4.4.1 (Spectral Density of an MA(1) Process). If

$$X_t = Z_t + \theta Z_{t-1}, \qquad \{Z_t\} \sim \text{WN}(0, \sigma^2),$$

then

$$f_X(\lambda) = \frac{\sigma^2}{2\pi}|1 + \theta e^{-i\lambda}|^2 = \frac{\sigma^2}{2\pi}(1 + 2\theta\cos\lambda + \theta^2), \qquad -\pi \le \lambda \le \pi.$$

The graph of $f_X(\lambda)$, $0 \le \lambda \le \pi$, is displayed in Figure 4.1 for each of the values $\theta = -.9$ and $\theta = .9$. Observe that for $\theta = .9$ the density is large for low frequencies and small for high frequencies. This is not unexpected since when $\theta = .9$ the process has a large lag one correlation which makes the series smooth with only a small contribution from high frequency components. For

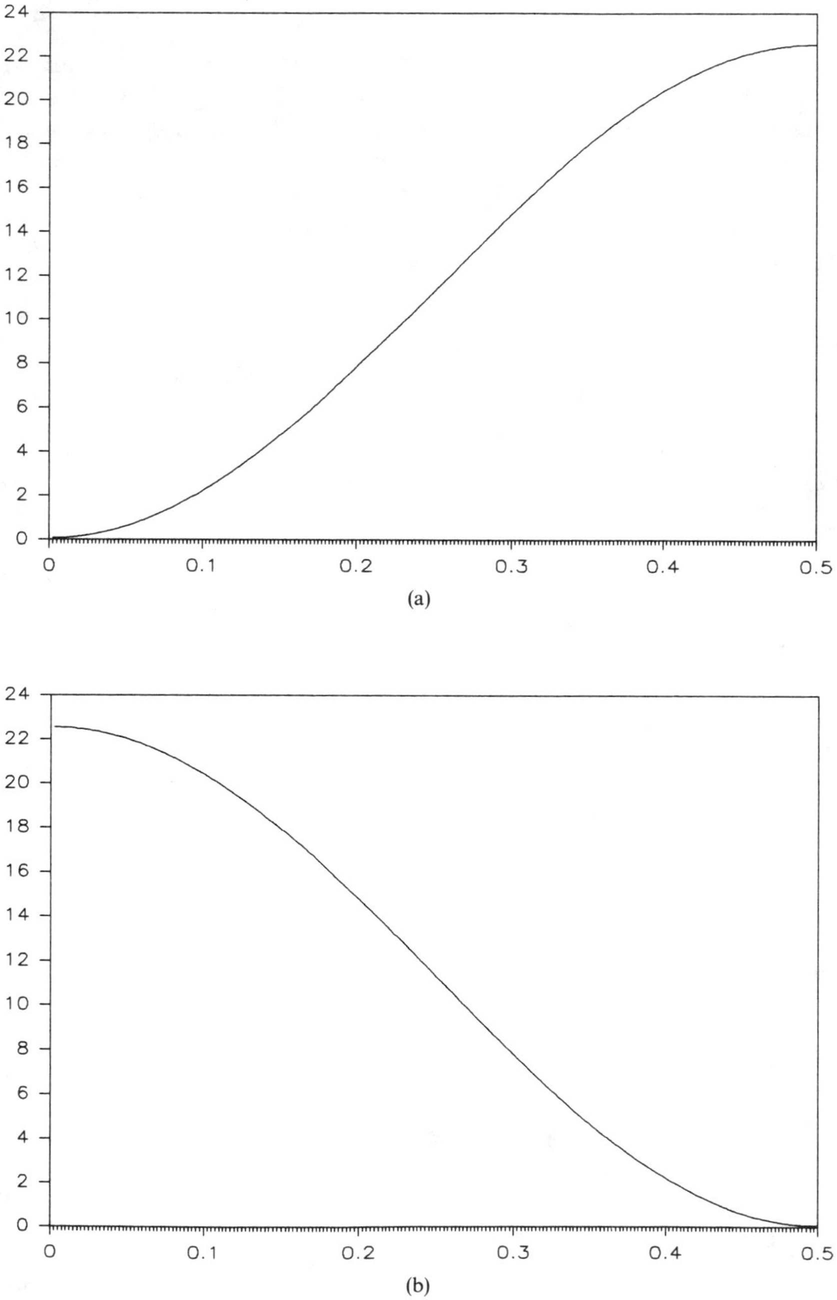

Figure 4.1. The spectral densities $f_X(2\pi c)$, $0 \le c \le \frac{1}{2}$, of $X_t = Z_t + \theta Z_{t-1}$, $\{Z_t\} \sim \text{WN}(0, 6.25)$, (a) when $\theta = -.9$ and (b) when $\theta = .9$.

$\theta = -.9$ the lag one correlation is large and negative, the series fluctuates rapidly about its mean value and, as expected, the spectral density is large for high frequencies and small for low frequencies. (See also Figures 1.18 and 1.19.) □

EXAMPLE 4.4.2 (The Spectral Density of an AR(1) Process). If

$$X_t - \phi X_{t-1} = Z_t, \qquad \{Z_t\} \sim \mathrm{WN}(0, \sigma^2),$$

then by Theorem 4.4.2, $\{X_t\}$ has spectral density

$$f_X(\lambda) = \frac{\sigma^2}{2\pi}|1 - \phi e^{-i\lambda}|^{-2} = \frac{\sigma^2}{2\pi}(1 - 2\phi\cos\lambda + \phi^2)^{-1}.$$

This function is shown in Figure 4.2 for each of the values $\phi = .7$ and $\phi = -.7$. Interpretations of the graphs analogous to those in Example 4.4.1 can again be made.

Causality, Invertibility and the Spectral Density

Consider the ARMA(p, q) process $\{X_t\}$ satisfying $\phi(B)X_t = \theta(B)Z_t$, where $\phi(z)\theta(z) \neq 0$ for all $z \in \mathbb{C}$ such that $|z| = 1$. Factorizing the polynomials $\phi(\cdot)$ and $\theta(\cdot)$ we can rewrite the defining equations in the form,

$$\prod_{j=1}^{p}(1 - a_j^{-1}B)X_t = \prod_{j=1}^{q}(1 - b_j^{-1}B)Z_t, \qquad \{Z_t\} \sim \mathrm{WN}(0, \sigma^2), \tag{4.4.7}$$

where

$$|a_j| > 1, \quad 1 \le j \le r, \quad |a_j| < 1, \quad r < j \le p,$$

and

$$|b_j| > 1, \quad 1 \le j \le s, \quad |b_j| < 1, \quad s < j \le q.$$

By Theorem 4.4.2, $\{X_t\}$ has spectral density

$$f_X(\lambda) = \frac{\sigma^2}{2\pi}\frac{\prod_{j=1}^{q}|1 - b_j^{-1}e^{-i\lambda}|^2}{\prod_{j=1}^{p}|1 - a_j^{-1}e^{-i\lambda}|^2}.$$

Now define

$$\tilde{\phi}(B) = \prod_{1 \le j \le r}(1 - a_j^{-1}B)\prod_{r < j \le p}(1 - \bar{a}_j B) \tag{4.4.8}$$

and

$$\tilde{\theta}(B) = \prod_{1 \le j \le s}(1 - b_j^{-1}B)\prod_{s < j \le q}(1 - \bar{b}_j B).$$

Then the ARMA(p, q) process $\{\tilde{X}_t\}$ defined by

$$\tilde{\phi}(B)\tilde{X}_t = \tilde{\theta}(B)Z_t$$

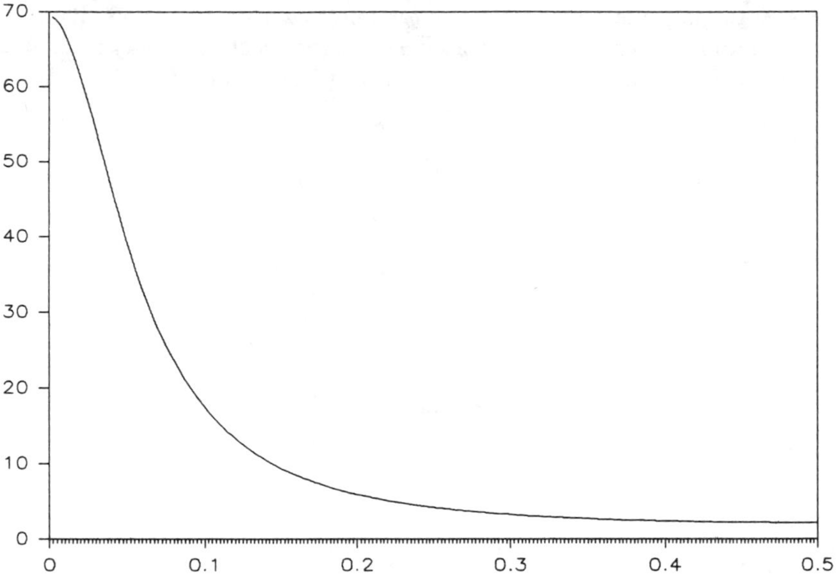

(a)

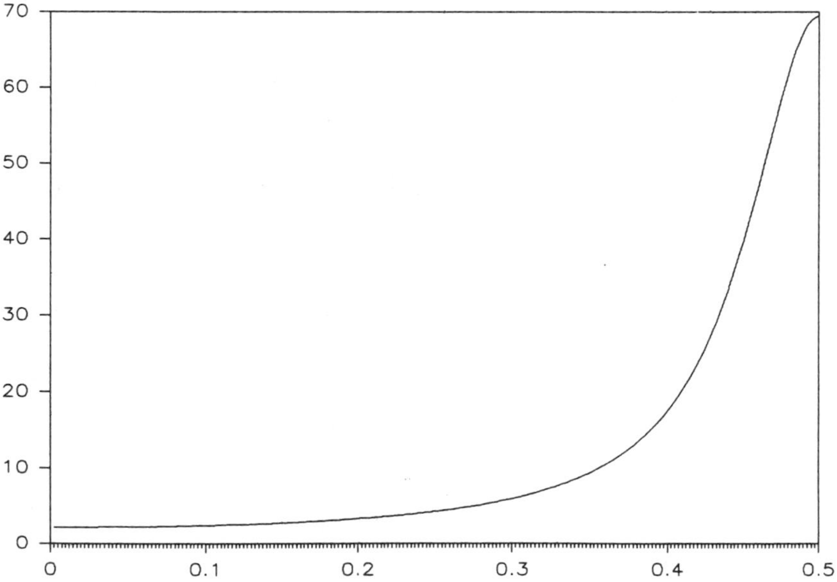

(b)

Figure 4.2. The spectral densities $f_X(2\pi c)$, $0 \le c \le \frac{1}{2}$, of $X_t - \phi X_{t-1} = Z_t$, $\{Z_t\} \sim \text{WN}(0, 6.25)$, (a) when $\phi = .7$ and (b) when $\phi = -.7$.

has spectral density

$$f_{\tilde{X}}(\lambda) = \frac{\sigma^2}{2\pi} \frac{|\tilde{\theta}(e^{-i\lambda})|^2}{|\tilde{\phi}(e^{-i\lambda})|^2}.$$

Since

$$|1 - \bar{b}_j e^{-i\lambda}| = |1 - b_j e^{i\lambda}| = |b_j||1 - b_j^{-1} e^{-i\lambda}|,$$

we can rewrite $f_{\tilde{X}}(\lambda)$ as

$$f_{\tilde{X}}(\lambda) = \frac{\sigma^2}{2\pi} \frac{\prod_{s<j\le q} |b_j|^2}{\prod_{r<j\le p} |a_j|^2} \frac{|\theta(e^{-i\lambda})|^2}{|\phi(e^{-i\lambda})|^2} = \frac{\prod_{s<j\le q} |b_j|^2}{\prod_{r<j\le p} |a_j|^2} f_X(\lambda).$$

Thus the ARMA(p, q) process $\{X_t^+\}$ defined by

$$\tilde{\phi}(B)X_t^+ = \tilde{\theta}(B)\tilde{Z}_t, \quad \{\tilde{Z}_t\} \sim \text{WN}\left(0, \sigma^2 \left(\prod_{r<j\le p} |a_j|\right)^2 \left(\prod_{s<j\le q} |b_j|\right)^{-2}\right),$$

is causal and invertible and has exactly the same spectral density (and hence autocovariance function) as the ARMA process (4.4.7). In fact $\{X_t\}$ itself has the causal invertible representation

$$\tilde{\phi}(B)X_t = \tilde{\theta}(B)Z_t^*$$

where $\{Z_t^*\}$ is white noise with the same variance as $\{\tilde{Z}_t\}$. This is easily checked by using the latter equation as the definition of $\{Z_t^*\}$. (See Proposition 3.5.1.)

Example 4.4.3. The ARMA process

$$X_t - 2X_{t-1} = Z_t + 4Z_{t-1}, \qquad \{Z_t\} \sim \text{WN}(0, \sigma^2),$$

is neither causal nor invertible. Introducing $\tilde{\phi}(z) = 1 - 0.5z$ and $\tilde{\theta}(z) = 1 + 0.25z$, we see that $\{X_t\}$ has the causal invertible representation

$$X_t - 0.5X_{t-1} = Z_t^* + 0.25Z_{t-1}^*, \qquad \{Z_t^*\} \sim \text{WN}(0, .25\sigma^2).$$

The case when the moving average polynomial $\theta(z)$ has zeroes on the unit circle is dealt with in the following propositions.

Proposition 4.4.1. *Let $\{X_t\}$ be an* ARMA(p, q) *process satisfying*

$$\phi(B)X_t = \theta(B)Z_t, \qquad \{Z_t\} \sim \text{WN}(0, \sigma^2),$$

where $\phi(z)$ and $\theta(z)$ have no common zeroes, $\phi(z) \ne 0$ for $|z| = 1$ and $\theta(z) \ne 0$ for $|z| < 1$. Then $Z_t \in \overline{\text{sp}}\{X_s, -\infty < s \le t\}$.

Proof. Factorize $\theta(z)$ as $\theta^\dagger(z)\theta^*(z)$, where

$$\theta^\dagger(z) = \prod_{1\le j\le s} (1 - b_j^{-1}z),$$

$$\theta^*(z) = \prod_{s<j\le q} (1 - b_j^{-1}z),$$

$|b_j| > 1$, $1 \le j \le s$ and $|b_j| = 1$, $s < j \le q$. Then consider the MA($q - s$) process,

$$Y_t = \theta^*(B)Z_t,$$

and note, since $\phi(B)X_t = \theta^\dagger(B)Y_t$, that

$$\overline{\text{sp}}\{Y_k, -\infty < k \le t\} \subseteq \overline{\text{sp}}\{X_k, -\infty < k \le t\}$$

for all t. Consequently it suffices to show that $Z_t \in \overline{\text{sp}}\{Y_k, -\infty < k \le t\}$. By Proposition 3.2.1,

$$Y_t = U_t + \alpha_1 U_{t-1} + \cdots + \alpha_{q-s} U_{t-q+s}, \qquad \{U_t\} \sim \text{WN}(0, \sigma_U^2),$$

where

$$U_t = Y_t - P_{\overline{\text{sp}}\{Y_k, -\infty < k < t\}} Y_t.$$

Using the two moving average representations for $\{Y_t\}$, we can write the spectral density f_Y of $\{Y_t\}$ as

$$f_Y(\lambda) = \frac{\sigma_U^2}{2\pi} |\alpha(e^{-i\lambda})|^2 = \frac{\sigma^2}{2\pi} |\theta^*(e^{-i\lambda})|^2.$$

Since $\theta^*(z)$ has all of its zeroes on the unit circle, $\alpha(z)$ and $\theta^*(z)$ must have the same zeroes. This in turn implies that

$$\theta(z) = \alpha(z) \quad \text{and} \quad \sigma^2 = \sigma_U^2.$$

It now follows that the two vectors, $(U_t, Y_t, \ldots, Y_{t-n})'$ and $(Z_t, Y_t, \ldots, Y_{t-n})'$, have the same covariance matrix and hence that

$$P_{\overline{\text{sp}}\{Y_k, t-n \le k \le t\}} Z_t = P_{\overline{\text{sp}}\{Y_k, t-n \le k \le t\}} U_t.$$

Taking mean square limits as $n \to \infty$ and using the fact that

$$U_t \in \overline{\text{sp}}\{Y_k, -\infty < k \le t\},$$

we find that

$$P_{\overline{\text{sp}}\{Y_k, -\infty < k \le t\}} Z_t = U_t.$$

Hence, since $\sigma^2 = \sigma_U^2$,

$$E(Z_t - P_{\overline{\text{sp}}\{Y_k, -\infty < k \le t\}} Z_t)^2 = EZ_t^2 - EU_t^2 = 0.$$

This implies that $Z_t = P_{\overline{\text{sp}}\{Y_k, -\infty < k \le t\}} Z_t$, or equivalently that

$$Z_t \in \overline{\text{sp}}\{Y_k, -\infty < k \le t\}$$

as was to be shown. □

Remark 1. If we extend the definition (Section 3.1) of invertibility of the ARMA equations, $\phi(B)X_t = \theta(B)Z_t$, by requiring only that

$$Z_t \in \overline{\text{sp}}\{X_k, -\infty < k \le t\},$$

Proposition 4.4.1 then states that invertibility is implied by the condition $\theta(z) \neq 0$ for $|z| < 1$. The converse is established below as Proposition 4.4.3.

Remark 2. If $\phi(z) \neq 0$ for all $|z| \leq 1$, then by projecting each side of the equation, $\phi(B)X_t = \theta(B)Z_t$, in Proposition 4.4.1 onto

$$\overline{\mathrm{sp}}\{X_s, -\infty < s \leq t-1\},$$

we see at once that Z_t is the innovation,

$$Z_t = X_t - \tilde{P}_{t-1}X_t,$$

where $\tilde{P}_{t-1}$ denotes projection onto $\overline{\mathrm{sp}}\{X_s, -\infty < s \leq t-1\}$.

Proposition 4.4.2. *Let $\{X_t\}$ be an* ARMA(p, q) *process satisfying*

$$\phi(B)X_t = \theta(B)Z_t, \qquad \{Z_t\} \sim \mathrm{WN}(0, \sigma^2),$$

where $\phi(z)$ and $\theta(z)$ have no common zeroes and $\phi(z) \neq 0$ for all $z \in \mathbb{C}$ such that $|z| = 1$. Then, if $\{b_j, 1 \leq j \leq q\}$ are the zeroes of $\theta(z)$, with $|b_j| \geq 1$, $1 \leq j \leq m$, and $|b_j| < 1$, $m < j \leq q$, there exists a white noise sequence $\{U_t\}$ such that $\{X_t\}$ is the unique stationary solution of the equations,

$$\tilde{\phi}(B)X_t = \tilde{\theta}(B)U_t, \qquad \{U_t\} \sim \mathrm{WN}(0, \tilde{\sigma}^2),$$

where $\tilde{\phi}(z)$ is the polynomial defined in (4.4.8) *and $\tilde{\theta}(z)$ is the polynomial,*

$$\tilde{\theta}(z) = \prod_{1 \leq j \leq m} (1 - b_j^{-1}z) \prod_{m < j \leq q} (1 - \bar{b}_j z).$$

The variance of U_t is given by

$$\tilde{\sigma}^2 = \sigma^2 \left(\prod_{r < j \leq p} |a_j|^2 \right) \left(\prod_{m < j \leq q} |b_j|^{-2} \right).$$

[*Remark* 2 *implies that* $U_t = X_t - \tilde{P}_{t-1}X_t$.]

Proof. By Problem 4.29, we know that there exists a white noise sequence $\{\tilde{Z}_t\}$ such that

$$\tilde{\phi}(B)X_t = \theta(B)\tilde{Z}_t.$$

The required white noise sequence $\{U_t\}$ is therefore the unique stationary solution of the equations,

$$\prod_{m < j \leq q} (1 - b_j B)U_t = \prod_{m < j \leq q} (1 - \bar{b}_j^{-1} B)\tilde{Z}_t. \qquad \square$$

Proposition 4.4.3. *If $\{X_t\}$ is defined as in Proposition* 4.4.2 *and the polynomial $\theta(z)$ has one or more zeros in the interior of the unit circle, then $Z_t \notin \overline{\mathrm{sp}}\{X_s, -\infty < s \leq t\}$.*

PROOF. By Proposition 4.4.2, we can express X_t in the form,

$$X_t = \sum_{j=0}^{\infty} \tilde{\psi}_j U_{t-j},$$

where $U_t = X_t - \tilde{P}_{t-1}X_t$, $\sum_{j=0}^{\infty} \tilde{\psi}_j z^j = \tilde{\theta}(z)/\tilde{\phi}(z)$ for $|z| \le 1$, and

$$\overline{\text{sp}}\{X_k, -\infty < k \le t\} = \overline{\text{sp}}\{U_k, -\infty < k \le t\}.$$

Suppose that the zeroes of $\theta(z)$ in the interior of the unit circle are $\{b_j: m < j \le q\}$ where $m < q$, and let

$$\theta(z) = \theta^*(z) \prod_{m<j\le q} (1 - b_j^{-1}z) = \theta^*(z)\theta^i(z).$$

From the equations $\tilde{\phi}(B)X_t = \tilde{\theta}(B)U_t$ and $\phi(B)X_t = \theta^*(B)\theta^i(B)Z_t$, it follows that

$$\tilde{\phi}(B)\theta^*(B)Z_t = \sum_{j=-\infty}^{\infty} \xi_j U_{t-j},$$

where ξ_j is the coefficient of z^j in the Laurent expansion of $\phi(z)\tilde{\theta}(z)/\theta^i(z)$, $|z - 1| < \varepsilon$, which is valid for some $\varepsilon > 0$. Since $\phi(z)\tilde{\theta}(z)$ and $\theta^i(z)$ have no zeroes in common and since $\theta^i(z)$ has all of its zeroes in the interior of the unit circle, it follows that $\xi_j \ne 0$ for some $j = -j_0 < 0$.

From $Z_t \in \overline{\text{sp}}\{X_k, -\infty < k \le t\}$, it would follow that

$$\tilde{\phi}(B)\theta^*(B)Z_t \in \overline{\text{sp}}\{U_k, -\infty < k \le t\}.$$

But this is impossible since

$$\langle U_{t+j_0}, \tilde{\phi}(B)\theta^*(B)Z_t\rangle = \xi_{-j_0} \operatorname{Var}(U_{t+j_0}) \ne 0.$$

We conclude therefore that $Z_t \notin \overline{\text{sp}}\{X_k, -\infty < k \le t\}$, as required. □

Rational Approximations for Spectral Densities

For any real-valued stationary process $\{X_t\}$ with continuous spectral density f, it is possible to find both a causal AR(p) process and an invertible MA(q) process whose spectral densities are arbitrarily close to f. This suggests that $\{X_t\}$ can be approximated in some sense by either an AR(p) or an MA(q) process. These results depend on Theorem 4.4.3 below. Recall that f is the spectral density of a real-valued stationary process if and only if f is symmetric, non-negative and integrable on $[-\pi, \pi]$.

Theorem 4.4.3. *If f is a symmetric continuous spectral density on $[-\pi, \pi]$, then for every $\varepsilon > 0$ there exists a non-negative integer p and a polynomial $a(z) = \prod_{j=1}^{p} (1 - \eta_j^{-1} z) = 1 + a_1 z + \cdots + a_p z^p$ with $|\eta_j| > 1, j = 1, \ldots, p$, and* *real-valued coefficients* $a_0, \ldots, a_p$, *such that*

$$|A|a(e^{-i\lambda})|^2 - f(\lambda)| < \varepsilon \quad \text{for all } \lambda \in [-\pi, \pi] \tag{4.4.9}$$

where $A = (1 + a_1^2 + \cdots + a_p^2)^{-1}(2\pi)^{-1}\int_{-\pi}^{\pi} f(v)\,dv$.

Proof. If $f(\lambda) \equiv 0$ the result is clearly true with $p = 0$. Assume therefore that $M = \sup_{-\pi \le \lambda \le \pi} f(\lambda) > 0$. Now for any $\varepsilon > 0$ let

$$\delta = \min\left\{M, \left[4\pi M\left(\int_{-\pi}^{\pi} f(v)\,dv\right)^{-1} + 2\right]^{-1}\varepsilon\right\}$$

and define

$$f^\delta(\lambda) = \max\{f(\lambda), \delta\}.$$

Clearly $f^\delta(\lambda)$ is also a symmetric continuous spectral density with $f^\delta(\lambda) \ge \delta$ and

$$0 \le f^\delta(\lambda) - f(\lambda) \le \delta \quad \text{for all } \lambda \in [-\pi, \pi]. \tag{4.4.10}$$

Now by Theorem 2.11.1 there exists an integer r such that

$$\left| r^{-1} \sum_{j=0}^{r-1} \sum_{|k| \le j} b_k e^{-ik\lambda} - f^\delta(\lambda) \right| < \delta \quad \text{for all } \lambda \in [-\pi, \pi], \tag{4.4.11}$$

where $b_k = (2\pi)^{-1}\int_{-\pi}^{\pi} f^\delta(v) e^{ivk}\,dv$. Interchanging the order of summation and using the fact that f^δ is a symmetric function, we have

$$r^{-1} \sum_{j=0}^{r-1} \sum_{|k| \le j} b_k e^{-ik\lambda} = \sum_{|k| < r} (1 - |k|/r) b_k e^{-ik\lambda}.$$

This function is strictly positive for all λ by (4.4.10) and the definition of $f^\delta(\lambda)$. Let

$$C(z) = \sum_{|k| < r} (1 - |k|/r) b_k z^k,$$

and observe that if $C(m) = 0$ then by symmetry $C(m^{-1}) = 0$. Hence, letting $p = \max\{k : b_k \ne 0\}$, we can write

$$z^p C(z) = K_1 \prod_{j=1}^{p} (1 - \eta_j^{-1} z)(1 - \eta_j z)$$

for some $K_1, \eta_1, \ldots, \eta_p$ such that $|\eta_j| > 1$, $j = 1, \ldots, p$. This equation can be rewritten in the form

$$K_2 a(z) a(z^{-1}) = C(z), \tag{4.4.12}$$

where $a(z)$ is the polynomial $1 + a_1 z + \cdots + a_p z^p = \prod_{j=1}^{p}(1 - \eta_j^{-1} z)$, and $K_2 = (-1)^p \eta_1 \ldots \eta_p K_1$. Equating the coefficients of z^0 on each side of (4.4.12) we find that

$$K_2 = b_0(1 + a_1^2 + \cdots + a_p^2)^{-1} = (2\pi)^{-1}(1 + a_1^2 + \cdots + a_p^2)^{-1} \int_{-\pi}^{\pi} f^\delta(v)\,dv.$$

Moreover from (4.4.11) we have

$$|K_2|a(e^{-i\lambda})|^2 - f^\delta(\lambda)| < \delta \quad \text{for all } \lambda. \tag{4.4.13}$$

From (4.4.13) and (4.4.10) we obtain the uniform bound

$$\begin{aligned}(1 + a_1^2 + \cdots + a_p^2)^{-1}|a(e^{-i\lambda})|^2 &\le (f^\delta(\lambda) + \delta)2\pi\left(\int_{-\pi}^{\pi} f^\delta(v)\,dv\right)^{-1} \\ &\le 4\pi M\left(\int_{-\pi}^{\pi} f(v)\,dv\right)^{-1}.\end{aligned}$$

Now with A defined as in the statement of the theorem

$$\begin{aligned}&|K_2|a(e^{-i\lambda})|^2 - A|a(e^{-i\lambda})|^2| \\ &\quad\le (2\pi)^{-1}\left(\int_{-\pi}^{\pi} (f^\delta(v) - f(v))\,dv\right)4\pi M\left(\int_{-\pi}^{\pi} f(v)\,dv\right)^{-1} \\ &\quad\le 4\pi M\delta\left(\int_{-\pi}^{\pi} f(v)\,dv\right)^{-1}.\end{aligned} \tag{4.4.14}$$

From the inequalities (4.4.10), (4.4.13) and (4.4.14) we obtain

$$\begin{aligned}|A|a(e^{-i\lambda})|^2 - f(\lambda)| &< \delta + \delta + 4\pi M\delta\left(\int_{-\pi}^{\pi} f(v)\,dv\right)^{-1} \\ &< \varepsilon,\end{aligned}$$

by the definition of δ. □

Corollary 4.4.1. *If f is a symmetric continuous spectral density and $\varepsilon > 0$, then there exists an invertible* MA(q) *process*

$$X_t = Z_t + a_1 Z_{t-1} + \cdots + a_q Z_{t-q}, \qquad \{Z_t\} \sim \text{WN}(0, \sigma^2),$$

such that

$$|f_X(\lambda) - f(\lambda)| < \varepsilon \quad \textit{for all } \lambda \in [-\pi, \pi],$$

where $\sigma^2 = (1 + a_1^2 + \cdots + a_q^2)^{-1} \int_{-\pi}^{\pi} f(v)\,dv$.

PROOF. Problem 4.14. □

Corollary 4.4.2. *If f is a symmetric continuous spectral density and $\varepsilon > 0$ then there exists a causal* AR(p) *process*

$$X_t + a_1 X_{t-1} + \cdots + a_p X_{t-p} = Z_t, \qquad \{Z_t\} \sim \text{WN}(0, \sigma^2),$$

such that

$$|f_X(\lambda) - f(\lambda)| < \varepsilon \quad \textit{for all } \lambda \in [-\pi, \pi].$$

PROOF. Let $f^\varepsilon(\lambda) = \max\{f(\lambda), \varepsilon/2\}$. Then $f^\varepsilon(\lambda) \geq \varepsilon/2$ and

$$0 \leq f^\varepsilon(\lambda) - f(\lambda) \leq \varepsilon/2 \quad \text{for all } \lambda \in [-\pi, \pi]. \tag{4.4.15}$$

Let $M = \max_\lambda f^\varepsilon(\lambda)$ and $\delta = \min\{(2M)^{-2}\varepsilon, (2M)^{-1}\}$. Applying Theorem 4.4.3 to the function $1/f^\varepsilon(\lambda)$, we have

$$|K|a(e^{-i\lambda})|^2 - 1/f^\varepsilon(\lambda)| < \delta \quad \text{for all } \lambda \in [-\pi, \pi], \tag{4.4.16}$$

where the polynomial $a(z) = 1 + a_1 z + \cdots + a_p z^p$ is non-zero for $|z| \leq 1$ and K is a positive constant. Moreover by our definition of δ, the inequality (4.4.16) yields the bound

$$K^{-1}|(a(e^{-i\lambda})|^{-2} \leq f^\varepsilon(\lambda)/(1 - \delta f^\varepsilon(\lambda)) \leq M/(1 - M\delta) \leq 2M.$$

Thus

$$\begin{aligned} |K^{-1}|a(e^{-i\lambda})|^{-2} - f^\varepsilon(\lambda)| &= |K|a(e^{-i\lambda})|^2 - 1/f^\varepsilon(\lambda)|\,[K^{-1}|a(e^{-i\lambda})|^{-2} f^\varepsilon(\lambda)] \\ &< 2M^2\delta \leq \varepsilon/2. \end{aligned} \tag{4.4.17}$$

Combining the inequalities (4.4.15) and (4.4.17) we get

$$|K^{-1}|a(e^{-i\lambda})|^{-2} - f(\lambda)| < \varepsilon \quad \text{for all } \lambda \in [-\pi, \pi]. \tag{4.4.18}$$

Now by Theorem 4.4.1 the causal AR(p) process

$$a(B)X_t = Z_t, \qquad \{Z_t\} \sim \text{WN}(0, 2\pi K^{-1})$$

has spectral density $K^{-1}|a(e^{-i\lambda})|^2$, which by (4.4.18) furnishes the required approximation to $f(\lambda)$. □

§4.5* Circulants and Their Eigenvalues

It is often desirable to be able to diagonalize a covariance matrix in a simple manner. By first diagonalizing a circulant matrix it is possible to obtain a relatively easy and useful asymptotic diagonalization of the covariance matrix of the first n observations from a stationary time series. We say that the $n \times n$ matrix $M = [m_{ij}]_{i,j=1}^n$ is a circulant matrix if there exists a function $m(\cdot)$ with period n such that $m_{ij} = m(j - i)$. That is

$$M = \begin{bmatrix} m(0) & m(1) & \cdots & m(n-1) \\ m(n-1) & m(0) & \cdots & m(n-2) \\ m(n-2) & m(n-1) & \cdots & m(n-3) \\ \vdots & \vdots & & \vdots \\ m(1) & m(2) & \cdots & m(0) \end{bmatrix}. \tag{4.5.1}$$

The eigenvalues and eigenvectors of M are easy to compute. Let

$$\omega_j = \frac{2\pi j}{n},$$

and

$$r_j = \exp(i\omega_j),$$

for $j = 0, 1, \ldots, n-1$.

Proposition 4.5.1. *The circulant matrix M has eigenvalues*

$$\lambda_j = \sum_{h=0}^{n-1} m(h) r_j^{-h}, \qquad j = 0, 1, \ldots, n-1,$$

with corresponding orthonormal left eigenvectors,

$$\mathbf{v}_j = n^{-1/2}[1, r_j, r_j^2, \ldots, r_j^{n-1}], \qquad j = 0, 1, \ldots, n-1.$$

PROOF. Straightforward calculations give

$$\begin{aligned}
\mathbf{v}_j M &= n^{-1/2}[m(0) + m(n-1)r_j + \cdots + m(1)r_j^{n-1}, m(1) + m(0)r_j + \cdots \\
&\qquad + m(2)r_j^{n-1}, \ldots, m(n-1) + m(n-2)r_j + \cdots + m(0)r_j^{n-1}] \\
&= \lambda_j n^{-1/2}[1, r_j, r_j^2, \ldots, r_j^{n-1}] \\
&= \lambda_j \mathbf{v}_j,
\end{aligned}$$

showing that $\mathbf{v}_j$ is a left eigenvector of M with corresponding eigenvalue λ_j, $j = 0, 1, \ldots, n-1$. Moreover, if $\mathbf{v}_k^*$ is the conjugate transpose of $\mathbf{v}_k$, then

$$\begin{aligned}
\mathbf{v}_j \mathbf{v}_k^* &= n^{-1}(1 + r_j r_k^{-1} + \cdots + r_j^{n-1} r_k^{-n+1}) \\
&= \begin{cases} n^{-1}[1 - (r_j/r_k)^n][1 - r_j/r_k]^{-1} = 0 & \text{if } j \neq k, \\ 1 & \text{if } j = k. \end{cases}
\end{aligned}$$

□

In order to diagonalize the matrix we now introduce the matrix

$$V = \begin{bmatrix} \mathbf{v}_0 \\ \mathbf{v}_1 \\ \vdots \\ \mathbf{v}_{n-1} \end{bmatrix},$$

observing from Proposition 4.5.1 that $VM = \Lambda V$ and hence that

$$VMV^{-1} = \Lambda, \tag{4.5.2}$$

where $\Lambda = \text{diag}\{\lambda_0, \lambda_1, \ldots, \lambda_{n-1}\}$ and $V^{-1} = [\mathbf{v}_0^*, \mathbf{v}_1^*, \ldots, \mathbf{v}_{n-1}^*]$.

Diagonalization of a Real Symmetric Circulant Matrix

If the circulant matrix M defined by (4.5.1) is also real and symmetric (i.e. if $m(j) = m(n-j) \in \mathbb{R}, j = 0, 1, \ldots, n-1$), then we can rewrite the eigenvalues λ_j of Proposition 4.5.1 in the form

$$\lambda_j = \begin{cases} \sum_{|h| \le [n/2]} m(h) r_j^{-h} & \text{if } n \text{ is odd,} \\ \sum_{|h| < [n/2]} m(h) r_j^{-h} + m(n/2 + 1) r_j^{-(n+2)/2} & \text{if } n \text{ is even,} \end{cases} \tag{4.5.3}$$

where $[n/2]$ is the integer part of $n/2$.

We first consider the case when n is odd. Since $m(\cdot)$ is an even function, we can express the n eigenvalues of M as

$$\lambda_0 = \sum_{|h| \le [n/2]} m(h)$$

$$\lambda_j = \sum_{|h| \le [n/2]} m(h) \exp(-i\omega_j h), \qquad j = 1, 2, \ldots, [n/2],$$

and

$$\lambda_{n-j} = \lambda_j, \qquad j = 1, 2, \ldots, [n/2].$$

Corresponding to the repeated eigenvalue $\lambda_j = \lambda_{n-j}$ $(1 \le j \le [n/2])$ of M there are two orthonormal left eigenvectors $\mathbf{v}_j$ and $\mathbf{v}_{n-j} = \bar{\mathbf{v}}_j$ as specified in Proposition 4.5.1. From these we can easily find a pair of *real* orthonormal eigenvectors corresponding to λ_j, viz.

$$\mathbf{c}_j = (\mathbf{v}_j + \mathbf{v}_{n-j})/\sqrt{2} = \sqrt{2/n}[1, \cos \omega_j, \cos 2\omega_j, \ldots, \cos(n-1)\omega_j]$$

and

$$\mathbf{s}_j = i(\mathbf{v}_{n-j} - \mathbf{v}_j)/\sqrt{2} = \sqrt{2/n}[0, \sin \omega_j, \sin 2\omega_j, \ldots, \sin(n-1)\omega_j].$$

Setting

$$\mathbf{c}_0 = \sqrt{1/n}[1, 1, 1, \ldots, 1]$$

and defining the real orthogonal matrix P by

$$P = \begin{bmatrix} \mathbf{c}_0 \\ \mathbf{c}_1 \\ \mathbf{s}_1 \\ \vdots \\ \mathbf{c}_{[n/2]} \\ \mathbf{s}_{[n/2]} \end{bmatrix} \tag{4.5.4}$$

we have $PM = \Lambda^{(s)}P$ and hence

$$PMP' = \Lambda^{(s)}, \tag{4.5.5}$$

where $\Lambda^{(s)} = \text{diag}\{\lambda_0, \lambda_1, \lambda_1, \ldots, \lambda_{[n/2]}, \lambda_{[n/2]}\}$.

For the case when n is even, both λ_0 and $\lambda_{n/2}$ have multiplicity 1. If we replace $c_{[n/2]}$ by $2^{-1/2}c_{n/2}$ in the definition of P and drop the last rows of the matrices P and $\Lambda^{(s)}$, then we again have $PMP' = \Lambda^{(s)}$.

Proposition 4.5.2. *Let $\gamma(\cdot)$ be an absolutely summable real autocovariance* *function, let $f(\cdot)$ be its spectral density*

$$f(\omega) = (2\pi)^{-1} \sum_{h=-\infty}^{\infty} \gamma(h)e^{-ih\omega},$$

and let D_n be the $n \times n$ matrix,

$$D_n = \begin{cases} \text{diag}\{f(0), f(\omega_1), f(\omega_1), \ldots, f(\omega_{[n/2]}), f(\omega_{[n/2]})\} & \text{if } n \text{ is odd}, \\ \text{diag}\{f(0), f(\omega_1), f(\omega_1), \ldots, f(\omega_{(n-2)/2}), f(\omega_{(n-2)/2}), f(\omega_{n/2})\} & \text{if } n \text{ is even}. \end{cases}$$

If P is the matrix defined by (4.5.4) and $\Gamma_n = [\gamma(i-j)]_{i,j=1}^n$, then the components $x_{ij}^{(n)}$ of the matrix

$$P\Gamma_n P' - 2\pi D_n,$$

converge to zero uniformly as $n \to \infty$ (i.e. $\sup_{1 \le i,j \le n} |x_{ij}^{(n)}| \to 0$).

Proof. Let $\mathbf{p}_i = [p_{i1}, p_{i2}, \ldots, p_{in}]$ denote the ith row of the matrix P and let $\Gamma_n^{(s)}$ denote the real symmetric circulant matrix

$$\Gamma_n^{(s)} = \begin{bmatrix} \gamma(0) & \gamma(1) & \gamma(2) & \cdots & \gamma(2) & \gamma(1) \\ \gamma(1) & \gamma(0) & \gamma(1) & \cdots & \gamma(3) & \gamma(2) \\ \gamma(2) & \gamma(1) & \gamma(0) & \cdots & \gamma(4) & \gamma(3) \\ \vdots & \vdots & \vdots & & \vdots & \vdots \\ \gamma(1) & \gamma(2) & \gamma(3) & \cdots & \gamma(1) & \gamma(0) \end{bmatrix}.$$

We know from (4.5.5) that $P\Gamma_n^{(s)}P' = \Lambda^{(s)}$. Moreover since the elements of the matrix $\Lambda^{(s)} - 2\pi D_n$ are bounded in absolute value by $\sum_{|h|>[n/2]} |\gamma(h)|$ which converges to zero as $n \to \infty$, it suffices to show that

$$|\mathbf{p}_i \Gamma_n^{(s)} \mathbf{p}_j' - \mathbf{p}_i \Gamma_n \mathbf{p}_j'| \to 0 \quad \text{uniformly in } i \text{ and } j. \tag{4.5.6}$$

But

$$|\mathbf{p}_i(\Gamma_n^{(s)} - \Gamma_n)\mathbf{p}_j'| = \left| \sum_{m=1}^{c} (\gamma(m) - \gamma(n-m)) \sum_{k=1}^{m} (p_{ik}p_{j,n-m+k} - p_{i,n-m+k}p_{jk}) \right|$$

where $c = [(n-1)/2]$. Since $|p_{ij}| \le (2/n)^{1/2}$ this expression is bounded by

$$4n^{-1}\left(2 \sum_{m=1}^{c} m|\gamma(m)| + 2 \sum_{m=1}^{c} m|\gamma(n-m)|\right) \le 8 \sum_{m=1}^{c} \frac{m}{n}|\gamma(m)| + 8 \sum_{m=n-c}^{n-1} \frac{c}{n}|\gamma(m)|.$$

The first term converges to zero as $n \to \infty$ by the dominated convergence theorem since the summand is dominated by $|\gamma(m)|$ and $\sum_{m=1}^{\infty} |\gamma(m)| < \infty$. The second term goes to zero since it is bounded by $\sum_{m=[n/2]}^{\infty} |\gamma(m)|$. Since both terms are independent of i and j, the proof of (4.5.6) is complete. □

Now let $\{X_t\}$ be a real-valued zero-mean stationary time series with autocovariance function $\gamma(\cdot)$ which is absolutely summable. Consider the transformed vector of random variables

$$\mathbf{Z} = V\mathbf{X} = \begin{bmatrix} \mathbf{v}_0 \\ \vdots \\ \mathbf{v}_{n-1} \end{bmatrix} \begin{bmatrix} X_0 \\ \vdots \\ X_{n-1} \end{bmatrix} = n^{-1/2} \begin{bmatrix} \sum_{h=0}^{n-1} r_0^h X_h \\ \vdots \\ \sum_{h=0}^{n-1} r_{n-1}^h X_h \end{bmatrix}, \tag{4.5.7}$$

with $r_j, \mathbf{v}_j, j = 0, \ldots, n-1$ defined as in Proposition 4.5.1. The components of $\mathbf{Z}$ are approximately uncorrelated for large n by Proposition 4.5.2. Moreover the matrix V, being orthogonal, is easily inverted to give

$$X_j = n^{-1/2} \sum_{h=0}^{n-1} Z_h \exp(-ij\omega_h).$$

Thus we have represented $X_0, X_1, \ldots, X_{n-1}$ as a sum of sinusoids with random coefficients which are asymptotically uncorrelated. This is one (albeit rough) interpretation of the spectral representation of the process $\{X_t\}$.

Another easily verified consequence of Proposition 4.5.2 is that with $\mathbf{Z}$ defined as in (4.5.7),

$$\sup_{0 \le k \le n-1} |E|Z_k|^2 - \sum_{h=-\infty}^{\infty} \gamma(h)\exp(-ih\omega_k)| \to 0 \quad \text{as } n \to \infty.$$

Let us consider now an *arbitrary* frequency $\omega \in (-\pi, \pi]$ and define, by analogy with (4.5.7),

$$Z_{\omega,n} = n^{-1/2} \sum_{h=0}^{n-1} X_h \exp(ih\omega).$$

Then

$$\begin{aligned} E|Z_{\omega,n}|^2 &= n^{-1} \sum_{k,l=0}^{n-1} \gamma(k-l)\exp[i\omega(k-l)] \\ &= n^{-1} \sum_{|h|<n} (n - |h|)\gamma(h)\exp(i\omega h) \\ &\to \sum_{h=-\infty}^{\infty} \gamma(h)\exp(i\omega h) = 2\pi f(\omega) \quad \text{as } n \to \infty. \end{aligned}$$

This shows that $(2\pi)^{-1}|Z_{\omega,n}|^2$ is an asymptotically unbiased estimator of $f(\omega)$. In Chapter 10 we shall show how this estimator can be modified in order to construct a consistent estimator of $f(\omega)$.

We conclude this section by deriving upper and lower bounds for the eigenvalues of the covariance matrix of $\mathbf{X}_n = (X_1, \ldots, X_n)'$ when $\{X_t\}$ is a stationary process with spectral density $f(\lambda)$, $-\pi \le \lambda \le \pi$.

Proposition 4.5.3. *Let $\{X_t\}$ be a stationary process with spectral density f such that*

$$m := \inf_\lambda f(\lambda) > 0 \quad \textit{and} \quad M := \sup_\lambda f(\lambda) < \infty,$$

and denote by $\lambda_1, \ldots, \lambda_n$ $(\lambda_1 \le \lambda_2 \le \cdots \le \lambda_n)$ *the eigenvalues of the covariance matrix* Γ_n *of* $(X_1, \ldots, X_n)'$. *Then*

$$2\pi m \le \lambda_1 \le \lambda_n \le 2\pi M.$$

PROOF. Let $\mathbf{x} = (x_1, \ldots, x_n)'$ be a non-zero right eigenvector of Γ_n corresponding to the eigenvalue λ_n. Then

$$\begin{aligned}
\lambda_n \mathbf{x}'\mathbf{x} &= \mathbf{x}'\Gamma_n \mathbf{x} \\
&= \int_{-\pi}^{\pi} \left| \sum_j x_j e^{-ijv} \right|^2 f(v)\, dv \\
&\le \int_{-\pi}^{\pi} \sum_j \sum_k x_j x_k e^{-i(j-k)v} M \, dv \\
&= 2\pi M \sum_j x_j^2,
\end{aligned}$$

showing that $\lambda_n \le 2\pi M$. A similar argument shows that $\lambda_1 \ge 2\pi m$. □

§4.6* Orthogonal Increment Processes on $[-\pi, \pi]$

In order to give a precise meaning to the spectral representation (4.2.5) mentioned earlier, it is necessary to introduce the concept of stochastic integration of a non-random function with respect to an orthogonal-increment process $\{Z(\lambda)\}$.

Definition 4.6.1. An orthogonal-increment process on $[-\pi, \pi]$ is a complex-valued stochastic process $\{Z(\lambda), -\pi \le \lambda \le \pi\}$ such that

$$\langle Z(\lambda), Z(\lambda) \rangle < \infty, \qquad -\pi \le \lambda \le \pi, \tag{4.6.1}$$

$$\langle Z(\lambda), 1 \rangle = 0, \qquad -\pi \le \lambda \le \pi, \tag{4.6.2}$$

and

$$\langle Z(\lambda_4) - Z(\lambda_3), Z(\lambda_2) - Z(\lambda_1) \rangle = 0, \qquad \text{if } (\lambda_1, \lambda_2] \cap (\lambda_3, \lambda_4] = \phi, \tag{4.6.3}$$

where the inner product is defined by $\langle X, Y \rangle = E(X\bar{Y})$.

The process $\{Z(\lambda), -\pi \le \lambda \le \pi\}$ will be called *right-continuous* if for all $\lambda \in [-\pi, \pi)$,

$$\|Z(\lambda + \delta) - Z(\lambda)\|^2 = E|Z(\lambda + \delta) - Z(\lambda)|^2 \to 0 \quad \text{as } \delta \downarrow 0.$$

It will be understood from now on that the term orthogonal-increment process will mean right-continuous orthogonal-increment process unless specifically indicated otherwise.

Proposition 4.6.1. *If $\{Z(\lambda), -\pi \le \lambda \le \pi\}$ is an orthogonal-increment process, then there is a unique distribution function F (i.e. a unique non-decreasing, right-continuous function) such that*

$$F(\lambda) = 0, \qquad \lambda \le -\pi,$$

$$F(\lambda) = F(\pi), \qquad \lambda \ge \pi,$$

and

$$F(\mu) - F(\lambda) = \|Z(\mu) - Z(\lambda)\|^2, \qquad -\pi \le \lambda \le \mu \le \pi. \tag{4.6.4}$$

Proof. For F to satisfy the prescribed conditions it is clear, on setting $\lambda = -\pi$, that

$$F(\mu) = \|Z(\mu) - Z(-\pi)\|^2, \qquad -\pi \le \mu \le \pi. \tag{4.6.5}$$

To check that the function so defined is non-decreasing, we use the orthogonality of $Z(\mu) - Z(\lambda)$ and $Z(\lambda) - Z(-\pi)$, $-\pi \le \lambda \le \mu \le \pi$, to write

$$\begin{aligned} F(\mu) &= \|Z(\mu) - Z(\lambda) + Z(\lambda) - Z(-\pi)\|^2 \\ &= \|Z(\mu) - Z(\lambda)\|^2 + \|Z(\lambda) - Z(-\pi)\|^2 \\ &\ge F(\lambda). \end{aligned}$$

The same calculation gives, for $-\pi \le \mu \le \mu + \delta \le \pi$,

$$\begin{aligned} F(\mu + \delta) - F(\mu) &= \|Z(\mu + \delta) - Z(\mu)\|^2 \\ &\to 0 \quad \text{as } \delta \downarrow 0, \end{aligned}$$

by the assumed right-continuity of $\{Z(\lambda)\}$. □

Remark. The distribution function F of Proposition 4.6.1, defined on $[-\pi, \pi]$ by (4.6.5) will be referred to as the distribution function associated with the orthogonal-increment process $\{Z(\lambda), -\pi \le \lambda \le \pi\}$. It is common practice in time series analysis to use the shorthand notation,

$$E(dZ(\lambda)\, \overline{dZ(\mu)}) = \delta_{\lambda,\mu}\, dF(\lambda),$$

for the equations (4.6.3) and (4.6.4).

Example 4.6.1. Brownian motion $\{B(\lambda), -\pi \le \lambda \le \pi\}$ with $EB(\lambda) = 0$ and $\text{Var}(B(\lambda)) = \sigma^2(\lambda + \pi)/2\pi$, $-\pi \le \lambda \le \pi$, is an orthogonal-increment process on $[-\pi, \pi]$. The associated distribution function satisfies $F(\lambda) = 0$, $\lambda \le -\pi$, $F(\lambda) = \sigma^2$, $\lambda \ge \pi$, and

$$F(\lambda) = \sigma^2(\lambda + \pi)/2\pi, \qquad -\pi \le \lambda \le \pi.$$

Example 4.6.2. If $\{N(\lambda), -\pi \le \lambda \le \pi\}$ is a Poisson process on $[-\pi, \pi]$ with constant intensity c then the process $Z(\lambda) = N(\lambda) - EN(\lambda)$, $-\pi \le \lambda \le \pi$, is an orthogonal-increment process with associated distribution function $F(\lambda) = 0$,

$\lambda \le -\pi$, $F(\lambda) = 2\pi c$, $\lambda \ge \pi$ and $F(\lambda) = c(\lambda + \pi)$, $-\pi \le \lambda \le \pi$. If c is chosen to be $\sigma^2/2\pi$ then $\{Z(\lambda)\}$ has exactly the same associated distribution function as $\{B(\lambda)\}$ in Example 4.6.1.

§4.7* Integration with Respect to an Orthogonal Increment Process

We now show how to define the stochastic integral

$$I(f) = \int_{(-\pi,\pi]} f(v)\, dZ(v),$$

where $\{Z(\lambda), -\pi \le \lambda \le \pi\}$ is an orthogonal-increment process defined on the probability space $(\Omega, \mathscr{F}, P)$ and f is any function on $[-\pi, \pi]$ which is square integrable with respect to the distribution function F associated with $Z(\lambda)$. We proceed step by step, first defining $I(f)$ for any f of the form

$$f(\lambda) = \sum_{i=0}^{n} f_i I_{(\lambda_i, \lambda_{i+1}]}(\lambda), \qquad -\pi = \lambda_0 < \lambda_1 < \cdots < \lambda_{n+1} = \pi, \tag{4.7.1}$$

as

$$I(f) = \sum_{i=0}^{n} f_i [Z(\lambda_{i+1}) - Z(\lambda_i)], \tag{4.7.2}$$

and then extending the mapping I to an isomorphism (see Section 2.9) of $L^2([-\pi, \pi], \mathscr{B}, F) \equiv L^2(F)$ onto a subspace of $L^2(\Omega, \mathscr{F}, P)$.

Let $\mathscr{D}$ be the class of all functions having the form (4.7.1) for some $n \in \{0, 1, 2, \ldots\}$. Then the definition (4.7.2) is consistent on $\mathscr{D}$ since for any given $f \in \mathscr{D}$ there is a *unique* representation of f,

$$f(\lambda) = \sum_{i=0}^{m} r_i I_{(v_i, v_{i+1}]}(\lambda), \qquad -\pi = v_0 < v_1 < \cdots < v_{m+1} = \pi,$$

in which $r_i \ne r_{i+1}$, $0 \le i < m$. All other representations of f having the form (4.7.1) are obtained by reexpressing one or more of the indicator functions $I_{(v_i, v_{i+1}]}$ as a sum of indicator functions of adjoining intervals. However this makes no difference to the value of $I(f)$, and hence the definition (4.7.2) is the same for all representations (4.7.1) of f. It is clear that (4.7.2) defines I as a linear mapping on $\mathscr{D}$. Moreover the mapping preserves inner products since if $f \in \mathscr{D}$ and $g \in \mathscr{D}$ then there exist representations

$$f(\lambda) = \sum_{i=0}^{n} f_i I_{(\lambda_i, \lambda_{i+1}]}(\lambda)$$

and

$$g(\lambda) = \sum_{i=0}^{n} g_i I_{(\lambda_i, \lambda_{i+1}]}(\lambda)$$

in terms of a single partition $-\pi = \lambda_0 < \lambda_1 < \cdots < \lambda_{n+1} = \pi$. Hence the inner-product of $I(f)$ and $I(g)$ in $L^2(\Omega, \mathscr{F}, P)$ is

$$\langle I(f), I(g)\rangle = \left\langle \sum_{i=0}^{n} f_i[Z(\lambda_{i+1}) - Z(\lambda_i)], \sum_{i=0}^{n} g_i[Z(\lambda_{i+1}) - Z(\lambda_i)]\right\rangle$$

$$= \sum_{i=0}^{n} f_i\bar{g}_i(F(\lambda_{i+1}) - F(\lambda_i)),$$

by the orthogonality of the increments of $\{Z(\lambda)\}$ and Proposition 4.6.1. But the last expression can be written as

$$\int_{(-\pi,\pi]} f(v)\bar{g}(v)\,dF(v) = \langle f, g\rangle_{L^2(F)},$$

the inner product in $L^2(F)$ of f and g. Hence the mapping I on $\mathscr{D}$ preserves inner products.

Now let $\bar{\mathscr{D}}$ denote the closure in $L^2(F)$ of the set $\mathscr{D}$. If $f \in \bar{\mathscr{D}}$ then there exists a sequence $\{f_n\}$ of elements of $\mathscr{D}$ such that $\|f_n - f\|_{L^2(F)} \to 0$. We therefore define $I(f)$ as the mean square limit

$$I(f) = \underset{n\to\infty}{\text{m.s.lim}}\, I(f_n), \tag{4.7.3}$$

after first checking (a) that the limit exists and (b) that the limit is the same for all sequences $\{f_n\}$ such that $\|f_n - f\|_{L^2(F)} \to 0$. To check (a) we simply observe that for $f_m, f_n \in \mathscr{D}$,

$$\|I(f_n) - I(f_m)\| = \|I(f_n - f_m)\|$$
$$= \|f_n - f_m\|_{L^2(F)},$$

so if $\|f_n - f\|_{L^2(F)} \to 0$, the sequence $\{I(f_n)\}$ is a Cauchy sequence and therefore convergent in $L^2(\Omega, \mathscr{F}, P)$. To check (b), suppose that $\|f_n - f\|_{L^2(F)} \to 0$ and $\|g_n - f\|_{L^2(F)} \to 0$ where $f_n, g_n \in \mathscr{D}$. Then the sequence $f_1, g_1, f_2, g_2, \ldots$, must be norm convergent and therefore the sequence $I(f_1), I(g_1), I(f_2), I(g_2), \ldots$, must converge in $L^2(\Omega, \mathscr{F}, P)$. However this is not possible unless the sub-sequences $\{I(f_n)\}$ and $\{I(g_n)\}$ have the same mean square limit. This completes the proof that the definition (4.7.3) is both meaningful and consistent for $f \in \bar{\mathscr{D}}$.

The mapping I on $\bar{\mathscr{D}}$ is linear and preserves inner products since if $f^{(i)} \in \bar{\mathscr{D}}$ and $\|f_n^{(i)} - f^{(i)}\|_{L^2(F)} \to 0$, $f_n^{(i)} \in \mathscr{D}$, $i = 1, 2$, then by the linearity of I on $\mathscr{D}$,

$$I(a_1 f^{(1)} + a_2 f^{(2)}) = \lim_{n\to\infty} I(a_1 f_n^{(1)} + a_2 f_n^{(2)})$$
$$= \lim_{n\to\infty} (a_1 I(f_n^{(1)}) + a_2 I(f_n^{(2)}))$$
$$= a_1 I(f^{(1)}) + a_2 I(f^{(2)})$$

and by the continuity of the inner product,

$$\begin{aligned}\langle I(f^{(1)}), I(f^{(2)})\rangle &= \lim_{n\to\infty} \langle I(f_n^{(1)}), I(f_n^{(2)})\rangle \\ &= \lim_{n\to\infty} \langle f_n^{(1)}, f_n^{(2)}\rangle_{L^2(F)} \\ &= \langle f^{(1)}, f^{(2)}\rangle_{L^2(F)}.\end{aligned}$$

It remains now only to show that $\bar{\mathscr{D}} = L^2(F)$. To do this we first observe that the continuous functions on $[-\pi, \pi]$ are dense in $L^2(F)$ since F is a bounded distribution function (see e.g. Ash (1972), p. 88). Moreover $\mathscr{D}$ is a dense subset (in the $L^2(F)$ sense) of the set of continuous functions on $[-\pi, \pi]$. Hence $\bar{\mathscr{D}} = L^2(F)$.

Equations (4.7.2) and (4.7.3) thus define I as a linear, inner-product preserving mapping of $\bar{\mathscr{D}} = L^2(F)$ into $L^2(\Omega, \mathscr{F}, P)$. The image $I(\bar{\mathscr{D}})$ of $\bar{\mathscr{D}}$ is clearly a closed linear subspace of $L^2(\Omega, \mathscr{F}, P)$, and the mapping I is an isormorphism (see Section 2.9) of $\bar{\mathscr{D}}$ onto $I(\bar{\mathscr{D}})$. The mapping I provides us with the required definition of the stochastic integral.

Definition 4.7.1 (The Stochastic Integral). If $\{Z(\lambda)\}$ is an orthogonal-increment process on $[-\pi, \pi]$ with associated distribution function F and if $f \in L^2(F)$, then the stochastic integral $\int_{(-\pi,\pi]} f(\lambda)\, dZ(\lambda)$ is defined as the random variable $I(f)$ constructed above, i.e.

$$\int_{(-\pi,\pi]} f(v)\, dZ(v) := I(f).$$

Properties of the Stochastic Integral

For any functions f and g in $L^2(F)$ we have established the properties

$$I(a_1 f + a_2 g) = a_1 I(f) + a_2 I(g), \qquad a_1, a_2 \in \mathbb{C}, \tag{4.7.4}$$

and

$$E(I(f)\overline{I(g)}) = \int_{(-\pi,\pi]} f(v)\overline{g(v)}\, dF(v). \tag{4.7.5}$$

Moreover if $\{f_n\}$ and $\{g_n\}$ are sequences in $L^2(F)$ such that $\|f_n - f\|_{L^2(F)} \to 0$ and $\|g_n - g\|_{L^2(F)} \to 0$, then by continuity of the inner product,

$$E(I(f_n)\overline{I(g_n)}) \to E(I(f)\overline{I(g)}) = \int_{(-\pi,\pi]} f(v)\overline{g(v)}\, dF(v). \tag{4.7.6}$$

From (4.7.2) it is clear that

$$E(I(f)) = 0 \tag{4.7.7}$$

for all $f \in \mathscr{D}$; if $f \in \bar{\mathscr{D}}$ then there is a sequence $\{f_n\}$, $f_n \in \mathscr{D}$, such that $f_n \xrightarrow{L^2(F)} f$ and $I(f_n) \xrightarrow{\text{m.s.}} I(f)$, so $E(I(f)) = \lim_{n\to\infty} E(I(f_n))$ and (4.7.7) remains valid. This argument is frequently useful for establishing properties of stochastic integrals.

Finally we note from (4.7.5) and (4.7.7) that if $\{Z(\lambda)\}$ is any orthogonal increment process on $[-\pi, \pi]$ with associated distribution function F, then

$$X_t = I(e^{it\cdot}) = \int_{(-\pi,\pi]} e^{it\nu}\, dZ(\nu), \qquad t \in \mathbb{Z}, \tag{4.7.8}$$

is a stationary process with mean zero and autocovariance function

$$E(X_{t+h}\bar{X}_t) = \int_{(-\pi,\pi]} e^{i\nu h}\, dF(\nu). \tag{4.7.9}$$

In the following section we establish a converse of this result, namely that if $\{X_t\}$ is *any* stationary process, then $\{X_t\}$ has the representation (4.7.8) for an appropriately chosen orthogonal increment process $\{Z(\lambda)\}$ whose associated distribution function is the same as the spectral distribution function of $\{X_t\}$.

§4.8* The Spectral Representation

Let $\{X_t\}$ be a zero mean stationary process with spectral distribution function F. To establish the spectral representation (4.2.5) of the process $\{X_t\}$ we first need to identify an appropriate orthogonal increment process $\{Z(\lambda), \lambda \in [-\pi, \pi]\}$. The identification of $\{Z(\lambda)\}$ and the proof of the representation will be achieved by defining a certain isomorphism between the subspaces $\bar{\mathscr{H}} = \overline{\text{sp}}\{X_t, t \in \mathbb{Z}\}$ of $L^2(\Omega, \mathscr{F}, P)$ and $\bar{\mathscr{K}} = \overline{\text{sp}}\{e^{it\cdot}, t \in \mathbb{Z}\}$ of $L^2(F)$. This isomorphism will provide a link between random variables in the "time domain" and functions on $[-\pi, \pi]$ in the "frequency domain".

Let $\mathscr{H} = \text{sp}\{X_t, t \in \mathbb{Z}\}$ and $\mathscr{K} = \text{sp}\{e^{it\cdot}, t \in \mathbb{Z}\}$ denote the (not necessarily closed) subspaces $\mathscr{H} \subset L^2(\Omega, \mathscr{F}, P)$ and $\mathscr{K} \subset L^2(F)$ consisting of finite linear combinations of X_t, $t \in \mathbb{Z}$, and $e^{it\cdot}$, $t \in \mathbb{Z}$, respectively. We first show that the mapping

$$T\left(\sum_{j=1}^{n} a_j X_{t_j}\right) = \sum_{j=1}^{n} a_j e^{it_j\cdot}, \tag{4.8.1}$$

defines an isomorphism between $\mathscr{H}$ and $\mathscr{K}$. To check that T is well-defined, suppose that $\|\sum_{j=1}^{n} a_j X_{t_j} - \sum_{k=1}^{m} b_k X_{t_k}\| = 0$. Then by definition of the $L^2(F)$ norm and Herglotz's theorem,

$$\begin{aligned}\left\| T\left(\sum_{j=1}^{n} a_j X_{t_j}\right) - T\left(\sum_{k=1}^{m} b_k X_{t_k}\right) \right\|^2_{L^2(F)} &= \int_{(-\pi,\pi]} \left| \sum_{j=1}^{n} a_j e^{it_j\nu} - \sum_{k=1}^{m} b_k e^{it_k\nu} \right|^2 dF(\nu) \\ &= E\left| \sum_{j=1}^{n} a_j X_{t_j} - \sum_{k=1}^{m} b_k X_{t_k} \right|^2 = 0,\end{aligned}$$

showing that (4.8.1) defines T consistently on $\mathscr{H}$. The linearity of T follows easily from this fact. In addition,

$$\begin{aligned}\left\langle T\left(\sum_{j=1}^{n} a_j X_{t_j}\right), T\left(\sum_{k=1}^{m} b_k X_{s_k}\right)\right\rangle &= \sum_{j=1}^{n}\sum_{k=1}^{m} a_j \bar{b}_k \langle e^{it_j \cdot}, e^{is_k \cdot}\rangle_{L^2(F)} \\ &= \sum_{j=1}^{n}\sum_{k=1}^{m} a_j \bar{b}_k \int_{(-\pi,\pi]} e^{i(t_j - s_k)v}\, dF(v) \\ &= \sum_{j=1}^{n}\sum_{k=1}^{m} a_j \bar{b}_k \langle X_{t_j}, X_{s_k}\rangle \\ &= \left\langle \sum_{j=1}^{n} a_j X_{t_j}, \sum_{k=1}^{m} b_k X_{s_k}\right\rangle,\end{aligned}$$

showing that T does in fact define an isormorphism between $\mathscr{H}$ and $\mathscr{K}$.

We show next that the mapping T can be extended uniquely to an isomorphism from $\bar{\mathscr{H}}$ onto $\bar{\mathscr{K}}$. If $Y \in \bar{\mathscr{H}}$ then there is a sequence $Y_n \in \mathscr{H}$ such that $\|Y_n - Y\| \to 0$. This implies that $\{Y_n\}$ is a Cauchy sequence and hence, since T is norm-preserving, the sequence $\{TY_n\}$ is Cauchy in $L^2(F)$. The sequence $\{TY_n\}$ therefore converges in norm to an element of $\bar{\mathscr{K}}$. If T is to be norm-preserving on $\bar{\mathscr{H}}$ we must define

$$TY = \underset{n\to\infty}{\text{m.s.lim}}\, TY_n.$$

This is a consistent definition of T on $\bar{\mathscr{H}}$ since if $\|\tilde{Y}_n - Y\| \to 0$ then the sequence $TY_1, T\tilde{Y}_1, TY_2, T\tilde{Y}_2, \ldots$ is convergent, implying that the subsequences $\{TY_n\}$ and $\{T\tilde{Y}_n\}$ have the same limit, namely TY. Moreover using the same argument as given in Section 4.7 it is easy to show that the mapping T extended to $\bar{\mathscr{H}}$ is linear and preserves inner products.

Finally, by Theorem 2.11.1, $\mathscr{K}$ is uniformly dense in the space of continuous functions ϕ on $[-\pi, \pi]$ with $\phi(\pi) = \phi(-\pi)$, which in turn is dense in $L^2(F)$ (see Ash (1972), p. 88). Hence $\bar{\mathscr{K}} = L^2(F)$. We have therefore established the following theorem.

Theorem 4.8.1. *If F is the spectral distribution function of the stationary process $\{X_t, t \in \mathbb{Z}\}$, then there is a unique isomorphism T of $\overline{\text{sp}}\{X_t, t \in \mathbb{Z}\}$ onto $L^2(F)$ such that*

$$TX_t = e^{it\cdot}, \qquad t \in \mathbb{Z}.$$

Theorem 4.8.1 is particularly useful in the theory of linear prediction (see Section 5.6). It is also the key to the identification of the orthogonal increment process $\{Z(\lambda), -\pi \le \lambda \le \pi\}$ appearing in the spectral representation (4.2.5). We introduce the process $\{Z(\lambda)\}$ in the following proposition.

Proposition 4.8.1. *If T is defined as in Theorem* 4.8.1 *then the process $\{Z(\lambda), -\pi \le \lambda \le \pi\}$ defined by*

$$Z(\lambda) = T^{-1}(I_{(-\pi,\lambda]}(\cdot)), \qquad -\pi \le \lambda \le \pi,$$

is an orthogonal increment process (see Definition 4.6.1). Moreover the distribution function associated with $\{Z(\lambda)\}$ (see Proposition 4.6.1) is exactly the spectral distribution function F of $\{X_t\}$.

PROOF. For each $\lambda \in [-\pi, \pi]$, $Z(\lambda)$ is a well-defined element of $\overline{sp}\{X_t, t \in \mathbb{Z}\}$ by Theorem 4.8.1. Hence $\langle Z(\lambda), Z(\lambda)\rangle < \infty$. Since $Z(\lambda) \in \overline{sp}\{X_t, t \in \mathbb{Z}\}$ there is a sequence $\{Y_n\}$ of elements of $\overline{sp}\{X_t, t \in \mathbb{Z}\}$ such that $\|Y_n - Z(\lambda)\| \to 0$ as $n \to \infty$. By the continuity of the inner product we have

$$\langle Z(\lambda), 1\rangle = \lim_{n\to\infty} \langle Y_n, 1\rangle = 0$$

since each X_t, and hence each Y_n, has zero mean. Finally if $-\pi \le \lambda_1 \le \lambda_2 \le \lambda_3 \le \lambda_4 \le \pi$,

$$\begin{aligned}\langle Z(\lambda_4) - Z(\lambda_3), Z(\lambda_2) - Z(\lambda_1)\rangle &= \langle TZ(\lambda_4) - TZ(\lambda_3), TZ(\lambda_2) - TZ(\lambda_1)\rangle \\ &= \langle I_{(\lambda_3,\lambda_4]}(\cdot), I_{(\lambda_1,\lambda_2]}(\cdot)\rangle_{L^2(F)} \\ &= \int_{(-\pi,\pi]} I_{(\lambda_3,\lambda_4]}(\nu) I_{(\lambda_1,\lambda_2]}(\nu)\, dF(\nu) \\ &= 0,\end{aligned}$$

completing the proof that $\{Z(\lambda)\}$ has orthogonal increments. A calculation which is almost identical to the previous one gives

$$\langle Z(\mu) - Z(\lambda), Z(\mu) - Z(\lambda)\rangle = F(\mu) - F(\lambda),$$

showing that $\{Z(\lambda)\}$ is right-continuous with associated distribution function F as claimed. □

It is now a simple matter to establish the spectral representation (4.2.5).

Theorem 4.8.2 (The Spectral Representation Theorem). *If $\{X_t\}$ is a stationary sequence with mean zero and spectral distribution function F, then there exists a right-continuous orthogonal-increment process $\{Z(\lambda), -\pi \le \lambda \le \pi\}$ such that*

$$\text{(i) } E|Z(\lambda) - Z(-\pi)|^2 = F(\lambda), \qquad -\pi \le \lambda \le \pi,$$

and

$$\text{(ii) } X_t = \int_{(-\pi,\pi]} e^{it\nu}\, dZ(\nu) \quad \textit{with probability one.}$$

PROOF. Let $\{Z(\lambda)\}$ be the process defined in Proposition 4.8.1 and let I be the isomorphism,

$$I(f) = \int_{(-\pi,\pi]} f(\nu)\, dZ(\nu),$$

from $\bar{\mathscr{D}} = L^2(F)$ onto $I(\bar{\mathscr{D}}) \subseteq L^2(\Omega, \mathscr{F}, P)$, which was discussed in Section 4.7. If $f \in \mathscr{D}$ has the representation (4.7.1) then

$$I(f) = \sum_{i=0}^{n} f_i(Z(\lambda_{i+1}) - Z(\lambda_i))$$
$$= T^{-1}(f).$$

This relationship remains valid for all $f \in \bar{\mathscr{D}} = L^2(F)$ since both I and T^{-1} are isomorphisms. Therefore we must have $I = T^{-1}$ (i.e. $TI(f) = f$ for all $f \in L^2(F)$) and hence from Theorem 4.8.1.

$$X_t = I(e^{it\cdot}) = \int_{(-\pi,\pi]} e^{itv}\, dZ(v),$$

giving the required representation for $\{X_t\}$. The first assertion of the theorem is an immediate consequence of Proposition 4.8.1. □

Corollary 4.8.1. *If $\{X_t\}$ is a zero-mean stationary sequence then there exists a right continuous orthogonal increment process $\{Z(\lambda), -\pi \le \lambda \le \pi\}$ such that $Z(-\pi) = 0$ and*

$$X_t = \int_{(-\pi,\pi]} e^{itv}\, dZ(v) \quad \textit{with probability one.}$$

If $\{Y(\lambda)\}$ and $\{Z(\lambda)\}$ are two such processes then

$$P(Y(\lambda) = Z(\lambda)) = 1 \quad \textit{for each } \lambda \in [-\pi, \pi].$$

Proof. If we denote by $\{Z^*(\lambda)\}$ the orthogonal-increment process defined by Proposition 4.8.1, then the process

$$Z(\lambda) = Z^*(\lambda) - Z^*(-\pi), \qquad -\pi \le \omega \le \pi,$$

not only satisfies $Z(-\pi) = 0$, but also has exactly the same increments as $\{Z^*(\lambda)\}$. Hence

$$X_t = \int_{(-\pi,\pi]} e^{itv}\, dZ^*(v) = \int_{(-\pi,\pi]} e^{itv}\, dZ(v).$$

Suppose now that $\{Y(\lambda)\}$ is another orthogonal-increment process such that $Y(-\pi) = 0$ and

$$X_t = \int_{(-\pi,\pi]} e^{itv}\, dY(v) = \int_{(-\pi,\pi]} e^{itv}\, dZ(v) \quad \text{with probability one.} \tag{4.8.2}$$

If we define for $f \in L^2(F)$,

$$I_Y(f) = \int_{(-\pi,\pi]} f(v)\, dY(v)$$

and

$$I_Z(f) = \int_{(-\pi,\pi]} f(v)\,dZ(v)$$

then we have from (4.8.2)

$$I_Y(e^{it\cdot}) = I_Z(e^{it\cdot}) \quad \text{for all } t \in \mathbb{Z}. \tag{4.8.3}$$

Since I_Y and I_Z are equal on $\operatorname{sp}\{e^{it\cdot}, t \in \mathbb{Z}\}$ which is dense in $L^2(F)$ (see the comment preceding Theorem 4.8.1), it follows that $I_Y(f) = I_Z(f)$ for all $f \in L^2(F)$. Choosing $f(v) = I_{(-\pi,\lambda]}(v)$ we obtain (with probability one)

$$Y(\lambda) = \int_{(-\pi,\pi]} f(v)\,dZ(v) = Z(\lambda), \qquad -\pi \le \lambda \le \pi. \qquad \square$$

Remark 1. In the course of the proof of Theorem 4.8.2, the following result was established: $Y \in \overline{\operatorname{sp}}\{X_t, t \in \mathbb{Z}\}$ if and only if there exists a function $f \in L^2(F)$ such that $Y = I(f) = \int_{(-\pi,\pi]} f(v)\,dZ(v)$. This means that I is an isomorphism of $L^2(F)$ onto $\overline{\operatorname{sp}}\{X_t, t \in \mathbb{Z}\}$ (with the property that $I(e^{it\cdot}) = X_t$).

Remark 2. The argument supplied for Theorem 4.8.2 is an existence proof which does not reveal in an explicit manner how $\{Z(\lambda)\}$ is constructed. In the next section, we give a formula for obtaining $Z(\lambda)$ from $\{X_t\}$.

Remark 3. The corollary states that the orthogonal-increment process in the spectral representation is unique if one uses the normalization $Z(-\pi) = 0$. Two different stationary processes may have the same spectral distribution function, for example the processes $X_t = \int_{(-\pi,\pi]} e^{it\lambda}\,dB(\lambda)$ and $Y_t = \int_{(-\pi,\pi]} e^{it\lambda}\,dN(\lambda)$ with $\{B(\lambda)\}$ and $\{N(\lambda)\}$ defined as in Examples 4.6.1 and 4.6.2. In such cases the processes must of course have the same autocovariance function.

EXAMPLE 4.8.1. Let $Z(\lambda) = B(\lambda)$ be Brownian motion on $[-\pi,\pi]$ as defined in Example 4.6.1 with $EZ(\lambda) = 0$ and $\operatorname{Var}(Z(\lambda)) = \sigma^2(\lambda + \pi)/2\pi$, $-\pi \le \lambda \le \pi$. For $t \in \mathbb{Z}$, set $g_t(v) = \sqrt{2}\cos(tv) I_{(-\pi,0]}(v) + \sqrt{2}\sin(tv) I_{(0,\pi]}(v)$ and

$$X_t = \int_{(-\pi,\pi]} g_t(v)\,dB(v) = \sqrt{2}\left(\int_{(-\pi,0]} \cos(tv)\,dB(v) + \int_{(0,\pi]} \sin(tv)\,dB(v)\right), \tag{4.8.4}$$

(cf. Problem 4.25). Then $EX_t = 0$ by (4.7.7), and by (4.7.5),

$$E(X_{t+h}X_h) = \frac{\sigma^2}{2\pi}\int_{(-\pi,\pi]} g_{t+h}(v) g_t(v)\,dv = \frac{\sigma^2}{2\pi} 2\int_0^{\pi} \cos(hv)\,dv. \tag{4.8.5}$$

Hence $E(X_{t+h}X_t) = \sigma^2 \delta_{h,0}$ and consequently $\{X_t\} \sim \operatorname{WN}(0, \sigma^2)$.

Since however $B(\lambda)$ is Gaussian we can go further and show that the random variables X_t, $t = 0, \pm 1, \ldots$, are independent with $X_t \sim N(0, \sigma^2)$. To prove this let $s_1, \ldots, s_k$ be any k distinct integers and for each fixed j let $\{f_j^{(n)}\}$ be a sequence of elements of $\mathscr{D}$, i.e. functions of the form (4.7.1), such that $f_j^{(n)} \to g_{s_j}(\cdot)$ in $L^2(F)$. Since the mapping I_B is an isomorphism of $\bar{\mathscr{D}} = L^2(F)$ onto $I_B(\bar{\mathscr{D}})$, we conclude from (4.8.4) that

$$\theta_1 I_B(f_1^{(n)}) + \cdots + \theta_k I_B(f_k^{(n)}) \xrightarrow{\text{m.s.}} \theta_1 X_{s_1} + \cdots + \theta_k X_{s_k}. \tag{4.8.6}$$

The left -hand side, $I_B(\sum_{j=1}^k \theta_j f_j^{(n)})$, is clearly normally distributed with mean zero and variance $\|\sum_{j=1}^k \theta_j f_j^{(n)}\|^2$. The characteristic function of $I_B(\sum_{j=1}^k \theta_j f_j^{(n)})$ is therefore (Section 1.6)

$$\phi_n(u) = \exp\left[-\tfrac{1}{2}u^2 \left\| \sum_{j=1}^k \theta_j f_j^{(n)} \right\|_{L^2(F)}^2\right].$$

By the continuity of the inner product in $L^2(F)$, as $n \to \infty$,

$$\left\| \sum_{j=1}^k \theta_j f_j^{(n)} \right\|_{L^2(F)}^2 \to \left\| \sum_{j=1}^k \theta_j e^{is_j \cdot} \right\|_{L^2(F)}^2 = \sigma^2 \sum_{j=1}^k \theta_j^2.$$

From (4.8.6) we conclude therefore that $\sum_{j=1}^k \theta_j X_{s_j}$ has the Gaussian characteristic function,

$$\phi(u) = \lim_{n\to\infty} \phi_n(u) = \exp\left[-\tfrac{1}{2}u^2\sigma^2 \sum_{j=1}^k \theta_j^2\right].$$

Since this is true for all choices of $\theta_1, \ldots, \theta_k$, we deduce that $X_{s_1}, \ldots, X_{s_k}$ are jointly normal. From the covariances (4.8.5) it then follows that the random variables X_t, $t = 0, \pm 1, \ldots$, are iid $N(0, \sigma^2)$.

Remark 4. If A is a Borel subset of $[-\pi, \pi]$, it will be convenient in the following proposition (and elsewhere) to define

$$\int_A f(v)\, dZ(v) = \int_{(-\pi,\pi]} f(v) I_A(v)\, dZ(v), \tag{4.8.7}$$

where the right-hand side has already been defined in Section 4.7.

Proposition 4.8.2. *Suppose the spectral distribution function F of the stationary process $\{X_t\}$ has a point of discontinuity at λ_0 where $-\pi < \lambda_0 < \pi$. Then with probability one,*

$$X_t = \int_{(-\pi,\pi]\setminus\{\lambda_0\}} e^{itv}\, dZ(v) + (Z(\lambda_0) - Z(\lambda_0^-))e^{it\lambda_0},$$

where the two terms on the right side are uncorrelated and

$$\operatorname{Var}(Z(\lambda_0) - Z(\lambda_0^-)) = F(\lambda_0) - F(\lambda_0^-).$$

PROOF. The left limit $Z(\lambda_0^-)$ is defined to be

$$Z(\lambda_0^-) = \underset{n\to\infty}{\text{m.s.lim}}\, Z(\lambda_n), \tag{4.8.8}$$

where λ_n is any sequence such that $\lambda_n \uparrow \lambda_0$. To check that (4.8.8) makes sense we note first that $\{Z(\lambda_n)\}$ is a Cauchy sequence since $\|Z(\lambda_n) - Z(\lambda_m)\|^2 = |F(\lambda_n) - F(\lambda_m)| \to 0$ as $m, n \to \infty$. Hence the limit in (4.8.8) exists. Moreover if $\nu_n \uparrow \lambda_0$ as $n \to \infty$ then $\|Z(\lambda_n) - Z(\nu_n)\|^2 = |F(\lambda_n) - F(\nu_n)| \to 0$ as $n \to \infty$, and hence the limit in (4.8.8) is the same for all non-decreasing sequences with limit λ_0.

For $\delta > 0$ define $\lambda_{\pm\delta} = \lambda_0 \pm \delta$. Now by the spectral representation, if $0 < \delta < \pi - |\lambda_0|$,

$$X_t = \int_{(-\pi,\pi]\setminus(\lambda_{-\delta},\lambda_\delta]} e^{it\nu}\, dZ(\nu) + \int_{(\lambda_{-\delta},\lambda_\delta]} e^{it\nu}\, dZ(\nu). \tag{4.8.9}$$

Note that the two terms are uncorrelated since the regions of integration are disjoint. Now as $\delta \to 0$ the first term converges in mean square to $\int_{(-\pi,\pi]\setminus\{\lambda_0\}} e^{it\nu}\, dZ(\nu)$ since

$$e^{it\cdot} I_{(-\pi,\pi]\setminus(\lambda_{-\delta},\lambda_\delta]} \to e^{it\cdot} I_{(-\pi,\pi]\setminus\{\lambda_0\}} \text{ in } L^2(F).$$

To see how the last term of (4.8.9) behaves as $\delta \to 0$ we use the inequality

$$\begin{aligned}
&\left\| \int_{(\lambda_{-\delta},\lambda_\delta]} e^{it\nu}\, dZ(\nu) - e^{it\lambda_0}(Z(\lambda_0) - Z(\lambda_0^-)) \right\| \\
&\quad \le \left\| \int_{(\lambda_{-\delta},\lambda_\delta]} e^{it\nu}\, dZ(\nu) - e^{it\lambda_0}(Z(\lambda_\delta) - Z(\lambda_{-\delta})) \right\| \\
&\quad + \|Z(\lambda_\delta) - Z(\lambda_{-\delta}) - (Z(\lambda_0) - Z(\lambda_0^-))\|.
\end{aligned} \tag{4.8.10}$$

As $\delta \to 0$ the second term on the right of (4.8.10) goes to zero by the right continuity of $\{Z(\lambda)\}$ and the definition of $Z(\lambda_0^-)$. The first term on the right side of (4.8.10) can be written as

$$\begin{aligned}
&\left\| \int_{(-\pi,\pi]} (e^{it\nu} - e^{it\lambda_0}) I_{(\lambda_{-\delta},\lambda_\delta]}(\nu)\, dZ(\nu) \right\| \\
&\quad = \|(e^{it\cdot} - e^{it\lambda_0}) I_{(\lambda_{-\delta},\lambda_\delta]}(\cdot)\|_{L^2(F)} \\
&\quad \le \left[\sup_{\lambda_{-\delta} \le \lambda \le \lambda_\delta} |e^{it\lambda} - e^{it\lambda_0}|^2 (F(\lambda_\delta) - F(\lambda_{-\delta})) \right]^{1/2} \\
&\quad \to 0 \quad \text{as } \delta \to 0,
\end{aligned}$$

by the continuity of the function $e^{it\cdot}$. Hence we deduce from (4.8.10) that

$$\int_{(\lambda_{-\delta},\lambda_\delta]} e^{it\nu}\, dZ(\nu) \xrightarrow{\text{m.s.}} e^{it\lambda_0}(Z(\lambda_0) - Z(\lambda_0^-)) \quad \text{as } \delta \to 0.$$

The continuity of the inner product and the orthogonality of the two integrals in (4.8.9) guarantee that their mean-square limits are also orthogonal.

Moreover

$$\operatorname{Var}(Z(\lambda_0) - Z(\lambda_0^-)) = \lim_{\lambda_n \uparrow \lambda_0} \operatorname{Var}(Z(\lambda_0) - Z(\lambda_n)) = F(\lambda_0) - F(\lambda_0^-). \qquad \square$$

If the spectral distribution function has k points of discontinuity at $\lambda_1, \ldots, \lambda_k$ then $\{X_t\}$ has the representation

$$X_t = \int_{(-\pi,\pi]\setminus\{\lambda_1,\ldots,\lambda_k\}} e^{it\nu}\,dZ(\nu) + \sum_{j=1}^{k} (Z(\lambda_j) - Z(\lambda_j^-))e^{it\lambda_j}, \qquad (4.8.11)$$

where the $(k+1)$ terms on the right are uncorrelated (this should be compared with the example in Section 4.2).

The importance of (4.8.11) in time series analysis is immense. The process $Y_t = (Z(\lambda_0) - Z(\lambda_0^-))e^{it\lambda_0}$ is said to be deterministic since Y_t is determined for all t if Y_{t_0} is known for some t_0. The existence of a discontinuity in the spectral distribution function at a given frequency λ_0 therefore indicates the presence in the time series of a deterministic sinusoidal component with frequency λ_0.

§4.9* Inversion Formulae

Using the isomorphism T of Theorem 4.8.1 and a Fourier approximation to $I_{(\nu,\omega]}(\cdot) \in L^2(F)$, it is possible to express directly in terms of $\{X_t\}$ the orthogonal-increment process $\{Z(\lambda)\}$ appearing in the spectral representation (4.2.5).

Recall that for $-\pi < \nu < \omega < \pi$

$$T(Z(\omega) - Z(\nu)) = I_{(\nu,\omega]}(\cdot)$$

and

$$TX_t = e^{it\cdot} \quad \text{for all } t \in \mathbb{Z}.$$

Consequently if

$$\sum_{|j| \le n} \alpha_j e^{ij\cdot} \xrightarrow{L^2(F)} I_{(\nu,\omega]}(\cdot), \qquad (4.9.1)$$

then by the isomorphism,

$$\sum_{|j| \le n} \alpha_j X_j \xrightarrow{\text{m.s.}} Z(\omega) - Z(\nu). \qquad (4.9.2)$$

An appropriate trigonometric polynomial satisfying (4.9.1) is given by the nth-order Fourier series approximation to $I_{(\nu,\omega]}(\cdot)$, viz.

$$h_n(\lambda) = \sum_{|j| \le n} \alpha_j e^{ij\lambda}, \qquad (4.9.3)$$

where

$$\alpha_j = \frac{1}{2\pi} \int_{-\pi}^{\pi} I_{(\nu,\omega]}(\lambda) e^{-ij\lambda}\,d\lambda. \qquad (4.9.4)$$

In Section 4.11 we establish the following essential properties of the sequence of approximants $\{h_n(\cdot)\}$:

$$\sup_{\lambda \in [-\pi,\pi]\backslash E} |h_n(\lambda) - I_{(\nu,\omega]}(\lambda)| \to 0 \quad \text{as } n \to \infty, \tag{4.9.5}$$

where E is any open set containing the points ν and ω, and

$$\sup_{\lambda \in [-\pi,\pi]} |h_n(\lambda)| \le M < \infty \quad \text{for some constant } M \text{ and all } n. \tag{4.9.6}$$

Proposition 4.9.1. *If F is the spectral distribution function of the stationary sequence $\{X_n\}$ and if ν and ω are continuity points of F such that $-\pi < \nu < \omega < \pi$, then*

$$h_n(\cdot) \xrightarrow{L^2(F)} I_{(\nu,\omega]}(\cdot).$$

PROOF. Problem 4.26. □

Theorem 4.9.1. *If $\{X_n\}$ is a stationary sequence with autocovariance function $\gamma(\cdot)$, spectral distribution function F, and spectral representation $X_t = \int_{(-\pi,\pi]} e^{it\lambda}\,dZ(\lambda)$, and if ν and ω $(-\pi < \nu < \omega < \pi)$ are continuity points of F, then as $n \to \infty$*

$$\frac{1}{2\pi}\sum_{|j|\le n} X_j \left(\int_\nu^\omega e^{-ij\lambda}\,d\lambda\right) \xrightarrow{\text{m.s.}} Z(\omega) - Z(\nu) \tag{4.9.7}$$

and

$$\frac{1}{2\pi}\sum_{|j|\le n} \gamma(j) \left(\int_\nu^\omega e^{-ij\lambda}\,d\lambda\right) \to F(\omega) - F(\nu). \tag{4.9.8}$$

PROOF. The left side of (4.9.7) is just $T^{-1}h_n$ where T is the isomorphism of Theorem 4.8.1 and h_n is defined by (4.9.3). By Proposition 4.9.1 we conclude that

$$T^{-1}h_n \xrightarrow{\text{m.s.}} T^{-1}I_{(\nu,\omega]} = Z(\omega) - Z(\nu),$$

which establishes (4.9.7). To find $Z(\theta) - Z(\theta^-)$, $-\pi < \theta \le \pi$, see Problem 4.32.

To prove (4.9.8) we note, from the spectral representation of $\gamma(\cdot)$ that the left-hand side of (4.9.8) can be written as

$$\begin{aligned}
&\frac{1}{2\pi}\sum_{|j|\le n}\left(\int_{-\pi}^{\pi} I_{(\nu,\omega]}(\lambda)e^{-ij\lambda}\,d\lambda\right)\left(\int_{(-\pi,\pi]} e^{ij\theta}\,dF(\theta)\right)\\
&\quad= \int_{(-\pi,\pi]} \frac{1}{2\pi}\sum_{|j|\le n}\left(\int_{-\pi}^{\pi} I_{(\nu,\omega]}(\lambda)e^{-ij\lambda}\,d\lambda\right)e^{ij\theta}\,dF(\theta)\\
&\quad= \int_{(-\pi,\pi]} h_n(\theta)\,dF(\theta).
\end{aligned}$$

By Proposition 4.9.1 and the Cauchy–Schwarz inequality,

$$|\langle h_n - I_{(\nu,\omega]}, 1\rangle| \le \|h_n - I_{(\nu,\omega]}\| F^{1/2}(\pi) \to 0 \quad \text{as } n \to \infty.$$

Hence

$$\int_{(-\pi,\pi]} h_n(\lambda)\, dF(\lambda) \to \int_{(-\pi,\pi]} I_{(\nu,\omega]}(\lambda)\, dF(\lambda) = F(\omega) - F(\nu),$$

as required. □

Although in this book we are primarily concerned with time series in discrete-time, there is a simple analogue of the above theorem for continuous-time stationary processes which are mean square continuous. This is stated below. The major differences are that sums are replaced by integrals and the range of integration is $\mathbb{R}$ instead of $(-\pi, \pi]$.

Theorem 4.9.2. *Let $\{X(t), t \in \mathbb{R}\}$ be a zero mean stationary process with autocovariance function $\gamma(\cdot)$ which is continuous at 0. Then there exists a spectral distribution function $F(t)$ and an orthogonal-increment process $Z(t)$ on $-\infty < t < \infty$ such that*

$$\gamma(s) = \int_{-\infty}^{\infty} e^{is\lambda}\, dF(\lambda)$$

and

$$X_t = \int_{-\infty}^{\infty} e^{it\lambda}\, dZ(\lambda).$$

Moreover if ν and ω $(-\infty < \nu < \omega < \infty)$ are continuity points of F, then as $T \to \infty$,

$$\frac{1}{2\pi} \int_{-T}^{T} \left(\int_{\nu}^{\omega} e^{-ity}\, dy \right) \gamma(t)\, dt \to F(\omega) - F(\nu)$$

and

$$\frac{1}{2\pi} \int_{-T}^{T} \left(\int_{\nu}^{\omega} e^{-ity}\, dy \right) X(t)\, dt \xrightarrow{\text{m.s.}} Z(\omega) - Z(\nu).$$

For a proof of this result see Hannan (1970).

§4.10* Time-Invariant Linear Filters

The process $\{Y_t, t = 0, \pm 1, \ldots\}$ is said to be obtained from $\{X_t, t = 0, \pm 1, \ldots\}$ by application of the linear filter $C = \{c_{t,k}, t, k = 0, \pm 1, \ldots\}$ if

$$Y_t = \sum_{k=-\infty}^{\infty} c_{t,k} X_k, \qquad t = 0, \pm 1, \ldots; \tag{4.10.1}$$

the coefficients $c_{t,k}$ are called the weights of the filter. The filter C is said to be time-invariant if $c_{t,k}$ depends only on $(t-k)$, i.e. if

$$c_{t,k} = h_{t-k}, \tag{4.10.2}$$

since then

$$\begin{aligned} Y_{t-s} &= \sum_{k=-\infty}^{\infty} c_{t-s,k} X_k \\ &= \sum_{k=-\infty}^{\infty} c_{t,s+k} X_k \\ &= \sum_{k=-\infty}^{\infty} c_{t,k} X_{k-s}, \end{aligned}$$

i.e. the time-shifted process $\{Y_{t-s}, t = 0, \pm 1, \ldots\}$ is obtained from $\{X_{t-s}, t = 0, \pm 1, \ldots\}$ by application of the same linear filter C. For the time-invariant linear filter $H = \{h_i, i = 0, \pm 1, \ldots\}$ we can rewrite (4.10.1) in the form

$$Y_t = \sum_{k=-\infty}^{\infty} h_k X_{t-k}. \tag{4.10.3}$$

The time-invariant linear filter (TLF) H is said to be causal if

$$h_k = 0 \quad \text{for } k < 0, \tag{4.10.4}$$

since then Y_t is expressible in terms only of X_s, $s \le t$.

EXAMPLE 4.10.1. The filter defined by

$$Y_t = aX_{-t}, \qquad t = 0, \pm 1, \ldots,$$

is linear but not time-invariant since the weights are $c_{t,k} = a\delta_{t,-k}$ which do not depend on $(t-k)$ only.

EXAMPLE 4.10.2. The filter

$$Y_t = a_{-1}X_{t+1} + a_0 X_t + a_1 X_{t-1}, \qquad t = 0, \pm 1, \ldots,$$

is a TLF with $h_j = a_j, j = -1, 0, 1$ and $h_j = 0$ otherwise. It is not causal unless $a_{-1} = 0$.

EXAMPLE 4.10.3. The causal ARMA(p, q) process,

$$\phi(B)X_t = \theta(B)Z_t, \qquad t = 0, \pm 1, \ldots,$$

can be written (by Theorem 3.1.1) in the form

$$X_t = \sum_{j=0}^{\infty} \psi_j Z_{t-j}$$

where $\sum_{j=0}^{\infty} \psi_j z^j = \theta(z)/\phi(z)$, $|z| \le 1$. Hence $\{X_t\}$ is obtained from $\{Z_t\}$ by application of the causal TLF $\{\psi_j, j = 0, 1, 2, \ldots\}$.

Definition 4.10.1. An absolutely summable time-invariant linear filter $H = \{h_j, j = 0, \pm 1, \ldots\}$ is a TLF such that

$$\sum_{j=-\infty}^{\infty} |h_j| < \infty.$$

If $\{X_t\}$ is a zero-mean stationary process and H is an absolutely summable TLF, then applying H to $\{X_t\}$ gives

$$Y_t = \sum_{j=-\infty}^{\infty} h_j X_{t-j}. \tag{4.10.5}$$

By Theorem 4.4.1 we know that $\{Y_t\}$ is stationary with zero mean and spectral distribution function

$$F_Y(\lambda) = \int_{(-\pi, \lambda]} |h(e^{-iv})|^2 \, dF_X(v), \tag{4.10.6}$$

where

$$h(e^{-iv}) = \sum_{j=-\infty}^{\infty} h_j e^{-ijv}.$$

In the following theorem, we show that (4.10.5) and (4.10.6) remain valid under conditions weaker than absolute summability of the TLF. The theorem also shows how the spectral representation of the process $\{Y_t\}$ itself is related to that of $\{X_t\}$.

Theorem 4.10.1. *Let $\{X_t\}$ be a zero-mean stationary process with spectral representation*

$$X_t = \int_{(-\pi, \pi]} e^{itv} \, dZ_X(v)$$

and spectral distribution function $F_X(\cdot)$. Suppose $H = \{h_j, j = 0, \pm 1, \ldots\}$ is a TLF *such that the series $\sum_{j=-n}^{n} h_j e^{-ij\cdot}$ converges in $L^2(F_X)$ norm to*

$$h(e^{-i\cdot}) = \sum_{j=-\infty}^{\infty} h_j e^{-ij\cdot} \tag{4.10.7}$$

as $n \to \infty$. Then the process

$$Y_t = \sum_{j=-\infty}^{\infty} h_j X_{t-j}$$

is stationary with zero mean, spectral distribution function

$$F_Y(\lambda) = \int_{(-\pi, \lambda]} |h(e^{-iv})|^2 \, dF_X(v) \tag{4.10.8}$$

and spectral representation

$$Y_t = \int_{(-\pi,\pi]} e^{itv} h(e^{-iv})\, dZ_X(v). \tag{4.10.9}$$

If $h(e^{-iv})$ is non-zero for $v \notin A$ where $\int_A dF_X(\lambda) = 0$, the filter H can be inverted in the sense that X_t can be expressed as

$$X_t = \int_{(-\pi,\pi]} g(e^{-iv}) e^{itv}\, dZ_Y(v), \tag{4.10.10}$$

where $g(e^{-iv}) = 1/h(e^{-iv})$ and $dZ_Y(v) = h(e^{-iv})\, dZ_X(v)$. From (4.10.10) and Remark 1 of Section 4.8 it then follows that $X_t \in \overline{\text{sp}}\{Y_s, -\infty < s < \infty\}$.

Proof. From the spectral representation of $\{X_t\}$ we have

$$\sum_{j=-n}^{n} h_j X_{t-j} = \int_{(-\pi,\pi]} e^{itv} \sum_{j=-n}^{n} h_j e^{-ijv}\, dZ_X(v) \tag{4.10.11}$$

and since $\sum_{j=-n}^{n} h_j e^{-ij\cdot}$ converges to $h(e^{-i\cdot})$ in $L^2(F_X)$, it follows that the left side of (4.10.11) converges in mean square and that

$$Y_t = \sum_{j=-\infty}^{\infty} h_j X_{t-j} = \int_{(-\pi,\pi]} e^{itv} h(e^{-iv})\, dZ_X(v).$$

Equation (4.10.8) then follows directly from (4.7.5).

Once (4.10.10) is established, it will follow immediately that

$$X_t \in \overline{\text{sp}}\{Y_s, -\infty < s < \infty\}.$$

Since $g(e^{-iv})h(e^{-iv}) = 1$ for $v \notin A$ and $\int_A dF_Y(\lambda) = 0$, it follows that $g \in L^2(F_Y)$ so that the stochastic integral in (4.10.10) is well-defined. Also,

$$\begin{aligned}\left\| X_t - \int_{(-\pi,\pi]} g(e^{-itv}) e^{itv}\, dZ_Y(v) \right\|^2 &= \left\| \int_{(-\pi,\pi]} e^{itv}(1 - g(e^{-itv})h(e^{-itv}))\, dZ_Y(v) \right\|^2 \\ &= \int_{(-\pi,\pi]} |1 - g(e^{-itv})h(e^{-itv})|^2\, dF_X(v) \\ &= 0,\end{aligned}$$

which establishes (4.10.10). □

Remark 1. An absolutely summable TLF satisfies the assumptions of the first part of the theorem since in this case, $\sum_{j=-n}^{n} h_j e^{-ij\cdot}$ converges uniformly to $h(e^{-i\cdot})$.

Remark 2. Heuristically speaking the spectral representation of $\{X_t\}$ decomposes X_t into a sum of sinusoids,

$$e^{itv}\, dZ_X(v), \qquad -\pi < v \leq \pi.$$

The effect of the TLF H is to produce corresponding components

$$h(e^{-i\nu})e^{it\nu}\,dZ_X(\nu), \qquad -\pi < \nu \le \pi,$$

which combine to form the filtered process $\{Y_t\}$. Consequently $|h(e^{-i\nu})|$ is often called the *amplitude gain* of the filter, $\arg h(e^{-i\nu})$ the *phase gain* and $h(e^{-i\nu})$ itself the *transfer function*. In view of (4.10.8) with $\lambda = \pi$, the quantity $|h(e^{-i\nu})|^2$ is referred to as the *variance* (*or power*) *gain* or as the *power transfer function* at frequency ν.

Remark 3. The spectral viewpoint is particularly convenient for linear filtering since techniques are available for producing physical devices with prescribed transfer functions. The analysis of the behaviour of networks of such devices is particularly simple in terms of spectral analysis. For example if $\{X_t\}$ is operated on sequentially by two absolutely summable TLF's H_1 and H_2 in series, then the output process $\{W_t\}$ will have spectral representation

$$W_t = \int_{(-\pi,\pi]} e^{it\nu} h_1(e^{-i\nu}) h_2(e^{-i\nu})\,dZ_X(\nu)$$

and spectral distribution function

$$F_W(\lambda) = \int_{(-\pi,\pi]} |h_1(e^{-i\nu}) h_2(e^{-i\nu})|^2\,dF_X(\nu).$$

Remark 4. Let $\{Y_t\}$ be the MA(1) process,

$$Y_t = Z_t - Z_{t-1}, \qquad \{Z_t\} \sim \mathrm{WN}(0, \sigma^2).$$

Then since $h(e^{-i\nu}) = 1 - e^{-i\nu}$ is non-zero for $\nu \ne 0$ and since F_Z is continuous at zero, it follows from Theorem 4.10.1 that

$$Z_t = \int_{(-\pi,\pi]} e^{it\nu}(1 - e^{-i\nu})^{-1}\,dZ_Y(\nu).$$

Although in this case $Z_t \in \overline{\mathrm{sp}}\{Y_s, -\infty < s \le t\}$ (see also Problem 3.8), Z_t does not have a representation of the form $\sum_{j=0}^{\infty} a_j Y_{t-j}$. More generally, the filter $(1 - B)$ can be inverted whenever the spectral distribution function of the input process is continuous at zero. To illustrate the possible non-invertibility of $(1 - B)$, let us apply it to the stationary process,

$$Z_t' = Z_t + A,$$

where A is uncorrelated with $\{Z_t\}$, $E(A) = 0$ and $\mathrm{Var}(A) = \sigma_A^2 > 0$. Since

$$Y_t = (1 - B)Z_t = (1 - B)Z_t',$$

it is clear that we cannot hope to recover $\{Z_t'\}$ from $\{Y_t\}$. In this example the transfer function, $h(e^{-i\nu}) = 1 - e^{-i\nu}$, is zero at $\nu = 0$, a frequency to which $F_{Z'}$ assigns positive measure, σ_A^2.

§4.11* Properties of the Fourier Approximation h_n to $I_{(\nu,\omega]}$

In this section we establish the properties (4.9.5) and (4.9.6) of the trigonometric polynomials

$$h_n(\theta) = \frac{1}{2\pi} \sum_{|j| \le n} e^{ij\theta} \int_{-\pi}^{\pi} I_{(\nu,\omega]}(\lambda) e^{-ij\lambda}\, d\lambda, \tag{4.11.1}$$

which were used in deriving the inversion formulae of Section 4.9.

Let $D_n(\cdot)$ be the Dirichlet kernel (see Section 2.11),

$$D_n(\lambda) = \sum_{|j| \le n} e^{ij\lambda} = \begin{cases} \dfrac{\sin[(n + \frac{1}{2})\lambda]}{\sin(\lambda/2)} & \text{if } \lambda \ne 0, \\ 2n + 1 & \text{if } \lambda = 0. \end{cases}$$

Proposition 4.11.1. *If $\delta \in (0, 2\pi)$ and $\{f_n\}$ is the sequence of functions defined by*

$$f_n(x) = \int_0^x D_n(\lambda)\, d\lambda,$$

then as $n \to \infty$, $f_n(x) \to \pi$ uniformly on the set $[\delta, 2\pi - \delta]$ and

$$\sup_{x \in [0, 2\pi - \delta]} |f_n(x)| \le M < \infty \quad \textit{for all } n \ge 1.$$

Proof. We have for $x \ge 0$,

$$\int_0^x D_n(\lambda)\, d\lambda = 2 \int_0^{x/2} \lambda^{-1} \sin((2n + 1)\lambda)\, d\lambda$$
$$+ 2 \int_0^{x/2} g(\lambda)\sin((2n + 1)\lambda)\, d\lambda, \tag{4.11.2}$$

where $g(\lambda) = [\sin^{-1}(\lambda) - \lambda^{-1}] = (\lambda - \sin \lambda)/(\lambda \sin \lambda)$. Straightforward calculations show that $g(\lambda) = \lambda/6 + o(\lambda)$, $g'(\lambda) = \frac{1}{6} + o(1)$ as $\lambda \to 0$ and that $g(\lambda)$ and $g'(\lambda)$ are uniformly bounded on $[0, \pi - \delta/2]$. Integrating the second term on the right side of (4.11.2) and using the fact that $g(0) = 0$, we obtain

$$2 \int_0^{x/2} g(\lambda)\sin((2n + 1)\lambda)\, d\lambda = -2(2n + 1)^{-1} g(x/2)\cos((2n + 1)x/2$$
$$+ 2(2n + 1)^{-1} \int_0^{x/2} g'(\lambda)\cos((2n + 1)\lambda)\, d\lambda.$$

Since $g(\lambda)$ and $g'(\lambda)$ are bounded for $\lambda \in [0, \pi - \delta/2]$ it follows easily that this expression converges to zero uniformly in x on the set $[0, 2\pi - \delta]$. On the other hand by a standard result in analysis (see Billingsley (1986), p. 239),

$$2\int_0^{x/2} \lambda^{-1}\sin((2n+1)\lambda)\,d\lambda = 2\int_0^{(2n+1)x/2} \lambda^{-1}\sin\lambda\,d\lambda$$
$$\to \pi \quad \text{as } n \to \infty. \tag{4.11.3}$$

It is also evident that the convergence in (4.11.3) is uniform in x on the set $[\delta, 2\pi - \delta]$ and hence that $f_n(x)$ converges uniformly to π on the same set. Moreover since the integral $2\int_0^y \lambda^{-1}\sin\lambda\,d\lambda$ is a continuous function on $[0, \infty)$ with finite limits at $y = 0$ and $y = \infty$, the integral in (4.11.3) is uniformly bounded in x (≥ 0) and n. Combined with the uniform convergence to zero of the second integral on the right of (4.11.2), this shows that $f_n(x)$ is uniformly bounded for $x \in [0, 2\pi - \delta]$ and $n \geq 1$. □

Proposition 4.11.2. *If $\{h_n\}$ is the sequence of functions defined in* (4.11.1), *then for $-\pi < \nu < \omega < \pi$,*

$$\sup_{\theta\in[-\pi,\pi]\backslash E} |h_n(\theta) - I_{(\nu,\omega]}(\theta)| \to 0 \quad \textit{as } n \to \infty,$$

where E is any open subset of $[-\pi, \pi]$ containing both ν and ω. Also

$$\sup_{\theta\in[-\pi,\pi]} |h_n(\theta)| \leq M < \infty \quad \textit{for all } n \geq 1.$$

PROOF.

$$\begin{aligned} h_n(\theta) &= \frac{1}{2\pi}\sum_{|j|\leq n} e^{ij\theta}\int_\nu^\omega e^{-ij\lambda}\,d\lambda \\ &= \frac{1}{2\pi}\int_\nu^\omega D_n(\theta-\lambda)\,d\lambda \\ &= \frac{1}{2\pi}\int_{\theta-\omega}^{\theta-\nu} D_n(\lambda)\,d\lambda \\ &= \frac{1}{2\pi}\left(\int_0^{\theta-\nu} D_n(\lambda)\,d\lambda - \int_0^{\theta-\omega} D_n(\lambda)\,d\lambda\right). \end{aligned} \tag{4.11.4}$$

Given the set E, there exists a $\delta > 0$ such that $\delta < |\theta - \nu| < 2\pi - \delta$ and $\delta < |\theta - \omega| < 2\pi - \delta$ for all $\theta \in [-\pi, \pi]\backslash E$. Since $D_n(\lambda)$ is an even function it follows from Proposition 4.11.1 that

$$\frac{1}{2\pi}\int_0^{\theta-\nu} D_n(\lambda)\,d\lambda \to \begin{cases} \frac{1}{2} & \text{if } \theta - \nu > 0, \\ -\frac{1}{2} & \text{if } \theta - \nu < 0, \end{cases}$$

and the convergence is uniform in θ on $[-\pi, \pi]\backslash E$. The same result holds with ν replaced by ω, and hence $h_n(\theta) \to I_{(\nu,\omega]}(\theta)$ uniformly on $[-\pi, \pi]\backslash E$.

The uniform boundedness of $h_n(\theta)$ follows on applying Proposition 4.11.1 to (4.11.4) and noting that $|\theta - \nu| < 2\pi$ and $|\theta - \omega| < 2\pi$ for all $\theta \in [-\pi, \pi]$. □

Problems

4.1. If $\gamma(\cdot)$ is the autocovariance function of a complex-valued stationary process, show that $\gamma(0) \geq 0$, $|\gamma(h)| \leq \gamma(0)$, $\gamma(h) = \overline{\gamma(-h)}$, and that $\gamma(\cdot)$ is non-negative definite (see (4.1.6)).

4.2. Establish whether or not the following function is the autocovariance function of a stationary process:

$$\gamma(h) = \begin{cases} 1 & \text{if } h = 0, \\ -.5 & \text{if } h = \pm 2, \\ -.25 & \text{if } h = \pm 3, \\ 0 & \text{otherwise.} \end{cases}$$

4.3. If $0 < a < \pi$, use equation (4.3.1) to show that

$$\gamma(h) = \begin{cases} h^{-1} \sin ah, & h = \pm 1, \pm 2, \ldots, \\ a, & h = 0, \end{cases}$$

is the autocovariance function of a stationary process $\{X_t, t = 0, \pm 1, \ldots\}$. What is the spectral density of $\{X_t\}$?

4.4. If $\{X_t\}$ is the process defined by equation (4.2.1), show that $\{X_t\}$ is real-valued if and only if $\lambda_j = -\lambda_{n-j}$ and $A(\lambda_j) = \overline{A(\lambda_{n-j})}$, $j = 1, \ldots, n-1$ and $A(\lambda_n)$ is real. Show that $\{X_t\}$ then satisfies equation (4.2.2).

4.5. Determine the autocovariance function of the process with spectral density $f(\lambda) = (\pi - |\lambda|)/\pi^2$, $-\pi \leq \lambda \leq \pi$.

4.6. Evaluate the spectral density of $\{Z_t\}$, where $\{Z_t\} \sim \text{WN}(0, \sigma^2)$.

4.7. If $\{X_t\}$ and $\{Y_t\}$ are uncorrelated stationary processes with spectral distribution functions $F_X(\cdot)$ and $F_Y(\cdot)$, show that the process $\{Z_t := X_t + Y_t\}$ is stationary and determine its spectral distribution function.

4.8. Let $\{X_t\}$ and $\{Y_t\}$ be stationary zero-mean processes with spectral densities f_X and f_Y. If $f_X(\lambda) \leq f_Y(\lambda)$ for all $\lambda \in [-\pi, \pi]$, show that
(a) $\Gamma_{n,Y} - \Gamma_{n,X}$ is a non-negative definite matrix, where $\Gamma_{n,Y}$ and $\Gamma_{n,X}$ are the covariance matrices of $\mathbf{Y} = (Y_1, \ldots, Y_n)'$ and $\mathbf{X} = (X_1, \ldots, X_n)'$ respectively, and
(b) $\text{Var}(\mathbf{b}'\mathbf{X}) \leq \text{Var}(\mathbf{b}'\mathbf{Y})$ for all $\mathbf{b} = (b_1, \ldots, b_n)' \in \mathbb{R}^n$.

4.9. Let $\{X_t\}$ be the process

$$X_t = A\cos(\pi t/3) + B\sin(\pi t/3) + Y_t$$

where $Y_t = Z_t + 2.5Z_{t-1}$, $\{Z_t\} \sim \text{WN}(0, \sigma^2)$, and A and B are uncorrelated $(0, v^2)$ random variables which are also uncorrelated with $\{Z_t\}$. Find the covariance function and the spectral distribution function of $\{X_t\}$.

4.10. Construct a process $\{X_t\}$ which has spectral distribution function,

$$F_X(\omega) = \begin{cases} \pi + \omega, & -\pi \leq \omega < -\pi/6, \\ 3\pi + \omega, & -\pi/6 \leq \omega < \pi/6, \\ 5\pi + \omega, & \pi/6 \leq \omega \leq \pi. \end{cases}$$

For which values of d does the differenced process $\nabla_d X_t = X_t - X_{t-d}$ have a spectral density? What is the significance of this result for deseasonalizing a time series by differencing?

4.11. Let $\{X_t\}$ be the ARMA(p, q) process defined by

$$\phi(B)X_t = \theta(B)Z_t, \qquad \{Z_t\} \sim \text{WN}(0, \sigma^2),$$

where $\phi(z) \neq 0$ for all $z \in \mathbb{C}$ such that $|z| = 1$. Recall from Example 3.5.1 that for some $r > 1$,

$$\sum_{k=-\infty}^{\infty} \gamma(k)z^k = \sigma^2\theta(z)\theta(z^{-1})/[\phi(z)\phi(z^{-1})], \qquad r^{-1} < |z| < r,$$

where the series converges absolutely in the region specified. Use this result in conjunction with Corollary 4.3.2 to deduce that $\{X_t\}$ has a spectral density and to express the density in terms of σ^2, $\theta(\cdot)$ and $\phi(\cdot)$.

4.12. Let $\{X_t\}$ denote the Wölfer sunspot numbers (Example 1.1.5) and let $\{Y_t\}$ denote the mean-corrected series, $Y_t = X_t - 46.93$, $t = 1, \ldots, 100$. The following AR(2) model for $\{Y_t\}$ is obtained by equating the theoretical and sample autocovariances at lags 0, 1 and 2:

$$Y_t - 1.317Y_{t-1} + .634Y_{t-2} = Z_t, \qquad \{Z_t\} \sim \text{WN}(0, 289.3).$$

(These estimated parameter values are called "Yule–Walker" estimates and can be found using the program PEST, option 3.)

Determine the spectral density of the fitted model and find the frequency at which it achieves its maximum value. What is the corresponding period? (The spectral density of any ARMA process can be computed numerically using the program PEST, option 5.)

4.13. If $\{X_t\}$ and $\{Y_t\}$ are stationary processes satisfying

$$X_t - \alpha X_{t-1} = W_t, \qquad \{W_t\} \sim \text{WN}(0, \sigma^2),$$

and

$$Y_t - \alpha Y_{t-1} = X_t + Z_t, \qquad \{Z_t\} \sim \text{WN}(0, \sigma^2),$$

where $|\alpha| < 1$ and $\{W_t\}$ and $\{Z_t\}$ are uncorrelated, find the spectral density of $\{Y_t\}$.

4.14. Prove Corollary 4.4.1.

4.15. Let $\{X_t\}$ be the MA(1) process,

$$X_t = Z_t - 2Z_{t-1}, \qquad \{Z_t\} \sim \text{WN}(0, \sigma^2).$$

Given $\varepsilon > 0$, find a positive integer $k(\varepsilon)$ and constants $a_0 = 1, a_1, \ldots, a_k$ such that the spectral density of the process

$$Y_t = \sum_{j=0}^{k} a_j X_{t-j}$$

satisfies $\sup_{-\pi \le \lambda \le \pi} |f_Y(\lambda) - \text{Var}(Y_t)/2\pi| < \varepsilon$.

4.16. Compute and sketch the spectral density $f(\lambda)$, $0 \le \lambda \le \pi$, of the stationary process $\{X_t\}$ defined by

$$X_t - .99X_{t-3} = Z_t, \qquad \{Z_t\} \sim \text{WN}(0, 1).$$

Does the spectral density suggest that the sample paths of $\{X_t\}$ will exhibit oscillatory behaviour? If so, then what is the approximate period of the oscillation? Compute the spectral density of the filtered process,

$$Y_t = \tfrac{1}{3}(X_{t-1} + X_t + X_{t+1}),$$

and compare the numerical values of the spectral densities of $\{X_t\}$ and $\{Y_t\}$ at frequency $\omega = 2\pi/3$ radians per unit time. What effect would you expect the filter to have on the oscillations of $\{X_t\}$?

4.17. The spectral density of a real-valued process $\{X_t\}$ is defined on $[0, \pi]$ by

$$f(\lambda) = \begin{cases} 100, & \pi/6 - .01 < \lambda < \pi/6 + .01, \\ 0, & \text{otherwise,} \end{cases}$$

and on $[-\pi, 0]$ by $f(\lambda) = f(-\lambda)$.

(a) Evaluate the covariance function of $\{X_t\}$ at lags 0 and 1.

(b) Find the spectral density of the process $\{Y_t\}$ where

$$Y_t := \nabla_{12} X_t = X_t - X_{t-12}.$$

(c) What is the variance of Y_t?

(d) Sketch the power transfer function of the filter ∇_{12} and use the sketch to explain the effect of the filter on sinusoids with frequencies (i) near zero and (ii) near $\pi/6$.

4.18. Let $\{X_t\}$ be any stationary series with continuous spectral density f such that $0 \le f(\lambda) \le K$ and $f(\pi) \ne 0$.
Let $f_n(\lambda)$ denote the spectral density of the differenced series $\{(1 - B)^n X_t\}$.

(a) Express $f_n(\lambda)$ in terms of $f_{n-1}(\lambda)$ and hence evaluate $f_n(\lambda)$.

(b) Show that $f_n(\lambda)/f_n(\pi) \to 0$ as $n \to \infty$ for each $\lambda \in [0, \pi)$.

(c) What does (b) suggest regarding the behaviour of the sample-paths of $\{(1 - B)^n X_t\}$ for large values of n?

(d) Plot $\{(1 - B)^n X_t\}$ for $n = 1, 2, 3$ and 4, where X_t, $t = 1, \ldots, 100$ are the Wölfer sunspot numbers (Example 1.1.5). Do the realizations exhibit the behaviour expected from (c)? Notice the dependence of the sample variance on the order of differencing. (The graphs and the sample variances can be found using the program PEST.)

4.19. Determine the power transfer function of the time-invariant linear filter with coefficients $\psi_0 = 1$, $\psi_1 = -2\alpha$, $\psi_2 = 1$ and $\psi_j = 0$ for $j \ne 0, 1$ or 2. If you wish to use the filter to suppress sinusoidal oscillations with period 6, what value of α should you use? If the filter is applied to the process defined in Problem 4.9, what is the spectral distribution function of the filtered process, $Y_t = X_t - 2\alpha X_{t-1} + X_{t-2}$?

4.20.* Suppose $\{Z(\lambda), -\pi \le \lambda \le \pi\}$ is an orthogonal-increment process which is not necessarily right-continuous. Show that for all $\lambda \in [-\pi, \pi)$, $Z(\lambda + \delta)$ converges in mean square as $\delta \downarrow 0$. Call the limit $Z(\lambda^+)$ and show that this new process is a right-continuous orthogonal-increment process which is equal to $Z(\lambda)$ with probability one except possibly at a countable number of values of λ in $[-\pi, \pi)$.

4.21.* (a) If $\{Z_t\}$ is an iid sequence of $N(0, 1)$ random variables, show that the associated orthogonal-increment process for $\{Z_t\}$ is given by

$$dZ(\lambda) = dB(\lambda) + dB(-\lambda) + i(dB(-\lambda) - dB(\lambda))$$

where $B(\lambda)$ is Brownian motion on $[-\pi, \pi]$ with $\sigma^2 = 1/4$ (see Example 4.6.1) and where integration relative to $dB(-\lambda)$ is defined by

$$\int_{[-\pi,\pi]} f(\lambda)\, dB(-\lambda) = \int_{[-\pi,\pi]} f(-\lambda)\, dB(\lambda)$$

for all $f \in L^2([-\pi, \pi])$.

(b) Let $X_t = A\cos(\omega t) + B\sin(\omega t) + Z_t$ where A, B and Z_t, $t = 0, \pm 1, \ldots$, are iid $N(0, 1)$ random variables. Give a complete description of the orthogonal-increment process associated with $\{X_t\}$.

4.22.* If $X_t = \int_{(-\pi,\pi]} e^{itv}\, dZ(v)$ where $\{Z(v), -\pi \le v \le \pi\}$ is an orthogonal increment process with associated distribution function $F(\cdot)$, and if $Y_t - \phi Y_{t-1} = X_t$, where $\phi \in (-1, 1)$, find a function $\psi(\cdot)$ such that

$$Y_t = \int_{(-\pi,\pi]} e^{itv}\psi(v)\, dZ(v).$$

Hence express $E(Y_{t+h}\bar{X}_t)$ as an integral with respect to F. Evaluate the integral in the special case when $F(v) = \sigma^2(v + \pi)/2\pi$, $-\pi \le v \le \pi$.

4.23.* Let $\{Z(v), -\pi \le v \le \pi\}$ be an orthogonal increment process with associated distribution function $F(\cdot)$ and suppose that $\psi \in L^2(F)$.

(a) Show that

$$W(v) = \int_{(-\pi,v]} \psi(\lambda)\, dZ(\lambda), \qquad -\pi \le v \le \pi,$$

is an orthogonal increment process with associated distribution function,

$$G(v) = \int_{(-\pi,v]} |\psi(\lambda)|^2\, dF(\lambda).$$

(b) Show that if $g \in L^2(G)$ then $g\psi \in L^2(F)$ and

$$\int_{(-\pi,\pi]} g(\lambda)\, dW(\lambda) = \int_{(-\pi,\pi]} g(\lambda)\psi(\lambda)\, dZ(\lambda).$$

(c) Show that if $|\psi| > 0$ (except possibly on a set of F-measure zero), then

$$Z(v) = \int_{(-\pi,v]} \frac{1}{\psi(\lambda)}\, dW(\lambda), \qquad -\pi \le v \le \pi.$$

4.24.* If $\{X_t\}$ is the stationary process with spectral representation,

$$X_t = \int_{(-\pi,\pi]} e^{itv}\, dZ_X(v), \qquad t = 0, \pm 1, \ldots,$$

where $E|dZ_X(\nu)|^2 = |\phi(\nu)|^2\,dF(\nu)$, F is a distribution function on $[-\pi, \pi]$, $\phi \neq 0$ almost everywhere relative to dF, and $\phi \in L^2(F)$, show that

$$Y_t = \int_{(-\pi,\pi]} e^{it\nu}\phi^{-1}(\nu)\,dZ_X(\lambda), \qquad t = 0, \pm 1, \ldots,$$

is a stationary process with spectral distribution function F.

4.25.* Suppose that $\{X_t\}$ is a real stationary process with mean zero, spectral distribution function F and spectral representation $X_t = \int_{[-\pi,\pi]} e^{it\nu}\,dZ(\nu)$. Let $U(\lambda) = \mathrm{Re}\{Z(\lambda)\}$ and $V(\lambda) = -\mathrm{Im}\{Z(\lambda)\}$ and assume that F is continuous at 0 and π.

(a) Show that $dF(\lambda) = dF(-\lambda)$, i.e. $\int_{[-\pi,\pi]} g(\lambda)\,dF(\lambda) = \int_{[-\pi,\pi]} g(\lambda)\,dF(-\lambda)$ for all $g \in L^2(F)$, where integration with respect to $F(-\lambda)$ is defined by

$$\int_{[-\pi,\pi]} g(\lambda)\,dF(-\lambda) = \int_{[-\pi,\pi]} g(-\lambda)\,dF(\lambda).$$

(b) Show that $dZ(\lambda) = \overline{dZ(-\lambda)}$, i.e. $\int_{[-\pi,\pi]} g(\lambda)\,dZ(\lambda) = \int_{[-\pi,\pi]} g(\lambda)\,\overline{dZ(-\lambda)}$ for all $g \in L^2(F)$, where integration with respect to $\overline{Z(-\lambda)}$ is defined by

$$\int_{[-\pi,\pi]} g(\lambda)\,\overline{dZ(-\lambda)} = \overline{\int_{[-\pi,\pi]} \overline{g(-\lambda)}\,dZ(\lambda)}.$$

Deduce that $dU(\lambda) = dU(-\lambda)$ and $dV(\lambda) = -dV(-\lambda)$.

(c) Show that $\{U(\lambda),\ 0 \le \lambda \le \pi\}$ and $\{V(\lambda),\ 0 \le \lambda \le \pi\}$ are orthogonal increment processes such that for all λ and μ,

$$E(dU(\lambda)\,dU(\mu)) = 2^{-1}\delta_{\lambda,\mu}\,dF(\lambda),$$
$$E(dV(\lambda)\,dV(\mu)) = 2^{-1}\delta_{\lambda,\mu}\,dF(\lambda),$$

and

$$E(dU(\lambda)\,dV(\mu)) = 0.$$

(d) Show that

$$X_t = \int_{[-\pi,\pi]} \cos(\nu t)\,dU(\nu) + \int_{[-\pi,\pi]} \sin(\nu t)\,dV(\nu)$$

and that

$$X_t = 2\int_{[0,\pi]} \cos(\nu t)\,dU(\nu) + 2\int_{[0,\pi]} \sin(\nu t)\,dV(\nu).$$

4.26.* Use the properties (4.9.5) and (4.9.6) to establish Proposition 4.9.1.

4.27.* Let $\{Z(\nu), -\pi \le \nu \le \pi\}$ be an orthogonal-increment process with $E|Z(\nu_2) - Z(\nu_1)|^2 = a(\nu_2 - \nu_1)$, $\nu_2 \ge \nu_1$. Show that

$$Y_t = \int_{(-\pi,\pi]} (\pi - |\nu|/2)^{-1} e^{i\nu t}\,dZ(\nu),$$

is a stationary process and determine its spectral density and variance. If

$$X_t = \int_{(-\pi,\pi]} e^{i\nu t}\, dZ(\nu),$$

find the coefficients of the time-invariant linear filter which, when applied to $\{Y_t\}$, gives $\{X_t\}$.

4.28.* Show that if $\phi(\cdot)$ and $\theta(\cdot)$ are polynomials with no common zeroes and if $\phi(z) = 0$ for some $z \in \mathbb{C}$ such that $|z| = 1$, then the ARMA equations

$$\phi(B)X_t = \theta(B)Z_t, \qquad \{Z_t\} \sim \text{WN}(0, \sigma^2),$$

have no stationary solution. (Assume the existence of a stationary solution and then use the relation between the spectral distributions of $\{X_t\}$ and $\{Z_t\}$ to derive a contradiction.)

4.29.* Let $\{X_t\}$ be the stationary solution of the ARMA(p, q) equations,

$$\phi(B)X_t = \theta(B)Z_t, \qquad \{Z_t\} \sim \text{WN}(0, \sigma^2),$$

where

$$\phi(z) = (1 - a_1^{-1}z)\cdots(1 - a_r^{-1}z)(1 - a_{r+1}^{-1}z)\cdots(1 - a_p^{-1}z),$$

and

$$|a_i| > 1,\ i = 1, \ldots, r; \qquad |a_i| < 1,\ i = r+1, \ldots, p.$$

Define

$$\tilde{\phi}(z) = (1 - a_1^{-1}z)\cdots(1 - a_r^{-1}z)(1 - \bar{a}_{r+1}z)\cdots(1 - \bar{a}_p z),$$

and show (by computing the spectral density) that $\{\tilde{\phi}(B)X_t\}$ has the same autocovariance function as $\{|a_{r+1}\cdots a_p|^2\theta(B)Z_t\}$. It follows (see Section 3.2) that there exists a white noise process $\{\tilde{Z}_t\}$ in terms of which $\{X_t\}$ has the causal representation,

$$\tilde{\phi}(B)X_t = \theta(B)\tilde{Z}_t.$$

What is the variance of $\tilde{Z}_t$?

4.30.* Prove Theorem 4.1.1. (To establish the sufficiency of condition (4.1.6), let K_1 and K_2 be the real and imaginary parts respectively of the Hermitian function K and let $L^{(n)}$ be the $2n \times 2n$-matrix,

$$L^{(n)} = \frac{1}{2}\begin{bmatrix} K_1^{(n)} & K_2^{(n)} \\ -K_2^{(n)} & K_1^{(n)} \end{bmatrix}, \qquad \text{where } K_r^{(n)} = [K_r(i-j)]_{i,j=1}^n,\ r = 1, 2.$$

Show that $L^{(n)}$ is a real symmetric non-negative definite matrix and let $(Y_1, \ldots, Y_n, Z_1, \ldots, Z_n)'$ be a Gaussian random vector with mean $\mathbf{0}$ and covariance matrix $L^{(n)}$. Define the family of distribution functions,

$$F_{t+1,\ldots,t+n}(y_1, z_1, \ldots, y_n, z_n) := P(Y_1 \le y_1, Z_1 \le z_1, \ldots, Y_n \le y_n, Z_n \le z_n),$$

$n \in \{1, 2, \ldots\}$, $t \in \mathbb{Z}$, and use Kolmogorov's theorem to deduce the existence of a bivariate Gaussian process $\{(Y_t, Z_t), t \in \mathbb{Z}\}$ with mean zero and covariances

$$E(Y_{t+h} Y_t) = E(Z_{t+h} Z_t) = \tfrac{1}{2} K_1(h),$$
$$E(Y_{t+h} Z_t) = -E(Z_{t+h} Y_t) = \tfrac{1}{2} K_2(h).$$

Conclude by showing that $\{X_t := Y_t - iZ_t, t \in \mathbb{Z}\}$ is a complex-valued process with autocovariance function $K(\cdot)$.)

4.31.* Let $\{B(\lambda), -\pi \leq \lambda \leq \pi\}$ be Brownian motion as described in Example 4.6.1. If $g \in L^2(d\lambda)$ is a real-valued symmetric function, show that the process defined by

$$X_t = \int_{(-\pi, 0]} \sqrt{2} \cos(tv) g(v)\, dB(v) + \int_{(0, \pi]} \sqrt{2} \sin(tv) g(v)\, dB(v), \qquad t = 0, \pm 1, \ldots,$$

is a stationary Gaussian process with spectral density function $g^2(\lambda)\sigma^2/(2\pi)$. Conversely, suppose $\{X_t\}$ is a real stationary Gaussian process with orthogonal-increment process $Z(\lambda)$ and spectral density function $f(\lambda)$ which is strictly positive. Show that the process,

$$B(\lambda) := \int_{(0, \lambda]} f^{-1/2}(v)[dZ(v) + \overline{dZ(v)}], \qquad \lambda \in [0, \pi],$$

is Brownian motion.

4.32.* Show that for any spectral c.d.f. F on $[-\pi, \pi]$, the function $D_n(\cdot - \theta)/(2n + 1)$ (see Section 2.11) converges in $L^2(F)$ to the indicator function of the set $\{\theta\}$. Use this result to deduce, in the notation of Theorem 4.9.1, that $Z(\theta) - Z(\theta^-)$, $-\pi < \theta \leq \pi$, is the mean square limit as $n \to \infty$ of

$$(2n + 1)^{-1} \sum_{|j| \leq n} X_j \exp(-ij\theta).$$

CHAPTER 5

Prediction of Stationary Processes

In this chapter we investigate the problem of predicting the values $\{X_t, t \geq n + 1\}$ of a stationary process in terms of $\{X_1, \ldots, X_n\}$. The idea is to utilize observations taken at or before time n to forecast the subsequent behaviour of $\{X_t\}$. Given any closed subspace $\mathscr{M}$ of $L^2(\Omega, \mathscr{F}, P)$, the best predictor in $\mathscr{M}$ of X_{n+h} is defined to be the element of $\mathscr{M}$ with minimum mean-square distance from X_{n+h}. This of course is not the only possible definition of "best", but for processes with finite second moments it leads to a theory of prediction which is simple, elegant and useful in practice. (In Chapter 13 we shall introduce alternative criteria which are needed for the prediction of processes with infinite second-order moments.) In Section 2.7, we showed that the projections $P_{\mathscr{M}(X_1,\ldots,X_n)} X_{n+h}$ and $P_{\overline{\text{sp}}\{1,X_1,\ldots,X_n\}} X_{n+h}$ are respectively the best function of $X_1, \ldots, X_n$ and the best linear combination of $1, X_1, \ldots, X_n$ for predicting X_{n+h}. For the reasons given in Section 2.7 we shall concentrate almost exclusively on predictors of the latter type (best linear predictors) instead of attempting to work with conditional expectations.

§5.1 The Prediction Equations in the Time Domain

Let $\{X_t\}$ be a stationary process with mean μ and autocovariance function $\gamma(\cdot)$. Then the process $\{Y_t\} = \{X_t - \mu\}$ is a zero-mean stationary process with autocovariance function $\gamma(\cdot)$ and it is not difficult to show (Problem 5.1) that

$$P_{\overline{\text{sp}}\{1,X_1,\ldots,X_n\}} X_{n+h} = \mu + P_{\overline{\text{sp}}\{Y_1,\ldots,Y_n\}} Y_{n+h}. \tag{5.1.1}$$

Throughout this chapter we shall assume therefore, without loss of generality, that $\mu = 0$. Under this assumption it is clear from (5.1.1) that

$$P_{\overline{\text{sp}}\{1,X_1,\ldots,X_n\}} X_{n+h} = P_{\overline{\text{sp}}\{X_1,\ldots,X_n\}} X_{n+h}. \tag{5.1.2}$$

Equations for the One-Step Predictors

Let $\mathscr{H}_n$ denote the closed linear subspace $\overline{\text{sp}}\{X_1,\dots,X_n\}$, $n \geq 1$, and let $\hat{X}_{n+1}$, $n \geq 0$, denote the one-step predictors, defined by

$$\hat{X}_{n+1} = \begin{cases} 0 & \text{if } n = 0, \\ P_{\mathscr{H}_n} X_{n+1} & \text{if } n \geq 1. \end{cases} \tag{5.1.3}$$

Since $\hat{X}_{n+1} \in \mathscr{H}_n$, $n \geq 1$, we can write

$$\hat{X}_{n+1} = \phi_{n1} X_n + \cdots + \phi_{nn} X_1, \qquad n \geq 1, \tag{5.1.4}$$

where $\phi_{n1}, \dots, \phi_{nn}$ satisfy the prediction equations (2.3.8), viz.

$$\left\langle \sum_{i=1}^{n} \phi_{ni} X_{n+1-i}, X_{n+1-j} \right\rangle = \langle X_{n+1}, X_{n+1-j} \rangle, \qquad j = 1, \dots, n,$$

with $\langle X, Y \rangle = E(XY)$. By the linearity of the inner product these equations can be rewritten in the form,

$$\sum_{i=1}^{n} \phi_{ni} \gamma(i-j) = \gamma(j), \qquad j = 1, \dots, n,$$

or equivalently

$$\Gamma_n \boldsymbol{\phi}_n = \boldsymbol{\gamma}_n, \tag{5.1.5}$$

where $\Gamma_n = [\gamma(i-j)]_{i,j=1,\dots,n}$, $\boldsymbol{\gamma}_n = (\gamma(1), \dots, \gamma(n))'$ and $\boldsymbol{\phi}_n = (\phi_{n1}, \dots, \phi_{nn})'$. The projection theorem (Theorem 2.3.1) guarantees that equation (5.1.5) has at least one solution since $\hat{X}_{n+1}$ must be expressible in the form (5.1.4) for some $\boldsymbol{\phi}_n \in \mathbb{R}^n$. Equations (5.1.4) and (5.1.5) are known as the one-step prediction equations. Although there may be many solutions of (5.1.5), every one of them, when substituted in (5.1.4), must give the same predictor $\hat{X}_{n+1}$ since we know (also from Theorem 2.3.1) that $\hat{X}_{n+1}$ is uniquely defined. There is exactly one solution of (5.1.5) if and only if Γ_n is non-singular, in which case the solution is

$$\boldsymbol{\phi}_n = \Gamma_n^{-1} \boldsymbol{\gamma}_n. \tag{5.1.6}$$

The conditions specified in the following proposition are sufficient to ensure that Γ_n is non-singular for every n.

Proposition 5.1.1. *If $\gamma(0) > 0$ and $\gamma(h) \to 0$ as $h \to \infty$ then the covariance matrix $\Gamma_n = [\gamma(i-j)]_{i,j=1,\dots,n}$ of $(X_1, \dots, X_n)'$ is non-singular for every n.*

PROOF. Suppose that Γ_n is singular for some n. Then since $EX_t = 0$ there exists an integer $r \geq 1$ and real constants $a_1, \dots, a_r$ such that Γ_r is non-singular and

$$X_{r+1} = \sum_{j=1}^{r} a_j X_j$$

(see Problem 1.17). By stationarity we then have

$$X_{r+h} = \sum_{j=1}^{r} a_j X_{j+h-1}, \quad \text{for all } h \geq 1,$$

and consequently for all $n \geq r + 1$ there exist real constants $a_1^{(n)}, \ldots, a_r^{(n)}$, such that

$$X_n = \mathbf{a}^{(n)\prime} \mathbf{X}_r, \tag{5.1.7}$$

where $\mathbf{X}_r = (X_1, \ldots, X_r)'$ and $\mathbf{a}^{(n)} = (a_1^{(n)}, \ldots, a_r^{(n)})'$. Now from (5.1.7)

$$\begin{aligned}\gamma(0) &= \mathbf{a}^{(n)\prime} \Gamma_r \mathbf{a}^{(n)} \\ &= \mathbf{a}^{(n)\prime} P \Lambda P' \mathbf{a}^{(n)},\end{aligned}$$

where (see Proposition 1.6.3) Λ is a diagonal matrix whose entries are the strictly positive eigenvalues $\lambda_1 \leq \lambda_2 \leq \cdots \leq \lambda_r$ of Γ_r, and PP' is the identity matrix. Hence

$$\begin{aligned}\gamma(0) &\geq \lambda_1 \mathbf{a}^{(n)\prime} P P' \mathbf{a}^{(n)} \\ &= \lambda_1 \sum_{j=1}^{r} (a_j^{(n)})^2,\end{aligned}$$

which shows that for each fixed j, $a_j^{(n)}$ is a bounded function of n.

We can also write $\gamma(0) = \operatorname{Cov}(X_n, \sum_{j=1}^{r} a_j^{(n)} X_j)$, from which it follows that

$$\gamma(0) \leq \sum_{j=1}^{r} |a_j^{(n)}| \, |\gamma(n-j)|.$$

In view of this inequality and the boundedness of $a_j^{(\cdot)}$, it is clearly not possible to have $\gamma(0) > 0$ and $\gamma(h) \to 0$ as $h \to \infty$ if Γ_n is singular for some n. This completes the proof. □

Corollary 5.1.1. *Under the conditions of Proposition* 5.1.1, *the best linear predictor* $\hat{X}_{n+1}$ *of* X_{n+1} *in terms of* $X_1, \ldots, X_n$ *is*

$$\hat{X}_{n+1} = \sum_{i=1}^{n} \phi_{ni} X_{n+1-i}, \qquad n = 1, 2, \ldots,$$

where $\boldsymbol{\phi}_n := (\phi_{n1}, \ldots, \phi_{nn})' = \Gamma_n^{-1} \boldsymbol{\gamma}_n$, $\boldsymbol{\gamma}_n := (\gamma(1), \ldots, \gamma(n))'$ *and* $\Gamma_n = [\gamma(i-j)]_{i,j=1,\ldots,n}$. *The mean squared error is* $v_n = \gamma(0) - \boldsymbol{\gamma}_n' \Gamma_n^{-1} \boldsymbol{\gamma}_n$.

Proof. The result is an immediate consequence of (5.1.5) and Proposition 5.1.1. □

Equations for the h-Step Predictors, $h \geq 1$

The best linear predictor of X_{n+h} in terms of $X_1, \ldots, X_n$ for any $h \geq 1$ can be found in exactly the same manner as $\hat{X}_{n+1}$. Thus

$$P_{\mathscr{H}_n} X_{n+h} = \phi_{n1}^{(h)} X_n + \cdots + \phi_{nn}^{(h)} X_1, \qquad n, h \geq 1, \tag{5.1.8}$$

where $\boldsymbol{\phi}_n^{(h)} = (\phi_{n1}^{(h)}, \dots, \phi_{nn}^{(h)})'$ is any solution (unique if Γ_n is non-singular) of

$$\Gamma_n \boldsymbol{\phi}_n^{(h)} = \boldsymbol{\gamma}_n^{(h)}, \tag{5.1.9}$$

where $\boldsymbol{\gamma}_n^{(h)} = (\gamma(h), \gamma(h+1), \dots, \gamma(n+h-1))'$.

§5.2 Recursive Methods for Computing Best Linear Predictors

In this section we establish two recursive algorithms for determining the one-step predictors $\hat{X}_{n+1}$, $n \geq 1$, defined by (5.1.3), and show how they can be used also to compute the h-step predictors $P_{\mathscr{H}_n} X_{n+h}$, $h \geq 1$. Recursive prediction is of great practical importance since direct computation of $P_{\mathscr{H}_n} X_{n+h}$ from (5.1.8) and (5.1.9) requires, for large n, the solution of a large system of linear equations. Moreover, each time the number of observations is increased, the whole procedure must be repeated. The algorithms to be described in this section however allow us to compute best predictors without having to perform any matrix inversions. Furthermore they utilize the predictors based on n observations to compute those based on $n+1$ observations, $n = 1, 2, \dots$. We shall also see in Chapter 8 how the second algorithm greatly facilitates the computation of the exact likelihood of $\{X_1, \dots, X_n\}$ when the process $\{X_t\}$ is Gaussian.

Recursive Prediction Using the Durbin–Levinson Algorithm

Since $\hat{X}_{n+1} = P_{\mathscr{H}_n} X_{n+1} \in \mathscr{H}_n$, $n \geq 1$, we can express $\hat{X}_{n+1}$ in the form,

$$\hat{X}_{n+1} = \phi_{n1} X_n + \cdots + \phi_{nn} X_1, \qquad n \geq 1. \tag{5.2.1}$$

The mean squared error of prediction will be denoted by v_n. Thus

$$v_n = E(X_{n+1} - \hat{X}_{n+1})^2, \qquad n \geq 1, \tag{5.2.2}$$

and clearly $v_0 = \gamma(0)$.

The algorithm specified in the following proposition, known as the Durbin or Levinson algorithm, is a recursive scheme for computing $\boldsymbol{\phi}_n = (\phi_{n1}, \dots, \phi_{nn})'$ and v_n for $n = 1, 2, \dots$.

Proposition 5.2.1 (The Durbin–Levinson Algorithm). *If $\{X_t\}$ is a zero mean stationary process with autocovariance function $\gamma(\cdot)$ such that $\gamma(0) > 0$ and $\gamma(h) \to 0$ as $h \to \infty$, then the coefficients ϕ_{nj} and mean squared errors v_n as defined by* (5.2.1) *and* (5.2.2) *satisfy $\phi_{11} = \gamma(1)/\gamma(0)$, $v_0 = \gamma(0)$,*

$$\phi_{nn} = \left[\gamma(n) - \sum_{j=1}^{n-1} \phi_{n-1,j}\gamma(n-j)\right] v_{n-1}^{-1}, \tag{5.2.3}$$

$$\begin{bmatrix} \phi_{n1} \\ \vdots \\ \phi_{n,n-1} \end{bmatrix} = \begin{bmatrix} \phi_{n-1,1} \\ \vdots \\ \phi_{n-1,n-1} \end{bmatrix} - \phi_{nn} \begin{bmatrix} \phi_{n-1,n-1} \\ \vdots \\ \phi_{n-1,1} \end{bmatrix} \tag{5.2.4}$$

and

$$v_n = v_{n-1}[1 - \phi_{nn}^2]. \tag{5.2.5}$$

Proof. By the definition of $P_{\mathscr{K}_1}$, $\mathscr{K}_1 = \overline{\text{sp}}\{X_2, \ldots, X_n\}$ and $\mathscr{K}_2 = \overline{\text{sp}}\{X_1 - P_{\mathscr{K}_1} X_1\}$ are orthogonal subspaces of $\mathscr{H}_n = \overline{\text{sp}}\{X_1, \ldots, X_n\}$. Moreover it is easy to see that for any $Y \in L^2(\Omega, \mathscr{F}, P)$, $P_{\mathscr{H}_n} Y = P_{\mathscr{K}_1} Y + P_{\mathscr{K}_2} Y$. Hence

$$\hat{X}_{n+1} = P_{\mathscr{K}_1} X_{n+1} + P_{\mathscr{K}_2} X_{n+1} = P_{\mathscr{K}_1} X_{n+1} + a(X_1 - P_{\mathscr{K}_1} X_1), \tag{5.2.6}$$

where

$$a = \langle X_{n+1}, X_1 - P_{\mathscr{K}_1} X_1 \rangle / \|X_1 - P_{\mathscr{K}_1} X_1\|^2. \tag{5.2.7}$$

Now by stationarity, $(X_1, \ldots, X_n)'$ has the same covariance matrix as both $(X_n, X_{n-1}, \ldots, X_1)'$ and $(X_2, \ldots, X_{n+1})'$, so that

$$P_{\mathscr{K}_1} X_1 = \sum_{j=1}^{n-1} \phi_{n-1,j} X_{j+1}, \tag{5.2.8}$$

$$P_{\mathscr{K}_1} X_{n+1} = \sum_{j=1}^{n-1} \phi_{n-1,j} X_{n+1-j}, \tag{5.2.9}$$

and

$$\|X_1 - P_{\mathscr{K}_1} X_1\|^2 = \|X_{n+1} - P_{\mathscr{K}_1} X_{n+1}\|^2 = \|X_n - \hat{X}_n\|^2 = v_{n-1}. \tag{5.2.10}$$

From equations (5.2.6), (5.2.8) and (5.2.9) we obtain

$$\hat{X}_{n+1} = aX_1 + \sum_{j=1}^{n-1} [\phi_{n-1,j} - a\phi_{n-1,n-j}] X_{n+1-j}, \tag{5.2.11}$$

where, from (5.2.7) and (5.2.8),

$$\begin{aligned} a &= \left(\langle X_{n+1}, X_1 \rangle - \sum_{j=1}^{n-1} \phi_{n-1,j} \langle X_{n+1}, X_{j+1} \rangle \right) v_{n-1}^{-1} \\ &= \left[\gamma(n) - \sum_{j=1}^{n-1} \phi_{n-1,j} \gamma(n-j) \right] v_{n-1}^{-1}. \end{aligned}$$

In view of (5.1.6) and Proposition 5.1.1, the assumption that $\gamma(h) \to 0$ as $h \to \infty$ guarantees that the representation

$$\hat{X}_{n+1} = \sum_{j=1}^{n} \phi_{nj} X_{n+1-j} \tag{5.2.12}$$

is unique. Comparing coefficients in (5.2.11) and (5.2.12) we therefore deduce that

$$\phi_{nn} = a \tag{5.2.13}$$

and

$$\phi_{nj} = \phi_{n-1,j} - a\phi_{n-1,n-j}, \qquad j = 1, \ldots, n-1, \tag{5.2.14}$$

in accordance with (5.2.3) and (5.2.4).

It remains only to establish (5.2.5). The mean squared error of the predictor $\hat{X}_{n+1}$ is

$$\begin{aligned} v_n &= \|X_{n+1} - \hat{X}_{n+1}\|^2 \\ &= \|X_{n+1} - P_{\mathscr{K}_1} X_{n+1} - P_{\mathscr{K}_2} X_{n+1}\|^2 \\ &= \|X_{n+1} - P_{\mathscr{K}_1} X_{n+1}\|^2 + \|P_{\mathscr{K}_2} X_{n+1}\|^2 - 2\langle X_{n+1} - P_{\mathscr{K}_1} X_{n+1}, P_{\mathscr{K}_2} X_{n+1}\rangle \\ &= v_{n-1} + a^2 v_{n-1} - 2a\langle X_{n+1}, X_1 - P_{\mathscr{K}_1} X_1\rangle, \end{aligned}$$

where we have used (5.2.10), the orthogonality of $\mathscr{K}_1$ and $\mathscr{K}_2$, and the fact that $P_{\mathscr{K}_2} X_{n+1} = a(X_1 - P_{\mathscr{K}_1} X_1)$. Finally from (5.2.7) we obtain

$$v_n = v_{n-1}(1 - a^2)$$

as required. □

In Section 3.4 we gave two definitions of the partial autocorrelation of $\{X_t\}$ at lag n, viz.

$$\alpha(n) = \operatorname{Corr}(X_{n+1} - P_{\overline{\operatorname{sp}}\{X_2,\ldots,X_n\}} X_{n+1}, X_1 - P_{\overline{\operatorname{sp}}\{X_2,\ldots,X_n\}} X_1)$$

and

$$\alpha(n) = \phi_{nn}.$$

In the following corollary we establish the equivalence of these two definitions under the conditions of Proposition 5.2.1.

Corollary 5.2.1 (The Partial Autocorrelation Function). *Under the assumptions of Proposition* 5.2.1

$$\phi_{nn} = \operatorname{Corr}(X_{n+1} - P_{\overline{\operatorname{sp}}\{X_2,\ldots,X_n\}} X_{n+1}, X_1 - P_{\overline{\operatorname{sp}}\{X_2,\ldots,X_n\}} X_1).$$

Proof. Since $P_{\mathscr{K}_1} X_{n+1} \perp (X_1 - P_{\mathscr{K}_1} X_1)$, equations (5.2.13), (5.2.7) and (5.2.10) give

$$\begin{aligned} \phi_{nn} &= \langle X_{n+1}, X_1 - P_{\mathscr{K}_1} X_1\rangle / \|X_1 - P_{\mathscr{K}_1} X_1\|^2 \\ &= \langle X_{n+1} - P_{\mathscr{K}_1} X_{n+1}, X_1 - P_{\mathscr{K}_1} X_1\rangle / \|X_1 - P_{\mathscr{K}_1} X_1\|^2 \\ &= \operatorname{Corr}(X_{n+1} - P_{\mathscr{K}_1} X_{n+1}, X_1 - P_{\mathscr{K}_1} X_1). \end{aligned}$$

□

Recursive Prediction Using the Innovations Algorithm

The central idea in the proof of Proposition 5.2.1 was the decomposition of $\mathscr{H}_n$ into the two orthogonal subspaces $\mathscr{K}_1$ and $\mathscr{K}_2$. The second recursion, established below as Proposition 5.2.2, depends on the decomposition of $\mathscr{H}_n$ into n orthogonal subspaces by means of the Gram–Schmidt procedure.

Proposition 5.2.2 is more generally applicable than Proposition 5.2.1 since we allow $\{X_t\}$ to be a possibly non-stationary process with mean zero and autocovariance function,

$$\kappa(i,j) = \langle X_i, X_j \rangle = E(X_i X_j).$$

As before, we define $\mathscr{H}_n = \overline{\text{sp}}\{X_1,\ldots,X_n\}$, $\hat{X}_{n+1}$ as in (5.1.3), and $v_n = \|X_{n+1} - \hat{X}_{n+1}\|^2$. Clearly (defining $\hat{X}_1 := 0$),

$$\mathscr{H}_n = \overline{\text{sp}}\{X_1 - \hat{X}_1, X_2 - \hat{X}_2,\ldots, X_n - \hat{X}_n\}, \qquad n \geq 1,$$

so that

$$\hat{X}_{n+1} = \sum_{j=1}^{n} \theta_{nj}(X_{n+1-j} - \hat{X}_{n+1-j}).$$

We now establish the recursive scheme for computing $\{\theta_{nj}, j = 1,\ldots,n; v_n\}$, $n = 1, 2, \ldots$.

Proposition 5.2.2 (The Innovations Algorithm). *If $\{X_t\}$ has zero mean and $E(X_i X_j) = \kappa(i,j)$, where the matrix $[\kappa(i,j)]_{i,j=1}^{n}$ is non-singular for each $n = 1, 2, \ldots$, then the one-step predictors $\hat{X}_{n+1}$, $n \geq 0$, and their mean squared errors v_n, $n \geq 1$, are given by*

$$\hat{X}_{n+1} = \begin{cases} 0 & \text{if } n = 0, \\ \sum_{j=1}^{n} \theta_{nj}(X_{n+1-j} - \hat{X}_{n+1-j}) & \text{if } n \geq 1, \end{cases} \tag{5.2.15}$$

and

$$\begin{cases} v_0 = \kappa(1,1), \\ \theta_{n,n-k} = v_k^{-1}\left(\kappa(n+1,k+1) - \sum_{j=0}^{k-1} \theta_{k,k-j}\theta_{n,n-j}v_j\right), \qquad k = 0, 1, \ldots, n-1, \\ v_n = \kappa(n+1,n+1) - \sum_{j=0}^{n-1} \theta_{n,n-j}^2 v_j. \end{cases} \tag{5.2.16}$$

(It is a trivial matter to solve (5.2.16) *recursively in the order v_0; θ_{11}, v_1; θ_{22}, θ_{21}, v_2; θ_{33}, θ_{32}, θ_{31}, v_3; ....)*

PROOF. The set $\{X_1 - \hat{X}_1, X_2 - \hat{X}_2,\ldots, X_n - \hat{X}_n\}$ is orthogonal since $(X_i - \hat{X}_i) \in \mathscr{H}_{j-1}$ for $i < j$ and $(X_j - \hat{X}_j) \perp \mathscr{H}_{j-1}$ by definition of $\hat{X}_j$. Taking the inner product on both sides of (5.2.15) with $X_{k+1} - \hat{X}_{k+1}$, $0 \leq k < n$, we have

$$\langle \hat{X}_{n+1}, X_{k+1} - \hat{X}_{k+1} \rangle = \theta_{n,n-k} v_k.$$

Since $(X_{n+1} - \hat{X}_{n+1}) \perp (X_{k+1} - \hat{X}_{k+1})$, the coefficients $\theta_{n,n-k}$, $k = 0, \ldots, n-1$ are given by

$$\theta_{n,n-k} = v_k^{-1}\langle X_{n+1}, X_{k+1} - \hat{X}_{k+1} \rangle. \tag{5.2.17}$$

Making use of the representation (5.2.15) with n replaced by k, we obtain

$$\theta_{n,n-k} = v_k^{-1}\left(\kappa(n+1,k+1) - \sum_{j=0}^{k-1} \theta_{k,k-j}\langle X_{n+1}, X_{j+1} - \hat{X}_{j+1}\rangle\right). \quad (5.2.18)$$

Since by (5.2.17), $\langle X_{n+1}, X_{j+1} - \hat{X}_{j+1}\rangle = v_j\theta_{n,n-j}$, $0 \le j < n$, we can rewrite (5.2.18) in the form

$$\theta_{n,n-k} = v_k^{-1}\left(\kappa(n+1,k+1) - \sum_{j=0}^{k-1} \theta_{k,k-j}\theta_{n,n-j}v_j\right),$$

as required. By the projection theorem and Proposition 2.3.2,

$$v_n = \|X_{n+1} - \hat{X}_{n+1}\|^2 = \|X_{n+1}\|^2 - \|\hat{X}_{n+1}\|^2 = \kappa(n+1,n+1) - \sum_{k=0}^{n-1} \theta_{n,n-k}^2 v_k,$$

completing the derivation of (5.2.16). □

Remark 1. While the Durbin–Levinson recursion gives the coefficients of $X_1, \ldots, X_n$ in the representation $\hat{X}_{n+1} = \sum_{j=1}^{n} \phi_{nj}X_{n+1-j}$, Proposition 5.2.2 gives the coefficients of the "innovations", $(X_j - \hat{X}_j)$, $j = 1, \ldots, n$, in the orthogonal expansion $\hat{X}_{n+1} = \sum_{j=1}^{n} \theta_{nj}(X_{n+1-j} - \hat{X}_{n+1-j})$. The latter expansion is extremely simple to use and, in the case of ARMA(p, q) processes, can be simplified still further as described in Section 5.3. Proposition 5.2.2 also yields an innovations representation of X_{n+1} itself. Thus, defining $\theta_{n0} = 1$, we can write

$$X_{n+1} = \sum_{j=0}^{n} \theta_{nj}(X_{n+1-j} - \hat{X}_{n+1-j}), \qquad n = 0, 1, 2, \ldots.$$

Example 5.2.1 (Prediction of an MA(1) Process Using the Innovations Algorithm). If $\{X_t\}$ is the process,

$$X_t = Z_t + \theta Z_{t-1}, \qquad \{Z_t\} \sim \text{WN}(0, \sigma^2),$$

then $\kappa(i,j) = 0$ for $|i - j| > 1$, $\kappa(i,i) = \sigma^2(1 + \theta^2)$ and $\kappa(i, i+1) = \theta\sigma^2$. From this it is easy to see, using (5.2.16), that

$$\theta_{nj} = 0, \qquad 2 \le j \le n,$$

$$\theta_{n1} = v_{n-1}^{-1}\,\theta\sigma^2,$$

$$v_0 = (1 + \theta^2)\sigma^2,$$

and

$$v_n = [1 + \theta^2 - v_{n-1}^{-1}\,\theta^2\sigma^2]\sigma^2.$$

If we define $r_n = v_n/\sigma^2$, then we can write

$$\hat{X}_{n+1} = \theta(X_n - \hat{X}_n)/r_{n-1}$$

where $r_0 = 1 + \theta^2$ and $r_{n+1} = 1 + \theta^2 - \theta^2/r_n$. Table 5.2.1 illustrates the use of these recursions in computing $\hat{X}_6$ from observations of $X_1, \ldots, X_5$ with $\theta = -.9$. Note that v_n is non-increasing in n and, since $\|X_n - \hat{X}_n - Z_n\| \to 0$ as $n \to \infty$, $v_n \to \sigma^2$ (see Problem 5.5). The convergence of v_n to σ^2 is quite rapid in the example shown in Table 5.2.1.

Table 5.2.1. Calculation of $\hat{X}_t$ and v_t from Five Observations of the MA(1) Process, $X_t = Z_t - .9Z_{t-1}$, $Z_t \sim N(0,1)$

t	X_{t+1}	$\hat{X}_{t+1}$	v_t
0	−2.58	0	1.810
1	1.62	1.28	1.362
2	−0.96	−0.22	1.215
3	2.62	0.55	1.144
4	−1.36	−1.63	1.102
5		−0.22	1.075

EXAMPLE 5.2.2 (Prediction of an MA(1) Process Using the Durbin–Levinson Algorithm). If we apply the Durbin–Levinson algorithm to the problem considered in Example 5.2.1 we obtain

$$\begin{array}{lllll}
v_0 = 1.810 & & & & \\
\phi_{11} = -.4972 & v_1 = 1.362 & & & \\
\phi_{22} = -.3285 & \phi_{21} = -.6605 & v_2 = 1.215 & & \\
\phi_{33} = -.2433 & \phi_{32} = -.4892 & \phi_{31} = -.7404 & v_3 = 1.144 & \\
\phi_{44} = -.1914 & \phi_{43} = -.3850 & \phi_{42} = -.5828 & \phi_{41} = -.7870 & v_4 = 1.102 \\
\phi_{55} = -.1563 & \phi_{54} = -.3144 & \phi_{53} = -.4761 & \phi_{52} = -.6430 & \phi_{51} = -.8169 \\
v_5 = 1.075, & & & &
\end{array}$$

giving $\hat{X}_6 = \phi_{55}X_1 + \cdots + \phi_{51}X_5 = -0.22$, in agreement with the much simpler calculation based on Proposition 5.2.2 and shown in Table 5.2.1. Note that the constants ϕ_{nn}, $n = 1, 2, \ldots, 5$, are the partial autocorrelations at lags $1, 2, \ldots, 5$ respectively.

Recursive Calculation of the h-Step Predictors, $h \geq 1$

Let us introduce the notation P_n for the projection operator $P_{\mathscr{H}_n}$. Then the h-step predictors $P_n X_{n+h}$ can easily be found with the aid of Proposition 5.2.2. By Proposition 2.3.2, for $h \geq 1$,

$$\begin{aligned}
P_n X_{n+h} &= P_n P_{n+h-1} X_{n+h} \\
&= P_n \hat{X}_{n+h} \\
&= P_n\left(\sum_{j=1}^{n+h-1} \theta_{n+h-1,j}(X_{n+h-j} - \hat{X}_{n+h-j})\right).
\end{aligned}$$

Since $(X_{n+h-j} - \hat{X}_{n+h-j}) \perp \mathscr{H}_n$ for $j < h$, it follows from Proposition 2.3.2 that

$$P_n X_{n+h} = \sum_{j=h}^{n+h-1} \theta_{n+h-1,j}(X_{n+h-j} - \hat{X}_{n+h-j}) \tag{5.2.19}$$

where the coefficients θ_{nj} are determined as before by (5.2.16). Moreover the mean squared error can be expressed as

$$\begin{aligned} E(X_{n+h} - P_n X_{n+h})^2 &= \|X_{n+h}\|^2 - \|P_n X_{n+h}\|^2 \\ &= \kappa(n+h, n+h) - \sum_{j=h}^{n+h-1} \theta_{n+h-1,j}^2 v_{n+h-j-1}. \end{aligned} \tag{5.2.20}$$

§5.3 Recursive Prediction of an ARMA(p, q) Process

Proposition 5.2.2 can of course be applied directly to the prediction of the causal ARMA process,

$$\phi(B)X_t = \theta(B)Z_t, \qquad \{Z_t\} \sim \mathrm{WN}(0, \sigma^2), \tag{5.3.1}$$

where as usual, $\phi(B) = 1 - \phi_1 B - \cdots - \phi_p B^p$ and $\theta(B) = 1 + \theta_1 B + \cdots + \theta_q B^q$. We shall see below however that a drastic simplification in the calculations can be achieved if, instead of applying Proposition 5.2.2 directly to $\{X_t\}$, we apply it to the transformed process (cf. Ansley (1979)),

$$\begin{cases} W_t = \sigma^{-1} X_t, & t = 1, \ldots, m, \\ W_t = \sigma^{-1}\phi(B) X_t, & t > m, \end{cases} \tag{5.3.2}$$

where

$$m = \max(p, q). \tag{5.3.3}$$

For notational convenience we define $\theta_0 = 1$ and assume that $p \geq 1$ and $q \geq 1$. (There is no loss of generality in these assumptions since in the analysis which follows we may take any of the coefficients ϕ_i and θ_i to be zero.)

With the subspaces $\mathscr{H}_n$ as defined in Section 5.1, we can write

$$\mathscr{H}_n = \overline{\mathrm{sp}}\{X_1, \ldots, X_n\} = \overline{\mathrm{sp}}\{W_1, \ldots, W_n\}, \qquad n \geq 1. \tag{5.3.4}$$

For $n \geq 1$, $\hat{X}_{n+1}$ and $\hat{W}_{n+1}$ will denote the projections on $\mathscr{H}_n$ of X_{n+1} and W_{n+1} respectively. As in (5.1.3) we also define $\hat{X}_1 = \hat{W}_1 = 0$.

The autocovariance function $\gamma_X(\cdot)$ of $\{X_t\}$ can easily be computed using any of the methods described in Section 3.3. The autocovariances $\kappa(i,j) = E(W_i W_j)$ are then found from

$$\kappa(i,j) = \begin{cases} \sigma^{-2}\gamma_X(i-j), & 1 \leq i, j \leq m, \\[4pt] \sigma^{-2}\left[\gamma_X(i-j) - \displaystyle\sum_{r=1}^{p} \phi_r \gamma_X(r - |i-j|)\right], & \min(i,j) \leq m < \max(i,j) \leq 2m, \\[4pt] \displaystyle\sum_{r=0}^{q} \theta_r \theta_{r+|i-j|}, & \min(i,j) > m, \\[4pt] 0, & \text{otherwise}, \end{cases} \tag{5.3.5}$$

where we have adopted the convention $\theta_j = 0$ for $j > q$.

Applying Proposition 5.2.2 to the process $\{W_t\}$ we obtain

$$\begin{cases} \hat{W}_{n+1} = \sum_{j=1}^{n} \theta_{nj}(W_{n+1-j} - \hat{W}_{n+1-j}), & 1 \le n < m, \\ \hat{W}_{n+1} = \sum_{j=1}^{q} \theta_{nj}(W_{n+1-j} - \hat{W}_{n+1-j}), & n \ge m, \end{cases} \tag{5.3.6}$$

where the coefficients θ_{nj} and mean squared errors $r_n = E(W_{n+1} - \hat{W}_{n+1})^2$ are found recursively from (5.2.16) with κ defined as in (5.3.5). The notable feature of the predictors (5.3.6) is the vanishing of θ_{nj} when both $n \ge m$ and $j > q$. This is a consequence of (5.2.16) and the fact that $\kappa(n,j) = 0$ if $n > m$ and $|n - j| > q$.

To find $\hat{X}_n$ from $\hat{W}_n$ we observe, by projecting each side of (5.3.2) onto $\mathscr{H}_{t-1}$, that

$$\begin{cases} \hat{W}_t = \sigma^{-1}\hat{X}_t, & t = 1, \dots, m, \\ \hat{W}_t = \sigma^{-1}[\hat{X}_t - \phi_1 X_{t-1} - \cdots - \phi_p X_{t-p}], & t > m, \end{cases} \tag{5.3.7}$$

which, together with (5.3.2), shows that

$$X_t - \hat{X}_t = \sigma[W_t - \hat{W}_t] \quad \text{for all } t \ge 1. \tag{5.3.8}$$

Replacing $(W_j - \hat{W}_j)$ by $\sigma^{-1}(X_j - \hat{X}_j)$ in (5.3.6) and then substituting into (5.3.7) we finally obtain,

$$\begin{cases} \hat{X}_{n+1} = \sum_{j=1}^{n} \theta_{nj}(X_{n+1-j} - \hat{X}_{n+1-j}), & 1 \le n < m, \\ \hat{X}_{n+1} = \phi_1 X_n + \cdots + \phi_p X_{n+1-p} + \sum_{j=1}^{q} \theta_{nj}(X_{n+1-j} - \hat{X}_{n+1-j}), & n \ge m, \end{cases} \tag{5.3.9}$$

and

$$E(X_{n+1} - \hat{X}_{n+1})^2 = \sigma^2 E(W_{n+1} - \hat{W}_{n+1})^2 = \sigma^2 r_n, \tag{5.3.10}$$

where θ_{nj} and r_n are found from (5.2.16) with κ as in (5.3.5). Equations (5.3.9) determine the one-step predictors $\hat{X}_2, \hat{X}_3, \dots,$ recursively.

Remark 1. The covariances $\kappa(i,j)$ of the transformed process $\{W_t\}$ depend only on $\phi_1, \dots, \phi_p, \theta_1, \dots, \theta_q$ and not on σ^2. The same is therefore true of θ_{nj} and r_n.

Remark 2. The representation (5.3.9) for $\hat{X}_{n+1}$ is particularly convenient from a practical point of view, not only because of the simple recursion relations for the coefficients, but also because for $n \ge m$ it requires the storage of at most p past observations $X_n, \dots, X_{n+1-p}$ and at most q past innovations $(X_{n+1-j} - \hat{X}_{n+1-j}), j = 1, \dots, q$, in order to predict X_{n+1}. Direct application of Proposition 5.2.2 to $\{X_t\}$ on the other hand leads to a representation of $\hat{X}_{n+1}$ in terms of all the n preceding innovations $(X_j - \hat{X}_j), j = 1, \dots, n$.

Remark 3. It can be shown (see Problem 5.6) that if $\{X_t\}$ is invertible then as $n \to \infty$, $r_n \to 1$ and $\theta_{nj} \to \theta_j$, $j = 1, \dots, q$.

EXAMPLE 5.3.1 (Prediction of an AR(p) Process). Applying (5.3.9) to the ARMA$(p, 1)$ process with $\theta_1 = 0$, we easily find that

$$\hat{X}_{n+1} = \phi_1 X_n + \cdots + \phi_p X_{n+1-p}, \qquad n \geq p.$$

EXAMPLE 5.3.2 (Prediction of an MA(q) Process). Applying (5.3.9) to the ARMA$(1, q)$ process with $\phi_1 = 0$, we obtain

$$\hat{X}_{n+1} = \sum_{j=1}^{\min(n,q)} \theta_{nj}(X_{n+1-j} - \hat{X}_{n+1-j}), \qquad n \geq 1,$$

where the coefficients θ_{nj} are found by applying the algorithm (5.2.16) to the covariances $\kappa(i,j)$ defined in (5.3.5). Since in this case the processes $\{W_t\}$ and $\{\sigma^{-1} X_t\}$ are identical, these covariances are simply

$$\kappa(i,j) = \sigma^{-2}\gamma_X(i-j) = \sum_{r=0}^{q-|i-j|} \theta_r \theta_{r+|i-j|}.$$

EXAMPLE 5.3.3 (Prediction of an ARMA(1, 1) Process). If

$$X_t - \phi X_{t-1} = Z_t + \theta Z_{t-1}, \qquad \{Z_t\} \sim \text{WN}(0, \sigma^2), \tag{5.3.11}$$

and $|\phi| < 1$, then equations (5.3.9) reduce to the single equation

$$\hat{X}_{n+1} = \phi X_n + \theta_{n1}(X_n - \hat{X}_n), \qquad n \geq 1. \tag{5.3.12}$$

To compute θ_{n1} we first use equations (3.3.8) with $k = 0$ and $k = 1$ to find that $\gamma_X(0) = \sigma^2(1 + 2\theta\phi + \theta^2)/(1 - \phi^2)$. Substituting in (5.3.5) then gives, for $i, j \geq 1$,

$$\kappa(i,j) = \begin{cases} (1 + 2\theta\phi + \theta^2)/(1 - \phi^2), & i = j = 1, \\ 1 + \theta^2, & i = j \geq 2, \\ \theta, & |i-j| = 1, i \geq 1, \\ 0, & \text{otherwise.} \end{cases}$$

With these values of $\kappa(i,j)$, the recursions (5.2.16) reduce to

$$\begin{cases} r_0 = (1 + 2\theta\phi + \theta^2)/(1 - \phi^2), \\ \theta_{n1} = \theta/r_{n-1}, \\ r_n = 1 + \theta^2 - \theta^2/r_{n-1}, \end{cases} \tag{5.3.13}$$

which are quite trivial to solve (see Problem 5.13).

In Table 5.3.1 we show simulated values of $X_1, \ldots, X_{10}$ for the process (5.3.11) with $Z_t \sim N(0, 1)$, $\phi_1 = \phi = 0.2$ and $\theta_1 = \theta = 0.4$. The table also shows the values of r_n and θ_{n1}, $n = 1, \ldots, 10$, computed from (5.3.13) and the corresponding predicted values $\hat{X}_{n+1}$, $n = 1, \ldots, 10$, as specified by (5.3.12). Since $\sigma^2 = 1$ in this case, the mean squared errors are

$$E(X_{n+1} - \hat{X}_{n+1})^2 = \sigma^2 r_n = r_n.$$

Table 5.3.1. Calculation of $\hat{X}_n$ for Data from the ARMA(1, 1) Process of Example 5.3.3

n	X_{n+1}	r_n	θ_{n1}	$\hat{X}_{n+1}$
0	−1.100	1.3750		0
1	0.514	1.0436	0.2909	−0.5340
2	0.116	1.0067	0.3833	0.5068
3	−0.845	1.0011	0.3973	−0.1321
4	0.872	1.0002	0.3996	−0.4539
5	−0.467	1.0000	0.3999	0.7046
6	−0.977	1.0000	0.4000	−0.5620
7	−1.699	1.0000	0.4000	−0.3614
8	−1.228	1.0000	0.4000	−0.8748
9	−1.093	1.0000	0.4000	−0.3869
10		1.0000	0.4000	−0.5010

EXAMPLE 5.3.4 (Prediction of an ARMA(2, 3) Process). Simulated values of $X_1, \ldots, X_{10}$ for the causal ARMA(2, 3) process

$$X_t - X_{t-1} + 0.24X_{t-2} = Z_t + 0.4Z_{t-1} + 0.2Z_{t-2} + 0.1Z_{t-3},$$
$$\{Z_t\} \sim \text{WN}(0, 1),$$

are shown in Table 5.3.2.

In order to find the one-step predictors $\hat{X}_n$, $n = 2, \ldots, 11$ we first need the covariances $\gamma_X(h)$, $h = 0, 1, 2$, which are easily found from equations (3.3.8) with $k = 0, 1, 2$, to be

$$\gamma_X(0) = 7.17133, \quad \gamma_X(1) = 6.44139 \quad \text{and} \quad \gamma_X(2) = 5.06027.$$

Substituting in (5.3.5), we find that the symmetric matrix $K = [\kappa(i,j)]_{i,j=1,2,\ldots}$ is given by

$$K = \begin{bmatrix} 7.17133 & & & & & & & \\ 6.44139 & 7.17133 & & & & & & \\ 5.06027 & 6.44139 & 7.17133 & & & & & \\ 0.10 & 0.34 & 0.816 & 1.21 & & & & \\ 0 & 0.10 & 0.34 & 0.50 & 1.21 & & & \\ 0 & 0 & 0.10 & 0.24 & 0.50 & 1.21 & & \\ \cdot & 0 & 0 & 0.10 & 0.24 & 0.50 & 1.21 & \\ \cdot & \cdot & 0 & 0 & \cdot & \cdot & \cdot & \cdot \\ \cdot & \cdot & \cdot & \cdot & \cdot & \cdot & \cdot & \cdot \\ \cdot & \cdot & \cdot & \cdot & \cdot & \cdot & \cdot & \cdot \end{bmatrix} \tag{5.3.14}$$

The next step is to solve the recursions (5.2.16) with $\kappa(i,j)$ as in (5.3.14) for θ_{nj} and $r_{n-1}, j = 1, \ldots, n; n = 1, \ldots, 10$. Then

$$\hat{X}_{n+1} = \sum_{j=1}^{n} \theta_{nj}(X_{n+1-j} - \hat{X}_{n+1-j}), \qquad n = 1, 2,$$

$$\hat{X}_{n+1} = X_n - 0.24X_{n-1} + \sum_{j=1}^{3} \theta_{nj}(X_{n+1-j} - \hat{X}_{n+1-j}), \qquad n = 3, 4, \ldots,$$

and

$$E(X_{n+1} - \hat{X}_{n+1})^2 = \sigma^2 r_n = r_n.$$

The results are shown in Table 5.3.2.

Table 5.3.2. Calculation of $\hat{X}_{n+1}$ for Data from the ARMA(2, 3) Process of Example 5.3.4

n	X_{n+1}	r_n	θ_{n1}	θ_{n2}	θ_{n3}	$\hat{X}_{n+1}$
0	1.704	7.1713				0
1	0.527	1.3856	0.8982			1.5305
2	1.041	1.0057	1.3685	0.7056		−0.1710
3	0.942	1.0019	0.4008	0.1806	0.0139	1.2428
4	0.555	1.0016	0.3998	0.2020	0.0722	0.7443
5	−1.002	1.0005	0.3992	0.1995	0.0994	0.3138
6	−0.585	1.0000	0.4000	0.1997	0.0998	−1.7293
7	0.010	1.0000	0.4000	0.2000	0.0998	−0.1688
8	−0.638	1.0000	0.4000	0.2000	0.0999	0.3193
9	0.525	1.0000	0.4000	0.2000	0.1000	−0.8731
10		1.0000	0.4000	0.2000	0.1000	1.0638
11		1.0000	0.4000	0.2000	0.1000	
12		1.0000	0.4000	0.2000	0.1000	

h-Step Prediction of an ARMA(p, q) Process, $h \geq 1$

As in Section 5.2 we shall use the notation P_n for the projection operator $P_{\mathscr{H}_n}$. Then from (5.2.19) we have

$$\begin{aligned} P_n W_{n+h} &= \sum_{j=h}^{n+h-1} \theta_{n+h-1,j}(W_{n+h-j} - \hat{W}_{n+h-j}) \\ &= \sigma^{-1} \sum_{j=h}^{n+h-1} \theta_{n+h-1,j}(X_{n+h-j} - \hat{X}_{n+h-j}). \end{aligned}$$

Using this result and applying the operator P_n to each side of the equations (5.3.2), we conclude that the h-step predictors $P_n X_{n+h}$ satisfy

$$P_n X_{n+h} = \begin{cases} \displaystyle\sum_{j=h}^{n+h-1} \theta_{n+h-1,j}(X_{n+h-j} - \hat{X}_{n+h-j}), & 1 \leq h \leq m - n, \\ \displaystyle\sum_{i=1}^{p} \phi_i P_n X_{n+h-i} + \sum_{h \leq j \leq q} \theta_{n+h-1,j}(X_{n+h-j} - \hat{X}_{n+h-j}), & h > m - n. \end{cases} \tag{5.3.15}$$

Once the predictors $\hat{X}_1, \ldots, \hat{X}_n$ have been computed from (5.3.9), it is a straightforward calculation, with n fixed, to determine the predictors $P_n X_{n+1}$, $P_n X_{n+2}, P_n X_{n+3}, \ldots$, recursively from (5.3.15).

Assuming that $n > m$, as is invariably the case in practical prediction problems, we have for $h \geq 1$,

$$P_n X_{n+h} = \sum_{i=1}^{p} \phi_i P_n X_{n+h-i} + \sum_{j=h}^{q} \theta_{n+h-1,j}(X_{n+h-j} - \hat{X}_{n+h-j}), \tag{5.3.16}$$

where the second term is zero if $h > q$. Expressing X_{n+h} as $\hat{X}_{n+h} + (X_{n+h} - \hat{X}_{n+h})$, we can also write,

$$X_{n+h} = \sum_{i=1}^{p} \phi_i X_{n+h-i} + \sum_{j=0}^{q} \theta_{n+h-1,j}(X_{n+h-j} - \hat{X}_{n+h-j}), \tag{5.3.17}$$

where $\theta_{n0} := 1$ for all n. Subtracting (5.3.16) from (5.3.17) gives

$$X_{n+h} - P_n X_{n+h} - \sum_{i=1}^{p} \phi_i(X_{n+h-i} - P_n X_{n+h-i}) = \sum_{j=0}^{h-1} \theta_{n+h-1,j}(X_{n+h-j} - \hat{X}_{n+h-j}),$$

and hence,

$$\Phi \begin{bmatrix} X_{n+1} - P_n X_{n+1} \\ \vdots \\ X_{n+h} - P_n X_{n+h} \end{bmatrix} = \Theta \begin{bmatrix} X_{n+1} - \hat{X}_{n+1} \\ \vdots \\ X_{n+h} - \hat{X}_{n+h} \end{bmatrix}, \tag{5.3.18}$$

where Φ and Θ are the lower triangular matrices,

$$\Phi = -[\phi_{i-j}]_{i,j=1}^{h} \quad (\phi_0 := -1, \phi_j := 0 \text{ if } j > p \text{ or } j < 0),$$

and

$$\Theta = [\theta_{n+i-1,i-j}]_{i,j=1}^{h} \quad (\theta_{n0} := 1, \theta_{nj} := 0 \text{ if } j > q \text{ or } j < 0).$$

From (5.3.18) we immediately find that the covariance matrix of the vector $(X_{n+1} - P_n X_{n+1}, \ldots, X_{n+h} - P_n X_{n+h})'$ of prediction errors is

$$C = \Phi^{-1}\Theta V \Theta' \Phi'^{-1}, \tag{5.3.19}$$

where $V = \operatorname{diag}(v_n, v_{n+1}, \ldots, v_{n+h-1})$. It is not difficult to show (Problem 5.7) that Φ^{-1} is the lower triangular matrix

$$\Phi^{-1} = [\chi_{i-j}]_{i,j=1}^{h} \quad (\chi_0 := 1, \chi_j := 0 \text{ if } j < 0), \tag{5.3.20}$$

whose components $\chi_j, j \geq 1$, are easily computed from the recursion relations,

$$\chi_j = \sum_{k=1}^{\min(p,j)} \phi_k \chi_{j-k}, \qquad j = 1, 2, \ldots. \tag{5.3.21}$$

[By writing down the recursion relations for the coefficients in the power series expansion of $1/\phi(z)$ (cf. (3.3.3)), we see in fact that

$$\sum_{j=0}^{\infty} \chi_j z^j = (1 - \phi_1 z - \cdots - \phi_p z^p)^{-1}, \qquad |z| \leq 1.]$$

The mean squared error of the h-step predictor $P_n X_{n+h}$ is then found from (5.3.19) to be

$$\sigma_n^2(h) := E(X_{n+h} - P_n X_{n+h})^2 = \sum_{j=0}^{h-1} \left(\sum_{r=0}^{j} \chi_r \theta_{n+h-r-1, j-r} \right)^2 v_{n+h-j-1}. \quad (5.3.22)$$

Assuming invertibility of the ARMA process, we can let $n \to \infty$ in (5.3.16) and (5.3.22) to get the large-sample approximations,

$$P_n X_{n+h} \simeq \sum_{i=1}^{p} \phi_i P_n X_{n+h-i} + \sum_{j=h}^{q} \theta_j (X_{n+h-j} - \hat{X}_{n+h-j}) \quad (5.3.23)$$

and

$$\sigma_n^2(h) \simeq \sigma^2 \sum_{j=0}^{h-1} \left(\sum_{r=0}^{j} \chi_r \theta_{j-r} \right)^2 = \sigma^2 \sum_{j=0}^{h-1} \psi_j^2, \quad (5.3.24)$$

where

$$\sum_{j=0}^{\infty} \psi_j z^j = \left(\sum_{j=0}^{\infty} \chi_j z^j \right) \left(\sum_{j=0}^{q} \theta_j z^j \right) = \theta(z)/\phi(z), \qquad |z| \le 1.$$

Example 5.3.5 (Two- and Three-Step Prediction of an ARMA(2, 3) Process). We illustrate the use of equations (5.3.16) and (5.3.22) by applying them to the data of Example 5.3.4 (see Table 5.3.2). From (5.3.16) we obtain

$$\begin{aligned} P_{10} X_{12} &= \sum_{i=1}^{2} \phi_i P_{10} X_{12-i} + \sum_{j=2}^{3} \theta_{11,j} (X_{12-j} - \hat{X}_{12-j}) \\ &= \phi_1 \hat{X}_{11} + \phi_2 X_{10} + .2(X_{10} - \hat{X}_{10}) + .1(X_9 - \hat{X}_9) \\ &= 1.1217 \end{aligned}$$

and

$$\begin{aligned} P_{10} X_{13} &= \sum_{i=1}^{2} \phi_i P_{10} X_{13-i} + \sum_{j=3}^{3} \theta_{12,j} (X_{13-j} - \hat{X}_{13-j}) \\ &= \phi_1 P_{10} X_{12} + \phi_2 \hat{X}_{11} + .1(X_{10} - \hat{X}_{10}) \\ &= 1.0062 \end{aligned}$$

For $k > 13$, $P_{10} X_k$ is easily found recursively from

$$P_{10} X_k = \phi_1 P_{10} X_{k-1} + \phi_2 P_{10} X_{k-2}.$$

To find the mean squared error of $P_n X_{n+h}$ we apply (5.3.22) with $\chi_0 = 1$, $\chi_1 = \phi_1 = 1$ and $\chi_2 = \phi_1 \chi_1 + \phi_2 = .76$. Using the values of θ_{nj} and v_j $(= r_j)$ in Table 5.3.2, we obtain

$$\sigma_{10}^2(2) = E(X_{12} - P_{10} X_{12})^2 = 2.960,$$

and

$$\sigma_{10}^2(3) = E(X_{13} - P_{10} X_{13})^2 = 4.810.$$

If we use the large-sample approximation (5.3.23) and (5.3.24), the predicted values $P_{10}X_{10+h}$ and mean squared errors $\sigma_{10}^2(h)$, $h \geq 1$, are unchanged since the coefficients $\theta_{nj}, j = 1, 2, 3$, and the one-step mean squared errors $v_n = r_n\sigma^2$ have attained their asymptotic values (to four decimal places) when $n = 10$.

§5.4 Prediction of a Stationary Gaussian Process; Prediction Bounds

Let $\{X_t\}$ be a zero-mean stationary Gaussian process (see Definition 1.3.4) with covariance function $\gamma(\cdot)$ such that $\gamma(0) > 0$ and $\gamma(h) \to 0$ as $h \to \infty$. By equation (5.1.8) the best linear predictor of X_{n+h} in terms of $\mathbf{X}_n = (X_1, \ldots, X_n)'$ is

$$P_n X_{n+h} = [\gamma(n+h-1), \gamma(n+h-2), \ldots, \gamma(h)]\Gamma_n^{-1}\mathbf{X}_n, \qquad h \geq 1. \quad (5.4.1)$$

(The calculation of $P_n X_{n+h}$ is most simply carried out recursively with the aid of (5.2.19) or, in the case of an ARMA(p, q) process, by using (5.3.15).) Since $(X_1, \ldots, X_{n+h})'$ has a multivariate normal distribution, it follows from Problem 2.20 that

$$P_n X_{n+h} = E_{\mathscr{M}(X_1, \ldots, X_n)} X_{n+h} = E(X_{n+h} | X_1, \ldots, X_n).$$

For a stationary Gaussian process it is clear that the prediction error, $\Delta_n(h) := X_{n+h} - P_n X_{n+h}$, is normally distributed with mean zero and variance

$$\sigma_n^2(h) = E\Delta_n(h)^2,$$

which can be calculated either from (5.2.20) in the general case, or from (5.3.22) if $\{X_t\}$ is an ARMA(p, q) process.

Denoting by $\Phi_{1-\alpha/2}$ the $(1 - \alpha/2)$-quantile of the standard normal distribution function, we conclude from the observations of the preceding paragraph that X_{n+h} lies between the bounds $P_n X_{n+h} \pm \Phi_{1-\alpha/2}\sigma_n(h)$ with probability $(1 - \alpha)$. These bounds are therefore called $(1 - \alpha)$-prediction bounds for X_{n+h}.

§5.5 Prediction of a Causal Invertible ARMA Process in Terms of X_j, $-\infty < j \leq n$

It is sometimes useful, primarily in order to approximate $P_n X_{n+h}$ for large n, to determine the projection of X_{n+h} onto $\mathscr{M}_n = \overline{\text{sp}}\{X_j, -\infty < j \leq n\}$. In this section we shall consider this problem in the case when $\{X_t\}$ is a causal invertible ARMA(p, q) process,

$$\phi(B)X_t = \theta(B)Z_t, \qquad \{Z_t\} \sim \text{WN}(0, \sigma^2). \quad (5.5.1)$$

In order to simplify notation we shall consider n to be a fixed positive integer and define

$$\tilde{X}_t := P_{\mathcal{M}_n} X_t \quad (= X_t \text{ for } t \le n). \tag{5.5.2}$$

We can then determine $\tilde{X}_{n+h}$ and $E(X_{n+h} - \tilde{X}_{n+h})^2$ from the following theorem. The quantity $E(X_{n+h} - \tilde{X}_{n+h})^2$ is useful for large n as an approximation to $E(X_{n+h} - P_n X_{n+h})^2$.

Theorem 5.5.1. *If X_t is the causal invertible* ARMA *process* (5.5.1) and $\tilde{X}_t$ *is defined by* (5.5.2) *then*

$$\tilde{X}_{n+h} = -\sum_{j=1}^{\infty} \pi_j \tilde{X}_{n+h-j} \tag{5.5.3}$$

and

$$\tilde{X}_{n+h} = \sum_{j=h}^{\infty} \psi_j Z_{n+h-j}, \tag{5.5.4}$$

where $\sum_{j=0}^{\infty} \pi_j z^j = \phi(z)/\theta(z)$ *and* $\sum_{j=0}^{\infty} \psi_j z^j = \theta(z)/\phi(z)$, $|z| \le 1$. *Moreover*

$$E(X_{n+h} - \tilde{X}_{n+h})^2 = \sigma^2 \sum_{j=0}^{h-1} \psi_j^2. \tag{5.5.5}$$

Proof. We know from Theorems 3.1.1 and 3.1.2 that

$$Z_{n+h} = X_{n+h} + \sum_{j=1}^{\infty} \pi_j X_{n+h-j} \tag{5.5.6}$$

and

$$X_{n+h} = \sum_{j=0}^{\infty} \psi_j Z_{n+h-j}. \tag{5.5.7}$$

Applying the operator $P_{\mathcal{M}_n}$ to each side of these equations and using the fact that Z_{n+k} is orthogonal to $\mathcal{M}_n$ for each $k \ge 1$, we obtain equations (5.5.3) and (5.5.4). Then subtracting (5.5.4) from (5.5.7) we find that

$$X_{n+h} - \tilde{X}_{n+h} = \sum_{j=0}^{h-1} \psi_j Z_{n+h-j}, \tag{5.5.8}$$

from which the result (5.5.5) follows at once. □

Remark 1. Equation (5.5.3) is the most convenient one for calculation of $\tilde{X}_{n+h}$. It can be solved recursively for $h = 1, 2, 3, \ldots$, using the conditions $\tilde{X}_t = X_t$, $t \le n$. Thus

$$\tilde{X}_{n+1} = -\sum_{j=1}^{\infty} \pi_j X_{n+1-j},$$

$$\tilde{X}_{n+2} = -\pi_1 \tilde{X}_{n+1} - \sum_{j=2}^{\infty} \pi_j X_{n+2-j},$$

etc.

For large n, a truncated solution $\tilde{X}_n^T$, obtained by setting $\sum_{j=n+h}^{\infty} \pi_j X_{n+h-j} = 0$ in (5.5.3) and solving the resulting equations

$$\tilde{X}_{n+h}^T = -\sum_{j=1}^{n+h-1} \pi_j \tilde{X}_{n+h-j}^T \quad \text{with } \tilde{X}_t^T = X_t,\ t = 1, \ldots, n,$$

is sometimes used as an approximation to $P_n X_{n+h}$. This procedure gives

$$\tilde{X}_{n+1}^T = -\sum_{j=1}^{n} \pi_j X_{n+1-j},$$

$$\tilde{X}_{n+2}^T = -\pi_1 \tilde{X}_{n+1}^T - \sum_{j=2}^{n+1} \pi_j X_{n+2-j},$$

etc.

The mean squared error of $\tilde{X}_{n+h}$ as specified by (5.5.5) is also sometimes used as an approximation to $E(X_{n+h} - P_n X_{n+h})^2$. The approximation (5.5.5) is in fact simply the large sample approximation (5.3.24) to the exact mean squared error (5.3.22) of $P_n X_{n+h}$.

Remark 2. For an AR(p) process, equation (5.5.3) leads to the expected result,

$$\tilde{X}_{n+1} = \phi_1 X_n + \cdots + \phi_p X_{n+1-p},$$

with mean squared error

$$E(X_{n+1} - \tilde{X}_{n+1})^2 = \sigma^2.$$

For an MA(1) process (5.5.3) gives

$$\tilde{X}_{n+1} = -\sum_{j=0}^{\infty} (-\theta_1)^{j+1} X_{n-j},$$

with mean squared error

$$E(X_{n+1} - \tilde{X}_{n+1})^2 = \sigma^2.$$

The "truncation" approximation to $P_n X_{n+1}$ for the MA(1) process is

$$\tilde{X}_{n+1}^T = -\sum_{j=0}^{n-1} (-\theta_1)^{j+1} X_{n-j},$$

which may be a poor approximation if $|\theta_1|$ is near one.

Remark 3. For fixed n, the prediction errors $X_{n+h} - \tilde{X}_{n+h}$, $h = 1, 2, \ldots$, are not uncorrelated. In fact it is clear from (5.5.8) that the covariance of the h-step and k-step prediction errors is

$$E[(X_{n+h} - \tilde{X}_{n+h})(X_{n+k} - \tilde{X}_{n+k})] = \sigma^2 \sum_{j=0}^{h-1} \psi_j \psi_{j+k-h} \quad \text{for } k \geq h. \tag{5.5.9}$$

The corresponding covariance of $(X_{n+h} - P_n X_{n+h})$ and $(X_{n+k} - P_n X_{n+k})$ is rather more complicated, but it can be found from (5.3.19).

§5.6* Prediction in the Frequency Domain

If $\{X_t, t \in \mathbb{Z}\}$ is a zero-mean stationary process with spectral distribution function F and associated orthogonal increment process $\{Z(\lambda), -\pi \le \lambda \le \pi\}$, then the mapping I defined by

$$I(g) = \int_{(-\pi,\pi]} g(\lambda)\, dZ(\lambda), \qquad g \in L^2(F), \tag{5.6.1}$$

is an isomorphism of $L^2(F)$ onto $\overline{\mathscr{H}} = \overline{\text{sp}}\{X_t, t \in \mathbb{Z}\}$ (see Remark 1 of Section 4.8), with the property that

$$I(e^{it\cdot}) = X_t, \qquad t \in \mathbb{Z}. \tag{5.6.2}$$

This isomorphism allows us to compute projections (i.e. predictors) in $\overline{\mathscr{H}}$ by computing projections in $L^2(F)$ and then applying the mapping I. For example the best linear predictor $P_{\mathscr{M}_n} X_{n+h}$ of X_{n+h} in $\mathscr{M}_n = \overline{\text{sp}}\{X_t, -\infty < t \le n\}$ can be expressed (see Section 2.9) as

$$P_{\mathscr{M}_n} X_{n+h} = I(P_{\overline{\text{sp}}\{\exp(it\cdot), -\infty < t \le n\}} e^{i(n+h)\cdot}). \tag{5.6.3}$$

The calculation of the projection on the right followed by application of the mapping I is illustrated in the following examples.

EXAMPLE 5.6.1. Suppose that $\{X_t\}$ has mean zero and spectral density f such that

$$f(\lambda) = 0, \qquad 1 < |\lambda| \le \pi. \tag{5.6.4}$$

It is simple to check that $|1 - e^{-i\lambda}| < 1$ for $\lambda \in E = [-1, 1]$ and hence that the series

$$\sum_{k=0}^{\infty} (1 - e^{-i\lambda})^k = \sum_{k=0}^{\infty} \sum_{j=0}^{k} \binom{k}{j} (-1)^j e^{-i\lambda j}, \tag{5.6.5}$$

converges uniformly on E to $[1 - (1 - e^{-i\lambda})]^{-1} = e^{i\lambda}$. Consequently

$$e^{i(n+1)\lambda} = \sum_{k=0}^{\infty} \sum_{j=0}^{k} \binom{k}{j} (-1)^j e^{i\lambda(n-j)}, \qquad \lambda \in E, \tag{5.6.6}$$

with the series on the right converging uniformly on E. By (5.6.4) the series (5.6.6) converges also in $L^2(F)$ to a limit which is clearly an element of $\overline{\text{sp}}\{\exp(it\cdot), -\infty < t \le n\}$. Projecting each side of (5.6.6) onto this subspace, we therefore obtain

$$P_{\overline{\text{sp}}\{\exp(it\cdot), -\infty < t \le n\}} e^{i(n+1)\cdot} = e^{i(n+1)\cdot} = \sum_{k=0}^{\infty} \sum_{j=0}^{k} \binom{k}{j} (-1)^j e^{i(n-j)\cdot}. \tag{5.6.7}$$

Applying the mapping I to this equation and using (5.6.2) and (5.6.3), we conclude that

$$P_{\mathscr{M}_n} X_{n+1} = X_{n+1} = \sum_{k=0}^{\infty} \sum_{j=0}^{k} \binom{k}{j} (-1)^j X_{n-j}. \tag{5.6.8}$$

Computation of this predictor using the time-domain methods of Section 5.1 is a considerably more difficult task.

EXAMPLE 5.6.2 (Prediction of an ARMA Process). Consider the causal invertible ARMA process,

$$\phi(B)X_t = \theta(B)Z_t, \qquad \{Z_t\} \sim \mathrm{WN}(0, \sigma^2),$$

with spectral density

$$f(\lambda) = a(\lambda)\overline{a(\lambda)}, \tag{5.6.9}$$

where

$$a(\lambda) = (2\pi)^{-1/2}\sigma \sum_{k=0}^{\infty} \psi_k e^{-ik\lambda} \tag{5.6.10}$$

and $\sum_{k=0}^{\infty} \psi_k z^k = \theta(z)/\phi(z)$, $|z| \le 1$. Convergence of the series in (5.6.10) is uniform on $[-\pi, \pi]$ since, by the causality assumption, $\sum_k |\psi_k| < \infty$.

The function $g(\cdot) = P_{\overline{\mathrm{sp}}\{\exp(it\cdot), -\infty < t \le n\}} e^{i(n+h)\cdot}$ must satisfy

$$\int_{-\pi}^{\pi} (e^{i(n+h)\lambda} - g(\lambda))e^{-im\lambda} a(\lambda)\overline{a(\lambda)}\, d\lambda = 0, \qquad m \le n. \tag{5.6.11}$$

This equation implies that $(e^{i(n+h)\cdot} - g(\cdot))a(\cdot)\overline{a(\cdot)}$ is an element of the subspace $\mathscr{M}_+ = \overline{\mathrm{sp}}\{\exp(im\cdot),\ m > n\}$ of $L^2([-\pi, \pi], \mathscr{B}, d\lambda)$. Noting from (5.6.10) that $1/\overline{a(\cdot)} \in \overline{\mathrm{sp}}\{\exp(im\cdot),\ m \ge 0\}$, we deduce that the function $(e^{i(n+h)\cdot} - g(\cdot))a(\cdot)$ is also an element of $\mathscr{M}_+$. Let us now write

$$e^{i(n+h)\lambda} a(\lambda) = g(\lambda)a(\lambda) + (e^{i(n+h)\lambda} - g(\lambda))a(\lambda), \tag{5.6.12}$$

observing that $g(\cdot)a(\cdot)$ is orthogonal to $\mathscr{M}_+$ (in $L^2(d\lambda)$). But from (5.6.10),

$$e^{i(n+h)\lambda} a(\lambda) = (2\pi)^{-1/2}\sigma e^{in\lambda} \sum_{k=-h}^{\infty} \psi_{k+h} e^{-ik\lambda}, \tag{5.6.13}$$

and since the element $e^{i(n+h)\cdot}a(\cdot)$ of $L^2(d\lambda)$ has a unique representation as a sum of two components, one in $\mathscr{M}_+$ and one orthogonal to $\mathscr{M}_+$, we can immediately make the identification,

$$g(\lambda)a(\lambda) = (2\pi)^{-1/2}\sigma e^{in\lambda} \sum_{k=0}^{\infty} \psi_{k+h} e^{-ik\lambda}.$$

Using (5.6.10) again we obtain

$$g(\lambda) = e^{in\lambda}[\phi(e^{-i\lambda})/\theta(e^{-i\lambda})] \sum_{k=0}^{\infty} \psi_{k+h} e^{-ik\lambda},$$

i.e.

$$g(\lambda) = \sum_{j=0}^{\infty} \alpha_j e^{i(n-j)\lambda}, \tag{5.6.14}$$

where $\sum_{j=0}^{\infty} \alpha_j z^j = [\phi(z)/\theta(z)] \sum_{k=0}^{\infty} \psi_{k+h} z^k$, $|z| \le 1$. Applying the mapping I to each side of (5.6.14) and using (5.6.2) and (5.6.3), we conclude that

$$P_{\mathscr{M}_n} X_{n+h} = \sum_{j=0}^{\infty} \alpha_j X_{n-j}. \tag{5.6.15}$$

It is not difficult to check (Problem 5.17) that this result is equivalent to (5.5.4).

§5.7* The Wold Decomposition

In Example 5.6.1, the values $X_{n+j}, j \geq 1$, of the process $\{X_t, t \in \mathbb{Z}\}$ were perfectly predictable in terms of elements of $\mathscr{M}_n = \overline{\text{sp}}\{X_t, -\infty < t \leq n\}$. Such processes are called deterministic. Any zero-mean stationary process $\{X_t\}$ which is not deterministic can be expressed as a sum $X_t = U_t + V_t$ of an MA(∞) process $\{U_t\}$ and a deterministic process $\{V_t\}$ which is uncorrelated with $\{U_t\}$. In the statement and proof of this decomposition (Theorem 5.7.1) we shall use the notation σ^2 for the one-step mean squared error,

$$\sigma^2 = E|X_{n+1} - P_{\mathscr{M}_n} X_{n+1}|^2,$$

and $\mathscr{M}_{-\infty}$ for the closed linear subspace,

$$\mathscr{M}_{-\infty} = \bigcap_{n=-\infty}^{\infty} \mathscr{M}_n,$$

of the Hilbert space $\mathscr{M} = \overline{\text{sp}}\{X_t, t \in \mathbb{Z}\}$. All subspaces and orthogonal complements should be interpreted as relative to $\mathscr{M}$. For orthogonal subspaces $\mathscr{S}_1$ and $\mathscr{S}_2$ we define $\mathscr{S}_1 \oplus \mathscr{S}_2 := \{x + y : x \in \mathscr{S}_1 \text{ and } y \in \mathscr{S}_2\}$.

Remark 1. The process $\{X_t\}$ is said to be deterministic if and only if $\sigma^2 = 0$, or equivalently if and only if $X_t \in \mathscr{M}_{-\infty}$ for each t (Problem 5.18).

Theorem 5.7.1 (The Wold Decomposition). *If $\sigma^2 > 0$ then X_t can be expressed as*

$$X_t = \sum_{j=0}^{\infty} \psi_j Z_{t-j} + V_t, \tag{5.7.1}$$

where

(i) $\psi_0 = 1$ *and* $\sum_{j=0}^{\infty} \psi_j^2 < \infty$,

(ii) $\{Z_t\} \sim \text{WN}(0, \sigma^2)$,

(iii) $Z_t \in \mathscr{M}_t$ *for each* $t \in \mathbb{Z}$,

(iv) $E(Z_t V_s) = 0$ *for all* $s, t \in \mathbb{Z}$,

(v) $V_t \in \mathscr{M}_{-\infty}$ *for each* $t \in \mathbb{Z}$,

and

(vi) $\{V_t\}$ *is deterministic.*

((v) *and* (vi) *are not the same since* $\mathcal{M}_{-\infty}$ *is defined in terms of* $\{X_t\}$, *not* $\{V_t\}$.) *The sequences* $\{\psi_j\}$, $\{Z_j\}$ *and* $\{V_j\}$ *are uniquely determined by* (5.7.1) *and the conditions* (i)–(vi).

PROOF. We first show that the sequences defined by

$$Z_t = X_t - P_{\mathcal{M}_{t-1}} X_t, \tag{5.7.2}$$

$$\psi_j = \langle X_t, Z_{t-j} \rangle / \sigma^2 \tag{5.7.3}$$

and

$$V_t = X_t - \sum_{j=0}^{\infty} \psi_j Z_{t-j}, \tag{5.7.4}$$

satisfy (5.7.1) and conditions (i)–(vi). The proof is then completed by establishing the uniqueness of the three sequences.

Clearly Z_t as defined by (5.7.2) is an element of $\mathcal{M}_t$ and is orthogonal to $\mathcal{M}_{t-1}$ by the definition of $P_{\mathcal{M}_{t-1}} X_t$. Hence

$$Z_t \in \mathcal{M}_{t-1}^{\perp} \subset \mathcal{M}_{t-2}^{\perp} \subset \cdots,$$

which shows that for $s < t$, $E(Z_s Z_t) = 0$. By Problem 5.19 this establishes (ii) and (iii). Now by Theorem 2.4.2(ii) we can write

$$P_{\overline{\text{sp}}\{Z_j, j \le t\}} X_t = \sum_{j=0}^{\infty} \psi_j Z_{t-j}, \tag{5.7.5}$$

where ψ_j is defined by (5.7.3) and $\sum_{j=0}^{\infty} \psi_j^2 < \infty$. The coefficients ψ_j are independent of t by stationarity and

$$\psi_0 = \sigma^{-2} \langle X_t, X_t - P_{\mathcal{M}_{t-1}} X_t \rangle = \sigma^{-2} \| X_t - P_{\mathcal{M}_{t-1}} X_t \|^2 = 1.$$

Equations (5.7.4) and (5.7.5) and the definition of $P_{\overline{\text{sp}}\{Z_j, j \le t\}} X_t$ imply that

$$\langle V_t, Z_s \rangle = 0 \quad \text{for } s \le t.$$

On the other hand if $s > t$, $Z_s \in \mathcal{M}_{s-1}^{\perp} \subset \mathcal{M}_t^{\perp}$, and since $V_t \in \mathcal{M}_t$ we conclude that

$$\langle V_t, Z_s \rangle = 0 \quad \text{for } s > t,$$

establishing (iv.) To establish (v) and (vi) it will suffice (by Remark 1) to show that

$$\overline{\text{sp}}\{V_j, j \le t\} = \mathcal{M}_{-\infty} \quad \text{for every } t. \tag{5.7.6}$$

Since $V_t \in \mathcal{M}_t = \mathcal{M}_{t-1} \oplus \overline{\text{sp}}\{Z_t\}$ and since $\langle V_t, Z_t \rangle = 0$, we conclude that $V_t \in \mathcal{M}_{t-1} = \mathcal{M}_{t-2} \oplus \overline{\text{sp}}\{Z_{t-1}\}$. But since $\langle V_t, Z_{t-1} \rangle = 0$ it then follows that $V_t \in \mathcal{M}_{t-2}$. Continuing with this argument we see that $V_t \in \mathcal{M}_{t-j}$ for each $j \ge 0$, whence $V_t \in \bigcap_{j=0}^{\infty} \mathcal{M}_{t-j} = \mathcal{M}_{-\infty}$. Thus

$$\overline{\text{sp}}\{V_j, j \le t\} \subseteq \mathcal{M}_{-\infty} \quad \text{for every } t. \tag{5.7.7}$$

Now by (5.7.4), $\mathcal{M}_t = \overline{\text{sp}}\{Z_j, j \le t\} \oplus \overline{\text{sp}}\{V_j, j \le t\}$. If $Y \in \mathcal{M}_{-\infty}$ then $Y \in \mathcal{M}_{s-1}$ for every s, so that $\langle Y, Z_s \rangle = 0$ for every s, and consequently $Y \in \overline{\text{sp}}\{V_j, j \le t\}$.

But this means that

$$\mathcal{M}_{-\infty} \subseteq \overline{\text{sp}}\{V_j, j \leq t\} \quad \text{for every } t, \tag{5.7.8}$$

which completes the proof of (5.7.6) and hence of (v) and (vi).

To establish uniqueness we observe from (5.7.1) that if $\{Z_t\}$ and $\{V_t\}$ are any sequences satisfying (5.7.1) and having the properties (i)–(vi), then $\mathcal{M}_{t-1} \subseteq \overline{\text{sp}}\{Z_j, j \leq t-1\} \oplus \overline{\text{sp}}\{V_j, j \leq t-1\}$ from which it follows, using (ii) and (iv), that Z_t is orthogonal to $\mathcal{M}_{t-1}$. Projecting each side of (5.7.1) onto $\mathcal{M}_{t-1}$ and subtracting the resulting equation from (5.7.1), we then find that the process $\{Z_t\}$ must satisfy (5.7.2). By taking inner products of each side of (5.7.1) with Z_{t-j} we see that ψ_j must also satisfy (5.7.3). Finally, if (5.7.1) is to hold, it is obviously necessary that V_t must be defined as in (5.7.4). □

In the course of the preceding proof we have established a number of results which are worth collecting together as a corollary.

Corollary 5.7.1

(a) $\overline{\text{sp}}\{V_j, j \leq t\} = \mathcal{M}_{-\infty}$ *for every* t.
(b) $\mathcal{M}_t = \overline{\text{sp}}\{Z_j, j \leq t\} \oplus \mathcal{M}_{-\infty}$.
(c) $\mathcal{M}_{-\infty}^{\perp} = \overline{\text{sp}}\{Z_j, j \in \mathbb{Z}\}$.
(d) $\overline{\text{sp}}\{U_j, j \leq t\} = \overline{\text{sp}}\{Z_j, j \leq t\}$, *where* $U_t = \sum_{j=0}^{\infty} \psi_j Z_{t-j}$.

PROOF.

(a) This is a restatement of (5.7.6).
(b) Use part (a) together with the relation, $\mathcal{M}_t = \overline{\text{sp}}\{Z_j, j \leq t\} \oplus \overline{\text{sp}}\{V_j, j \leq t\}$.
(c) Observe that $\mathcal{M} = \overline{\text{sp}}\{X_t, t \in \mathbb{Z}\} = \overline{\text{sp}}\{Z_t, t \in \mathbb{Z}\} \oplus \mathcal{M}_{-\infty}$.
(d) This follows from the fact that $\mathcal{M}_t = \overline{\text{sp}}\{U_j, j \leq t\} \oplus \mathcal{M}_{-\infty}$. □

In view of part (b) of the corollary it is now possible to interpret the representation (5.7.1) as the decomposition of the subspace $\mathcal{M}_t$ into two orthogonal subspaces $\overline{\text{sp}}\{Z_j, j \leq t\}$ and $\mathcal{M}_{-\infty}$.

A stationary process is said to be *purely non-deterministic* if and only if $\mathcal{M}_{-\infty} = \{0\}$. In this case the Wold decomposition has no deterministic component, and the process can be represented as an MA(∞), $X_t = \sum_{j=0}^{\infty} \psi_j Z_{t-j}$. Many of the time series dealt with in this book (e.g. ARMA processes) are purely non-deterministic.

Observe that the h-step predictor for the process (5.7.1) is

$$P_{\mathcal{M}_t} X_{t+h} = \sum_{j=h}^{\infty} \psi_j Z_{t+h-j} + V_{t+h},$$

since $Z_j \perp \mathcal{M}_t$ for all $j < t$, and $V_{t+h} \in \mathcal{M}_t$. The corresponding mean squared error is

$$\|X_{t+h} - P_{\mathcal{M}_t} X_{t+h}\|^2 = \text{Var}\left(\sum_{j=0}^{h-1} \psi_j Z_{t+h-j}\right) = \sigma^2 \sum_{j=0}^{h-1} \psi_j^2,$$

which should be compared with the result (5.5.5). For a purely non-deterministic process it is clear that the h-step prediction mean squared error converges as $h \to \infty$ to the variance of the process. In general we have from part (d) of Corollary 5.7.1,

$$P_{\overline{\text{sp}}\{U_j, j \le t\}} U_{t+h} = P_{\overline{\text{sp}}\{Z_j, j \le t\}} U_{t+h} = \sum_{j=h}^{\infty} \psi_j Z_{t+h-j},$$

which shows that the h-step prediction error for the $\{U_t\}$ sequence coincides with that of the $\{X_t\}$ process. This is not unexpected since the purely deterministic component does not contribute to the prediction error.

EXAMPLE 5.7.1. Consider the stationary process $X_t = Z_t + Y$, where $\{Z_t\} \sim \text{WN}(0, \sigma^2)$, $\{Z_t\}$ is uncorrelated with the random variable Y and Y has mean zero and variance σ^2. Since

$$\frac{1}{n}\sum_{j=0}^{n-1} X_{t-j} = \frac{1}{n}\sum_{j=0}^{n-1} Z_{t-j} + Y \xrightarrow{\text{m.s.}} Y,$$

it follows that $Y \in \mathcal{M}_t$ for every t. Also $Z_t \perp \mathcal{M}_s$ for $s < t$ so $Z_t \perp \mathcal{M}_{-\infty}$. Hence $Y = P_{\mathcal{M}_{-\infty}} X_t$ is the deterministic component of the Wold decomposition and $Z_t = X_t - Y$ is the purely non-deterministic component.

For a stationary process $\{X_t\}$ satisfying the hypotheses of Theorem 5.7.1, the spectral distribution function F_X is the sum of two spectral distribution functions F_U and F_V corresponding to the two components $U_t = \sum_{j=0}^{\infty} \psi_j Z_{t-j}$ and V_t appearing in (5.7.1) (see Problem 4.7). From Chapter 4, F_U is absolutely continuous with respect to Lebesgue measure and has the spectral density

$$f_U(\lambda) = |\psi(e^{-i\lambda})|^2 \sigma^2/(2\pi), \qquad \text{where } \psi(e^{-i\lambda}) = \sum_{j=0}^{\infty} \psi_j e^{-ij\lambda}. \tag{5.7.9}$$

On the other hand, the spectral distribution F_V has no absolutely continuous component (see Doob (1953)). Consequently the Wold decomposition of a stationary process is analogous to the Lebesgue decomposition of the spectral measure into its absolutely continuous and singular parts. We state this as a theorem.

Theorem 5.7.2. *If $\sigma^2 > 0$, then*

$$F_X = F_U + F_V$$

where F_U and F_V are respectively the absolutely continuous and singular components in the Lebesgue decomposition of F_X. The density function associated with F_U is defined by (5.7.9).

The requirement $\sigma^2 > 0$ is critical in the above theorem. In other words it is possible for a deterministic process to have an absolutely continuous spectral distribution function. This is illustrated by Example 5.6.1. In the next section, a formula for σ^2 will be given in terms of the derivative of F_X which is valid even in the case $\sigma^2 = 0$. This immediately yields a necessary and sufficient criterion for a stationary process to be deterministic.

§5.8* Kolmogorov's Formula

Let $\{X_t\}$ be a real-valued zero-mean stationary process with spectral distribution function F_X and let f denote the derivative of F_X (defined everywhere on $[-\pi, \pi]$ except possibly on a set of Lebesgue measure zero). We shall assume, to simplify the proof of the following theorem, that f is continuous on $[-\pi, \pi]$ and is bounded away from zero. Since $\{X_t\}$ is real, we must have $f(\lambda) = f(-\lambda)$, $0 \le \lambda \le \pi$. For a general proof, see Hannan (1970) or Ash and Gardner (1975).

Theorem 5.8.1 (Kolmogorov's Formula). *The one-step mean square prediction error of the stationary process $\{X_t\}$ is*

$$\sigma^2 = 2\pi \exp\left\{\frac{1}{2\pi}\int_{-\pi}^{\pi} \ln f(\lambda)\, d\lambda\right\}. \tag{5.8.1}$$

Proof. Using a Taylor series expansion of $\ln(1 - z)$ for $|z| < 1$ and the identity $\int_{-\pi}^{\pi} e^{ik\lambda}\, d\lambda = 0$, $k \neq 0$, we have for $|a| < 1$,

$$\begin{aligned}\int_{-\pi}^{\pi} \ln|1 - ae^{-i\lambda}|^2\, d\lambda &= \int_{-\pi}^{\pi} \ln(1 - ae^{-i\lambda})(1 - \bar{a}e^{i\lambda})\, d\lambda \\ &= -\int_{-\pi}^{\pi}\left(\sum_{j=1}^{\infty}\frac{a^j e^{-ij\lambda}}{j} + \sum_{k=1}^{\infty}\frac{\bar{a}^k e^{ik\lambda}}{k}\right)d\lambda \\ &= 0. \end{aligned} \tag{5.8.2}$$

If $\{X_t\}$ is an AR(p) process satisfying $\phi(B)X_t = Z_t$, where $\{Z_t\} \sim \text{WN}(0, \sigma^2)$ and $\phi(z) = 1 - \phi_1 z - \cdots - \phi_p z^p \neq 0$ for $|z| \le 1$, then $\{X_t\}$ has spectral density,

$$g(\lambda) = \frac{\sigma^2}{2\pi}|1 - \phi_1 e^{-i\lambda} - \cdots - \phi_p e^{-ip\lambda}|^{-2} = \frac{\sigma^2}{2\pi}\prod_{j=1}^{p}|1 - a_j e^{-i\lambda}|^{-2},$$

where $|a_j| < 1$, $j = 1, \ldots, p$. Hence

$$\int_{-\pi}^{\pi} \ln g(\lambda)\, d\lambda = \int_{-\pi}^{\pi} \ln\frac{\sigma^2}{2\pi}\, d\lambda - \sum_{j=1}^{p}\int_{-\pi}^{\pi} \ln|1 - a_j e^{-i\lambda}|^2\, d\lambda = 2\pi \ln\frac{\sigma^2}{2\pi},$$

establishing Kolmogorov's formula for causal AR processes.

Under the assumptions made on f, it is clear that $\min_{-\pi\le\lambda\le\pi} f(\lambda) > 0$. Moreover, it is easily shown from Corollary 4.4.2 that for any $\varepsilon \in (0, \min f(\lambda))$, there exist causal AR processes with spectral densities $g_\varepsilon^{(1)}$ and $g_\varepsilon^{(2)}$ such that

$$f(\lambda) - \varepsilon \le g_\varepsilon^{(1)}(\lambda) \le f(\lambda) \le g_\varepsilon^{(2)}(\lambda) \le f(\lambda) + \varepsilon. \tag{5.8.3}$$

Now define

$$\begin{aligned}\sigma_n^2(f) &= E[(X_t - P_{\overline{\mathrm{sp}}\{X_{t-1},\ldots,X_{t-n}\}}X_t)^2]\\ &= \min_{c_1,\ldots,c_n} E(X_t - c_1X_{t-1} - \cdots - c_nX_{t-n})^2\\ &= \min_{c_1,\ldots,c_n} \int_{-\pi}^{\pi} |1 - c_1e^{-i\lambda} - \cdots - c_ne^{-in\lambda}|^2 f(\lambda)\,d\lambda.\end{aligned}$$

By (5.8.3) and the definition of $\sigma_n^2(\cdot)$,

$$\sigma_n^2(g_\varepsilon^{(1)}) \le \sigma_n^2(f) \le \sigma_n^2(g_\varepsilon^{(2)}).$$

Since, by Problem 2.18, $\sigma_n^2(f) \to \sigma^2(f) := E[(X_t - P_{\overline{\mathrm{sp}}\{X_s,\, -\infty<s<t\}}X_t)^2]$,

$$\sigma^2(g_\varepsilon^{(1)}) \le \sigma^2(f) \le \sigma^2(g_\varepsilon^{(2)}). \tag{5.8.4}$$

However we have already established that

$$\sigma^2(g_\varepsilon^{(i)}) = 2\pi \exp\left\{\frac{1}{2\pi}\int_{-\pi}^{\pi} \ln g_\varepsilon^{(i)}(\lambda)\,d\lambda\right\}, \qquad i = 1, 2.$$

If follows therefore from (5.8.4) that $\sigma^2(f)$ must equal the common limit, as $\varepsilon \to 0$, of $\sigma^2(g_\varepsilon^{(1)})$ and $\sigma^2(g_\varepsilon^{(2)})$, i.e.

$$\sigma^2(f) = 2\pi \exp\left\{\frac{1}{2\pi}\int_{-\pi}^{\pi} \ln f(\lambda)\,d\lambda\right\}. \qquad \square$$

Remark 1. Notice that $-\infty \le \int_{-\pi}^{\pi} \ln f(\lambda)\,d\lambda < \infty$ since $\ln f(\lambda) \le f(\lambda)$. If $\int_{-\pi}^{\pi} \ln f(\lambda)\,d\lambda = -\infty$, the theorem is still true with $\sigma^2 = 0$. Thus

$$\sigma^2 > 0 \quad \text{if and only if} \quad \int_{-\pi}^{\pi} \ln f(\lambda)\,d\lambda > -\infty,$$

and in this case $f(\lambda) > 0$ almost everywhere.

Remark 2. Equation (5.8.1) was first derived by Szegö in the absolutely continous case and was later extended by Kolmogorov to the general case. In the literature however it is usually referred to as Kolmogorov's formula.

EXAMPLE 5.8.1. For the process defined in Example 5.7.1, $F_U(d\lambda) = \sigma^2\,d\lambda/2\pi$ and $F_V(d\lambda) = \sigma^2\delta_0(d\lambda)$ where δ_0 is the unit mass at the origin. Not surprisingly, the one-step mean square prediction error is therefore

$$2\pi \exp\left\{\frac{1}{2\pi}\int_{-\pi}^{\pi} \ln\left(\frac{\sigma^2}{2\pi}\right)d\lambda\right\} = \sigma^2.$$

Problems

5.1. Let $\{X_t\}$ be a stationary process with mean μ. Show that

$$P_{\overline{\mathrm{sp}}\{1,X_1,\ldots,X_n\}}X_{n+h} = \mu + P_{\overline{\mathrm{sp}}\{Y_1,\ldots,Y_n\}}Y_{n+h},$$

where $\{Y_t\} = \{X_t - \mu\}$.

5.2. Suppose that $\{\mathscr{H}_n, n = 1, 2, \dots\}$ is a sequence of subspaces of a Hilbert space $\mathscr{H}$ with the property that $\mathscr{H}_n \subseteq \mathscr{H}_{n+1}$, $n = 1, 2, \dots$. Let $\mathscr{H}_\infty$ be the smallest closed subspace of $\mathscr{H}$ containing $\bigcup \mathscr{H}_n$, and let X be an element of $\mathscr{H}$. If $P_n X$ and $P_\infty X$ are the projections of X onto $\mathscr{H}_n$ and $\mathscr{H}_\infty$ respectively, show that
(a) $P_1 X, (P_2 - P_1)X, (P_3 - P_2)X, \dots$, are orthogonal,
(b) $\sum_{j=1}^{\infty} \|(P_{j+1} - P_j)X\|^2 < \infty$,
and
(c) $P_n X \to P_\infty X$.

5.3. Show that the converse of Proposition 5.1.1 is not true by constructing a stationary process $\{X_t\}$ such that Γ_n is non-singular for all n and $\gamma(h) \not\to 0$ as $h \to \infty$.

5.4. Suppose that $\{X_t\}$ is a stationary process with mean zero and spectral density

$$f_X(\lambda) = (\pi - |\lambda|)/\pi^2, \qquad -\pi \le \lambda \le \pi.$$

Find the coefficients $\{\theta_{ij}, j = 1, \dots, i; i = 1, \dots, 5\}$ and the mean squared errors $\{v_i, i = 0, \dots, 5\}$.

5.5. Let $\{X_t\}$ be the MA(1) process of Example 5.2.1. If $|\theta| < 1$, show that as $n \to \infty$,
(a) $\|X_n - \hat{X}_n - Z_n\| \to 0$,
(b) $v_n \to \sigma^2$,
and
(c) $\theta_{n1} \to \theta$. (Note that $\theta = E(X_{n+1} Z_n)\sigma^{-2}$ and $\theta_{n1} = v_{n-1}^{-1} E(X_{n+1}(X_n - \hat{X}_n))$.)

5.6. Let $\{X_t\}$ be the invertible MA(q) process

$$X_t = Z_t + \theta_1 Z_{t-1} + \cdots + \theta_q Z_{t-q}, \qquad \{Z_t\} \sim \text{WN}(0, \sigma^2).$$

Show that as $n \to \infty$,
(a) $\|X_n - \hat{X}_n - Z_n\| \to 0$,
(b) $v_n \to \sigma^2$,
and that
(c) there exist constants $K > 0$ and $c \in (0, 1)$ such that $|\theta_{nj} - \theta_j| \le K c^n$ for all n.

5.7. Verify equations (5.3.20) and (5.3.21).

5.8. The values .644, −.442, −.919, −1.573, .852, −.907, .686, −.753, −.954, .576, are simulated values of $X_1, \dots, X_{10}$ where $\{X_t\}$ is the ARMA(2, 1) process,

$$X_t - .1X_{t-1} - .12X_{t-2} = Z_t - .7Z_{t-1}, \qquad \{Z_t\} \sim \text{WN}(0, 1).$$

(a) Compute the forecasts $P_{10}X_{11}$, $P_{10}X_{12}$ and $P_{10}X_{13}$ and the corresponding mean squared errors.
(b) Assuming that $Z_t \sim N(0, 1)$, construct 95% prediction bounds for X_{11}, X_{12} and X_{13}.
(c) Using the method of Problem 5.15, compute $\tilde{X}_{11}^T$, $\tilde{X}_{12}^T$ and $\tilde{X}_{13}^T$ and compare these values with those obtained in (a).
[The simulated values of X_{11}, X_{12} and X_{13} were in fact .074, 1.097 and −.187 respectively.]

5.9. Repeat parts (a)–(c) of Problem 5.8 for the simulated values −1.222, 1.707, .049, 1.903, −3.341, 3.041, −1.012, −.779, 1.837, −3.693 of $X_1, \dots, X_{10}$, where $\{X_t\}$ is the MA(2) process

$$X_t = Z_t - 1.1Z_{t-1} + .28Z_{t-2}, \qquad \{Z_t\} \sim \text{WN}(0, 1).$$

[The simulated values of X_{11}, X_{12} and X_{13} in this case were 3.995, −3.859 3.746.]

5.10. If $\{X_1,\dots,X_n\}$ are observations of the AR(p) process,

$$X_t - \phi_1 X_{t-1} - \cdots - \phi_p X_{t-p} = Z_t, \qquad \{Z_t\} \sim \text{WN}(0,\sigma^2),$$

show that the mean squared error of the predictor $P_n X_{n+h}$ is

$$\sigma_n^2(h) = \sigma^2 \sum_{j=0}^{h-1} \psi_j^2 \qquad \text{for } n \geq p,\ h \geq 1,$$

where $\psi(z) = \sum_{j=0}^{\infty} \psi_j z^j = 1/\phi(z)$. This means that the asymptotic approximation (5.3.24) is exact for an autoregressive process when $n \geq p$.

5.11. Use the model defined in Problem 4.12 to find the best linear predictors of the Wölfer sunspot numbers $X_{101},\dots,X_{105}$ (being careful to take into account the non-zero mean of the series). Assuming that the series is Gaussian, find 95% prediction bounds for each value. (The observed values of $X_{101},\dots,X_{105}$ are in fact 139, 111, 102, 66, 45.) How do the predicted values $P_{100}X_{100+h}$ and their mean squared errors behave for large h?

5.12. Let $\{X_t\}$ be the ARMA(2, 1) process,

$$X_t - .5X_{t-1} + .25X_{t-2} = Z_t + 2Z_{t-1}, \qquad \{Z_t\} \sim \text{WN}(0,1),$$

and let

$$Y_t = \begin{cases} X_t, & t \leq 2, \\ X_t - .5X_{t-1} + .25X_{t-2}, & t > 2. \end{cases}$$

(a) Find the covariance matrix of $(Y_1, Y_2, Y_3)'$ and hence find the coefficients θ_{11} and θ_{21} in the representations

$$\hat{X}_2 = \theta_{11}(X_1 - \hat{X}_1),$$
$$\hat{X}_3 = .5X_2 - .25X_1 + \theta_{21}(X_2 - \hat{X}_2).$$

(b) Use the mean squared errors of the predictors $\hat{X}_1$, $\hat{X}_2$ and $\hat{X}_3$ to evaluate the determinant of the covariance matrix of $(X_1, X_2, X_3)'$.

(c) Find the limits as $n \to \infty$ of the coefficients θ_{n1} and of the one-step mean-square prediction errors v_n.

(d) Given that $X_{199} = 6.2$, $X_{200} = -2.2$ and $\hat{X}_{200} = .5$, use the limiting values found in (c) to compute the best predictor $\hat{X}_{201}$ and its mean squared error.

(e) What is the value of $\lim_{h\to\infty} E(X_{n+h} - P_n X_{n+h})^2$?

5.13. The coefficients θ_{nj} and one-step mean squared errors $v_n = r_n\sigma^2$ can be determined for the general causal ARMA(1, 1) process (5.3.11) by solving the equations (5.3.13) as follows:

(a) Show that if $y_n := r_n/(r_n - 1)$, then the last of the equations (5.3.13), can be rewritten in the form,

$$y_n = \theta^{-2} y_{n-1} + 1, \qquad n \geq 1.$$

(b) Deduce that $y_n = \theta^{-2n} y_0 + \sum_{j=1}^{n} \theta^{-2(j-1)}$ and hence determine r_n and θ_{n1}, $n = 1, 2, \dots$.

(c) Evaluate the limits as $n \to \infty$ of r_n and θ_{n1} in the two cases $|\theta| < 1$ and $|\theta| \geq 1$.

5.14. Let $\{X_t\}$ be the MA(1) process

$$X_t = Z_t + \theta Z_{t-1}, \qquad \{Z_t\} \sim \mathrm{WN}(0, \sigma^2)$$

with $|\theta| < 1$.

(a) Show that $v_n := E|X_{n+1} - \hat{X}_{n+1}|^2 = \sigma^2(1 - \theta^{2n+4})/(1 - \theta^{2n+2})$.

(b) If $\tilde{X}^T_{n+1} = -\sum_{j=1}^{n}(-\theta)^j X_{n+1-j}$ is the truncation approximation to $P_n X_{n+1}$, show that $E|X_{n+1} - \tilde{X}^T_{n+1}|^2 = (1 + \theta^{2n+2})\sigma^2$ and compare this value with v_n for $|\theta|$ near one.

5.15. Let $\{X_t\}$ be a causal invertible ARMA(p, q) process

$$\phi(B)X_t = \theta(B)Z_t, \qquad \{Z_t\} \sim \mathrm{WN}(0, \sigma^2).$$

Given the sample $\{X_1, \ldots, X_n\}$, we define

$$Z_t^* = \begin{cases} 0 & \text{if } t \le 0 \text{ or } t > n, \\ \phi(B)X_t - \theta_1 Z_{t-1}^* - \cdots - \theta_q Z_{t-q}^* & \text{if } t = 1, \ldots, n, \end{cases}$$

where we set $X_t = 0$ for $t \le 0$.

(a) Show that $\phi(B)X_t = \theta(B)Z_t^*$ for all $t \le n$ (with the understanding that $X_t = 0$ for $t \le 0$) and hence that $Z_t^* = \pi(B)X_t$ where $\pi(z) = \phi(z)/\theta(z)$.

(b) If $\tilde{X}^T_{n+1} = -\sum_{j=1}^{n} \pi_j X_{n+1-j}$ is the truncation approximation to $P_n X_{n+1}$ (see Remark 1 in Section 5.5), show that

$$\tilde{X}^T_{n+1} = \phi_1 X_n + \cdots + \phi_p X_{n+1-p} + \theta_1 Z_n^* + \cdots + \theta_q Z_{n+1-q}^*.$$

(c) Generalize (b) to show that for all $h \ge 1$

$$\tilde{X}^T_{n+h} = \phi_1 \tilde{X}^T_{n+h-1} + \cdots + \phi_p \tilde{X}^T_{n+h-p} + \sum_{j=h}^{q} \theta_j Z_{n+h-j}^*$$

where $\tilde{X}^T_j = X_j$ if $j = 1, \ldots, n$.

5.16.* Consider the process $X_t = A\cos(Bt + U)$, $t = 0, \pm 1, \ldots$, where A, B and U are random variables such that (A, B) and U are independent, and U is uniformly distributed on $(0, 2\pi)$.

(a) Show that $\{X_t\}$ is stationary and determine its mean and covariance function.

(b) Show that the joint distribution of A and B can be chosen in such a way that $\{X_t\}$ has the autocovariance function of the MA(1) process, $Y_t = Z_t + \theta Z_{t-1}$, $\{Z_t\} \sim \mathrm{WN}(0, \sigma^2)$, $|\theta| \le 1$.

(c) Suppose that A and B have the joint distribution found in (b) and let X_{t+h}^* and $\hat{X}_{t+h}$ be the best and best linear predictors respectively of X_{t+h} in terms of $\{X_j, -\infty < j \le t\}$. Find the mean squared errors of X_{t+h}^* and $\hat{X}_{t+h}$, $h \ge 2$.

5.17.* Check that equation (5.6.15) is equivalent to (5.5.4).

5.18.* If σ^2 is the one-step mean-square prediction error for a stationary process $\{X_t\}$ show that $\sigma^2 = 0$ if and only if $X_t \in \mathcal{M}_{-\infty}$ for every t.

5.19.* Suppose that $\{X_t\}$ is a stationary process with mean zero. Define $\mathcal{M}_t = \overline{\mathrm{sp}}\{X_s, s \le t\}$ and $Z_t = X_t - P_{\mathcal{M}_{t-1}} X_t$.

(a) Show that $\sigma^2 = E|X_{t+1} - P_{\mathcal{M}_t} X_{t+1}|^2$ does not depend on t.

(b) Show that $\psi_j = E(X_t Z_{t-j})/\sigma^2$ does not depend on t.

5.20.* Let $\{Y_t\}$ be the MA(1) process,

$$Y_t = Z_t + 2.5Z_{t-1}, \qquad \{Z_t\} \sim \mathrm{WN}(0, \sigma^2),$$

and define $X_t = A\cos(\omega t) + B\sin(\omega t) + Y_t$ where A and B are uncorrelated (0, σ_1^2) random variables which are uncorrelated with $\{Y_t\}$.

(a) Show that $\{X_t\}$ is non-deterministic.

(b) Determine the Wold decomposition of $\{X_t\}$.

(c) What are the components of the spectral distribution function of $\{X_t\}$ corresponding to the deterministic and purely non-deterministic components of the Wold decomposition?

5.21.* Let $\{X_n, n = 0, \pm 1, \ldots\}$ be a stationary Markov chain with states ± 1 and transition probabilities $P(X_{n+1} = j | X_n = i) = p$ if $i = j$, $(1 - p)$ if $i \neq j$. Find the white noise sequence $\{Z_n\}$ and coefficients a_j such that

$$Z_n \in \overline{\mathrm{sp}}\{X_t, -\infty < t \leq n\}$$

and

$$X_n = \sum_{j=0}^{\infty} a_j Z_{n-j}, \qquad n = 0, \pm 1, \ldots.$$

5.22. Suppose that

$$X_t = A\cos(\pi t/3) + B\sin(\pi t/3) + Z_t + .5Z_{t-1}, \qquad t = 0, \pm 1, \ldots,$$

where $\{Z_t\} \sim \mathrm{WN}(0, 1)$, A and B are uncorrelated random variables with mean zero, variance 4 and $E(AZ_t) = E(BZ_t) = 0$, $t = 0, \pm 1, \ldots$. Find the best linear predictor of X_{t+1} based on X_t and X_{t-1}. What is the mean squared error of the best linear predictor of X_{t+1} based on $\{X_j, -\infty < j \leq t\}$?

5.23.* Let $\{X_t\}$ be the moving average

$$X_t = \sum_{j=-\infty}^{\infty} \psi_j Z_{t-j}, \qquad \{Z_t\} \sim \mathrm{WN}(0, \sigma^2)$$

where $\psi_k = \left(\frac{1}{\pi}\right)\left(\frac{\sin k}{k}\right)$.

(a) Find the spectral density of $\{X_t\}$.

(b) Is the process purely non-deterministic, non-deterministic, or deterministic?

5.24.* If the zero-mean stationary process $\{X_n\}$ has autocovariance function,

$$\gamma(h) = \begin{cases} 1 & \text{if } h = 0, \\ \rho & \text{if } h \neq 0, \end{cases}$$

where $0 < \rho < 1$,

(a) show that the mean square limit as $n \to \infty$ of $n^{-1} \sum_{j=1}^{n} X_{-j}$ exists,

(b) show that X_n can be represented as

$$X_n = Z + Y_n,$$

where $\{Z, Y_j, j = 0, \pm 1, \ldots\}$ are zero-mean uncorrelated random variables with $EZ^2 = \rho$ and $EY_j^2 = 1 - \rho$, $j = 0, \pm 1, \ldots$,

(c) find the spectral distribution function of $\{X_n\}$,

(d) determine the components in the Wold decomposition of X_n, and

(e) find the mean squared error of the one-step predictor $P_{\overline{\mathrm{sp}}\{X_j,\,-\infty<j\leq n\}}X_{n+1}$.

5.25. Suppose that $\{U_t\}$ and $\{V_t\}$ are two stationary processes having the same autocovariance functions. Without appealing to Kolmogorov's formula, show that the two processes have the same one-step mean-square prediction errors.

5.26.* Under the assumptions made in our proof of Kolmogorov's formula (Theorem 5.8.1), show that the mean squared error of the two-step linear predictor $\tilde{X}_{t+2} := P_{\mathrm{sp}\{X_j,\,-\infty<j\leq t\}}X_{t+2}$ is

$$\left[1 + (2\pi)^{-2}\left(\int_{-\pi}^{\pi} e^{-i\lambda} \ln f(\lambda)\, d\lambda\right)^2\right]\sigma^2$$

with σ^2 as in (5.8.1).

5.27. Let $\{X_t\}$ be the causal AR(1) process,

$$X_t - \phi X_{t-1} = Z_t, \qquad \{Z_t\} \sim \mathrm{WN}(0, \sigma^2),$$

and let $\hat{X}_{n+1}$ be the best linear predictor of X_{n+1} based on $X_1, \ldots, X_n$. Defining $\theta_{n0} = 1$ and $\hat{X}_1 = 0$, find $\theta_{n1}, \ldots, \theta_{nn}$ such that

$$X_{n+1} = \sum_{j=0}^{n} \theta_{nj}(X_{n+1-j} - \hat{X}_{n+1-j}).$$

5.28.* Suppose that $X_t = \sum_{j=0}^{\infty} \psi_j Z_{t-j}$, $\{Z_t\} \sim \mathrm{WN}(0, 1)$ and $\sum_{j=0}^{\infty} \psi_j^2 < \infty$. Show that the h-step mean-square prediction error,

$$\sigma^2(h) := E(X_{t+h} - P_{\overline{\mathrm{sp}}\{X_s,\,-\infty<s\leq t\}}X_{t+h})^2,$$

satisfies

$$\sigma^2(h) \geq \psi_0^2 + \cdots + \psi_{h-1}^2.$$

Conclude that $\{X_t\}$ is purely non-deterministic.

CHAPTER 6*

Asymptotic Theory

In order to carry out statistical inference for time series it is necessary to be able to derive the distributions of various statistics used for the estimation of parameters from the data. For finite n the exact distribution of such a statistic $f_n(X_1, \ldots, X_n)$ is usually (even for Gaussian processes) prohibitively complicated. In such cases, we can still however base the inference on large-sample approximations to the distribution of the statistic in question. The mathematical tools for deriving such approximations are developed in this chapter. A comprehensive treatment of asymptotic theory is given in the book of Serfling (1980). Chapter 5 of the book by Billingsley (1986) is also strongly recommended.

§6.1 Convergence in Probability

We first define convergence in probability and the related order concepts which, as we shall see, are closely analogous to their deterministic counterparts. With these tools we can then develop convergence in probability analogues of Taylor expansions which will be used later to derive the large-sample asymptotic distributions of estimators of our time series parameters.

Let $\{a_n, n = 1, 2, \ldots\}$ be a sequence of strictly positive real numbers and let $\{X_n, n = 1, 2, \ldots\}$ be a sequence of random variables all defined on the same probability space.

Definition 6.1.1 (Convergence in Probability to Zero). We say that X_n converges in probability to zero, written $X_n = o_p(1)$ or $X_n \xrightarrow{P} 0$, if for every $\varepsilon > 0$,

$$P(|X_n| > \varepsilon) \to 0 \quad \text{as } n \to \infty.$$

Definition 6.1.2 (Boundedness in Probability). We say that the sequence $\{X_n\}$ is bounded in probability (or tight), written $X_n = O_p(1)$, if for every $\varepsilon > 0$ there exists $\delta(\varepsilon) \in (0, \infty)$ such that

$$P(|X_n| > \delta(\varepsilon)) < \varepsilon \quad \text{for all } n.$$

The relation between these two concepts is clarified by the following equivalent characterization of convergence in probability to zero, viz. $X_n = o_p(1)$ if and only if for every $\varepsilon > 0$ there exists a sequence $\delta_n(\varepsilon) \downarrow 0$ such that

$$P(|X_n| > \delta_n(\varepsilon)) < \varepsilon \quad \text{for all } n,$$

(see Problem 6.3). The definitions should also be compared with their non-random counterparts, viz. $x_n = o(1)$ if $x_n \to 0$ and $x_n = O(1)$ if $\{x_n\}$ is bounded.

Definition 6.1.3 (Convergence in Probability and Order in Probability).

(i) X_n converges in probability to the random variable X, written $X_n \xrightarrow{P} X$, if and only if $X_n - X = o_p(1)$.
(ii) $X_n = o_p(a_n)$ if and only if $a_n^{-1} X_n = o_p(1)$.
(iii) $X_n = O_p(a_n)$ if and only if $a_n^{-1} X_n = O_p(1)$.

Notice that if we drop the subscripts p in Definitions 6.1.3 (ii) and (iii) we recover the usual definitions of $o(\cdot)$ and $O(\cdot)$ for non-random sequences. In fact most of the rules governing the manipulation of $o(\cdot)$ and $O(\cdot)$ carry over to $o_p(\cdot)$ and $O_p(\cdot)$. In particular we have the following results.

Proposition 6.1.1. *If X_n and Y_n, $n = 1, 2, \ldots$, are random variables defined on the same probability space and $a_n > 0$, $b_n > 0$, $n = 1, 2, \ldots$, then*

(i) *if $X_n = o_p(a_n)$ and $Y_n = o_p(b_n)$, we have*
$X_n Y_n = o_p(a_n b_n)$,
$X_n + Y_n = o_p(\max(a_n, b_n))$,

and

$|X_n|^r = o_p(a_n^r)$, *for* $r > 0$;

(ii) *if $X_n = o_p(a_n)$ and $Y_n = O_p(b_n)$, we have*
$X_n Y_n = o_p(a_n b_n)$.

Moreover

(iii) *the statement* (i) *remains valid if o_p is everywhere replaced by O_p.*

PROOF. (i) If $|X_n Y_n|/(a_n b_n) > \varepsilon$ then either $|Y_n|/b_n \le 1$ and $|X_n|/a_n > \varepsilon$ or $|Y_n|/b_n > 1$ and $|X_n Y_n|/(a_n b_n) > \varepsilon$. Hence

$$P(|X_n Y_n|/(a_n b_n) > \varepsilon) \le P(|X_n|/a_n > \varepsilon) + P(|Y_n|/b_n > 1)$$
$$\to 0 \quad \text{as } n \to \infty.$$

If $|X_n + Y_n|/\max(a_n, b_n) > \varepsilon$ then either $|X_n|/a_n > \varepsilon/2$ or $|Y_n|/b_n > \varepsilon/2$. Hence

$$P(|X_n + Y_n|/\max(a_n, b_n) > \varepsilon) \le P(|X_n|/a_n > \varepsilon/2) + P(|Y_n|/b_n > \varepsilon/2)$$
$$\to 0 \quad \text{as } n \to \infty.$$

For the last part of (i) we simply observe that

$$P(|X_n|^r/a_n^r > \varepsilon) = P(|X_n|/a_n > \varepsilon^{1/r}) \to 0 \quad \text{as } n \to \infty.$$

Parts (ii) and (iii) are left as exercises for the reader. □

The Definitions 6.1.1–6.1.3 extend in a natural way to sequences of random vectors. Suppose now that $\{\mathbf{X}_n, n = 1, 2, \ldots\}$ is a sequence of random vectors, all defined on the same probability space and such that $\mathbf{X}_n$ has k components $X_{n1}, X_{n2}, \ldots, X_{nk}$, $n = 1, 2, \ldots$.

Definition 6.1.4 (Order in Probability for Random Vectors).

(i) $\mathbf{X}_n = o_p(a_n)$ if and only if $X_{nj} = o_p(a_n)$, $j = 1, \ldots, k$.
(ii) $\mathbf{X}_n = O_p(a_n)$ if and only if $X_{nj} = O_p(a_n)$, $j = 1, \ldots, k$.
(iii) $\mathbf{X}_n$ converges in probability to the random vector $\mathbf{X}$, written $\mathbf{X}_n \xrightarrow{P} \mathbf{X}$, if and only if $\mathbf{X}_n - \mathbf{X} = o_p(1)$.

Convergence in probability of $\mathbf{X}_n$ to $\mathbf{X}$ can also be conveniently characterized in terms of the Euclidean distance $|\mathbf{X}_n - \mathbf{X}| = [\sum_{j=1}^k (X_{nj} - X_j)^2]^{1/2}$.

Proposition 6.1.2. $\mathbf{X}_n - \mathbf{X} = o_p(1)$ *if and only if* $|\mathbf{X}_n - \mathbf{X}| = o_p(1)$.

PROOF. If $\mathbf{X}_n - \mathbf{X} = o_p(1)$ then for each $\varepsilon > 0$, $\lim_{n\to\infty} P(|X_{nj} - X_j|^2 > \varepsilon/k) = 0$ for each $j = 1, \ldots, k$. But

$$P\left(\sum_{j=1}^k |X_{nj} - X_j|^2 > \varepsilon\right) \le \sum_{j=1}^k P(|X_{nj} - X_j|^2 > \varepsilon/k) \tag{6.1.1}$$

since $\sum_{j=1}^k |X_{nj} - X_j|^2 > \varepsilon$ implies that at least one summand exceeds ε/k. Since the right side of (6.1.1) converges to zero so too does the left side and hence $|\mathbf{X}_n - \mathbf{X}|^2 = o_p(1)$. By Proposition 6.1.1 this implies that $|\mathbf{X}_n - \mathbf{X}| = o_p(1)$.

Conversely if $|\mathbf{X}_n - \mathbf{X}| = o_p(1)$ we have $|X_{ni} - X_i|^2 \le |\mathbf{X}_n - \mathbf{X}|^2$ whence $P(|X_{ni} - X_i| > \varepsilon) \le P(|\mathbf{X}_n - \mathbf{X}|^2 > \varepsilon^2) \to 0$. □

Proposition 6.1.3. *If* $\mathbf{X}_n - \mathbf{Y}_n \xrightarrow{P} \mathbf{0}$ *and* $\mathbf{Y}_n \xrightarrow{P} \mathbf{Y}$ *then* $\mathbf{X}_n \xrightarrow{P} \mathbf{Y}$.

PROOF. $|\mathbf{X}_n - \mathbf{Y}| \le |\mathbf{X}_n - \mathbf{Y}_n| + |\mathbf{Y}_n - \mathbf{Y}| = o_p(1)$, by Propositions 6.1.1 and 6.1.2. □

Proposition 6.1.4. *If* $\{\mathbf{X}_n\}$ *is a sequence of k-dimensional random vectors such that* $\mathbf{X}_n \xrightarrow{P} \mathbf{X}$ *and if* $g: \mathbb{R}^k \to \mathbb{R}^m$ *is a continuous mapping, then* $g(\mathbf{X}_n) \xrightarrow{P} g(\mathbf{X})$.

PROOF. Let K be a positive real number. Then given any $\varepsilon > 0$ we have

$$P(|g(\mathbf{X}_n) - g(\mathbf{X})| > \varepsilon) \le P(|g(\mathbf{X}_n) - g(\mathbf{X})| > \varepsilon, |\mathbf{X}| \le K, |\mathbf{X}_n| \le K) + P(\{|\mathbf{X}| > K\} \cup \{|\mathbf{X}_n| > K\}).$$

Since g is uniformly continuous on $\{\mathbf{x} : |\mathbf{x}| \le K\}$, there exists $\gamma(\varepsilon) > 0$ such that for all n,

$$\{|g(\mathbf{X}_n) - g(\mathbf{X})| > \varepsilon, |\mathbf{X}| \le K, |\mathbf{X}_n| \le K\} \subseteq \{|\mathbf{X}_n - \mathbf{X}| > \gamma(\varepsilon)\}.$$

Hence

$$\begin{aligned} P(|g(\mathbf{X}_n) - g(\mathbf{X})| > \varepsilon) &\le P(|\mathbf{X}_n - \mathbf{X}| > \gamma(\varepsilon)) + P(|\mathbf{X}| > K) + P(|\mathbf{X}_n| > K) \\ &\le P(|\mathbf{X}_n - \mathbf{X}| > \gamma(\varepsilon)) + P(|\mathbf{X}| > K) \\ &\quad + P(|\mathbf{X}| > K/2) + P(|\mathbf{X}_n - \mathbf{X}| > K/2). \end{aligned}$$

Now given any $\delta > 0$ we can choose K to make the second and third terms each less than $\delta/4$. Then since $|\mathbf{X}_n - \mathbf{X}| \xrightarrow{P} 0$, the first and fourth terms will each be less than $\delta/4$ for all n sufficiently large. Consequently $g(\mathbf{X}_n) \xrightarrow{P} g(\mathbf{X})$. □

Taylor Expansions in Probability

If g is continuous at $\mathbf{a}$ and $\mathbf{X}_n = \mathbf{a} + o_p(1)$ then the argument of Proposition 6.1.4 tells us that $g(\mathbf{X}_n) = g(\mathbf{a}) + o_p(1)$. If we strengthen the assumptions on g to include the existence of derivatives, then it is possible to derive probabilistic analogues of the Taylor expansions of non-random functions about a given point $\mathbf{a}$. Some of these analogues which will be useful in deriving asymptotic distributions are given below.

Proposition 6.1.5. *Let $\{X_n\}$ be a sequence of random variables such that $X_n = a + O_p(r_n)$ where $a \in \mathbb{R}$ and $0 < r_n \to 0$ as $n \to \infty$. If g is a function with s derivatives at a then*

$$g(X_n) = \sum_{j=0}^{s} \frac{g^{(j)}(a)}{j!}(X_n - a)^j + o_p(r_n^s),$$

where $g^{(j)}$ is the jth derivative of g and $g^{(0)} = g$.

Proof. Let

$$h(x) = \left[g(x) - \sum_{j=0}^{s} \frac{g^{(j)}(a)}{j!}(x - a)^j\right] \Bigg/ \left[\frac{(x - a)^s}{s!}\right], \qquad x \ne a,$$

and $h(a) = 0$. Then the function h is continuous at a so that $h(X_n) = h(a) + o_p(1)$. This implies that $h(X_n) = o_p(1)$ and so by Proposition 6.1.1 (ii),

$$(X_n - a)^s h(X_n) = o_p(r_n^s),$$

which proves the result. □

Example 6.1.1. Suppose that $\{X_t\} \sim \mathrm{IID}(\mu, \sigma^2)$ with $\mu > 0$. If $\bar{X}_n = n^{-1} \sum_{t=1}^n X_t$, then by Chebychev's inequality (see Proposition 6.2.1 below), $P(n^{1/2}|\bar{X}_n - \mu| > \varepsilon) \le \sigma^2 \varepsilon^{-2}$, and hence

$$\bar{X}_n - \mu = O_p(n^{-1/2}).$$

Since $\ln x$ has a derivative at μ, the conditions of Proposition 6.1.5 are satisfied and we therefore obtain the expansion,

$$\ln \bar{X}_n = \ln \mu + \mu^{-1}(\bar{X}_n - \mu) + o_p(n^{-1/2}).$$

We conclude this section with a multivariate analogue of Proposition 6.1.5.

Proposition 6.1.6. *Let $\{\mathbf{X}_n\}$ be a sequence of random $k \times 1$ vectors such that*

$$\mathbf{X}_n - \mathbf{a} = O_p(r_n),$$

where $\mathbf{a} \in \mathbb{R}^k$ and $r_n \to 0$ as $n \to \infty$. If g is a function from $\mathbb{R}^k$ into $\mathbb{R}$ such that the derivatives $\partial g/\partial x_i$ are continuous in a neighborhood $N(\mathbf{a})$ of $\mathbf{a}$, then

$$g(\mathbf{X}_n) = g(\mathbf{a}) + \sum_{i=1}^k \frac{\partial g}{\partial x_i}(\mathbf{a})(X_{ni} - a_i) + o_p(r_n).$$

Proof. From the usual Taylor expansion for a function of several variables (see for example Seeley (1970), p. 160) we have, as $\mathbf{x} \to \mathbf{a}$,

$$g(\mathbf{x}) = g(\mathbf{a}) + \sum_{i=1}^k \frac{\partial g}{\partial x_i}(\mathbf{a})(x_i - a_i) + o(|\mathbf{x} - \mathbf{a}|).$$

Defining

$$h(\mathbf{x}) = \left[g(\mathbf{x}) - g(\mathbf{a}) - \sum_{i=1}^k \frac{\partial g}{\partial x_i}(\mathbf{a})(x_i - a_i)\right] \Big/ |\mathbf{x} - \mathbf{a}|, \qquad \mathbf{x} \ne \mathbf{a},$$

and $h(\mathbf{a}) = 0$, we deduce that h is continuous at $\mathbf{a}$ and hence that $h(\mathbf{X}_n) = o_p(1)$ as $n \to \infty$. By Proposition 6.1.1 this implies that $h(\mathbf{X}_n)|\mathbf{X}_n - \mathbf{a}| = o_p(r_n)$, which proves the result. □

§6.2 Convergence in r^{th} Mean, $r > 0$

Mean square convergence was introduced in Section 2.7 where we discussed the space L^2 of square integrable random variables on a probability space $(\Omega, \mathcal{F}, P)$. In this section we consider a generalization of this concept, convergence in r^{th} mean, and discuss some of its properties. It reduces to mean-square convergence when $r = 2$.

Definition 6.2.1 (Convergence in r^{th} Mean, $r > 0$). The sequence of random variables $\{X_n\}$ is said to converge in r^{th} mean to X, written $X_n \xrightarrow{r} X$, if $E|X_n - X|^r \to 0$ as $n \to \infty$.

Proposition 6.2.1 (Chebychev's Inequality). *If* $E|X|^r < \infty$, $r \geq 0$ *and* $\varepsilon > 0$, *then*

$$P(|X| \geq \varepsilon) \leq \varepsilon^{-r} E|X|^r.$$

PROOF.

$$\begin{aligned} P(|X| \geq \varepsilon) &= P(|X|^r \varepsilon^{-r} \geq 1) \\ &= EI_{[1,\infty)}(|X|^r \varepsilon^{-r}), \quad \text{where } I_A(x) = \begin{cases} 1 & \text{if } x \in A, \\ 0 & \text{if } x \notin A, \end{cases} \\ &\leq E[|X|^r \varepsilon^{-r} I_{[1,\infty)}(|X|^r \varepsilon^{-r})] \\ &\leq \varepsilon^{-r} E|X|^r. \end{aligned}$$

□

The following three propositions provide useful connections between the behaviour of moments and order in probability.

Proposition 6.2.2. *If* $X_n \xrightarrow{r} X$ *then* $X_n \xrightarrow{P} X$.

PROOF. By Chebychev's inequality we have for any $\varepsilon > 0$,

$$P(|X_n - X| > \varepsilon) \leq \varepsilon^{-r} E|X_n - X|^r \to 0 \quad \text{as } n \to \infty.$$

□

Proposition 6.2.3. *If* $a_n > 0$, $n = 1, 2, \ldots$, *and* $E(X_n^2) = O(a_n^2)$, *then* $X_n = O_p(a_n)$.

PROOF. Applying Chebychev's inequality again, we have for any $M > 0$,

$$\begin{aligned} P(a_n^{-1}|X_n| > M) &\leq a_n^{-2} E|X_n|^2/M^2 \\ &\leq C/M^2 \quad \text{where } C = \sup_n (a_n^{-2} E|X_n|^2) < \infty. \end{aligned}$$

Defining $\delta(\varepsilon) = 2(C/\varepsilon)^{1/2}$ if $C > 0$ and any positive constant if $C = 0$, we see from Definition 6.1.2 that $a_n^{-1}|X_n| = O_p(1)$. □

Proposition 6.2.4. *If* $EX_n \to \mu$ *and* $\operatorname{Var}(X_n) \to 0$ *then* $X_n \xrightarrow{\text{m.s.}} \mu$ *(and* $X_n \xrightarrow{P} \mu$ *by Proposition* 6.2.2).

PROOF.

$$\begin{aligned} E|X_n - \mu|^2 &= E|X_n - EX_n|^2 + |EX_n - \mu|^2 \\ &\to 0 \quad \text{as } n \to \infty. \end{aligned}$$

□

§6.3 Convergence in Distribution

The statements $X_n \xrightarrow{\text{m.s.}} X$ and $X_n \xrightarrow{P} X$ are meaningful only when the random variables $X, X_1, X_2, \ldots,$ are all defined on the same probability space. The notion of convergence in distribution however depends only on the distribution functions of $X, X_1, X_2, \ldots,$ and is meaningful even if $X, X_1, X_2, \ldots,$ are all defined on different probability spaces. We shall show in Proposition 6.3.2 that convergence in distribution of a sequence $\{X_n\}$ is implied by convergence in probability. We begin with a definition.

Definition 6.3.1 (Convergence in Distribution). The sequence $\{\mathbf{X}_n\}$ of random k-vectors with distribution functions $\{F_{\mathbf{X}_n}(\cdot)\}$ is said to converge in distribution if there exists a random k-vector $\mathbf{X}$ such that

$$\lim_{n\to\infty} F_{\mathbf{X}_n}(\mathbf{x}) = F_{\mathbf{X}}(\mathbf{x}) \quad \text{for all } \mathbf{x} \in C, \tag{6.3.1}$$

where C is the set of continuity points of the distribution function $F_{\mathbf{X}}(\cdot)$ of $\mathbf{X}$. If (6.3.1) holds we shall say that $\mathbf{X}_n$ converges in distribution to $\mathbf{X}$. Such convergence will be denoted by $\mathbf{X}_n \Rightarrow \mathbf{X}$ or $F_{\mathbf{X}_n} \Rightarrow F_{\mathbf{X}}$.

If $\mathbf{X}_n \Rightarrow \mathbf{X}$ then the distribution of $\mathbf{X}_n$ can be well approximated for large n by the distribution of $\mathbf{X}$. This observation is extremely useful since $F_{\mathbf{X}}$ is often easier to compute than $F_{\mathbf{X}_n}$.

A proof of the equivalence of the following characterizations of convergence in distribution can be found in Billingsley (1986), Chapter 5.

Theorem 6.3.1 (Characterizations of Convergence in Distribution). *If* $F_0, F_1, F_2, \ldots$ *are distribution functions on* $\mathbb{R}^k$ *with corresponding characteristic functions* $\phi_n(\mathbf{t}) = \int_{\mathbb{R}^k} \exp(i\mathbf{t}'\mathbf{x})\, dF_n(\mathbf{x})$, $n = 0, 1, 2, \ldots,$ *then the following statements are equivalent:*

(i) $F_n \Rightarrow F_0$,
(ii) $\int_{\mathbb{R}^k} g(\mathbf{x})\, dF_n(\mathbf{x}) \to \int_{\mathbb{R}^k} g(\mathbf{x})\, dF_0(\mathbf{x})$ *for every bounded continuous function* g,
(iii) $\lim_{n\to\infty} \phi_n(\mathbf{t}) = \phi_0(\mathbf{t})$ *for every* $\mathbf{t} = (t_1, \ldots, t_k)' \in \mathbb{R}^k$.

Proposition 6.3.1 (The Cramer–Wold Device). *Let* $\{\mathbf{X}_n\}$ *be a sequence of random* k*-vectors. Then* $\mathbf{X}_n \Rightarrow \mathbf{X}$ *if and only if* $\boldsymbol{\lambda}'\mathbf{X}_n \Rightarrow \boldsymbol{\lambda}'\mathbf{X}$ *for all* $\boldsymbol{\lambda} = (\lambda_1, \ldots, \lambda_k)' \in \mathbb{R}^k$.

PROOF. First assume that $\mathbf{X}_n \Rightarrow \mathbf{X}$. Then for any fixed $\boldsymbol{\lambda} \in \mathbb{R}^k$, Theorem 6.3.1 (iii) gives

$$\phi_{\boldsymbol{\lambda}'\mathbf{X}_n}(t) = E\exp(it\boldsymbol{\lambda}'\mathbf{X}_n) = \phi_{\mathbf{X}_n}(t\boldsymbol{\lambda}) \to \phi_{\mathbf{X}}(t\boldsymbol{\lambda}) = \phi_{\boldsymbol{\lambda}'\mathbf{X}}(t),$$

showing that $\boldsymbol{\lambda}'\mathbf{X}_n \Rightarrow \boldsymbol{\lambda}'\mathbf{X}$.

Now suppose that $\boldsymbol{\lambda}'\mathbf{X}_n \Rightarrow \boldsymbol{\lambda}'\mathbf{X}$ for each $\boldsymbol{\lambda} \in \mathbb{R}^k$. Then using Theorem 6.3.1 again, we have for any $\boldsymbol{\lambda} \in \mathbb{R}^k$,

$$\phi_{\mathbf{X}_n}(\boldsymbol{\lambda}) = E\exp(i\boldsymbol{\lambda}'\mathbf{X}_n) = \phi_{\boldsymbol{\lambda}'\mathbf{X}_n}(1) \to \phi_{\boldsymbol{\lambda}'\mathbf{X}}(1) = \phi_{\mathbf{X}}(\boldsymbol{\lambda})$$

which shows that $\mathbf{X}_n \Rightarrow \mathbf{X}$. □

Remark 1. If $\mathbf{X}_n \Rightarrow \mathbf{X}$ then the Cramer–Wold device with $\lambda_i = 1$ and $\lambda_j = 0$, $j \neq i$, shows at once that $X_{ni} \Rightarrow X_i$ where X_{ni} and X_i are the ith components of $\mathbf{X}_n$ and $\mathbf{X}$ respectively. If on the other hand $X_{ni} \Rightarrow X_i$ for each i, then it is not necessarily true that $\mathbf{X}_n \Rightarrow \mathbf{X}$ (see Problem 6.8).

Proposition 6.3.2. *If* $\mathbf{X}_n \xrightarrow{P} \mathbf{X}$ *then*

(i) $E|\exp(i\mathbf{t}'\mathbf{X}_n) - \exp(i\mathbf{t}'\mathbf{X})| \to 0$ *as* $n \to \infty$ *for every* $\mathbf{t} \in \mathbb{R}^k$
and
(ii) $\mathbf{X}_n \Rightarrow \mathbf{X}$.

PROOF. Given $\mathbf{t} \in \mathbb{R}^k$ and $\varepsilon > 0$, choose $\delta(\varepsilon) > 0$ such that

$$|\exp(i\mathbf{t}'\mathbf{x}) - \exp(i\mathbf{t}'\mathbf{y})| = |1 - \exp(i\mathbf{t}'(\mathbf{y} - \mathbf{x}))| < \varepsilon \quad \text{if } |\mathbf{x} - \mathbf{y}| < \delta. \tag{6.3.2}$$

We then have

$$\begin{aligned} E|\exp(i\mathbf{t}'\mathbf{X}_n) - \exp(i\mathbf{t}'\mathbf{X})| &= E|1 - \exp(i\mathbf{t}'(\mathbf{X}_n - \mathbf{X}))| \\ &= E[|1 - \exp(i\mathbf{t}'(\mathbf{X}_n - \mathbf{X}))| I_{\{|X_n - X| < \delta\}}] \\ &\quad + E[|1 - \exp(i\mathbf{t}'(\mathbf{X}_n - \mathbf{X}))| I_{\{|X_n - X| \geq \delta\}}]. \end{aligned}$$

The first term is less than ε by (6.3.2) and the second term is bounded above by $2P(|\mathbf{X}_n - \mathbf{X}| \geq \delta)$ which goes to zero as $n \to \infty$ since $\mathbf{X}_n \xrightarrow{P} \mathbf{X}$. This proves (i).

To establish the result (ii) we first note that

$$|E\exp(i\mathbf{t}'\mathbf{X}_n) - E\exp(i\mathbf{t}'\mathbf{X})| \leq E|\exp(i\mathbf{t}'\mathbf{X}_n) - \exp(i\mathbf{t}'\mathbf{X})| \to 0 \quad \text{as } n \to \infty,$$

and then use Theorem 6.3.1 (iii). □

Proposition 6.3.3. *If* $\{\mathbf{X}_n\}$ *and* $\{\mathbf{Y}_n\}$ *are two sequences of random k-vectors such that* $\mathbf{X}_n - \mathbf{Y}_n = o_p(1)$ and $\mathbf{X}_n \Rightarrow \mathbf{X}$, then $\mathbf{Y}_n \Rightarrow \mathbf{X}$.

PROOF. By Theorem 6.3.1 (iii), it suffices to show that

$$|\phi_{\mathbf{Y}_n}(\mathbf{t}) - \phi_{\mathbf{X}_n}(\mathbf{t})| \to 0 \text{ as } n \to \infty \quad \text{for each } \mathbf{t} \in \mathbb{R}^k, \tag{6.3.3}$$

since then

$$|\phi_{\mathbf{Y}_n}(\mathbf{t}) - \phi_{\mathbf{X}}(\mathbf{t})| \leq |\phi_{\mathbf{Y}_n}(\mathbf{t}) - \phi_{\mathbf{X}_n}(\mathbf{t})| + |\phi_{\mathbf{X}_n}(\mathbf{t}) - \phi_{\mathbf{X}}(\mathbf{t})| \to 0.$$

But

$$\begin{aligned} |\phi_{\mathbf{Y}_n}(\mathbf{t}) - \phi_{\mathbf{X}_n}(\mathbf{t})| &= |E(\exp(i\mathbf{t}'\mathbf{Y}_n) - \exp(i\mathbf{t}'\mathbf{X}_n))| \\ &\leq E|1 - \exp(i\mathbf{t}'(\mathbf{X}_n - \mathbf{Y}_n))| \\ &\to 0 \quad \text{as } n \to \infty, \end{aligned}$$

by Proposition 6.3.2. □

Proposition 6.3.4. *If $\{\mathbf{X}_n\}$ is a sequence of random k-vectors such that $\mathbf{X}_n \Rightarrow \mathbf{X}$ and if $h: \mathbb{R}^k \to \mathbb{R}^m$ is a continuous mapping, then $h(\mathbf{X}_n) \Rightarrow h(\mathbf{X})$.*

PROOF. For a fixed $\mathbf{t} \in \mathbb{R}^m$, $e^{i\mathbf{t}'h(\mathbf{X})}$ is a bounded continuous function of $\mathbf{X}$ so that by Theorem 6.3.1 (ii), $\phi_{h(\mathbf{X}_n)}(\mathbf{t}) \to \phi_{h(\mathbf{X})}(\mathbf{t})$. Theorem 6.3.1 (iii) then implies that $h(\mathbf{X}_n) \Rightarrow h(\mathbf{X})$. □

In the special case when $\{\mathbf{X}_n\}$ converges in distribution to a constant random vector $\mathbf{b}$, it is also true that $\{\mathbf{X}_n\}$ converges in probability to $\mathbf{b}$, as shown in the following proposition. (Notice that convergence in probability to $\mathbf{b}$ is meaningful even when $\mathbf{X}_1, \mathbf{X}_2, \ldots,$ are all defined on different probability spaces.)

Proposition 6.3.5. *If $\mathbf{X}_n \Rightarrow \mathbf{b}$ where $\mathbf{b}$ is a constant k-vector, then $\mathbf{X}_n \xrightarrow{P} \mathbf{b}$.*

PROOF. We first prove the result for random variables (i.e. in the case $k = 1$). If $X_n \Rightarrow b$ then $F_{X_n}(x) \to I_{[b,\infty)}(x)$ for all $x \neq b$. Hence for any $\varepsilon > 0$,

$$\begin{aligned} P(|X_n - b| \le \varepsilon) &= P(b - \varepsilon \le X_n \le b + \varepsilon) \\ &\to I_{[b,\infty)}(b + \varepsilon) - I_{[b,\infty)}(b - \varepsilon) \\ &= 1, \end{aligned}$$

showing that $X_n \xrightarrow{P} b$.

To establish the result in the general case, $k \ge 1$, we observe that if $\mathbf{X}_n \Rightarrow \mathbf{b}$ then $X_{nj} \Rightarrow b_j$ for each $j = 1, \ldots, k$ by Remark 1. From the result of the preceding paragraph we deduce that $X_{nj} \xrightarrow{P} b_j$ for each $j = 1, \ldots, k$ and hence by Definition 6.1.4 that $\mathbf{X}_n \xrightarrow{P} \mathbf{b}$. □

Proposition 6.3.6 (The Weak Law of Large Numbers). *If $\{X_n\}$ is an* iid *sequence of random variables with a finite mean μ, then*

$$\bar{X}_n \xrightarrow{P} \mu$$

where $\bar{X}_n := (X_1 + \cdots + X_n)/n$.

PROOF. Since $\bar{X}_n - \mu = ((X_1 - \mu) + \cdots + (X_n - \mu))/n$, it suffices to prove the result for zero-mean sequences. Assuming that $\mu = 0$, and using the independence of $X_1, X_2, \ldots,$ we have

$$\begin{aligned} \phi_{\bar{X}_n}(t) &= Ee^{it\bar{X}_n} \\ &= (\phi_{X_1}(n^{-1}t))^n. \end{aligned}$$

From the inequality $|1 - y^n| \le n|1 - y|$, $|y| \le 1$, and the assumption that $EX_1 = 0$ it follows that

$$\begin{aligned} |1 - \phi_{\bar{X}_n}(t)| &\le n|1 - \phi_{X_1}(n^{-1}t)| \\ &= n|E(1 + itn^{-1}X_1 - e^{itn^{-1}X_1})| \\ &\le E|n(1 + itn^{-1}X_1 - e^{itn^{-1}X_1})|. \end{aligned}$$

A Taylor series approximation to $\cos x$ and $\sin x$ then gives

$$\begin{aligned}|1 + iy - e^{iy}| &= |1 + iy - \cos y - i\sin y| \\ &\le |1 - \cos y| + |y - \sin y| \\ &\le \min(2|y|, |y|^2)\end{aligned}$$

for all real y. Replacing y by $tn^{-1}x$ in this bound we see that for every x

$$|n(1 + itn^{-1}x - e^{itn^{-1}x})| \le 2|t|\,|x|, \qquad n = 1, 2, \ldots,$$

and

$$|n(1 + itn^{-1}x - e^{itn^{-1}x})| \to 0 \quad \text{as } n \to \infty.$$

Since $E|X_1| < \infty$ by assumption, $E|n(1 + itn^{-1}X_1 - e^{itn^{-1}X_1})| \to 0$ by the dominated convergence theorem. Hence $\phi_{\bar{X}_n}(t) \to 1$ for every t and since 1 is the characteristic function of the zero random variable we conclude from Propositions 6.3.1 (iii) and 6.3.5 that $\bar{X}_n \xrightarrow{P} 0$. □

Proposition 6.3.7. *If $\{\mathbf{X}_n\}$ and $\{\mathbf{Y}_n\}$ are sequences of random k- and m-vectors respectively and if $\mathbf{X}_n \Rightarrow \mathbf{X}$ and $\mathbf{Y}_n \Rightarrow \mathbf{b}$ where $\mathbf{b}$ is a constant vector, then*

$$\begin{bmatrix}\mathbf{X}_n \\ \mathbf{Y}_n\end{bmatrix} \Rightarrow \begin{bmatrix}\mathbf{X} \\ \mathbf{b}\end{bmatrix}. \tag{6.3.4}$$

(Note that (6.3.4) is not necessarily true if $\mathbf{Y}_n$ converges in distribution to a non-constant random vector.)

PROOF. If we define $\mathbf{Z}_n = [\mathbf{X}_n', \mathbf{b}']'$, then from Proposition 6.3.5 we have $\mathbf{Z}_n - [\mathbf{X}_n', \mathbf{Y}_n']' = o_p(1)$. It is clear that $\mathbf{Z}_n \Rightarrow [\mathbf{X}', \mathbf{b}']'$ and so (6.3.4) follows from Proposition 6.3.3. □

The following proposition is stated without proof since it follows at once from Propositions 6.3.4 and 6.3.7.

Proposition 6.3.8. *If $\{\mathbf{X}_n\}$ and $\{\mathbf{Y}_n\}$ are sequences of random k-vectors such that $\mathbf{X}_n \Rightarrow \mathbf{X}$ and $\mathbf{Y}_n \Rightarrow \mathbf{b}$ where $\mathbf{b}$ is constant, then*

$$\text{(i)}\ \ \mathbf{X}_n + \mathbf{Y}_n \Rightarrow \mathbf{X} + \mathbf{b}$$

and

$$\text{(ii)}\ \ \mathbf{Y}_n'\mathbf{X}_n \Rightarrow \mathbf{b}'\mathbf{X}.$$

The next proposition will prove to be very useful in establishing asymptotic normality of the sample mean and sample autocovariance function for a wide class of time series models.

Proposition 6.3.9. *Let $\mathbf{X}_n$, $n = 1, 2, \ldots,$ and $\mathbf{Y}_{nj}$, $j = 1, 2, \ldots; n = 1, 2, \ldots,$ be random k-vectors such that*

(i) $\mathbf{Y}_{nj} \Rightarrow \mathbf{Y}_j$ *as* $n \to \infty$ *for each* $j = 1, 2, \ldots,$
(ii) $\mathbf{Y}_j \Rightarrow \mathbf{Y}$ *as* $j \to \infty$, *and*
(iii) $\lim_{j\to\infty} \limsup_{n\to\infty} P(|\mathbf{X}_n - \mathbf{Y}_{nj}| > \varepsilon) = 0$ *for every* $\varepsilon > 0$.

Then

$$\mathbf{X}_n \Rightarrow \mathbf{Y} \quad \textit{as } n \to \infty.$$

PROOF. By Theorem 6.3.1, it suffices to show that for each $\mathbf{t} \in \mathbb{R}^k$

$$|\phi_{\mathbf{X}_n}(\mathbf{t}) - \phi_{\mathbf{Y}}(\mathbf{t})| \to 0 \quad \text{as } n \to \infty.$$

The triangle inequality gives the bound

$$\begin{aligned} |\phi_{\mathbf{X}_n}(\mathbf{t}) - \phi_{\mathbf{Y}}(\mathbf{t})| \le |\phi_{\mathbf{X}_n}(\mathbf{t}) - \phi_{\mathbf{Y}_{nj}}(\mathbf{t})| + |\phi_{\mathbf{Y}_{nj}}(\mathbf{t}) - \phi_{\mathbf{Y}_j}(\mathbf{t})| \\ + |\phi_{\mathbf{Y}_j}(\mathbf{t}) - \phi_{\mathbf{Y}}(\mathbf{t})|. \end{aligned} \tag{6.3.5}$$

From (iii) it follows, by an argument similar to the proof of Proposition 6.3.2 (i), that $\limsup_{n\to\infty} |\phi_{\mathbf{X}_n}(\mathbf{t}) - \phi_{\mathbf{Y}_{nj}}(\mathbf{t})| \to 0$ as $j \to \infty$. Assumption (ii) guarantees that the last term in (6.3.5) also goes to zero as $j \to \infty$. For any positive δ we can therefore choose j so that the upper limits as $n \to \infty$ of the first and third terms on the right side of (6.3.5) are both less than $\delta/2$. For this fixed value of j, $\lim_{n\to\infty} |\phi_{\mathbf{Y}_{nj}}(\mathbf{t}) - \phi_{\mathbf{Y}_j}(\mathbf{t})| = 0$ by assumption (i). Consequently $\limsup_{n\to\infty} |\phi_{\mathbf{X}_n}(\mathbf{t}) - \phi_{\mathbf{Y}}(\mathbf{t})| < \frac{1}{2}\delta + \frac{1}{2}\delta = \delta$, and since δ was chosen arbitrarily, $\limsup_{n\to\infty} |\phi_{\mathbf{X}_n}(\mathbf{t}) - \phi_{\mathbf{Y}}(\mathbf{t})| = 0$ as required. □

Proposition 6.3.10 (The Weak Law of Large Numbers for Moving Averages). *Let* $\{X_t\}$ *be the two-sided moving average*

$$X_t = \sum_{j=-\infty}^{\infty} \psi_j Z_{t-j}$$

where $\{Z_t\}$ *is* iid *with mean* μ *and* $\sum_{j=-\infty}^{\infty} |\psi_j| < \infty$. *Then*

$$\bar{X}_n \xrightarrow{P} \left(\sum_{j=-\infty}^{\infty} \psi_j \right) \mu.$$

(*Note that the variance of* Z_t *may be infinite.*)

PROOF. First note that the series $\sum_{j=-\infty}^{\infty} \psi_j Z_{t-j}$ converges absolutely with probability one since

$$E\left(\sum_{j=-\infty}^{\infty} |\psi_j Z_{t-j}| \right) \le \sum_{j=-\infty}^{\infty} |\psi_j| E|Z_1| < \infty.$$

Now for each j, we have from the weak law of large numbers,

$$n^{-1} \sum_{t=1}^{n} Z_{t-j} \xrightarrow{P} \mu.$$

Defining $Y_{nk} := n^{-1} \sum_{t=1}^{n} \sum_{|j| \le k} \psi_j Z_{t-j} = \sum_{|j| \le k} \psi_j (n^{-1} \sum_{t=1}^{n} Z_{t-j})$, we see from

Proposition 6.1.4 that

$$Y_{nk} \xrightarrow{P} \left(\sum_{|j| \le k} \psi_j \right) \mu.$$

If we define $Y_k = (\sum_{|j| \le k} \psi_j)\mu$, then since $Y_k \to Y := (\sum_{j=-\infty}^{\infty} \psi_j)\mu$, it suffices to show by Proposition 6.3.9 that

$$\lim_{k \to \infty} \limsup_{n \to \infty} P(|\bar{X}_n - Y_{nk}| > \varepsilon) = 0 \quad \text{for every } \varepsilon > 0. \tag{6.3.6}$$

Applying Proposition 6.2.1 with $r = 1$, we have

$$\begin{aligned} P(|\bar{X}_n - Y_{nk}| > \varepsilon) &= P\left(\left| n^{-1} \sum_{t=1}^{n} \sum_{|j|>k} \psi_j Z_{t-j} \right| > \varepsilon \right) \\ &\le E\left| \sum_{|j|>k} \psi_j Z_{1-j} \right| \Big/ \varepsilon \\ &\le \left(\sum_{|j|>k} |\psi_j| \right) E|Z_1| \Big/ \varepsilon, \end{aligned}$$

which implies (6.3.6). □

§6.4 Central Limit Theorems and Related Results

Many of the estimators used in time series analysis turn out to be asymptotically normal as the number of observations goes to infinity. In this section we develop some of the standard techniques to be used for establishing asymptotic normality.

Definition 6.4.1. A sequence of random variables $\{X_n\}$ is said to be asymptotically normal with "mean" μ_n and "standard deviation" σ_n if $\sigma_n > 0$ for n sufficiently large and

$$\sigma_n^{-1}(X_n - \mu_n) \Rightarrow Z, \qquad \text{where } Z \sim \mathrm{N}(0, 1).$$

In the notation of Serfling (1980) we shall write this as

$$X_n \text{ is AN}(\mu_n, \sigma_n^2).$$ □

Remark 1. If X_n is AN(μ_n, σ_n^2) it is not necessarily the case that $\mu_n = EX_n$ or that $\sigma_n^2 = \mathrm{Var}(X_n)$. See Example 6.4.1 below.

Remark 2. In order to prove that X_n is AN(μ_n, σ_n^2) it is often simplest to establish the result in the equivalent form (see Theorem 6.3.1 (iii)),

$$\phi_{Z_n}(t) \to \exp(-t^2/2),$$

where $\phi_{Z_n}(\cdot)$ is the characteristic function of $Z_n = \sigma_n^{-1}(X_n - \mu_n)$. This approach works especially well when X_n is a sum of independent random variables as in the following theorem.

Theorem 6.4.1 (The Central Limit Theorem). *If* $\{X_n\} \sim \text{IID}(\mu, \sigma^2)$ *and* $\bar{X}_n = (X_1 + \cdots + X_n)/n$, *then*

$$\bar{X}_n \text{ is } \text{AN}(\mu, \sigma^2/n).$$

PROOF. Define the iid sequence $\{Y_t\}$ with mean zero and variance one by $Y_t = (X_t - \mu)/\sigma$ and set $\bar{Y}_n = n^{-1}\sum_{i=1}^n Y_i$. By Remark 2, it suffices to show that $\phi_{n^{1/2}\bar{Y}_n}(t) \to e^{-t^2/2}$. By independence, we have

$$\begin{aligned}\phi_{n^{1/2}\bar{Y}_n}(t) &= E\exp\left[itn^{-1/2}\sum_{j=1}^n Y_j\right]\\ &= [\phi_{Y_1}(tn^{-1/2})]^n.\end{aligned}$$

First we need the inequality, $|x^n - y^n| \le n|x - y|$ for $|x| \le 1$ and $|y| \le 1$, which can be proved easily by induction on n. This implies that for $n \ge t^2/4$,

$$\begin{aligned}|[\phi_{Y_1}(tn^{-1/2})]^n - (1 - t^2/(2n))^n| &\le n|\phi_{Y_1}(tn^{-1/2}) - (1 - t^2/(2n))|\\ &= n|E(e^{itn^{-1/2}Y_1} - (1 + itn^{-1/2}Y_1 - t^2Y_1^2/(2n)))|.\end{aligned} \tag{6.4.1}$$

Using a Taylor series expansion of e^{ix} in a neighborhood of $x = 0$ we have

$$n|e^{itn^{-1/2}x} - (1 + itn^{-1/2}x - t^2x^2/(2n))| \to 0 \quad \text{as } n \to \infty$$

and

$$n|e^{itn^{-1/2}x} - (1 + itn^{-1/2}x - t^2x^2/(2n))| \le (tx)^2 \quad \text{for all } n \text{ and } x.$$

Thus, by the dominated convergence theorem, the right-hand side of (6.4.1) converges to zero as $n \to \infty$ and since $(1 - t^2/(2n))^n \to e^{-t^2/2}$ we obtain $\phi_{n^{1/2}\bar{Y}_n}(t) \to e^{-t^2/2}$ as required. □

Remark 3. The assumption of identical distributions in Theorem 6.4.1 can be replaced by others such as the Lindeberg condition (see Billingsley, 1986) which is a restriction on the truncated variances of the random variables X_n. However the assumptions of Theorem 6.4.1 will suffice for our purposes.

Proposition 6.4.1. *If* X_n *is* $\text{AN}(\mu, \sigma_n^2)$ *where* $\sigma_n \to 0$ *as* $n \to \infty$, *and if* g *is a function which is differentiable at* μ, *then*

$$g(X_n) \text{ is } \text{AN}(g(\mu), g'(\mu)^2\sigma_n^2).$$

PROOF. Since $Z_n = \sigma_n^{-1}(X_n - \mu) \Rightarrow Z$ where $Z \sim \text{N}(0, 1)$, we may conclude from Problem 6.7 that $Z_n = O_p(1)$ as $n \to \infty$. Hence $X_n = \mu + O_p(\sigma_n)$. By Proposition 6.1.5 we therefore have

$$\sigma_n^{-1}[g(X_n) - g(\mu)] = \sigma_n^{-1}g'(\mu)[X_n - \mu] + o_p(1),$$

which with Proposition 6.3.3 proves the result. □

EXAMPLE 6.4.1. Suppose that $\{X_n\} \sim \text{IID}(\mu, \sigma^2)$ where $\mu \neq 0$ and $0 < \sigma < \infty$. If $\bar{X}_n = n^{-1}(X_1 + \cdots + X_n)$ then by Theorem 6.4.1

$$\bar{X}_n \text{ is AN}(\mu, \sigma^2/n),$$

and by Proposition 6.4.1,

$$\bar{X}_n^{-1} \text{ is AN}(\mu^{-1}, \mu^{-4}\sigma^2/n).$$

Depending on the distribution of X_n, it is possible that the mean of $\bar{X}_n^{-1}$ may not exist (see Problem 6.17).

We now extend the notion of asymptotic normality to random k-vectors, $k \geq 1$. Recall from Proposition 1.5.5 that $\mathbf{X}$ is multivariate normal if and only if every linear combination $\boldsymbol{\lambda}'\mathbf{X}$ is univariate normal. This fact, in conjunction with the Cramer–Wold device, motivates the following definition (see Serfling (1980)) of asymptotic multivariate normality.

Definition 6.4.2. The sequence $\{\mathbf{X}_n\}$ of random k-vectors is asymptotically normal with "mean vector" $\boldsymbol{\mu}_n$ and "covariance matrix" Σ_n if

(i) Σ_n has no zero diagonal elements for all sufficiently large n, and
(ii) $\boldsymbol{\lambda}'\mathbf{X}_n$ is $\text{AN}(\boldsymbol{\lambda}'\boldsymbol{\mu}_n, \boldsymbol{\lambda}'\Sigma_n\boldsymbol{\lambda})$ for every $\boldsymbol{\lambda} \in \mathbb{R}^k$ such that $\boldsymbol{\lambda}'\Sigma_n\boldsymbol{\lambda} > 0$ for all sufficient large n.

Proposition 6.4.2. *If $\mathbf{X}_n$ is $\text{AN}(\boldsymbol{\mu}_n, \Sigma_n)$ and B is any non-zero $m \times k$ matrix such that the matrices $B\Sigma_n B'$, $n = 1, 2, \ldots$, have no zero diagonal elements then*

$$B\mathbf{X}_n \text{ is AN}(B\boldsymbol{\mu}_n, B\Sigma_n B').$$

PROOF. Problem 6.21. □

The following proposition is the multivariate analogue of Proposition 6.4.1.

Proposition 6.4.3. *Suppose that $\mathbf{X}_n$ is $\text{AN}(\boldsymbol{\mu}, c_n^2\Sigma)$ where Σ is a symmetric non-negative definite matrix and $c_n \to 0$ as $n \to \infty$. If $\mathbf{g}(\mathbf{X}) = (g_1(\mathbf{X}), \ldots, g_m(\mathbf{X}))'$ is a mapping from $\mathbb{R}^k$ into $\mathbb{R}^m$ such that each $g_i(\cdot)$ is continuously differentiable in a neighborhood of $\boldsymbol{\mu}$, and if $D\Sigma D'$ has all of its diagonal elements non-zero, where D is the $m \times k$ matrix $[(\partial g_i/\partial x_j)(\boldsymbol{\mu})]$, then*

$$\mathbf{g}(\mathbf{X}_n) \text{ is AN}(\mathbf{g}(\boldsymbol{\mu}), c_n^2 D\Sigma D').$$

PROOF. First we show that $X_{nj} = \mu_j + O_p(c_n)$. Applying Proposition 6.4.2 with $B = (\delta_{j1}, \delta_{j2}, \ldots, \delta_{jk})$ we find that $X_{nj} = B\mathbf{X}$ is $\text{AN}(\mu_j, c_n^2\sigma_{jj})$ where σ_{jj} is the jth diagonal element of Σ and $\sigma_{jj} > 0$ by Definition 6.4.2. Since $c_n^{-1}(X_{nj} - \mu_j)$ converges in distribution we may conclude that it is bounded in probability (Problem 6.7) and hence that $X_{nj} = \mu_j + O_p(c_n)$.

Now applying Proposition 6.1.6 we can write, for $i = 1, \ldots, m$,

$$g_i(\mathbf{X}_n) = g_i(\boldsymbol{\mu}) + \sum_{j=1}^{k} \frac{\partial g_i}{\partial x_j}(\boldsymbol{\mu})(X_{nj} - \mu_j) + o_p(c_n),$$

or equivalently,

$$\mathbf{g}(\mathbf{X}_n) - \mathbf{g}(\boldsymbol{\mu}) = D(\mathbf{X}_n - \boldsymbol{\mu}) + o_p(c_n).$$

Dividing both sides by c_n we obtain

$$c_n^{-1}[\mathbf{g}(\mathbf{X}_n) - \mathbf{g}(\boldsymbol{\mu})] = c_n^{-1}D(\mathbf{X}_n - \boldsymbol{\mu}) + o_p(1),$$

and since $c_n^{-1}D(\mathbf{X}_n - \boldsymbol{\mu})$ is $\text{AN}(\mathbf{0}, D\Sigma D')$, we conclude from Proposition 6.3.3 that the same is true of $c_n^{-1}[\mathbf{g}(\mathbf{X}_n) - \mathbf{g}(\boldsymbol{\mu})]$. □

Example 6.4.2 (The Sample Coefficient of Variation). Suppose that $\{X_n\} \sim \text{IID}(\mu, \sigma^2)$, $\sigma > 0$, $EX_n^4 = \mu_4 < \infty$, $EX_n^3 = \mu_3$, $EX_n^2 = \mu_2 = \mu^2 + \sigma^2$ and $EX_n = \mu_1 = \mu \neq 0$. The sample coefficient of variation is defined as $Y_n = s_n/\bar{X}_n$ where $\bar{X}_n = n^{-1}(X_1 + \cdots + X_n)$ and $s_n^2 = n^{-1}\sum_{i=1}^n (X_i - \bar{X}_n)^2$. It is easy to verify (Problem 6.22) that

$$\begin{bmatrix} \bar{X}_n \\ n^{-1}\sum_{i=1}^{n} X_i^2 \end{bmatrix} \text{ is } \text{AN}\left(\begin{bmatrix} \mu_1 \\ \mu_2 \end{bmatrix}, n^{-1}\Sigma\right) \tag{6.4.2}$$

where Σ is the matrix with components

$$\Sigma_{ij} = \mu_{i+j} - \mu_i\mu_j, \qquad i, j = 1, 2.$$

Now $Y_n = g(\bar{X}_n, n^{-1}\sum_{i=1}^n X_i^2)$ where $g(x, y) = x^{-1}(y - x^2)^{1/2}$. Applying Proposition 6.4.3 with

$$D = \left[\frac{\partial g}{\partial x}(\boldsymbol{\mu}), \frac{\partial g}{\partial y}(\boldsymbol{\mu})\right] = [-\mu_2/(\sigma\mu_1^2), 1/(2\sigma\mu_1)],$$

we find at once that

$$Y_n \text{ is } \text{AN}(\sigma/\mu_1, n^{-1}D\Sigma D').$$

We shall frequently have need for a central limit theorem which applies to sums of dependent random variables. It will be sufficient for our purposes to have a theorem which applies to m-dependent strictly stationary sequences, defined as follows.

Definition 6.4.3 (m-Dependence). A strictly stationary sequence of random variables $\{X_t\}$ is said to be m-dependent (where m is a non-negative integer) if for each t the two sets of random variables $\{X_j, j \leq t\}$ and $\{X_j, j \geq t + m + 1\}$ are independent.

Remark 4. In checking for m-dependence of a strictly stationary sequence $\{X_t, t = 0, \pm 1, \pm 2, \ldots\}$ it is clearly sufficient to check the independence of the two sets $\{X_j, j \le 0\}$ and $\{X_j, j \ge m+1\}$ since they have the same joint distributions as $\{X_j, j \le t\}$ and $\{X_j, j \ge t+m+1\}$ respectively.

Remark 5. The property of m-dependence generalizes that of independence in a natural way. Observations of an m-dependent process are independent provided they are separated in time by more than m time units. In the special case when $m = 0$, m-dependence reduces to independence. The MA(q) processes introduced in Section 3.1 are m-dependent with $m = q$.

The following result, due originally to Hoeffding and Robbins (1948), extends the classical central limit theorem (Theorem 6.4.1) to m-dependent sequences.

Theorem 6.4.2 (The Central Limit Theorem for Strictly Stationary m-Dependent Sequences). *If $\{X_t\}$ is a strictly stationary m-dependent sequence of random variables with mean zero and autocovariance function $\gamma(\cdot)$, and if $v_m = \gamma(0) + 2\sum_{j=1}^{m}\gamma(j) \ne 0$, then*

(i) $\lim_{n\to\infty} n\,\mathrm{Var}(\bar{X}_n) = v_m$ *and*
(ii) $\bar{X}_n$ *is* $\mathrm{AN}(0, v_m/n)$.

PROOF. (i) $n\,\mathrm{Var}(\bar{X}_n) = n^{-1}\sum_{i=1}^{n}\sum_{j=1}^{n}\gamma(i-j)$

$$= \sum_{|j|<n}(1 - n^{-1}|j|)\gamma(j)$$

$$= \sum_{|j|\le m}(1 - n^{-1}|j|)\gamma(j) \quad \text{for } n > m$$

$$\to v_m \quad \text{as } n \to \infty.$$

(ii) For each integer k such that $k > 2m$, let $Y_{nk} = n^{-1/2}[(X_1 + \cdots + X_{k-m}) + (X_{k+1} + \cdots + X_{2k-m}) + \cdots + (X_{(r-1)k+1} + \cdots + X_{rk-m})]$ where $r = [n/k]$, the integer part of n/k. Observe that $n^{1/2}Y_{nk}$ is a sum of r iid random variables each having mean zero and variance,

$$R_{k-m} = \mathrm{Var}(X_1 + \cdots + X_{k-m}) = \sum_{|j|<k-m}(k - m - |j|)\gamma(j).$$

Applying the central limit theorem (Theorem 6.4.1), we have

$$Y_{nk} \Rightarrow Y_k \quad \text{where } Y_k \sim N(0, k^{-1}R_{k-m}).$$

Moreover, since $k^{-1}R_{k-m} \to v_m$ as $k \to \infty$, we may conclude (Problem 6.16) that

$$Y_k \Rightarrow Y \quad \text{where } Y \sim N(0, v_m).$$

It remains only to show that

$$\lim_{k\to\infty}\limsup_{n\to\infty} P(|n^{1/2}\bar{X}_n - Y_{nk}| > \varepsilon) = 0 \quad \text{for every } \varepsilon > 0, \tag{6.4.3}$$

since the second conclusion of the theorem will then follow directly from Proposition 6.3.9.

In order to establish (6.4.3) we write $(n^{1/2}\bar{X}_n - Y_{nk})$ as a sum of $r = [n/k]$ independent terms, viz.

$$n^{1/2}\bar{X}_n - Y_{nk} = n^{-1/2}\sum_{j=1}^{r-1}(X_{jk-m+1} + X_{jk-m+2} + \cdots + X_{jk})$$
$$+ n^{-1/2}(X_{rk-m+1} + \cdots + X_n).$$

Making use of this independence and the stationarity of $\{X_t\}$, we find that

$$\operatorname{Var}(n^{1/2}\bar{X}_n - Y_{nk}) = n^{-1}[([n/k] - 1)R_m + R_{h(n)}],$$

where $R_m = \operatorname{Var}(X_1 + \cdots + X_m)$, $R_{h(n)} = \operatorname{Var}(X_1 + \cdots + X_{h(n)})$ and $h(n) = n - k[n/k] + m$. Now R_m is independent of n and $R_{h(n)}$ is a bounded function of n since $0 \le h(n) \le k + m$. Hence $\limsup_{n\to\infty} \operatorname{Var}(n^{1/2}\bar{X}_n - Y_{nk}) = k^{-1}R_m$, and so by Chebychev's inequality condition (6.4.3) is satisfied. □

Remark 6. Recalling Definition 6.4.1, we see that the condition $v_m \neq 0$ is essential for conclusion (ii) of Theorem 6.4.2 to be meaningful. In cases where $v_m = 0$ it is not difficult to show that $n^{1/2}\bar{X}_n \xrightarrow{P} 0$ and $n\operatorname{Var}(\bar{X}_n) \to 0$ as $n \to \infty$ (see Problem 6.6). The next example illustrates this point.

Example 6.4.3. The strictly stationary MA(1) process,

$$X_t = Z_t - Z_{t-1}, \qquad \{Z_t\} \sim \operatorname{IID}(0, \sigma^2),$$

is m-dependent with $m = 1$, and

$$v_m = \gamma(0) + 2\gamma(1) = 0.$$

For this example $\bar{X}_n = n^{-1}(Z_n - Z_0)$, which shows directly that $n\bar{X}_n \Rightarrow Z_1 - Z_0$, $n^{1/2}\bar{X}_n \xrightarrow{P} 0$ and $n\operatorname{Var}(\bar{X}_n) \to 0$ as $n \to \infty$.

Example 6.4.4 (Asymptotic Behaviour of $\bar{X}_n$ for the MA(q) Process with $\sum_{j=0}^{q}\theta_j \neq 0$). The MA($q$) process,

$$X_t = \sum_{j=0}^{q}\theta_j Z_{t-j}, \qquad \{Z_t\} \sim \operatorname{IID}(0, \sigma^2), \qquad \sum_{j=0}^{q}\theta_j \neq 0, \qquad \theta_0 = 1,$$

is a q-dependent strictly stationary sequence with

$$v_q = \sum_{j=-q}^{q}\gamma(j) = \sigma^2\left(\sum_{j=0}^{q}\theta_j\right)^2 = 2\pi f(0),$$

where $f(\cdot)$ is the spectral density of $\{X_t\}$ (see Theorem 4.4.2). A direct application of Theorem 6.4.2 shows that

$$\bar{X}_n \text{ is AN}\left(0, \sigma^2\left(\sum_{j=0}^{q}\theta_j\right)^2 \Big/ n\right). \tag{6.4.4}$$

Problems

6.1. Show that a finite set of random variables $\{X_1, \dots, X_n\}$ is bounded in probability.

6.2. Prove parts (ii) and (iii) of Proposition 6.1.1.

6.3. Show that $X_n = o_p(1)$ if and only if for every $\varepsilon > 0$, there exists a sequence $\delta_n(\varepsilon) \downarrow 0$ such that $P(|X_n| > \delta_n(\varepsilon)) < \varepsilon$ for all n.

6.4. Let $X_1, X_2, \dots,$ be iid random variables with distribution function F. If $M_n := \max(X_1, \dots, X_n)$ and $m_n := \min(X_1, \dots, X_n)$, show that $M_n/n \xrightarrow{P} 0$ if $x(1 - F(x)) \to 0$ as $x \to \infty$ and $m_n/n \xrightarrow{P} 0$ if $xF(-x) \to 0$ as $x \to \infty$.

6.5. If $X_n = O_p(1)$, is it true that there exists a subsequence $\{X_{n_k}\}$ and a constant $K \in (0, \infty)$ such that $P(|X_{n_k}| < K, k = 1, 2, \dots) = 1$?

6.6. Let $\{X_t\}$ be a stationary process with mean zero and an absolutely summable autocovariance function $\gamma(\cdot)$ such that $\sum_{h=-\infty}^{\infty} \gamma(h) = 0$. Show that $n \operatorname{Var}(\bar{X}_n) \to 0$ and hence that $n^{1/2}\bar{X}_n \xrightarrow{P} 0$.

6.7. If $\{X_n\}$ is a sequence of random variables such that $X_n \Rightarrow X$, show that $\{X_n\}$ is also bounded in probability.

6.8. Give an example of two sequences of random variables $\{X_n, n = 0, 1, \dots\}$ and $\{Y_n, n = 0, 1, \dots\}$ such that $X_n \Rightarrow X_0$ and $Y_n \Rightarrow Y_0$ while $(X_n, Y_n)'$ does not converge in distribution.

6.9. Suppose that the random vectors $\mathbf{X}_n$ and $\mathbf{Y}_n$ are independent for each n and that $\mathbf{X}_n \Rightarrow \mathbf{X}$ and $\mathbf{Y}_n \Rightarrow \mathbf{Y}$. Show that $[\mathbf{X}_n', \mathbf{Y}_n']' \Rightarrow [\mathbf{X}', \mathbf{Y}']'$ where $\mathbf{X}$ and $\mathbf{Y}$ are independent.

6.10. Show that if $X_n \Rightarrow X$, $Y_n \Rightarrow Y$ and X_n is independent of Y_n for each n, then $X_n + Y_n \Rightarrow X + Y$ where X and Y are independent.

6.11. Let $\{X_n\}$ be a sequence of random variables such that $EX_n = m$ and $\operatorname{Var}(X_n) = \sigma_n^2 > 0$ for all n, where $\sigma_n^2 \to 0$ as $n \to \infty$. Define

$$Z_n = \sigma_n^{-1}(X_n - m),$$

and let f be a function with non-zero derivative $f'(m)$ at m.
(a) Show that $Z_n = O_p(1)$ and $X_n = m + o_p(1)$.
(b) If $Y_n = [f(X_n) - f(m)]/[\sigma_n f'(m)]$, show that $Y_n - Z_n = o_p(1)$.
(c) Show that if Z_n converges in probability or in distribution then so does Y_n.
(d) If S_n is binomially distributed with parameters n and p, and $f'(p) \neq 0$, use the preceding results to determine the asymptotic distribution of $f(S_n/n)$.

6.12. Suppose that X_n is $\mathrm{AN}(\mu, \sigma_n^2)$ where $\sigma_n^2 \to 0$. Show that $X_n \xrightarrow{P} \mu$.

6.13. Suppose that X_n is $\mathrm{AN}(\mu, \sigma_n^2)$ and $Y_n = a_n + o_p(a_n)$. If $\sigma_n/a_n \to c$, where $0 < c < \infty$, show that $(X_n - \mu)/Y_n$ is $\mathrm{AN}(0, c^2)$.

6.14. If $\mathbf{X}_n = (X_{n1}, \dots, X_{nm})' \Rightarrow N(\mathbf{0}, \Sigma)$ and $\Sigma_n \xrightarrow{P} \Sigma$ where Σ is non-singular, show that $\mathbf{X}_n'\Sigma_n^{-1}\mathbf{X}_n \Rightarrow \chi^2(m)$.

6.15. If X_n is $\mathrm{AN}(\mu_n, \sigma_n^2)$, show that
(a) X_n is $\mathrm{AN}(\tilde{\mu}_n, \tilde{\sigma}_n^2)$ if and only if $\tilde{\sigma}_n/\sigma_n \to 1$ and $(\tilde{\mu}_n - \mu_n)/\sigma_n \to 0$, and

(b) $a_n X_n + b_n$ is $\text{AN}(\mu_n, \sigma_n^2)$ if and only if $a_n \to 1$ and $(\mu_n(a_n - 1) + b_n)/\sigma_n \to 0$.
(c) If X_n is $\text{AN}(n, 2n)$, show that $(1 - n^{-1})X_n$ is $\text{AN}(n, 2n)$ but that $(1 - n^{-1/2})X_n$ is not $\text{AN}(n, 2n)$.

6.16. Suppose that $X_n \sim N(\mu_n, v_n)$ where $\mu_n \to \mu$, $v_n \to v$ and $0 < v < \infty$. Show that $X_n \Rightarrow X$, where $X \sim N(\mu, v)$.

6.17. Suppose that $\{X_t\} \sim \text{IID}(\mu, \sigma^2)$ where $0 < \sigma^2 < \infty$. If $\bar{X}_n = n^{-1}(X_1 + \cdots + X_n)$ has a probability density function $f(x)$ which is continuous and positive at $x = 0$, show that $E|\bar{X}_n^{-1}| = \infty$. What is the limit distribution of $\bar{X}_n^{-1}$ when $\mu = 0$?

6.18. If $X_1, X_2, \ldots,$ are iid normal random variables with mean μ and variance σ^2, find the asymptotic distributions of $\bar{X}_n^2 = (n^{-1} \sum_{j=1}^n X_j)^2$
(a) when $\mu \neq 0$, and
(b) when $\mu = 0$.

6.19. Define

$$\ln^+(x) = \begin{cases} \ln(x) & \text{if } x > 0, \\ 0 & x \leq 0. \end{cases}$$

If X_n is $\text{AN}(\mu, \sigma_n^2)$ where $\mu > 0$ and $\sigma_n \to 0$, show that $\ln^+(X_n)$ is $\text{AN}(\ln(\mu), \mu^{-2}\sigma_n{}^2)$.

6.20. Let $f(x) = 3x^{-2} - 2x^{-3}$ for $x \neq 0$. If X_n is $\text{AN}(1, \sigma_n^2)$ find the limit distribution of $(f(X_n) - 1)/\sigma_n^2$ assuming that $0 < \sigma_n \to 0$.

6.21. Prove Proposition 6.4.2.

6.22. Verify (6.4.2) in Example 6.4.2. If $\mu \neq 0$, what is the limit distribution of $n^{-1/2} Y_n$?

6.23. Let $X_1, X_2, \ldots,$ be iid positive stable random variables with support $[0, \infty)$, exponent $\alpha \in (0, 1)$ and scale parameter $c^{1/\alpha}$ where $c > 0$. This means that

$$Ee^{-\theta X_1} = \exp(-c\theta^\alpha), \qquad \theta \geq 0.$$

The parameters c and α can be estimated by solving the two "moment" equations

$$n^{-1} \sum_{j=1}^n e^{-\theta_i X_j} = \exp(-c\theta_i^\alpha), \qquad i = 1, 2,$$

where $0 < \theta_1 < \theta_2$, for c and α. Find the asymptotic joint distribution of the estimators.

6.24. Suppose $\{Z_t\} \sim \text{IID}(0, \sigma^2)$.
(a) For $h \geq 1$ and $k \geq 1$, show that $Z_t Z_{t+h}$ and $Z_s Z_{s+k}$ are uncorrelated for all $s \neq t$, $s \geq 1$, $t \geq 1$.
(b) For a fixed $h \geq 1$, show that

$$\sigma^{-2} n^{-1/2} \sum_{t=1}^n (Z_t Z_{t+1}, \ldots, Z_t Z_{t+h})' \Rightarrow (N_1, \ldots, N_h)'$$

where $N_1, N_2, \ldots, N_h$ are iid N(0, 1) random variables. (Note that the sequence $\{Z_t Z_{t+h}, t = 1, 2, \ldots\}$ is h-dependent and is also $\text{WN}(0, \sigma^4)$.)
(c) Show that for each $h \geq 1$,

$$n^{-1/2}\left(\sum_{t=1}^n Z_t Z_{t+h} - \sum_{t=1}^{n-h} (Z_t - \bar{Z}_n)(Z_{t+h} - \bar{Z}_n) \right) \xrightarrow{P} 0$$

where

$$\bar{Z}_n = n^{-1}(Z_1 + \cdots + Z_n).$$

(d) Noting by the weak law of large numbers that $n^{-1}\sum_{t=1}^{n} Z_t^2 \xrightarrow{P} \sigma^2$, conclude from (b) and (c) that

$$n^{1/2}(\hat{\rho}(1), \ldots, \hat{\rho}(h))' \Rightarrow (N_1, \ldots, N_h)'$$

where

$$\hat{\rho}(h) = \sum_{t=1}^{n-h} (Z_t - \bar{Z}_n)(Z_{t+h} - \bar{Z}_n) \Big/ \sum_{t=1}^{n} (Z_t - \bar{Z}_n)^2.$$

CHAPTER 7

Estimation of the Mean and the Autocovariance Function

If $\{X_t\}$ is a real-valued stationary process, then from a second-order point of view it is characterized by its mean μ and its autocovariance function $\gamma(\cdot)$. The estimation of μ, $\gamma(\cdot)$ and the autocorrelation function $\rho(\cdot) = \gamma(\cdot)/\gamma(0)$ from observations of $X_1, \ldots, X_n$, therefore plays a crucial role in problems of inference and in particular in the problem of constructing an appropriate model for the data. In this chapter we consider several estimators which will be used and examine some of their properties.

§7.1 Estimation of μ

A natural unbiased estimator of the mean μ of the stationary process $\{X_t\}$ is the sample mean

$$\bar{X}_n = n^{-1}(X_1 + X_2 + \cdots + X_n). \tag{7.1.1}$$

We first examine the behavior of the mean squared error $E(\bar{X}_n - \mu)^2$ for large n.

Theorem 7.1.1. *If $\{X_t\}$ is stationary with mean μ and autocovariance function $\gamma(\cdot)$, then as $n \to \infty$,*

$$\operatorname{Var}(\bar{X}_n) = E(\bar{X}_n - \mu)^2 \to 0 \quad \textit{if } \gamma(n) \to 0,$$

and

$$nE(\bar{X}_n - \mu)^2 \to \sum_{h=-\infty}^{\infty} \gamma(h) \quad \textit{if } \sum_{h=-\infty}^{\infty} |\gamma(h)| < \infty.$$

PROOF.

$$\begin{aligned} n\operatorname{Var}(\bar{X}_n) &= \frac{1}{n}\sum_{i,j=1}^{n} \operatorname{Cov}(X_i, X_j) \\ &= \sum_{|h|<n}\left(1 - \frac{|h|}{n}\right)\gamma(h) \\ &\le \sum_{|h|<n} |\gamma(h)|. \end{aligned}$$

If $\gamma(n) \to 0$ as $n \to \infty$ then $\lim_{n\to\infty} n^{-1}\sum_{|h|<n}|\gamma(h)| = 2\lim_{n\to\infty}|\gamma(n)| = 0$, whence $\operatorname{Var}(\bar{X}_n) \to 0$. If $\sum_{h=-\infty}^{\infty}|\gamma(h)| < \infty$ then the dominated convergence theorem gives

$$\lim_{n\to\infty} n\operatorname{Var}(\bar{X}_n) = \lim_{n\to\infty}\sum_{|h|<n}\left(1 - \frac{|h|}{n}\right)\gamma(h) = \sum_{h=-\infty}^{\infty}\gamma(h). \qquad \square$$

Remark 1. If $\sum_{h=-\infty}^{\infty}|\gamma(h)| < \infty$, then $\{X_t\}$ has a spectral density $f(\cdot)$ and, by Corollary 4.3.2,

$$n\operatorname{Var}(\bar{X}_n) \to \sum_{h=-\infty}^{\infty}\gamma(h) = 2\pi f(0).$$

Remark 2. If $X_t = \mu + \sum_{j=-\infty}^{\infty}\psi_j Z_{t-j}$ with $\sum_{j=-\infty}^{\infty}|\psi_j| < \infty$, then $\sum_{h=-\infty}^{\infty}|\gamma(h)| < \infty$ (see Problem 3.9) and

$$n\operatorname{Var}(\bar{X}_n) \to \sum_{h=-\infty}^{\infty}\gamma(h) = 2\pi f(0) = \sigma^2\left(\sum_{j=-\infty}^{\infty}\psi_j\right)^2.$$

Remark 3. Theorem 7.1.1 shows that if $\gamma(n) \to 0$ as $n \to \infty$, then $\bar{X}_n$ converges in mean square (and hence in probability) to the mean μ. Moreover under the stronger condition $\sum_{h=-\infty}^{\infty}|\gamma(h)| < \infty$ (which is satisfied by all ARMA(p,q) processes) $\operatorname{Var}(\bar{X}_n) \sim n^{-1}\sum_{h=-\infty}^{\infty}\gamma(h)$. This suggests that under suitable conditions it might be true that $\bar{X}_n$ is $\mathrm{AN}(\mu, n^{-1}\sum_{h=-\infty}^{\infty}\gamma(h))$. One set of assumptions which guarantees the asymptotic normality is given in the next theorem.

Theorem 7.1.2. *If $\{X_t\}$ is the stationary process,*

$$X_t = \mu + \sum_{j=-\infty}^{\infty}\psi_j Z_{t-j}, \qquad \{Z_t\} \sim \mathrm{IID}(0, \sigma^2),$$

where $\sum_{j=-\infty}^{\infty}|\psi_j| < \infty$ and $\sum_{j=-\infty}^{\infty}\psi_j \neq 0$, then

$$\bar{X}_n \text{ is } \mathrm{AN}(\mu, n^{-1}v),$$

where $v = \sum_{h=-\infty}^{\infty}\gamma(h) = \sigma^2(\sum_{j=-\infty}^{\infty}\psi_j)^2$, and $\gamma(\cdot)$ is the autocovariance function of $\{X_t\}$.

PROOF. See Section 7.3. $\square$

Theorem 7.1.2 is useful for finding approximate large-sample confidence intervals for μ. If the process $\{X_t\}$ is not only stationary but also Gaussian, then from the second line of the proof of Theorem 7.1.1 we can go further and write down the exact distribution of $\bar{X}_n$ for finite n, viz.

$$n^{1/2}(\bar{X}_n - \mu) \sim \mathrm{N}\left(0, \sum_{|h|<n} \left(1 - \frac{|h|}{n}\right)\gamma(h)\right),$$

a result which gives exact confidence bounds for μ if $\gamma(\cdot)$ is known, and approximate bounds if it is necessary to estimate $\gamma(\cdot)$ from the observations.

Although we have concentrated here on $\bar{X}_n$ as an estimator of μ, there are other possibilities. If for example we assume a particular model for the data such as $\phi(B)(X_t - \mu) = \theta(B)Z_t$, then it is possible to compute the best linear unbiased estimator $\hat{\mu}_n$ of μ in terms of $X_1, \ldots, X_n$ (see Problem 7.2). However even with this more elaborate procedure, there is little to be gained asymptotically as $n \to \infty$ since it can be shown (see Grenander and Rosenblatt (1957), Section 7.3) that for processes $\{X_t\}$ with piecewise continuous spectral densities (and in particular for ARMA processes)

$$\lim_{n\to\infty} n \operatorname{Var}(\hat{\mu}_n) = \lim_{n\to\infty} n \operatorname{Var}(\bar{X}_n).$$

We shall use the simple estimator $\bar{X}_n$.

§7.2 Estimation of $\gamma(\cdot)$ and $\rho(\cdot)$

The estimators which we shall use for $\gamma(h)$ and $\rho(h)$ are

$$\hat{\gamma}(h) = n^{-1} \sum_{t=1}^{n-h} (X_t - \bar{X}_n)(X_{t+h} - \bar{X}_n), \qquad 0 \le h \le n-1, \tag{7.2.1}$$

and

$$\hat{\rho}(h) = \hat{\gamma}(h)/\hat{\gamma}(0), \tag{7.2.2}$$

respectively. The estimator (7.2.1) is biassed but its asymptotic distribution (as $n \to \infty$) has mean $\gamma(h)$ under the conditions of Proposition 7.3.4 below. The estimators $\hat{\gamma}(h)$, $h = 0, \ldots, n-1$, also have the desirable property that for each $n \ge 1$ the matrix

$$\hat{\Gamma}_n = \begin{bmatrix} \hat{\gamma}(0) & \hat{\gamma}(1) & \cdots & \hat{\gamma}(n-1) \\ \hat{\gamma}(1) & \hat{\gamma}(0) & \cdots & \hat{\gamma}(n-2) \\ \vdots & & & \\ \hat{\gamma}(n-1) & \hat{\gamma}(n-2) & \cdots & \hat{\gamma}(0) \end{bmatrix} \tag{7.2.3}$$

is non-negative definite. To see this we write

$$\hat{\Gamma}_n = n^{-1} T T',$$

where T is the $n \times 2n$ matrix,

$$T = \begin{bmatrix} 0 & \cdots & & 0 & Y_1 & Y_2 & \cdots & Y_n \\ 0 & \cdots & 0 & Y_1 & Y_2 & \cdots & Y_n & 0 \\ \vdots & & & & & & & \vdots \\ 0 & Y_1 & Y_2 & \cdots & Y_n & 0 & \cdots & 0 \end{bmatrix}$$

and $Y_i = X_i - \bar{X}_n$, $i = 1, \ldots, n$. Then for any real $n \times 1$ vector $\mathbf{a}$ we have

$$\mathbf{a}'\hat{\Gamma}_n\mathbf{a} = n^{-1}(\mathbf{a}'T)(\mathbf{a}'T)' \geq 0,$$

and consequently the sample autocovariance matrix $\hat{\Gamma}_n$ and sample autocorrelation matrix,

$$\hat{R}_n = \hat{\Gamma}_n/\hat{\gamma}(0), \tag{7.2.4}$$

are both non-negative definite. The factor n^{-1} is sometimes replaced by $(n-h)^{-1}$ in the definition of $\hat{\gamma}(h)$, but the matrices $\hat{\Gamma}_n$ and $\hat{R}_n$ may not then be non-negative definite. We shall therefore always use the definitions (7.2.1) and (7.2.2) of $\hat{\gamma}(h)$ and $\hat{\rho}(h)$. Note that $\det \hat{\Gamma}_n > 0$ if $\hat{\gamma}(0) > 0$ (Problem 7.11).

From $X_1, \ldots, X_n$ it is of course impossible without further information to estimate $\gamma(k)$ and $\rho(k)$ for $k \geq n$, and for k slightly smaller than n we should expect that any estimators will be unreliable since there are so few pairs (X_t, X_{t+k}) available (only one if $k = n - 1$). Box and Jenkins (1976), p. 33, suggest that useful estimates of correlation $\rho(k)$ can only be made if n is roughly 50 or more and $k \leq n/4$.

It will be important in selecting an appropriate ARMA model for a given set of observations to be able to recognize when sample autocorrelations are significantly different from zero. In order to do this we use the following theorem which gives the asymptotic joint distribution for fixed h of $\hat{\rho}(1), \ldots, \hat{\rho}(h)$ as $n \to \infty$.

Theorem 7.2.1. *If $\{X_t\}$ is the stationary process,*

$$X_t - \mu = \sum_{j=-\infty}^{\infty} \psi_j Z_{t-j}, \qquad \{Z_t\} \sim \text{IID}(0, \sigma^2),$$

where $\sum_{j=-\infty}^{\infty} |\psi_j| < \infty$ and $EZ_t^4 < \infty$, then for each $h \in \{1, 2, \ldots\}$ we have

$$\hat{\boldsymbol{\rho}}(h) \text{ is } \text{AN}(\boldsymbol{\rho}(h), n^{-1}W),$$

where

$$\hat{\boldsymbol{\rho}}(h)' = [\hat{\rho}(1), \hat{\rho}(2), \ldots, \hat{\rho}(h)],$$

$$\boldsymbol{\rho}(h)' = [\rho(1), \rho(2), \ldots, \rho(h)],$$

and W is the covariance matrix whose (i,j)-element is given by Bartlett's formula,

$$w_{ij} = \sum_{k=-\infty}^{\infty} \{\rho(k+i)\rho(k+j) + \rho(k-i)\rho(k+j) + 2\rho(i)\rho(j)\rho^2(k)$$
$$- 2\rho(i)\rho(k)\rho(k+j) - 2\rho(j)\rho(k)\rho(k+i)\}.$$

PROOF. See Section 7.3. □

In the following theorem, the finite fourth moment assumption is relaxed at the expense of a slightly stronger assumption on the sequence $\{\psi_j\}$.

Theorem 7.2.2. *If $\{X_t\}$ is the stationary process*

$$X_t - \mu = \sum_{j=-\infty}^{\infty} \psi_j Z_{t-j}, \qquad \{Z_t\} \sim \text{IID}(0, \sigma^2),$$

where $\sum_{j=-\infty}^{\infty} |\psi_j| < \infty$ and $\sum_{j=-\infty}^{\infty} \psi_j^2 |j| < \infty$, then for each $h \in \{1, 2, \ldots\}$

$$\hat{\boldsymbol{\rho}}(h) \text{ is } \text{AN}(\boldsymbol{\rho}(h), n^{-1} W),$$

where $\hat{\boldsymbol{\rho}}(h)$, $\boldsymbol{\rho}(h)$ and W are defined as in Theorem 7.2.1.

PROOF. See Section 7.3.

Remark 1. Simple algebra shows that

$$\begin{aligned} w_{ij} = \sum_{k=1}^{\infty} &\{\rho(k+i) + \rho(k-i) - 2\rho(i)\rho(k)\} \\ &\times \{\rho(k+j) + \rho(k-j) - 2\rho(j)\rho(k)\}, \end{aligned} \tag{7.2.5}$$

which is a more convenient form of w_{ij} for computational purposes. This formula also shows that the asymptotic distribution of $n^{1/2}(\hat{\boldsymbol{\rho}}(h) - \boldsymbol{\rho}(h))$ is the same as that of the random vector $(Y_1, \ldots, Y_h)'$, where

$$Y_i = \sum_{k=1}^{\infty} (\rho(k+i) + \rho(k-i) - 2\rho(i)\rho(k)) N_k, \qquad i = 1, \ldots, h, \tag{7.2.6}$$

and $N_1, N_2, \ldots$ are iid N(0, 1) random variables. The proof of Theorem 7.2.2 shows in fact that the limit distribution of $n^{1/2}(\hat{\boldsymbol{\rho}}(h) - \boldsymbol{\rho}(h))$ is completely determined by the limit distribution of the random variables $\sigma^{-2} n^{-1/2} \sum_{t=1}^{n} Z_t Z_{t+i}$, $i = 1, 2, \ldots$ which are asymptotically iid N(0, 1) (see Problem 6.24).

Remark 2. Before considering some applications of Theorem 7.2.2 we note that its conditions are satisfied by every ARMA(p, q) process driven by an iid sequence $\{Z_t\}$ with zero mean and finite variance. The assumption of identical distributions in Theorems 7.1.2 and 7.2.1 can also be replaced by the boundedness of $E|Z_t|^3$ and $E|Z_t|^6$ respectively (or by other conditions which permit the use in the proofs of a central limit theorem for non-identically distributed random variables). This should be kept in mind in applying the results.

EXAMPLE 7.2.1 (Independent White Noise). If $\{X_t\} \sim \text{IID}(0, \sigma^2)$, then $\rho(l) = 0$ if $|l| > 0$, so from (7.2.5) we obtain

$$w_{ij} = \begin{cases} 1 & \text{if } i = j, \\ 0 & \text{otherwise.} \end{cases}$$

For large n therefore $\hat{\rho}(1), \ldots, \hat{\rho}(h)$ are approximately independent and identically distributed normal random variables with mean 0 and variance n^{-1}. If we plot the sample autocorrelation function $\hat{\rho}(k)$ as a function of k, approximately .95 of the sample autocorrelations should lie between the bounds $\pm 1.96n^{-1/2}$. This can be used as a check that the observations truly are from an IID process. In Figure 7.1 we have plotted the sample autocorrelation $\hat{\rho}(k)$, $k = 1, \ldots, 40$ for a sample of 200 independent observations from the distribution N(0, 1). It can be seen that all but one of the autocorrelations lie between the bounds $\pm 1.96n^{-1/2}$. If we had been given the data with no prior information, inspection of the sample autocorrelation function would have given us no grounds on which to reject the simple hypothesis that the data is a realization of a white noise process.

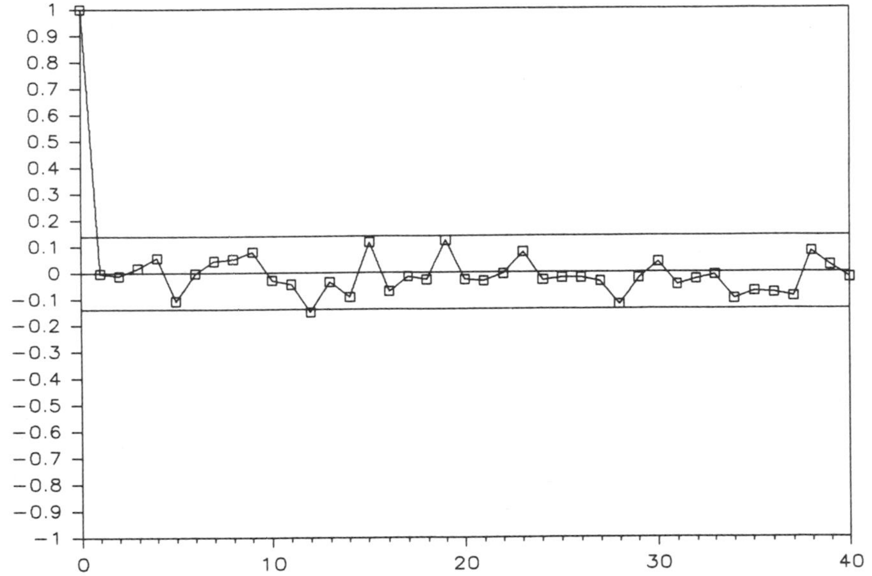

Figure 7.1. The sample autocorrelation function of $n = 200$ observations of Gaussian white noise, showing the bounds $\pm 1.96n^{-1/2}$.

Example 7.2.2 (Moving Average of Order q). If

$$X_t = Z_t + \theta_1 Z_{t-1} + \cdots + \theta_q Z_{t-q}, \qquad \{Z_t\} \sim \text{IID}(0, \sigma^2),$$

then from Bartlett's formula (7.2.5) we have

$$w_{ii} = [1 + 2\rho^2(1) + 2\rho^2(2) + \cdots + 2\rho^2(q)], \qquad i > q,$$

as the variance of the asymptotic distribution of $n^{1/2}\hat{\rho}(i)$ as $n \to \infty$. In Figure

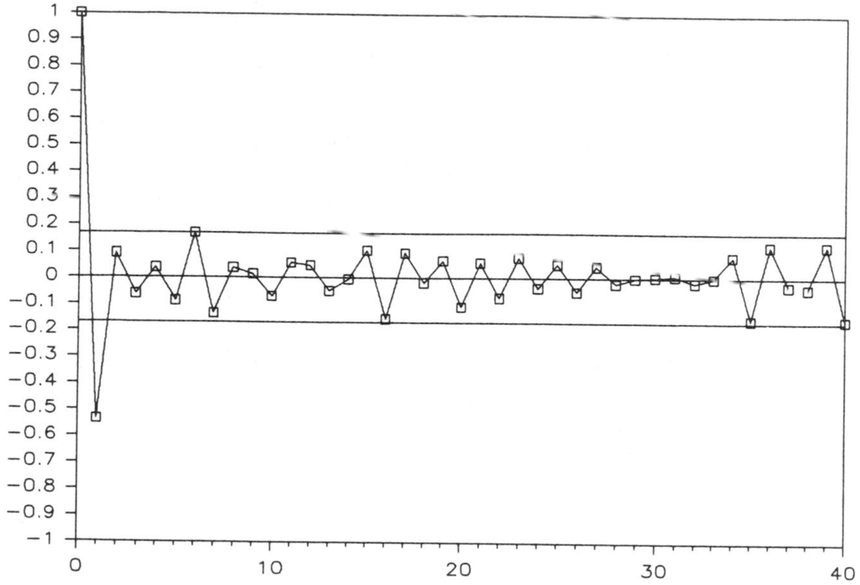

Figure 7.2. The sample autocorrelation function of $n = 200$ observations of the Gaussian MA(1) process, $X_t = Z_t - .8Z_{t-1}$, $\{Z_t\} \sim \text{WN}(0, 1)$, showing the bounds $\pm 1.96n^{-1/2}[1 + 2\rho^2(1)]^{1/2}$.

7.2 we have plotted the sample autocorrelation function $\hat{\rho}(k)$, $k = 0, 1, \ldots, 40$, for 200 observations from the Gaussian MA(1) process

$$X_t = Z_t - .8Z_{t-1}, \qquad \{Z_t\} \sim \text{IID}(0, 1). \tag{7.2.6}$$

The lag-one sample autocorrelation is found to be $\hat{\rho}(1) = -.5354 = -7.57n^{-1/2}$, which would cause us (in the absence of our prior knowledge of $\{X_t\}$) to reject the hypothesis that the data is a sample from a white noise process. The fact that $|\hat{\rho}(k)| < 1.96n^{-1/2}$ for $k \in \{2, \ldots, 40\}$ strongly suggests that the data is from a first-order moving average process. In Figure 7.2 we have plotted the bounds $\pm 1.96n^{-1/2}[1 + 2\rho^2(1)]^{1/2}$ where $\rho(1) = -.8/1.64 = -.4878$. The sample autocorrelations $\hat{\rho}(2), \ldots, \hat{\rho}(40)$ all lie within these bounds, indicating the compatibility of the data with the model (7.2.6). Since however $\rho(1)$ is not normally known in advance, the autocorrelations $\hat{\rho}(2), \ldots, \hat{\rho}(40)$ would in practice have been compared with the more stringent bounds $\pm 1.96n^{-1/2}$ or with the bounds $\pm 1.96n^{-1/2}[1 + 2\hat{\rho}^2(1)]^{1/2}$ in order to check the hypothesis that the data is generated by a moving average process of order 1.

EXAMPLE 7.2.3 (Autoregressive Process of Order 1). Applying Bartlett's formula to the causal AR(1) process,

$$X_t - \phi X_{t-1} = Z_t, \qquad \{Z_t\} \sim \text{IID}(0, \sigma^2),$$

and using the result (see Section 3.1) that $\rho(i) = \phi^{|i|}$, we find that the asymptotic variance of $n^{1/2}(\hat{\rho}(i) - \rho(i))$ is

$$\begin{aligned} w_{ii} &= \sum_{k=1}^{i} \phi^{2i}(\phi^{-k} - \phi^{k})^2 + \sum_{k=i+1}^{\infty} \phi^{2k}(\phi^{-i} - \phi^{i})^2 \\ &= (1 - \phi^{2i})(1 + \phi^2)(1 - \phi^2)^{-1} - 2i\phi^{2i}, \qquad i = 1, 2, \ldots, \\ &\simeq (1 + \phi^2)/(1 - \phi^2) \quad \text{for } i \text{ large.} \end{aligned}$$

The result is not of the same importance in model identification as the corresponding result for moving average processes, since autoregressive processes are more readily identified from the vanishing of the partial autocorrelation function at lags greater than the order of the autoregression. We shall return to the general problem of identifying an appropriate model for a given time series in Chapter 9.

§7.3* Derivation of the Asymptotic Distributions

This section is devoted to the proofs of Theorems 7.1.2, 7.2.1 and 7.2.2. For the statements of these we refer the reader to Sections 7.1 and 7.2. The proof of Theorem 7.1.2, being a rather straightforward application of the techniques of Chapter 6, is given first. We then proceed in stages through Propositions 7.3.1–7.3.4 to the proof of Theorem 7.2.1 and Propositions 7.3.5–7.3.8 to the proof of Theorem 7.2.2.

PROOF OF THEOREM 7.1.2. We first define

$$X_{tm} = \mu + \sum_{j=-m}^{m} \psi_j Z_{t-j}$$

and

$$Y_{nm} = \bar{X}_{nm} = \left(\sum_{t=1}^{n} X_{tm}\right)\bigg/ n.$$

By Example 6.4.4, as $n \to \infty$,

$$n^{1/2}(Y_{nm} - \mu) \Rightarrow Y_m \quad \text{where } Y_m \sim \mathrm{N}\left(0, \sigma^2\left(\sum_{j=-m}^{m} \psi_j\right)^2\right). \tag{7.3.1}$$

Now as $m \to \infty$, $\sigma^2(\sum_{j=-m}^{m} \psi_j)^2 \to \sigma^2(\sum_{j=-\infty}^{\infty} \psi_j)^2$, and so by Problem 6.16,

$$Y_m \Rightarrow Y \quad \text{where } Y \sim \mathrm{N}\left(0, \sigma^2\left(\sum_{j=-\infty}^{\infty} \psi_j\right)^2\right). \tag{7.3.2}$$

By Remark 2 of Section 7.1,

$$\operatorname{Var}(n^{1/2}(\bar{X}_n - Y_{nm})) = n \operatorname{Var}\left(n^{-1} \sum_{t=1}^{n} \sum_{|j|>m} \psi_j Z_{t-j}\right)$$

$$\to \left(\sum_{|j|>m} \psi_j\right)^2 \sigma^2 \quad \text{as } n \to \infty.$$

Hence

$$\lim_{m\to\infty} \limsup_{n\to\infty} \operatorname{Var}(n^{1/2}(\bar{X}_n - Y_{nm})) = 0,$$

which, in conjunction with Chebychev's inequality, implies that condition (iii) of Proposition 6.3.9 is satisfied. In view of (7.3.1) and (7.3.2) we can therefore apply the Proposition to conclude that $n^{1/2}(\bar{X}_n - \mu) \Rightarrow Y$. □

The asymptotic multivariate normality of the sample autocorrelations (Theorem 7.2.1) will be established by first examining the asymptotic behavior of the sample autocovariances $\hat{\gamma}(h)$ defined by (7.2.1). In order to do this it is simplest to work in terms of the function

$$\gamma^*(h) = n^{-1} \sum_{t=1}^{n} X_t X_{t+h}, \qquad h = 0, 1, 2, \ldots,$$

which, as we shall see in Proposition 7.3.4, has the same asymptotic properties as the sample autocovariance function.

Proposition 7.3.1. *Let $\{X_t\}$ be the two-sided moving average,*

$$X_t = \sum_{j=-\infty}^{\infty} \psi_j Z_{t-j}, \qquad \{Z_t\} \sim \mathrm{IID}(0, \sigma^2),$$

where $EZ_t^4 = \eta\sigma^4 < \infty$ and $\sum_{j=-\infty}^{\infty} |\psi_j| < \infty$. Then if $p \geq 0$ and $q \geq 0$,

$$\lim_{n\to\infty} n \operatorname{Cov}(\gamma^*(p), \gamma^*(q))$$

$$= (\eta - 3)\gamma(p)\gamma(q) + \sum_{k=-\infty}^{\infty} [\gamma(k)\gamma(k - p + q) + \gamma(k + q)\gamma(k - p)], \tag{7.3.3}$$

where $\gamma(\cdot)$ is the autocovariance function of $\{X_t\}$.

Proof. First observe that

$$E(Z_s Z_t Z_u Z_v) = \begin{cases} \eta\sigma^4 & \text{if } s = t = u = v, \\ \sigma^4 & \text{if } s = t \neq u = v, \\ 0 & \text{if } s \neq t, s \neq u \text{ and } s \neq v. \end{cases} \tag{7.3.4}$$

Now

$$E(X_t X_{t+p} X_{t+h+p} X_{t+h+p+q})$$

$$= \sum_i \sum_j \sum_k \sum_l \psi_i \psi_{j+p} \psi_{k+h+p} \psi_{l+h+p+q} E(Z_{t-i} Z_{t-j} Z_{t-k} Z_{t-l})$$

and the sum can be rewritten, using (7.3.4), in the form

$$(\eta - 3)\sigma^4 \sum_i \psi_i\psi_{i+p}\psi_{i+h+p}\psi_{i+h+p+q} + \gamma(p)\gamma(q)$$
$$+ \gamma(h+p)\gamma(h+q) + \gamma(h+p+q)\gamma(h).$$

It follows that

$$E\gamma^*(p)\gamma^*(q) = n^{-2}E\left(\sum_{s=1}^{n}\sum_{t=1}^{n} X_tX_{t+p}X_sX_{s+q}\right)$$
$$= n^{-2}\sum_{s=1}^{n}\sum_{t=1}^{n}\Big[\gamma(p)\gamma(q) + \gamma(s-t)\gamma(s-t-p+q)$$
$$+ \gamma(s-t+q)\gamma(s-t-p)$$
$$+ (\eta-3)\sigma^4\sum_i \psi_i\psi_{i+p}\psi_{i+s-t}\psi_{i+s-t+q}\Big].$$

Letting $k = s - t$, interchanging the order of summation and subtracting $\gamma(p)\gamma(q)$, we find that

$$\operatorname{Cov}(\gamma^*(p), \gamma^*(q)) = n^{-1}\sum_{|k|<n}(1 - n^{-1}|k|)T_k, \tag{7.3.5}$$

where

$$T_k = \gamma(k)\gamma(k-p+q) + \gamma(k+q)\gamma(k-p) + (\eta-3)\sigma^4\sum_i \psi_i\psi_{i+p}\psi_{i+k}\psi_{i+k+q}.$$

The absolute summability of $\{\psi_j\}$ implies that $\{T_k\}$ is also absolutely summable. We can therefore apply the dominated convergence theorem in (7.3.5) to deduce that

$$\lim_{n\to\infty} n\operatorname{Cov}(\gamma^*(p), \gamma^*(q))$$
$$= \sum_{k=-\infty}^{\infty} T_k$$
$$= (\eta-3)\gamma(p)\gamma(q) + \sum_{k=-\infty}^{\infty}[\gamma(k)\gamma(k-p+q) + \gamma(k+q)\gamma(k-p)]. \quad \square$$

Proposition 7.3.2. *If $\{X_t\}$ is the moving average,*

$$X_t = \sum_{j=-m}^{m} \psi_j Z_{t-j}, \qquad \{Z_t\} \sim \text{IID}(0, \sigma^2), \tag{7.3.6}$$

where $EZ_t^4 = \eta\sigma^4 < \infty$, and if $\gamma(\cdot)$ is the autocovariance function of $\{X_t\}$, then for any non-negative integer h,

$$\begin{bmatrix} \gamma^*(0) \\ \vdots \\ \gamma^*(h) \end{bmatrix} \text{ is AN}\left(\begin{bmatrix} \gamma(0) \\ \vdots \\ \gamma(h) \end{bmatrix}, n^{-1}V\right),$$

where V is the covariance matrix,

$$V = \left[(\eta - 3)\gamma(p)\gamma(q) + \sum_{k=-\infty}^{\infty} (\gamma(k)\gamma(k - p + q) \right.$$
$$\left. + \gamma(k + q)\gamma(k - p)) \right]_{p,q=0,\dots,h}.$$

PROOF. We define a sequence of random $(h + 1)$-vectors $\{\mathbf{Y}_t\}$ by

$$\mathbf{Y}_t' = (X_t X_t, X_t X_{t+1}, \dots, X_t X_{t+h}).$$

Then $\{\mathbf{Y}_t\}$ is a strictly stationary $(2m + h)$-dependent sequence and

$$n^{-1} \sum_{t=1}^{n} \mathbf{Y}_t = (\gamma^*(0), \dots, \gamma^*(h))'.$$

We therefore need to show that as $n \to \infty$,

$$n^{-1} \sum_{t=1}^{n} \boldsymbol{\lambda}'\mathbf{Y}_t \quad \text{is} \quad \text{AN}\left(\boldsymbol{\lambda}' \begin{bmatrix} \gamma(0) \\ \vdots \\ \gamma(h) \end{bmatrix}, n^{-1}\boldsymbol{\lambda}'V\boldsymbol{\lambda} \right), \tag{7.3.7}$$

for all vectors $\boldsymbol{\lambda} \in \mathbb{R}^{h+1}$ such that $\boldsymbol{\lambda}'V\boldsymbol{\lambda} > 0$. For any such $\boldsymbol{\lambda}$, the sequence $\{\boldsymbol{\lambda}'\mathbf{Y}_t\}$ is $(2m + h)$-dependent and since, by Proposition 7.3.1,

$$\lim_{n \to \infty} n^{-1} \operatorname{Var}\left(\sum_{t=1}^{n} \boldsymbol{\lambda}'\mathbf{Y}_t \right) = \boldsymbol{\lambda}'V\boldsymbol{\lambda} > 0,$$

we conclude from Remark 6 of Section 6.4 that $\{\boldsymbol{\lambda}'\mathbf{Y}_t\}$ satisfies the hypotheses of Theorem 6.4.2. Application of the theorem immediately gives the required result (7.3.7). □

The next step is to extend Proposition 7.3.2 to MA(∞) processes.

Proposition 7.3.3. *Proposition* 7.3.2 *remains true if we replace* (7.3.6) *by*

$$X_t = \sum_{j=-\infty}^{\infty} \psi_j Z_{t-j}, \qquad \{Z_t\} \sim \text{IID}(0, \sigma^2), \tag{7.3.8}$$

where $\sum_{j=-\infty}^{\infty} |\psi_j| < \infty$ *and* $EZ_t^4 = \eta\sigma^4 < \infty$.

PROOF. The idea of the proof is to apply Proposition 7.3.2 to the truncated sequence

$$X_{tm} = \sum_{j=-m}^{m} \psi_j Z_{t-j},$$

and then to derive the result for $\{X_t\}$ by letting $m \to \infty$. For $0 \le p \le h$ we define

$$\gamma_m^*(p) = n^{-1} \sum_{t=1}^{n} X_{tm} X_{(t+p)m}.$$

Then by Proposition 7.3.2

$$n^{1/2}\begin{bmatrix}\gamma_m^*(0)-\gamma_m(0)\\ \vdots \\ \gamma_m^*(h)-\gamma_m(h)\end{bmatrix} \Rightarrow \mathbf{Y}_m,$$

where $\gamma_m(\cdot)$ is the autocovariance function of $\{X_{tm}\}$, $\mathbf{Y}_m \sim \mathrm{N}(\mathbf{0}, V_m)$ and

$$V_m = \Big[(\eta-3)\gamma_m(p)\gamma_m(q) + \sum_{k=-\infty}^{\infty}(\gamma_m(k)\gamma_m(k-p+q)$$
$$+ \gamma_m(k+q)\gamma_m(k-p))\Big]_{p,q=0,\ldots,h}.$$

Now as $m \to \infty$,

$$V_m \to V,$$

where V is defined like V_m with $\gamma_m(\cdot)$ replaced by $\gamma(\cdot)$. Hence

$$\mathbf{Y}_m \Rightarrow \mathbf{Y} \quad \text{where } \mathbf{Y} \sim \mathrm{N}(\mathbf{0}, V).$$

The proof can now be completed by an application of Proposition 6.3.9 provided we can show that

$$\lim_{m\to\infty}\limsup_{n\to\infty} P(n^{1/2}|\gamma_m^*(p)-\gamma_m(p)-\gamma^*(p)+\gamma(p)| > \varepsilon) = 0, \tag{7.3.9}$$

for $p = 0, 1, \ldots, h$.

The probability in (7.3.9) is bounded by $\varepsilon^{-2} n \operatorname{Var}(\gamma_m^*(p) - \gamma^*(p)) = \varepsilon^{-2}[n\operatorname{Var}(\gamma_m^*(p)) + n\operatorname{Var}(\gamma^*(p)) - 2n\operatorname{Cov}(\gamma_m^*(p), \gamma^*(p))]$. From the calculations of Proposition 7.3.1 and the preceding paragraph,

$$\lim_{m\to\infty}\lim_{n\to\infty} n\operatorname{Var}(\gamma_m^*(p)) = \lim_{n\to\infty} n\operatorname{Var}(\gamma^*(p))$$
$$= v_{pp},$$

where v_{pq} is the (p, q)-element of V. Moreover by a calculation similar to that given in the proof of Proposition 7.3.1, it can be shown that

$$\lim_{m\to\infty}\lim_{n\to\infty} n\operatorname{Cov}(\gamma_m^*(p), \gamma^*(p)) = v_{pp}, \tag{7.3.10}$$

whence

$$\lim_{m\to\infty}\limsup_{n\to\infty} \varepsilon^{-2} n\operatorname{Var}(\gamma_m^*(p) - \gamma^*(p)) = 0. \tag{7.3.11}$$

This establishes (7.3.9). □

Next we show that, under the conditions of Proposition 7.3.3, the vectors $[\gamma^*(0), \ldots, \gamma^*(h)]'$ and $[\hat{\gamma}(0), \ldots, \hat{\gamma}(h)]'$ have the same asymptotic distribution.

Proposition 7.3.4. *If* $\{X_t\}$ *is the moving average process,*

$$X_t = \sum_{j=-\infty}^{\infty} \psi_j Z_{t-j}, \qquad \{Z_t\} \sim \text{IID}(0, \sigma^2),$$

where $\sum_{j=-\infty}^{\infty} |\psi_j| < \infty$ *and* $EZ_t^4 = \eta\sigma^4 < \infty$, *and if* $\gamma(\cdot)$ *is the autocovariance function of* $\{X_t\}$, *then for any non-negative integer* h,

$$\begin{bmatrix} \hat{\gamma}(0) \\ \vdots \\ \hat{\gamma}(h) \end{bmatrix} \text{ is AN}\left(\begin{bmatrix} \gamma(0) \\ \vdots \\ \gamma(h) \end{bmatrix}, n^{-1}V \right) \tag{7.3.12}$$

where V *is the covariance matrix,*

$$V = \left[(\eta - 3)\gamma(p)\gamma(q) + \sum_{k=-\infty}^{\infty} (\gamma(k)\gamma(k - p + q) + \gamma(k + q)\gamma(k - p)) \right]_{p,q=0,\ldots,h}. \tag{7.3.13}$$

PROOF. Simple algebra gives, for $0 \le p \le h$,

$$n^{1/2}(\gamma^*(p) - \hat{\gamma}(p)) = n^{1/2}\bar{X}_n \left[n^{-1} \sum_{t=1}^{n-p} X_{t+p} + n^{-1} \sum_{t=1}^{n-p} X_t + (1 - n^{-1}p)\bar{X}_n \right] + n^{-1/2} \sum_{t=n-p+1}^{n} X_t X_{t+p}.$$

The last term is $o_p(1)$ since $n^{-1/2}E|\sum_{t=n-p+1}^{n} X_t X_{t+p}| \le n^{-1/2}p\gamma(0)$ and $n^{-1/2}p\gamma(0) \to 0$ as $n \to \infty$. By Theorem 7.1.2 we also know that

$$n^{1/2}\bar{X}_n \Rightarrow Y \quad \text{where } Y \sim \text{N}\left(0, \sigma^2 \left(\sum_{j=-\infty}^{\infty} \psi_j \right)^2 \right),$$

which implies that $n^{1/2}\bar{X}_n$ is $O_p(1)$. Moreover by the weak law of large numbers (cf. Proposition 6.3.10),

$$\left[n^{-1} \sum_{t=1}^{n-p} X_{t+p} + n^{-1} \sum_{t=1}^{n-p} X_t + (1 - n^{-1}p)\bar{X}_n \right] \xrightarrow{P} 0.$$

From these observations we conclude that

$$n^{1/2}(\gamma^*(p) - \hat{\gamma}(p)) = o_p(1) \quad \text{as } n \to \infty,$$

and the conclusion of the proposition then follows from Propositions 6.3.3 and 7.3.3. □

Remark 1. If $\{Y_t\}$ is a stationary process with mean μ, then Propositions 7.3.1–7.3.4 apply to the process $\{X_t\} = \{Y_t - \mu\}$, provided of course the specified conditions are satisfied by $\{Y_t - \mu\}$. In particular if

$$Y_t = \mu + \sum_{j=-\infty}^{\infty} \psi_j Z_{t+j}, \qquad \{Z_t\} \sim \text{IID}(0, \sigma^2),$$

where $\sum_{j=-\infty}^{\infty} |\psi_j| < \infty$ and $EZ_t^4 = \eta\sigma^4 < \infty$ and if $\gamma(\cdot)$ is the autocovariance function of $\{Y_t\}$, then for any non-negative integer h,

$$\begin{bmatrix} \hat{\gamma}(0) \\ \vdots \\ \hat{\gamma}(h) \end{bmatrix} \text{ is AN}\left(\begin{bmatrix} \gamma(0) \\ \vdots \\ \gamma(h) \end{bmatrix}, n^{-1}V\right),$$

where V is defined by (7.3.13) and $\hat{\gamma}(p) = n^{-1}\sum_{j=1}^{n-h}(Y_j - \bar{Y}_n)(Y_{j+h} - \bar{Y}_n)$.

We are now in a position to prove the asymptotic joint normality of the sample autocorrelations.

Proof of Theorem 7.2.1. Let $g(\cdot)$ be the function from $\mathbb{R}^{h+1}$ into $\mathbb{R}^h$ defined by

$$\mathbf{g}([x_0, x_1, \ldots, x_h]') = [x_1/x_0, \ldots, x_h/x_0]', \qquad x_0 \neq 0.$$

If $\gamma(\cdot)$ is the autocovariance function of $\{X_t\}$, then by Proposition 6.4.3 and Remark 1 above,

$$\hat{\boldsymbol{\rho}}(h) = \mathbf{g}([\hat{\gamma}(0), \ldots, \hat{\gamma}(h)]') \quad \text{is AN}(\mathbf{g}([\gamma(0), \ldots, \gamma(h)]'), n^{-1}DVD'),$$

i.e. $\hat{\boldsymbol{\rho}}(h)$ is AN$(\boldsymbol{\rho}(h), n^{-1}DVD')$, where V is defined by (7.3.13) and D is the matrix of partial derivatives,

$$D = \gamma(0)^{-1}\begin{bmatrix} -\rho(1) & 1 & 0 & \cdots & 0 \\ -\rho(2) & 0 & 1 & & 0 \\ \vdots & & & \ddots & \\ -\rho(h) & 0 & 0 & \cdots & 1 \end{bmatrix}.$$

Denoting by v_{ij} and w_{ij} the (i,j)-elements of V and DVD' respectively, we find that

$$\begin{aligned} w_{ij} &= v_{ij} - \rho(i)v_{0j} - \rho(j)v_{i0} + \rho(i)\rho(j)v_{00} \\ &= \sum_{k=-\infty}^{\infty}\Big[\rho(k)\rho(k-i+j) + \rho(k-i)\rho(k+j) + 2\rho(i)\rho(j)\rho^2(k) \\ &\qquad - 2\rho(i)\rho(k)\rho(k+j) - 2\rho(j)\rho(k)\rho(k-i)\Big]. \end{aligned}$$

Noting that $\sum_k \rho(k)\rho(k-i+j) = \sum_k \rho(k+i)\rho(k+j)$ and that $\sum_k \rho(j)\rho(k)\rho(k-i) = \sum_k \rho(j)\rho(k+i)\rho(k)$, we see that w_{ij} is exactly as specified in the statement of Theorem 7.2.1. □

We next turn to the proof of Theorem 7.2.2 which is broken up into a series of propositions.

Proposition 7.3.5. *If $\{X_t\}$ is the moving average process*

$$X_t = \sum_{j=-\infty}^{\infty} \psi_j Z_{t-j}, \qquad \{Z_t\} \sim \text{IID}(0, \sigma^2),$$

where $\sum_{j=-\infty}^{\infty} |\psi_j| < \infty$ *and* $\sum_{j=-\infty}^{\infty} \psi_j^2 |j| < \infty$, *then for* $h \geq 0$,

$$\gamma^*(h) \xrightarrow{P} \left(\sum_{j=-\infty}^{\infty} \psi_j \psi_{j+h} \right) \sigma^2 = \gamma(h).$$

PROOF. We give the proof for $h = 0$. The general case is similar. Now

$$\begin{aligned} \gamma^*(0) &= n^{-1} \sum_{t=1}^{n} \sum_{i,j} \psi_i \psi_j Z_{t-i} Z_{t-j} \\ &= n^{-1} \sum_{t=1}^{n} \sum_{i} \psi_i^2 Z_{t-i}^2 + Y_n, \end{aligned}$$

where $Y_n = \sum \sum_{i \neq j} \psi_i \psi_j n^{-1} \sum_{t=1}^{n} Z_{t-i} Z_{t-j}$. By the weak law of large numbers for moving averages (Proposition 6.3.10), the first term converges in probability to $(\sum_i \psi_i^2)\sigma^2$. So it suffices to show that $Y_n \xrightarrow{P} 0$. For $i \neq j$, $\{Z_{t-i} Z_{t-j}, t = 0, \pm 1, \ldots\} \sim \text{WN}(0, \sigma^4)$ and hence

$$\text{Var}\left(n^{-1} \sum_{t=1}^{n} Z_{t-i} Z_{t-j} \right) = n^{-1} \sigma^4 \to 0.$$

Thus for each positive integer k

$$Y_{nk} = \sum\sum_{|i| \leq k, |j| \leq k, i \neq j} \psi_i \psi_j n^{-1} \sum_{t=1}^{n} Z_{t-i} Z_{t-j} \xrightarrow{P} 0,$$

and

$$\begin{aligned} \lim_{k \to \infty} \limsup_{n \to \infty} E|Y_n - Y_{nk}| &\leq \lim_{k \to \infty} \limsup_{n \to \infty} \sum_{|i| > k} \sum_{|j| > k} |\psi_i \psi_j| E|Z_1 Z_2| \\ &= 0. \end{aligned}$$

Now appealing to Proposition 6.3.9, we deduce that $Y_n \xrightarrow{P} 0$. □

Proposition 7.3.6. *Let* $\{X_t\}$ *be as defined in Proposition* 7.3.5 *and set*

$$\rho^*(h) = \frac{\gamma^*(h)}{\gamma^*(0)} \quad \textit{for } h = 1, 2, \ldots.$$

Then

$$n^{1/2} \left(\rho^*(h) - \rho(h) - (\gamma^*(0))^{-1} n^{-1/2} \sum_{t=1}^{n} \sum_{j \neq 0} \sum_{i} a_{ij} Z_{t-i} Z_{t-i+j} \right) \xrightarrow{P} 0 \qquad (7.3.14)$$

where

$$a_{ij} = \psi_i(\psi_{i-j+h} - \rho(h)\psi_{i-j}), \qquad i = 0, \pm 1, \ldots; \quad j = \pm 1, \pm 2, \ldots.$$

PROOF. We have

$$\rho^*(h)-\rho(h)=(\gamma^*(0))^{-1}(\gamma^*(h)-\rho(h)\gamma^*(0))$$
$$=(\gamma^*(0))^{-1}n^{-1}\sum_{t=1}^{n}\left(\sum_i\sum_j\psi_i\psi_jZ_{t-i}Z_{t+h-j}-\rho(h)\sum_i\sum_j\psi_i\psi_jZ_{t-i}Z_{t-j}\right)$$
$$=(\gamma^*(0))^{-1}n^{-1}\sum_{t=1}^{n}\sum_i\sum_j\psi_i(\psi_{i-j+h}-\rho(h)\psi_{i-j})Z_{t-i}Z_{t-i+j},$$

so that the left side of (7.3.14) is

$$(\gamma^*(0))^{-1}n^{-1/2}\sum_i\left(\psi_i(\psi_{i+h}-\rho(h)\psi_i)\sum_{t=1}^{n}Z_{t-i}^2\right)$$
$$=(\gamma^*(0))^{-1}n^{-1/2}\sum_i\left[\psi_i(\psi_{i+h}-\rho(h)\psi_i)\left(\sum_{t=1}^{n}Z_t^2+U_{ni}\right)\right], \tag{7.3.15}$$

where $U_{ni}=\sum_{t=1-i}^{n-i}Z_t^2-\sum_{t=1}^{n}Z_t^2$ is a sum of at most $2|i|$ random variables. Since $\sum_i\psi_i(\psi_{i+h}-\rho(h)\psi_i)=0$ and $\gamma^*(0)\xrightarrow{P}(\sum_i\psi_i^2)\sigma^2$, the proof will be complete by Proposition 6.1.1 once we show that

$$\sum_i\psi_i(\psi_{i+h}-\rho(h)\psi_i)U_{ni}=O_p(1). \tag{7.3.16}$$

But,

$$\limsup_{n\to\infty}E\left|\sum_i\psi_i(\psi_{i+h}-\rho(h)\psi_i)U_{ni}\right|$$
$$\le\sum_i\left(|\psi_i\psi_{i+h}|+|\psi_i|^2\right)(2|i|)\sigma^2$$
$$\le 2\sigma^2\left(\sum_i\psi_i^2|i|\right)^{1/2}\left(\sum_i\psi_{i+h}^2|i|\right)^{1/2}+2\sigma^2\sum_i\psi_i^2|i|$$
$$<\infty$$

and this inequality implies (7.3.16) as required. □

Proposition 7.3.7. *Let X_t and a_{ij} be as defined in Proposition 7.3.6. Then for each positive integer j,*

$$n^{-1/2}\left[\sum_i a_{ij}\sum_{t=1}^{n}Z_{t-i}Z_{t-i+j}-\sum_i a_{ij}\sum_{t=1}^{n}Z_tZ_{t+j}\right]\xrightarrow{P}0 \tag{7.3.17}$$

and

$$n^{-1/2}\left[\sum_i a_{i,-j}\sum_{t=1}^{n}Z_{t-i}Z_{t-i-j}-\sum_i a_{i,-j}\sum_{t=1}^{n}Z_tZ_{t+j}\right]\xrightarrow{P}0. \tag{7.3.18}$$

PROOF. The left side of (7.3.17) is equal to $n^{-1/2}\sum_i a_{ij}U_{ni}$ where $U_{ni}=\sum_{t=1-i}^{n-i}Z_tZ_{t+j}-\sum_{t=1}^{n}Z_tZ_{t+j}$ is a sum of at most $2|i|$ products Z_tZ_{t+j}. Moreover,

$$\begin{aligned}\limsup_{n\to\infty} E\left|\sum_i a_{ij}U_{ni}\right| &\le \sum_i \left(|\psi_i\psi_{i-j+h}| + |\psi_i\psi_{i-j}|\right)2|i|\sigma^2\\ &\le 2\sigma^2\left[\left(\sum_i \psi_i^2|i|\right)^{1/2}\left(\sum_i \psi_{i-j+h}^2|i|\right)^{1/2}\right.\\ &\quad\left.+\left(\sum_i \psi_i^2|i|\right)^{1/2}\left(\sum_i \psi_{i-j}^2|i|\right)^{1/2}\right]\\ &< \infty.\end{aligned}$$

Consequently $n^{-1/2}\sum_i a_{ij}U_{ni} \xrightarrow{P} 0$. The proof of (7.3.18) is practically identical and is therefore omitted. □

Proposition 7.3.8. *Let $\{X_t\}$ be the moving average process defined in Proposition 7.3.5. Then for every positive integer h,*

$$n^{1/2}(\boldsymbol{\rho}^*(h) - \boldsymbol{\rho}(h))' \Rightarrow (Y_1, \ldots, Y_h)'$$

where $\boldsymbol{\rho}^*(h) = (\rho^*(1), \ldots, \rho^*(h))$,

$$Y_k = \sum_{j=1}^{\infty} (\rho(k+j) + \rho(k-j) - 2\rho(j)\rho(k))N_j,$$

and $N_1, N_2, \ldots$ are iid N(0, 1) *random variables.*

PROOF. By Proposition 7.3.6,

$$n^{1/2}(\rho^*(h) - \rho(h)) = (\gamma^*(0))^{-1}n^{-1/2}\sum_{t=1}^{n}\sum_{j\neq 0}\sum_i a_{ij}Z_{t-i}Z_{t-i+j} + o_p(1). \quad (7.3.19)$$

Also by Problem 6.24, we have for each fixed positive integer m,

$$\sigma^{-2}n^{-1/2}\left(\sum_{t=1}^{n} Z_tZ_{t+1}, \ldots, \sum_{t=1}^{n} Z_tZ_{t+m}\right) \Rightarrow (N_1, \ldots, N_m)'$$

where $N_1, \ldots, N_m$ are iid N(0, 1) random variables. It then follows from Propositions 7.3.7 and 6.3.4 that

$$\sigma^{-2}n^{-1/2}\sum_{0<|j|\le m}\sum_{t=1}^{n}\sum_i a_{ij}Z_{t-i}Z_{t-i+j} \Rightarrow \sum_{j=1}^{m}\left(\sum_i (a_{ij} + a_{i,-j})\right)N_j. \quad (7.3.20)$$

We next show that (7.3.20) remains valid with m replaced by ∞. By Proposition 6.3.9, m may be replaced by ∞ provided

$$\sum_{j=1}^{m}\sum_i (a_{ij} + a_{i,-j})N_j \Rightarrow \sum_{j=1}^{\infty}\sum_i (a_{ij} + a_{i,-j})N_j \quad \text{as } m \to \infty \quad (7.3.21)$$

and

$$\lim_{m\to\infty}\limsup_{n\to\infty} \operatorname{Var}\left(n^{-1/2}\sum_{|j|>m}\sum_{t=1}^{n}\left(\sum_i a_{ij}Z_{t-i}Z_{t-i+j}\right)\right) = 0. \quad (7.3.22)$$

Now (7.3.21) is clear (in fact we have convergence in probability). To prove (7.3.22), we write $\operatorname{Var}(n^{-1/2}\sum_{|j|>m}\sum_{t=1}^{n}\sum_i a_{ij}Z_{t-i}Z_{t-i+j})$ as

$$\begin{aligned}
n^{-1}&\sum_{s=1}^{n}\sum_{t=1}^{n}\sum_{i}\sum_{|j|>m}\sum_{k}\sum_{|l|>m} a_{ij}a_{kl}E(Z_{t-i}Z_{t-i+j}Z_{s-k}Z_{s-k+l})\\
&= n^{-1}\sum_{s=1}^{n}\sum_{t=1}^{n}\sum_{i}\sum_{|j|>m} a_{ij}(a_{s-t+i,j}+a_{s-t+i-j,-j})\sigma^4\\
&\le \sum_{|t|<n}\sum_{i}\sum_{|j|>m} |a_{ij}||a_{t+i,j}+a_{t+i-j,-j}|\sigma^4\\
&\le \sum_{i}\sum_{|j|>m} |a_{ij}|\left(\sum_{t}(|a_{t+i,j}|+|a_{t+i-j,-j}|)\right)\sigma^4\\
&= 2\sum_{|j|>m}\left(\sum_{i}|a_{ij}|\right)\left(\sum_{t}|a_{tj}|\right)\sigma^4\\
&= 2\sum_{|j|>m}\left(\sum_{i}|a_{ij}|\right)^2\sigma^4.
\end{aligned}$$

This bound is independent of n. Using the definition of a_{ij} it is easy to show that the bound converges to zero as $m \to \infty$, thus verifying (7.3.22). Consequently (7.3.20) is valid with $m = \infty$. Since $\gamma^*(0) \xrightarrow{P} (\sum_j \psi_j^2)\sigma^2$, it follows from (7.3.19) and Proposition 6.3.8 that

$$\begin{aligned}
n^{1/2}(\rho^*(h)-\rho(h)) &\Rightarrow \sum_{j=1}^{\infty}\left(\sum_i a_{ij}+a_{i,-j}\right)N_j\bigg/\left(\sum_i \psi_i^2\right)\\
&= Y_h.
\end{aligned}$$

Finally the proof of the joint convergence of the components of the vector $n^{1/2}(\boldsymbol{\rho}^*(h)-\boldsymbol{\rho}(h))$ can be carried out by writing vector analogues of the preceding equations. □

Proof of Theorem 7.2.2. As in the proof of Theorem 7.2.1, (see Remark 1) we may assume without loss of generality that $\mu = 0$. By Proposition 6.3.3, it suffices to show that

$$n^{1/2}(\rho^*(h)-\hat\rho(h)) \xrightarrow{P} 0 \quad \text{for } h = 1, 2, \ldots.$$

As in the proof of Proposition 7.3.4 (and assuming only that $\{Z_t\} \sim \text{IID}(0,\sigma^2)$) we have for $h \ge 0$,

$$n^{1/2}(\gamma^*(h)-\hat\gamma(h)) = o_p(1).$$

By Proposition 7.3.5,

$$\gamma^*(h) \xrightarrow{P} \gamma(h) \quad \text{for } h \ge 0$$

and hence

$$\hat\gamma(h) \xrightarrow{P} \gamma(h) \quad \text{for } h \ge 0.$$

Thus by Proposition 6.1.1,

$$\begin{aligned} n^{1/2}(\rho^*(h) - \hat{\rho}(h)) &= n^{1/2}(\gamma^*(h) - \hat{\gamma}(h))/\gamma^*(0) \\ &\quad + n^{1/2}(\hat{\gamma}(0) - \gamma^*(0))\hat{\gamma}(h)/(\gamma^*(0)\hat{\gamma}(0)) \\ &= o_p(1). \end{aligned}$$ □

Problems

7.1. If $\{X_t\}$ is a causal AR(1) process with mean μ, show that $\bar{X}_n$ is AN(μ, $\sigma^2(1-\phi)^{-2}n^{-1}$). In a sample of size 100 from an AR(1) process with $\phi = .6$ and $\sigma^2 = 2$, we obtain $\bar{X}_n = .271$. Construct an approximate 95% confidence interval for the mean μ. Does the data suggest that $\mu = 0$?

7.2. Let $\{X_t\}$ be a stationary process with mean μ. Show that the best linear unbiased estimator $\hat{\mu}$ of μ is given by $\hat{\mu}_n = (\mathbf{1}'\Gamma_n^{-1}\mathbf{1})^{-1}\mathbf{1}'\Gamma_n^{-1}\mathbf{X}_n$ where Γ_n is the covariance matrix of $\mathbf{X}_n = (X_1, \ldots, X_n)'$ and $\mathbf{1} = (1, \ldots, 1)'$. Show also that $\operatorname{Var}(\hat{\mu}_n) = (\mathbf{1}'\Gamma_n^{-1}\mathbf{1})^{-1}$. [$\hat{\mu}_n$ is said to be the best linear unbiased estimator of μ if $E(\hat{\mu}_n - \mu)^2 = \min E|Y - \mu|^2$ where the minimum is taken over all $Y \in \overline{\text{sp}}\{X_1, \ldots, X_n\}$ with $EY = \mu$.]

7.3. Show that for any series $\{x_1, \ldots, x_n\}$, the sample autocovariances satisfy $\sum_{|h|<n} \hat{\gamma}(h) = 0$.

7.4. Use formula (7.2.5) to compute the asymptotic covariance matrix of $\hat{\rho}(1), \ldots, \hat{\rho}(h)$ for an MA(1) process. For which values of j and k in $\{1, 2, \ldots\}$ are $\hat{\rho}(j)$ and $\hat{\rho}(k)$ asymptotically independent?

7.5. Use formula (7.2.5) to compute the asymptotic covariance of $\hat{\rho}(1)$ and $\hat{\rho}(2)$ for an AR(1) process. What is the behaviour of the asymptotic correlation of $\hat{\rho}(1)$ and $\hat{\rho}(2)$ as $\phi \to \pm 1$?

7.6. For an AR(1) process the sample autocorrelation $\hat{\rho}(1)$ is AN(ϕ, $(1-\phi^2)n^{-1}$). Show that $n^{1/2}(\hat{\rho}(1) - \phi)/(1 - \hat{\rho}^2(1))^{1/2}$ is AN(0, 1). If a sample of size 100 from an AR(1) process gives $\hat{\rho}(1) = .638$, construct a 95% confidence interval for ϕ. Is the data consistent with the hypothesis that $\phi = .7$?

7.7. In Problem 7.6, suppose that we estimate ϕ by $(\hat{\rho}(3))^{1/3}$. Show that $(\hat{\rho}(3))^{1/3}$ is AN(ϕ, $n^{-1}v$) and express v in terms of ϕ. Compare the asymptotic variances of the estimators $\hat{\rho}(1)$ and $(\hat{\rho}(3))^{1/3}$ as ϕ varies between -1 and 1.

7.8. Suppose that $\{X_t\}$ is the AR(1) process,

$$X_t - \mu = \phi(X_t - \mu) + Z_t, \qquad \{Z_t\} \sim \text{IID}(0, \sigma^2),$$

where $|\phi| < 1$. Find constants $a_n > 0$ and b_n such that $\exp(\bar{X}_n)$ is AN(b_n, a_n).

7.9. Find the asymptotic distribution of $\hat{\rho}(2)/\hat{\rho}(1)$ for the Gaussian MA(1) process,

$$X_t = Z_t + \theta Z_{t-1}, \qquad \{Z_t\} \sim \text{IID}(0, v),$$

where $0 < |\theta| < 1$.

7.10. If $\{X_t\}$ is the MA(1) process in Problem 7.9, the moment estimators $\hat{\theta}$ and $\hat{v}$ of θ and v based on the observations $\{X_1, \ldots, X_n\}$ are obtained by equating the sample and theoretical autocovariances at lags 0 and 1. Thus

$$\hat{v}(1 + \hat{\theta}^2) = \hat{\gamma}(0),$$

and

$$\hat{\theta}/(1 + \hat{\theta}^2) = \hat{\rho}(1).$$

Use the asymptotic joint distribution of $(\hat{\gamma}(0), \hat{\rho}(1))$

(a) to estimate the probability that these equations have a solution when $\theta = .6$ and $n = 200$ ($\hat{\theta}$ must be real), and

(b) to determine the asymptotic joint distribution of $(\hat{v}, \hat{\theta})'$.

7.11. If $X_1, \ldots, X_n$ are n observations of a stationary time series, define

$$\hat{\gamma}(h) = \begin{cases} n^{-1} \sum_{t=1}^{n-|h|} (X_t - \bar{X}_n)(X_{t+h} - \bar{X}_n) & \text{if } |h| < n, \\ 0 & \text{if } |h| \geq n. \end{cases}$$

Show that the function $\hat{\gamma}(\cdot)$ is non-negative definite and hence, by Theorem 1.5.1, that $\hat{\gamma}(\cdot)$ is the autocovariance function of some stationary process $\{Y_t\}$. From Proposition 3.2.1 it then follows at once that $\{Y_t\}$ is an MA$(n-1)$ process. (Show that $\hat{\Gamma}_{n+h}$ is non-negative definite for all $h \geq 0$ by setting $Y_{n+1} = Y_{n+2} = \cdots = Y_{n+h} = 0$ in the argument of Section 7.2.) Conclude from Proposition 5.1.1 that if $\hat{\gamma}(0) > 0$, then $\hat{\Gamma}_n$ is non-singular for every n.

CHAPTER 8

Estimation for ARMA Models

The determination of an appropriate ARMA(p, q) model to represent an observed stationary time series involves a number of inter-related problems. These include the choice of p and q (order selection), and estimation of the remaining parameters, i.e. the mean, the coefficients $\{\phi_i, \theta_j : i = 1, \ldots, p; j = 1, \ldots, q\}$ and the white noise variance σ^2, for given values of p and q. Goodness of fit of the model must also be checked and the estimation procedure repeated with different values of p and q. Final selection of the most appropriate model depends on a variety of goodness of fit tests, although it can be systematized to a large degree by use of criteria such as the AICC statistic discussed in Chapter 9.

This chapter is devoted to the most straightforward part of the modelling procedure, namely the estimation, for fixed values of p and q, of the parameters $\boldsymbol{\phi} = (\phi_1, \ldots, \phi_p)'$, $\boldsymbol{\theta} = (\theta_1, \ldots, \theta_q)'$ and σ^2. It will be assumed throughout that the data has been adjusted by subtraction of the mean, so our problem becomes that of fitting a zero-mean ARMA model to the adjusted data $x_1, \ldots, x_n$. If the model fitted to the adjusted data is

$$X_t - \phi_1 X_{t-1} - \cdots - \phi_p X_{t-p} = Z_t + \theta_1 Z_{t-1} + \cdots + \theta_q Z_{t-q},$$
$$\{Z_t\} \sim \text{WN}(0, \sigma^2),$$

then the corresponding model for the original stationary series $\{Y_t\}$ is found by substituting $Y_j - \bar{y}$ for X_j, $j = t, \ldots, t - p$, where $\bar{y} = n^{-1} \sum_{j=1}^{n} y_j$ is the sample mean of the original data, treated as a fixed constant.

In the case $q = 0$ a good estimate of $\boldsymbol{\phi}$ can be obtained by the simple device of equating the sample and theoretical autocovariances at lags 0, 1, ..., p. This is the Yule–Walker estimator discussed in Sections 8.1 and 8.2. When $q > 0$ the corresponding procedure, i.e. equating sample and theoretical autocovariances at lags $0, \ldots, p + q$, is neither simple nor efficient. In Sections 8.3 and 8.4 we discuss a simple method, based on the innovations algorithm (Proposition 5.2.2), for obtaining more efficient preliminary estimators of the coefficients when $q > 0$. These are still not as efficient as least squares or maximum likelihood estimators, and serve primarily as initial values for the non-linear optimization procedure required for computing these more efficient estimators.

Calculation of the exact Gaussian likelihood of an arbitrary second order process and in particular of an ARMA process is greatly simplified by use of the innovations algorithm. We make use of this simplification in our discussion of maximum likelihood and least squares estimation for ARMA processes in Section 8.7. The asymptotic properties of the estimators and the determination of large-sample confidence intervals for the parameters are discussed in Sections 8.8, 8.9, 8.11 and 10.8.

§8.1 The Yule–Walker Equations and Parameter Estimation for Autoregressive Processes

Let $\{X_t\}$ be the zero-mean causal autoregressive process,

$$X_t - \phi_1 X_{t-1} - \cdots - \phi_p X_{t-p} = Z_t, \qquad \{Z_t\} \sim \text{WN}(0, \sigma^2). \tag{8.1.1}$$

Our aim is to find estimators of the coefficient vector $\boldsymbol{\phi} = (\phi_1, \ldots, \phi_p)'$ and the white noise variance σ^2 based on the observations $X_1, \ldots, X_n$.

The causality assumption allows us to write X_t in the form

$$X_t = \sum_{j=0}^{\infty} \psi_j Z_{t-j}, \tag{8.1.2}$$

where by Theorem 3.1.1, $\psi(z) = \sum_{j=0}^{\infty} \psi_j z^j = 1/\phi(z)$, $|z| \le 1$. Multiplying each side of (8.1.1) by X_{t-j}, $j = 0, \ldots, p$, taking expectations, and using (8.1.2) to evaluate the right-hand sides, we obtain the Yule–Walker equations,

$$\Gamma_p \boldsymbol{\phi} = \boldsymbol{\gamma}_p, \tag{8.1.3}$$

and

$$\sigma^2 = \gamma(0) - \boldsymbol{\phi}' \boldsymbol{\gamma}_p, \tag{8.1.4}$$

where Γ_p is the covariance matrix $[\gamma(i-j)]_{i,j=1}^{p}$ and $\boldsymbol{\gamma}_p = (\gamma(1), \gamma(2), \ldots, \gamma(p))'$. These equations can be used to determine $\gamma(0), \ldots, \gamma(p)$ from σ^2 and $\boldsymbol{\phi}$.

On the other hand, if we replace the covariances $\gamma(j)$, $j = 0, \ldots, p$, appearing in (8.1.3) and (8.1.4) by the corresponding sample covariances $\hat{\gamma}(j)$, we obtain a set of equations for the so-called Yule–Walker estimators $\hat{\boldsymbol{\phi}}$ and $\hat{\sigma}^2$ of $\boldsymbol{\phi}$ and σ^2, namely

$$\hat{\Gamma}_p \hat{\boldsymbol{\phi}} = \hat{\boldsymbol{\gamma}}_p, \tag{8.1.5}$$

and

$$\hat{\sigma}^2 = \hat{\gamma}(0) - \hat{\boldsymbol{\phi}}'\hat{\boldsymbol{\gamma}}_p, \tag{8.1.6}$$

where $\hat{\Gamma}_p = [\hat{\gamma}(i-j)]_{i,j=1}^p$ and $\hat{\boldsymbol{\gamma}}_p = (\hat{\gamma}(1), \hat{\gamma}(2), \ldots, \hat{\gamma}(p))'$.

If $\hat{\gamma}(0) > 0$, then by Problem 7.11, $\hat{\Gamma}_p$ is non-singular. Dividing each side of (8.1.5) by $\hat{\gamma}(0)$, we therefore obtain

$$\hat{\boldsymbol{\phi}} - \hat{R}_p^{-1}\hat{\boldsymbol{\rho}}_p \tag{8.1.7}$$

and

$$\hat{\sigma}^2 = \hat{\gamma}(0)[1 - \hat{\boldsymbol{\rho}}_p'\hat{R}_p^{-1}\hat{\boldsymbol{\rho}}_p], \tag{8.1.8}$$

where $\hat{\boldsymbol{\rho}}_p = (\hat{\rho}(1), \ldots, \hat{\rho}(p))' = \hat{\boldsymbol{\gamma}}_p/\hat{\gamma}(0)$.

With $\hat{\boldsymbol{\phi}}$ as defined by (8.1.7), it can be shown that $1 - \hat{\phi}_1 z - \cdots - \hat{\phi}_p z^p \neq 0$ for $|z| \leq 1$ (see Problem 8.3). Hence the fitted model,

$$X_t - \hat{\phi}_1 X_{t-1} - \cdots - \hat{\phi}_p X_{t-p} = Z_t, \qquad \{Z_t\} \sim \mathrm{WN}(0, \hat{\sigma}^2),$$

is causal. The autocovariances $\gamma_F(h)$, $h = 0, \ldots, p$ of the fitted model must therefore satisfy the $p + 1$ linear equations (cf. (8.1.3) and (8.1.4))

$$\gamma_F(h) - \hat{\phi}_1\gamma_F(h-1) - \cdots - \hat{\phi}_p\gamma_F(h-p) = \begin{cases} 0, & h = 1, \ldots, p, \\ \hat{\sigma}^2, & h = 0. \end{cases}$$

However, from (8.1.5) and (8.1.6) we see that the solution of these equations is $\gamma_F(h) = \hat{\gamma}(h)$, $h = 0, \ldots, p$ so that the autocovariances of the fitted model at lags $0, \ldots, p$ coincide with the corresponding sample autocovariances.

The argument of the preceding paragraph shows that for *every* non-singular covariance matrix $\Gamma_{p+1} = [\gamma(i-j)]_{i,j=1}^{p+1}$ there is an AR(p) process whose autocovariances at lags $0, \ldots, p$ are $\gamma(0), \ldots, \gamma(p)$. (The required coefficients and white noise variance are found from (8.1.7) and (8.1.8) on replacing $\hat{\rho}(j)$ by $\gamma(j)/\gamma(0)$, $j = 0, \ldots, p$, and $\hat{\gamma}(0)$ by $\gamma(0)$.) There may not however be an MA(p) process with this property. For example if $\gamma(0) = 1$ and $\gamma(1) = \gamma(-1) = \beta$, the matrix Γ_2 is a non-singular covariance matrix for all $\beta \in (-1, 1)$. Consequently there is an AR(1) process with autocovariances 1 and β at lags 0 and 1 for all $\beta \in (-1, 1)$. However there is an MA(1) process with autocovariances 1 and β at lags 0 and 1 if and only if $|\beta| \leq 1/2$. (See Example 1.5.1.)

It is often the case that moment estimators, i.e. estimators which (like $\hat{\boldsymbol{\phi}}$) are obtained by equating theoretical and sample moments, are far less efficient than estimators obtained by alternative methods such as least squares or maximum likelihood. For example, estimation of the coefficient of an MA(1) process by equating the theoretical and sample autocorrelations at lag 1 is very inefficient (see Section 8.5). However for an AR(p) process, we shall see that the Yule–Walker estimator, $\hat{\boldsymbol{\phi}}$, has the same asymptotic distribution as $n \to \infty$ as the maximum likelihood estimator of $\boldsymbol{\phi}$ to be discussed in Sections 8.7 and 8.8.

Theorem 8.1.1. *If $\{X_t\}$ is the causal* AR(p) *process* (8.1.1) *with* $\{Z_t\} \sim \mathrm{IID}(0, \sigma^2)$, and $\hat{\boldsymbol{\phi}}$ *is the Yule–Walker estimator of* $\boldsymbol{\phi}$, *then*

$$n^{1/2}(\hat{\boldsymbol{\phi}} - \boldsymbol{\phi}) \Rightarrow N(\mathbf{0}, \sigma^2 \Gamma_p^{-1}),$$

where Γ_p *is the covariance matrix* $[\gamma(i-j)]_{i,j=1}^{p}$. *Moreover,*

$$\hat{\sigma}^2 \xrightarrow{P} \sigma^2.$$

PROOF. See Section 8.10. □

Theorem 8.1.1 enables us in particular to specify large-sample confidence regions for $\boldsymbol{\phi}$ and for each of its components. This is illustrated in Example 8.2.1.

In fitting autoregressive models to data, the order p will usually be unknown. If the true order is p and we attempt to fit a process of order m, we should expect the estimated coefficient vector $\hat{\boldsymbol{\phi}}_m = (\hat{\phi}_{m1}, \ldots, \hat{\phi}_{mm})'$ to have a small value of $\hat{\phi}_{mm}$ for each $m > p$. Although the exact distribution of $\hat{\boldsymbol{\phi}}_{mm}$ for $m > p$ is not known even in the Gaussian case, the following asymptotic result is extremely useful in helping us to identify the appropriate order of the process to be fitted.

Theorem 8.1.2. *If* $\{X_t\}$ *is the causal* AR(p) *process* (8.1.1) *with* $\{Z_t\} \sim \text{IID}(0, \sigma^2)$, *and if* $\hat{\boldsymbol{\phi}}_m = (\hat{\phi}_{m1}, \ldots, \hat{\phi}_{mm})' = \hat{R}_m^{-1} \hat{\boldsymbol{\rho}}_m$, $m > p$, *then*

$$n^{1/2}(\hat{\boldsymbol{\phi}}_m - \boldsymbol{\phi}_m) \Rightarrow \text{N}(\mathbf{0}, \sigma^2 \Gamma_m^{-1}),$$

where $\boldsymbol{\phi}_m$ *is the coefficient vector of the best linear predictor* $\boldsymbol{\phi}_m' \mathbf{X}_m$ *of* X_{m+1} *based on* $\mathbf{X}_m = (X_m, \ldots, X_1)'$, *i.e.* $\boldsymbol{\phi}_m = R_m^{-1} \boldsymbol{\rho}_m$. *In particular for* $m > p$,

$$n^{1/2} \hat{\phi}_{mm} \Rightarrow \text{N}(0, 1).$$

PROOF. See Section 8.10. □

The application of Theorem 8.1.2 to order selection will be discussed in Section 8.2 in connection with the recursive fitting of autoregressive models.

§8.2 Preliminary Estimation for Autoregressive Processes Using the Durbin–Levinson Algorithm

Suppose we have observations $x_1, \ldots, x_n$ of a zero-mean stationary time series. Provided $\hat{\gamma}(0) > 0$ we can fit an autoregressive process of order $m < n$ to the data by means of the Yule–Walker equations. The fitted AR(m) process is

$$X_t - \hat{\phi}_{m1} X_{t-1} - \cdots - \hat{\phi}_{mm} X_{t-m} = Z_t, \qquad \{Z_t\} \sim \text{WN}(0, \hat{v}_m), \quad (8.2.1)$$

where from (8.1.7) and (8.1.8),

$$\hat{\boldsymbol{\phi}}_m := (\hat{\phi}_{m1}, \ldots, \hat{\phi}_{mm})' = \hat{R}_m^{-1} \hat{\boldsymbol{\rho}}_m, \tag{8.2.2}$$

and

$$\hat{v}_m = \hat{\gamma}(0)[1 - \hat{\boldsymbol{\rho}}_m' \hat{R}_m^{-1} \hat{\boldsymbol{\rho}}_m]. \tag{8.2.3}$$

Now if we compare (8.2.2) and (8.2.3) with the statement of Corollary 5.1.1, we see that $\hat{\boldsymbol{\phi}}_m$ and $\hat{v}_m$ are related to the sample autocovariances in the same way that $\boldsymbol{\phi}_m$ and v_m are related to the autocovariances of the underlying process $\{X_t\}$. (As in Theorem 8.1.2, $\boldsymbol{\phi}_m$ is defined as the coefficient vector of the best linear predictor $\boldsymbol{\phi}_m'\mathbf{X}_m$ of X_{m+1} based on $\mathbf{X}_m = (X_m, \ldots, X_1)'$; v_m is the corresponding mean squared error.)

Consequently (if $\hat{\gamma}(0) > 0$ so that $\hat{R}_1, \hat{R}_2, \ldots$ are non-singular) we can use the Durbin–Levinson algorithm to fit autoregressive models of successively increasing orders 1, 2, ..., to the data. The estimated coefficient vectors $\hat{\boldsymbol{\phi}}_1$, $\hat{\boldsymbol{\phi}}_2, \ldots$, and white noise variances $\hat{v}_1, \hat{v}_2, \ldots$, are computed recursively from the sample covariances just as we computed $\boldsymbol{\phi}_1, \boldsymbol{\phi}_2, \ldots$, and $v_1, v_2, \ldots$, from the covariances in Chapter 5. Restated in terms of the estimates $\hat{\boldsymbol{\phi}}_m$, $\hat{v}_m$, the algorithm becomes:

Proposition 8.2.1 (The Durbin–Levinson Algorithm for Fitting Autoregressive Models). *If* $\hat{\gamma}(0) > 0$ *then the fitted autoregressive models* (8.2.1) *for* $m = 1, 2, \ldots, n-1$, *can be determined recursively from the relations*, $\hat{\phi}_{11} = \hat{\rho}(1)$, $\hat{v}_1 = \hat{\gamma}(0)[1 - \hat{\rho}^2(1)]$,

$$\hat{\phi}_{mm} = \left[\hat{\gamma}(m) - \sum_{j=1}^{m-1} \hat{\phi}_{m-1,j}\hat{\gamma}(m-j)\right] \Big/ \hat{v}_{m-1}, \tag{8.2.4}$$

$$\begin{bmatrix} \hat{\phi}_{m1} \\ \vdots \\ \hat{\phi}_{m,m-1} \end{bmatrix} = \hat{\boldsymbol{\phi}}_{m-1} - \hat{\phi}_{mm} \begin{bmatrix} \hat{\phi}_{m-1,m-1} \\ \vdots \\ \hat{\phi}_{m-1,1} \end{bmatrix} \tag{8.2.5}$$

and

$$\hat{v}_m = \hat{v}_{m-1}(1 - \hat{\phi}_{mm}^2). \tag{8.2.6}$$

Use of these recursions bypasses the matrix inversion required in the direct computation of $\hat{\boldsymbol{\phi}}_m$ and $\hat{v}_m$ from (8.1.7) and (8.1.8). It also provides us with estimates $\hat{\phi}_{11}, \hat{\phi}_{22}, \ldots$, of the partial autocorrelation function at lags 1, 2, ... These estimates are extremely valuable, first for deciding on the appropriateness of an autoregressive model, and then for choosing an appropriate order for the model to be fitted.

We already know from Section 3.4 that for an AR(p) process the partial autocorrelations $\alpha(m) = \phi_{mm}$, $m > p$, are zero. Moreover we know from Theorem 8.1.2 that for an AR(p) process the estimator $\hat{\phi}_{mm}$, is, for large n and each $m > p$, approximately normally distributed with mean 0 and variance

$1/n$. If an autoregressive model is appropriate for the data there should consequently be a finite lag beyond which the observed values $\hat{\phi}_{mm}$ are compatible with the distribution $N(0, 1/n)$. In particular if the order of the process is p then for $m > p$, $\hat{\phi}_{mm}$ will fall between the bounds $\pm 1.96n^{-1/2}$ with probability close to .95. This suggests using as a preliminary estimator of p the smallest value of r such that $|\hat{\phi}_{mm}| < 1.96n^{-1/2}$ for $m > r$. (A more systematic approach to order selection based on the AICC will be discussed in Section 9.2.) Once a value for p has been selected, the fitted process is specified by (8.2.1), (8.2.2) and (8.2.3) with $m = p$.

Asymptotic confidence regions for the true coefficient vector $\boldsymbol{\phi}_p$ and for its individual components ϕ_{pj} can be found with the aid of Theorem 8.1.1. Thus, if $\chi^2_{1-\alpha}(p)$ denotes the $(1 - \alpha)$ quantile of the chi-squared distribution with p degrees of freedom, then for large sample size n, the region

$$\{\boldsymbol{\phi} \in \mathbb{R}^p : (\boldsymbol{\phi} - \hat{\boldsymbol{\phi}}_p)' \hat{\Gamma}_p (\boldsymbol{\phi} - \hat{\boldsymbol{\phi}}_p) \le n^{-1} \hat{v}_p \chi^2_{1-\alpha}(p)\} \tag{8.2.7}$$

contains $\boldsymbol{\phi}_p$ with probability close to $(1 - \alpha)$. (See Problems 1.16 and 6.14.) Similarly, if $\Phi_{1-\alpha}$ denotes the $(1 - \alpha)$ quantile of the standard normal distribution and $\hat{v}_{jj}$ is the jth diagonal element of $\hat{v}_p \hat{\Gamma}_p^{-1}$, then for large n the interval

$$\{\phi \in \mathbb{R} : |\phi - \hat{\phi}_{pj}| \le n^{-1/2} \Phi_{1-\alpha/2} \hat{v}_{jj}^{1/2}\} \tag{8.2.8}$$

contains ϕ_{pj} with probability close to $(1 - \alpha)$.

Example 8.2.1. One thousand observations $x_1, \ldots, x_{1000}$ of a zero-mean stationary process gave sample autocovariances $\hat{\gamma}(0) = 3.6840$, $\hat{\gamma}(1) = 2.2948$ and $\hat{\gamma}(2) = 1.8491$.

Applying the Durbin–Levinson algorithm to fit successively higher order autoregressive processes to the data, we obtain

$$\hat{\phi}_{11} = \hat{\rho}(1) = .6229,$$

$$\hat{v}_1 = \hat{\gamma}(0)(1 - \hat{\rho}^2(1)) = 2.2545,$$

$$\hat{\phi}_{22} = [\hat{\gamma}(2) - \hat{\phi}_{11}\hat{\gamma}(1)]/\hat{v}_1 = .1861,$$

$$\hat{\phi}_{21} = \hat{\phi}_{11} - \hat{\phi}_{22}\hat{\phi}_{11} = .5070,$$

$$\hat{v}_2 = \hat{v}_1(1 - \hat{\phi}_{22}^2) = 2.1764.$$

The computer program PEST can be used to apply the recursions (8.2.4)–(8.2.6) for increasing values of n, and hence to determine the sample partial autocorrelation function $\hat{\phi}_{jj}$, shown with the sample autocorrelation function $\hat{\rho}(j)$ in Figure 8.1. The bounds plotted on both graphs are the values $\pm 1.96n^{-1/2}$.

Inspection of the graph of $\hat{\phi}_{jj}$ strongly suggests that the appropriate model for this data is an AR(2) process. Using the Yule–Walker estimates $\hat{\phi}_{21}$, $\hat{\phi}_{22}$ and $\hat{v}_2$ computed above, we obtain the fitted process,

$$X_t - .5070X_{t-1} - .1861X_{t-2} = Z_t, \qquad \{Z_t\} \sim \text{WN}(0, 2.1764).$$

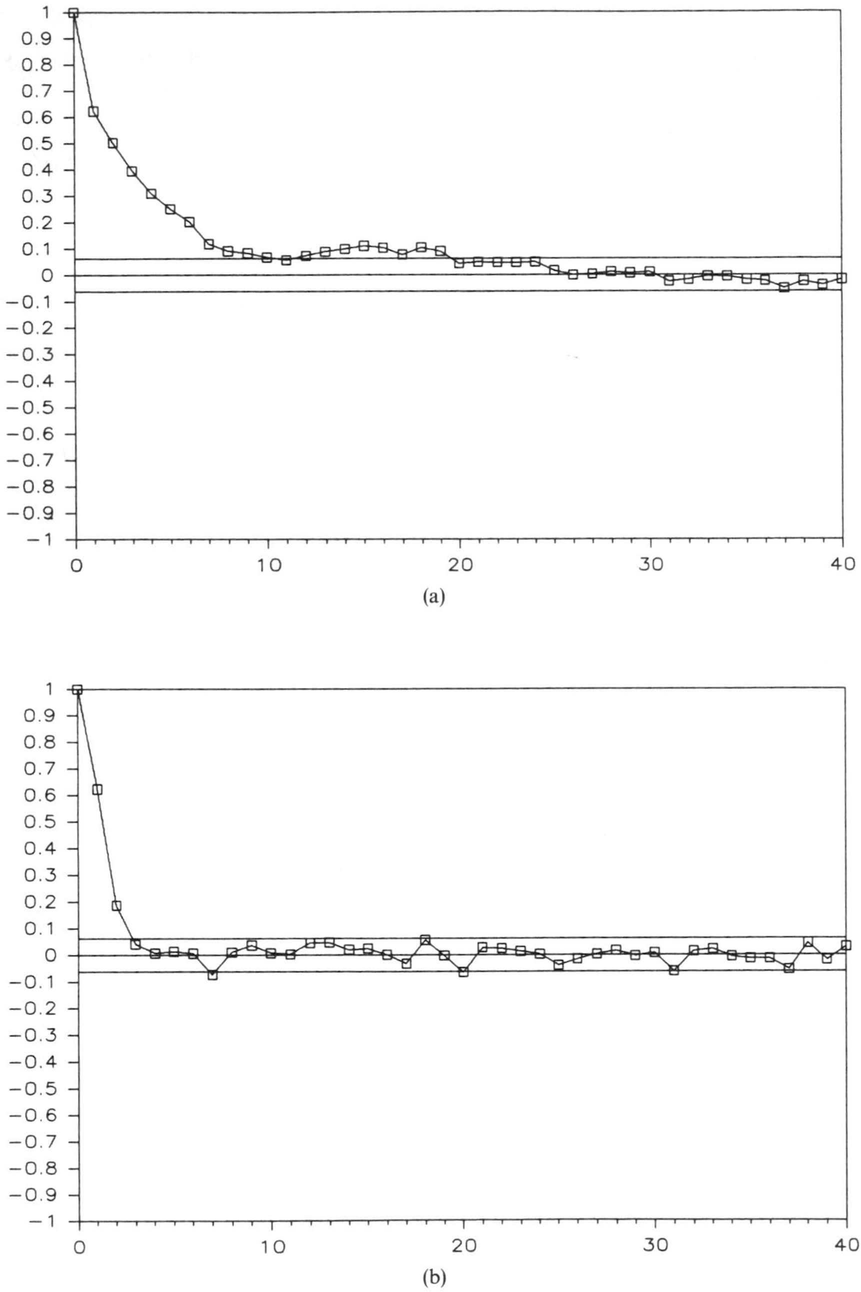

Figure 8.1. The sample ACF (a) and PACF (b) for the data of Example 8.2.1, showing the bounds $\pm 1.96n^{-1/2}$.

From Theorem 8.1.1, the error vector $\hat{\boldsymbol{\phi}} - \boldsymbol{\phi}$ is approximately normally distributed with mean $\mathbf{0}$ and covariance matrix,

$$n^{-1}\hat{v}_2\hat{\Gamma}_2^{-1} = n^{-1}\left[1 - \sum_{j=1}^{2} \hat{\rho}(j)\hat{\phi}_{2j}\right]\begin{bmatrix} 1 & \hat{\rho}(1) \\ \hat{\rho}(1) & 1 \end{bmatrix}^{-1}$$
$$= \begin{bmatrix} .000965 & -.000601 \\ -.000601 & .000965 \end{bmatrix}.$$

From (8.2.8) we obtain the approximate .95 confidence bounds, $\hat{\phi}_i \pm 1.96(.000965)^{1/2}$ for ϕ_i, $i = 1, 2$. These are $.5070 \pm .0609$ for ϕ_1 and $.1861 \pm .0609$ for ϕ_2.

The data for this example came from a simulated AR(2) process with coefficients $\phi_1 = .5$, $\phi_2 = .2$ and white noise variance 2.25. The true coefficients thus lie between the confidence bounds computed in the preceding paragraph.

§8.3 Preliminary Estimation for Moving Average Processes Using the Innovations Algorithm

Just as we can fit autoregressive models of orders 1, 2, ..., to the data $x_1, \ldots, x_n$ by applying the Durbin–Levinson algorithm to the sample autocovariances, we can also fit moving average models,

$$X_t = Z_t + \hat{\theta}_{m1}Z_{t-1} + \cdots + \hat{\theta}_{mm}Z_{t-m}, \qquad \{Z_t\} \sim \mathrm{WN}(0, \hat{v}_m), \tag{8.3.1}$$

of orders $m = 1, 2, \ldots$, by means of the innovations algorithm (Proposition 5.2.2). The estimated coefficient vectors $\hat{\boldsymbol{\theta}}_m := (\hat{\theta}_{m1}, \ldots, \hat{\theta}_{mm})'$, and white noise variances $\hat{v}_m$, $m = 1\ 2, \ldots$, are specified in the following definition. (The justification for using estimators defined in this way is contained in Theorem 8.3.1.)

Definition 8.3.1 (Innovation Estimates of Moving Average Parameters). If $\hat{\gamma}(0) > 0$, we define the innovation estimates $\hat{\boldsymbol{\theta}}_m$, $\hat{v}_m$ appearing in (8.3.1) for $m = 1, 2, \ldots, n-1$, by the recursion relations, $\hat{v}_0 = \hat{\gamma}(0)$,

$$\hat{\theta}_{m,m-k} = \hat{v}_k^{-1}\left[\hat{\gamma}(m-k) - \sum_{j=0}^{k-1} \hat{\theta}_{m,m-j}\hat{\theta}_{k,k-j}\hat{v}_j\right], \quad k = 0, \ldots, m-1, \tag{8.3.2}$$

and

$$\hat{v}_m = \hat{\gamma}(0) - \sum_{j=0}^{m-1} \hat{\theta}_{m,m-j}^2\hat{v}_j. \tag{8.3.3}$$

Theorem 8.3.1 (The Asymptotic Behavior of $\hat{\boldsymbol{\theta}}_m$). *Let $\{X_t\}$ be the causal invertible* ARMA *process $\phi(B)X_t = \theta(B)Z_t$, $\{Z_t\} \sim \text{IID}(0, \sigma^2)$, $EZ_t^4 < \infty$, and let $\psi(z) = \sum_{j=0}^{\infty} \psi_j z^j = \theta(z)/\phi(z)$, $|z| \le 1$, (with $\psi_0 = 1$ and $\psi_j = 0$ for $j < 0$). Then for any sequence of positive integers $\{m(n), n = 1, 2, \dots\}$ such that $m < n$, $m \to \infty$ and $m = o(n^{1/3})$ as $n \to \infty$, we have for each k,*

$$n^{1/2}(\hat{\theta}_{m1} - \psi_1, \hat{\theta}_{m2} - \psi_2, \dots, \hat{\theta}_{mk} - \psi_k)' \Rightarrow N(\mathbf{0}, A),$$

where $A = [a_{ij}]_{i,j=1}^k$ and

$$a_{ij} = \sum_{r=1}^{\min(i,j)} \psi_{i-r}\psi_{j-r}.$$

Moreover,

$$\hat{v}_m \xrightarrow{P} \sigma^2.$$

Proof. See Brockwell and Davis (1988b). □

Remark. Although the recursive fitting of moving average models using the innovations algorithm is closely analogous to the recursive fitting of autoregressive models using the Durbin–Levinson algorithm, there is one important distinction. For an AR(p) process the Yule–Walker estimator $\hat{\boldsymbol{\phi}}_p = (\hat{\phi}_{p1}, \dots, \hat{\phi}_{pp})'$ is consistent for $\boldsymbol{\phi}_p$ (i.e. $\hat{\boldsymbol{\phi}}_p \xrightarrow{P} \boldsymbol{\phi}_p$) as the sample size $n \to \infty$. However for an MA(q) process the estimator $\hat{\boldsymbol{\theta}}_q = (\hat{\theta}_{q1}, \dots, \hat{\theta}_{qq})'$ is not consistent for the true parameter vector $\boldsymbol{\theta}_q$ as $n \to \infty$. For consistency it is necessary to use the estimators $(\hat{\theta}_{m1}, \dots, \hat{\theta}_{mq})'$ of $\boldsymbol{\theta}_q$ with $\{m(n)\}$ satisfying the conditions of Theorem 8.3.1. The choice of m for any fixed sample size can be made by increasing m until the vector $(\hat{\theta}_{m1}, \dots, \hat{\theta}_{mq})'$ stabilizes. It is found in practice that there is a large range of values of m for which the fluctuations in $\hat{\theta}_{mj}$ are small compared with the estimated asymptotic standard deviation $n^{-1/2}(\sum_{k=0}^{j-1} \hat{\theta}_{mk}^2)^{1/2}$ as given by Theorem 8.3.1.

We know from Section 3.3 that for an MA(q) process the autocorrelations $\rho(m)$, $m > q$, are zero. Moreover we know from Bartlett's formula (see Example 7.2.2) that the sample autocorrelation $\hat{\rho}(m)$, $m > q$, is approximately normally distributed with mean $\rho(m) = 0$ and variance $n^{-1}[1 + 2\rho^2(1) + \cdots + 2\rho^2(q)]$. This result enables us to use the graph of $\hat{\rho}(m)$, $m = 1, 2, \dots$, both to decide whether or not a given set of data can be plausibly modelled by a moving average process and also to obtain a preliminary estimate of the order q. This procedure was described in Example 7.2.2.

If, in addition to examining $\hat{\rho}(m)$, $m = 1, 2, \dots$, we examine the coefficient vectors $\hat{\boldsymbol{\theta}}_m$, $m = 1, 2, \dots$, we are able not only to assess the appropriateness of a moving average model and estimate its order q, but also to obtain preliminary estimates $\hat{\theta}_{m1}, \dots, \hat{\theta}_{mq}$ of the coefficients. We plot the values $\hat{\theta}_{m1}, \dots, \hat{\theta}_{mm}, 0, 0, \dots$ for $m = 1, 2, \dots$, increasing m until the values stabilize (until the fluctuations in each component are of order $n^{-1/2}$, the asymptotic standard deviation of $\hat{\theta}_{m1}$). Since from Theorem 8.3.1 the asymptotic variance of $\hat{\theta}_{mj}$ is $\sigma_j^2(\theta_1, \ldots, \theta_{j-1}) = n^{-1}\sum_{k=0}^{j-1}\theta_k^2$, we also plot the bounds $\pm 1.96\hat{\sigma}_j$ where $\hat{\sigma}_j = \sigma_j(\hat{\theta}_{m1}, \ldots, \hat{\theta}_{m,j-1})$. A value of $\hat{\theta}_{mj}$ outside these bounds suggests that the corresponding coefficient θ_j is non-zero. The estimate of θ_j is $\hat{\theta}_{mj}$ and the largest lag for which $\hat{\theta}_{mj}$ lies outside the bounds $\pm 1.96\hat{\sigma}_j$ is the estimate of the order q of the moving average process. (A more systematic approach to order selection using the AICC will be discussed in Section 9.2.)

Asymptotic confidence regions for the coefficient vector $\boldsymbol{\theta}_q$ and for its individual components can be found with the aid of Theorem 8.3.1. For example an approximate .95 confidence interval for θ_j is given by

$$\left\{\theta \in \mathbb{R} : |\theta - \hat{\theta}_{mj}| \le 1.96 n^{-1/2}\left(\sum_{k=0}^{j-1}\hat{\theta}_{mk}^2\right)^{1/2}\right\}. \tag{8.3.4}$$

EXAMPLE 8.3.1. One thousand observations $x_1, \ldots, x_{1000}$ of a zero-mean stationary process gave sample autocovariances $\hat{\gamma}(0) = 7.5541$, $\hat{\gamma}(1) = -5.1241$ and $\hat{\gamma}(2) = 1.3805$.

The sample autocorrelations and partial autocorrelations for lags up to 40 are shown in Figure 8.2. They strongly suggest a moving average model of order 2 for the data. Although five sample autocorrelations at lags greater than 2 are outside the bounds $\pm 1.96 n^{-1/2}$, none are outside the bounds $\pm 1.96 n^{-1/2}[1 + 2\hat{\rho}^2(1) + 2\hat{\rho}^2(2)]^{1/2}$.

Applying the innovations algorithm to fit successively higher moving average processes to the data, we obtain $\hat{v}_0 = 7.5541$,

$$\begin{aligned}
\hat{\theta}_{11} &= \hat{\rho}(1) = -.67832,\\
\hat{v}_1 &= \hat{\gamma}(0) - \hat{\theta}_{11}^2\hat{v}_0 = 4.0785,\\
\hat{\theta}_{22} &= \hat{v}_0^{-1}\hat{\gamma}(2) = .18275,\\
\hat{\theta}_{21} &= \hat{v}_1^{-1}[\hat{\gamma}(1) - \hat{\theta}_{22}\hat{\theta}_{11}\hat{v}_0] = -1.0268,\\
\hat{v}_2 &= \hat{\gamma}(0) - \hat{\theta}_{22}^2\hat{v}_0 - \hat{\theta}_{21}^2\hat{v}_1 = 3.0020.
\end{aligned}$$

Option 3 of the program PEST can be used to apply the recursions (8.3.2) and (8.3.3) for larger values of m. The estimated values $\hat{\theta}_{mj}, j = 1, \ldots, 10$ and $\hat{v}_m$ are shown in Table 8.1 for $m = 1, \ldots, 10, 20, 50$ and 100. It is clear from the table that the fluctuations in the coefficients from $m = 7$ up to 100 are of order $1000^{-1/2} = .032$. The values of $\hat{\theta}_{7j}, j = 1, \ldots, 7$, plotted in Figure 8.3 confirm the MA(2) model suggested by the sample autocorrelation function.

The model fitted to the data on the basis of $\hat{\boldsymbol{\theta}}_7$ is

$$X_t = Z_t - 1.41 Z_{t-1} + .60 Z_{t-2}, \qquad \{Z_t\} \sim \mathrm{WN}(0, 2.24). \tag{8.3.5}$$

In fact from Table 8.1 we see that the estimated coefficients show very little change as m varies between 7 and 100.

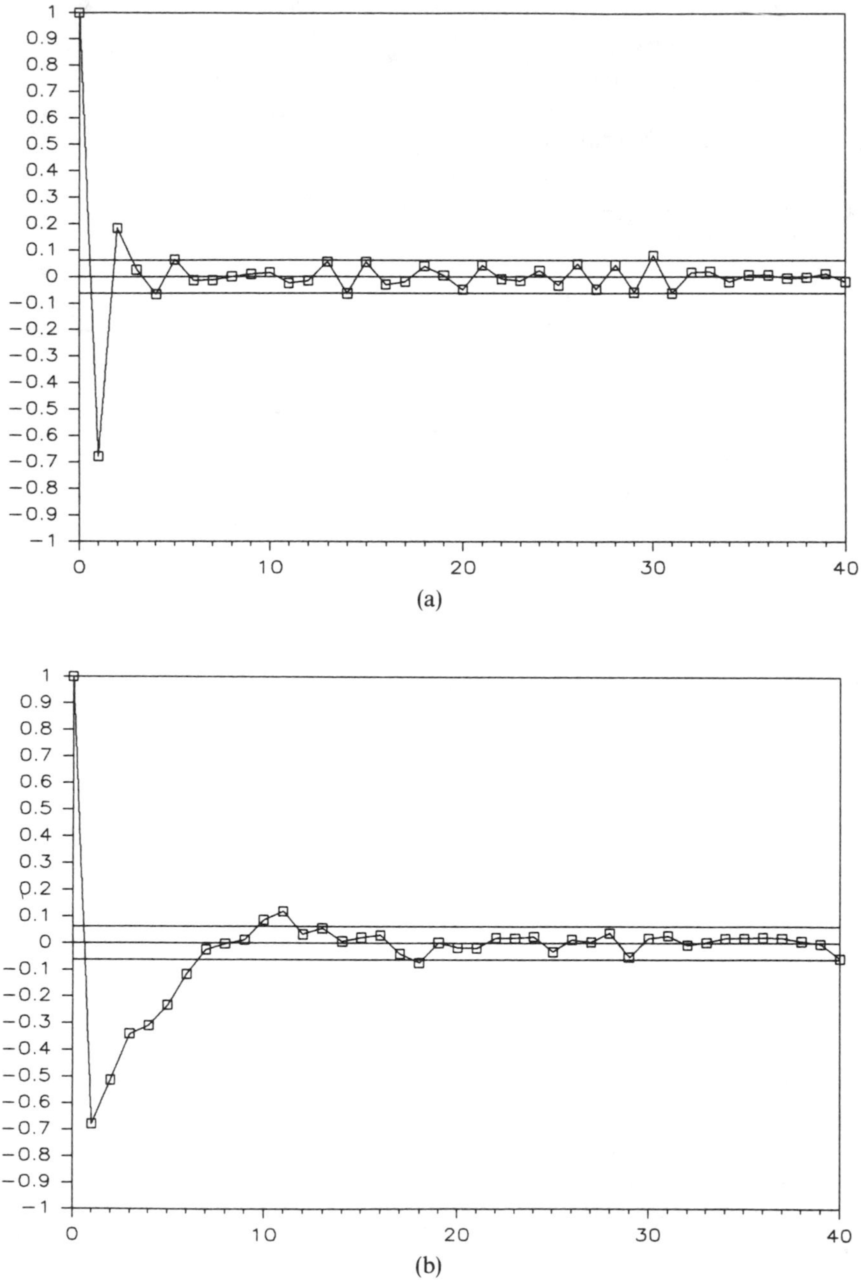

Figure 8.2. The sample ACF (a) and PACF (b) for the data of Example 8.3.1, showing the bounds $\pm 1.96n^{-1/2}$.

Table 8.1. $\hat{\theta}_{mj}, j = 1, \ldots, 10$, and $\hat{v}_m$ for the Data of Example 8.3.1

	$\hat{\theta}_{mj}$										$\hat{v}_m$
m \ j	1	2	3	4	5	6	7	8	9	10	
1	−0.68										4.08
2	−1.03	.18									3.00
3	−1.20	.37	.03								2.65
4	−1.31	.44	−.04	.07							2.40
5	−1.38	.51	−.03	−.04	.06						2.27
6	−1.41	.57	−.01	−.02	.10	−.02					2.24
7	−1.41	.60	−.01	−.02	.10	−.05	−.01				2.24
8	−1.41	.61	−.02	−.03	.10	−.07	−.02	.00			2.24
9	−1.41	.61	−.02	−.03	.10	−.08	−.02	.01	.01		2.24
10	−1.41	.61	−.02	−.02	.12	−.07	.00	.05	.04	.02	2.22
20	−1.43	.63	−.03	−.02	.11	−.08	.00	.03	.02	−.03	2.16
50	−1.43	.62	−.02	−.02	.12	−.08	.00	.03	.02	−.03	2.10
100	−1.43	.62	−.03	−.01	.11	−.08	−.01	.04	.01	−.03	2.00

An alternative method for obtaining preliminary estimates of the coefficients (once q has been determined) is to equate the theoretical and sample autocorrelations at lags $1, \ldots, q$ and solve the resulting non-linear equations for $\theta_1, \ldots, \theta_q$. Using the algorithm of Wilson (1969) to determine the solution

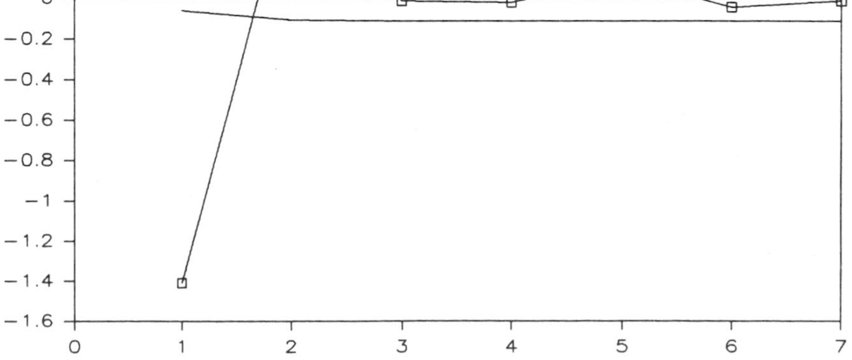

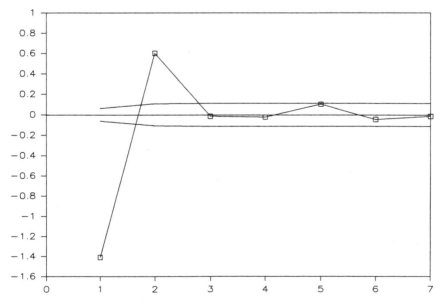

Figure 8.3. The estimates $\hat{\theta}_{7j}, j = 1, \ldots, 7$, for the data of Example 8.3.1, showing the bounds $\pm 1.96(\sum_{k=0}^{j-1} \hat{\theta}_{7k}^2)^{1/2} n^{-1/2}$.

for (θ_1, θ_2) such that $1 + \theta_1 z + \theta_2 z^2 \neq 0$ for $|z| < 1$, we arrive at the model,

$$X_t = Z_t - 1.49Z_{t-1} + .67Z_{t-2}, \qquad \{Z_t\} \sim \text{WN}(0, 2.06).$$

The actual process used to generate the data in this example was the Gaussian moving average,

$$X_t = Z_t - 1.40Z_{t-1} + .60Z_{t-2}, \qquad \{Z_t\} \sim \text{WN}(0, 2.25).$$

It is very well approximated by the preliminary model (8.3.5).

§8.4 Preliminary Estimation for ARMA(p, q) Processes

Let $\{X_t\}$ be the zero-mean causal ARMA(p, q) process,

$$X_t - \phi_1 X_{t-1} - \cdots - \phi_p X_{t-p} = Z_t + \theta_1 Z_{t-1} + \cdots + \theta_q Z_{t-q}, \qquad \{Z_t\} \sim \text{WN}(0, \sigma^2). \tag{8.4.1}$$

The causality assumption ensures that

$$X_t = \sum_{j=0}^{\infty} \psi_j Z_{t-j},$$

where by (3.3.3) and (3.3.4), the coefficients ψ_j satisfy

$$\begin{cases} \psi_0 = 1, \\ \psi_j = \theta_j + \sum_{i=1}^{\min(j,p)} \phi_i \psi_{j-i}, \qquad j = 1, 2, \ldots \end{cases} \tag{8.4.2}$$

and by convention, $\theta_j = 0$ for $j > q$ and $\phi_j = 0$ for $j > p$. To estimate $\psi_1, \ldots, \psi_{p+q}$, we can use the innovation estimates $\hat{\theta}_{m1}, \ldots, \hat{\theta}_{m,p+q}$, whose asymptotic behaviour is specified in Theorem 8.3.1. Replacing ψ_j by $\hat{\theta}_{mj}$ in (8.4.2) and solving the resulting equations,

$$\hat{\theta}_{mj} = \theta_j + \sum_{i=1}^{\min(j,p)} \phi_i \hat{\theta}_{m,j-i}, \qquad j = 1, 2, \ldots, p + q, \tag{8.4.3}$$

for $\boldsymbol{\phi}$ and $\boldsymbol{\theta}$, we obtain initial parameter estimates $\hat{\boldsymbol{\phi}}$ and $\hat{\boldsymbol{\theta}}$. From equations (8.4.3) with $j = q + 1, \ldots, q + p$, we see that $\hat{\boldsymbol{\phi}}$ should satisfy the equation,

$$\begin{bmatrix} \hat{\theta}_{m,q+1} \\ \hat{\theta}_{m,q+2} \\ \vdots \\ \hat{\theta}_{m,q+p} \end{bmatrix} = \begin{bmatrix} \hat{\theta}_{m,q} & \hat{\theta}_{m,q-1} & \cdots & \hat{\theta}_{m,q+1-p} \\ \hat{\theta}_{m,q+1} & \hat{\theta}_{m,q} & \cdots & \hat{\theta}_{m,q+2-p} \\ \vdots & \vdots & & \vdots \\ \hat{\theta}_{m,q+p-1} & \hat{\theta}_{m,q+p-2} & \cdots & \hat{\theta}_{m,q} \end{bmatrix} \begin{bmatrix} \phi_1 \\ \phi_2 \\ \vdots \\ \phi_p \end{bmatrix}. \tag{8.4.4}$$

Having solved (8.4.4) for $\hat{\boldsymbol{\phi}}$ (which may not be causal), the estimate of $\boldsymbol{\theta}$ is then easily found from

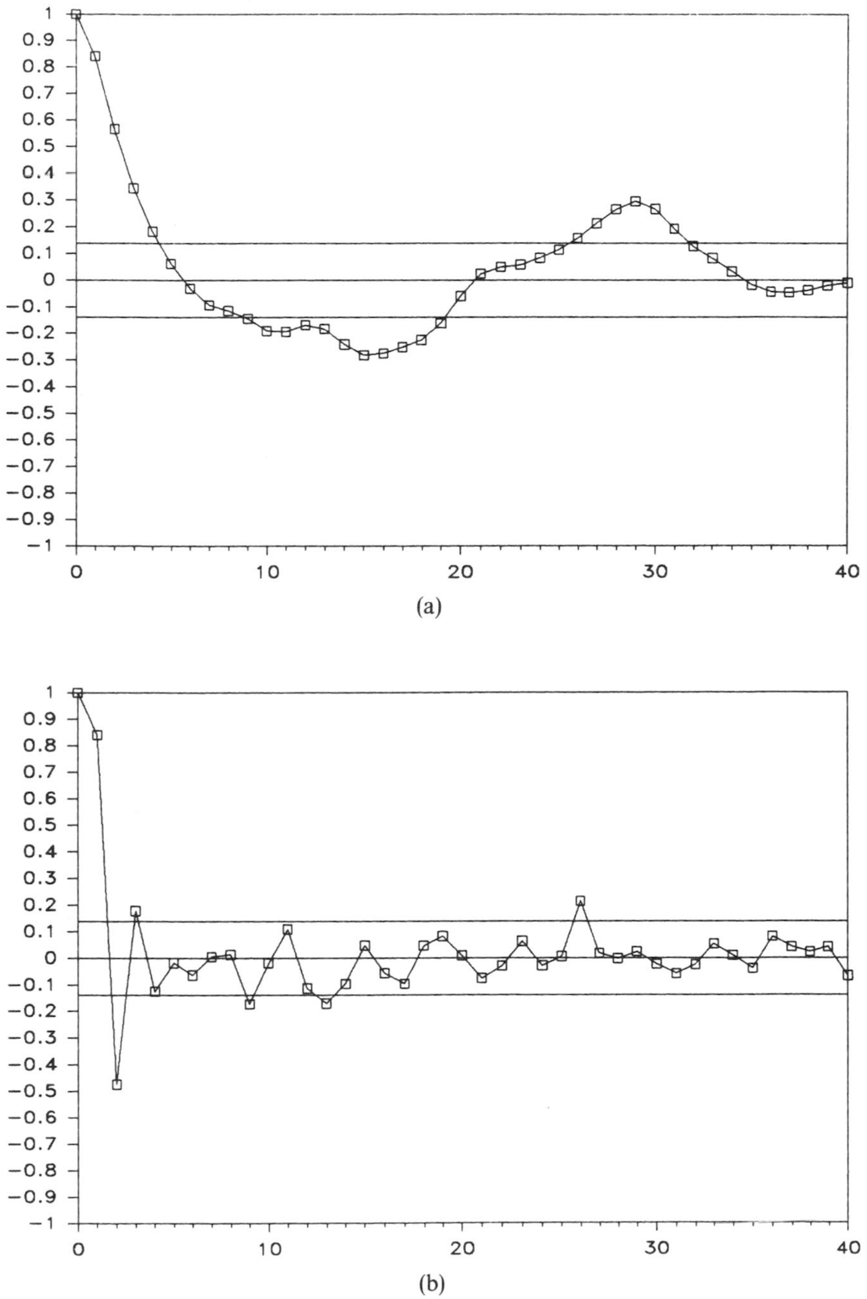

Figure 8.4. The sample ACF (a) and PACF (b) for the data of Example 8.4.1, showing the bounds $\pm 1.96n^{-1/2}$.

$$\hat{\theta}_j = \hat{\theta}_{mj} - \sum_{i=1}^{\min(j,p)} \hat{\phi}_i \hat{\theta}_{m,j-i}, \qquad j = 1, 2, \ldots, q. \tag{8.4.5}$$

Finally the white noise variance σ^2 is estimated by

$$\hat{\sigma}^2 = \hat{v}_m.$$

In the case of a pure moving average process, $p = 0$ and the method reduces to the one described in Section 8.3.

EXAMPLE 8.4.1. The sample autocorrelation function and partial autocorrelation function of a zero-mean time series of length 200 are shown in Figure 8.4. Identification of an appropriate model is much less obvious than in Examples 8.2.1 and 8.3.1. However we can proceed as follows. First use program PEST, Option 3, to fit a moving average model (8.3.1), with m chosen so as to give the smallest AICC value. (The AICC is a measure of goodness of fit, defined and discussed later in Section 9.3.) For this example the minimum occurs when $m = 8$ and the corresponding moving average model has coefficients as follows:

Table 8.2. $\hat{\theta}_{8,j}, j = 1, \ldots, 8$, for the Data of Example 8.4.1

j	1	2	3	4	5	6	7	8
θ_j	1.341	1.019	.669	.423	.270	.129	.011	−.115

The next step is to search for an ARMA(p, q) process, with p and q small, such that the equations (8.4.3) are satisfied with $m = 8$. For any given p and q (with $p + q \leq 8$), the equations (8.4.3) can be solved for $\boldsymbol{\phi}$ and $\boldsymbol{\theta}$ using Option 3 of PEST with m set equal to 8. At the same time the program computes the AICC value for the fitted model. The procedure is repeated for values of p and q such that $p + q \leq 8$ and models with small AICC value are noted as potentially useful preliminary models. In this particular example the AICC is minimized when $p = q = 1$ and the corresponding preliminary model is

$$X_t - .760X_{t-1} = Z_t + .582Z_{t-1}, \qquad \{Z_t\} \sim \text{WN}(0, 1.097).$$

This has a close resemblance to the true model, $X_t - .8X_{t-1} = Z_t + .6Z_{t-1}$, with $\{Z_t\} \sim \text{WN}(0, 1)$, which was used to generate the data. In general the resemblance will not be so close, so it is essential that preliminary estimation be followed by application of a more efficient procedure (see Section 8.5). For larger values of p and q, it is preferable to carry out the search procedure using maximum likelihood estimation (Option 8 of PEST) without preliminary estimation. Thus we can fit maximum likelihood models with $p + q = 1$, then $p + q = 2$, $p + q = 3, \ldots$, using lower order models with appended zero coefficients as initial models for the likelihood maximization. (See Sections 8.7 and 9.2.)

§8.5 Remarks on Asymptotic Efficiency

The preliminary estimates $(\hat{\boldsymbol{\phi}}, \hat{\boldsymbol{\theta}}, \hat{\sigma}^2)$ of the parameters in the ARMA(p, q) model discussed in Section 8.4 are weakly consistent in the sense that

$$\hat{\boldsymbol{\phi}} \xrightarrow{P} \boldsymbol{\phi}, \hat{\boldsymbol{\theta}} \xrightarrow{P} \boldsymbol{\theta} \quad \text{and} \quad \hat{\sigma}^2 \xrightarrow{P} \sigma^2 \quad \text{as } n \to \infty.$$

This is because (with $m(n)$ satisfying the conditions of Theorem 8.3.1) $\hat{\theta}_{mj} \xrightarrow{P} \psi_j$ and $\hat{v}_m \xrightarrow{P} \sigma^2$. Hence $(\hat{\boldsymbol{\phi}}, \hat{\boldsymbol{\theta}})$ must converge in probability to a solution of (8.4.2), i.e. to $(\boldsymbol{\phi}, \boldsymbol{\theta})$. In fact using Theorem 8.3.1, it may be shown (see Problem 8.22 and Brockwell and Davis (1988a)) that

$$\hat{\boldsymbol{\phi}} = \boldsymbol{\phi} + O_p(n^{-1/2}) \quad \text{and} \quad \hat{\boldsymbol{\theta}} = \boldsymbol{\theta} + O_p(n^{-1/2}).$$

In the next section we discuss a more efficient estimation procedure (strictly more efficient if $q \geq 1$) of $(\boldsymbol{\phi}, \boldsymbol{\theta})$ based on maximization of the Gaussian likelihood. We first introduce, through an example, the concept of relative efficiency of two competing estimators. Consider the MA(1) process $X_t = Z_t + \theta Z_{t-1}$ where $|\theta| < 1$ and $\{Z_t\} \sim \text{IID}(0, \sigma^2)$. If $\hat{\theta}_n^{(1)}$ and $\hat{\theta}_n^{(2)}$ are two estimators of θ based on the observations $X_1, \ldots, X_n$ such that $\hat{\theta}_n^{(i)}$ is $\text{AN}(\theta, \sigma_i^2(\theta)/n)$, $i = 1, 2$, then the asymptotic efficiency of $\hat{\theta}_n^{(1)}$ relative to $\hat{\theta}_n^{(2)}$ is defined to be

$$e(\theta, \hat{\theta}^{(1)}, \hat{\theta}^{(2)}) = \frac{\sigma_2^2(\theta)}{\sigma_1^2(\theta)}.$$

(This notion of efficiency extends in an obvious way to more general estimation problems.) If $e(\theta, \hat{\theta}^{(1)}, \hat{\theta}^{(2)}) \leq 1$ for all $\theta \in (-1, 1)$ then we say that $\hat{\theta}_n^{(2)}$ is a more efficient estimator of θ than $\hat{\theta}_n^{(1)}$ (strictly more efficient if in addition $e(\theta, \hat{\theta}^{(1)}, \hat{\theta}^{(2)}) < 1$ for some $\theta \in (-1, 1)$). For the MA(1) process let $\hat{\theta}_n^{(1)}$ denote the moment estimator of θ obtained by solving the equations $\hat{\gamma}(0) = \hat{\sigma}^2(1 + \hat{\theta}^2)$ and $\hat{\gamma}(1) = \hat{\sigma}^2\hat{\theta}$ for $\hat{\sigma}$ and $\hat{\theta}$. If $|\hat{\rho}(1)| > \frac{1}{2}$ there is no real solution $\hat{\theta}$ so we define $\hat{\theta} = \text{sgn}(\hat{\rho}(1))$. If $|\hat{\rho}(1)| \leq \frac{1}{2}$ then

$$\hat{\rho}(1) = \hat{\theta}_n^{(1)}/(1 + (\hat{\theta}_n^{(1)})^2).$$

In general therefore we can write,

$$\hat{\theta}_n^{(1)} = g(\hat{\rho}(1))$$

where

$$g(x) = \begin{cases} -1 & \text{if } x < -\frac{1}{2}, \\ (1 - (1 - 4x^2)^{1/2})/2x & \text{if } |x| \leq \frac{1}{2}, \\ 1 & \text{if } x > \frac{1}{2}. \end{cases}$$

From Theorem 7.2.2, $\hat{\rho}(1)$ is $\text{AN}(\rho(1), (1 - 3\rho^2(1) + 4\rho^4(1))/n)$, and so by Proposition 6.4.1,

$$\hat{\theta}_n^{(1)} \text{ is AN}(g(\rho(1)), \sigma_1^2(\theta)/n),$$

where

$$\sigma_1^2(\theta) = [g'(\rho(1))]^2[1 - 3\rho^2(1) + 4\rho^4(1)]$$
$$= (1 + \theta^2 + 4\theta^4 + \theta^6 + \theta^8)/(1 - \theta^2)^2.$$

If we now define $\hat{\theta}_n^{(2)} = \hat{\theta}_{m1}$, the estimator obtained from the innovations algorithm, then by Theorem 8.3.1,

$$\hat{\theta}_n^{(2)} \text{ is } \mathrm{AN}(\theta, n^{-1}).$$

Thus $e(\theta, \hat{\theta}^{(1)}, \hat{\theta}^{(2)}) = \sigma_1^{-2}(\theta) \leq 1$ for all $|\theta| < 1$, with strict inequality when $\theta \neq 0$. In particular

$$e(\theta, \hat{\theta}^{(1)}, \hat{\theta}^{(2)}) = \begin{cases} .82, & \theta = .25, \\ .37, & \theta = .50, \\ .06, & \theta = .75, \end{cases}$$

demonstrating the superiority of $\hat{\theta}_n^{(2)}$ over $\hat{\theta}_n^{(1)}$. We shall see in Example 8.8.2 that the maximum likelihood estimator $\hat{\theta}_n^{(3)}$ is $\mathrm{AN}(\theta, (1 - \theta^2)/n)$. Hence

$$e(\theta, \hat{\theta}^{(2)}, \hat{\theta}^{(3)}) = \begin{cases} .94, & \theta = .25, \\ .75, & \theta = .50, \\ .44, & \theta = .75. \end{cases}$$

While $\hat{\theta}_n^{(3)}$ is more efficient, $\hat{\theta}_n^{(2)}$ has reasonably good efficiency except when $|\theta|$ is close to 1. The superiority of maximum likelihood estimators from the point of view of asymptotic efficiency holds for a very large class of time-series models.

§8.6 Recursive Calculation of the Likelihood of an Arbitrary Zero-Mean Gaussian Process

In this section $\{X_t\}$ is assumed to be a Gaussian process with mean zero and covariance function $\kappa(i,j) = EX_iX_j$. Let $\mathbf{X}_n = (X_1, \ldots, X_n)'$ and let $\hat{\mathbf{X}}_n = (\hat{X}_1, \ldots, \hat{X}_n)'$ where $\hat{X}_1 = 0$ and $\hat{X}_j = E(X_j|X_1, \ldots, X_{j-1}) = P_{\overline{\mathrm{sp}}\{X_1, \ldots, X_{j-1}\}}X_j$, $j \geq 2$. Let Γ_n denote the covariance matrix, $\Gamma_n = E(\mathbf{X}_n\mathbf{X}_n')$, and assume that Γ_n is non-singular.

The likelihood of $\mathbf{X}_n$ is

$$L(\Gamma_n) = (2\pi)^{-n/2}(\det \Gamma_n)^{-1/2}\exp(-\tfrac{1}{2}\mathbf{X}_n'\Gamma_n^{-1}\mathbf{X}_n). \tag{8.6.1}$$

The direct calculation of $\det \Gamma_n$ and Γ_n^{-1} can be avoided by expressing this in terms of the one-step predictors $\hat{X}_j$, and their mean squared errors v_{j-1}, $j = 1, \ldots, n$, both of which are easily calculated recursively from the innovations algorithm, Proposition 5.2.2.

Let θ_{ij}, $j = 1, \ldots, i$; $i = 1, 2, \ldots$, denote the coefficients obtained when Proposition 5.2.2 is applied to the covariance function κ of $\{X_t\}$, and let $\theta_{i0} = 1$, $\theta_{ij} = 0$ for $j < 0$, $i = 0, 1, 2, \ldots$. Now define the $n \times n$ lower triangular matrix,

$$C = [\theta_{i,i-j}]_{i,j=0}^{n-1}, \tag{8.6.2}$$

and the $n \times n$ diagonal matrix,

$$D = \operatorname{diag}(v_0, v_1, \dots, v_{n-1}). \tag{8.6.3}$$

The innovations representation (5.2.15) of $\hat{X}_j, j = 1, \dots, n$, can then be written in the form,

$$\hat{\mathbf{X}}_n = (C - I)(\mathbf{X}_n - \hat{\mathbf{X}}_n),$$

where I is the $n \times n$ identity matrix. Hence

$$\mathbf{X}_n = \mathbf{X}_n - \hat{\mathbf{X}}_n + \hat{\mathbf{X}}_n = C(\mathbf{X}_n - \hat{\mathbf{X}}_n). \tag{8.6.4}$$

Since D is the covariance matrix of $(\mathbf{X}_n - \hat{\mathbf{X}}_n)$, it follows that

$$\Gamma_n = CDC' \tag{8.6.5}$$

(from which the Cholesky factorization $\Gamma_n = UU'$, with U lower triangular, can easily be deduced).

From (8.6.4) and (8.6.5), we obtain

$$\mathbf{X}_n'\Gamma_n^{-1}\mathbf{X}_n = (\mathbf{X}_n - \hat{\mathbf{X}}_n)'D^{-1}(\mathbf{X}_n - \hat{\mathbf{X}}_n) = \sum_{j=1}^{n} (X_j - \hat{X}_j)^2/v_{j-1}, \tag{8.6.6}$$

and

$$\det \Gamma_n = (\det C)^2(\det D) = v_0 v_1 \cdots v_{n-1}. \tag{8.6.7}$$

The likelihood (8.6.1) of the vector $\mathbf{X}_n$ therefore reduces to

$$L(\Gamma_n) = (2\pi)^{-n/2}(v_0 \cdots v_{n-1})^{-1/2} \exp\left\{-\frac{1}{2}\sum_{j=1}^{n} (X_j - \hat{X}_j)^2/v_{j-1}\right\}. \tag{8.6.8}$$

Applying Proposition 5.2.2 to the covariance function κ gives $\hat{X}_1, \hat{X}_2, \dots, v_0, v_1, \dots$, and hence $L(\Gamma_n)$.

If Γ_n is expressible in terms of a finite number of unknown parameters $\beta_1, \dots, \beta_r$, as for example when $\{X_t\}$ is an ARMA(p, q) process and $r = p + q + 1$, it is usually necessary to estimate the parameters from the data $\mathbf{X}_n$. A standard statistical procedure in such situations (see e.g. Lehmann (1983)) is to maximize the likelihood $L(\beta_1, \dots, \beta_r)$ with respect to $\beta_1, \dots, \beta_r$. In the case when $X_1, X_2, \dots$ are independently and identically distributed, it is known that under rather general conditions the maximum likelihood estimators are consistent as $n \to \infty$ and asymptotically normal with variances as small or smaller than those of any other asymptotically normal estimators. A natural estimation procedure for Gaussian processes therefore is to maximize (8.6.8) with respect to $\beta_1, \dots, \beta_r$. The dependence of the sequence $\{X_n\}$ must however be kept in mind when studying the asymptotic behaviour of the estimators. (See Sections 8.8, 8.11 and 10.8 below.)

Even if $\{X_t\}$ is not Gaussian, it makes sense to regard (8.6.8) as a measure of the goodness of fit of the covariance matrix $\Gamma_n(\beta_1, \dots, \beta_r)$ to the data, and still to choose the parameters $\beta_1, \ldots, \beta_r$ in such a way as to maximize (8.6.8). We shall always refer to the estimators $\hat{\beta}_1, \ldots, \hat{\beta}_r$ so obtained as "maximum likelihood" estimators, even when $\{X_t\}$ is not Gaussian. Regardless of the joint distribution of $X_1, \ldots, X_n$, we shall also refer to (8.6.1) (and its algebraic equivalent (8.6.8)) as the "Gaussian likelihood" of $X_1, \ldots, X_n$.

§8.7 Maximum Likelihood and Least Squares Estimation for ARMA Processes

Suppose now that $\{X_t\}$ is the causal ARMA(p, q) process,

$$X_t = \phi_1 X_{t-1} + \cdots + \phi_p X_{t-p} + \theta_0 Z_t + \cdots + \theta_q Z_{t-q}, \qquad \{Z_t\} \sim \text{WN}(0, \sigma^2), \tag{8.7.1}$$

where $\theta_0 = 1$. The causality assumption means that $1 - \phi_1 z - \cdots - \phi_p z^p \neq 0$ for $|z| \leq 1$. To avoid ambiguity we shall assume also that the coefficients θ_i and white noise variance σ^2 have been adjusted (without affecting the autocovariance function of $\{X_t\}$) to ensure that $\theta(z) = 1 + \theta_1 z + \cdots + \theta_q z^q \neq 0$ for $|z| < 1$. Our first problem is to find maximum likelihood estimates of the parameter vectors $\boldsymbol{\phi} = (\phi_1, \ldots, \phi_p)'$, $\boldsymbol{\theta} = (\theta_1, \ldots, \theta_q)'$ and of the white noise variance σ^2.

In Section 5.3 we showed that the one-step predictors $\hat{X}_{i+1}$ and their mean squared errors are given by,

$$\begin{cases} \hat{X}_{i+1} = \sum_{j=1}^{i} \theta_{ij}(X_{i+1-j} - \hat{X}_{i+1-j}), & 1 \leq i < m = \max(p, q), \\ \hat{X}_{i+1} = \phi_1 X_i + \cdots + \phi_p X_{i+1-p} + \sum_{j=1}^{q} \theta_{ij}(X_{i+1-j} - \hat{X}_{i+1-j}), & i \geq m, \end{cases} \tag{8.7.2}$$

and

$$E(X_{i+1} - \hat{X}_{i+1})^2 = \sigma^2 r_i, \tag{8.7.3}$$

where θ_{ij} and r_i are obtained by applying Proposition 5.2.2 to the covariance function (5.3.5). We recall also that θ_{ij} and r_i are independent of σ^2.

Substituting in the general expression (8.6.8), we find that the Gaussian likelihood of the vector of observations $\mathbf{X}_n = (X_1, \ldots, X_n)'$ is

$$L(\boldsymbol{\phi}, \boldsymbol{\theta}, \sigma^2) = (2\pi\sigma^2)^{-n/2}(r_0 \cdots r_{n-1})^{-1/2} \exp\left[-\tfrac{1}{2}\sigma^{-2} \sum_{j=1}^{n} (X_j - \hat{X}_j)^2 / r_{j-1}\right]. \tag{8.7.4}$$

Differentiating $\ln L(\boldsymbol{\phi}, \boldsymbol{\theta}, \sigma^2)$ partially with respect to σ^2 and noting that $\hat{X}_j$ and r_j are independent of σ^2, we deduce (Problem 8.11) that the maximum likelihood estimators $\hat{\boldsymbol{\phi}}$, $\hat{\boldsymbol{\theta}}$ and $\hat{\sigma}^2$ satisfy

$$\hat{\sigma}^2 = n^{-1} S(\hat{\boldsymbol{\phi}}, \hat{\boldsymbol{\theta}}), \tag{8.7.5}$$

where

$$S(\hat{\boldsymbol{\phi}}, \hat{\boldsymbol{\theta}}) = \sum_{j=1}^{n} (X_j - \hat{X}_j)^2 / r_{j-1}, \tag{8.7.6}$$

and $\hat{\boldsymbol{\phi}}$, $\hat{\boldsymbol{\theta}}$ are the values of $\boldsymbol{\phi}$, $\boldsymbol{\theta}$ which minimize

$$l(\boldsymbol{\phi}, \boldsymbol{\theta}) = \ln(n^{-1} S(\boldsymbol{\phi}, \boldsymbol{\theta})) + n^{-1} \sum_{j=1}^{n} \ln r_{j-1}. \tag{8.7.7}$$

We shall refer to $l(\boldsymbol{\phi}, \boldsymbol{\theta})$ as the "reduced likelihood". The calculation of $l(\boldsymbol{\phi}, \boldsymbol{\theta})$ can easily be carried out using Proposition 5.2.2 which enables us to compute $\theta_{i-1,j}$, r_{i-1} and $\hat{X}_i$ recursively for any prescribed pair of parameter vectors $\boldsymbol{\phi}$, $\boldsymbol{\theta}$. A non-linear minimization program is used in the computer program PEST, in conjunction with the innovations algorithm, to search for the values of $\boldsymbol{\phi}$ and $\boldsymbol{\theta}$ which minimize $l(\boldsymbol{\phi}, \boldsymbol{\theta})$. These are the maximum likelihood estimates of $\boldsymbol{\phi}$ and $\boldsymbol{\theta}$ respectively. The maximum likelihood estimator of σ^2 is then found from (8.7.5).

The search procedure may be greatly accelerated if we begin with parameter values $\boldsymbol{\phi}_0$, $\boldsymbol{\theta}_0$ which are close to the minimum of l. It is for this reason that simple, reasonably good preliminary estimates of $\boldsymbol{\phi}$ and $\boldsymbol{\theta}$, such as those described in Sections 8.2, 8.3 and 8.4, are important. It is essential to begin the search with a *causal* parameter vector $\boldsymbol{\phi}_0$ since causality is assumed in the computation of $l(\boldsymbol{\phi}, \boldsymbol{\theta})$. Failure to do so will result in an error message from the program. The estimate of $\boldsymbol{\phi}$ returned by the program is constrained to be causal. The estimate of $\boldsymbol{\theta}$ is not constrained to be invertible, although if the initial vector $\boldsymbol{\theta}_0$ satisfies the condition $1 + \theta_{01} z + \cdots + \theta_{0q} z^q \neq 0$ for $|z| < 1$ and if $(\boldsymbol{\phi}_0, \boldsymbol{\theta}_0)$ is close to the minimum, then it is likely that the value of $\hat{\boldsymbol{\theta}}$ returned by the program will also satisfy $1 + \hat{\theta}_1 z + \cdots + \hat{\theta}_q z^q \neq 0$ for $|z| < 1$. If not, it is a simple matter to adjust the estimates of σ^2 and $\boldsymbol{\theta}$ in order to satisfy the condition without altering the value of the likelihood function (see Section 4.4). Since we specified in (8.7.1) that $\theta(z) \neq 0$ for $|z| < 1$, the estimates $\hat{\boldsymbol{\theta}}$ and $\hat{\sigma}^2$ are chosen as those which satisfy the condition $\hat{\theta}(z) \neq 0$ for $|z| < 1$. Note however that this constraint is not always desirable (see Example 9.2.2).

An intuitively appealing alternative estimation procedure is to minimize the weighted sum of squares

$$S(\boldsymbol{\phi}, \boldsymbol{\theta}) = \sum_{j=1}^{n} (X_j - \hat{X}_j)^2 / r_{j-1}, \tag{8.7.8}$$

with respect to $\boldsymbol{\phi}$ and $\boldsymbol{\theta}$. The estimators obtained in this way will be referred to as the "least squares" estimators $\tilde{\boldsymbol{\phi}}$ and $\tilde{\boldsymbol{\theta}}$ of $\boldsymbol{\phi}$ and $\boldsymbol{\theta}$. In view of the close relationship (8.7.7) between $l(\boldsymbol{\phi}, \boldsymbol{\theta})$ and $S(\boldsymbol{\phi}, \boldsymbol{\theta})$, the least squares estimators can easily be found (if required) using the same computer program PEST. For the minimization of $S(\boldsymbol{\phi}, \boldsymbol{\theta})$ however, it is necessary not only to restrict $\boldsymbol{\phi}$ to be causal, but also to restrict $\boldsymbol{\theta}$ to be invertible. Without the latter constraint there will in general be no finite $(\boldsymbol{\phi}, \boldsymbol{\theta})$ at which S achieves its minimum value (see Problem 8.13). If $n^{-1}\sum_{j=1}^{n} \ln r_{j-1}$ is asymptotically negligible compared with $\ln S(\boldsymbol{\phi}, \boldsymbol{\theta})$, as is the case when $\boldsymbol{\theta}$ is constrained to be invertible (since $r_n \to 1$), then from (8.7.7), minimization of S will be equivalent to minimization of l and the least squares and maximum likelihood estimators will have similar asymptotic properties. The least squares estimator $\tilde{\sigma}_{\text{LS}}^2$ is found from

$$\tilde{\sigma}_{\text{LS}}^2 = (n - p - q)^{-1} S(\tilde{\boldsymbol{\phi}}, \tilde{\boldsymbol{\theta}}), \tag{8.7.9}$$

where the divisor $(n - p - q)$ is used (as in standard linear regression theory) since $\sigma^{-2} S(\tilde{\boldsymbol{\phi}}, \tilde{\boldsymbol{\theta}})$ is distributed approximately as chi-squared with $(n - p - q)$ degrees of freedom (see Section 8.9).

§8.8 Asymptotic Properties of the Maximum Likelihood Estimators

If $\{X_t\}$ is the causal invertible process,

$$X_t - \phi_1 X_{t-1} - \cdots - \phi_p X_{t-p} = Z_t + \theta_1 Z_{t-1} + \cdots + \theta_q Z_{t-q}, \qquad \{Z_t\} \sim \text{IID}(0, \sigma^2), \tag{8.8.1}$$

and if $\phi(\cdot)$ and $\theta(\cdot)$ have no common zeroes, then the maximum likelihood estimator $\hat{\boldsymbol{\beta}}' = (\hat{\phi}_1, \dots, \hat{\phi}_p, \hat{\theta}_1, \dots, \hat{\theta}_q) = (\hat{\boldsymbol{\phi}}', \hat{\boldsymbol{\theta}}')$ is defined to be the causal invertible value of $\boldsymbol{\beta}' = (\boldsymbol{\phi}', \boldsymbol{\theta}')$ which minimizes the reduced likelihood $l(\boldsymbol{\phi}, \boldsymbol{\theta})$ defined by (8.7.7). The program PEST can be used to determine $\hat{\boldsymbol{\phi}}$, $\hat{\boldsymbol{\theta}}$ numerically. It also gives the maximum likelihood estimate $\hat{\sigma}^2$ of the white noise variance determined by (8.7.5).

The least squares estimators $\tilde{\boldsymbol{\phi}}$, $\tilde{\boldsymbol{\theta}}$ are the causal invertible values of $\boldsymbol{\phi}$ and $\boldsymbol{\theta}$ which minimize $\ln(n^{-1} S(\boldsymbol{\phi}, \boldsymbol{\theta})) = l(\boldsymbol{\phi}, \boldsymbol{\theta}) - n^{-1}\sum_{j=1}^{n} \ln r_{j-1}$. Because of the invertibility the term $n^{-1}\sum_{j=1}^{n} \ln r_{j-1}$ is asymptotically negligible as $n \to \infty$ and the estimators $\tilde{\boldsymbol{\phi}}$ and $\tilde{\boldsymbol{\theta}}$ have the same asymptotic properties as $\hat{\boldsymbol{\phi}}$ and $\hat{\boldsymbol{\theta}}$. It follows, (see Theorem 10.8.2), that if $\{Z_t\} \sim \text{IID}(0, \sigma^2)$ and $\phi(\cdot)$ and $\theta(\cdot)$ are causal and invertible with no common zeroes, then

$$n^{1/2}(\hat{\boldsymbol{\beta}} - \boldsymbol{\beta}) \Rightarrow N(\mathbf{0}, V(\boldsymbol{\beta})), \tag{8.8.2}$$

where the asymptotic covariance matrix $V(\boldsymbol{\beta})$ can be computed explicitly from (8.11.14) (see also (10.8.30)). Specifically for $p \geq 1$ and $q \geq 1$,

$$V(\boldsymbol{\beta}) = \sigma^2 \begin{bmatrix} E\mathbf{U}_t\mathbf{U}_t' & E\mathbf{U}_t\mathbf{V}_t' \\ E\mathbf{V}_t\mathbf{U}_t' & E\mathbf{V}_t\mathbf{V}_t' \end{bmatrix}^{-1}, \tag{8.8.3}$$

where $\mathbf{U}_t = (U_t, \dots, U_{t+1-p})'$, $\mathbf{V}_t = (V_t, \dots, V_{t+1-q})'$ and $\{U_t\}$, $\{V_t\}$ are the autoregressive processes,

$$\phi(B)U_t = Z_t, \tag{8.8.4}$$

and

$$\theta(B)V_t = Z_t. \tag{8.8.5}$$

(For $p = 0$, $V(\boldsymbol{\beta}) = \sigma^2[E\mathbf{V}_t\mathbf{V}_t']^{-1}$, and for $q = 0$, $V(\boldsymbol{\beta}) = \sigma^2[E\mathbf{U}_t\mathbf{U}_t']^{-1}$.)

We now compute the asymptotic distributions for several special cases of interest.

EXAMPLE 8.8.1 (AR(p)). From (8.8.3),

$$V(\boldsymbol{\phi}) = \sigma^2[E\mathbf{U}_t\mathbf{U}_t']^{-1},$$

where $\phi(B)U_t = Z_t$. Hence

$$V(\boldsymbol{\phi}) = \sigma^2\Gamma_p^{-1},$$

where $\Gamma_p = E(\mathbf{U}_t\mathbf{U}_t') = [EX_iX_j]_{i,j=1}^p$, and

$$\hat{\boldsymbol{\phi}} \text{ is AN}(\boldsymbol{\phi}, n^{-1}\sigma^2\Gamma_p^{-1}).$$

In the special cases $p = 1$ and $p = 2$ it is easy to express Γ_p^{-1} in terms of $\boldsymbol{\phi}$, giving the results,

$$\text{AR}(1): \hat{\phi} \text{ is AN}(\phi, n^{-1}(1-\phi^2)),$$

$$\text{AR}(2): \begin{bmatrix} \hat{\phi}_1 \\ \hat{\phi}_2 \end{bmatrix} \text{ is AN}\left(\begin{bmatrix} \phi_1 \\ \phi_2 \end{bmatrix}, n^{-1}\begin{bmatrix} 1-\phi_2^2 & -\phi_1(1+\phi_2) \\ -\phi_1(1+\phi_2) & 1-\phi_2^2 \end{bmatrix}\right).$$

EXAMPLE 8.8.2 (MA(q)). From (8.8.3)

$$V(\boldsymbol{\theta}) = \sigma^2[E\mathbf{V}_t\mathbf{V}_t']^{-1},$$

where $\theta(B)V_t = Z_t$. Hence

$$V(\boldsymbol{\theta}) = \sigma^2[\Gamma_q^*]^{-1},$$

where Γ_q^* is the covariance matrix $[EV_iV_j]_{i,j=1}^q$ of the autoregressive process

$$V_t + \theta_1 V_{t-1} + \cdots + \theta_q V_{t-q} = Z_t.$$

Inspection of the results of Example 8.8.1 yields, for the cases MA(1) and MA(2),

$$\text{MA}(1): \hat{\theta} \text{ is AN}(\theta, n^{-1}(1-\theta^2)),$$

$$\text{MA}(2): \begin{bmatrix} \hat{\theta}_1 \\ \hat{\theta}_2 \end{bmatrix} \text{ is AN}\left(\begin{bmatrix} \theta_1 \\ \theta_2 \end{bmatrix}, n^{-1}\begin{bmatrix} 1-\theta_2^2 & \theta_1(1-\theta_2) \\ \theta_1(1-\theta_2) & 1-\theta_2^2 \end{bmatrix}\right).$$

EXAMPLE 8.8.3 (ARMA(1, 1)). In this case

$$V(\phi, \theta) = \sigma^2\begin{bmatrix} EU_t^2 & EU_tV_t \\ EU_tV_t & EV_t^2 \end{bmatrix}^{-1},$$

where $U_t - \phi U_{t-1} = Z_t$ and $V_t + \theta V_{t-1} = Z_t$. A simple calculation gives,

$$V(\phi, \theta) = \begin{bmatrix} (1-\phi^2)^{-1} & (1+\phi\theta)^{-1} \\ (1+\phi\theta)^{-1} & (1-\theta^2)^{-1} \end{bmatrix}^{-1},$$

whence

$$\begin{bmatrix} \hat{\phi} \\ \hat{\theta} \end{bmatrix} \text{ is AN}\left(\begin{bmatrix} \phi \\ \theta \end{bmatrix}, n^{-1}\frac{1+\phi\theta}{(\phi+\theta)^2}\begin{bmatrix} (1-\phi^2)(1+\phi\theta) & -(1-\theta^2)(1-\phi^2) \\ -(1-\theta^2)(1-\phi^2) & (1-\theta^2)(1+\phi\theta) \end{bmatrix}\right).$$

These asymptotic distributions provide us with a general technique for computing asymptotic confidence regions for $\boldsymbol{\phi}$ and $\boldsymbol{\theta}$ from the maximum likelihood or least squares estimates. This is discussed in more detail, together with an alternative technique based on the likelihood surface, in Section 8.9.

§8.9 Confidence Intervals for the Parameters of a Causal Invertible ARMA Process

Large-sample confidence regions for the coefficients $\boldsymbol{\phi}$ and $\boldsymbol{\theta}$ of a causal invertible ARMA process can be derived from the asymptotic distribution of the maximum likelihood estimators in exactly the same way as those derived from the asymptotic distribution of the Yule–Walker estimator of $\boldsymbol{\phi}$ in Section 8.2. For the process (8.8.1) let $\boldsymbol{\beta}' = (\boldsymbol{\phi}', \boldsymbol{\theta}')$ and let $\hat{\boldsymbol{\beta}}$ be the maximum likelihood estimator of $\boldsymbol{\beta}$. Then defining $V(\boldsymbol{\beta})$ by (8.8.3) we obtain the approximate $(1-\alpha)$ confidence region for $\boldsymbol{\beta}$,

$$\{\boldsymbol{\beta} \in \mathbb{R}^{p+q} : (\boldsymbol{\beta} - \hat{\boldsymbol{\beta}})' V^{-1}(\hat{\boldsymbol{\beta}})(\boldsymbol{\beta} - \hat{\boldsymbol{\beta}}) \le n^{-1}\chi^2_{1-\alpha}(p+q)\}. \tag{8.9.1}$$

Writing $v_{jj}(\hat{\boldsymbol{\beta}})$ for the jth diagonal element of $V(\hat{\boldsymbol{\beta}})$, we also have the approximate $(1-\alpha)$ confidence region for β_j, i.e.

$$\{\beta \in \mathbb{R} : |\beta - \hat{\beta}_j| \le n^{-1/2}\Phi_{1-\alpha/2} v_{jj}^{1/2}(\hat{\boldsymbol{\beta}})\}. \tag{8.9.2}$$

An alternative approach, based on the shape of the reduced likelihood surface, $l(\boldsymbol{\beta}) = l(\boldsymbol{\phi}, \boldsymbol{\theta})$, near its minimum can also be used. We shall assume for the remainder of this section that $\{X_t\}$ is Gaussian. For large n, the invertibility assumption allows us to approximate $n\exp(l(\boldsymbol{\beta}))$ (since $n^{-1}\sum_{i=0}^{n-1}\ln r_i \to 0$) by

$$S(\boldsymbol{\beta}) = \mathbf{X}_n' G_n^{-1}(\boldsymbol{\beta})\mathbf{X}_n, \tag{8.9.3}$$

where

$$G_n(\boldsymbol{\beta}) = \sigma^{-2}\Gamma_n, \tag{8.9.4}$$

and the maximum likelihood estimator $\hat{\boldsymbol{\beta}}$ by the value of $\boldsymbol{\beta}$ which minimizes $S(\boldsymbol{\beta})$, i.e. by the least squares estimator.

The behavior of $\hat{\boldsymbol{\beta}}$ can then be investigated by making a further approximation which reduces the problem to a standard one in the theory of the general linear model. To do this we define $T = \sigma^{-1} CD^{1/2}$, where C and D were defined in Section 8.6, and let

$$\mathbf{W}_n(\boldsymbol{\beta}) = T^{-1}\mathbf{X}_n.$$

Then by (8.6.5), $TT' = G_n(\boldsymbol{\beta})$ and by (8.6.4), the ith component W_{ni} of $W_n(\boldsymbol{\beta})$ is $(X_i - \hat{X}_i)/r_{i-1}^{1/2}$. The problem of finding $\hat{\boldsymbol{\beta}}$ is thus equivalent to minimizing $S(\boldsymbol{\beta}) = \mathbf{W}_n(\boldsymbol{\beta})'\mathbf{W}_n(\boldsymbol{\beta})$ with respect to $\boldsymbol{\beta}$. Now make the approximation that for each j, $\partial \mathbf{W}_n/\partial \beta_j$ is constant in a small neighborhood of the true parameter vector $\boldsymbol{\beta}^*$, and let

$$X = \left[-\frac{\partial W_{ni}}{\partial \beta_j}(\beta^*) \right]_{i,j=1}^{p+q}.$$

If $\boldsymbol{\beta}_0$ is some fixed vector in this neighborhood, we can then write

$$\mathbf{W}_n(\boldsymbol{\beta}_0) = \mathbf{W}_n(\boldsymbol{\beta}^*) + X(\boldsymbol{\beta}^* - \boldsymbol{\beta}_0),$$

i.e.

$$\mathbf{Y}_n = X\boldsymbol{\beta}^* + \mathbf{W}_n(\boldsymbol{\beta}^*), \tag{8.9.5}$$

where $\mathbf{Y}_n$ is the transformed vector of observations, $\mathbf{W}_n(\boldsymbol{\beta}_0) + X\boldsymbol{\beta}_0$, and the components of $\mathbf{W}_n(\boldsymbol{\beta}^*)$ are independent with the distribution $\mathrm{N}(0, \sigma^2)$.

Equation (8.9.5) is a linear model of standard type, and our estimator $\hat{\boldsymbol{\beta}}$ is the value of $\boldsymbol{\beta}$ which minimizes the squared Euclidean distance,

$$S(\boldsymbol{\beta}) = \|\mathbf{Y}_n - X\boldsymbol{\beta}\|^2 = \mathbf{W}_n(\boldsymbol{\beta})'\mathbf{W}_n(\boldsymbol{\beta}) = \mathbf{X}_n' G_n(\boldsymbol{\beta})^{-1}\mathbf{X}_n. \tag{8.9.6}$$

The standard theory of least-squares estimation for the general linear model (see Section 2.6 and Problem 2.19) suggests the approximations, $S(\hat{\boldsymbol{\beta}}) \sim \sigma^2\chi^2(n-p-q)$ and $S(\boldsymbol{\beta}^*) - S(\hat{\boldsymbol{\beta}}) \sim \sigma^2\chi^2(p+q)$, with $S(\hat{\boldsymbol{\beta}})$ and $S(\boldsymbol{\beta}^*) - S(\hat{\boldsymbol{\beta}})$ approximately independent. These observations yield the $(1-\alpha)$ confidence regions:

$$\boldsymbol{\beta}^* : \left\{ \boldsymbol{\beta} \in \mathbb{R}^{p+q} : \frac{S(\boldsymbol{\beta}) - S(\hat{\boldsymbol{\beta}})}{S(\hat{\boldsymbol{\beta}})} \leq \frac{p+q}{n-p-q} F_{1-\alpha}(p+q, n-p-q) \right\}, \tag{8.9.7}$$

$$\sigma^2 : \left\{ v \in \mathbb{R} : \frac{S(\hat{\boldsymbol{\beta}})}{\chi^2_{1-\alpha/2}(n-p-q)} \leq v \leq \frac{S(\hat{\boldsymbol{\beta}})}{\chi^2_{\alpha/2}(n-p-q)} \right\}, \tag{8.9.8}$$

where F_α and χ^2_α denote the α-quantiles of the F and χ^2 distributions. Since the function S can be computed for any value of $\boldsymbol{\beta}$ using the program PEST, these regions can be determined numerically. Marginal confidence intervals for the components of $\boldsymbol{\beta}^*$ can be found analogously. These are:

$$\beta_j^* : \left\{ \beta_j \in \mathbb{R} : \frac{S(\boldsymbol{\beta}) - S(\hat{\boldsymbol{\beta}})}{S(\hat{\boldsymbol{\beta}})} \leq \frac{t^2_{1-\alpha/2}(n-p-q)}{n-p-q} \text{ for some } \boldsymbol{\beta} \in \mathbb{R}^{p+q} \text{ with } j^{\text{th}} \text{ component } \beta_j \right\}. \tag{8.9.9}$$

§8.10* Asymptotic Behavior of the Yule–Walker Estimates

Throughout this section, it is assumed that $\{X_t\}$ is a causal AR(p) process

$$X_t - \phi_1 X_{t-1} - \cdots - \phi_p X_{t-p} = Z_t, \tag{8.10.1}$$

where $\{Z_t\} \sim \text{IID}(0, \sigma^2)$. The Yule–Walker estimates of $\boldsymbol{\phi}$ and σ^2 are given by equations (8.1.3) and (8.1.4), or equivalently by

$$\hat{\boldsymbol{\phi}} = \hat{\Gamma}_p^{-1} \hat{\boldsymbol{\gamma}}_p$$

and

$$\hat{\sigma}^2 = \hat{\gamma}(0) - \hat{\boldsymbol{\gamma}}_p' \hat{\boldsymbol{\phi}}.$$

It will be convenient to express (8.10.1) in the form

$$\mathbf{Y} = X\boldsymbol{\phi} + \mathbf{Z} \tag{8.10.2}$$

where $\mathbf{Y} = (X_1, \ldots, X_n)'$, $\mathbf{Z} = (Z_1, \ldots, Z_n)'$ and X is the $n \times p$ design matrix,

$$X = \begin{bmatrix} X_0 & X_{-1} & \cdots & X_{1-p} \\ X_1 & X_0 & \cdots & X_{2-p} \\ \vdots & \vdots & & \\ X_{n-1} & X_{n-2} & \cdots & X_{n-p} \end{bmatrix}.$$

Because of the similarity between (8.10.2) and the general linear model (see Section 2.6 and Problem 2.19), we introduce the "linear regression estimate" $\boldsymbol{\phi}^*$ of $\boldsymbol{\phi}$ defined by

$$\boldsymbol{\phi}^* = (X'X)^{-1} X'\mathbf{Y}. \tag{8.10.3}$$

The vector $\boldsymbol{\phi}^*$ is not an estimator in the usual sense since it depends on the values $X_{1-p}, X_{2-p}, \ldots, X_n$ and not only on the observed values $X_1, \ldots, X_n$. Nevertheless, as we shall see, $\boldsymbol{\phi}^*$ and $\hat{\boldsymbol{\phi}}$ have similar properties. This is because $n^{-1}(X'X)$ is the matrix whose (i,j)-element is equal to $n^{-1} \sum_{k=1-i}^{n-i} X_k X_{k+|i-j|}$ and $n^{-1} X'\mathbf{Y}$ is the vector whose i^{th} component is $n^{-1} \sum_{k=1-i}^{n-i} X_k X_{k+i}$. Consequently (see Proposition 7.3.5),

$$n^{-1}(X'X) \xrightarrow{P} \Gamma_p \quad \text{and} \quad n^{-1} X'\mathbf{Y} \xrightarrow{P} \boldsymbol{\gamma}_p. \tag{8.10.4}$$

The proof of Theorem 8.1.1 is divided into two parts. First we establish the limit distribution for $\boldsymbol{\phi}^*$ and then show that $\boldsymbol{\phi}^*$ and $\hat{\boldsymbol{\phi}}$ must have the same limit law.

Proposition 8.10.1. *With $\boldsymbol{\phi}^*$ defined as in* (8.10.3)

$$n^{1/2}(\boldsymbol{\phi}^* - \boldsymbol{\phi}) \Rightarrow \text{N}(0, \sigma^2 \Gamma_p^{-1}).$$

PROOF. From (8.10.1) and (8.10.2) we have

$$n^{1/2}(\boldsymbol{\phi}^* - \boldsymbol{\phi}) = n^{1/2}((X'X)^{-1}X'(X\boldsymbol{\phi} + \mathbf{Z}) - \boldsymbol{\phi})$$
$$= n(X'X)^{-1}(n^{-1/2}X'\mathbf{Z}).$$

By setting $\mathbf{U}_t = (X_{t-1}, \ldots, X_{t-p})'Z_t$, $t \geq 1$ we have

$$n^{-1/2}X'\mathbf{Z} = n^{-1/2}\sum_{t=1}^{n}\mathbf{U}_t.$$

Observe that $E\mathbf{U}_t = \mathbf{0}$ and that

$$E\mathbf{U}_t\mathbf{U}_{t+h}' = \begin{cases} \sigma^2\Gamma_p, & h = 0, \\ O_{p\times p}, & h \neq 0, \end{cases}$$

since Z_t is independent of $(X_{t-1}, \ldots, X_{t-p})$. Let $X_t = \sum_{j=0}^{\infty}\psi_j Z_{t-j}$ be the causal representation of X_t. For a fixed positive integer m set $X_t^{(m)} = \sum_{j=0}^{m}\psi_j Z_{t-j}$ and $\mathbf{U}_t^{(m)} = (X_{t-1}^{(m)}, \ldots, X_{t-p}^{(m)})'Z_t$ and let $\boldsymbol{\lambda}$ be a fixed element of $\mathbb{R}^p$. Then $\boldsymbol{\lambda}'\mathbf{U}_t^{(m)}$ is a strictly stationary $(m+p)$-dependent white noise sequence with variance given by $\sigma^2\boldsymbol{\lambda}'\Gamma_p^{(m)}\boldsymbol{\lambda}$ where $\Gamma_p^{(m)}$ is the covariance matrix of $(X_{t-1}^{(m)}, \ldots, X_{t-p}^{(m)})$. Hence by Theorem 6.4.2, we have

$$n^{-1/2}\sum_{t=1}^{n}\boldsymbol{\lambda}'\mathbf{U}_t^{(m)} \Rightarrow \boldsymbol{\lambda}'\mathbf{V}^{(m)} \quad \text{where } \mathbf{V}^{(m)} \sim \mathrm{N}(\mathbf{0}, \sigma^2\Gamma_p^{(m)}).$$

Since $\sigma^2\Gamma_p^{(m)} \to \sigma^2\Gamma_p$ as $m \to \infty$, we have $\boldsymbol{\lambda}'\mathbf{V}^{(m)} \Rightarrow \mathbf{V}$ where $\mathbf{V} \sim \mathrm{N}(\mathbf{0}, \sigma^2\Gamma_p)$. Also it is easy to check that

$$n^{-1}\operatorname{Var}\left(\boldsymbol{\lambda}'\sum_{t=1}^{n}(\mathbf{U}_t^{(m)} - \mathbf{U}_t)\right) = \boldsymbol{\lambda}'E[(\mathbf{U}_t - \mathbf{U}_t^{(m)})(\mathbf{U}_t - \mathbf{U}_t^{(m)})']\boldsymbol{\lambda}$$
$$\to 0 \quad \text{as } m \to \infty.$$

Since $\mathbf{X}_t^{(m)} \xrightarrow{\text{m.s.}} \mathbf{X}_t$ as $m \to \infty$, application of Proposition 6.3.9 and the Cramer–Wold device gives

$$n^{-1/2}X'\mathbf{Z} \Rightarrow \mathrm{N}(\mathbf{0}, \sigma^2\Gamma_p).$$

It then follows from (8.10.4) that $n(X'X)^{-1} \xrightarrow{P} \Gamma_p^{-1}$, from which we conclude by Propositions 6.3.8 and 6.4.2 that

$$\boldsymbol{\phi}^* \text{ is } \mathrm{AN}(\boldsymbol{\phi}, n^{-1}\sigma^2\Gamma_p^{-1}). \qquad \square$$

Proof of Theorem 8.1.1. In view of the above proposition and Proposition 6.3.3, it suffices to show that

$$n^{1/2}(\hat{\boldsymbol{\phi}} - \boldsymbol{\phi}^*) = o_p(1).$$

We have

$$n^{1/2}(\hat{\boldsymbol{\phi}} - \boldsymbol{\phi}^*) = n^{1/2}(\hat{\Gamma}_p^{-1}\hat{\boldsymbol{\gamma}}_p - (X'X)^{-1}X'\mathbf{Y}),$$

i.e.

$$\begin{aligned} n^{1/2}(\hat{\boldsymbol{\phi}} - \boldsymbol{\phi}^*) &= n^{1/2}\hat{\Gamma}_p^{-1}(\hat{\boldsymbol{\gamma}}_p - n^{-1}X'\mathbf{Y}) \\ &\quad + n^{1/2}(\hat{\Gamma}_p^{-1} - n(X'X)^{-1})n^{-1}X'\mathbf{Y}. \end{aligned} \tag{8.10.5}$$

The ith component of $n^{1/2}(\hat{\gamma}_p - n^{-1}X'\mathbf{Y})$ is

$$\begin{aligned} n^{-1/2}&\left(\sum_{k=1}^{n-i}(X_k - \bar{X}_n)(X_{k+i} - \bar{X}_n) - \sum_{k=1-i}^{n-i} X_k X_{k+i}\right) \\ &= n^{-1/2}\sum_{k=1-i}^{0} X_k X_{k+i} + n^{1/2}\bar{X}_n((1 - n^{-1}i)\bar{X}_n - n^{-1}\sum_{k=1}^{n-i}(X_k + X_{k+i})), \end{aligned} \tag{8.10.6}$$

which by Theorem 7.1.2 and Proposition 6.1.1 is $o_p(1)$. Next we show that

$$n^{1/2}\|\hat{\Gamma}_p^{-1} - n(X'X)^{-1}\| = o_p(1), \tag{8.10.7}$$

where $\|A\|$ is defined for a $p \times p$ matrix A to be the Euclidean length of the p^2 dimensional vector consisting of all the components of the matrix. A simple calculation gives

$$\begin{aligned} n^{1/2}\|\hat{\Gamma}_p^{-1} - n(X'X)^{-1}\| &= n^{1/2}\|\hat{\Gamma}_p^{-1}(n^{-1}(X'X) - \hat{\Gamma}_p)n(X'X)^{-1}\| \\ &\le n^{1/2}\|\hat{\Gamma}_p^{-1}\|\,\|n^{-1}(X'X) - \hat{\Gamma}_p\|\,\|n(X'X)^{-1}\|. \end{aligned}$$

Equation (8.10.6) implies that $n^{1/2}\|n^{-1}(X'X) - \hat{\Gamma}_p\| = o_p(1)$, and since $\hat{\Gamma}_p^{-1} \xrightarrow{P} \Gamma_p^{-1}$ and $n(X'X)^{-1} \xrightarrow{P} \Gamma_p^{-1}$, (8.10.7) follows. Combining (8.10.7) with the fact that $n^{-1}X'\mathbf{Y} \xrightarrow{P} \boldsymbol{\gamma}_p$, gives the desired conclusion that $n^{1/2}(\hat{\boldsymbol{\phi}} - \boldsymbol{\phi}^*) = o_p(1)$. Since $\hat{\boldsymbol{\gamma}}_p \xrightarrow{P} \boldsymbol{\gamma}_p$ and $\hat{\boldsymbol{\phi}} \xrightarrow{P} \boldsymbol{\phi}$, it follows that $\hat{\sigma}^2 \xrightarrow{P} \sigma^2$. □

Proof of Theorem 8.1.2. The same ideas used in the proof of Theorem 8.1.1 can be adapted to prove Theorem 8.1.2. Fix an integer $m > p$ and note that the linear model in (8.10.2) can be written as

$$\begin{bmatrix} X_1 \\ X_2 \\ \vdots \\ X_n \end{bmatrix} = \begin{bmatrix} X_0 & X_{-1} & \cdots & X_{1-m} \\ X_1 & X_0 & \cdots & X_{2-m} \\ & & & \vdots \\ X_{n-1} & X_{n-2} & \cdots & X_{n-m} \end{bmatrix} \begin{bmatrix} \phi_{m1} \\ \phi_{m2} \\ \vdots \\ \phi_{mm} \end{bmatrix} + \begin{bmatrix} Z_1 \\ Z_2 \\ \vdots \\ Z_n \end{bmatrix},$$

where, since $\{X_t\}$ is an AR(p) process, $\boldsymbol{\phi}_m' = (\phi_{m1}, \ldots, \phi_{mm}) := \Gamma_m^{-1}\gamma_m = (\phi_1, \ldots, \phi_p, 0, \ldots, 0)'$. The linear regression estimate of $\boldsymbol{\phi}_m$ in the model $\mathbf{Y} = X\boldsymbol{\phi}_m + \mathbf{Z}$ is then

$$\boldsymbol{\phi}_m^* = (X'X)^{-1}X'\mathbf{Y},$$

which differs by $o_p(n^{-1/2})$ from the Yule–Walker estimate

$$\hat{\boldsymbol{\phi}}_m = \hat{\Gamma}_m^{-1}\hat{\gamma}_m.$$

It follows from the proof of Proposition 8.10.1 that $\boldsymbol{\phi}_m^*$ is AN$(\boldsymbol{\phi}_m, \sigma^2\Gamma_m^{-1})$ and hence that

$$n^{1/2}(\hat{\boldsymbol{\phi}}_m - \boldsymbol{\phi}_m) \Rightarrow \text{N}(\mathbf{0}, \sigma^2\Gamma_m^{-1}).$$

In particular,

$$\hat{\phi}_{mm} \text{ is AN}(0, n^{-1}),$$

since the (m, m) component of Γ_m^{-1} is (see Problem 8.15)

$$(\det \Gamma_{m-1})/(\det \Gamma_m) = (\det \Gamma_{m-1})/(\sigma^2 \det \Gamma_{m-1}) = \sigma^{-2}.$$ □

§8.11* Asymptotic Normality of Parameter Estimators

In this section we discuss, for a causal invertible ARMA(p, q) process, the asymptotic normality of an estimator of the coefficient vector which has the same asymptotic distribution as the least squares and maximum likelihood estimators. The asymptotic distribution of the maximum likelihood and least squares estimators will be derived in Section 10.8.

Recall that the least squares estimators minimize the sum of squares,

$$S(\boldsymbol{\phi}, \boldsymbol{\theta}) = \sum_{t=1}^{n} (X_t - \hat{X}_t)^2 / r_{t-1}.$$

However we shall consider the following approximation to $S(\boldsymbol{\phi}, \boldsymbol{\theta})$. First we approximate the "standardized innovations" $(X_t - \hat{X}_t)/(r_{t-1})^{1/2}$ by $Z_t(\boldsymbol{\phi}, \boldsymbol{\theta})$ where

$$\begin{cases} Z_1(\boldsymbol{\phi}, \boldsymbol{\theta}) = X_1, \\ Z_2(\boldsymbol{\phi}, \boldsymbol{\theta}) = X_2 - \phi_1 X_1 - \theta_1 Z_1(\boldsymbol{\phi}, \boldsymbol{\theta}), \\ \vdots \\ Z_n(\boldsymbol{\phi}, \boldsymbol{\theta}) = X_n - \phi_1 X_{n-1} - \cdots - \phi_p X_{n-p} - \theta_1 Z_{n-1}(\boldsymbol{\phi}, \boldsymbol{\theta}) - \cdots - \theta_q Z_{n-q}(\boldsymbol{\phi}, \boldsymbol{\theta}). \end{cases} \tag{8.11.1}$$

By the assumed invertibility we can write Z_t in the form,

$$Z_t = X_t + \sum_{j=1}^{\infty} \pi_j X_{t-j},$$

and then (8.11.1) corresponds to setting (see Problem 5.15)

$$Z_t(\boldsymbol{\phi}, \boldsymbol{\theta}) = X_t + \sum_{j=1}^{t-1} \pi_j X_{t-j}.$$

Using the relations (see Problem 8.21)

$$\|Z_t(\boldsymbol{\phi}, \boldsymbol{\theta}) - Z_t\| \le \sum_{j=t}^{\infty} |\pi_j| \, \|X_1\|,$$

$$\|X_t - \hat{X}_t\|^2 = r_{t-1}\sigma^2 \le \|Z_t(\boldsymbol{\phi}, \boldsymbol{\theta})\|^2,$$

and

$$\|Z_t - (X_t - \hat{X}_t)\|^2 = (r_{t-1} - 1)\sigma^2,$$

we can show that

$$\begin{cases} \|Z_t(\boldsymbol{\phi}, \boldsymbol{\theta}) - Z_t\| \le c_1 a^t, \\ \|Z_t(\boldsymbol{\phi}, \boldsymbol{\theta}) - (X_t - \hat{X}_t)/(r_{t-1})^{1/2}\| \le c_2 a^t, \\ \|Z_t(\boldsymbol{\phi}, \boldsymbol{\theta})\| \le k, \end{cases} \tag{8.11.2}$$

for all t where a, c_1, c_2 and k are constants with $0 < a < 1$. It is useful to make one further approximation to $(X_t - \hat{X}_t)/(r_{t-1})^{1/2}$ by linearizing $Z_t(\boldsymbol{\phi}, \boldsymbol{\theta})$ about an initial estimate $(\boldsymbol{\phi}_0, \boldsymbol{\theta}_0)$ of $(\boldsymbol{\phi}, \boldsymbol{\theta})$. Thus, if $\boldsymbol{\beta}' = (\phi_1, \ldots, \phi_p, \theta_1, \ldots, \theta_q)$ and $\boldsymbol{\beta}'_0 = (\boldsymbol{\phi}'_0, \boldsymbol{\theta}'_0)$, we approximate $Z_t(\boldsymbol{\beta})$ by

$$Z_t(\boldsymbol{\beta}_0) - \mathbf{D}_t'(\boldsymbol{\beta} - \boldsymbol{\beta}_0), \tag{8.11.3}$$

where $\mathbf{D}_t' = (D_{t,1}(\boldsymbol{\beta}_0), \ldots, D_{t,p+q}(\boldsymbol{\beta}_0))$, with

$$D_{t,i}(\boldsymbol{\beta}) = -\frac{\partial Z_t(\boldsymbol{\beta})}{\partial \beta_i}, \qquad i = 1, \ldots, p + q.$$

Then by minimizing the sum of squares

$$\sum_{t=1}^{n} (Z_t(\boldsymbol{\beta}_0) - \mathbf{D}_t'(\boldsymbol{\beta} - \boldsymbol{\beta}_0))^2$$

(which by (8.11.2) and (8.11.3) is a reasonably good approximation to $S(\boldsymbol{\phi}, \boldsymbol{\theta})$), we obtain an estimator $\boldsymbol{\beta}^\dagger$ of $\boldsymbol{\beta}$ which has the same asymptotic properties as the least squares estimator $\tilde{\boldsymbol{\beta}}$. The estimator $\boldsymbol{\beta}^\dagger$ is easy to compute from the methods of Section 2.6. Specifically, if we let $\mathbf{Z}(\boldsymbol{\beta}_0) = (Z_1(\boldsymbol{\beta}_0), \ldots, Z_n(\boldsymbol{\beta}_0))'$ and write D for the $n \times (p + q)$ design matrix $(\mathbf{D}_1, \ldots, \mathbf{D}_n)'$, then the linear regression estimate of $\Delta\boldsymbol{\beta} = \boldsymbol{\beta} - \boldsymbol{\beta}_0$ is

$$\widehat{\Delta\boldsymbol{\beta}} = (D'D)^{-1} D'\mathbf{Z}(\boldsymbol{\beta}_0)$$

so that

$$\boldsymbol{\beta}^\dagger = \boldsymbol{\beta}_0 + \widehat{\Delta\boldsymbol{\beta}}.$$

The asymptotic normality of this estimator is established in the following theorem.

Theorem 8.11.1. *Let $\{X_t\}$ be the causal invertible* ARMA(p, q) *process*

$$X_t - \phi_1 X_{t-1} - \cdots - \phi_p X_{t-p} = Z_t + \theta_1 Z_{t-1} + \cdots + \theta_q Z_{t-q},$$

where $\{Z_t\} \sim \mathrm{IID}(0, \sigma^2)$ and where $\phi(z)$ and $\theta(z)$ have no common zeros. Suppose that $\boldsymbol{\beta}_0 = (\beta_{01}, \ldots, \beta_{0,p+q})'$ is a preliminary estimator of $\boldsymbol{\beta} = (\phi_1, \ldots, \phi_p, \theta_1, \ldots, \theta_q)'$ such that $\boldsymbol{\beta}_0 - \boldsymbol{\beta} = o_p(n^{-1/4})$, and that $\boldsymbol{\beta}^\dagger$ is the estimator constructed from $\boldsymbol{\beta}_0$ as described above. Then

$$\text{(i)}\ n^{-1} D'D \xrightarrow{P} \sigma^2 V^{-1}(\boldsymbol{\beta})$$

where $V(\boldsymbol{\beta})$ is a $(p + q) \times (p + q)$ nonsingular matrix and

$$\text{(ii)}\ n^{1/2}(\boldsymbol{\beta}^\dagger - \boldsymbol{\beta}) \Rightarrow \mathrm{N}(\mathbf{0}, V(\boldsymbol{\beta})).$$

In addition for the least squares estimator $\tilde{\boldsymbol{\beta}}$, we have

$$\text{(iii)}\ n^{1/2}(\tilde{\boldsymbol{\beta}} - \boldsymbol{\beta}) \Rightarrow \mathrm{N}(\mathbf{0}, V(\boldsymbol{\beta})).$$

(*A formula for computing $V(\boldsymbol{\beta})$ is given in* (8.11.14).)

Sketch of Proof. We shall give only an outline of the proofs of (i) and (ii). The result (iii) is discussed in Section 10.8. Expanding $Z_t(\boldsymbol{\beta})$ in a Taylor series about $\boldsymbol{\beta} = \boldsymbol{\beta}_0$, we have

$$Z_t(\boldsymbol{\beta}) = Z_t(\boldsymbol{\beta}_0) - \mathbf{D}_t'(\boldsymbol{\beta} - \boldsymbol{\beta}_0) + H_t, \tag{8.11.4}$$

where

$$H_t = \frac{1}{2}\sum_{i=1}^{p+q}\sum_{j=1}^{p+q}\frac{\partial^2 Z_t}{\partial\beta_i\partial\beta_j}(\boldsymbol{\beta}_t^*)(\beta_i - \beta_{0i})(\beta_j - \beta_{0j})$$

and $\boldsymbol{\beta}_t^*$ is between $\boldsymbol{\beta}$ and $\boldsymbol{\beta}_0$. Rearranging (8.11.4) and combining the equations for $t = 1, \ldots, n$, we obtain the matrix equation,

$$\mathbf{Z}(\boldsymbol{\beta}_0) = D(\boldsymbol{\beta} - \boldsymbol{\beta}_0) - \mathbf{H} + \mathbf{Z}(\boldsymbol{\beta}),$$

where $\mathbf{Z}(\boldsymbol{\beta}) = (Z_1(\boldsymbol{\beta}), \ldots, Z_n(\boldsymbol{\beta}))'$, $D = (\mathbf{D}_1, \ldots, \mathbf{D}_n)'$ and $\mathbf{H} = (H_1, \ldots, H_n)'$. Hence

$$\begin{aligned} n^{1/2}(\boldsymbol{\beta}^\dagger - \boldsymbol{\beta}) &= n^{1/2}(\boldsymbol{\beta}_0 + \widehat{\Delta\boldsymbol{\beta}} - \boldsymbol{\beta}) \\ &= n^{1/2}(\boldsymbol{\beta}_0 + (D'D)^{-1}D'\mathbf{Z}(\boldsymbol{\beta}_0) - \boldsymbol{\beta}), \end{aligned}$$

i.e.

$$n^{1/2}(\boldsymbol{\beta}^\dagger - \boldsymbol{\beta}) = n^{1/2}(D'D)^{-1}D'\mathbf{Z}(\boldsymbol{\beta}) - n^{1/2}(D'D)^{-1}D'\mathbf{H}. \tag{8.11.5}$$

The idea of the proof is to show that

$$n^{-1}D'D \xrightarrow{P} \sigma^2 V^{-1}(\boldsymbol{\beta}), \tag{8.11.6}$$

$$n^{-1/2}D'Z(\boldsymbol{\beta}) \Rightarrow \mathrm{N}(\mathbf{0}, \sigma^4 V^{-1}(\boldsymbol{\beta})), \tag{8.11.7}$$

and

$$n^{-1/2}D'\mathbf{H} = o_p(1). \tag{8.11.8}$$

Once these results are established, the conclusion of the theorem follows from Propositions 6.1.1, 6.3.3 and 6.3.8.

From (8.11.1) we have for $t > \max(p, q)$,

$$D_{t,i}(\boldsymbol{\beta}) = -\frac{\partial Z_t}{\partial\phi_i}(\boldsymbol{\beta}) = X_{t-i} - \theta_1 D_{t-1,i}(\boldsymbol{\beta}) - \cdots - \theta_q D_{t-q,i}(\boldsymbol{\beta}), \qquad i = 1, \ldots, p,$$

and

$$D_{t,i+p}(\boldsymbol{\beta}) = -\frac{\partial Z_t}{\partial\theta_i}(\boldsymbol{\beta}) = Z_{t-i}(\boldsymbol{\beta}) - \theta_1 D_{t-1,i+p}(\boldsymbol{\beta}) - \cdots - \theta_q D_{t-q,i+p}(\boldsymbol{\beta}),$$

$$i = 1, \ldots, q,$$

so that for $t > \max(p, q)$, $D_{t,i}(\boldsymbol{\beta})$ satisfies the difference equations,

$$\begin{cases} \theta(B)D_{t,i}(\boldsymbol{\beta}) = X_{t-i}, & i = 1, \ldots, p, \\ \theta(B)D_{t,i+p}(\boldsymbol{\beta}) = Z_{t-i}(\boldsymbol{\beta}), & i = 1, \ldots, q. \end{cases} \tag{8.11.9}$$

If we define the two autoregressive processes,

$$U_t = \theta^{-1}(B)X_t = \phi^{-1}(B)Z_t,$$

and

$$V_t = \theta^{-1}(B)Z_t,$$

then it follows from (8.11.9) and (8.11.2) that $D_{t,i}(\boldsymbol{\beta})$ can be approximated by

$$\begin{cases} B^i U_t = U_{t-i} & \text{for } i = 1, \ldots, p, \\ B^{i-p} V_t = V_{t-i+p} & \text{for } i = p+1, \ldots, p+q. \end{cases} \tag{8.11.10}$$

Set $\mathbf{U}_t = (U_{t-1}, \ldots, U_{t-p})'$, $\mathbf{V}_t = (V_{t-1}, \ldots, V_{t-q})'$ and $\mathbf{W}_t' = (\mathbf{U}_t', \mathbf{V}_t')$.

The limit in (8.11.6) is established by first showing that

$$n^{-1}\left(\sum_{t=1}^{n} D_{t,i}(\boldsymbol{\beta}_0)D_{t,j}(\boldsymbol{\beta}_0) - \sum_{t=1}^{n} D_{t,i}(\boldsymbol{\beta})D_{i,j}(\boldsymbol{\beta})\right) = o_p(1),$$

using the assumption $(\boldsymbol{\beta}_0 - \boldsymbol{\beta}) = o(n^{-1/4})$, and then expanding $D_{t,i}(\boldsymbol{\beta}_0)D_{t,j}(\boldsymbol{\beta}_0)$ in a Taylor series about $\boldsymbol{\beta}$. Next, from the approximation (8.11.10) we have

$$n^{-1}\left(\sum_{t=1}^{n} D_{t,i}(\boldsymbol{\beta})D_{t,j}(\boldsymbol{\beta}) - \sum_{t=1}^{n} W_{ti}W_{tj}\right) = o_p(1).$$

If $EZ_t^4 < \infty$, then by applying Theorem 7.1.1 to the individual components of $\mathbf{W}_t\mathbf{W}_t'$, we obtain

$$n^{-1}\sum_{t=1}^{n} \mathbf{W}_t\mathbf{W}_t' \xrightarrow{P} E(\mathbf{W}_1\mathbf{W}_1'), \tag{8.11.11}$$

from which we identify $V(\boldsymbol{\beta})$ as

$$V(\boldsymbol{\beta}) = \sigma^2[E(\mathbf{W}_1\mathbf{W}_1')]^{-1}. \tag{8.11.12}$$

However if we assume only that $EZ_t^2 < \infty$, then (8.11.11) also holds by the ergodic theorem (see Hannan, 1970).

The verification of (8.11.7) is completed in the following steps. By expanding $D_{t,1}(\boldsymbol{\beta}_0)$ in a Taylor series about $\boldsymbol{\beta}$ and using (8.11.2) to approximate $Z_t(\boldsymbol{\beta})$ by $\mathbf{Z}_t$, it can be shown that

$$\left|n^{-1/2}\left(\sum_{t=1}^{n} D_{t,i}(\boldsymbol{\beta}_0)Z_t(\boldsymbol{\beta}) - \sum_{t=1}^{n} W_{t,i}Z_t\right)\right| = o_p(1). \tag{8.11.13}$$

Since for each $i = 1, \ldots, p+q$, $W_{t,i}$, $t = \ldots, -1, 0, 1, \ldots$, is a moving average of $Z_{t-1}, Z_{t-2}, \ldots$, the sequence of random vectors $\mathbf{W}_t Z_t$, $t \geq 1$, is uncorrelated with mean zero and covariance matrix,

$$\begin{aligned} E[(\mathbf{W}_t Z_t)(\mathbf{W}_t Z_t)'] &= \sigma^2 E(\mathbf{W}_t\mathbf{W}_t') \\ &= \sigma^4 V^{-1}(\boldsymbol{\beta}). \end{aligned}$$

Using the argument given in the proof of Proposition 8.10.1, it follows that

$$n^{-1/2}\sum_{t=1}^{n} \mathbf{W}_t Z_t \Rightarrow \mathrm{N}(\mathbf{0}, \sigma^4 V^{-1}(\boldsymbol{\beta})),$$

which, with (8.11.13), establishes (8.11.7).

Finally to prove (8.11.8), it suffices to show that

$$n^{-1}\sum_{t=1}^{n}\left|D_{t,k}(\boldsymbol{\beta}_0)\frac{\partial^2 Z_t(\boldsymbol{\beta}_t^*)}{\partial\beta_i\partial\beta_j}\right| = O_p(1), \qquad i, j, k = 1, \ldots, p+q,$$

since $(\beta_i - \beta_{0i})(\beta_j - \beta_{0j}) = o_p(n^{-1/2})$. This term is handled by first showing that $\boldsymbol{\beta}_0$ and $\boldsymbol{\beta}_t^*$ may be replaced by $\boldsymbol{\beta}$ and then that the resulting expression has an expectation which is bounded in n. □

Note that the expression for $V(\boldsymbol{\beta})$ simplifies to

$$V(\boldsymbol{\beta}) = \sigma^2 \begin{bmatrix} E\mathbf{U}_1\mathbf{U}_1' & E\mathbf{U}_1\mathbf{V}_1' \\ E\mathbf{V}_1\mathbf{U}_1' & E\mathbf{V}_1\mathbf{V}_1' \end{bmatrix}^{-1} \tag{8.11.14}$$

where $\mathbf{U}_t$ and $\mathbf{V}_t$ were defined in the course of the proof. The application of (8.11.14) was illustrated for several low-order ARMA models in Section 8.8.

Problems

8.1. The Wölfer sunspot numbers $\{X_t, t = 1, \ldots, 100\}$ of Example 1.1.5 have sample autocovariances $\hat{\gamma}(0) = 1382.2$, $\hat{\gamma}(1) = 1114.4$, $\hat{\gamma}(2) = 591.72$ and $\hat{\gamma}(3) = 96.215$. Find the Yule–Walker estimates of ϕ_1, ϕ_2 and σ^2 in the model

$$Y_t = \phi_1 Y_{t-1} + \phi_2 Y_{t-2} + Z_t, \qquad \{Z_t\} \sim \mathrm{WN}(0, \sigma^2),$$

for the mean-corrected series $Y_t = X_t - 46.93$, $t = 1, \ldots, 100$. Use Theorem 8.1.1 to find 95% confidence intervals for ϕ_1 and ϕ_2.

8.2. Use the Durbin–Levinson algorithm to compute the sample partial autocorrelations $\hat{\phi}_{11}$, $\hat{\phi}_{22}$ and $\hat{\phi}_{33}$ of the Wölfer sunspot numbers. Is the value of $\hat{\phi}_{33}$ compatible with the hypothesis that the data is generated by an AR(2) process? (Use Theorem 8.1.2 and significance level .05.)

8.3. Let $(X_1, \ldots, X_{p+1})'$ be a random vector with mean $\mathbf{0}$ and non-singular covariance matrix $\Gamma_{p+1} = [\gamma(i-j)]_{i,j=1}^{p+1}$. Note that $P_{\overline{\mathrm{sp}}\{X_1, \ldots, X_p\}} X_{p+1} = \phi_1 X_p + \cdots + \phi_p X_1$ where $\boldsymbol{\phi} = \Gamma_p^{-1} \boldsymbol{\gamma}_p$ (see (8.1.3)). Show that $\phi(z) = 1 - \phi_1 z - \cdots - \phi_p z^p \neq 0$ for $|z| \leq 1$. (If $\phi(z) = (1 - az)\xi(z)$, with $|a| \geq 1$, set $\tilde{\phi}(z) = (1 - \rho z)\xi(z)$ where $\rho = \mathrm{Corr}(Y_{p+1}, Y_p)$ and $Y_j = \xi(B) X_j$. Then $E|\tilde{\phi}(B) X_{p+1}|^2 = E|Y_{p+1} - \rho Y_p|^2 \leq E|Y_{p+1} - a Y_p|^2 = E|\phi(B) X_{p+1}|^2$ with equality holding if and only if $\rho = a$.)

8.4. Show that the zero-mean stationary Gaussian process $\{X_t\}$ with spectral density

$$f(\lambda) = \sigma^2(\pi - |\lambda|)/\pi^2, \qquad -\pi \leq \lambda \leq \pi,$$

has the autocovariance function

$$\gamma(h) = \begin{cases} \sigma^2 & \text{if } h = 0, \\ 4\sigma^2/(\pi|h|)^2 & \text{if } |h| = 1, 3, 5, \ldots, \\ 0 & \text{otherwise.} \end{cases}$$

Hence find the coefficients $\theta_{41}, \ldots, \theta_{44}$ in the innovation representation,

$$\hat{X}_5 = \sum_{j=1}^{4} \theta_{4j}(X_{5-j} - \hat{X}_{5-j}).$$

Find an explicit expression, in terms of X_i and $\hat{X}_i$, $i = 1, \ldots, 5$, for the maximum likelihood estimator of σ^2 based on $X_1, \ldots, X_5$.

8.5. Use the program PEST to simulate and file 20 realizations of length 200 of the Gaussian ARMA(1, 1) process,

$$X_t - \phi X_{t-1} = Z_t + \theta Z_{t-1}, \qquad \{Z_t\} \sim \mathrm{WN}(0, 1),$$

with $\phi = \theta = .6$. Use the program PEST as in Example 8.4.1 to find preliminary models for each series.

8.6. Use the program PEST to simulate and file 20 realizations of length 200 of the Gaussian MA(1) process

$$X_t = Z_t + \theta Z_{t-1}, \qquad \{Z_t\} \sim \mathrm{WN}(0, 1),$$

with $\theta = .6$.

(a) For each series find the moment estimate $\hat{\theta}_M$ of θ (see Section 8.5), recording the number of times the sample autocorrelation $\hat{\rho}(1)$ falls outside the interval $[-\frac{1}{2}, \frac{1}{2}]$.

(b) For each series use the program PEST to find the innovations estimate $\hat{\theta}_I$ of θ (choosing m to minimize the preliminary AICC value).

(c) Use the program PEST to compute the least squares estimate $\hat{\theta}_{LS}$ for each series.

(d) Use the program PEST to compute the maximum likelihood estimate $\hat{\theta}_{ML}$ for each series.

Compare the performances of the four estimators with each other and with the behavior expected from their asymptotic distributions. Compare the number of series for which $|\hat{\rho}(1)| > \frac{1}{2}$ with the expected number based on the asymptotic probability computed in Problem 7.10.

8.7. Use equation (8.7.4) to show that if $n > p$, the likelihood of the observations $\{X_1, \ldots, X_n\}$ of the causal AR(p) process,

$$X_t - \phi_1 X_{t-1} - \cdots - \phi_p X_{t-p} = Z_t, \qquad \{Z_t\} \sim \mathrm{WN}(0, \sigma^2),$$

is

$$L(\boldsymbol{\phi}, \sigma^2) = (2\pi\sigma^2)^{-n/2}(\det G_p)^{-1/2}$$

$$\times \exp\left\{-\frac{1}{2}\sigma^{-2}\left[\mathbf{X}_p' G_p^{-1} \mathbf{X}_p + \sum_{t=p+1}^{n} (X_t - \phi_1 X_{t-1} - \cdots - \phi_p X_{t-p})^2\right]\right\},$$

where $\mathbf{X}_p = (X_1, \ldots, X_p)'$ and $G_p = \sigma^{-2}\Gamma_p = \sigma^{-2}E(\mathbf{X}_p\mathbf{X}_p')$.

8.8. Use the result of Problem 8.7 to derive a pair of linear equations for the least squares estimates of ϕ_1 and ϕ_2 for a causal AR(2) process. Compare your equations with those for the Yule–Walker estimates.

8.9. Given two observations x_1 and x_2 from the causal AR(1) process

$$X_t - \phi X_{t-1} = Z_t, \qquad \{Z_t\} \sim \mathrm{WN}(0, \sigma^2),$$

such that $|x_1| \neq |x_2|$, find the maximum likelihood estimates of ϕ and σ^2.

8.10. Derive a cubic equation for the maximum likelihood estimate of ϕ_1 for a causal AR(1) process.

8.11. Verify that the maximum likelihood estimators $\hat{\boldsymbol{\phi}}$ and $\hat{\boldsymbol{\theta}}$ are those values of $\boldsymbol{\phi}$ and $\boldsymbol{\theta}$ which minimize $l(\boldsymbol{\phi}, \boldsymbol{\theta})$ in equation (8.7.7). Also show that the maximum likelihood estimator of σ^2 is $n^{-1}S(\hat{\boldsymbol{\phi}}, \hat{\boldsymbol{\theta}})$.

8.12. For a causal ARMA process, determine the limit of $(1/n)\sum_{j=1}^{n} \ln r_{j-1}$. When is the limit non-zero?

8.13. In Section 8.6, suppose that the covariance matrix Γ_n depends on the parameter $\boldsymbol{\beta}$. Further assume that the n values $v_0, \ldots, v_{n-1}$ are unbounded functions of $\boldsymbol{\beta}$. Show that the function $S(\boldsymbol{\beta}) = \mathbf{X}_n'\Gamma_n^{-1}(\boldsymbol{\beta})\mathbf{X}_n$ can be made arbitrarily close to zero for a suitable choice of $\boldsymbol{\beta}$.

8.14. Specialize Problem 8.13 to the case when Γ_n is the covariance matrix of an MA(1) process with θ equal to any real number. Show that $S(\theta) = \sum_{j=1}^{n} (X_j - \hat{X}_j)^2/r_{j-1}$ can be made arbitrarily small by choosing θ sufficiently large.

8.15. For an AR(p) process, show that $\det \Gamma_m = (\det \Gamma_p)\sigma^{2(m-p)}$ for all $m > p$. Conclude that the (m, m) component of Γ_m^{-1} is $(\det \Gamma_{m-1})/(\det \Gamma_m) = \sigma^{-2}$.

8.16. *Simulation of a Gaussian process.* Show that n consecutive observations $\{X_k, k = 1, \ldots, n\}$ of a zero-mean Gaussian process with autocovariance function $\kappa(i,j)$ can be generated from n iid N(0, 1) random variables $Z_1, \ldots, Z_n$, by setting

$$X_k = v_{k-1}^{1/2} Z_k + \sum_{j=1}^{k-1} \theta_{k-1,j} v_{k-j-1}^{1/2} Z_{k-j}, \qquad k = 1, \ldots, n,$$

where $\theta_{kj}, j = 1, \ldots, k$ and v_{k-1} are computed from the innovations algorithm. (Use equation (8.6.4) to show that $(X_1, \ldots, X_n)'$ has covariance matrix $[\kappa(i,j)]_{i,j=1}^{n}$.)

8.17. *Simulation of an* ARMA(p,q) *process.* Show that a Gaussian ARMA(p,q) process $\{X_t, t = 1, 2, \ldots\}$ can be generated from iid $N(0,1)$ random variables Z_1, $Z_2, \ldots$ by first defining

$$W_k = \begin{cases} v_{k-1}^{1/2} Z_k + \sum_{j=1}^{k-1} \theta_{k-1,j} v_{k-1-j}^{1/2} Z_{k-j}, & k \le m = \max(p,q), \\[2ex] v_{k-1}^{1/2} Z_k + \sum_{j=1}^{q} \theta_{k-1,j} v_{k-1-j}^{1/2} Z_{k-j}, & k > m, \end{cases}$$

where $\theta_{kj}, j = 1, \ldots, k$ and v_{k-1} are found from the innovations algorithm with $\kappa(i,j)$ as defined in (5.3.5). The simulated values of the ARMA process $\{X_t\}$ are then found recursively from

$$X_k = \sigma W_k, \qquad k \le m,$$

$$X_k = \sigma W_k + \phi_1 X_{k-1} + \cdots + \phi_p X_{k-p}, \qquad k > m.$$

8.18. Verify the calculation of $V(\phi, \theta)$ in Example 8.8.3.

8.19. Verify the calculation of $V(\phi_1, \phi_2)$ for the AR(2) process in Example 8.8.1.

8.20. Using (8.9.1) and one of the series generated in Problem 8.5, plot the boundary of an approximate 95% confidence region for (ϕ, θ).

8.21.* Verify the relations (8.11.2).

8.22.* If $\hat{\boldsymbol{\phi}}$ and $\hat{\boldsymbol{\theta}}$ are the preliminary estimates of $\boldsymbol{\phi}$ and $\boldsymbol{\theta}$ obtained from equations (8.4.4) and (8.4.5), show that $\hat{\boldsymbol{\phi}} = \boldsymbol{\phi} + O_p(n^{-1/2})$ and $\hat{\boldsymbol{\theta}} = \boldsymbol{\theta} + O_p(n^{-1/2})$.

8.23.* Let $\tilde{l}(G)$ be the function

$$\tilde{l}(G) = \ln(n^{-1}\mathbf{X}'G^{-1}\mathbf{X}) + n^{-1}\ln(\det G)$$

where G is a positive definite matrix. Show that $\tilde{l}(aG) = \tilde{l}(G)$ where a is any positive constant. Conclude that for an MA(1) process, the reduced likelihood $l(\theta)$ given in (8.7.7) satisfies $l(\theta) = l(\theta^{-1})$ and that $l(\cdot)$ has either a local maximum or minimum at $\theta = 1$.

CHAPTER 9

Model Building and Forecasting with ARIMA Processes

In this chapter we shall examine the problem of selecting an appropriate model for a given set of observations $\{X_t, t = 1, \ldots, n\}$. If the data (a) exhibits no apparent deviations from stationarity and (b) has a rapidly decreasing autocorrelation function, we shall seek a suitable ARMA process to represent the mean-corrected data. If not, then we shall first look for a transformation of the data which generates a new series with the properties (a) and (b). This can frequently be achieved by differencing, leading us to consider the class of ARIMA (autoregressive-integrated moving average) processes which is introduced in Section 9.1. Once the data has been suitably transformed, the problem becomes one of finding a satisfactory ARMA(p, q) model, and in particular of choosing (or identifying) p and q. The sample autocorrelation and partial autocorrelation functions and the preliminary estimators $\hat{\boldsymbol{\phi}}_m$ and $\hat{\boldsymbol{\theta}}_m$ of Sections 8.2 and 8.3 can provide useful guidance in this choice. However our prime criterion for model selection will be the AICC, a modified version of Akaike's AIC, which is discussed in Section 9.3. According to this criterion we compute maximum likelihood estimators of $\boldsymbol{\phi}$, $\boldsymbol{\theta}$ and σ^2 for a variety of competing p and q values and choose the fitted model with smallest AICC value. Other techniques, in particular those which use the R and S arrays of Gray et al. (1978), are discussed in the recent survey of model identification by de Gooijer et al. (1985). If the fitted model is satisfactory, the residuals (see Section 9.4) should resemble white noise. A number of tests designed to check this are described in Section 9.4, and these should be applied to the minimum-AICC model to make sure that the residuals are consistent with their expected behaviour under the model. If they are not, then competing models (models with AICC-value close to the minimum) should be checked until we find one which passes the goodness of fit tests. In some cases a small difference in AICC-value (say less than 2) between two satisfactory models may be ignored in the interest of model simplicity. In Section 9.5 we consider the prediction of ARIMA processes, which can be treated as an extension of the techniques developed for ARMA processes in Section 5.3. Finally we examine the fitting and prediction of seasonal ARIMA (SARIMA) models, whose analysis, except for certain aspects of model identification, is quite analogous to that of ARIMA processes.

§9.1 ARIMA Models for Non-Stationary Time Series

We have already discussed the importance of the class of ARMA models for representing stationary series. A generalization of this class, which incorporates a wide range of non-stationary series, is provided by the ARIMA processes, i.e. processes which, after differencing finitely many times, reduce to ARMA processes.

Definition 9.1.1 (The ARIMA(p,d,q) Process). If d is a non-negative integer, then $\{X_t\}$ is said to be an ARIMA(p,d,q) process if $Y_t := (1-B)^d X_t$ is a causal ARMA(p,q) process.

This definition means that $\{X_t\}$ satisfies a difference equation of the form

$$\phi^*(B)X_t \equiv \phi(B)(1-B)^d X_t = \theta(B)Z_t, \qquad \{Z_t\} \sim \mathrm{WN}(0,\sigma^2), \qquad (9.1.1)$$

where $\phi(z)$ and $\theta(z)$ are polynomials of degrees p and q respectively and $\phi(z) \neq 0$ for $|z| \leq 1$. The polynomial $\phi^*(z)$ has a zero of order d at $z = 1$. The process $\{X_t\}$ is stationary if and only if $d = 0$, in which case it reduces to an ARMA(p,q) process.

Notice that if $d \geq 1$ we can add an arbitrary polynomial trend of degree $(d-1)$ to $\{X_t\}$ without violating the difference equation (9.1.1). ARIMA models are therefore useful for representing data with trend (see Sections 1.4 and 9.2). It should be noted however that ARIMA processes can also be appropriate for modelling series with no trend. Except when $d = 0$, the mean of $\{X_t\}$ is not determined by equation (9.1.1) and it can in particular be zero. Since for $d \geq 1$, equation (9.1.1) determines the second order properties of $\{(1-B)^d X_t\}$ but not those of $\{X_t\}$ (Problem 9.1), estimation of $\boldsymbol{\phi}$, $\boldsymbol{\theta}$ and σ^2 will be based on the observed differences $(1-B)^d X_t$. Additional assumptions are needed for prediction (see Section 9.5).

EXAMPLE 9.1.1. $\{X_t\}$ is an ARIMA(1, 1, 0) process if for some $\phi \in (-1, 1)$,

$$(1-\phi B)(1-B)X_t = Z_t, \qquad \{Z_t\} \sim \mathrm{WN}(0,\sigma^2).$$

We can then write

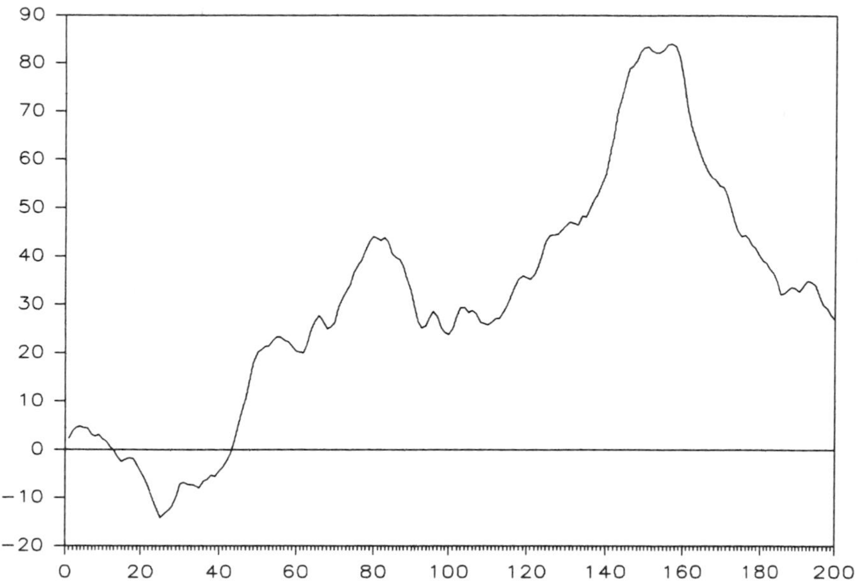

(a)

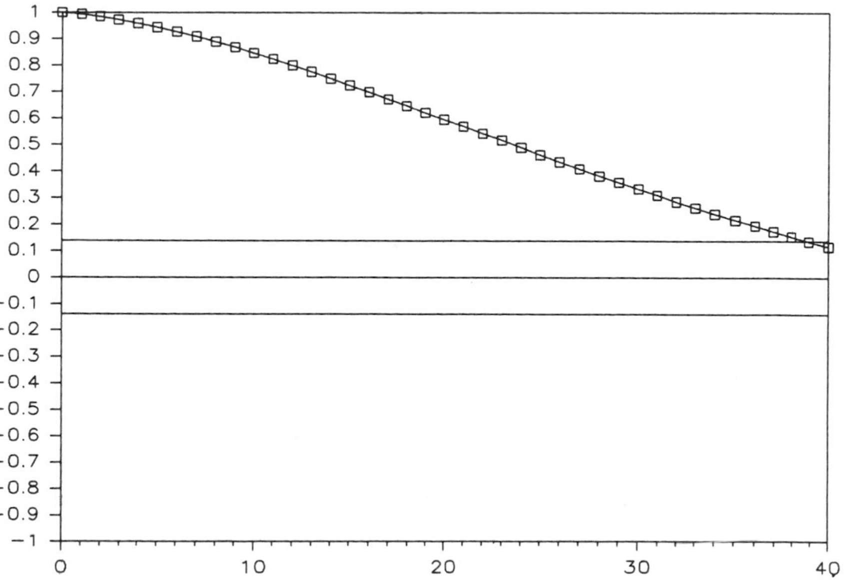

(b)

Figure 9.1. (a) A realization of $\{X_1, \ldots, X_{200}\}$ for the ARIMA process of Example 9.1.1, (b) the sample ACF and (c) the sample PACF.

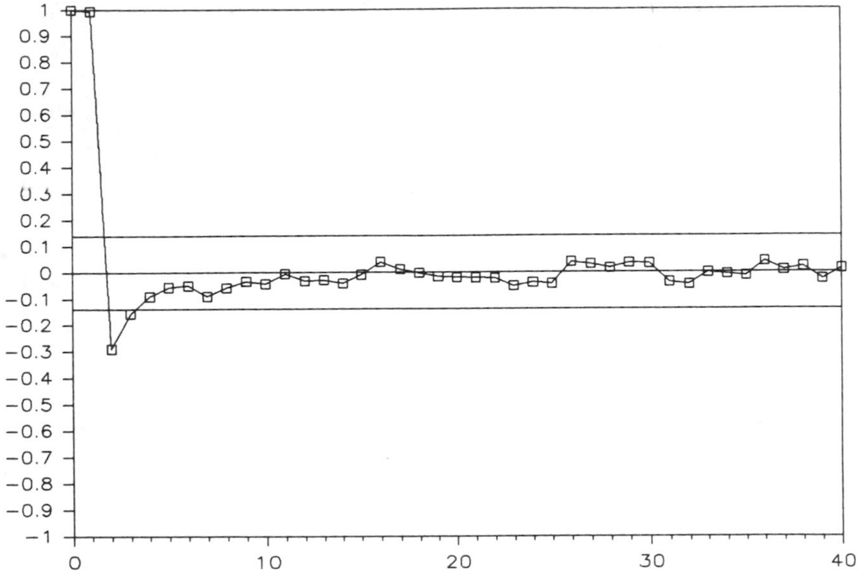

(c)

Figure 9.1. (continued)

$$X_t = X_0 + \sum_{j=1}^{t} Y_j, \qquad t \geq 1,$$

where

$$Y_t = (1 - B)X_t = \sum_{j=0}^{\infty} \phi^j Z_{t-j}.$$

A realization of $\{X_1, \ldots, X_{200}\}$ with $X_0 = 0$, $\phi = .8$ and $\sigma = 1$ is shown in Figure 9.1 together with the sample autocorrelation and partial autocorrelation functions.

A distinctive feature of the data which suggests the appropriateness of an ARIMA model is the slowly decaying positive sample autocorrelation function seen in Figure 9.1. If therefore we were given only the data and wished to find an appropriate model it would be natural to apply the operator $\nabla = 1 - B$ repeatedly in the hope that for some j, $\{\nabla^j X_t\}$ will have a rapidly decaying sample autocorrelation function compatible with that of an ARMA process with no zeroes of the autoregressive polynomial near the unit circle. For the particular time series in this example, one application of the operator ∇ produces the realization shown in Figure 9.2, whose sample autocorrelation and partial autocorrelation functions suggest an AR(1) model for $\{\nabla X_t\}$. The maximum likelihood estimates of ϕ and σ^2 obtained from PEST (under the assumption that $E(\nabla X_t) = 0$) are .808 and .978 respectively, giving the model,

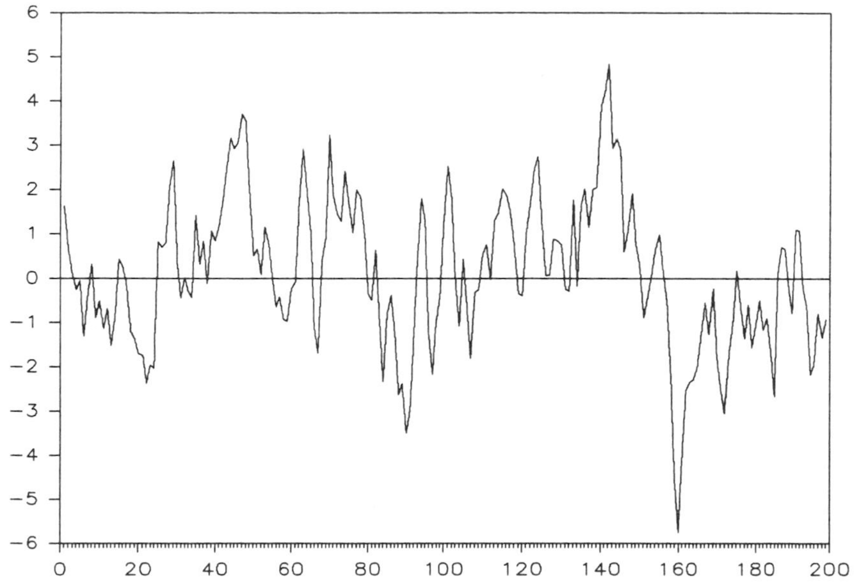

(a)

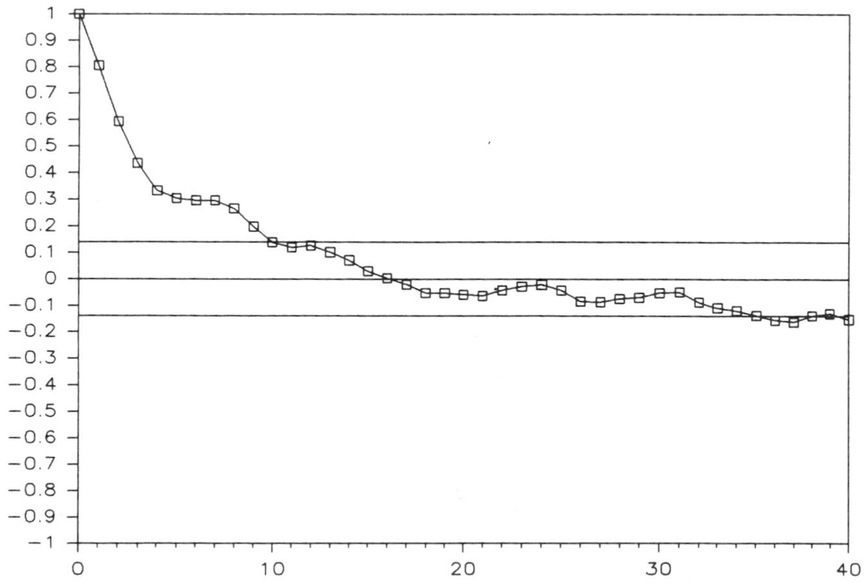

(b)

Figure 9.2. (a) The differenced series, $Y_t = X_{t+1} - X_t$, $t = 1, \ldots, 199$, of Example 9.1.1, (b) the sample ACF of $\{Y_t\}$ and (c) the sample PACF of $\{Y_t\}$.

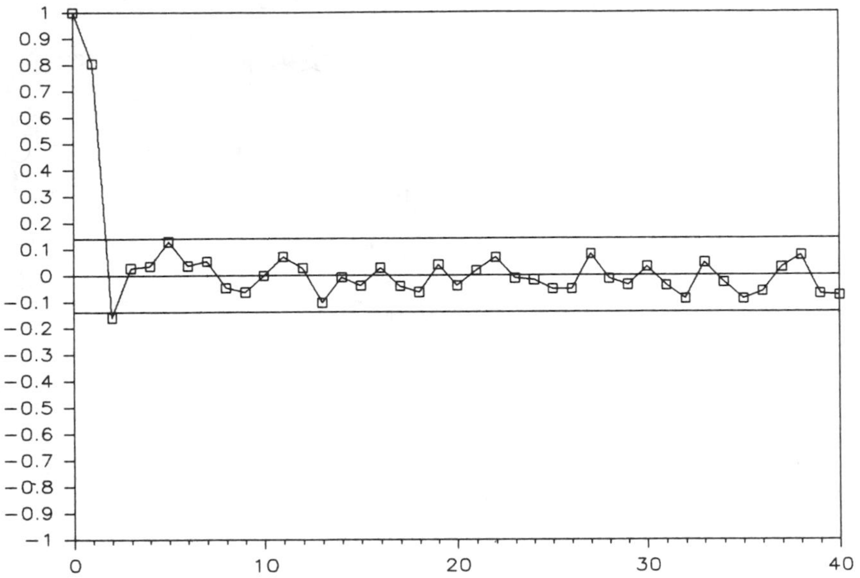

(c)

Figure 9.2. (continued)

$$(1 - .808B)(1 - B)X_t = Z_t, \qquad \{Z_t\} \sim \text{WN}(0, .978), \tag{9.1.2}$$

which bears a close resemblance to the true underlying process,

$$(1 - .8B)(1 - B)X_t = Z_t, \qquad \{Z_t\} \sim \text{WN}(0, 1). \tag{9.1.3}$$

Instead of differencing the series in Figure 9.1 we could proceed more directly by attempting to fit an AR(2) process as suggested by the sample partial autocorrelation function. Maximum likelihood estimation, carried out using the program PEST and assuming that $EX_t = 0$, gives the model,

$$\begin{aligned}(1 - 1.804B + .806B^2)X_t = (1 - .815B)(1 - .989B)X_t = Z_t,& \\ \{Z_t\} \sim \text{WN}(0, .970),&\end{aligned} \tag{9.1.4}$$

which, although stationary, has coefficients which closely resemble those of the true non-stationary process (9.1.3).

From a sample of finite length it will be extremely difficult to distinguish between a non-stationary process such as (9.1.3) for which $\phi^*(1) = 0$, and a process such as (9.1.4), which has very similar coefficients but for which ϕ^* has all of its zeroes outside the unit circle. In either case however, if it is possible by differencing to generate a series with rapidly decaying sample autocorrelation function, then the differenced data can be fitted by a low order ARMA process whose autoregressive polynomial ϕ^* has zeroes which are comfortably outside the unit circle. This means that the fitted parameters will be well away from the boundary of the allowable parameter set. This is desirable for numerical computation of parameter estimates and can be quite critical for some methods of estimation. For example if we apply the Yule–Walker equations to fit an AR(2) model to the data in Figure 9.1, we obtain the model

$$(1 - 1.282B + .290B^2)X_t = Z_t, \qquad \{Z_t\} \sim \text{WN}(0, 6.435), \qquad (9.1.5)$$

which bears little resemblance to either the maximum likelihood model (9.1.4) or the true model (9.1.3). In this case the matrix $\hat{R}_2$ appearing in (8.1.7) is nearly singular.

An obvious limitation in fitting an ARIMA(p, d, q) process $\{X_t\}$ to data is that $\{X_t\}$ is permitted to be non-stationary only in a very special way, i.e. by allowing the polynomial $\phi^*(B)$ in the representation $\phi^*(B)X_t = \theta(B)Z_t$ to have a zero of positive multiplicity d at the point 1 on the unit circle. Such models are appropriate when the sample autocorrelation function of the data is a slowly decaying positive function as in Figure 9.1, since sample autocorrelation functions of this form are associated with models $\phi^*(B)X_t = \theta(B)Z_t$ in which ϕ^* has a zero either at or close to 1.

Sample autocorrelations with slowly decaying oscillatory behavior as in Figures 9.3 and 9.4 are associated with models $\phi^*(B)X_t = \theta(B)Z_t$ in which ϕ^* has a zero close to $e^{i\theta}$ for some $\theta \in (-\pi, \pi]$ other than $\theta = 0$. Figure 9.3 was obtained from a sample of 200 simulated observations from the process,

$$X_t + .99X_{t-1} = Z_t, \qquad \{Z_t\} \sim \text{WN}(0, 1),$$

for which ϕ^* has a zero near $e^{i\pi}$. Figure 9.4 shows the sample autocorrelation function of 200 observations from the process,

$$X_t - X_{t-1} + .99X_{t-2} = Z_t, \qquad \{Z_t\} \sim \text{WN}(0, 1),$$

for which ϕ^* has zeroes near $e^{\pm i\pi/3}$. In such cases the sample autocorrelations can be made to decay more rapidly by applying the operator $[1 - (2\cos\theta)B + B^2] = (1 - e^{i\theta}B)(1 - e^{-i\theta}B)$ to the data, instead of the operator $(1 - B)$ as in the previous paragraph. If $2\pi/\theta$ is close to some integer s then the sample autocorrelation function will be nearly periodic with period s and the operator $\nabla_s = (1 - B^s)$ (with zeroes near $B = e^{\pm i\theta}$) can also be applied to produce a series with more rapidly decaying autocorrelation function (see also Section 9.6). The sample autocorrelation functions in Figures 9.3 and 9.4 are nearly periodic with periods 2 and 6 respectively. Applying the operators $(1 - B^2)$ to the first series and $(1 - B^6)$ to the second gives two new series with the much more rapidly decaying sample autocorrelation functions shown in Figure 9.5 and 9.6 respectively. For the new series it is then not difficult to fit an ARMA model $\phi(B)X_t = \theta(B)Z_t$ for which the zeroes of ϕ are all well outside the unit circle. Techniques for identifying and determining such ARMA models will be discussed in subsequent sections.

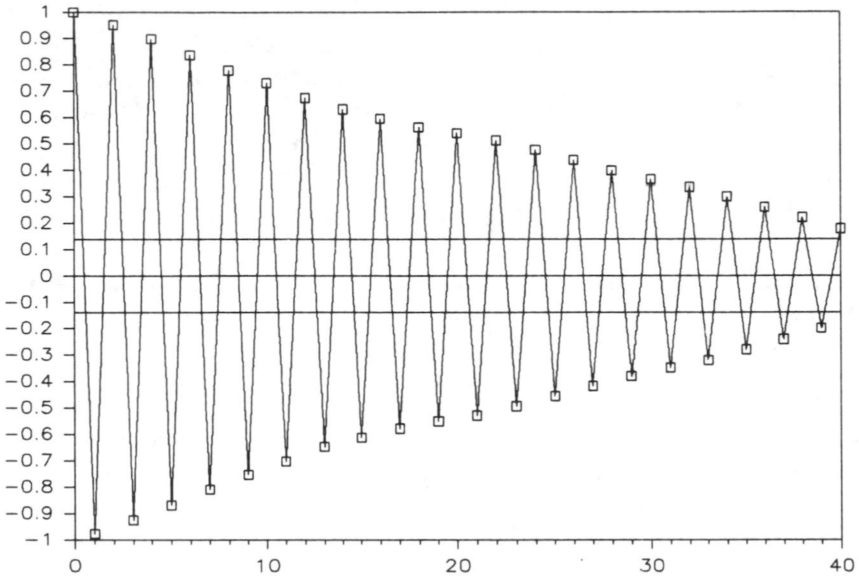

(a)

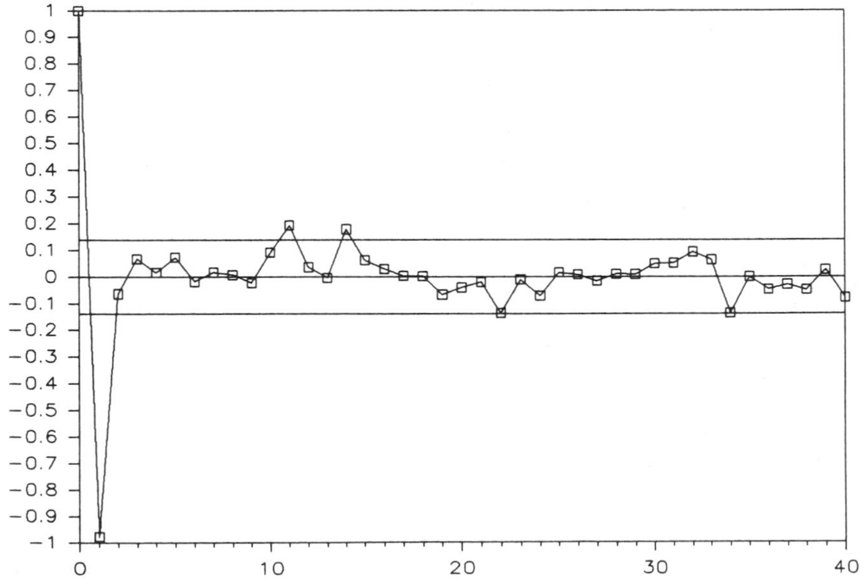

(b)

Figure 9.3. The sample ACF (a) and PACF (b) of a realization of length 200 of the process $X_t + .99X_{t-1} = Z_t$, $\{Z_t\} \sim \text{WN}(0, 1)$.

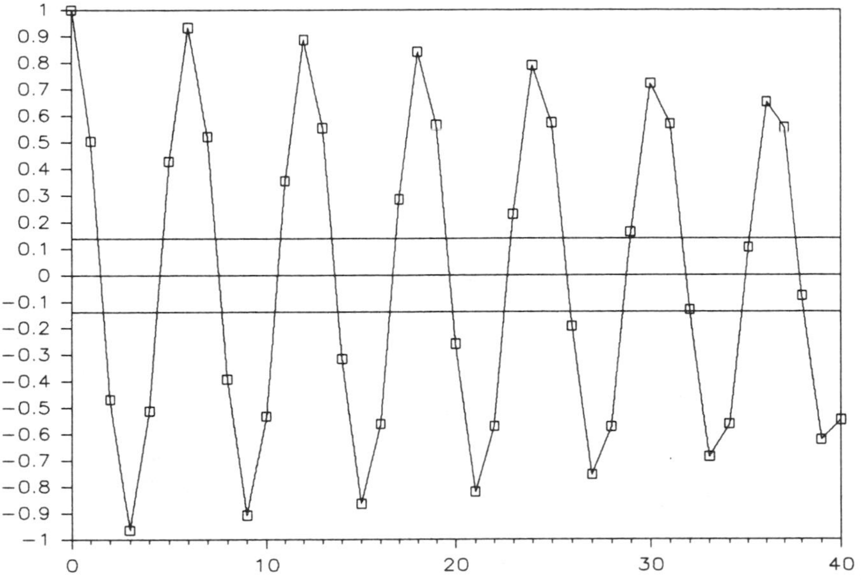

(a)

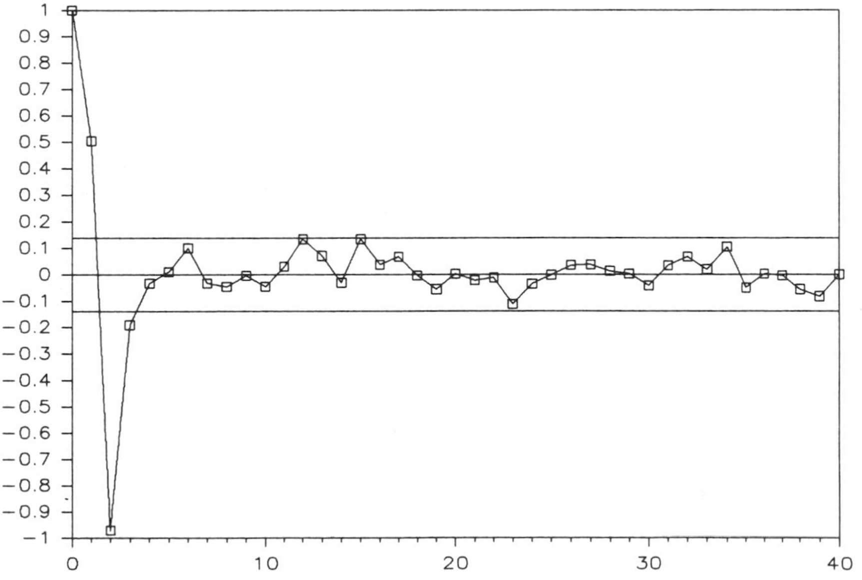

(b)

Figure 9.4. The sample ACF (a) and PACF (b) of a realization of length 200 of the process $X_t - X_{t-1} + 0.99X_{t-2} = Z_t$, $\{Z_t\} \sim \text{WN}(0, 1)$.

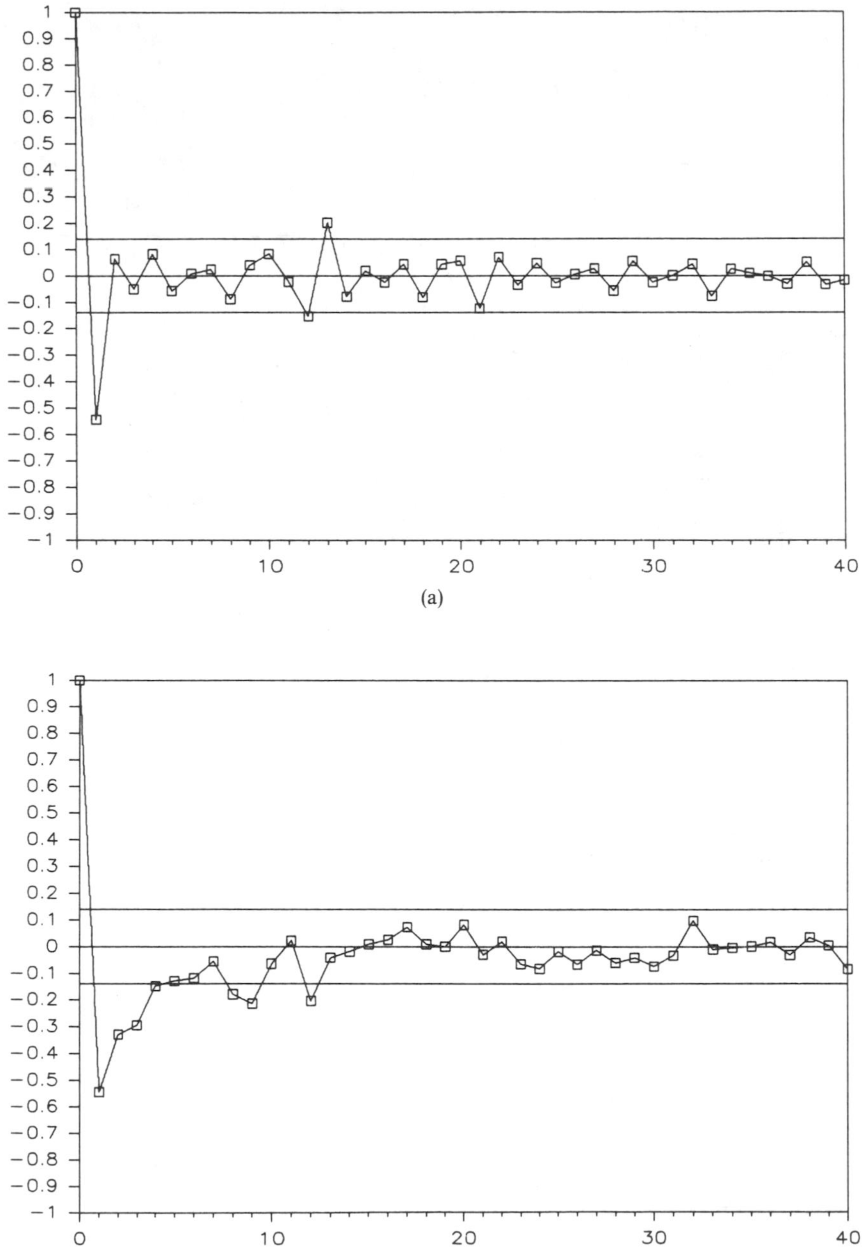

(b)

Figure 9.5. The sample ACF (a) and PACF (b) of $\{(1 - B^2)X_t\}$ where $\{X_t\}$ is the series whose sample ACF and PACF are shown in Figure 9.3.

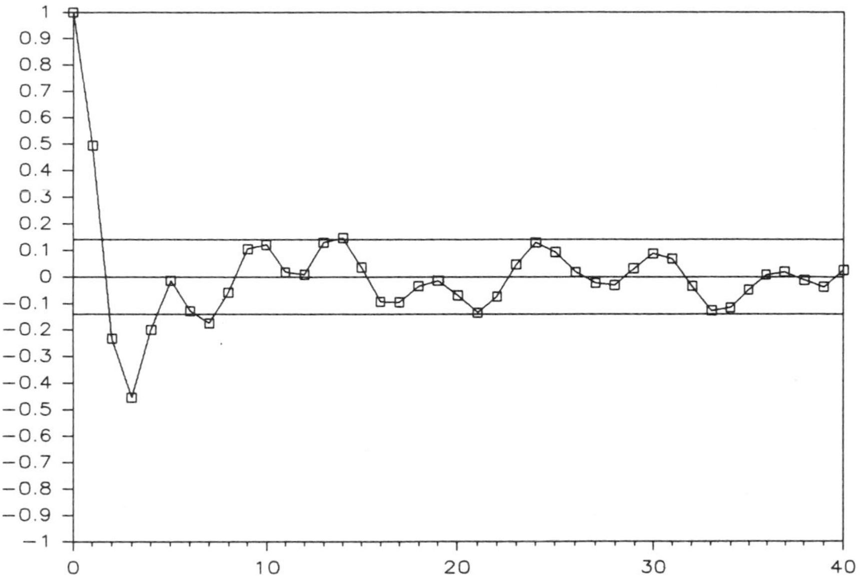

(a)

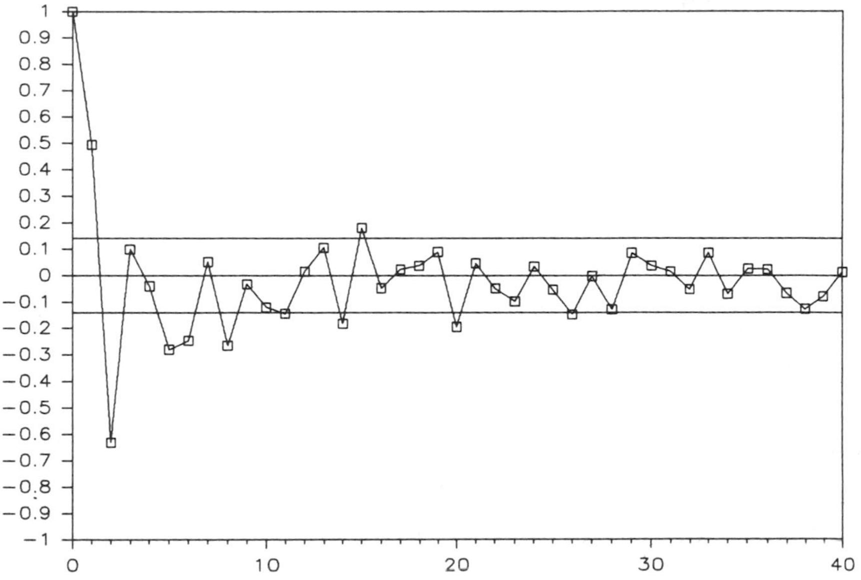

(b)

Figure 9.6. The sample ACF (a) and PACF (b) of $\{(1 - B^6)X_t\}$ where $\{X_t\}$ is the series whose sample ACF and PACF are shown in Figure 9.4.

§9.2 Identification Techniques

(a) *Preliminary Transformations.* The estimation methods described in Chapter 8 enable us to find, for given values of p and q, an ARMA(p, q) model to fit a given series of data. For this procedure to be meaningful it must be at least plausible that the data is in fact a realization of an ARMA process and in particular that it is a realization of a stationary process. If the data display characteristics suggesting non-stationarity (e.g. trend and seasonality), then it may be necessary to make a transformation so as to produce a new series which is more compatible with the assumption of stationarity.

Deviations from stationarity may be suggested by the graph of the series itself or by the sample autocorrelation function or both.

Inspection of the graph of the series will occasionally reveal a strong dependence of variability on the level of the series, in which case the data should first be transformed to reduce or eliminate this dependence. For example, Figure 9.7 shows the International Airline Passenger Data $\{U_t, t = 1, \ldots, 144\}$ of Box and Jenkins (1976), p. 531. It is clear from the graph that the variability increases as U_t increases. On the other hand the transformed series $V_t = \ln U_t$, shown in Figure 9.8, displays no increase in variability with V_t. The logarithmic transformation used here is in fact appropriate whenever $\{U_t\}$ is a series whose standard deviation increases linearly with the mean. For a systematic account of a general class of variance-stabilizing transformations, we refer the reader to Box and Cox (1964). The defining equation for the general Box–Cox transformation f_λ is

$$f_\lambda(U_t) = \begin{cases} \lambda^{-1}(U_t^\lambda - 1), & U_t \ge 0, \lambda > 0, \\ \ln U_t, & U_t > 0, \lambda = 0, \end{cases}$$

and the program PEST provides the option of applying f_λ (with $(0 \le \lambda \le 1.5)$ prior to the elimination of trend and/or seasonality from the data. In practice, if a Box–Cox transformation is necessary, it is often the case that either f_0 or $f_{1/2}$ is adequate.

Trend and seasonality are usually detected by inspecting the graph of the (possibly transformed) series. However they are also characterized by sample autocorrelation functions which are slowly decaying and nearly periodic respectively. The elimination of trend and seasonality was discussed in Section 1.4 where we described two methods:

(i) "classical decomposition" of the series into a trend component, a seasonal component, and a random residual component, and
(ii) differencing.

The program PEST(Option 1) offers a choice between these techniques. Both methods were applied to the transformed Airline Data $V_t = \ln U_t$ of the preceding paragraph. Figures 9.9 and 9.10 show respectively the two series found from PEST by (i) estimating and removing from $\{V_t\}$ a linear trend component and a seasonal component of period 12, and (ii) applying the

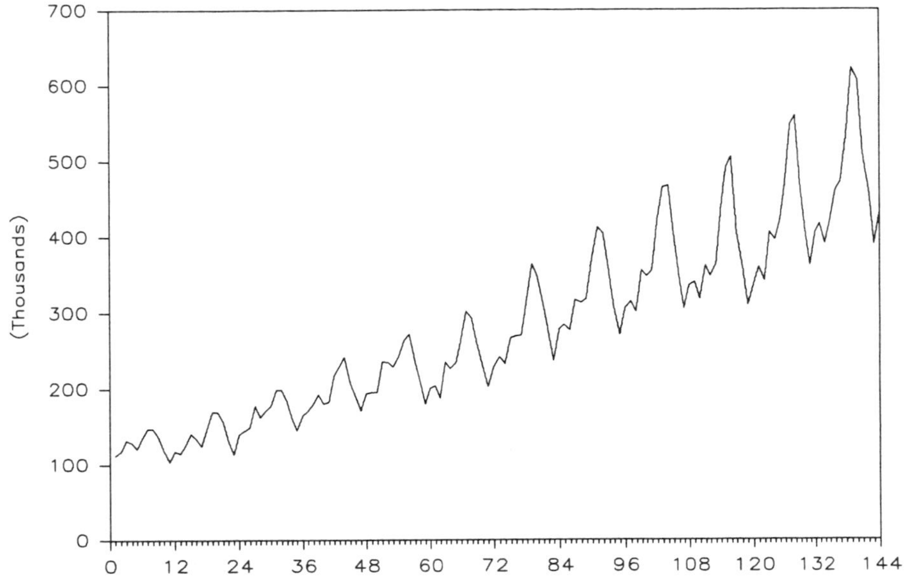

Figure 9.7. International airline passengers; monthly totals in thousands of passengers $\{U_t, t = 1, \ldots, 144\}$ from January 1949 to December 1960 (Box and Jenkins (1970)).

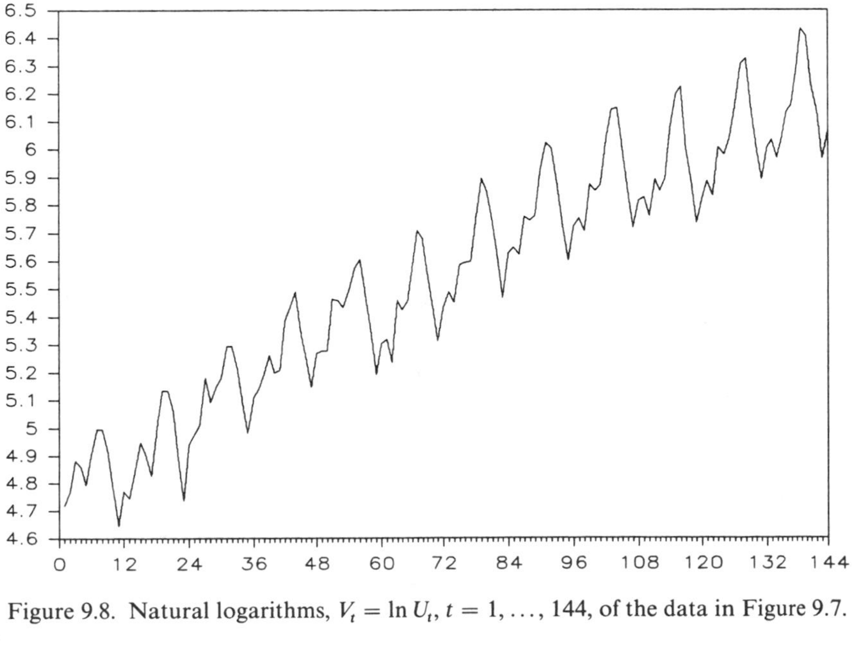

Figure 9.8. Natural logarithms, $V_t = \ln U_t$, $t = 1, \ldots, 144$, of the data in Figure 9.7.

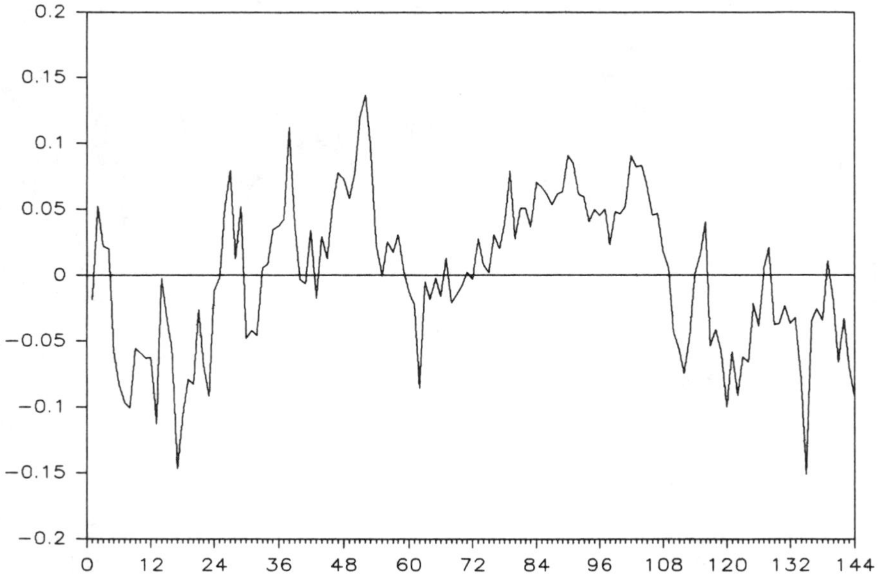

Figure 9.9. Residuals after removing a linear trend and seasonal component from the data $\{V_t\}$ of Figure 9.8.

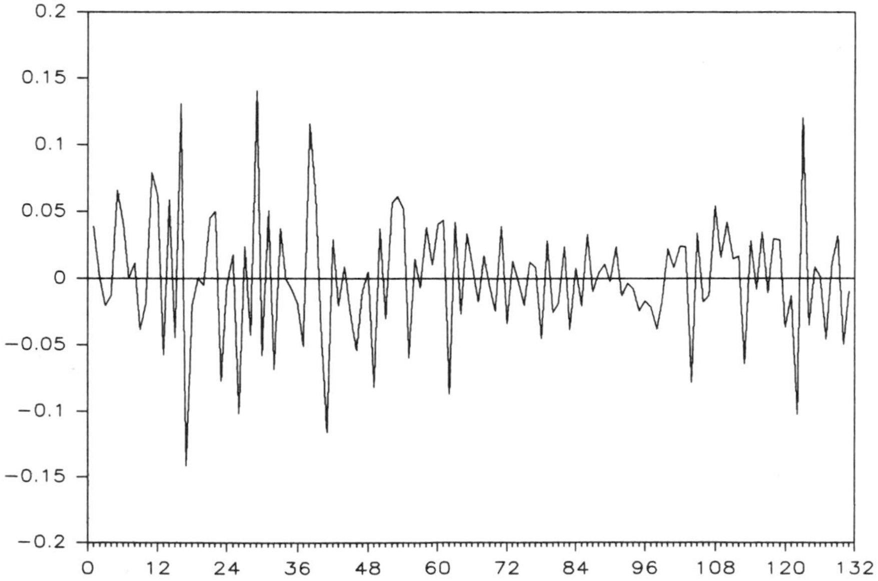

Figure 9.10. The differenced series $\{\nabla\nabla_{12} V_{t+13}\}$ where $\{V_t\}$ is the data shown in Figure 9.8.

difference operator $(1 - B)(1 - B^{12})$ to $\{V_t\}$. Neither of the two resulting series display any apparent deviations from stationarity, nor do their sample autocorrelation functions (the sample autocorrelation function of $\{\nabla\nabla_{12} V_t\}$ is shown in Figure 9.11).

After the elimination of trend and seasonality it is still possible that the sample autocorrelation function may appear to be that of a non-stationary or nearly non-stationary process, in which case further differencing as described in Section 9.1 may be carried out.

(b) *The Identification Problem.* Let $\{X_t\}$ denote the mean-corrected transformed series, found as described in (a). The problem now is to find the most satisfactory ARMA(p, q) model to represent $\{X_t\}$. If p and q were known in advance this would be a straightforward application of the estimation techniques developed in Chapter 8. However this is usually not the case, so that it becomes necessary also to identify appropriate values for p and q.

It might appear at first sight that the higher the values of p and q chosen, the better the fitted model will be. For example, if we fit a sequence of AR(p) processes, $p = 1, 2, \ldots$, the maximum likelihood estimate, $\hat{\sigma}^2$, of σ^2 generally decreases monotonically as p increases (see e.g. Table 9.2). However we must beware of the danger of overfitting, i.e. of tailoring the fit too closely to the particular numbers observed. An extreme case of overfitting (in a somewhat different context) occurs if we fit a polynomial of degree 99 to 100 observations generated from the model $Y_t = a + bt + Z_t$, where $\{Z_t\}$ is an independent sequence of standard normal random variables. The fit will be perfect for the given data set, but use of the model to predict future values may result in gross errors.

Criteria have been developed, in particular Akaike's AIC criterion and Parzen's CAT criterion, which attempt to prevent overfitting by effectively assigning a cost to the introduction of each additional parameter. In Section 9.3 we discuss a bias-corrected form of the AIC, defined for an ARMA(p, q) model with coefficient vectors $\boldsymbol{\phi}$ and $\boldsymbol{\theta}$, by

$$\text{AICC}(\boldsymbol{\phi}, \boldsymbol{\theta}) = -2 \ln L(\boldsymbol{\phi}, \boldsymbol{\theta}, S(\boldsymbol{\phi}, \boldsymbol{\theta})/n) + 2(p + q + 1)n/(n - p - q - 2), \tag{9.2.1}$$

where $L(\boldsymbol{\phi}, \boldsymbol{\theta}, \sigma^2)$ is the likelihood of the data under the Gaussian ARMA model with parameters $(\boldsymbol{\phi}, \boldsymbol{\theta}, \sigma^2)$ and $S(\boldsymbol{\phi}, \boldsymbol{\theta})$ is the residual sum of squares defined in Section 8.7. On the basis of the analysis given in Section 9.3, the model selected is the one which minimizes the value of AICC. Intuitively one can think of $2(p + q + 1)n/(n - p - q - 2)$ in (9.2.1) as a penalty term to discourage over-parameterization. Once a model has been found which minimizes the AICC value, it must then be checked for goodness of fit (essentially by checking that the residuals are like white noise) as discussed in Section 9.4.

Introduction of the AICC (or analogous) statistic reduces model identification to a well-defined problem. However the search for a model which minimizes the AICC can be very lengthy without some idea of the class

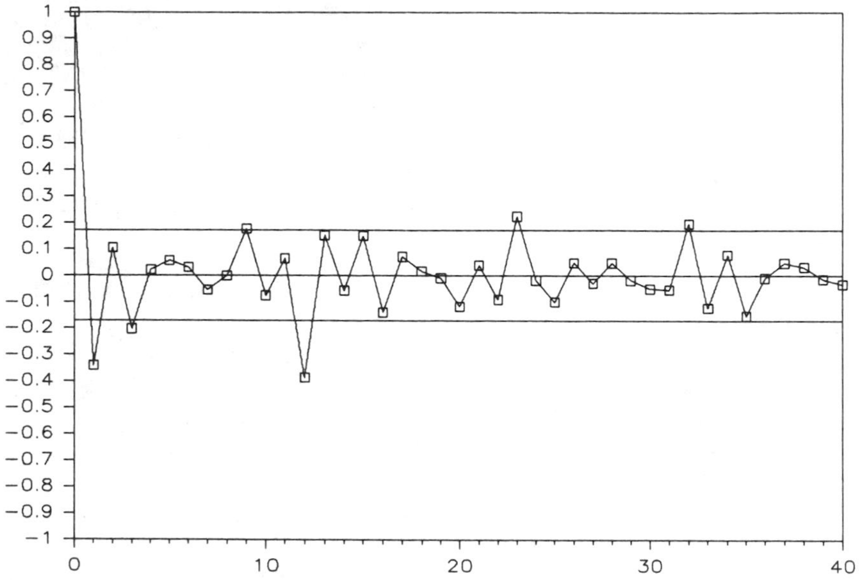

(a)

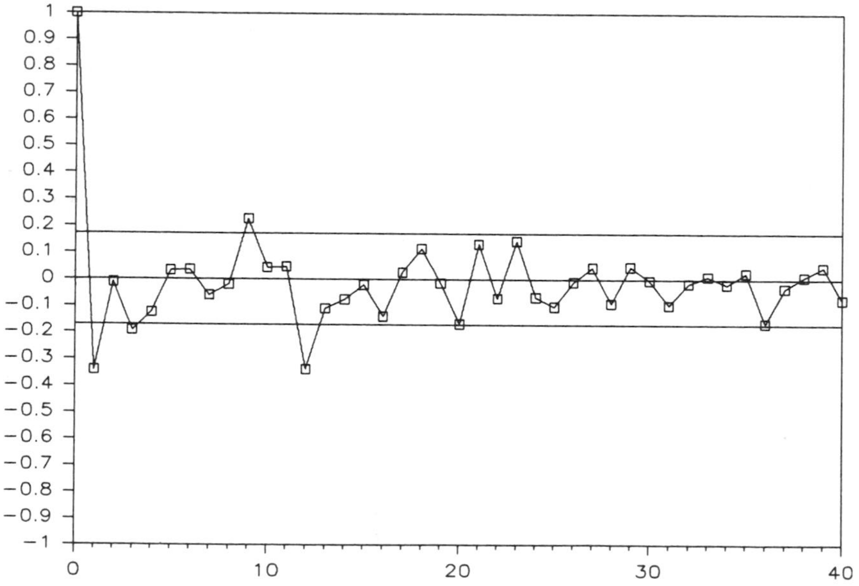

(b)

Figure 9.11. The sample ACF (a) and PACF (b) of the series $\{\nabla\nabla_{12}V_{t+13}\}$ shown in Figure 9.10.

of models to be explored. A variety of techniques can be used to accelerate the search by providing us with preliminary estimates of p and q, and possibly also preliminary estimates of the coefficients.

The primary tools used as indicators of p and q are the sample autocorrelation and partial autocorrelation functions and the preliminary estimators $\hat{\boldsymbol{\phi}}_m$ and $\hat{\boldsymbol{\theta}}_m$, $m = 1, 2, \ldots$, discussed in Sections 8.2 and 8.3 respectively. From these it is usually easy to judge whether a low order autoregressive or moving average model will prove satisfactory. If so then we can proceed by successively fitting models of orders 1, 2, 3, ..., until we find a minimum value of the AICC. (Mixed models should also be considered before making a final selection.)

EXAMPLE 9.2.1. Figure 9.12 shows the sample autocorrelation and partial autocorrelation functions of a series of 200 observations from a zero-mean stationary process. They suggest an autoregressive model of order 2 (or perhaps 3) for the data. This suggestion is supported by the Yule–Walker estimators $\hat{\boldsymbol{\phi}}_m$, $m = 1, 2, \ldots$, of the coefficient vectors of autoregressive models of order m. The Yule–Walker estimates $\hat{\boldsymbol{\phi}}_{mj}$, $j = 1, \ldots, m$; $m = 1, \ldots, 5$ are shown in Table 9.1 with the corresponding ratios,

$$r_{mj} = \hat{\phi}_{mj}/(1.96\hat{\sigma}_{mj}), \tag{9.2.2}$$

where $\hat{\sigma}_{mj}^2$ is the jth diagonal element of $\hat{\sigma}^2\Gamma_m^{-1}(\hat{\boldsymbol{\phi}}_m)/n$, the estimated version of the asymptotic covariance matrix of $\hat{\boldsymbol{\phi}}_m$ appearing in Theorem 8.1.2. A value of r_{mj} with absolute value greater than 1 causes us to reject, at approximate level .05, the hypothesis that ϕ_{mj} is zero (assuming that the true underlying process is an AR(p) process with $p \leq m$).

The next step is to fit autoregressive models of orders 1, 2, ..., by maximum likelihood, using the Yule–Walker estimates as initial values for the maximization algorithm. The maximum likelihood estimates for the mean-corrected data are shown in Table 9.2 together with the corresponding AICC values.

Table 9.1. The Yule–Walker Estimates $\hat{\phi}_{mj}, j = 1, \ldots, m$; $m = 1, \ldots, 5$, and the Ratios r_{mj} (in Parentheses) for the Data of Example 9.2.1

m \ j	1	2	3	4	5
1	.878				
	(13.255)				
2	1.410	−.606			
	(12.785)	(−5.490)			
3	1.301	−.352	−.180		
	(9.545)	(−1.595)	(−1.318)		
4	1.293	−.369	−.119	−.047	
	(9.339)	(−1.632)	(−.526)	(−.338)	
5	1.295	−.362	−.099	−.117	.054
	(9.361)	(−1.602)	(−.428)	(−.516)	(.391)

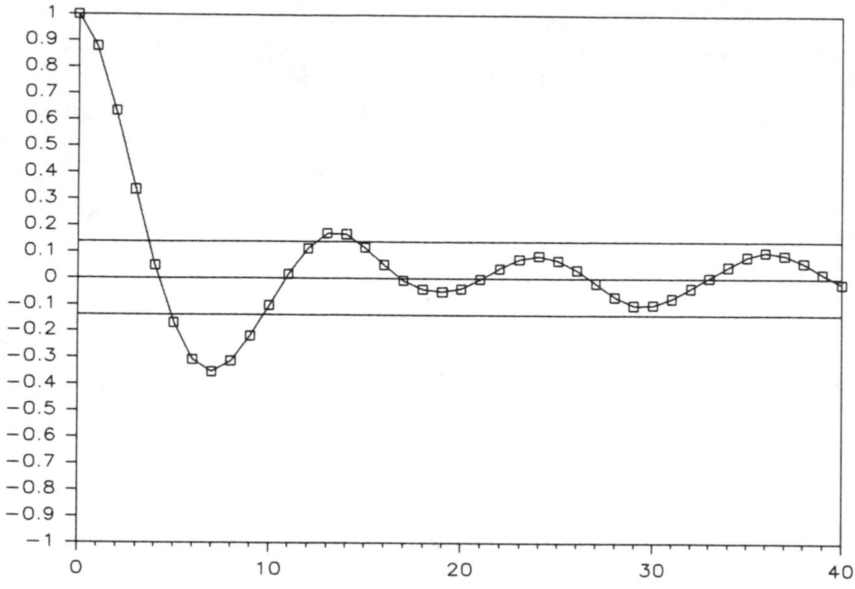

(a)

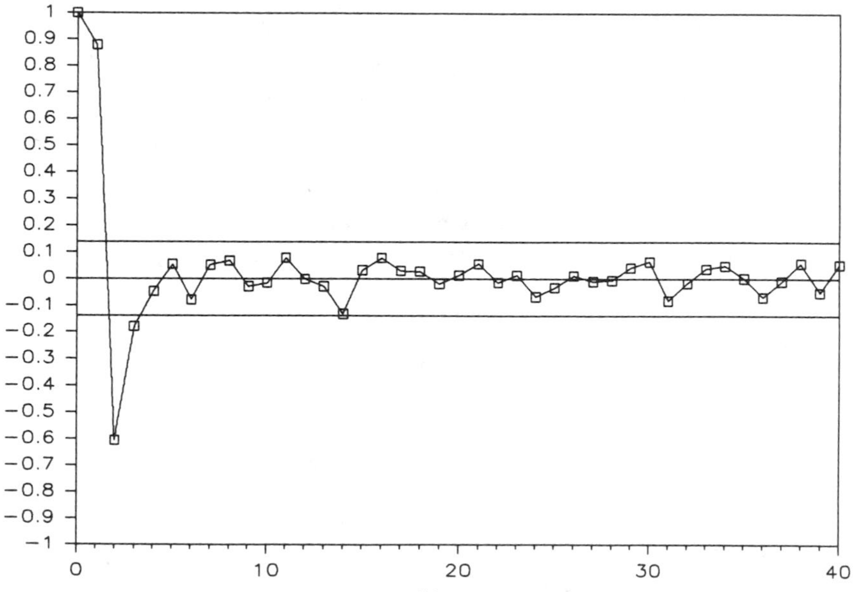

(b)

Figure 9.12. The sample ACF (a) and PACF (b) for the data of Example 9.2.1.

Table 9.2 The Maximum Likelihood Estimates $\hat{\phi}_{mj}$, $\hat{\sigma}_m^2$, $j = 1, \ldots, m$; $m = 1, \ldots, 5$, and the Corresponding AICC, BIC and FPE Values for the Data of Example 9.2.1

$m \backslash j$	1	2	3	4	5	$\hat{\sigma}_m^2$	AICC	BIC	FPE
1	.892					1.547	660.44	662.40	1.562
2	1.471	−.656				.885	551.94	558.29	.903
3	1.387	−.471	−.127			.871	550.86	561.49	.897
4	1.383	−.486	−.081	−.033		.870	552.75	567.31	.905
5	1.383	−.484	−.072	−.059	.019	.870	554.81	573.01	.914

The BIC and FPE statistics (which are analogous to the AICC but with different penalties for the introduction of additional parameters) are also shown. All three statistics are discussed in Section 9.3.

From Table 9.2 we see that the autoregressive model selected by the AICC criterion for the mean-corrected data $\{X_t\}$ is

$$X_t - 1.387X_{t-1} + .471X_{t-2} + .127X_{t-3} = Z_t, \qquad \{Z_t\} \sim \text{WN}(0, .871). \tag{9.2.3}$$

Application of the goodness of fit tests to be described in Section 9.4 shows that this model is indeed satisfactory. (If the residuals for the model (9.2.3) had turned out to be incompatible with white noise, it would be necessary to modify the model. The model modification technique described below in (d) is frequently useful for this purpose.)

Approximate confidence intervals for the coefficients can be found from the asymptotic distribution of the maximum likelihood estimators given in Section 8.8. The program PEST approximates the covariance matrix $V(\hat{\boldsymbol{\beta}})$ of (8.8.3) by $2H^{-1}(\hat{\boldsymbol{\beta}})$, where $H(\hat{\boldsymbol{\beta}})$ is the Hessian matrix of the reduced likelihood evaluated at $\hat{\boldsymbol{\beta}}$. From this we obtain the asymptotic .95 confidence bounds $\hat{\beta}_j \pm 1.96[v_{jj}(\hat{\boldsymbol{\beta}})/n]^{1/2}$ for β_j, where $v_{jj}(\hat{\boldsymbol{\beta}})$ is the jth diagonal element of $V(\hat{\boldsymbol{\beta}})$. This gives the following bounds for the coefficients ϕ_1, ϕ_2, ϕ_3.

$$\phi_1\text{: } 1.387 \pm .136,$$

$$\phi_2\text{: } -.471 \pm .226,$$

$$\phi_3\text{: } -.127 \pm .138.$$

The confidence bounds for ϕ_3 suggest that perhaps an AR(2) model should have been fitted to this data since 0 falls between the bounds for ϕ_3. In fact if we had minimized the BIC rather than the AICC (see Table 9.2) we would have chosen the AR(2) model. The BIC is a Bayesian modification of the AIC criterion which was introduced by Akaike to correct the tendency of the latter to overestimate the number of parameters. The true model in this example was

$$X_t - 1.4X_{t-1} + .6X_{t-2} = Z_t, \qquad \{Z_t\} \sim \text{WN}(0, 1). \tag{9.2.4}$$

Example 9.2.2. Inspection of the sample autocorrelation and partial autocorrelation functions of the logged and differenced airline data $\{\nabla\nabla_{12} V_t\}$ shown in Figure 9.11 suggests the possibility of either a moving average model of order 12 (or perhaps 23) with a large number of zero coefficients, or alternatively of an autoregressive model of order 12.

To explore these possibilities further, the program PEST (Option 3) was used to compute the preliminary estimates $\hat{\boldsymbol{\theta}}_m$ and $\hat{\boldsymbol{\phi}}_m$, $m = 15, 25, 30$, as described in Sections 8.2 and 8.3 respectively. These are shown, with the ratios r_{mj} of each estimated coefficient to 1.96 times its standard error, in Tables 9.3 and 9.4 respectively. For $\hat{\boldsymbol{\phi}}_m$, r_{mj} was defined by equation (9.2.2). For $\hat{\boldsymbol{\theta}}_m$,

$$r_{mj} = \hat{\theta}_{mj}/(1.96\hat{\sigma}_{mj}),$$

where by Theorem 8.3.1, $\hat{\sigma}_{mj}^2 = n^{-1}(1 + \hat{\theta}_{m1}^2 + \cdots + \hat{\theta}_{m,j-1}^2)$, $j > 1$, and $\hat{\sigma}_{m1}^2 = n^{-1}$.

For the preliminary moving average model of order 30 we have plotted the ratios r_{mj}, $j = 1, \ldots, 30$, with boundaries at the critical value 1, in Figure 9.13. The graph suggests that we consider models with non-zero coefficients at lags 1, 12, 23, possibly 3, and possibly also 5, 9 and 13. Of the models with non-zero coefficients at one or more of the lags 1, 3, 12 and 23, it is found that the one with smallest AICC value (-486.04) is (for $X_t = \nabla\nabla_{12} V_t - .00029$)

$$X_t = Z_t - .355Z_{t-1} - .201Z_{t-3} - .524Z_{t-12} + .241Z_{t-23}, \qquad (9.2.5)$$

where $\{Z_t\} \sim \text{WN}(0, .00125)$. If we expand the class of models considered to include non-zero coefficients at one or more of the lags 5, 9 and 13 suggested

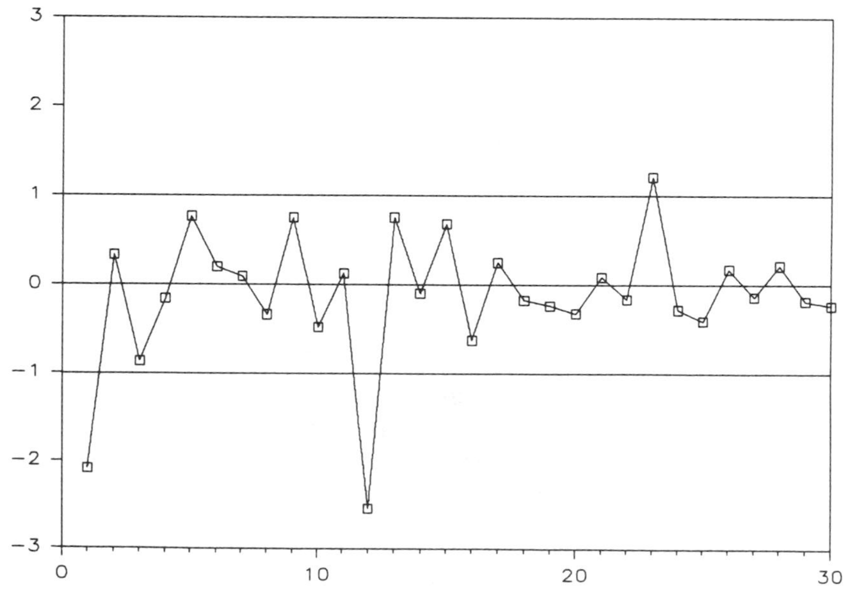

Figure 9.13. Ratio of the estimated coefficient $\hat{\theta}_{30,j}$ to 1.96 times its standard error, $j = 1, \ldots, 30$ (from Table 9.3).

Table 9.3. The Innovation Estimates $\hat{\boldsymbol{\theta}}_m$, $\hat{v}_m$, $m = 15, 25, 30$ for Ex. 9.2.2

$m = 15$

White Noise Variance
.0014261
MA Coefficients

−.40660	.07100	−.15392	.02185	.08851
.04968	−.04885	.03474	.09606	−.07203
−.07263	−.36395	.15123	−.00745	.14956

Ratio of Coefficients to (1.96*Standard Error)

−2.37442	.38408	−.83086	.11679	.47294
.26458	−.25991	.18465	.51031	−.38122
−.38355	−1.91782	.75712	−.03700	.74253

$m = 25$

White Noise Variance
.0012638
MA Coefficients

−.36499	.05981	−.14812	−.02483	.12247
.05701	−.01327	−.03646	.12737	−.09385
.01909	−.47123	.13557	−.03196	.12591
−.11667	.06124	−.01722	−.05076	−.06775
.00908	−.04050	.24405	−.05955	−.10028

Ratio of Coefficients to (1.96*Standard Error)

−2.13140	.32813	−.81127	−.13471	.66421
.30723	−.07141	−.19619	.68497	−.50128
.10160	−2.50732	.66284	−.15528	.61149
−.56352	.29444	−.08271	−.24370	−.32501
.04352	−.19399	1.16818	−.27954	−.47016

$m = 30$

White Noise Variance
.0012483
MA Coefficients

−.35719	.06006	−.15764	−.02979	.14018
.03632	.01689	−.06313	.13925	−.09032
.02333	−.47895	.15424	−.02092	.14133
−.13103	.05216	−.03701	−.04987	−.06822
.01851	−.03513	.25435	−.06067	−.08874
.03687	−.02951	.04555	−.04012	−.05106

Ratio of Coefficients to (1.96*Standard Error)

−2.08588	.33033	−.86556	−.16179	.76108
.19553	.09089	−.33967	.74792	−.48116
.12387	−2.54232	.75069	−.10099	.68213
−.62815	.24861	−.17626	−.23740	−.32447
.08791	−.16683	1.20720	−.28201	−.41199
.17077	−.13662	.21083	−.18554	−.23605

Table 9.4. The Yule–Walker Estimates $\hat{\boldsymbol{\phi}}_m$, $\hat{v}_m$, $m = 15, 25, 30$ for Example 9.2.2

$m = 15$

White Noise Variance

.0014262

AR Coefficients

−.40660	−.09364	−.16261	−.09185	.06259
.09282	−.00421	.04216	.15873	.01347
−.09957	−.38601	−.14160	−.08563	−.02174

Ratio of Coefficients to (1.96*Standard Error)

−2.37494	−.50827	−.88695	−.53016	.36155
.53517	−.02452	.24526	.92255	.07765
−.57517	−2.22798	−.77237	−.46481	−.12701

$m = 25$

White Noise Variance

.0012638

AR Coefficients

−.36498	−.07087	−.15643	−.13215	.06052
.11335	.03122	−.00683	.15464	.01640
−.03815	−.44895	−.19180	−.07800	−.04537
−.12301	.05101	.10160	−.04611	−.09381
.08851	−.03933	.10959	−.10408	−.10264

Ratio of Coefficients to (1.96*Standard Error)

−2.14268	−.39253	−.86904	−.72665	.33137
.62205	.17052	−.03747	.84902	.08970
−.20886	−2.46259	−.98302	−.42788	−.24836
−.67279	.28007	.55726	−.25190	−.51488
.48469	−.21632	.60882	−.57639	−.60263

$m = 30$

White Noise Variance

.0012483

AR Coefficients

−.35718	−.06759	−.15995	−.14058	.07806
.09844	.04452	−.01653	.15146	.02239
−.03322	−.46045	−.18454	−.07941	−.03957
−.15279	.06951	.09498	−.03316	−.10113
.08865	−.03566	.11481	−.11330	−.10948
.00673	.01324	−.07332	.04514	−.00489

Ratio of Coefficients to (1.96*Standard Error)

−2.08586	−.37210	−.88070	−.76545	.42148
.53284	.24132	−.09005	.82523	.12120
−.18061	−2.50310	−.92514	−.39393	−.19750
−.76248	.34477	.47615	−.18026	−.54978
.47991	−.19433	.62531	−.61407	−.59267
.03638	.07211	−.40372	.24853	−.02860

Table 9.5. Moving Average Models for Example 9.2.2

j	1	3	5	12	13	23	AICC	$\hat{\sigma}^2$	$\hat{v}_{130}$
Model (9.2.5)									
$\hat{\theta}_j$	−.355	−.201	0	−.524	0	.241	−486.04	.00125	.00125
Model (9.2.6)									
$\hat{\theta}_j$	−.433	−.306	.238	−.656	0	.352	−489.95	.00103	.00117
Model (9.2.9)									
$\hat{\theta}_j$	−.396	0	0	−.614	.243	0	−483.38	.00134	.00134

by Figure 9.13, we find that there is a model with even smaller AICC value than (9.2.5), namely

$$\begin{aligned} X_t = Z_t &- .433Z_{t-1} - .306Z_{t-3} + .238Z_{t-5} \\ &- .656Z_{t-12} + .352Z_{t-23}, \end{aligned} \tag{9.2.6}$$

with $\{Z_t\} \sim \text{WN}(0, .00103)$ and AICC $= -489.95$. Since the process defined by (9.2.6) passes the goodness of fit tests in Section 9.4, we choose it as our moving average model for the data.

The substantial reduction in white noise variance achieved by (9.2.6) must be interpreted carefully since (9.2.5) is an invertible model and (9.2.6) is not. Thus for (9.2.6) the asymptotic one-step linear predictor variance (the white noise variance of the equivalent invertible version of the model) is not σ^2 but $\sigma^2/|b_1 \cdots b_j|^2$ (see Section 4.4), where $b_1, \ldots, b_j$ are the zeroes of the moving average polynomial $\theta(z)$ inside the unit circle. For the model (9.2.6), $j = 4$ and $|b_1 \cdots b_j| = .939$, so the asymptotic one-step predictor variance is .00117, which is still noticeably smaller than the value .00125 for (9.2.5). The maximum likelihood program PEST also computes the estimated mean squared error of prediction, v_{n-1}, for the last observation based on the first $(n - 1)$. This is simply r_{n-1} times the maximum likelihood estimator of σ^2 (see Section 8.7). It can be seen in Table 9.5 that v_{n-1} is quite close to $\hat{\sigma}^2$ for each of the invertible models (9.2.5) and (9.2.9).

The model (9.2.6) does of course have an invertible version with the same likelihood (which can be found by using the program PEST), however it will have small non-zero coefficients at lags other than 1, 3, 5, 12 and 23. If we constrain the model to be invertible and to have zero coefficients except at lags 1, 3, 5, 12 and 23, the likelihood is maximized for parameter values precisely on the boundary of the invertible region and the maximum is strictly less than the likelihood of the model (9.2.6). Thus in the presence of lag constraints, insistence on invertibility can make it impossible to achieve the maximum value of the likelihood.

A similar analysis of the data, starting from Table 9.4 and fitting autoregressive rather than moving average models, leads first to the model,

$$X_t + .374X_{t-1} + .464X_{t-12} + .157X_{t-13} = Z_t, \tag{9.2.7}$$

with $\{Z_t\} \sim \text{WN}(0, .00146)$ and AICC $= -472.53$. Allowing non-zero coefficients also at lags 3, 4, 9 and 16, we obtain the improved model,

$$X_t + .365X_{t-1} + .467X_{t-12} + .179X_{t-13} + .129X_{t-16} = Z_t, \quad (9.2.8)$$

with $\{Z_t\} \sim \text{WN}(0, .00142)$ and AICC $= -472.95$. However neither (9.2.7) nor (9.2.8) comes close to the moving average model (9.2.6) from the point of view of the AICC value.

It is interesting to compare the model (9.2.6) with the multiplicative model for $\{\nabla\nabla_{12}V_t\}$ fitted by Box and Jenkins (1976), i.e. with $X_t^* = \nabla\nabla_{12}V_t$,

$$X_t^* = (1 - .396B)(1 - .614B^{12})Z_t, \qquad \{Z_t\} \sim \text{WN}(0, .00134). \quad (9.2.9)$$

The AICC value for this model is -483.38, making it preferable to (9.2.8) but inferior to both (9.2.5) and to our chosen model (9.2.6). Characteristics of the three moving average models can be compared by examining Table 9.5.

(c) *Identification of Mixed Models.* The identification of a pure autoregressive or moving average process is reasonably straightforward using the sample autocorrelation and partial autocorrelation functions, the preliminary estimators $\hat{\boldsymbol{\phi}}_m$ and $\hat{\boldsymbol{\theta}}_m$ and the AICC. On the other hand, for ARMA(p, q) processes with p and q both non-zero, the sample ACF and PACF are much more difficult to interpret. We therefore search directly for values of p and q such that the AICC defined by (9.2.1) is minimum. The search can be carried out in a variety of ways, e.g. by trying all (p, q) values such that $p + q = 1$, then $p + q = 2$, etc., or alternatively by using the following steps.

(i) Use maximum likelihood estimation (program PEST) to fit ARMA processes of orders $(1, 1), (2, 2), \ldots,$ to the data, selecting the model which gives the smallest value of the AICC. [Initial parameter estimates for PEST can be found using Option 3 to fit ARMA(p, p) models as described in Example 8.4.1, or by appending zero coefficients to fitted maximum likelihood models of lower order.]
(ii) Starting from the minimum-AICC ARMA(p, p) model, eliminate one or more coefficients (guided by the standard errors of the estimated coefficients), maximize the likelihood for each reduced model and compute the AICC value.
(iii) Select the model with smallest AICC value (subject to its passing the goodness of fit tests in Section 9.4).

The procedure is illustrated in the following example.

EXAMPLE 9.2.3. The sample autocorrelation and partial autocorrelation functions of 200 observations of a stationary series are shown in Figure 9.14. They suggest an AR(4) model for the data, or perhaps a mixed model with fewer coefficients. We shall explore both possibilities, first fitting a mixed model in accordance with the procedure outlined above.

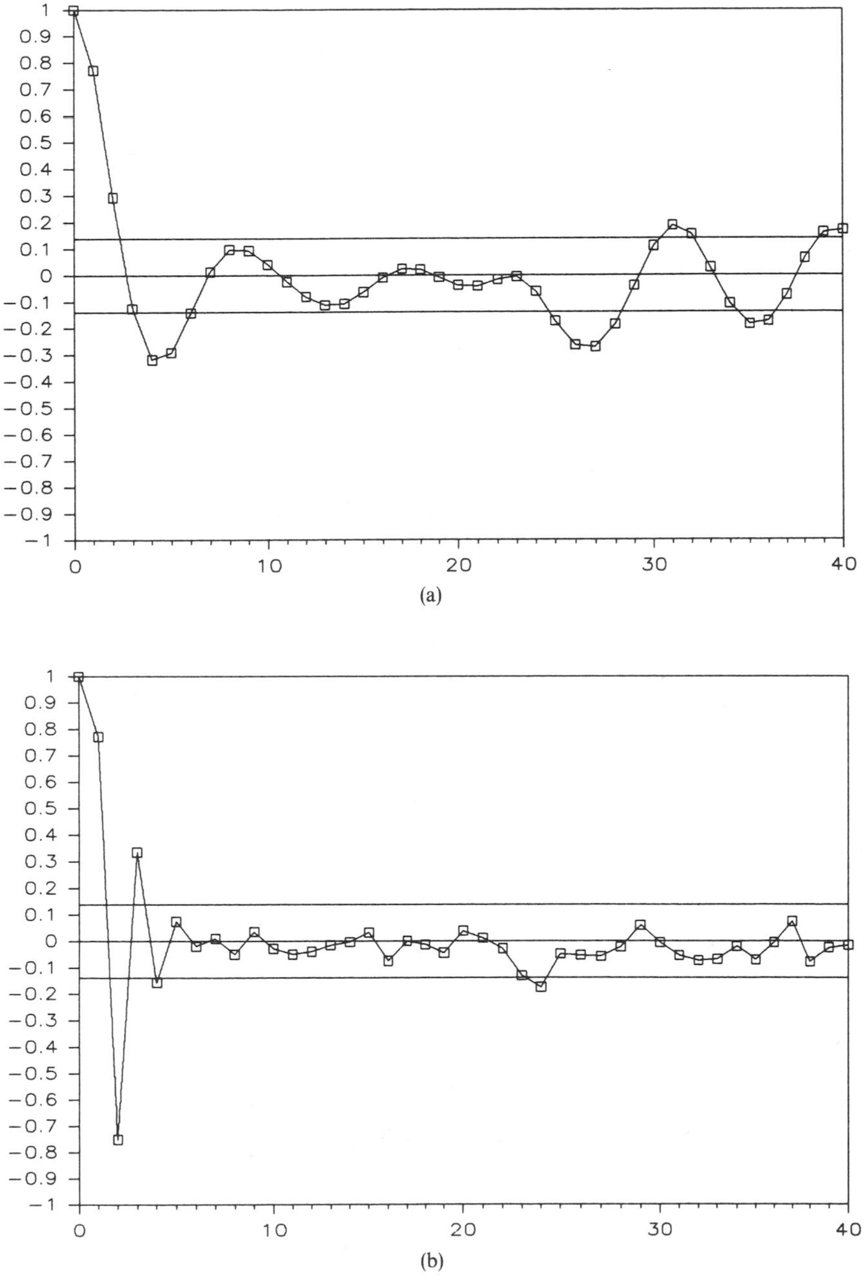

Figure 9.14. The sample ACF (a) and PACF (b) for the data of Example 9.2.3.

Table 9.6. Parameter Estimates for ARMA(p, p) Models, Example 9.2.3

(a) Preliminary Estimates (from PEST) with $m = 9$

p	$\tilde{\phi}_1$	$\tilde{\phi}_2$	$\tilde{\phi}_3$	$\tilde{\theta}_1$	$\tilde{\theta}_2$	$\tilde{\theta}_3$	AICC
1	.803			.868			656.61
2	1.142	−.592		.528	.025		591.43
3	−2.524	3.576	−2.156	4.195	1.982	.109	Non-causal

(b) Maximum Likelihood Estimates (from PEST)

p	$\hat{\phi}_1$	$\hat{\phi}_2$	$\hat{\phi}_3$	$\hat{\phi}_4$	$\hat{\theta}_1$	$\hat{\theta}_2$	$\hat{\theta}_3$	$\hat{\theta}_4$	$\hat{\sigma}^2$	AICC	BIC
1	.701				.892				1.458	652.33	657.36
2	1.118	−.580			.798	.103			.982	578.27	591.85
3	1.122	−.555	−.020		.792	.059	−.042		.982	582.39	603.17
4	1.016	−1.475	1.012	−.525	.889	1.207	.897	.216	.930	579.98	603.67

Table 9.6(a) shows the preliminary parameter estimates $\tilde{\boldsymbol{\phi}}, \tilde{\boldsymbol{\theta}}$ for ARMA(p, p) models with $p = 1$, 2 and 3 ($p = 3$ gives a non-causal model) and $m = 9$, obtained from PEST as described in Example 8.4.1. On the basis of the AICC values in Table 9.6(a), the ARMA(2, 2) model is the most promising. Since the preliminary ARMA(3, 3) model is not causal, it cannot be used to initialize the search for the maximum likelihood ARMA(3, 3) model. Instead, we use the maximum likelihood ARMA(2, 2) model with appended coefficients $\phi_3 = \theta_3 = 0$. The maximum likelihood results are shown in Table 9.6(b). The AICC values have a clearly defined minimum at $p = 2$. Comparing each coefficient of the maximum likelihood ARMA(2, 2) model with its standard error we obtain the results shown in Table 9.7, which suggest dropping the coefficient θ_2 and fitting an ARMA(2, 1) process. Maximum likelihood estimation then gives the model (for the mean-corrected data),

$$X_t - 1.185X_{t-1} + .624X_{t-2} = Z_t + .703Z_{t-1}, \qquad \{Z_t\} \sim \text{WN}(0, .986), \tag{9.2.10}$$

with AICC value 576.88 and BIC value 586.48.

Table 9.7. Comparison of $\hat{\phi}_1$, $\hat{\phi}_2$, $\hat{\theta}_1$ and $\hat{\theta}_2$ with Their Standard Errors (Obtained from the Program PEST)

	$\hat{\phi}_1$	$\hat{\phi}_2$	$\hat{\theta}_1$	$\hat{\theta}_2$
Estimated coefficient	1.118	−.580	.798	.103
Estimated coefficient / 1.96 * (Standard error)	5.811	−3.605	3.604	.450

If now we fit AR(p) models of order $p = 2, \ldots, 6$ we obtain the results shown in Table 9.8. The smallest AICC and BIC values are both achieved when $p = 5$, but the values are substantially larger than the corresponding values

Table 9.8. Maximum Likelihood AR(p) Models for Example 9.2.3

p	$\hat{\phi}_1$	$\hat{\phi}_2$	$\hat{\phi}_3$	$\hat{\phi}_4$	$\hat{\phi}_5$	$\hat{\phi}_6$	$\hat{\sigma}^2$	AICC	BIC	FPE
2	1.379	−.773					1.380	640.83	646.50	1.408
3	1.712	−1.364	.428				1.121	602.03	611.77	1.155
4	1.839	−1.760	.919	−.284			1.029	587.36	600.98	1.071
5	1.891	−1.932	1.248	−.627	.186		.992	582.35	599.66	1.043
6	1.909	−1.991	1.365	−.807	.362	−.092	.984	582.77	603.56	1.044

for (9.2.10). We therefore select the ARMA(2, 1) model, subject to its passing the goodness of fit tests to be discussed in Section 9.4.

The data for this example were in fact generated by the Gaussian process,

$$X_t - 1.3X_{t-1} + .7X_{t-2} = Z_t + .7Z_{t-1}, \qquad \{Z_t\} \sim \text{WN}(0, 1). \quad (9.2.11)$$

(d) *Use of the Residuals for Model Modification.* When an ARMA model $\phi(B)X_t = \theta(B)Z_t$ is fitted to a given series, an essential part of the procedure is to examine the residuals, which should, if the model is satisfactory, have the appearance of a realization of white noise. If the autocorrelations and partial autocorrelations of the residuals suggest that they come from some other clearly identifiable process, then this more complicated model for the residuals can be used to suggest a more appropriate model for the original data.

If the residuals appear to come from an ARMA process with coefficient vectors $\boldsymbol{\phi}_Z$ and $\boldsymbol{\theta}_Z$, this indicates that $\{Z_t\}$ in our fitted model should satisfy $\phi_Z(B)Z_t = \theta_Z(B)W_t$ where $\{W_t\}$ is white noise. Applying the operator $\phi_Z(B)$ to each side of the equation defining $\{X_t\}$, we obtain,

$$\phi_Z(B)\phi(B)X_t = \phi_Z(B)\theta(B)Z_t = \theta_Z(B)\theta(B)W_t, \quad (9.2.12)$$

where $\{W_t\}$ is white noise. The modified model for $\{X_t\}$ is thus an ARMA process with autoregressive and moving average operators $\phi_Z(B)\phi(B)$ and $\theta_Z(B)\theta(B)$ respectively.

EXAMPLE 9.2.4. Consider the AR(2) model in Table 9.8,

$$(1 - 1.379B + .773B^2)X_t = Z_t, \quad (9.2.13)$$

which was fitted to the data of Example 9.2.3. This is an unsatisfactory model, both for its high AICC value and the non-whiteness of its residuals. The sample autocorrelation and partial autocorrelation functions of its residuals are shown in Figure 9.15. They suggest an MA(2) model for $\{Z_t\}$, i.e.

$$Z_t = (1 + \theta_1 B + \theta_2 B^2)W_t, \qquad \{W_t\} \sim \text{WN}(0, \sigma^2). \quad (9.2.14)$$

From (9.2.13) and (9.2.14) we obtain an ARMA(2, 2) process as the modified model for $\{X_t\}$. Fitting an ARMA(2, 2) process by maximum likelihood and allowing subsets of the coefficients to be zero leads us to the same model for the data as was found in Example 9.2.3.

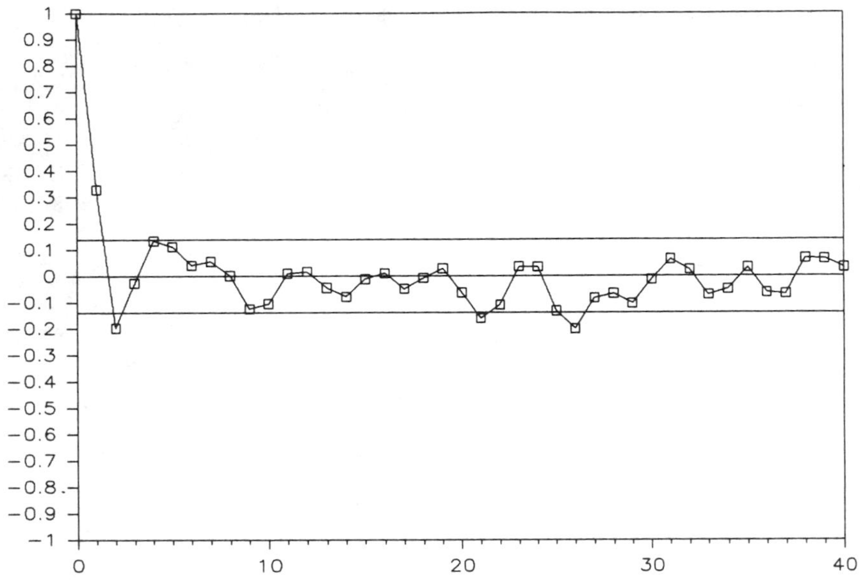

(a)

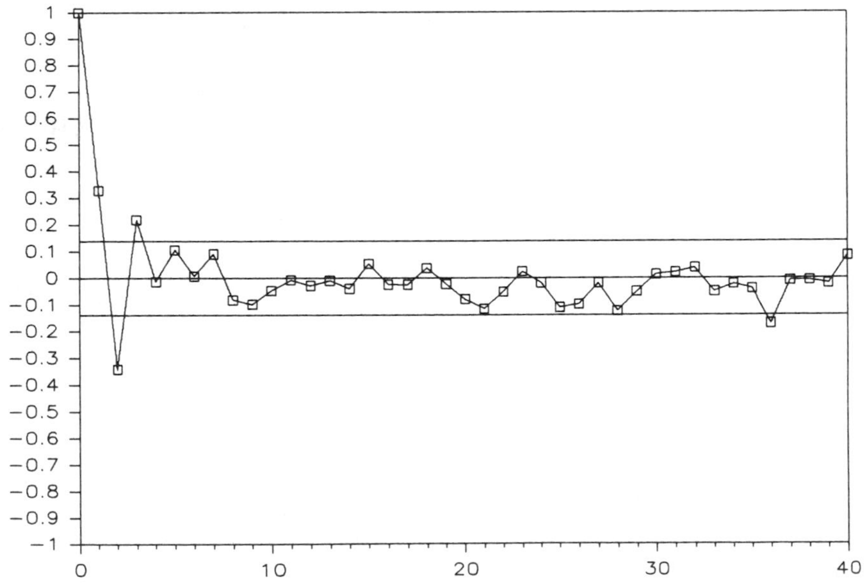

(b)

Figure 9.15. The sample ACF (a) and PACF (b) of the residuals when the data of Example 9.2.3 is fitted with the AR(2) model (9.2.13).

It is fortunate in Example 9.2.4 that the fitted AR(2) model has an autoregressive polynomial similar to that of the best-fitting ARMA(2, 1) process (9.2.10). It frequently occurs, when an AR(p) model is fitted to an ARMA(p, q) process, that the autoregressive polynomials for the two processes are totally different. The residuals from the AR(p) model in such a case are not likely to have a simple form such as the moving average form encountered in Example 9.2.4.

§9.3 Order Selection

In Section 9.2 we referred to the problem of overfitting and the need to avoid it by imposing a cost for increasing the number of parameters in the fitted model. One way in which this can be done for pure autoregressive models is to minimize the final prediction error (FPE) of Akaike (1969). The FPE is an estimate of the one-step prediction mean squared error for a realization of the process independent of the one observed. If we fit autoregressive processes of steadily increasing order p to the observed data, the maximum likelihood estimate of the white noise variance will usually decrease with p, however the estimation errors in the expanding set of fitted parameters will eventually cause the FPE to increase. According to the FPE criterion we then choose the order of the fitted process to be the value of p for which the FPE is minimum. To apply this criterion it remains only to express the FPE in terms of the data $X_1, \dots, X_n$.

Assume that $\{X_1, \dots, X_n\}$ is a realization of an AR(p) process with coefficients $\phi_1, \dots, \phi_p$, $(p < n)$, and let $\{Y_1, \dots, Y_n\}$ be an independent realization of the same process. If $\hat{\phi}_1, \dots, \hat{\phi}_p$ are the maximum likelihood estimators of the coefficients based on $\{X_1, \dots, X_n\}$ and if we use these to compute the one-step predictor $\hat{\phi}_1 Y_n + \cdots + \hat{\phi}_p Y_{n+1-p}$ of Y_{n+1}, then the mean-square prediction error is

$$\begin{aligned}
&E(Y_{n+1} - \hat{\phi}_1 Y_n - \cdots - \hat{\phi}_p Y_{n+1-p})^2 \\
&\quad = E[Y_{n+1} - \phi_1 Y_n - \cdots - \phi_p Y_{n+1-p} \\
&\qquad - (\hat{\phi}_1 - \phi_1) Y_n - \cdots - (\hat{\phi}_p - \phi_p) Y_{n+1-p}]^2 \\
&\quad = \sigma^2 + E[(\hat{\boldsymbol{\phi}}_p - \boldsymbol{\phi}_p)' [Y_{n+1-i} Y_{n+1-j}]_{i,j=1}^p (\hat{\boldsymbol{\phi}}_p - \boldsymbol{\phi}_p)],
\end{aligned}$$

where $\boldsymbol{\phi}_p' = (\phi_1, \dots, \phi_p)$, $\hat{\boldsymbol{\phi}}_p' = (\hat{\phi}_1, \dots, \hat{\phi}_p)$ and σ^2 is the white noise variance of the AR(p) model. Writing the last term in the preceding equation as the expectation of the conditional expectation given $X_1, \dots, X_n$, and using the independence of $\{X_1, \dots, X_n\}$ and $\{Y_1, \dots, Y_n\}$, we obtain

$$\begin{aligned}
&E(Y_{n+1} - \hat{\phi}_1 Y_n - \cdots - \hat{\phi}_p Y_{n+1-p})^2 \\
&\quad = \sigma^2 + E[(\hat{\boldsymbol{\phi}}_p - \boldsymbol{\phi}_p)' \Gamma_p (\hat{\boldsymbol{\phi}}_p - \boldsymbol{\phi}_p)],
\end{aligned}$$

where $\Gamma_p = E[Y_i Y_j]_{i,j=1}^p$. We can approximate the last term by assuming that

$n^{1/2}(\hat{\boldsymbol{\phi}}_p - \boldsymbol{\phi}_p)$ has its asymptotic distribution $N(\mathbf{0}, \sigma^2\Gamma_p^{-1})$ from Theorem 8.1.1. This gives (see Problem 1.16)

$$E(Y_{n+1} - \hat{\phi}_1 Y_n - \cdots - \hat{\phi}_p Y_{n+1-p})^2 \simeq \sigma^2\left(1 + \frac{p}{n}\right). \tag{9.3.1}$$

If $\hat{\sigma}^2$ is the maximum likelihood estimator of σ^2 then for large n, $n\hat{\sigma}^2/\sigma^2$ is distributed approximately as chi-squared with $(n - p)$ degrees of freedom (see Section 8.9). We therefore replace σ^2 in (9.3.1) by the estimator $n\hat{\sigma}^2/(n - p)$ to get the estimated mean square prediction error of Y_{n+1},

$$\text{FPE} = \hat{\sigma}^2 \frac{n+p}{n-p}. \tag{9.3.2}$$

Inspection of Table 9.2 shows how the FPE decreases to a minimum then increases as p is increased in Example 9.2.1. The same table shows the non-increasing behaviour of $\hat{\sigma}^2$.

A more generally applicable criterion for model-selection is the information criterion of Akaike (1973), known as the AIC. This was designed to be an approximately unbiased estimate of the Kullback–Leibler index of the fitted model relative to the true model (defined below). Here we use a bias-corrected version of the AIC, referred to as the AICC, suggested by Hurvich and Tsai (1989).

If $\mathbf{X}$ is an n-dimensional random vector whose probability density belongs to the family $\{f(\cdot; \psi), \psi \in \Psi\}$, the Kullback–Leibler discrepancy between $f(\cdot; \psi)$ and $f(\cdot; \theta)$ is defined as

$$d(\psi | \theta) = \Delta(\psi | \theta) - \Delta(\theta | \theta),$$

where

$$\begin{aligned}\Delta(\psi | \theta) &= E_\theta(-2 \ln f(\mathbf{X}; \psi)) \\ &= \int_{\mathbb{R}^n} -2 \ln(f(\mathbf{x}; \psi)) f(\mathbf{x}; \theta)\, d\mathbf{x},\end{aligned}$$

is the Kullback–Leibler index of $f(\cdot; \psi)$ relative to $f(\cdot; \theta)$. (In general $\Delta(\psi | \theta) \neq \Delta(\theta | \psi)$.) Applying Jensen's inequality, we see that

$$\begin{aligned}d(\psi | \theta) &= \int_{\mathbb{R}^n} -2 \ln\left(\frac{f(\mathbf{x}; \psi)}{f(\mathbf{x}; \theta)}\right) f(\mathbf{x}; \theta)\, d\mathbf{x} \\ &\geq -2 \ln\left(\int_{\mathbb{R}^n} \frac{f(\mathbf{x}, \psi)}{f(\mathbf{x}; \theta)} f(\mathbf{x}; \theta)\, d\mathbf{x}\right) \\ &= -2 \ln\left(\int_{\mathbb{R}^n} f(\mathbf{x}; \psi)\, d\mathbf{x}\right) \\ &= 0\end{aligned}$$

with equality holding if and only if $f(\mathbf{x}; \psi) = f(\mathbf{x}; \theta)$ a.e. $[f(\cdot, \theta)]$.

Given observations $X_1, \ldots, X_n$ of an ARMA process with unknown parameters $\theta = (\boldsymbol{\beta}, \sigma^2)$, the true model could be identified if it were possible to compute the Kullback–Leibler discrepancy between all candidate models and the true model. Since this is not possible we *estimate* the Kullback–Leibler discrepancies and choose the model whose estimated discrepancy (or index) is minimum. In order to do this, we assume that the true model and the alternatives are all Gaussian. (See the Remark below for further comments on this point.) Then for any given $\theta = (\boldsymbol{\beta}, \sigma^2)$, $f(\cdot; \theta)$ is the probability density of $(Y_1, \ldots, Y_n)'$, where $\{Y_t\}$ is a Gaussian ARMA(p, q) process with coefficient vector $\boldsymbol{\beta}$ and white noise variance σ^2. (The dependence of θ on p and q is through the dimension of the autoregressive and moving average coefficient vectors in $\boldsymbol{\beta}$.)

Suppose therefore that our observations $X_1, \ldots, X_n$ are from a Gaussian ARMA process with parameter vector $\theta = (\boldsymbol{\beta}, \sigma^2)$ and assume for the moment that the true order is (p, q). Let $\hat{\theta} = (\hat{\boldsymbol{\beta}}, \hat{\sigma}^2)$ be the maximum likelihood estimator of θ based on $X_1, \ldots, X_n$ and let $Y_1, \ldots, Y_n$ be an independent realization of the true process (with parameter θ). Then

$$-2 \ln L_Y(\hat{\boldsymbol{\beta}}, \hat{\sigma}^2) = -2 \ln L_X(\hat{\boldsymbol{\beta}}, \hat{\sigma}^2) + \hat{\sigma}^{-2} S_Y(\hat{\boldsymbol{\beta}}) - n,$$

so that

$$\begin{aligned} E_\theta(\Delta(\hat{\theta} | \theta)) &= E_{\boldsymbol{\beta},\sigma^2}(-2 \ln L_Y(\hat{\boldsymbol{\beta}}, \hat{\sigma}^2)) \\ &= E_{\boldsymbol{\beta},\sigma^2}(-2 \ln L_X(\hat{\boldsymbol{\beta}}, \hat{\sigma}^2)) + E_{\boldsymbol{\beta},\sigma^2}\left(\frac{S_Y(\hat{\boldsymbol{\beta}})}{\hat{\sigma}^2}\right) - n. \end{aligned} \tag{9.3.3}$$

Making the local linearity approximation used in Section 8.11, we can write, for large n,

$$\begin{aligned} S_Y(\hat{\boldsymbol{\beta}}) &\simeq S_Y(\boldsymbol{\beta}) + (\hat{\boldsymbol{\beta}} - \boldsymbol{\beta}) \frac{\partial S_Y}{\partial \boldsymbol{\beta}}(\boldsymbol{\beta}) + \frac{1}{2}(\hat{\boldsymbol{\beta}} - \boldsymbol{\beta})' \left[\frac{\partial^2 S_Y(\boldsymbol{\beta})}{\partial \beta_i \partial \beta_j}\right]_{i,j=1}^n (\hat{\boldsymbol{\beta}} - \boldsymbol{\beta}) \\ &\simeq S_Y(\boldsymbol{\beta}) + (\hat{\boldsymbol{\beta}} - \boldsymbol{\beta}) 2 \sum_{t=1}^n \frac{\partial Z_t}{\partial \boldsymbol{\beta}}(\boldsymbol{\beta}) Z_t(\boldsymbol{\beta}) + (\hat{\boldsymbol{\beta}} - \boldsymbol{\beta})' D' D (\hat{\boldsymbol{\beta}} - \boldsymbol{\beta}). \end{aligned}$$

From Section 8.11, we know that $n^{-1} D'D \xrightarrow{P} \sigma^2 V^{-1}(\boldsymbol{\beta})$, $\hat{\boldsymbol{\beta}}$ is AN$(\boldsymbol{\beta}, n^{-1} V(\boldsymbol{\beta}))$, and that $(\partial Z_t / \partial \boldsymbol{\beta})(\boldsymbol{\beta}) Z_t(\boldsymbol{\beta})$ has mean 0. Replacing $D'D$ by $n\sigma^2 V^{-1}(\boldsymbol{\beta})$ and assuming that $n^{1/2}(\hat{\boldsymbol{\beta}} - \boldsymbol{\beta})$ has covariance matrix $V(\boldsymbol{\beta})$, we obtain

$$\begin{aligned} E_{\boldsymbol{\beta},\sigma^2}[S_Y(\hat{\boldsymbol{\beta}})] &\simeq E_{\boldsymbol{\beta},\sigma^2}[S_Y(\boldsymbol{\beta})] + \sigma^2 E_{\boldsymbol{\beta},\sigma^2}[(\hat{\boldsymbol{\beta}} - \boldsymbol{\beta})' V^{-1}(\boldsymbol{\beta})(\hat{\boldsymbol{\beta}} - \boldsymbol{\beta})] \\ &\simeq \sigma^2 n + \sigma^2 (p + q), \end{aligned}$$

since $(\partial Z_t / \partial \boldsymbol{\beta})(\boldsymbol{\beta}) Z_t(\boldsymbol{\beta})$ is independent of $\hat{\boldsymbol{\beta}} - \boldsymbol{\beta}$ and $E(\mathbf{U}'\Sigma^{-1}\mathbf{U}) = \text{trace}(\Sigma\Sigma^{-1}) = k$ for any zero-mean random k-vector $\mathbf{U}$ with nonsingular covariance matrix Σ. From the argument given in Section 8.9, $n\hat{\sigma}^2 = S_X(\hat{\boldsymbol{\beta}})$ is distributed approximately as $\sigma^2 \chi^2(n - p - q)$ for large n and is asymptotically independent of $\hat{\boldsymbol{\beta}}$. With the independence of $\{X_1, \ldots, X_n\}$ and $\{Y_1, \ldots, Y_n\}$, this implies that $\hat{\sigma}^2$ is asymptotically independent of $S_Y(\hat{\boldsymbol{\beta}})$.

Consequently,

$$E_{\boldsymbol{\beta},\sigma^2}\left(\frac{S_Y(\hat{\boldsymbol{\beta}})}{\hat{\sigma}^2}\right) - n \simeq \sigma^2(n+p+q)(E_{\boldsymbol{\beta},\sigma^2}\hat{\sigma}^{-2}) - n$$

$$\simeq \sigma^2(n+p+q)\left(\sigma^2\frac{n-p-q-2}{n}\right)^{-1} - n$$

$$= \frac{2(p+q+1)n}{n-p-q-2}.$$

Thus the quantity, $-2\ln L_X(\hat{\boldsymbol{\beta}}, \hat{\sigma}^2) + 2(p+q+1)n/(n-p-q-2)$, is an approximately unbiased estimate of the expected Kullback–Leibler index $E_\theta(\Delta(\hat{\theta}|\theta))$ in (9.3.3). Since the preceding calculations (and the maximum likelihood estimators $\hat{\boldsymbol{\beta}}$ and $\hat{\sigma}^2$) are based on the assumption that the true order is (p, q), we therefore select the values of p and q for our fitted model to be those which minimize AICC($\hat{\boldsymbol{\beta}}$), where

$$\text{AICC}(\boldsymbol{\beta}) := -2\ln L_X(\boldsymbol{\beta}, S_X(\boldsymbol{\beta})/n) + 2(p+q+1)n/(n-p-q-2). \tag{9.3.4}$$

The AIC statistic, defined as

$$\text{AIC}(\boldsymbol{\beta}) := -2\ln L_X(\boldsymbol{\beta}, S_X(\boldsymbol{\beta})/n) + 2(p+q+1),$$

can be used in the same way. Both AICC($\boldsymbol{\beta}, \sigma^2$) and AIC($\boldsymbol{\beta}, \sigma^2$) can be defined for arbitrary σ^2 by replacing $S_X(\boldsymbol{\beta})/n$ in the preceding definitions by σ^2, however we shall use AICC($\boldsymbol{\beta}$) and AIC($\boldsymbol{\beta}$) as defined above since both AICC and AIC are minimized for any given $\boldsymbol{\beta}$ by setting $\sigma^2 = S_X(\boldsymbol{\beta})/n$.

For fitting autoregressive models, Monte Carlo studies (Jones (1975), Shibata (1976)) suggest that the AIC has a tendency to overestimate p. The penalty factors, $2(p+q+1)n/(n-p-q-2)$ and $2(p+q+1)$, for the AICC and AIC statistics are asymptotically equivalent as $n \to \infty$. The AICC statistic however has a more extreme penalty for large-order models which counteracts the overfitting tendency of the AIC. The BIC is another criterion which attempts to correct the overfitting nature of the AIC. For a zero-mean causal invertible ARMA(p, q) process, it is defined (Akaike (1978)) to be,

$$\begin{aligned}\text{BIC} = (n-p-q)\ln[n\hat{\sigma}^2/(n-p-q)] + n(1+\ln\sqrt{2\pi})\\ + (p+q)\ln\left[\left(\sum_{t=1}^{n} X_t^2 - n\hat{\sigma}^2\right)\Big/(p+q)\right],\end{aligned} \tag{9.3.5}$$

where $\hat{\sigma}^2$ is the maximum likelihood estimate of the white noise variance.

The BIC is a consistent order selection procedure in the sense that if the data $\{X_1, \ldots, X_n\}$ *are* in fact observations of an ARMA(p, q) process, and if $\hat{p}$ and $\hat{q}$ are the estimated orders found by minimizing the BIC, then $\hat{p} \to p$ and $\hat{q} \to q$ with probability one as $n \to \infty$ (Hannan (1980)). This property is not shared by the AICC or AIC. On the other hand, order selection by minimization of the AICC, AIC or FPE is asymptotically efficient for autoregressive models, while order selection by BIC minimization is not

(Shibata (1980), Hurvich and Tsai (1989)). Efficiency in this context is defined as follows. Suppose that $\{X_t\}$ is a causal AR(∞) process satisfying

$$\sum_{j=0}^{\infty} \pi_j X_{t-j} = Z_t, \qquad \{Z_t\} \sim \text{WN}(0, \sigma^2),$$

(where $\pi_0 = 1$) and let $(\hat{\phi}_{p1}, \ldots, \hat{\phi}_{pp})'$ be the Yule–Walker estimates of the coefficients of an AR(p) model fitted to the data $\{X_1, \ldots, X_n\}$ (see (8.2.2)). The one-step mean-square prediction error for an independent realization $\{Y_t\}$ of $\{X_t\}$, based on the AR(p) model fitted to $\{X_t\}$, is

$$\begin{aligned} &E_X(Y_{n+1} - \hat{\phi}_{p1} Y_n - \cdots - \hat{\phi}_{pp} Y_{n+1-p})^2 \\ &\quad = E_X(Z^*_{n+1} - (\hat{\phi}_{p1} + \pi_1)Y_n - \cdots - (\hat{\phi}_{pp} + \pi_p)Y_{n+1-p} - \sum_{j=p+1}^{\infty} \pi_j Y_{n+1-j})^2 \\ &\quad = \sigma^2 + (\hat{\boldsymbol{\phi}}_{p,\infty} + \boldsymbol{\pi}_\infty)'\Gamma_\infty(\hat{\boldsymbol{\phi}}_{p,\infty} + \boldsymbol{\pi}_\infty) \\ &\quad =: H(p) \end{aligned}$$

where E_X denotes expectation conditional on $X_1, \ldots, X_n$. Here $\{Z^*_t\}$ is the white noise associated with the $\{Y_t\}$ process, Γ_∞ is the infinite-dimensional covariance matrix of $\{Y_t\}$, and $\hat{\boldsymbol{\phi}}_{p,\infty}$ and $\boldsymbol{\pi}_\infty$ are the infinite-dimensional vectors, $(\hat{\phi}_{p1}, \ldots, \hat{\phi}_{pp}, 0, 0, \ldots)'$ and $(\pi_1, \pi_2, \ldots)'$. Now if p^*_n is the value of p which minimizes $H(p)$, $0 \le p \le k_n$, and k_n is a sequence of constants converging to infinity at a suitable rate, then an order selection procedure is said to be efficient if the estimated order $\hat{p}_n$ satisfies

$$\frac{H(p^*_n)}{H(\hat{p}_n)} \xrightarrow{P} 1$$

as $n \to \infty$. In other words, an efficient order selection procedure chooses an AR model which achieves the optimal rate of convergence of the mean-square prediction error.

Of course in the modelling of real data there is rarely such a thing as the "true order". For the process $X_t = \sum_{j=0}^{\infty} \psi_j Z_{t-j}$ there may be many polynomials $\theta(z)$, $\phi(z)$ such that the coefficients of z^j in $\theta(z)/\phi(z)$ closely approximate ψ_j for moderately small values of j. Correspondingly there may be many ARMA processes with properties similar to $\{X_t\}$. This problem of identifiability becomes much more serious for multivariate processes. The AICC criterion does however provide us with a rational criterion for choosing between competing models. It has been suggested (Duong (1981)) that models with AIC values within c of the minimum value should be considered competitive (with $c = 2$ as a typical value). Selection from amongst the competitive models can then be based on such factors as whiteness of the residuals (Section 9.4) and model simplicity.

Remark. In the course of the derivation of the AICC, it was assumed that the observations $\{X_1, \ldots, X_n\}$ were from a Gaussian ARMA(p, q) process. However, even if $(X_1, \ldots, X_n)$ has a non-Gaussian distribution, the argument given above shows that the AICC is an approximately unbiased estimator of

$$E(-2 \ln L_Y(\hat{\boldsymbol{\beta}}, \hat{\sigma}^2)), \tag{9.3.6}$$

where the expectation is now taken relative to the true (possibly non-Gaussian) distribution of $(X_1, \ldots, X_n)'$ and $(Y_1, \ldots, Y_n)'$ and L_Y is the Gaussian likelihood based on $(Y_1, \ldots, Y_n)'$. The quantity in (9.3.6) can be interpreted as the expected Kullback–Leibler index of the maximum likelihood Gaussian model relative to the true distribution of the process.

The AICC for Subset Models. We frequently have occasion, particularly in analyzing seasonal data, to fit ARMA(p, q) models in which all except $m\ (\leq p + q)$ of the coefficients are constrained to be zero (see Example 9.2.2). In such cases the definition (9.3.4) is replaced by,

$$\text{AICC}(\boldsymbol{\beta}) = -2 \ln L_X(\boldsymbol{\beta}, S_X(\boldsymbol{\beta})/n) + 2(m + 1)n/(n - m - 2). \tag{9.3.7}$$

§9.4 Diagnostic Checking

Typically the goodness of fit of a statistical model to a set of data is judged by comparing the observed values with the corresponding predicted values obtained from the fitted model. If the fitted model is appropriate, then the residuals should behave in a manner which is consistent with the model.

When we fit an ARMA(p, q) model to a given series we first find the maximum likelihood estimators $\hat{\boldsymbol{\phi}}$, $\hat{\boldsymbol{\theta}}$ and $\hat{\sigma}^2$ of the parameters $\boldsymbol{\phi}$, $\boldsymbol{\theta}$ and σ^2. In the course of this procedure the predicted values $\hat{X}_t(\hat{\boldsymbol{\phi}}, \hat{\boldsymbol{\theta}})$ of X_t based on $X_1, \ldots, X_{t-1}$ are computed for the fitted model. The residuals are then defined, in the notation of Section 8.7, by

$$\hat{W}_t = (X_t - \hat{X}_t(\hat{\boldsymbol{\phi}}, \hat{\boldsymbol{\theta}}))/(r_{t-1}(\hat{\boldsymbol{\phi}}, \hat{\boldsymbol{\theta}}))^{1/2}, \qquad t = 1, \ldots, n. \tag{9.4.1}$$

If we were to assume that the maximum likelihood ARMA(p, q) model is the *true* process generating $\{X_t\}$, then we could say that $\{\hat{W}_t\} \sim \text{WN}(0, \hat{\sigma}^2)$. However to check the appropriateness of an ARMA(p, q) model for the data, we should assume only that $X_1, \ldots, X_n$ is generated by an ARMA(p, q) process with unknown parameters $\boldsymbol{\phi}$, $\boldsymbol{\theta}$ and σ^2, whose maximum likelihood *estimators* are $\hat{\boldsymbol{\phi}}$, $\hat{\boldsymbol{\theta}}$ and $\hat{\sigma}^2$ respectively. Then it is not true that $\{\hat{W}_t\}$ is white noise. Nonetheless $\hat{W}_t$, $t = 1, \ldots, n$ should have properties which are similar to those of the white noise sequence,

$$W_t(\boldsymbol{\phi}, \boldsymbol{\theta}) = (X_t - \hat{X}_t(\boldsymbol{\phi}, \boldsymbol{\theta}))/(r_{t-1}(\boldsymbol{\phi}, \boldsymbol{\theta}))^{1/2}, \qquad t = 1, \ldots, n.$$

Moreover by (8.11.2), $E(W_t(\boldsymbol{\phi}, \boldsymbol{\theta}) - Z_t)^2$ is small for large t, so that properties of the residuals $\{\hat{W}_t\}$ should reflect those of the white noise sequence $\{Z_t\}$ generating the underlying ARMA(p, q) process. In particular the sequence $\{\hat{W}_t\}$ should be approximately (i) uncorrelated if $\{Z_t\} \sim \text{WN}(0, \sigma^2)$, (ii) independent if $\{Z_t\} \sim \text{IID}(0, \sigma^2)$, and (iii) normally distributed if $Z_t \sim \text{N}(0, \sigma^2)$.

Remark. There are several other candidates for the title "residuals" of a fitted ARMA process. One choice for example is (see Problem 5.15(a))

$$\hat{Z}_t = \hat{\theta}^{-1}(B)\hat{\phi}(B)X_t, \qquad t = 1, \ldots, n,$$

where $\hat{\phi}(z) := 1 - \hat{\phi}_1 z - \cdots - \hat{\phi}_p z^p$, $\hat{\theta}(z) := 1 + \hat{\theta}_1 z + \cdots + \hat{\theta}_q z^q$ and $X_t := 0$, $t \le 0$. However we prefer to use thc definition (9.4.1) because of its direct interpretation as a scaled difference between an observed and a predicted value, and because it is computed for each t in the course of determining the maximum likelihood estimates.

The Graph of $\{\hat{W}_t, t = 1, \ldots, n\}$. If the fitted model is appropriate, then the graph of $\hat{W}_t, t = 1, \ldots, n$, should resemble that of a white noise sequence. While it is difficult to identify the correlation structure of $\{\hat{W}_t\}$ (or any time series for that matter) from its graph, deviations of the mean from zero are sometimes clearly indicated by a trend or cyclic component, and non-constancy of the variance by fluctuations in $\hat{W}_t$ whose magnitude depends strongly on t.

The residuals obtained from fitting an AR(3) model to the data in Example 9.2.1 are displayed in Figure 9.16. The residuals $\hat{W}_t$ have been rescaled; i.e. divided by the estimated standard deviation $\hat{\sigma}$, so that most of the values should lie between ± 1.96. The graph gives no indication of a non-zero mean or non-constant variance, so on this basis there is no reason to doubt the compatibility of $\hat{W}_1, \ldots, \hat{W}_n$ with white noise.

The next step is to check that the sample autocorrelation function of $\hat{W}_1, \ldots, \hat{W}_n$ behaves as it should under the assumption that the fitted model is appropriate.

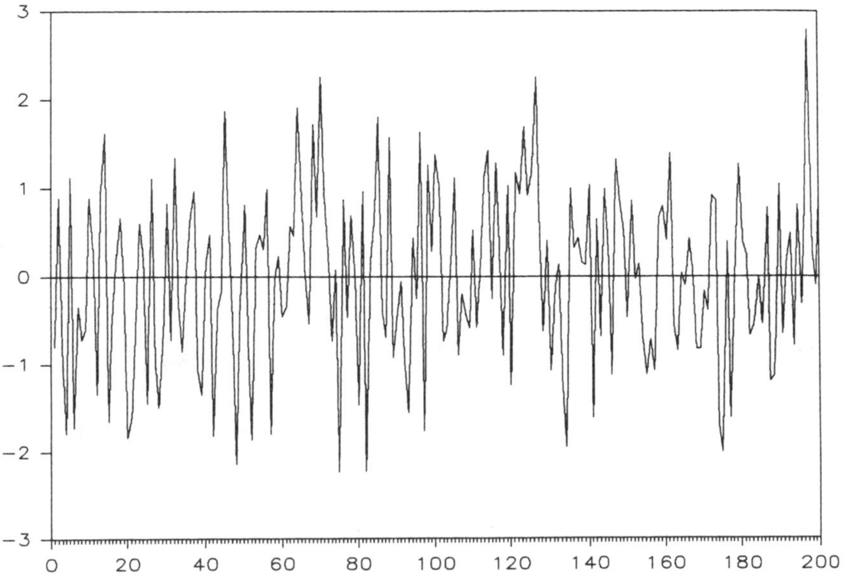

Figure 9.16. Rescaled residuals from the AR(3) model for the data of Example 9.2.1.

The Sample Autocorrelation Function of $\hat{W}_t$. The sample autocorrelations of an iid sequence $Z_1, \ldots, Z_n$ with $E(Z_t^2) < \infty$ are for large n approximately iid with distribution $N(0, 1/n)$ (see Example 7.2.1). Assuming therefore that we have fitted an appropriate ARMA model to our data and that the ARMA model is generated by an iid white noise sequence, the same approximation should be valid for the sample autocorrelation function of $\hat{W}_t$, $t = 1, \ldots, n$, defined by

$$\hat{\rho}_W(h) = \sum_{t=1}^{n-h} [(\hat{W}_t - \overline{W})(\hat{W}_{t+h} - \overline{W})] \Big/ \sum_{t=1}^{n} (\hat{W}_t - \overline{W})^2, \qquad h = 1, 2, \ldots$$

where $\overline{W} = n^{-1} \sum_{t=1}^{n} \hat{W}_t$. However, because each $\hat{W}_t$ is a function of the maximum likelihood estimator $(\hat{\boldsymbol{\phi}}, \hat{\boldsymbol{\theta}})$, $\hat{W}_1, \ldots, \hat{W}_n$ is not an iid sequence and the distribution of $\hat{\rho}_W(h)$ is not quite the same as in the iid case. In fact $\hat{\rho}_W(h)$ has an asymptotic variance which for small lags is less than $1/n$ and which for large lags is close to $1/n$. The asymptotic distribution of $\hat{\rho}_W(h)$ is discussed below.

Let $\hat{\boldsymbol{\rho}}_W = (\hat{\rho}_W(1), \ldots, \hat{\rho}_W(h))'$ where h is a fixed positive integer. If $\{X_t\}$ is the causal invertible ARMA process $\phi(B)X_t = \theta(B)Z_t$, define

$$\tilde{\phi}(z) = \phi(z)\theta(z) = 1 - \tilde{\phi}_1 z - \cdots - \tilde{\phi}_{p+q} z^{p+q}, \tag{9.4.2}$$

and

$$a(z) = (\tilde{\phi}(z))^{-1} = \sum_{j=0}^{\infty} a_j z^j. \tag{9.4.3}$$

It will be convenient also to define $a_j = 0$ for $j < 0$.

Assuming $h \geq p + q$, set

$$T_h = [a_{i-j}]_{1 \leq i \leq h,\, 1 \leq j \leq p+q},$$

$$\tilde{\Gamma}_{p+q} = \left[\sum_{k=0}^{\infty} a_k a_{k+|i-j|} \right]_{i,j=1}^{p+q}, \tag{9.4.4}$$

and

$$Q = T_h \tilde{\Gamma}_{p+q}^{-1} T_h' = [q_{ij}]_{i,j=1}^{h}. \tag{9.4.5}$$

Note that $\tilde{\Gamma}_{p+q}$ is the covariance matrix of $(Y_1, \ldots, Y_{p+q})$ where $\{Y_t\}$ is an AR$(p+q)$ process with autoregressive polynomial given by $\tilde{\phi}(z)$ in (9.4.2) and with $\sigma^2 = 1$. Then using the argument given in Box and Pierce (1970), it can be shown that

$$\hat{\boldsymbol{\rho}}_W \text{ is } \mathrm{AN}(0, n^{-1}(I_h - Q)), \tag{9.4.6}$$

where I_h is the $h \times h$ identity matrix. The asymptotic variance of $\hat{\rho}_W(i)$ is thus $n^{-1}(1 - q_{ii})$.

Example 9.4.1 (AR(1)). In this case $\tilde{\Gamma}_{p+q} = (1 - \phi^2)^{-1}$ and

$$q_{ii} = q_{ii}(\phi) = \phi^{2(i-1)}(1 - \phi^2), \qquad i = 1, 2, \ldots.$$

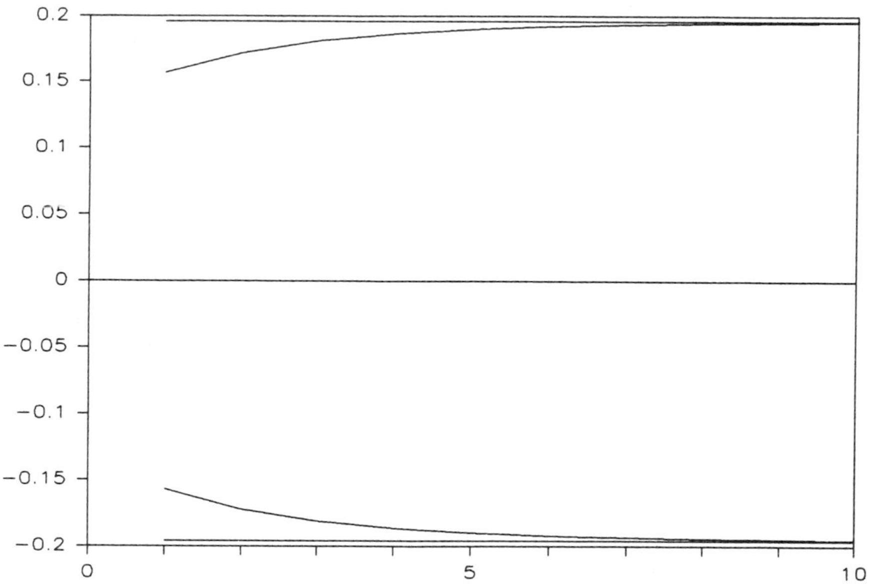

Figure 9.17. The bounds $\pm 1.96n^{-1/2}(1 - q_{ii}(\phi))^{1/2}$ of Example 9.4.1 with $n = 100$ and $\phi = 0$ (outer), $\phi = .8$ (inner).

The bounds $\pm 1.96(1 - q_{ii}(\phi))^{1/2}n^{-1/2}$ are plotted in Figure 9.17 for two values of ϕ. In applications, since the true value of ϕ is unknown, the bounds $\pm 1.96(1 - q_{ii}(\hat{\phi}))^{1/2}n^{-1/2}$ are plotted. A value of $\hat{\rho}_W(h)$ lying outside these bounds suggests possible inconsistency of the residuals, $\hat{W}_t$, $t = 1, \ldots, n$, with the fitted model. However it is essential to bear in mind that approximately 5 percent of the values of $\hat{\rho}_W(h)$ can be expected to fall outside the bounds, even if the fitted model is correct.

EXAMPLE 9.4.2 (AR(2)). A straightforward calculation yields

$$q_{11} = 1 - \phi_2^2, \qquad q_{12} = -\phi_1\phi_2(1 + \phi_2), \qquad q_{22} = 1 - \phi_2^2 - \phi_1^2(1 + \phi_2)^2.$$

Since the sequence $\{a_j\}$ in (9.4.3) satisfies the recursion relations,

$$a_j - \phi_1 a_{j-1} - \phi_2 a_{j-2} = 0, \qquad j \geq 2,$$

it follows from (9.4.5) that

$$q_{ij} - \phi_1 q_{i,j-1} - \phi_2 q_{i,j-2} = 0, \tag{9.4.7}$$

and hence that

$$q_{ii} = \phi_1 q_{i,i-1} + \phi_2 q_{i,i-2}.$$

The asymptotic variance $(1 - q_{ii}(\boldsymbol{\phi}))n^{-1}$ can thus easily be computed using the recursion (9.4.7) and the initial values q_{11}, q_{12} and q_{22}. The autocorrelations of the estimated residuals from the fitted AR(2) model in Example

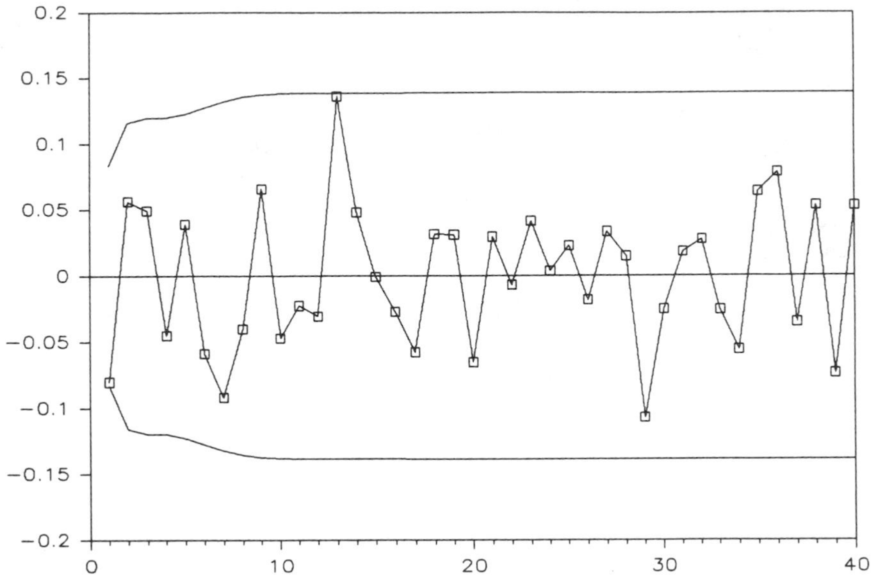

Figure 9.18. The autocorrelations of the residuals $\{\hat{W}_t\}$ from the AR(2) model, $X_t - 1.458X_{t-1} + .6X_{t-2} = Z_t$, fitted to the data in Example 9.2.1. The bounds are computed as described in Example 9.4.2.

9.2.1 and the bounds $\pm 1.96(1 - q_{ii}(\hat{\phi}_1, \hat{\phi}_2))^{1/2} n^{-1/2}$ are plotted in Figure 9.18. With the exception of $\hat{\rho}_W(13)$, the correlations are well within the confidence bounds.

The limit distribution of $\hat{\boldsymbol{\rho}}_W$ for MA(1) and MA(2) processes is the same as in Examples 9.4.1 and 9.4.2 with $\boldsymbol{\phi}$ replaced by $-\boldsymbol{\theta}$. Moreover the ARMA(1, 1) bounds can be found from the AR(2) bounds by setting $\phi_1 = (\phi - \theta)$ and $\phi_2 = -\phi\theta$ where ϕ and θ are the respective parameters in the ARMA(1, 1) model.

The Portmanteau Test. Instead of checking to see if each $\hat{\rho}_W(i)$ falls within the confidence bounds $\pm 1.96(1 - q_{ii})^{1/2} n^{-1/2}$, it is possible to consider instead a single statistic which depends on $\hat{\rho}_W(i)$, $1 \le i \le h$. Throughout this discussion h is assumed to depend on the sample size n in such a way that (i) $h_n \to \infty$ as $n \to \infty$, and (ii) the conditions of Box and Pierce (1970) are satisfied, namely

(a) $\psi_j = O(n^{-1/2})$ for $j \ge h_n$ where ψ_j, $j = 0, 1, \ldots$ are the coefficients in the expansion $X_t = \sum_{j=0}^{\infty} \psi_j Z_{t-j}$, and
(b) $h_n = O(n^{1/2})$.

Then since $h_n \to \infty$, the matrix $\tilde{\Gamma}_{p+q}$ may be approximated by $T_h' T_h$ and so the matrix Q in (9.4.5) and (9.4.6) may be approximated by the projection matrix (see Remark 2 of Section 2.5),

$$T_h(T_h'T_h)^{-1}T_h',$$

which has rank $p + q$. Thus if the model is appropriate, the distribution of $\hat{\boldsymbol{\rho}}_W = (\hat{\rho}_W(1), \dots, \hat{\rho}_W(h))'$ is approximately $N(\mathbf{0}, n^{-1}(I_k - T_h(T_h'T_h)^{-1}T_h'))$. It then follows from Problem 2.19 that the distribution of

$$Q_W = n\hat{\boldsymbol{\rho}}_W'\hat{\boldsymbol{\rho}}_W = n\sum_{j=1}^{h}\hat{\rho}_W^2(j)$$

is approximately chi-squared with $h - (p + q)$ degrees of freedom. The adequacy of the model is therefore rejected at level α if

$$Q_W > \chi_{1-\alpha}^2(h - p - q).$$

Applying this test to the residuals from the fitted AR(3) model in Example 9.2.1 with $h = 25$, we obtain $n\sum_{j=1}^{25}\hat{\rho}_W^2(j) = 11.995$, which is less than $\chi_{.95}^2(22) = 33.9$. Thus on the basis of this test, there is no reason to doubt the adequacy of the fitted model. For the airline data in Example 9.2.2, we have $n\sum_{j=1}^{25}\hat{\rho}_W^2(j) = 12.104$ for the fitted moving average model with non-zero coefficients at lags 1, 3, 5, 12 and 23. Comparing this value with $\chi_{.95}^2(25 - 5) = 31.4$, we see that the residuals pass the portmanteau test. Note that the number of coefficients fitted in the model is 5. For the residuals from the AR(2) model fitted to the data of Example 9.2.4, we obtain $n\sum_{j=1}^{25}\hat{\rho}_W^2(j) = 56.615$ which is larger than $\chi_{.95}^2(23) = 35.2$. Hence, as observed earlier, this model is not a good fit to the data.

Ljung and Box (1978) suggest replacing the statistic Q_W in the above test procedure with

$$\tilde{Q}_W = n(n + 2)\sum_{j=1}^{h}\hat{\rho}_W^2(j)/(n - j).$$

They argue that under the hypothesis of model adequacy, the cutoff value given by $\chi_{1-\alpha}^2(h - p - q)$ is closer to the true $(1 - \alpha)$-quantile of the distribution of $\tilde{Q}_W$ than to that of Q_W. However, as pointed out by Davies, Triggs and Newbold (1977) the variance of $\tilde{Q}_W$ may exceed that of a χ^2 distribution with $h - p - q$ degrees of freedom. The values of $\tilde{Q}_W$ with $h = 25$ for Examples 9.2.1 and 9.2.2 are 12.907 and 13.768, respectively. Hence the residuals pass this test of model adequacy.

Examination of the squared residuals may often suggest departures of the data from the fitted model which could not otherwise be detected from the residuals themselves. Granger and Anderson (1978) have found examples where the residuals were uncorrelated while the squared residuals were correlated. We can test the squared residuals for correlation in the same way that we test the residuals themselves. Let

$$\hat{\rho}_{WW}(h) = \frac{\sum_{t=1}^{n-h}(\hat{W}_t^2 - \overline{W^2})(\hat{W}_{t+h}^2 - \overline{W^2})}{\sum_{t=1}^{n}(\hat{W}_t^2 - \overline{W^2})}, \qquad h \geq 1$$

be the sample autocorrelation function of the squared residuals where $\overline{W^2} =$

$n^{-1}\sum_{i=1}^{n}\hat{W}_t^2$. Then McLeod and Li (1983) show that

$$\tilde{Q}_{WW} = n(n+2)\sum_{j=1}^{h}\hat{\rho}_{WW}^2(j)/(n-j)$$

has an approximate $\chi^2(h)$ distribution under the assumption of model adequacy. Consequently, the adequacy of the model is rejected at level α if

$$\tilde{Q}_{WW} > \chi^2_{1-\alpha}(h).$$

For Examples 9.2.1 and 9.2.2 with $h = 25$ we obtain the values $\tilde{Q}_{WW} = 26.367$ and $\tilde{Q}_{WW} = 16.356$, respectively. Since $\chi^2_{.95}(25) = 37.7$, the squared residuals for these two examples pass this portmanteau test.

An advantage of portmanteau tests is that they pool information from the correlations $\hat{\rho}_W(i)$, $i = 1, \ldots, h$ at different lags. A distinct disadvantage however, is that they frequently fail to reject poorly fitting models. In practice portmanteau tests are more useful for disqualifying unsatisfactory models from consideration than for selecting the best-fitting model among closely competing candidates.

Tests of Randomness. In addition to the tests based on the sample autocorrelation function of $\{\hat{W}_t\}$ which we have already described, there are a number of other tests available for checking the hypothesis of "randomness" of $\{\hat{W}_t\}$, i.e. the hypothesis that $\{\hat{W}_t\}$ is an iid sequence. Three of these tests are described below. For further details and for additional tests of randomness, see Kendall and Stuart (1976).

(a) *A Test Based on Turning Points.* If $y_1, \ldots, y_n$ is a sequence of observations, then we say that the data has a turning point at time i, $1 < i < n$, if $y_{i-1} < y_i$ and $y_i > y_{i+1}$ or if $y_{i-1} > y_i$ and $y_i < y_{i+1}$. Define T to be the number of turning points of the sequence $y_1, \ldots, y_n$. If $y_1, \ldots, y_n$ are observations of a random (iid) sequence, then the probability of a turning point at time i is $\frac{2}{3}$. The expected number of turning points is therefore

$$\mu_T = ET = 2(n-2)/3.$$

It can also be shown that the variance is

$$\sigma_T^2 = \operatorname{Var}(T) = (16n - 29)/90.$$

A large value of $T - \mu_T$ indicates that the series is fluctuating more rapidly than expected for a random series. On the other hand a value of $T - \mu_T$ much smaller than zero indicates a positive correlation between neighboring observations. It can be shown that for an iid sequence

$$T \text{ is } \mathrm{AN}(\mu_T, \sigma_T^2),$$

so the assumption that $y_1, \ldots, y_n$ are observations from a random sequence is rejected if $|T - \mu_T|/\sigma_T > \Phi_{1-\alpha/2}$ where $\Phi_{1-\alpha/2}$ is the $1 - \alpha/2$ percentage point of a standard normal distribution. The values of T for the residuals in

Examples 9.2.1–9.2.3 are displayed in Table 9.9. Inspecting the $|T - \mu_T|/\sigma_T$ column of the table we see that the three sets of residuals safely pass this test of randomness.

(b) *The Difference-Sign Test.* For this test we count the number of values of i such that $y_i > y_{i-1}$, $i = 2, \ldots, n$ or equivalently the number of times the differenced series $y_i - y_{i-1}$ is positive. If we denote this number by S, it is clear that under the random sequence assumption,

$$\mu_S = ES = \tfrac{1}{2}(n - 1).$$

It can also be shown, under the same assumption, that

$$\sigma_S^2 = \text{Var}(S) = (n + 1)/12,$$

and that

$$S \text{ is AN}(\mu_S, \sigma_S^2).$$

A large positive (or negative) value of $S - \mu_S$ indicates the presence of an increasing (or decreasing) trend in the data. We therefore reject the assumption of no trend in the data if $|S - \mu_S|/\sigma_S > \Phi_{1-\alpha/2}$. Table 9.9 contains the results of this test applied to the residuals of Examples 9.2.1–9.2.3. In all three cases, the residuals easily pass this test of randomness.

The difference-sign test as a test of randomness must be used with caution. A set of observations exhibiting a strong cyclic component will pass the difference-sign test for randomness since roughly half of the observations will be points of increase.

(c) *The Rank Test.* The rank test is particularly useful for detecting a linear trend in the data. Define P to be the number of pairs (i,j) such that $y_j > y_i$, $j > i$, $i = 1, \ldots, n - 1$. There is a total of $\binom{n}{2} = \frac{1}{2}n(n - 1)$ pairs (i,j) such that $j > i$, and for each pair the event $\{y_j > y_i\}$ has probability $\frac{1}{2}$ if $\{y_j\}$ is a random sequence. The mean of P is therefore $\mu_P = \frac{1}{4}n(n - 1)$. It can also be shown that the variance of P is $\sigma_P^2 = n(n - 1)(2n + 5)/8$ and that P is $\text{AN}(\mu_P, \sigma_P^2)$ (see Kendall and Stuart, 1976). A large positive (negative) value of $P - \mu_P$ indicates the presence of an increasing (decreasing) trend in the data. The assumption of randomness of $\{y_j\}$ is therefore rejected at level α if $|P - \mu_P|/\sigma_P > \Phi_{1-\alpha/2}$. From Table 9.9 we see that the residuals from Examples 9.2.1–9.2.3 easily pass this test of randomness.

Table 9.9. Tests of Randomness Applied to Residuals in Examples 9.2.1–9.2.3

	T	μ_T	$\|T - \mu_T\|/\sigma_T$	S	μ_S	$\|S - \mu_S\|/\sigma_S$	P	$\|P - \mu_P\|/\sigma_P$
Example 9.2.1	132	132	0	99	99.5	.12	10465	.36
Example 9.2.2	87	86	.21	65	65	0	3929	.44
Example 9.2.3	131	132	.10	104	99.5	1.10	10086	.10

Checking for Normality. If it can be assumed that the white noise process $\{Z_t\}$ generating an ARMA(p, q) process is Gaussian, then stronger conclusions can be drawn from the fitted model. For example, not only is it then possible to specify an estimated mean squared error for predicted values, but asymptotic prediction confidence bounds can also be computed (see Section 5.4). We now consider a test of the hypothesis that $\{Z_t\}$ is Gaussian.

Let $Y_{(1)} < Y_{(2)} < \cdots < Y_{(n)}$ be the order statistics of a random sample $Y_1, \ldots, Y_n$ from the distribution $\mathrm{N}(\mu, \sigma^2)$. If $X_{(1)} < X_{(2)} < \cdots < X_{(n)}$ are the order statistics from a $N(0, 1)$ sample of size n, then

$$EY_{(j)} = \mu + \sigma m_j,$$

where $m_j = EX_{(j)}, j = 1, \ldots, n$. Thus a plot of the points $(m_1, Y_{(1)}), \ldots, (m_n, Y_{(n)})$ should be approximately linear. However if the sample values Y_i are not normally distributed, then the plot should be non-linear. Consequently, the squared correlation of the points $(m_i, Y_{(i)})$, $i = 1, \ldots, n$ should be near one if the normal assumption is correct. The assumption of normality is therefore rejected if the squared correlation R^2 is sufficiently small. If we approximate m_i by $\Phi^{-1}((i - .5)/n)$ (see Mage (1982) for some alternative approximations), then R^2 reduces to

$$R^2 = \frac{\left(\sum_{i=1}^{n} (Y_{(i)} - \bar{Y})\Phi^{-1}\left(\frac{i - .5}{n}\right)\right)^2}{\sum_{i=1}^{n} (Y_{(i)} - \bar{Y})^2 \sum_{i=1}^{n} \left(\Phi^{-1}\left(\frac{i - .5}{n}\right)\right)^2},$$

where $\bar{Y} = n^{-1}(Y_1 + \cdots + Y_n)$. Percentage points for the distribution of R^2, assuming normality of the sample values, are given by Shapiro and Francia (1972) for sample sizes $n < 100$. For $n = 200$, $P(R^2 < .987) = .05$ and $P(R^2 < .989) = .10$; for $n = 131$, the corresponding quantiles are .980 and .983.

In Figure 9.19, we have plotted $(\Phi^{-1}((i - .5)/n), \hat{W}_{(i)})$, $i = 1, \ldots, n$ for the three sets of residuals obtained in Examples 9.2.1–9.2.3. The respective R^2 values are .992, .984 and .990. Based on the graphs and the R^2 values, the hypothesis that the residuals $\{\hat{W}_t\}$, and hence $\{Z_t\}$, are normally distributed is not rejected, even at level .10.

§9.5 Forecasting ARIMA Models

In this section we demonstrate how the methods of Section 5.3 can be adapted to forecast the future values of an ARIMA(p, d, q) process $\{X_t\}$. (The required numerical calculations can be carried out using the program PEST.) If $d \geq 1$ the first and second moments EX_t and $E(X_{t+h}X_t)$ are not determined by the difference equations (9.1.1). We cannot expect therefore to determine best linear predictors for $\{X_t\}$ without further assumptions.

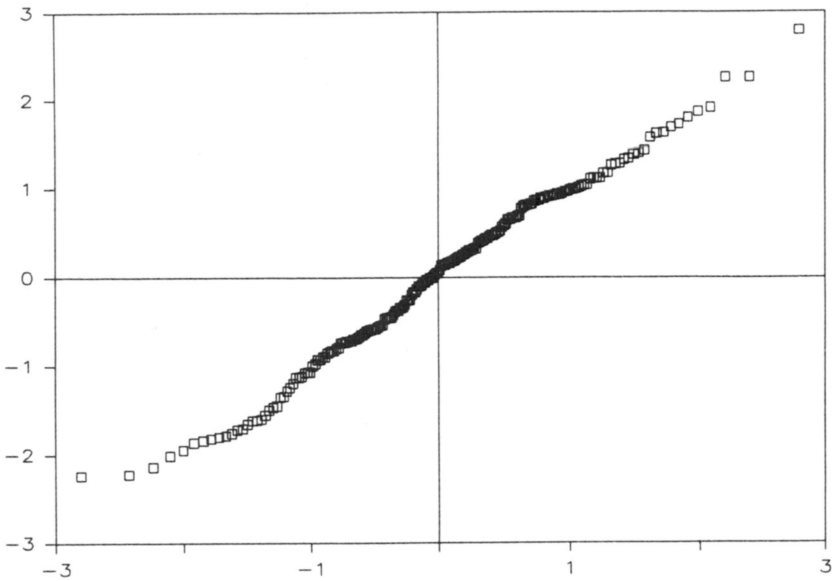

(a)

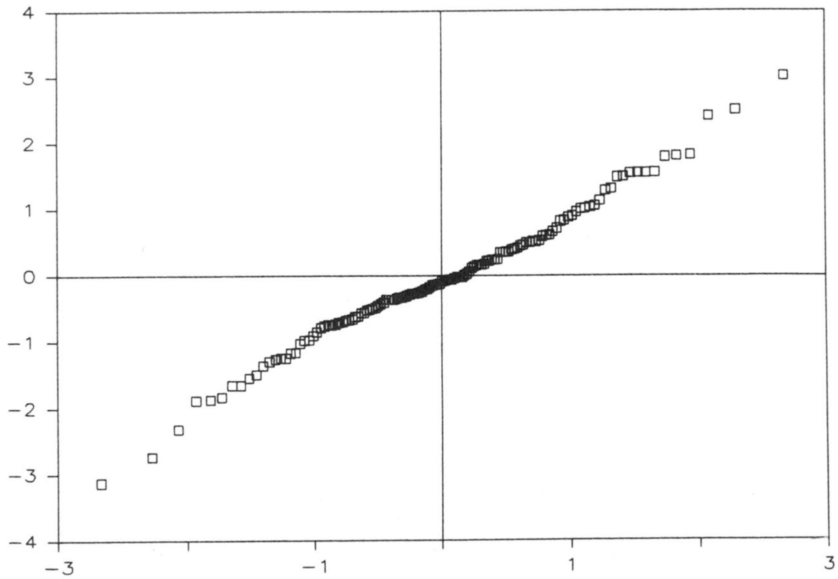

(b)

Figure 9.19. Scatter plots of the points $(\Phi^{-1}((i-.5)/n), \hat{W}_{(i)})$, $i = 1, \ldots, n$, for (a) Example 9.2.1, (b) Example 9.2.2 and (c) Example 9.2.3.

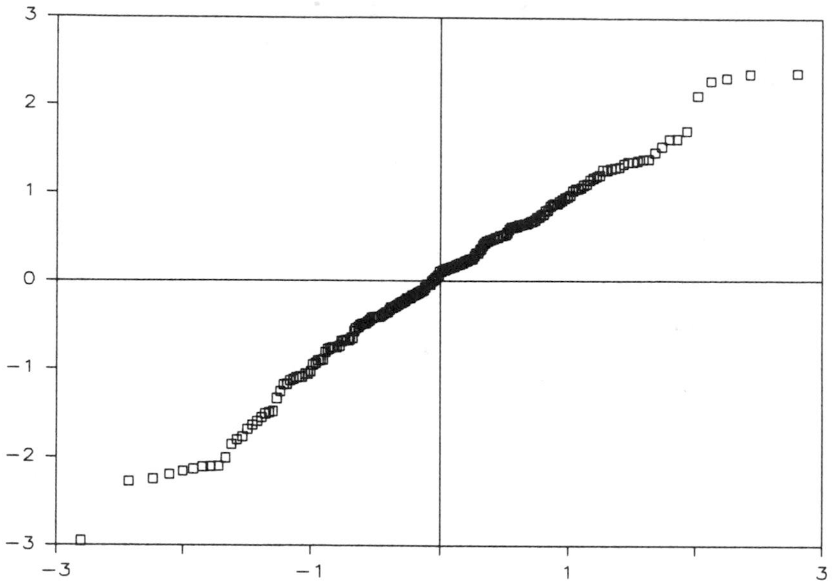

(c)

Figure 9.19. (continued)

For example, suppose that $\{Y_t\}$ is a causal ARMA(p, q) process and that X_0 is any random variable. Define

$$X_t = X_0 + \sum_{j=1}^{t} Y_j, \qquad t = 1, 2, \ldots.$$

Then $\{X_t, t \geq 0\}$ is an ARIMA$(p, 1, q)$ process with mean $EX_t = EX_0$ and autocovariances $E(X_{t+h}X_t) - (EX_0)^2$ depending on $\mathrm{Var}(X_0)$ and $\mathrm{Cov}(X_0, Y_j)$, $j = 1, 2, \ldots$. The best linear predictor of X_{n+1} based on $X_0, X_1, \ldots, X_n$ is the projection $P_{S_n}X_{n+1}$ where

$$S_n = \overline{\mathrm{sp}}\{X_0, X_1, \ldots, X_n\} = \overline{\mathrm{sp}}\{X_0, Y_1, \ldots, Y_n\}.$$

Thus

$$P_{S_n}X_{n+1} = P_{S_n}(X_0 + Y_1 + \cdots + Y_{n+1}) = X_n + P_{S_n}Y_{n+1}.$$

To evaluate this projection it is necessary in general to know $E(X_0 Y_j)$, $j = 1, \ldots, n + 1$, and EX_0^2. However if we assume that X_0 is uncorrelated with Y_j, $j = 1, 2, \ldots$, then $P_{S_n}Y_{n+1}$ is simply the projection of Y_{n+1} onto $\overline{\mathrm{sp}}\{Y_1, \ldots, Y_n\}$ which can be computed as described in Section 5.3. The assumption that X_0 is uncorrelated with $Y_1, Y_2, \ldots$, therefore suffices to determine the best linear predictor $P_{S_n}X_{n+1}$ in this case.

Turning now to the general case, we shall assume that our observed process $\{X_t\}$ satisfies the difference equations,

$$(1 - B)^d X_t = Y_t, \qquad t = 1, 2, \ldots,$$

where $\{Y_t\}$ is a causal ARMA(p, q) process, and that the vector $(X_{1-d}, \ldots, X_0)$ is uncorrelated with Y_t, $t > 0$. The difference equations can be rewritten in the form

$$X_t = Y_t - \sum_{j=1}^{d} \binom{d}{j} (-1)^j X_{t-j}, \qquad t = 1, 2, \ldots. \tag{9.5.1}$$

It is convenient, by relabelling the time axis if necessary, to assume that we observe $X_{1-d}, X_{2-d}, \ldots, X_n$. (The observed values of $\{Y_t\}$ are then $Y_1, \ldots, Y_n$.)

Our goal is to compute the best linear predictor of X_{n+h} based on $X_{1-d}, \ldots, X_n$, i.e.

$$P_{S_n} X_{n+h} := P_{\overline{\text{sp}}\{X_{1-d}, \ldots, X_n\}} X_{n+h}.$$

In the notation of Section 5.2 we shall write

$$P_n Y_{n+h} = P_{\overline{\text{sp}}\{Y_1, \ldots, Y_n\}} Y_{n+h}$$

and

$$\hat{Y}_{n+1} = P_n Y_{n+1}.$$

Since

$$S_n = \overline{\text{sp}}\{X_{1-d}, \ldots, X_0, Y_1, \ldots, Y_n\},$$

and since by assumption,

$$\overline{\text{sp}}\{X_{1-d}, \ldots, X_0\} \perp \overline{\text{sp}}\{Y_1, \ldots, Y_n\},$$

we have

$$P_{S_n} Y_{n+h} = P_{S_0} Y_{n+h} + P_n Y_{n+h} = P_n Y_{n+h}. \tag{9.5.2}$$

Hence if we apply the operator P_{S_n} to both sides of (9.5.1) with $t = n + h$, we obtain

$$P_{S_n} X_{n+h} = P_n Y_{n+h} - \sum_{j=1}^{d} \binom{d}{j} (-1)^j P_{S_n} X_{n+h-j}. \tag{9.5.3}$$

Since the predictors $P_n Y_{n+1}, P_n Y_{n+2}, \ldots$, can be found from (5.3.16), the predictors $P_{S_n} X_{n+1}, P_{S_n} X_{n+2}, \ldots$, are then easily computed recursively from (9.5.3).

In order to find the mean squared error of prediction it is convenient to express $P_n Y_{n+h}$ in terms of $\{X_j\}$. For $t \geq 0$ define

$$X_{t+1}^* = P_{S_t} X_{t+1}.$$

Then from (9.5.1) and (9.5.3) with $n = t$ we have

$$X_{t+1} - X_{t+1}^* = Y_{t+1} - \hat{Y}_{t+1}, \qquad t \geq 0,$$

and consequently for $n > m = \max(p, q)$ and $h \geq 1$,

$$P_n Y_{n+h} = \sum_{i=1}^{p} \phi_i P_n Y_{n+h-i} + \sum_{j=h}^{q} \theta_{n+h-1,j} (X_{n+h-j} - X_{n+h-j}^*). \tag{9.5.4}$$

Setting $\phi^*(z) = (1 - z)^d \phi(z) = 1 - \phi_1^* z - \cdots - \phi_{p+d}^* z^{p+d}$, we find from (9.5.2), (9.5.3) and (9.5.4) that for $n > m$ and $h \geq 1$,

$$P_{S_n} X_{n+h} = \sum_{j=1}^{p+d} \phi_j^* P_{S_n} X_{n+h-j} + \sum_{j=h}^{q} \theta_{n+h-1,j}(X_{n+h-j} - X_{n+h-j}^*), \quad (9.5.5)$$

which is analogous to the h-step prediction formula (5.3.16) for an ARMA process. The same argument which led to (5.3.22) shows that the mean squared error of the h-step predictor is (Problem 9.9)

$$\sigma_n^2(h) = E(X_{n+h} - P_{S_n} X_{n+h})^2 = \sum_{j=0}^{h-1} \left(\sum_{r=0}^{j} \chi_r \theta_{n+h-r-1,j-r} \right)^2 v_{n+h-j-1}, \quad (9.5.6)$$

where $\theta_{n0} = 1$,

$$\chi(z) = \sum_{r=0}^{\infty} \chi_r z^r = (1 - \phi_1^* z - \cdots - \phi_{p+d}^* z^{p+d})^{-1}, \qquad |z| < 1,$$

and

$$v_{n+h-j-1} = E(X_{n+h-j} - X_{n+h-j}^*)^2 = E(Y_{n+h-j} - \hat{Y}_{n+h-j})^2.$$

The coefficients χ_j can be found from the recursions (5.3.21) with ϕ_j^* replacing ϕ_j. For large n we can approximate (9.5.6), provided $\theta(\cdot)$ is invertible, by

$$\sigma_n^2(h) = \sum_{j=0}^{h-1} \psi_j^2 \sigma^2, \quad (9.5.7)$$

where

$$\psi(z) = \sum_{j=0}^{\infty} \psi_j z^j = (\phi^*(z))^{-1} \theta(z), \qquad |z| < 1.$$

EXAMPLE 9.5.1. Consider the ARIMA(1, 2, 1) model,

$$(1 - \phi B)(1 - B)^2 X_t = (1 + \theta B) Z_t, \qquad t = 1, 2, \ldots,$$

where (X_{-1}, X_0) is assumed to be uncorrelated with the ARMA(1, 1) process, $Y_t = (1 - B)^2 X_t$, $t = 1, 2, \ldots$. From (5.3.12) we have

$$P_n Y_{n+1} = \phi Y_n + \theta_{n1}(Y_n - \hat{Y}_n)$$

and

$$P_n Y_{n+h} = \phi P_n Y_{n+h-1} = \phi^{h-1} P_n Y_{n+1} \quad \text{for } h > 1.$$

Since in this case $\phi^*(z) = (1 - z)^2(1 - \phi z) = 1 - (\phi + 2)z + (2\phi + 1)z^2 - \phi z^3$, we find from (9.5.5) that

$$\begin{cases} P_{S_n} X_{n+1} = (\phi + 2) X_n - (2\phi + 1) X_{n-1} + \phi X_{n-2} + \theta_{n1}(Y_n - \hat{Y}_n), \\ P_{S_n} X_{n+h} = (\phi + 2) P_{S_n} X_{n+h-1} - (2\phi + 1) P_{S_n} X_{n+h-2} + \phi P_{S_n} X_{n+h-3} \\ \qquad \text{for } h > 1. \end{cases} \quad (9.5.8)$$

If for the moment we regard n as fixed and define the sequence $\{g(h)\}$ by

$$g(h) = P_{S_n} X_{n+h},$$

then $\{g(h)\}$ satisfies the difference equations

$$\phi^*(B)g(h) \equiv g(h) - (\phi + 2)g(h-1) + (2\phi + 1)g(h-2) - \phi g(h-3) = 0, \tag{9.5.9}$$

$h > 1$, with initial conditions,

$$\begin{cases} g(1) = P_{S_n} X_{n+1}, \\ g(0) = X_n, \\ g(-1) = X_{n-1}. \end{cases} \tag{9.5.10}$$

Using the results of Section 3.6, we can write the solution of the difference equation (9.5.9) in the form

$$g(h) = a_0 + a_1 h + a_2 \phi^h,$$

where a_0, a_1 and a_2 are determined by the initial conditions (9.5.10).

Table 9.10 shows the results of predicting the values X_{199}, X_{200} and X_{201} of an ARIMA(1, 2, 1) process with $\phi = .9$, $\theta = .8$ and $\sigma^2 = 1$, based on 200 observations $\{X_{-1}, X_0, \dots, X_{198}\}$. By running the program PEST to compute the likelihood of the observations $Y_t = (1 - B)^2 X_t$, $t = 1, \dots, 198$, under the model,

$$(1 - .9B)Y_t = (1 + .8B)Z_t, \qquad \{Z_t\} \sim \text{WN}(0, 1),$$

we find that $Y_{198} - \hat{Y}_{198} = -1.953$, $\theta_{197,1} = .800$ and $v_{197} = 1.000$. Since $\theta_{197,1} = \lim_{n\to\infty} \theta_{n,1}$ and $v_{197} = \lim_{n\to\infty} v_n$ to three decimal places, we use the large-sample approximation (9.5.7) to compute $\sigma_{198}^2(h)$. Thus

$$\sigma_{198}^2(h) = \sum_{j=0}^{h-1} \psi_j^2 \sigma^2 = \sum_{j=0}^{h-1} \psi_j^2,$$

where

$$\begin{aligned} \psi(z) &= \theta(z)/\phi^*(z) \\ &= (1 + .8z)(1 - 2.9z + 2.8z^2 - .9z^3)^{-1} \\ &= 1 + 3.7z + 7.93z^2 + 13.537z^3 + \cdots, \qquad |z| < 1. \end{aligned}$$

Since $X_{196} = -22195.57$, $X_{197} = -22335.07$, $X_{198} = -22474.41$ and

$$X_{198} - X_{198}^* = Y_{198} - \hat{Y}_{198} = -1.95,$$

equation (9.5.8) gives,

$$\begin{aligned} P_{S_{198}} X_{199} &= 2.9X_{198} - 2.8X_{197} + .9X_{196} + .8(X_{198} - X_{198}^*) \\ &= -22615.17. \end{aligned}$$

Table 9.10. Predicted Values Based on 200 Observations $\{X_{-1}, X_0, \ldots, X_{198}\}$ of the ARIMA(1, 2, 1) Process in Example 9.5.1 (the Standard Deviation of the Prediction Error Is Also Shown)

h	-1	0	1	2	3
$P_{S_{198}} X_{198+h}$	-22335.07	-22474.41	-22615.17	-22757.21	-22900.41
$\sigma_{198}(h)$	0	0	1	3.83	8.81

These predicted values and their mean squared errors can be found from PEST. The coefficients a_0, a_1 and a_2 in the function,

$$g(h) = P_{S_{198}} X_{198+h} = a_0 + a_1 h + a_2 (.9)^h, \qquad h \geq -1,$$

can now be determined from the initial conditions (9.5.10) with $n = 198$. These give $g(h) = -22346.61 - 153.54h - 127.8(.9)^h$. Predicted values $P_{S_{198}} X_{198+h}$ for any positive h can be computed directly from $g(h)$.

More generally, for an arbitrary ARIMA(p, d, q) process, the function defined by

$$g(h) = P_{S_n} X_{n+h}$$

satisfies the $(p + d)^{\text{th}}$-order difference equation,

$$\phi^*(B) g(h) = 0 \quad \text{for } h > q,$$

with initial conditions

$$g(h) = P_{S_n} X_{n+h}, \qquad h = q, q - 1, \ldots, q + 1 - p - d.$$

The solution $g(h)$ can be expressed for $d \geq 1$ as a polynomial of degree $(d - 1)$ plus a linear combination of geometrically decreasing terms corresponding to the reciprocals of the roots of $\phi(z) = 0$ (see Section 3.6). The presence of the polynomial term for $d \geq 1$ distinguishes the forecasts of an ARIMA process from those of a stationary ARMA process.

§9.6 Seasonal ARIMA Models

Seasonal series are characterized by a strong serial correlation at the seasonal lag (and possibly multiples thereof). For example, the correlation function in Figure 9.4 strongly suggests a seasonal series with six seasons. In Section 1.4, we discussed the classical decomposition of the time series $X_t = m_t + s_t + Y_t$ where m_t is the trend component, s_t is the seasonal component, and Y_t is the random noise component. However in practice it may not be reasonable to assume that the seasonality component repeats itself precisely in the same way cycle after cycle. Seasonal ARIMA models allow for randomness in the seasonal pattern from one cycle to the next.

Suppose we have r years of monthly data which we tabulate as follows:

	Month			
Year	1	2	···	12
1	X_1	X_2	···	X_{12}
2	X_{13}	X_{14}	···	X_{24}
3	X_{25}	X_{26}	···	X_{36}
⋮	⋮			⋮
r	$X_{1+12(r-1)}$	$X_{2+12(r-1)}$	···	$X_{12+12(r-1)}$

Each column in this table may itself be viewed as a realization of a time series. Suppose that each one of these twelve time series is generated by the same ARMA(P, Q) model, or more specifically that the series corresponding to the jth month, X_{j+12t}, $t = 0, \ldots, r-1$, satisfies a difference equation of the form,

$$\begin{aligned} X_{j+12t} = \Phi_1 X_{j+12(t-1)} + \cdots + \Phi_P X_{j+12(t-P)} + U_{j+12t} \\ + \Theta_1 U_{j+12(t-1)} + \cdots + \Theta_Q U_{j+12(t-Q)}, \end{aligned} \tag{9.6.1}$$

where

$$\{U_{j+12t}, t = \ldots, -1, 0, 1, \ldots\} \sim \text{WN}(0, \sigma_U^2). \tag{9.6.2}$$

Then since the same ARMA(P, Q) model is assumed to apply to each month, (9.6.1) can be rewritten for all t as

$$X_t = \Phi_1 X_{t-12} + \cdots + \Phi_P X_{t-12P} + U_t + \Theta_1 U_{t-12} + \cdots + \Theta_Q U_{t-12Q},$$

where (9.6.2) holds for each $j = 1, \ldots, 12$. (Notice however that $E(U_t U_{t+h})$ is not necessarily zero except when h is an integer multiple of 12.) We can thus write (9.6.1) in the compact form,

$$\Phi(B^{12}) X_t = \Theta(B^{12}) U_t, \tag{9.6.3}$$

where $\Phi(z) = 1 - \Phi_1 z - \cdots - \Phi_P z^P$, $\Theta(z) = 1 + \Theta_1 z + \cdots + \Theta_Q z^Q$, and $\{U_{j+12t}, t = \ldots, -1, 0, 1, \ldots\} \sim \text{WN}(0, \sigma_U^2)$ for each j. We refer to the model (9.6.3) as the between-year model.

EXAMPLE 9.6.1. Suppose $P = 0$, $Q = 1$ and $\Theta_1 = -.4$. Then the series of observations for any particular month is a moving average of order 1. If $E(U_t U_{t+h}) = 0$ for all h, i.e. if the white noise sequences for different months are uncorrelated with each other, then the columns themselves are uncorrelated. The correlation function for such a process is displayed in Figure 9.20.

EXAMPLE 9.6.2. Suppose $P = 1$, $Q = 0$ and $\Phi_1 = .7$. In this case the 12 series (one for each month) are AR(1) processes which are uncorrelated if the white noise sequences for different months are uncorrelated. A graph of the correlation function of this process is shown in Figure 9.20.

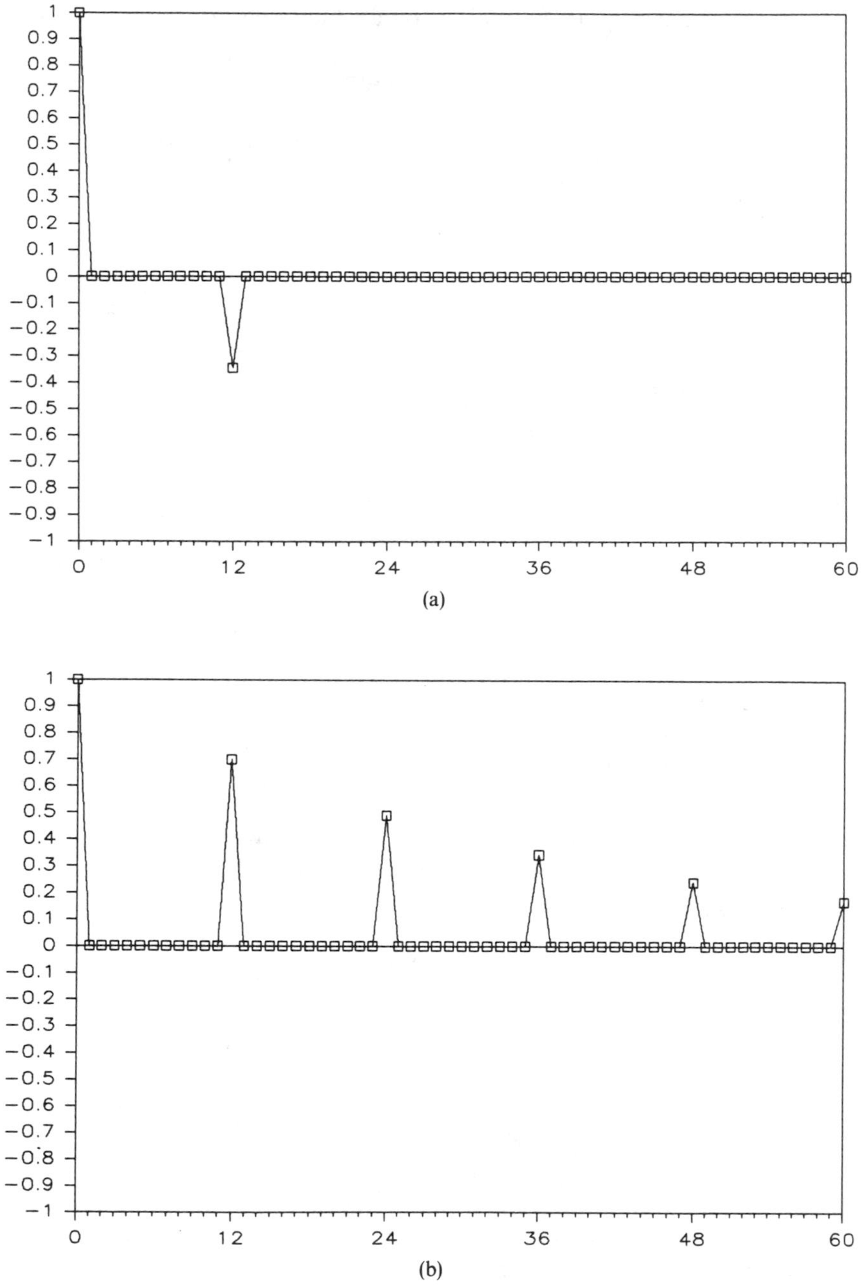

Figure 9.20. The autocorrelation functions of $\{X_t\}$ when (a) $X_t = U_t - .4U_{t-12}$ and (b) $X_t - .7X_{t-12} = U_t$ (see Examples 9.6.1 and 9.6.2).

It is unlikely that the 12 series corresponding to the different months are uncorrelated as in Examples 9.6.1 and 9.6.2. To incorporate dependence between these series we assume now that the $\{U_t\}$ sequence in (9.6.3) follows an ARMA(p, q) model,

$$\phi(B)U_t = \theta(B)Z_t, \qquad \{Z_t\} \sim \text{WN}(0, \sigma^2). \tag{9.6.4}$$

This assumption not only implies possible non-zero correlation between consecutive values of U_t, but also within the twelve sequences $\{U_{j+12t}, t = \ldots, -1, 0, 1, \ldots\}$, each of which was previously assumed to be uncorrelated. In this case (9.6.2) may no longer hold, however the coefficients in (9.6.4) will frequently have values such that $E(U_t U_{t+12j})$ is small for $j = \pm 1, \pm 2, \ldots$. Combining the two models (9.6.3) and (9.6.4), and allowing for differencing leads us to the definition of the general seasonal multiplicative ARIMA process.

Definition 9.6.1 (The SARIMA$(p, d, q) \times (P, D, Q)_s$ Process). If d and D are non-negative integers, then $\{X_t\}$ is said to be a seasonal ARIMA$(p, d, q) \times (P, D, Q)_s$ process with period s if the differenced process $Y_t := (1 - B)^d (1 - B^s)^D X_t$ is a causal ARMA process,

$$\phi(B)\Phi(B^s)Y_t = \theta(B)\Theta(B^s)Z_t, \qquad \{Z_t\} \sim \text{WN}(0, \sigma^2),$$

where $\phi(z) = 1 - \phi_1 z - \cdots - \phi_p z^p$, $\Phi(z) = 1 - \Phi_1 z - \cdots - \Phi_P z^P$, $\theta(z) = 1 + \theta_1 z + \cdots + \theta_z z^q$ and $\Theta(z) = 1 + \Theta_1 z + \cdots \Theta_Q z^Q$.

Note that the process $\{Y_t\}$ is causal if and only if $\phi(z) \neq 0$ and $\Phi(z) \neq 0$ for $|z| \leq 1$. In applications, D is rarely more than one and P and Q are typically less than three.

Because of the interaction between the two models describing the between-year and the between-season dependence structure, the covariance function for a SARIMA process can be quite complicated. Here we provide general guidelines for identifying SARIMA models from the sample correlation function of the data. First, we find d and D so as to make the differenced observations

$$Y_t = (1 - B)^d (1 - B^s)^D X_t$$

stationary in appearance (see Section 9.2). Next, we examine the sample autocorrelation and partial autocorrelation functions of $\{Y_t\}$ at lags which are multiples of s in order to identify the orders P and Q in the model (9.6.3). If $\hat{\rho}(\cdot)$ is the sample autocorrelation function of $\{Y_t\}$ then P and Q should be chosen so that $\hat{\rho}(ks)$, $k = 1, 2, \ldots$, is compatible with the autocorrelation function of an ARMA(P, Q) process. The orders p and q are then selected by attempting to match $\hat{\rho}(1), \ldots, \hat{\rho}(s - 1)$ with the autocorrelation function of an ARMA(p, q) process. Ultimately the AICC criterion (Section 9.3) and the goodness of fit tests (Section 9.4) are used to identify the best SARIMA model among competing alternatives.

For given values of p, d, q, P, D and Q, the parameters $\boldsymbol{\phi}$, $\boldsymbol{\theta}$, $\boldsymbol{\Phi}$, $\boldsymbol{\Theta}$ and σ^2 can be found using the maximum likelihood procedure of Section 8.7. The differences $Y_t = (1 - B)^d(1 - B^s)^D X_t$ constitute an ARMA$(p + sP, q + sQ)$ process in which some of the coefficients are zero and the rest are functions of the $(p + P + q + Q)$-dimensional vector $\boldsymbol{\beta}' = (\boldsymbol{\phi}', \boldsymbol{\Phi}', \boldsymbol{\theta}', \boldsymbol{\Theta}')$. For any fixed $\boldsymbol{\beta}$ the reduced likelihood $l(\boldsymbol{\beta})$ of the differences $Y_{1+d+sD}, \ldots, Y_n$, is easily computed as described in Section 8.7. The maximum likelihood estimate of $\boldsymbol{\beta}$ is the value which minimizes $l(\boldsymbol{\beta})$ and the maximum likelihood estimate of σ^2 is given by (8.7.5). The estimates can be found using the program PEST by specifying the required multiplicative relationships between the coefficients.

The forecasting methods described in Section 9.5 for ARIMA processes can also be applied to seasonal models. We first predict future values of the ARMA process $\{Y_t\}$ using (5.3.16) and then expand the operator $(1 - B)^d(1 - B^s)^D$ to derive the analogue of equation (9.5.3) which determines the best predictors recursively. The large sample approximation to the h-step mean squared error for prediction of $\{X_t\}$ is $\sigma_n^2(h) = \sigma^2 \sum_{j=0}^{h-1} \psi_j^2$, where σ^2 is the white noise variance and $\sum_{j=0}^{\infty} \psi_j z^j = \theta(z)\Theta(z^s)/[\phi(z)\Phi(z^s)(1 - z)^d(1 - z^s)^D]$, $|z| < 1$. Invertibility is required for the validity of this approximation.

The goodness of fit of a SARIMA model can be assessed by applying the same techniques and tests described in Section 9.4 to the residuals of the fitted model. In the following example we fit a SARIMA model to the series $\{X_t\}$ of monthly accidental deaths in the U.S.A. (Example 1.1.6).

EXAMPLE 9.6.3. The accidental death series $X_1, \ldots, X_{72}$ is plotted in Figure 1.6. Application of the operator $(1 - B)(1 - B^{12})$ generates a new series $\{Y_t\}$ with no apparent deviations from stationarity as seen in Figure 1.17. The sample autocorrelation function $\hat{\rho}(\cdot)$ of $\{Y_t\}$ is displayed in Figure 9.21. The values $\hat{\rho}(12) = -.333$, $\hat{\rho}(24) = -.099$ and $\hat{\rho}(36) = .013$ suggest a moving average of order 1 for the between-year model (i.e. $P = 0$, $Q = 1$). Moreover inspection of $\hat{\rho}(1), \ldots, \hat{\rho}(11)$ suggests that $\rho(1)$ is the only short-term correlation different from zero, so we also choose a moving average of order 1 for the between-month model (i.e. $p = 0$, $q = 1$). Taking into account the sample mean (28.831) of the differences $Y_t = (1 - B)(1 - B^{12})X_t$, we therefore arrive at the model,

$$Y_t = 28.831 + (1 + \theta_1 B)(1 + \Theta_1 B^{12})Z_t, \qquad \{Z_t\} \sim \text{WN}(0, \sigma^2),$$

for the series $\{Y_t\}$. The maximum likelihood estimates of the parameters are,

$$\hat{\theta}_1 = -.479,$$

$$\hat{\Theta}_1 = -.591,$$

and

$$\hat{\sigma}^2 = 94240,$$

with AICC value 855.53. The fitted model for $\{X_t\}$ is thus the

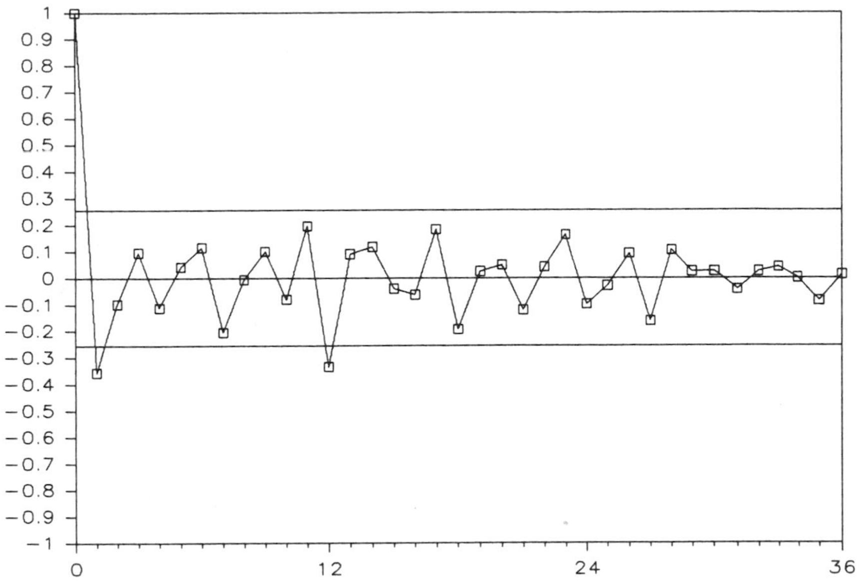

Figure 9.21. The sample ACF of the differenced accidental deaths $\{\nabla\nabla_{12}X_t\}$.

$\hat{\rho}(1) - \hat{\rho}(12)$:	−.36	−.10	.10	−.11	.04	.11	−.20	−.01	.10	−.08	.20	−.33
$\hat{\rho}(13) - \hat{\rho}(24)$:	.09	.12	−.04	−.06	.18	−.19	.02	.05	−.12	.04	.16	−.10
$\hat{\rho}(25) - \hat{\rho}(36)$:	−.03	.09	−.16	.11	.02	.03	−.04	.03	.04	.00	−.09	.01

SARIMA$(0, 1, 1) \times (0, 1, 1)_{12}$ process

$$(1 - B)(1 - B^{12})X_t = Y_t = 28.831 + (1 - .479B)(1 - .591B^{12})Z_t, \qquad (9.6.5)$$

where $\{Z_t\} \sim \text{WN}(0, 94240)$.

Notice that the model (9.6.5) has a slightly more general form than in Definition 9.6.1 owing to the presence of the constant term, 28.831. Predicted values and their mean squared errors can however still be computed as described in Section 9.5 with minor modifications. Thus predicted values of $\{Y_t\}$ are obtained by adding 28.831 to the corresponding predicted values of the ARMA process $\{Y_t - 28.831\}$. From (9.6.5) it is easy to write out the analogue of (9.5.1), from which the predicted values of $\{X_t\}$ are then found recursively as in Section 9.5. The mean-squared errors are given as before by (9.5.6), i.e. by ignoring the constant term in (9.6.5). Thus for large n the mean-squared h-step prediction error is approximately (from (9.5.7)),

$$\sigma_n^2(h) = \sigma^2 \sum_{j=0}^{h-1} \psi_j^2, \quad \text{where } \sigma^2 = 94240,$$

and $\sum_{j=0}^{\infty} \psi_j z^j = (1 - .479z)(1 - .591z^{12})(1 - z)^{-1}(1 - z^{12})^{-1}, |z| < 1$.

Table 9.11. Predicted Values of the Accidental Deaths Series for $t = 73, \ldots, 78$, the Standard Deviations σ_t of the Prediction Errors, and the Observed Values X_t

t	73	74	75	76	77	78
Model (9.6.5)						
Predictors	8441	7706	8550	8886	9844	10281
σ_t	307	346	381	414	443	471
Model (9.6.6)						
Predictors	8347	7620	8358	8743	9796	10180
σ_t	287	322	358	394	432	475
Observed values						
X_t	7798	7406	8363	8460	9217	9316

Instead of fitting a SARIMA model to the series $\{X_t\}$, we could look for the best-fitting moving average model for $\{\nabla\nabla_{12}X_t\}$ as we did in Example 9.2.2. This procedure leads to the model

$$\begin{aligned}\nabla\nabla_{12}X_t = Y_t = 28.831 + Z_t - .596Z_{t-1} - .405Z_{t-6} \\ - .685Z_{t-12} + .458Z_{t-13},\end{aligned} \tag{9.6.6}$$

where $\{Z_t\} \sim \text{WN}(0, 71370)$. The residuals for the models (9.6.5) and (9.6.6) both pass the goodness of fit tests in Section 9.4. The AICC value for (9.6.6) is 855.61, virtually the same as for (9.6.5).

The program PEST can be used to compute the best h-step linear predictors and their mean squared errors for any ARIMA (or SARIMA) process. The asymptotic form (9.5.7) of the mean squared error (with σ^2 replaced by v_n) is used if the model is invertible. If not then PEST computes the mean squared errors by converting the model to an invertible one. In Table 9.11 we show the predictors of the accidental deaths for the first six months of 1979 together with the standard deviations of the prediction errors and the observed numbers of accidental deaths for the same period. Both of the models (9.6.5) and (9.6.6) are illustrated in the table. The second of these is not invertible.

Problems

9.1. Suppose that $\{X_t\}$ is an ARIMA(p, d, q) process, satisfying the difference equations,

$$\phi(B)(1 - B)^d X_t = \theta(B)Z_t, \qquad \{Z_t\} \sim \text{WN}(0, \sigma^2).$$

Show that these difference equations are also satisfied by the process $W_t = X_t + A_0 + A_1 t + \cdots + A_{d-1}t^{d-1}$, where $A_0, \ldots, A_{d-1}$ are arbitrary random variables.

9.2. The model fitted to a data set $x_1, \ldots, x_{100}$ is

$$X_t + .4X_{t-1} = Z_t, \qquad \{Z_t\} \sim \text{WN}(0, 1).$$

The sample acf and pacf of the residuals are shown in the accompanying table. Are these values compatible with whiteness of the residuals? If not, suggest a better model for $\{X_t\}$, giving estimates of the coefficients.

Lag	1	2	3	4	5	6	7	8	9	10	11	12
ACF	.799	.412	.025	−.228	−.316	−.287	−.198	−.111	−.056	−.009	.048	.133
PACF	.799	−.625	−.044	.038	−.020	−.077	−.007	−.061	−.042	.089	.052	.125

9.3. Suppose $\{X_t\}$ is an MA(2) process, $X_t = Z_t + \theta_1 Z_{t-1} + \theta_2 Z_{t-2}$, $\{Z_t\} \sim \text{WN}(0, \sigma^2)$. If the AR(1) process, $(1 - \phi B)X_t = Y_t$, is mistakenly fitted to $\{X_t\}$, determine the autocovariance function of $\{Y_t\}$.

9.4. The following table shows the sample acf and pacf of the series,

$$Y_t = \nabla X_t,\ t = 1, \ldots, 399, \quad \text{with} \quad \textstyle\sum_{t=1}^{399} Y_t = 0 \quad \text{and} \quad \hat{\gamma}_Y(0) = 8.25.$$

(a) Specify a suitable ARMA model for $\{Y_t\}$, giving estimates of all the parameters. Explain your choice of model.

(b) Given that $X_{395} = 102.6$, $X_{396} = 105.3$, $X_{397} = 108.2$, $X_{398} = 110.5$ and $X_{399} = 113.9$, use your model to find the best mean square estimates of X_{400} and X_{401} and estimate the mean squared errors of your predictors.

Lag	1	2	3	4	5	6	7	8	9	10
ACF	.808	.654	.538	.418	.298	.210	.115	.031	.007	−.010
PACF	.808	.006	.023	−.068	−.080	.003	−.083	−.045	.091	.003
Lag	11	12	13	14	15	16	17	18	19	20
ACF	−.031	−.069	−.096	−.111	−.126	−.115	−.116	−.116	−.105	−.083
PACF	−.016	−.091	−.034	.001	−.034	.051	−.031	−.005	.008	.001

9.5. Consider the process $Y_t = \alpha t + \beta + Z_t$, $t = 0, 1, 2, \ldots$, $\{Z_t\} \sim \text{IID}(0, \sigma^2)$, where α and β are known. Observed values of $Y_0, \ldots, Y_n$ are available. Let $W_t = Y_t - Y_{t-1} - \alpha$, $t = 1, 2, \ldots$.

(a) Find the mean squared error of the best linear predictor $\hat{W}_{n+1} = P_{\overline{\text{sp}}\{W_1, \ldots, W_n\}} W_{n+1} = \theta_{n1}(W_n - \hat{W}_n)$ (use Problem 5.13).

(b) Find the mean squared error of the predictor of Y_{n+h} given by $Y_n + \hat{W}_{n+1} + h\alpha$, $h = 1, 2, \ldots$.

(c) Compare the mean squared error computed in (b) with that of the best predictor $E(Y_{n+h} | Y_0, \ldots, Y_n)$.

(d)* Compute the mean squared error of the predictor in (c) when α and β are replaced by the least squares estimators $\hat{\alpha}$ and $\hat{\beta}$ found from $Y_0, \ldots, Y_n$.

9.6. Series A (Appendix A) consists of the lake levels in feet (reduced by 570) of Lake Huron for July of each year from 1875 through 1972. In the class of ARIMA models, choose the model which you believe fits the data best. Your analysis should include:

(i) a logical explanation of the steps taken to find the chosen model,

(ii) approximate 95% confidence bounds for the components of $\boldsymbol{\phi}$ and $\boldsymbol{\theta}$,

(iii) an examination of the residuals to check for whiteness as described in Section 9.4.

9.7. The following observations are the values $X_0, \ldots, X_9$ of an ARIMA(0, 1, 2) process, $\nabla X_t = Z_t - 1.1Z_{t-1} + .28Z_{t-2}$, $\{Z_t\} \sim \text{WN}(0, 1)$: 2.83, 2.16, .85, −1.04, .35, −.90, .10, −.89, −1.57, −.42.

(a) Find an explicit formula for the function $g(h) = P_{S_9}X_{9+h}$, $h \geq 0$.

(b) Compute $\sigma_9^2(h)$ for $h = 1, \ldots, 5$.

9.8. Let $\{X_t\}$ be the ARIMA(2, 1, 0) process,

$$(1 - .8B + .25B^2)\nabla X_t = Z_t, \qquad \{Z_t\} \sim \text{WN}(0, 1).$$

Determine the function $g(h) = P_{S_n}X_{n+h}$ for $h > 0$. Assuming that n is large, compute $\sigma_n^2(h)$ for $h = 1, \ldots, 5$.

9.9. Verify equation (9.5.6).

9.10. Let $\{X_t\}$ be the seasonal process

$$(1 - .7B^2)X_t = (1 + .3B^2)Z_t, \qquad \{Z_t\} \sim \text{WN}(0, 1).$$

(a) Find the coefficients $\{\psi_j\}$ in the representation $X_t = \sum_{j=0}^{\infty} \psi_j Z_{t-j}$.

(b) Find the coefficients $\{\pi_j\}$ in the representation $Z_t = \sum_{j=0}^{\infty} \pi_j X_{t-j}$.

(c) Graph the autocorrelation function of $\{X_t\}$.

(d) Find an expression for $P_{10}X_{11}$ and $P_{10}X_{12}$ in terms of $X_1, \ldots, X_{10}$ and the innovations $X_t - \hat{X}_t$, $t = 1, \ldots, 10$.

(e) Find an explicit expression for $g(h) = P_{10}X_{10+h}$, $h \geq 1$, in terms of $g(1)$ and $g(2)$.

9.11. Let $\{X_t\}$ be the seasonal process,

$$X_t = (1 + .2B)(1 - .8B^{12})Z_t, \qquad \{Z_t\} \sim \text{WN}(0, \sigma^2).$$

(a) Determine the coefficients $\{\pi_j\}$ in the representation $Z_t = \sum_{j=0}^{\infty} \pi_j X_{t-j}$.

(b) Graph the autocorrelation function of $\{X_t\}$.

9.12. Monthly observations $\{D_t, -11 \leq t \leq n\}$ are deseasonalized by differencing at lag 12. The resulting differences $X_t = D_t - D_{t-12}$, $t = 1, \ldots, n$, are then found to be well fitted by the ARMA model,

$$X_t - 1.3X_{t-1} + .5X_{t-2} = Z_t + .5Z_{t-1}, \qquad \{Z_t\} \sim \text{WN}(0, 3.85).$$

Assume in the following questions that n is large and $\{D_t, -11 \leq t \leq 0\}$ is uncorrelated with $\{X_t, t \geq 1\}$; P_n denotes projection onto $\overline{\text{sp}}\{X_t, 1 \leq t \leq n\}$ and P_{S_n} denotes projection onto $\overline{\text{sp}}\{D_t, -11 \leq t \leq n\}$.

(a) Express P_nX_{n+1} and P_nX_{n+2} in terms of $X_1, \ldots, X_n$ and the innovations $(X_j - P_{j-1}X_j)$, $j = 1, \ldots, n$.

(b) Express $P_{S_n}D_{n+1}$ and $P_{S_n}D_{n+2}$ in terms of $\{D_t, -11 \le t \le n\}$, P_nX_{n+1} and P_nX_{n+2}.

(c) Find the mean squared errors of the predictors $P_{S_n}D_{n+1}$ and $P_{S_n}D_{n+2}$.

9.13. For each of the time series B–F in Appendix A find an ARIMA (or ARMA) model to represent the series obtained by deleting the last six observations. Explain and justify your choice of model in each case, giving approximate confidence bounds for estimated coefficients. Use each fitted model to obtain predicted values of the six observations deleted and the mean squared errors of the predictors. Compare the predicted and observed values. (Use PEST to carry out maximum likelihood estimation for each model and to generate the approximate variances of the estimators.)

CHAPTER 10

Inference for the Spectrum of a Stationary Process

In this chapter we consider problems of statistical inference for time series based on frequency-domain properties of the series. The fundamental tool used is the periodogram, which is defined in Section 10.1 for any time series $\{x_1, \ldots, x_n\}$. Section 10.2 deals with statistical tests for the presence of "hidden periodicities" in the data. Several tests are discussed, corresponding to various different models and hypotheses which we may wish to test. Spectral analysis for stationary time series, and in particular the estimation of the spectral density, depends very heavily on the asymptotic distribution as $n \to \infty$ of the periodogram ordinates of the series $\{X_1, \ldots, X_n\}$. The essential results are contained in Theorem 10.3.2. Under rather general conditions, the periodogram ordinates $I_n(\lambda_i)$ at any set of frequencies $\lambda_1, \ldots, \lambda_m$, $0 < \lambda_1 < \cdots < \lambda_m < \pi$, are asymptotically independent exponential random variables with means $2\pi f(\lambda_i)$, were f is the spectral density of $\{X_t\}$. Consequently the periodogram I_n is not a consistent estimator of $2\pi f$. Consistent estimators can however be constructed by applying linear smoothing filters to the periodogram. The asymptotic behaviour of the resulting discrete spectral average estimators can be derived from the asymptotic behaviour of the periodogram as shown in Section 10.4. Lag-window estimators of the form $(2\pi)^{-1} \sum_{|h| \le r} w(h/r)\hat{\gamma}(h)e^{-ih\omega}$, where $w(x)$, $-1 \le x \le 1$, is a suitably chosen weight function, are also discussed in Section 10.4 and compared with discrete spectral average estimators. Approximate confidence intervals for the spectral density are given in Section 10.5. An alternative approach to spectral density estimation, based on fitting an ARMA model to the data and computing the spectral density of the fitted process, is discussed in Section 10.6. An important role in the development of spectral analysis has been played by the fast Fourier transform algorithm, which makes possible the rapid calculation of the periodogram for very large data sets. An introduction to the algorithm and its application to the computation of autocovariances is given in Section 10.7. The chapter concludes with a discussion of the asymptotic behaviour of the maximum likelihood estimators of the coefficients of an ARMA(p, q) process.

§10.1 The Periodogram

Consider an arbitrary set of (possibly complex-valued) observations $x_1, \dots, x_n$ made at times $1, \dots, n$ respectively. The vector

$$\mathbf{x} := (x_1, \dots, x_n)',$$

belongs to the n-dimensional complex space $\mathbb{C}^n$. If $\mathbf{u}$ and $\mathbf{v}$ are two elements of $\mathbb{C}^n$, we define the inner product of $\mathbf{u}$ and $\mathbf{v}$ as in (2.1.2), i.e.

$$\langle \mathbf{u}, \mathbf{v} \rangle = \sum_{i=1}^{n} u_i \bar{v}_i. \tag{10.1.1}$$

By imagining the data $x_1, \dots, x_n$ to be the values at $1, \dots, n$ of a function with period n, we might expect (as is shown in Proposition 10.1.1 below) that each x_t can be expressed as a linear combination of harmonics,

$$x_t = n^{-1/2} \sum_{-\pi < \omega_j \le \pi} a_j e^{it\omega_j}, \qquad t = 1, \dots, n, \tag{10.1.2}$$

where the frequencies $\omega_j = 2\pi j/n$ are the integer multiples of the fundamental frequency $2\pi/n$ which fall in the interval $(-\pi, \pi]$. (Harmonics $e^{it\omega_j}$ with frequencies $2\pi j/n$ outside this interval cannot be distinguished on the basis of observations at integer times only.) The frequencies $\omega_j = 2\pi j/n$, $-\pi < \omega_j \le \pi$, are called the *Fourier frequencies* of the series $\{x_1, \dots, x_n\}$. The representation (10.1.2) can be rewritten in vector form as

$$\mathbf{x} = \sum_{j \in F_n} a_j \mathbf{e}_j, \tag{10.1.3}$$

where

$$\mathbf{e}_j = n^{-1/2}(e^{i\omega_j}, e^{i2\omega_j}, \dots, e^{in\omega_j})', \qquad j \in F_n, \tag{10.1.4}$$

$$F_n = \{ j \in \mathbb{Z} : -\pi < \omega_j \equiv 2\pi j/n \le \pi \} = \{ -[(n-1)/2], \dots, [n/2] \}, \tag{10.1.5}$$

and $[x]$ denotes the integer part of x. Notice that F_n contains n integers. The validity and uniqueness of the representation (10.1.3) and the values of the coefficients a_j are simple consequences of the following proposition.

Proposition 10.1.1. *The vectors* $\mathbf{e}_j, j \in F_n$, *defined by* (10.1.4) *constitute an orthonormal basis for* $\mathbb{C}^n$.

PROOF.
$$\langle \mathbf{e}_j, \mathbf{e}_k \rangle = n^{-1} \sum_{r=1}^{n} e^{ir(\omega_j - \omega_k)}$$
$$= \begin{cases} 1 & \text{if } j = k, \\ n^{-1} e^{i(\omega_j - \omega_k)} \dfrac{1 - e^{in(\omega_j - \omega_k)}}{1 - e^{i(\omega_j - \omega_k)}} = 0 & \text{if } j \neq k. \end{cases}$$ □

Corollary 10.1.1. *For any* $\mathbf{x} \in \mathbb{C}^n$,

$$\mathbf{x} = \sum_{j \in F_n} a_j \mathbf{e}_j, \tag{10.1.6}$$

where

$$a_j = \langle \mathbf{x}, \mathbf{e}_j \rangle = n^{-1/2} \sum_{t=1}^{n} x_t e^{-it\omega_j}. \tag{10.1.7}$$

PROOF. Take inner products of each side of (10.1.6) with $\mathbf{e}_j$, $j \in F_n$. □

Definition 10.1.1. The discrete Fourier transform of $\mathbf{x} \in \mathbb{C}^n$ is the sequence $\{a_j, j \in F_n\}$ defined by (10.1.7).

Definition 10.1.2 (The Periodogram of $\mathbf{x} \in \mathbb{C}^n$). The value $I(\omega_j)$ of the periodogram of $\mathbf{x}$ at frequency $\omega_j = 2\pi j/n$, $j \in F_n$, is defined in terms of the discrete Fourier transform $\{a_j\}$ of $\mathbf{x}$ by,

$$I(\omega_j) := |a_j|^2 = |\langle \mathbf{x}, \mathbf{e}_j \rangle|^2 = n^{-1} \left| \sum_{t=1}^{n} x_t e^{-it\omega_j} \right|^2. \tag{10.1.8}$$

Notice that the periodogram decomposes $\|\mathbf{x}\|^2$ into a sum of components $|\langle \mathbf{x}, \mathbf{e}_j \rangle|^2$ associated with the Fourier frequencies, ω_j, $j \in F_n$. Thus

$$\|\mathbf{x}\|^2 = \sum_{j \in F_n} I(\omega_j). \tag{10.1.9}$$

This decomposition can be neatly expressed as the "analysis of variance" shown in Table 10.1. ($[y]$ denotes the integer part of y.)

Table 10.1. Decomposition of $\|\mathbf{x}\|^2$ into Components Corresponding to the Harmonic Decomposition (10.1.6) of $\mathbf{x}$

Source	Degrees of freedom	Sum of squares
Frequency $\omega_{-[(n-1)/2]}$	1	$\lvert a_{-[(n-1)/2]}\rvert^2$
⋮	⋮	⋮
Frequency ω_0 (mean)	1	$\lvert a_0\rvert^2 = n^{-1}\lvert\sum_{t=1}^{n} x_t\rvert^2$
⋮	⋮	⋮
Frequency $\omega_{[n/2]}$	1	$\lvert a_{[n/2]}\rvert^2$
Total	n	$\|\mathbf{x}\|^2$

If $\mathbf{x} \in \mathbb{R}^n$ and if ω_j $(=2\pi j/n)$ and $-\omega_j$ are both in $(-\pi, \pi]$, it follows from (10.1.7) that $a_j = \bar{a}_{-j}$ and $I(\omega_j) = I(-\omega_j)$. We can therefore rewrite (10.1.6) in the form

$$\mathbf{x} = a_0\mathbf{e}_0 + \sum_{j=1}^{[(n-1)/2]} (a_j\mathbf{e}_j + \bar{a}_j\mathbf{e}_{-j}) + a_{n/2}\mathbf{e}_{n/2}, \tag{10.1.10}$$

where the last term is defined to be zero if n is odd. Writing a_j in its polar form, $a_j = r_j \exp(i\theta_j)$, we can reexpress (10.1.10) as

$$\mathbf{x} = a_0\mathbf{e}_0 + \sum_{j=1}^{[(n-1)/2]} 2^{1/2} r_j(\mathbf{c}_j \cos\theta_j - \mathbf{s}_j \sin\theta_j) + a_{n/2}\mathbf{e}_{n/2}, \tag{10.1.11}$$

where

$$\mathbf{c}_j = (2/n)^{1/2}(\cos\omega_j, \cos 2\omega_j, \ldots, \cos n\omega_j)'$$

and

$$\mathbf{s}_j = (2/n)^{1/2}(\sin\omega_j, \sin 2\omega_j, \ldots, \sin n\omega_j)'.$$

Now $\{\mathbf{e}_0, \mathbf{c}_1, \mathbf{s}_1, \ldots, \mathbf{c}_{[(n-1)/2]}, \mathbf{s}_{[(n-1)/2]}, \mathbf{e}_{n/2}\}$, with the last vector excluded if n is odd, is an orthonormal basis for $\mathbb{R}^n$. We can therefore decompose the sum of squares $\sum_{i=1}^n x_i^2$ into components corresponding to each vector in the set. For $1 \le j \le [(n-1)/2]$, the components corresponding to $\mathbf{c}_j$ and $\mathbf{s}_j$ are usually lumped together to produce a "frequency ω_j" component as in Table 10.2. This is just the squared length of the projection of $\mathbf{x}$ onto the two-dimensional subspace $\overline{\text{sp}}\{\mathbf{c}_j, \mathbf{s}_j\}$ of $\mathbb{R}^n$.

Notice that for $\mathbf{x} \in \mathbb{R}^n$ the same decomposition is obtained by pooling the contributions from frequencies ω_j and $-\omega_j$ in Table 10.1.

We have seen how the periodogram generates a decomposition of $\|\mathbf{x}\|^2$ into components associated with the Fourier frequencies $\omega_j = 2\pi j/n \in (-\pi, \pi]$.

Table 10.2. Decomposition of $\|\mathbf{x}\|^2$, $\mathbf{x} \in \mathbb{R}^n$, into Components Corresponding to the Harmonic Decomposition (10.1.11)

Source	Degrees of freedom	Sum of squares
Frequency ω_0 (mean)	1	$a_0^2 = n^{-1}\left(\sum_{t=1}^n x_t\right)^2 = I(0)$
Frequency ω_1	2	$2r_1^2 = 2\|a_1\|^2 = 2I(\omega_1)$
⋮	⋮	⋮
Frequency ω_k	2	$2r_k^2 = 2\|a_k\|^2 = 2I(\omega_k)$
⋮	⋮	⋮
Frequency $\omega_{n/2} = \pi$ (if n is even)	1	$a_{n/2}^2 = I(\pi)$
Total	n	$\sum_{t=1}^n x_t^2$

It is also closely related to the sample autocovariance function $\hat{\gamma}(k)$, $|k| < n$, as demonstrated in the following proposition.

Proposition 10.1.2 (The Periodogram of $\mathbf{x} \in \mathbb{C}^n$ in Terms of the Sample Autocovariance Function). *If ω_j is any non-zero Fourier frequency, then*

$$I(\omega_j) = \sum_{|k|<n} \hat{\gamma}(k)e^{-ik\omega_j}, \tag{10.1.12}$$

where $\hat{\gamma}(k) := n^{-1}\sum_{t=1}^{n-k}(x_{t+k} - m)(\bar{x}_t - \bar{m})$, $k \geq 0$, $m := n^{-1}\sum_{t=1}^{n} x_t$ *and* $\hat{\gamma}(k) = \overline{\hat{\gamma}(-k)}$, $k < 0$. [*If* $m = 0$, *or if we replace* $\hat{\gamma}(k)$ *in* (10.1.12) *by* $\tilde{\gamma}(k)$, *where* $\tilde{\gamma}$ *is defined like* $\hat{\gamma}$ *with m replaced by zero, the following proof shows that* (10.1.12) *is then valid for* all *Fourier frequencies,* $\omega_j \in (-\pi, \pi]$.]

PROOF. By Definition 10.1.2, we can write

$$I(\omega_j) = n^{-1}\sum_{s=1}^{n} x_s e^{-is\omega_j}\sum_{t=1}^{n}\bar{x}_t e^{it\omega_j}.$$

Now $\sum_{s=1}^{n} e^{is\omega_j} = \sum_{t=1}^{n} e^{-it\omega_j} = 0$ if $\omega_j \neq 0$, and hence

$$\begin{aligned} I(\omega_j) &= n^{-1}\sum_{s=1}^{n}\sum_{t=1}^{n}(x_s - m)(\bar{x}_t - \bar{m})e^{-i(s-t)\omega_j} \\ &= \sum_{|k|<n}\hat{\gamma}(k)e^{-ik\omega_j}. \end{aligned}$$

□

Remark. The striking resemblance between (10.1.12) and the expression $f(\omega) = (2\pi)^{-1}\sum_{k=-\infty}^{\infty}\gamma(k)e^{-ik\omega}$ for the spectral density of a stationary process with $\sum|\gamma(k)| < \infty$, suggests the potential value of the periodogram for spectral density estimation. This aspect of the periodogram will be taken up in Section 10.3.

§10.2 Testing for the Presence of Hidden Periodicities

In this section we shall consider a variety of tests (based on the periodogram) which can be used to test the null hypothesis H_0 that the data $\{X_1, \ldots, X_n\}$ is generated by a Gaussian white noise sequence, against the alternative hypothesis H_1 that the data is generated by a Gaussian white noise sequence with a superimposed deterministic periodic component. The form of the test will depend on the way in which the periodic component is specified. The data is assumed from now on to be real.

(a) *Testing for the Presence of a Sinusoid with Specified Frequency.* The model for the data is

$$X_t = \mu + A\cos\omega t + B\sin\omega t + Z_t, \tag{10.2.1}$$

where $\{Z_t\}$ is Gaussian white noise with variance σ^2, A and B are non-random constants and ω is a specified frequency. The null and alternative hypotheses are

$$H_0 : A = B = 0, \tag{10.2.2}$$

and

$$H_1 : A \text{ and } B \text{ are not both zero.} \tag{10.2.3}$$

If ω is one of the Fourier frequencies $\omega = 2\pi k/n \in (0, \pi)$, then the analysis of variance (Table 10.2) provides us with an easy test. The model (10.2.1) can be written, in the notation of (10.1.11), as

$$\mathbf{X} = n^{1/2}\mu\mathbf{e}_0 + (n/2)^{1/2}A\mathbf{c}_k + (n/2)^{1/2}B\mathbf{s}_k + \mathbf{Z}, \qquad \mathbf{Z} \sim \mathrm{N}(\mathbf{0}, \sigma^2 I_n). \tag{10.2.4}$$

We therefore reject H_0 in favour of H_1 if the frequency ω_k sum of squares in Table 10.2, i.e. $2I(\omega_k)$, is sufficiently large. To determine how large, we observe that under H_0 (see Problem 2.19),

$$2I(\omega_k) = \|P_{\overline{\mathrm{sp}}\{\mathbf{c}_k, \mathbf{s}_k\}}\mathbf{X}\|^2 = \|P_{\overline{\mathrm{sp}}\{\mathbf{c}_k, \mathbf{s}_k\}}\mathbf{Z}\|^2 \sim \sigma^2\chi^2(2),$$

and that $I(\omega_k)$ is independent of

$$\|\mathbf{X} - P_{\overline{\mathrm{sp}}\{\mathbf{e}_0, \mathbf{c}_k, \mathbf{s}_k\}}\mathbf{X}\|^2 = \sum_{i=1}^n X_i^2 - I(0) - 2I(\omega_k) \sim \sigma^2\chi^2(n-3).$$

We therefore reject H_0 in favor of H_1 at level α if

$$(n-3)I(\omega_k)\Bigg/\left[\sum_{i=1}^n X_i^2 - I(0) - 2I(\omega_k)\right] > F_{1-\alpha}(2, n-3).$$

An obvious modification of the above test can also be used if $\omega = \pi$. However if ω is not a Fourier frequency, the analysis is a little more complicated since the vectors

$$\mathbf{c} = (2/n)^{1/2}(\cos\omega, \cos 2\omega, \ldots, \cos n\omega)',$$
$$\mathbf{s} = (2/n)^{1/2}(\sin\omega, \sin 2\omega, \ldots, \sin n\omega)',$$

and $\mathbf{e}_0$ are not orthogonal. In principle however the test is quite analogous. The model now is

$$\mathbf{X} = n^{1/2}\mu\mathbf{e}_0 + (n/2)^{1/2}A\mathbf{c} + (n/2)^{1/2}B\mathbf{s} + \mathbf{Z}, \qquad \mathbf{Z} \sim \mathrm{N}(\mathbf{0}, \sigma^2 I_n),$$

and the two hypotheses H_0 and H_1 are again defined by (10.2.2) and (10.2.3). In this case we reject H_0 in favor of H_1 if

$$2I^*(\omega) := \|P_{\overline{\mathrm{sp}}\{\mathbf{e}_0, \mathbf{c}, \mathbf{s}\}}\mathbf{X} - P_{\overline{\mathrm{sp}}\{\mathbf{e}_0\}}\mathbf{X}\|^2$$

is large. Now under H_0,

$$2I^*(\omega) \sim \sigma^2\chi^2(2),$$

and $I^*(\omega)$ is independent of

$$\|\mathbf{X} - P_{\overline{\mathrm{sp}}\{\mathbf{e}_0, \mathbf{c}, \mathbf{s}\}}\mathbf{X}\|^2 \sim \sigma^2\chi^2(n-3).$$

We therefore reject H_0 in favour of H_1 at level α if

$$(n-3)I^*(\omega)/\|\mathbf{X} - P_{\overline{\text{sp}}\{\mathbf{e}_0,\mathbf{c},\mathbf{s}\}}\mathbf{X}\|^2 > F_{1-\alpha}(2, n-3).$$

To evaluate the test statistic we have

$$P_{\text{sp}\{\mathbf{e}_0\}}\mathbf{X} = n^{-1/2}\sum_{i=1}^{n} X_i\mathbf{e}_0,$$

and (see Section 2.6)

$$P_{\text{sp}\{\mathbf{e}_0,\mathbf{c},\mathbf{s}\}}\mathbf{X} = n^{1/2}\hat{\mu}\mathbf{e}_0 + (n/2)^{1/2}\hat{A}\mathbf{c} + (n/2)^{1/2}\hat{B}\mathbf{s},$$

where $\hat{\mu}$, $\hat{A}$ and $\hat{B}$ are least squares estimators satisfying

$$W'W(\hat{\mu}, \hat{A}, \hat{B})' = W'\mathbf{X},$$

and W is the $(n \times 3)$-matrix $[n^{1/2}\mathbf{e}_0, (n/2)^{1/2}\mathbf{c}, (n/2)^{1/2}\mathbf{s}]$.

(b) *Testing for the Presence of a Non-Sinusoidal Periodic Component with Specified Integer-Valued Period, $p < n$.* If f is any function with values f_t, $t \in \mathbb{Z}$, and with period $p \in (1, n)$, then the same argument which led to (10.1.11) shows that f has the representation,

$$f_t = \mu + \sum_{k=1}^{[(p-1)/2]} [A_k \cos(2\pi kt/p) + B_k \sin(2\pi kt/p)] + A_{p/2}(-1)^t, \qquad (10.2.5)$$

where $A_{p/2} := 0$ if p is odd. Our model for the data is therefore

$$X_t = f_t + Z_t, \qquad t = 1, \ldots, n \qquad (10.2.6)$$

where $\{Z_t\}$ is Gaussian white noise with variance σ^2 and f_t is defined by (10.2.5). The null hypothesis is

$$H_0 : A_j = B_j = 0 \quad \text{for all } j, \qquad (10.2.7)$$

and the alternative hypothesis is

$$H_1 : H_0 \text{ is false.} \qquad (10.2.8)$$

Define the n-component column vectors,

$$\mathbf{e}_0 = (1/n)^{1/2}(1, 1, \ldots, 1)',$$
$$\boldsymbol{\gamma}_j = (2/n)^{1/2}(\cos\psi_j, \cos 2\psi_j, \ldots, \cos n\psi_j)'$$

and

$$\boldsymbol{\sigma}_j = (2/n)^{1/2}(\sin\psi_j, \sin 2\psi_j, \ldots, \sin n\psi_j)'$$

where $\psi_j = 2\pi j/p$, $j = 1, 2, \ldots, [p/2]$. Now let $S(p)$ be the span of the p vectors $\mathbf{e}_0, \boldsymbol{\gamma}_1, \boldsymbol{\sigma}_1, \boldsymbol{\gamma}_2, \boldsymbol{\sigma}_2, \ldots,$ (the last is $\mathbf{e}_{p/2}$ if p is even, $\boldsymbol{\sigma}_{(p-1)/2}$ if p is odd) and let W be the $(n \times p)$-matrix

$$W = [\mathbf{e}_0, \boldsymbol{\gamma}_1, \boldsymbol{\sigma}_1, \boldsymbol{\gamma}_2, \cdots].$$

The projection of $\mathbf{X} = (X_1, \ldots, X_n)'$ onto $S(p)$ is then (see Section 2.6)

$$P_{S(p)}\mathbf{X} = W(W'W)^{-1}W'\mathbf{X}. \tag{10.2.9}$$

From (10.2.5) and (10.2.6),

$$\|\mathbf{X} - P_{S(p)}\mathbf{X}\|^2 = \|\mathbf{Z} - P_{S(p)}\mathbf{Z}\|^2 \sim \sigma^2\chi^2(n-p), \tag{10.2.10}$$

since $\mathbf{Z} := (Z_1, \ldots, Z_n)' \sim N(0, \sigma^2 I_n)$. Moreover under H_0,

$$\|P_{S(p)}\mathbf{X} - P_{\overline{\mathrm{sp}}\{\mathbf{e}_0\}}\mathbf{X}\|^2 = \|P_{S(p)}\mathbf{Z} - P_{\overline{\mathrm{sp}}\{\mathbf{e}_0\}}\mathbf{Z}\|^2 \sim \sigma^2\chi^2(p-1), \tag{10.2.11}$$

and is independent of $\|\mathbf{X} - P_{S(p)}\mathbf{X}\|^2$.

We reject H_0 in favour of H_1 if $\|P_{S(p)}\mathbf{X} - P_{\overline{\mathrm{sp}}\{\mathbf{e}_0\}}\mathbf{X}\|$ is sufficiently large. From (10.2.10) and (10.2.11), we obtain a size α test if we reject H_0 when

$$\frac{\|P_{S(p)}\mathbf{X} - \bar{X}\mathbf{1}\|^2/(p-1)}{\|\mathbf{X} - P_{S(p)}\mathbf{X}\|^2/(n-p)} > F_{1-\alpha}(p-1, n-p), \tag{10.2.12}$$

where $\bar{X} = \sum_{i=1}^n X_i/n$, $\mathbf{1} := (1, \ldots, 1)'$ and $P_{S(p)}\mathbf{X}$ is found from (10.2.9).

In the special case when n is an integer multiple of the period p, say $n = rp$, the calculations simplify dramatically as a result of the orthogonality of the p vectors $\mathbf{e}_0, \boldsymbol{\gamma}_1, \boldsymbol{\sigma}_1, \boldsymbol{\gamma}_2, \boldsymbol{\sigma}_2, \ldots$. In fact, in the notation of (10.1.11),

$$\boldsymbol{\gamma}_j = \mathbf{c}_{rj} \quad \text{and} \quad \boldsymbol{\sigma}_j = \mathbf{s}_{rj}, \qquad j = 0, \ldots, [p/2].$$

Hence, using Table 10.2,

$$\begin{aligned}\|P_{S(p)}\mathbf{X}\|^2 &= \sum_{j=0}^{[p/2]} \left[\|P_{\overline{\mathrm{sp}}\{\mathbf{c}_{rj}\}}\mathbf{X}\|^2 + \|P_{\overline{\mathrm{sp}}\{\mathbf{s}_{rj}\}}\mathbf{X}\|^2\right] \\ &= I(0) + 2\sum_{1 \le j < p/2} I(\omega_{rj}) + \delta_p I(\pi),\end{aligned}$$

where $\delta_p = 1$ if p is even, 0 if p is odd. The rejection criterion (10.2.12) therefore reduces to

$$\frac{2\sum_{1 \le j < p/2} I(\omega_{rj}) + \delta_p I(\pi)}{\sum X_i^2 - I(0) - \delta_p I(\pi) - 2\sum_{1 \le j < p/2} I(\omega_{rj})} > \frac{p-1}{n-p} F_{1-\alpha}(p-1, n-p), \tag{10.2.13}$$

where, as usual, $\omega_{rj} = 2\pi rj/n$.

(c) *Testing for Hidden Periodicities of Unspecified Frequency: Fisher's Test.* If $\{X_t\}$ is Gaussian white noise with variance σ^2 and $\mathbf{X} = (X_1, \ldots, X_n)'$, then, since $2I(\omega_k) = \|P_{\overline{\mathrm{sp}}\{\mathbf{c}_k, \mathbf{s}_k\}}\mathbf{X}\|^2$, $k = 1, \ldots, [(n-1)/2]$, we conclude from Problem 2.19 that

$$V_k := I(\omega_k)/\sigma^2 \sim \chi^2(2)/2, \qquad k = 1, \ldots, q, \tag{10.2.14}$$

where

$$q := [(n-1)/2],$$

and that $V_1, \ldots, V_q$ are independent. Since from (10.2.14) the density function of V_k is $e^{-x}I_{[0,\infty)}(x)$, we deduce that the joint density function of $V_1, \ldots, V_q$ is

$$f_{V_1\ldots V_q}(v_1,\ldots,v_q) = \prod_{i=1}^{q} e^{-v_i} I_{[0,\infty)}(v_i). \tag{10.2.15}$$

This is the key result used in the proof of the following proposition.

Proposition 10.2.1. *If $\{X_t\}$ is Gaussian white noise, then the random variables,*

$$Y_i := \frac{\sum_{k=1}^{i} V_k}{\sum_{k=1}^{q} V_k} = \frac{\sum_{k=1}^{i} I(\omega_k)}{\sum_{k=1}^{q} I(\omega_k)}, \qquad i = 1, \ldots, q-1,$$

are distributed as the order statistics of a sample of $(q-1)$ *independent random variables, each one uniformly distributed on the interval* $[0, 1]$.

PROOF. Let $S_i = \sum_{j=1}^{i} V_j$, $i = 1, \ldots, q$. Then from (10.2.15), the joint density function of $S_1, \ldots, S_q$ is (see e.g. Mood, Graybill and Boes (1974))

$$\begin{aligned} f_{S_1\ldots S_q}(s_1,\ldots,s_q) &= \exp[-s_1 - (s_2 - s_1) - \cdots - (s_q - s_{q-1})] \\ &= \exp(-s_q), \qquad 0 \le s_1 \le \cdots \le s_q. \end{aligned} \tag{10.2.16}$$

The marginal density function of S_q is the probability density function of the sum of q independent standard exponential random variables. Thus

$$f_{S_q}(s_q) = \frac{s_q^{q-1}}{(q-1)!} \exp(-s_q), \qquad s_q \ge 0. \tag{10.2.17}$$

From (10.2.16) and (10.2.17), the conditional density of $(S_1, \ldots, S_{q-1})$ given S_q is

$$f_{S_1\ldots S_{q-1}|S_q}(s_1,\ldots,s_{q-1}|s_q) = (q-1)!\, s_q^{-q+1}, \qquad 0 \le s_1 \le \cdots \le s_{q-1} \le s_q.$$

Since by definition $Y_i = S_i/S_q$, $i = 1, \ldots, q-1$, the conditional density of $Y_1, \ldots, Y_{q-1}$ given S_q is

$$f_{Y_1\ldots Y_{q-1}|S_q}(y_1,\ldots,y_{q-1}|s_q) = (q-1)!, \qquad 0 \le y_1 \le \cdots \le y_{q-1} \le 1,$$

and since this does not depend on s_q, we can write the unconditional joint density of $Y_1, \ldots, Y_{q-1}$ as,

$$f_{Y_1\ldots Y_{q-1}}(y_1,\ldots,y_{q-1}) = (q-1)!, \qquad 0 \le y_1 \le \cdots \le y_{q-1} \le 1. \tag{10.2.18}$$

This is precisely the joint density of the order statistics of a random sample of size $(q-1)$ from the uniform distribution on $(0, 1)$. □

Corollary 10.2.1. *Under the conditions of Proposition* 10.2.1, *the cumulative distribution function with jumps of size* $(q-1)^{-1}$ *at* Y_i, $i = 1, \ldots, q-1$, *is the empirical distribution function of a sample of size* $(q-1)$ *from the uniform distribution on* $(0, 1)$.

Corollary 10.2.2. *If we define* $Y_0 := 0$, $Y_q := 1$ *and*

$$M_q := \max_{1 \le i \le q} (Y_i - Y_{i-1}) = \max_{1 \le i \le q} \frac{I(\omega_i)}{\sum_{i=1}^{q} I(\omega_i)},$$

then under the conditions of Proposition 10.2.1,

$$P(M_q \le a) = \sum_{j=0}^{q} (-1)^j \binom{q}{j} (1 - ja)_+^{q-1}, \tag{10.2.19}$$

where $x_+ = \max(x, 0)$.

PROOF. It is clear from Proposition 10.2.1 that M_q is distributed as the length of the largest subinterval of $(0, 1)$ obtained when the interval is randomly partitioned by $(q - 1)$ points independently and uniformly distributed on $(0, 1)$. The distribution function of this length is shown by Feller (1971), p. 29, to have the form (10.2.19). □

Fisher's Test for Hidden Periodicities. Corollary 10.2.2 was used by Fisher to construct a test of the null hypothesis that $\{X_t\}$ is Gaussian white noise against the alternative hypothesis that $\{X_t\}$ contains an added deterministic periodic component of unspecified frequency. The idea is to reject the null hypothesis if the periodogram contains a value substantially larger than the average value, i.e. (recalling that $q = [(n - 1)/2]$) if

$$\xi_q := \left[\max_{1 \le i \le q} I(\omega_i)\right] \Big/ \left[q^{-1} \sum_{i=1}^{q} I(\omega_i)\right] = qM_q, \tag{10.2.20}$$

is sufficiently large. To apply the test, we compute the realized value x of ξ_q from the data $X_1, \ldots, X_n$ and then use (10.2.19) to compute

$$P(\xi_q \ge x) = 1 - \sum_{j=0}^{q} (-1)^j \binom{q}{j} (1 - jx/q)_+^{q-1}. \tag{10.2.21}$$

If this probability is less than α, we reject the null hypothesis at level α.

EXAMPLE 10.2.1. Figure 10.1 shows a realization of $\{X_1, \ldots, X_{100}\}$ together with the periodogram ordinates $I(\omega_j), j = 1, \ldots, 50$. In this case $q = [99/2] = 49$ and the realized value of ξ_{49} is $x = 9.4028/1.1092 = 8.477$. From (10.2.21),

$$P(\xi_{49} > 8.477) = .0054,$$

and consequently we reject the null hypothesis at level .01. [The data was in fact generated by the process

$$X_t = \cos(\pi t/3) + Z_t, \qquad t = 1, \ldots, 100,$$

where $\{Z_t\}$ is Gaussian white noise with variance 1. This explains the peak in the periodogram at $\omega_{17} = .34\pi$.]

The Kolmogorov–Smirnov Test. Corollary 10.2.1 suggests another test of the null hypothesis that $\{X_t\}$ is Gaussian white noise. We simply plot the empirical distribution function defined in the corollary and check its compatibility with the uniform distribution function $F(x) = x, 0 \le x \le 1$, using the Kolmogorov–

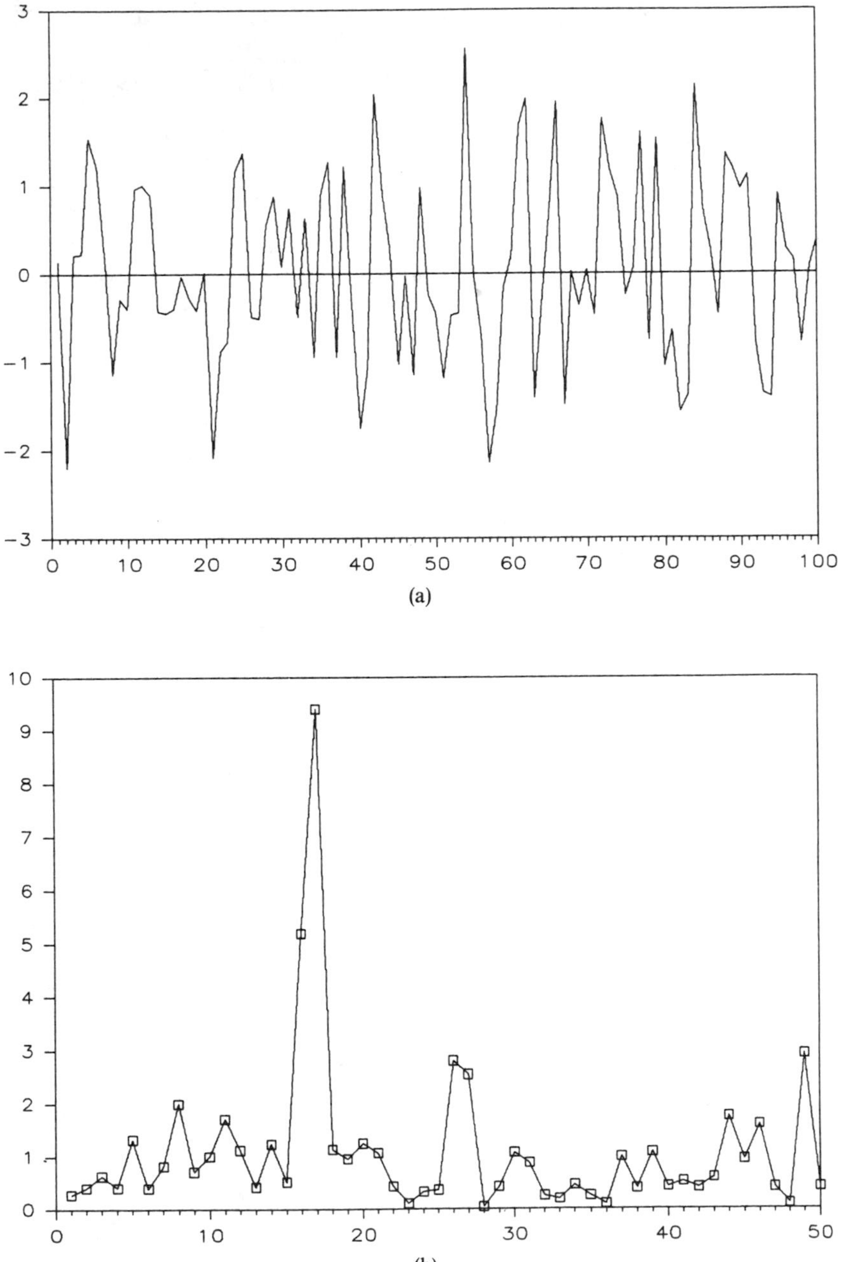

Figure 10.1. (a) The series $\{X_1, \ldots, X_{100}\}$ of Example 10.2.1 and (b) the corresponding periodogram ordinates $I(2\pi j/100), j = 1, \ldots, 50$.

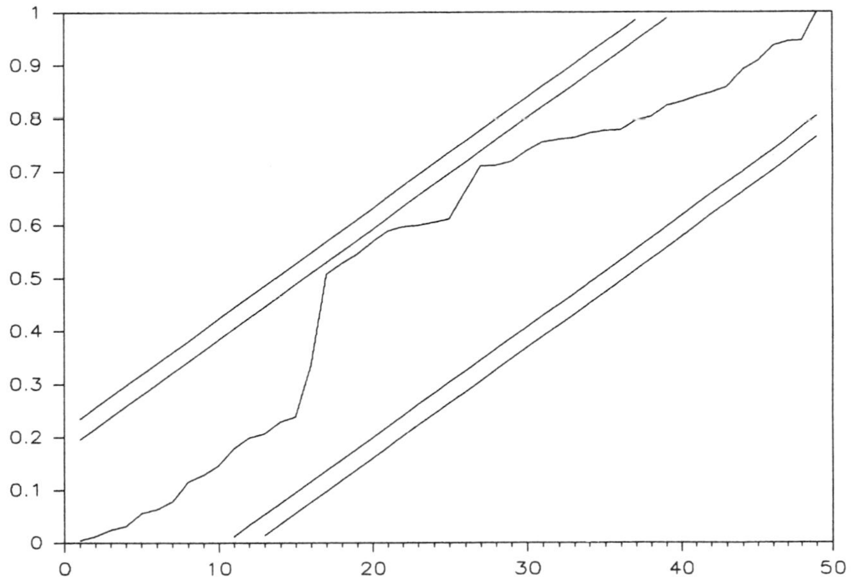

Figure 10.2. The standardized cumulative periodogram $C(x)$ for Example 10.2.1 showing the Kolmogorov–Smirnov bounds for $\alpha = .05$ (inner) and $\alpha = .01$ (outer).

Smirnov test. For $q > 30$ (i.e. for sample size $n > 62$), a good approximation to the level-α Kolmogorov-Smirnov test is to reject the null hypothesis if the empirical distribution function exits from the bounds

$$y = x \pm k_\alpha (q-1)^{-1/2}, \qquad 0 < x < 1,$$

where $k_{.05} = 1.36$ and $k_{.01} = 1.63$.

This procedure is precisely equivalent to plotting the standardized cumulative periodogram,

$$C(x) = \begin{cases} 0, & x < 1, \\ Y_i, & i \le x < i+1, i = 1, \ldots, q-1, \\ 1, & x \ge q, \end{cases} \tag{10.2.22}$$

and rejecting the null hypothesis at level α if for any x in $[1, q]$, the function C exits from the boundaries,

$$y = \frac{x-1}{q-1} \pm k_\alpha (q-1)^{-1/2}. \tag{10.2.23}$$

EXAMPLE 10.2.2. Figure 10.2 shows the cumulative periodogram and Kolmogorov–Smirnov boundaries for the data of Example 10.2.1 with $\alpha = .05$ and $\alpha = .01$. We do not reject the null hypothesis even at level .05 using this test. The Fisher test however rejected the null hypothesis at level

.01 since it is specifically designed to detect departures from the null hypothesis of the kind encountered in this example.

Generalization of the Fisher and Kolmogorov–Smirnov Tests. The null hypothesis assumed for both these tests was that $\{X_t\}$ is Gaussian white noise. However when n is large the tests can also be used to test the null hypothesis that $\{X_t\}$ has spectral density f by replacing $I(\omega_k)$ by $I(\omega_k)/f(\omega_k)$ in the definitions of Y_i and ξ_q.

§10.3 Asymptotic Properties of the Periodogram

In this section we shall consider the asymptotic properties of the periodogram of $X_1, \ldots, X_n$ when $\{X_t\}$ is a stationary time series with mean μ and absolutely summable autocovariance function $\gamma(\cdot)$. Under these conditions $\{X_t\}$ has a continuous spectral density (Corollary 4.3.2) given by

$$f(\omega) = (2\pi)^{-1} \sum_{k=-\infty}^{\infty} \gamma(k) e^{-ik\omega}, \qquad \omega \in [-\pi, \pi]. \tag{10.3.1}$$

The periodogram of $\{X_1, \ldots, X_n\}$ is defined at the Fourier frequencies $\omega_j = 2\pi j/n$, $\omega_j \in [-\pi, \pi]$, by

$$I_n(\omega_j) = n^{-1} \left| \sum_{t=1}^{n} X_t e^{-it\omega_j} \right|^2.$$

By Proposition 10.1.2, this definition is equivalent to

$$\begin{cases} I_n(0) = n|\bar{X}|^2, \\ I_n(\omega_j) = \displaystyle\sum_{|k|<n} \hat{\gamma}(k) e^{-ik\omega_j} \quad \text{if } \omega_j \neq 0, \end{cases} \tag{10.3.2}$$

where $\hat{\gamma}(k) = n^{-1} \sum_{t=1}^{n-|k|} (X_t - \bar{X})(X_{t+|k|} - \bar{X})$ and $\bar{X} = n^{-1} \sum_{t=1}^{n} X_t$. In deriving the asymptotic properties of I_n it will be convenient to use the alternative representation,

$$I_n(\omega_j) = \sum_{|k|<n} n^{-1} \left(\sum_{t=1}^{n-|k|} (X_t - \mu)(X_{t+|k|} - \mu) \right) e^{-ik\omega_j}, \qquad \omega_j \neq 0, \tag{10.3.3}$$

which can be established by the same argument used in the proof of Proposition 10.1.2.

In view of (10.3.2) a natural estimate of $f(\omega_j)$ for $\omega_j \neq 0$ is $I_n(\omega_j)/(2\pi)$. We now extend the domain of I_n to the whole interval $[-\pi, \pi]$ in order to estimate $f(\omega)$ for arbitrary non-zero frequencies in the interval $[-\pi, \pi]$. This can be done in various ways, e.g. by replacing ω_j in (10.3.2) by ω and allowing ω to take any value in $[-\pi, \pi]$. However we shall follow Fuller (1976) in defining the periodogram on $[-\pi, \pi]$ as a piecewise constant function which coincides with (10.3.2) at the Fourier frequencies $\omega_j \in [-\pi, \pi]$.

Definition 10.3.1 (Extension of the Periodogram). For any $\omega \in [-\pi, \pi]$ the periodogram is defined as follows:

$$I_n(\omega) = \begin{cases} I_n(\omega_k) & \text{if } \omega_k - \pi/n < \omega \le \omega_k + \pi/n \text{ and } 0 \le \omega \le \pi, \\ I_n(-\omega) & \text{if } \omega \in [-\pi, 0). \end{cases}$$

Clearly this definition implies that I_n is an even function which coincides with (10.3.2) at all integer multiples of $2\pi/n$. For $\omega \in [0, \pi]$, let $g(n, \omega)$ be the multiple of $2\pi/n$ closest to ω (the smaller one if there are two) and for $\omega \in [-\pi, 0)$ let $g(n, \omega) = g(n, -\omega)$. Then Definition 10.3.1 can be rewritten as

$$I_n(\omega) = I_n(g(n, \omega)). \tag{10.3.4}$$

The following proposition establishes the asymptotic unbiasedness of the periodogram estimate $I_n(\omega)/(2\pi)$ of $f(\omega)$ for $\omega \ne 0$.

Proposition 10.3.1. *If $\{X_t\}$ is stationary with mean μ and absolutely summable autocovariance function $\gamma(\cdot)$,* then

$$\text{(i)}\quad EI_n(0) - n\mu^2 \to 2\pi f(0)$$

and

$$\text{(ii)}\quad EI_n(\omega) \to 2\pi f(\omega) \text{ if } \omega \ne 0.$$

(If $\mu = 0$ then $EI_n(\omega)$ converges uniformly to $2\pi f(\omega)$ on $[-\pi, \pi]$.)

Proof. By Theorem 7.1.1,

$$EI_n(0) - n\mu^2 = n\operatorname{Var}(\bar{X}_n) \to \sum_{n=-\infty}^{\infty} \gamma(n) = 2\pi f(0).$$

Now if $\omega \in (0, \pi]$ then, for n sufficiently large, $g(n, \omega) \ne 0$. Hence, from (10.3.3) and (10.3.4)

$$\begin{aligned} EI_n(\omega) &= \sum_{|k|<n} n^{-1} \sum_{t=1}^{n-|k|} E[(X_t - \mu)(X_{t+|k|} - \mu)]e^{-ikg(n,\omega)} \\ &= \sum_{|k|<n} (1 - |k|/n)\gamma(k)e^{-ikg(n,\omega)}. \end{aligned}$$

However, since $\gamma(\cdot)$ is absolutely summable, $\sum_{|k|<n}(1 - |k|/n)\gamma(k)e^{-ik\lambda}$ converges uniformly to $2\pi f(\lambda)$ and therefore (since $g(n, \omega) \to \omega$) we have

$$EI_n(\omega) \to 2\pi f(\omega).$$

The uniform convergence of $EI_n(\omega)$ to $2\pi f(\omega)$ when $\mu = 0$ is easy to check using the uniform continuity of f on $[-\pi, \pi]$ and the uniform convergence of $g(n, \omega)$ to ω on $[0, \pi]$. □

As indicated earlier, the vectors $\{\mathbf{c}_j, \mathbf{s}_j; j = 1, \ldots, q = [(n-1)/2]\}$ in equation (10.1.11) are orthonormal. Consequently if $\{X_t\}$ is Gaussian white noise with variance σ^2, then the random variables,

$$\begin{cases} \alpha(\omega_j) := \langle \mathbf{X}, \mathbf{c}_j \rangle = (2/n)^{1/2} \sum_{t=1}^{n} X_t \cos(\omega_j t), & j = 1, \ldots, q, \\ \beta(\omega_j) := \langle \mathbf{X}, \mathbf{s}_j \rangle = (2/n)^{1/2} \sum_{t=1}^{n} X_t \sin(\omega_j t), & j = 1, \ldots, q, \end{cases} \tag{10.3.5}$$

are independent with distribution $N(0, \sigma^2)$. Consequently, as observed in Section 10.2, the periodogram ordinates,

$$I_n(\omega_j) = [\alpha^2(\omega_j) + \beta^2(\omega_j)]/2, \qquad j = 1, \ldots, q,$$

are independently and exponentially distributed with means $\sigma^2 = 2\pi f_X(\omega_j)$, where $f_X(\cdot)$ is the spectral density of $\{X_t\}$. An analogous asymptotic result (Theorem 10.3.2) can be established for linear processes. First however we shall consider the case when $\{X_t\} \sim \text{IID}(0, \sigma^2)$.

Proposition 10.3.2. *Suppose that* $\{Z_t\} \sim \text{IID}(0, \sigma^2)$ *and let* $I_n(\omega)$, $-\pi \le \omega \le \pi$, *denote the periodogram of* $\{Z_1, \ldots, Z_n\}$ *as defined by* (10.3.4).

(i) *If* $0 < \lambda_1 < \cdots < \lambda_m < \pi$ *then the random vector* $(I_n(\lambda_1), \ldots, I_n(\lambda_m))'$ *converges in distribution as* $n \to \infty$ *to a vector of independent and exponentially distributed random variables, each with mean* σ^2.

(ii) *If* $EZ_1^4 = \eta\sigma^4 < \infty$ *and* $\omega_j = 2\pi j/n \in [0, \pi]$, *then*

$$\text{Var}(I_n(\omega_j)) = \begin{cases} n^{-1}(\eta - 3)\sigma^4 + 2\sigma^4 & \text{if } \omega_j = 0 \text{ or } \pi, \\ n^{-1}(\eta - 3)\sigma^4 + \sigma^4 & \text{if } 0 < \omega_j < \pi, \end{cases} \tag{10.3.6}$$

and

$$\text{Cov}(I_n(\omega_j), I_n(\omega_k)) = n^{-1}(\eta - 3)\sigma^4 \quad \text{if } \omega_j \ne \omega_k. \tag{10.3.7}$$

(*If* Z_1 *is normally distributed, then* $\eta - 3 = 0$ *so that* $I_n(\omega_j)$ *and* $I_n(\omega_k)$ *are uncorrelated for* $j \ne k$, *as pointed out in Section* 10.2(c).)

Proof. (i) For an arbitrary frequency $\lambda \in (0, \pi)$ define

$$\alpha(\lambda) := \alpha(g(n, \lambda)) \quad \text{and} \quad \beta(\lambda) := \beta(g(n, \lambda)),$$

where $\alpha(\omega_j)$ and $\beta(\omega_j)$ are given by (10.3.5) with Z_t replacing X_t. Since $I_n(\lambda_j) = (\alpha^2(\lambda_j) + \beta^2(\lambda_j))/2$, it suffices to show that

$$(\alpha(\lambda_1), \beta(\lambda_1), \ldots, \alpha(\lambda_m), \beta(\lambda_m))' \text{ is } \text{AN}(\mathbf{0}, \sigma^2 I_{2m}), \tag{10.3.8}$$

where I_{2m} is the $2m \times 2m$ identity matrix.

Now if λ is a fixed frequency in $(0, \pi)$ then for all sufficiently large n, $g(n, \lambda) \in (0, \pi)$ and hence by the independence of the sequence $\{Z_t\}$

$$\begin{aligned} \text{Var}(\alpha(\lambda)) &= \text{Var}(\alpha(g(n, \lambda)) \\ &= \sigma^2(2/n) \sum_{t=1}^{n} \cos^2(g(n, \lambda)t) \\ &= \sigma^2. \end{aligned}$$

Moreover for any $\varepsilon > 0$,

$$\begin{aligned}
n^{-1} \sum_{t=1}^{n} E(\cos^2(g(n,\lambda)t)Z_t^2 I_{[|\cos(g(n,\lambda)t)Z_t| > \varepsilon n^{1/2}\sigma]}) \\
&\le n^{-1} \sum_{t=1}^{n} E(Z_t^2 I_{[|Z_t| > \varepsilon n^{1/2}\sigma]}) \\
&= E(Z_1^2 I_{[|Z_1| > \varepsilon n^{1/2}\sigma]}) \\
&\to 0 \quad \text{as } n \to \infty,
\end{aligned}$$

implying that $\alpha(\lambda)$ is $\text{AN}(0, \sigma^2)$ by the Lindeberg condition (see Billingsley (1986)). Finally, for all sufficiently large n, $g(n, \lambda_i) \in (0, \pi)$, $i = 1, \ldots, m$, and since the covariance matrix of $(\alpha(\lambda_1), \beta(\lambda_1), \ldots, \alpha(\lambda_m), \beta(\lambda_m))'$ is $\sigma^2 I_{2m}$, the joint convergence in (i) is easily established using the Cramer–Wold device.

(ii). By definition of $I_n(\omega_j)$, we have

$$I_n(\omega_j) = n^{-1} \sum_{s=1}^{n} \sum_{t=1}^{n} Z_s Z_t e^{i\omega_j(t-s)},$$

and hence,

$$EI_n(\omega_j)I_n(\omega_k) = n^{-2} \sum_{s=1}^{n} \sum_{t=1}^{n} \sum_{u=1}^{n} \sum_{v=1}^{n} E(Z_s Z_t Z_u Z_v) e^{i\omega_j(t-s)} e^{i\omega_k(v-u)}.$$

By (7.3.4), this expression can be rewritten as

$$n^{-1}(\eta - 3)\sigma^4 + \sigma^4\left(1 + n^{-2}\left|\sum_{t=1}^{n} e^{i(\omega_j+\omega_k)t}\right|^2 + n^{-2}\left|\sum_{t=1}^{n} e^{i(\omega_k-\omega_j)t}\right|^2\right),$$

and since $EI_n(\omega_j) = n^{-1}\sum_{t=1}^{n} EZ_t^2 = \sigma^2$, it follows that

$$\text{Cov}(I_n(\omega_j), I_n(\omega_k)) = n^{-1}(\eta - 3)\sigma^4 + n^{-2}\sigma^4\left(\left|\sum_{t=1}^{n} e^{i(\omega_j+\omega_k)t}\right|^2 + \left|\sum_{t=1}^{n} e^{i(\omega_k-\omega_j)t}\right|^2\right).$$

The relations (10.3.6) and (10.3.7) are immediate consequences of this equation. □

We next extend Proposition 10.3.2 to the linear process

$$X_t = \sum_{j=-\infty}^{\infty} \psi_j Z_{t-j}, \qquad \{Z_t\} \sim \text{IID}(0, \sigma^2), \tag{10.3.9}$$

where $\sum_{j=-\infty}^{\infty} |\psi_j| < \infty$. The spectral density of this process is related (see (4.4.3)) to the spectral density of the white noise sequence $\{Z_t\}$ by

$$f_X(\lambda) = |\psi(e^{-i\lambda})|^2 f_Z(\lambda), \qquad -\pi \le \lambda \le \pi,$$

where $\psi(e^{-i\lambda}) = \sum_{j=-\infty}^{\infty} \psi_j e^{-ij\lambda}$ (and $f_Z(\lambda) = \sigma^2/2\pi$). Since $I_n(\lambda)/2\pi$ can be thought of as a sample version of the spectral density function, we might expect a similar relationship to exist between the respective periodograms of $\{X_t\}$ and $\{Z_t\}$. This is the content of the following theorem.

Theorem 10.3.1. *Let $\{X_t\}$ be the linear process defined by (10.3.9) and let $I_{n,X}(\lambda)$ and $I_{n,Z}(\lambda)$ denote the periodograms of $\{X_1, \ldots, X_n\}$ and $\{Z_1, \ldots, Z_n\}$ respectively. Then, if $\omega_k = 2\pi k/n \in [0, \pi]$, we can write*

$$I_{n,X}(\omega_k) = |\psi(e^{-i\omega_k})|^2 I_{n,Z}(\omega_k) + R_n(\omega_k), \tag{10.3.10}$$

where $\max_{\omega_k \in [0,\pi]} E|R_n(\omega_k)| \to 0$ as $n \to \infty$. If in addition, $\sum_{j=-\infty}^{\infty} |\psi_j| |j|^{1/2} < \infty$ and $E|Z_1|^4 < \infty$, then $\max_{\omega_k \in [0,\pi]} E|R_n(\omega_k)|^2 = O(n^{-1})$.

Remark 1. Observe that we can rewrite (10.3.10) as

$$I_{n,X}(\lambda) = |\psi(e^{-ig(n,\lambda)})|^2 I_{n,Z}(\lambda) + R_n(g(n,\lambda)), \tag{10.3.11}$$

where $\sup_{\lambda \in [-\pi,\pi]} E|R_n(g(n,\lambda))| \to 0$. In particular, $R_n(g(n,\lambda)) \xrightarrow{P} 0$ for every $\lambda \in [-\pi, \pi]$.

PROOF. Set $\lambda = \omega_k \in [0, \pi]$ and let $J_X(\lambda)$ and $J_Z(\lambda)$ denote the discrete Fourier transforms of $\{X_t\}$ and $\{Z_t\}$ respectively. Then

$$\begin{aligned}
J_X(\lambda) &= n^{-1/2} \sum_{t=1}^{n} X_t e^{-i\lambda t} \\
&= n^{-1/2} \sum_{j=-\infty}^{\infty} \psi_j e^{-i\lambda j} \left(\sum_{t=1}^{n} Z_{t-j} e^{-i\lambda(t-j)} \right) \\
&= n^{-1/2} \sum_{j=-\infty}^{\infty} \psi_j e^{-i\lambda j} \left(\sum_{t=1-j}^{n-j} Z_t e^{-i\lambda t} \right) \\
&= n^{-1/2} \sum_{j=-\infty}^{\infty} \psi_j e^{-i\lambda j} \left(\sum_{t=1}^{n} Z_t e^{-i\lambda t} + U_{nj} \right),
\end{aligned}$$

i.e.

$$J_X(\lambda) = \psi(e^{-i\lambda}) J_Z(\lambda) + Y_n(\lambda), \tag{10.3.12}$$

where $U_{nj} = \sum_{t=1-j}^{n-j} Z_t e^{-i\lambda t} - \sum_{t=1}^{n} Z_t e^{-i\lambda t}$ and $Y_n(\lambda) = n^{-1/2} \sum_{j=-\infty}^{\infty} \psi_j e^{-i\lambda j} U_{nj}$. Note that if $|j| < n$, then U_{nj} is a sum of $2|j|$ independent random variables, whereas if $|j| \geq n$, U_{nj} is a sum of $2n$ independent random variables. It follows that

$$E|U_{nj}|^2 \leq 2 \min(|j|, n) \sigma^2,$$

and hence that

$$E|Y_n(\lambda)|^2 \leq \left(n^{-1/2} \sum_{j=-\infty}^{\infty} |\psi_j| (E|U_{nj}|^2)^{1/2} \right)^2.$$

Thus

$$E|Y_n(\lambda)|^2 \leq 2\sigma^2 \left(n^{-1/2} \sum_{j=-\infty}^{\infty} |\psi_j| \min(|j|, n)^{1/2} \right)^2. \tag{10.3.13}$$

Now if m is a fixed positive integer we have for $n > m$,

$$n^{-1/2} \sum_{j=-\infty}^{\infty} |\psi_j| \min(|j|, n)^{1/2} \le n^{-1/2} \sum_{|j| \le m} |\psi_j| |j|^{1/2} + \sum_{|j|>m} |\psi_j|,$$

whence

$$\overline{\lim_{n\to\infty}}\, n^{-1/2} \sum_{j=-\infty}^{\infty} |\psi_j| \min(|j|, n)^{1/2} \le \sum_{|j|>m} |\psi_j|.$$

Since m is arbitrary it follows that the bound in (10.3.13) converges to zero as $n \to \infty$. Recalling that $I_{n,X}(\omega_k) = J_X(\omega_k)J_X(-\omega_k)$, we deduce from (10.3.10) and (10.3.12) that

$$R_n(\lambda) = \psi(e^{-i\lambda})J_Z(\lambda)Y_n(-\lambda) + \psi(e^{i\lambda})J_Z(-\lambda)Y_n(\lambda) + |Y_n(\lambda)|^2.$$

Now $|\psi(e^{-i\lambda})| \le \sum_{j=-\infty}^{\infty} |\psi_j| < \infty$ and $E|J_Z(\lambda)|^2 = EI_{n,Z}(\lambda) = \sigma^2$. Moreover we have shown that the bound in (10.3.13) does not depend on λ. Application of the Cauchy–Schwarz inequality therefore gives

$$\max_{\omega_k \in [0,\pi]} E|R_n(\omega_k)| \to 0 \quad \text{as } n \to \infty.$$

Finally if $E|Z_1|^4 < \infty$ and $\sum_{j=-\infty}^{\infty} |\psi_j| |j|^{1/2} < \infty$, then (see Problem 10.14)

$$E|U_{nj}|^4 \le 2|j| E|Z_1|^4 + 3(2|j|\sigma^2)^2,$$

so that

$$E|Y_n(\lambda)|^4 \le n^{-2}\left(\sum_{j=-\infty}^{\infty} |\psi_j|(2|j| E|Z_1|^4 + 12|j|^2\sigma^4)^{1/4}\right)^4$$

$$= O(n^{-2}).$$

Hence by applying the Cauchy–Schwarz inequality and Proposition 10.3.2 to each of the terms in $R_n^2(\lambda)$, we obtain

$$\max_{\omega_k \in [0,\pi]} E|R_n(\omega_k)|^2 = O(n^{-1})$$

as desired. □

Theorem 10.3.2. *Let $\{X_t\}$ be the linear process,*

$$X_t = \sum_{j=-\infty}^{\infty} \psi_j Z_{t-j}, \qquad \{Z_t\} \sim \text{IID}(0, \sigma^2),$$

where $\sum_{j=-\infty}^{\infty} |\psi_j| < \infty$. Let $I_n(\lambda)$ denote the periodogram of $\{X_1, \ldots, X_n\}$ and let $f(\lambda)$ be the spectral density of $\{X_t\}$.

(i) If $f(\lambda) > 0$ for all $\lambda \in [-\pi, \pi]$ and if $0 < \lambda_1 < \cdots < \lambda_m < \pi$, then the random vector $(I_n(\lambda_1), \ldots, I_n(\lambda_m))'$ converges in distribution to a vector of independent and exponentially distributed random variables, the ith component of which has mean $2\pi f(\lambda_i)$, $i = 1, \ldots, m$.

(ii) If $\sum_{j=-\infty}^{\infty} |\psi_j| |j|^{1/2} < \infty$, $EZ_1^4 = \eta\sigma^4 < \infty$, $\omega_j = 2\pi j/n \ge 0$ and $\omega_k = 2\pi k/n \ge 0$, then

$$\operatorname{Cov}(I_n(\omega_j), I_n(\omega_k)) = \begin{cases} 2(2\pi)^2 f^2(\omega_j) + O(n^{-1/2}) & \text{if } \omega_j = \omega_k = 0 \text{ or } \pi, \\ (2\pi)^2 f^2(\omega_j) + O(n^{-1/2}) & \text{if } 0 < \omega_j = \omega_k < \pi, \\ O(n^{-1}) & \text{if } \omega_j \neq \omega_k, \end{cases}$$

where the terms $O(n^{-1/2})$ and $O(n^{-1})$ can be bounded uniformly in j and k by $c_1 n^{-1/2}$ and $c_2 n^{-1}$ respectively, for some positive constants c_1 and c_2.

PROOF. From Theorem 10.3.1, we have

$$I_n(\lambda_j) = I_n(g(n, \lambda_j)) = 2\pi f(g(n, \lambda_j))\sigma^{-2} I_{n,Z}(\lambda_j) + R_n(g(n, \lambda_j)).$$

Since $f(g(n, \lambda_j)) \to f(\lambda_j)$ and $R_n(g(n, \lambda_j)) \xrightarrow{P} 0$, the result (i) follows immediately from Propositions 10.3.2 and 6.3.8.

Now if $\sum_{j=-\infty}^{\infty} |\psi_j| |j|^{1/2} < \infty$ and $EZ_1^4 < \infty$ then from (10.3.11) we have

$$\begin{aligned} \operatorname{Var}(I_n(\omega_k)) &= (2\pi f(\omega_k)/\sigma^2)^2 \operatorname{Var}(I_{n,Z}(\omega_k)) + \operatorname{Var}(R_n(\omega_k)) \\ &\quad + 2(2\pi f(\omega_k))/\sigma^2) \operatorname{Cov}(I_{n,Z}(\omega_k), R_n(\omega_k)). \end{aligned}$$

Since $\operatorname{Var}(R_n(\omega_k)) \leq E|R_n(\omega_k)|^2 = O(n^{-1})$ and since $\operatorname{Var}(I_{n,Z}(\omega_k))$ is bounded uniformly in ω_k, the Cauchy–Schwarz inequality implies that $\operatorname{Cov}(I_{n,Z}(\omega_k), R_n(\omega_k)) = O(n^{-1/2})$. It therefore follows from (10.3.6) and Proposition 10.3.2 that

$$\operatorname{Var}(I_n(\omega_k)) = \begin{cases} 2(2\pi)^2 f^2(\omega_k) + O(n^{-1/2}) & \text{if } \omega_k = 0 \text{ or } \pi, \\ (2\pi)^2 f^2(\omega_k) + O(n^{-1/2}) & \text{if } 0 < \omega_k < \pi. \end{cases}$$

A similar argument also gives

$$\operatorname{Cov}(I_n(\omega_j), I_n(\omega_k)) = O(n^{-1/2}) \quad \text{if } \omega_j \neq \omega_k.$$

In order to improve the bound from $O(n^{-1/2})$ to $O(n^{-1})$ in this last relation we follow the argument of Fuller (1976).

Set $\omega = \omega_j$ and $\lambda = \omega_k$ with $\lambda \neq \omega$. Then by the definition of the periodogram, we have

$$\begin{aligned} &E(I_n(\omega)I_n(\lambda)) - EI_n(\omega)EI_n(\lambda) \\ &= E\left\{n^{-2} \sum_{s=1}^{n} \sum_{t=1}^{n} \sum_{u=1}^{n} \sum_{v=1}^{n} X_s X_t X_u X_v e^{-i\omega(s-t)} e^{-i\lambda(u-v)}\right\} \\ &\quad - \left(n^{-1} \sum_{|h|<n} (n - |h|)\gamma(h) e^{-i\omega h}\right)\left(n^{-1} \sum_{|k|<n} (n - |k|)\gamma(k) e^{-i\lambda k}\right). \end{aligned}$$

By the same steps taken in the proof of Proposition 7.3.1, the above expression may be written as the sum of the following three terms:

$$n^{-2}(\eta - 3)\sigma^4 \sum_{s=1}^{n} \sum_{t=1}^{n} \sum_{u=1}^{n} \sum_{v=1}^{n} \sum_{j=-\infty}^{\infty} \psi_j \psi_{t-s+j} \psi_{u-s+j} \psi_{v-s+j} e^{i\omega(t-s)} e^{i\lambda(v-u)}, \tag{10.3.14}$$

$$\left(n^{-1} \sum_{s=1}^{n} \sum_{u=1}^{n} \gamma(u - s) e^{-i\omega s} e^{-i\lambda u}\right)\left(n^{-1} \sum_{t=1}^{n} \sum_{v=1}^{n} \gamma(v - t) e^{i\omega t} e^{i\lambda v}\right) \tag{10.3.15}$$

and

$$\left(n^{-1}\sum_{s=1}^{n}\sum_{v=1}^{n}\gamma(v-s)e^{-i\omega s}e^{i\lambda v}\right)\left(n^{-1}\sum_{t=1}^{n}\sum_{u=1}^{n}\gamma(u-t)e^{i\omega t}e^{-i\lambda u}\right). \tag{10.3.16}$$

By interchanging the order of summation, we see that the first term is bounded by

$$n^{-2}(\eta-3)\sigma^4\left(\sum_{j=-\infty}^{\infty}|\psi_j|\right)^4 = O(n^{-2}). \tag{10.3.17}$$

Now the first factor of (10.3.15) can be written as

$$n^{-1}\sum_{s=1}^{n}\sum_{u=1}^{n}\gamma(u-s)e^{-i\omega(s-u)}e^{-i(\lambda+\omega)u} = n^{-1}\sum_{u=1}^{n}\sum_{s=1-u}^{n-u}\gamma(s)e^{-i\omega s}e^{-i(\omega+\lambda)u},$$

from which it follows that

$$\begin{aligned} &n^{-1}\sum_{s=1}^{n}\sum_{u=1}^{n}\gamma(u-s)e^{-i\omega s}e^{-i\lambda u} \\ &= n^{-1}\left(\sum_{s=0}^{n-1}\gamma(s)e^{-i\omega s}\sum_{u=1}^{n-s}e^{-i(\omega+\lambda)u} + \sum_{s=-n+1}^{-1}\gamma(s)e^{-i\omega s}\sum_{u=1-s}^{n}e^{-i(\omega+\lambda)u}\right). \end{aligned} \tag{10.3.18}$$

However, since $\omega + \lambda = 2\pi(j+k)/n \neq 0$ or 2π, we have for $0 \le s \le n-1$,

$$\begin{aligned} \left|\sum_{u=1}^{n-s}e^{i(\omega+\lambda)u}\right| &= \left|\sum_{u=1}^{n}e^{i(\omega+\lambda)u} - \sum_{u=n-s+1}^{n}e^{i(\omega+\lambda)u}\right| \\ &= \left|0 - \sum_{u=n-s+1}^{n}e^{i(\omega+\lambda)u}\right| \\ &\le s. \end{aligned}$$

Similarly,

$$\left|\sum_{u=1-s}^{n}e^{-i(\omega+\lambda)u}\right| \le |s|, \qquad -n+1 \le s \le -1.$$

These inequalities show that the right side of (10.3.18) is bounded by

$$\begin{aligned} n^{-1}\sum_{|s|<n}|s||\gamma(s)| &\le n^{-1}\sum_{|s|<n}\sum_{j=-\infty}^{\infty}|s||\psi_j\psi_{j+s}| \\ &\le n^{-1/2}\sum_{|s|<n}\sum_{j=-\infty}^{\infty}|s|^{1/2}|\psi_j\psi_{j+s}| \\ &\le n^{-1/2}\left(\sum_{s=-\infty}^{\infty}\sum_{j=-\infty}^{\infty}|s+j|^{1/2}|\psi_{s+j}\psi_j|\right. \\ &\qquad \left. + \sum_{s=-\infty}^{\infty}\sum_{j=-\infty}^{\infty}|j|^{1/2}|\psi_j\psi_{s+j}|\right) \\ &= 2n^{-1/2}\left(\sum_{s=-\infty}^{\infty}|s|^{1/2}|\psi_s|\right)\left(\sum_{j=-\infty}^{\infty}|\psi_j|\right) \\ &= O(n^{-1/2}). \end{aligned}$$

Hence

$$\left| n^{-1} \sum_{s=1}^{n} \sum_{u=1}^{n} \gamma(u-s)e^{-i\omega s}e^{-i\lambda u} \right| = O(n^{-1/2}). \tag{10.3.19}$$

The relation (10.3.19) remains valid if ω is replaced by $-\omega$ or if λ is replaced by $-\lambda$ or both. The terms (10.3.15) and (10.3.16) are therefore of order $O(n^{-1})$. Taking into account (10.3.17) we deduce that $\text{Cov}(I_n(\omega), I_n(\lambda)) = O(n^{-1})$ as desired. □

A good estimator $\hat{\theta}_n$ of any parameter θ should be at least consistent, i.e. should converge in probability to θ as $n \to \infty$. However Theorem 10.3.2 shows that $I_n(\lambda)/2\pi$ is not a consistent estimator of $f(\lambda)$. Since for large n the periodogram ordinates are approximately uncorrelated with variances changing only slightly over small frequency intervals, we might hope to construct a consistent estimator of $f(\lambda)$ by averaging the periodogram ordinates in a small neighborhood of λ (just as we obtain a consistent estimator of a population mean by averaging the observed values in a random sample of size n). The number of Fourier frequencies in a given interval increases approximately linearly with n. By averaging the periodogram ordinates over a suitably increasing number of frequencies in a neighborhood of λ, we can indeed construct consistent spectral density estimators as shown in the following section.

§10.4 Smoothing the Periodogram

Let $\{X_t\}$ be the linear process

$$X_t = \sum_{j=-\infty}^{\infty} \psi_j Z_{t-j}, \qquad \{Z_t\} \sim \text{IID}(0, \sigma^2),$$

where $\sum_{j=-\infty}^{\infty} |\psi_j| |j|^{1/2} < \infty$. If $I_n(\omega_j)$, $j \in F_n$, is the periodogram based on $X_1, \ldots, X_n$, then we may write

$$(2\pi)^{-1} I_n(\omega_j) = f(\omega_j) + U_j f(\omega_j), \qquad j = 1, \ldots, [(n-1)/2],$$

where f is defined by (10.3.1) and the sequence $\{U_j\}$ (by Theorem 10.3.2) is approximately WN(0, 1) for large n. In other words, we may think of $(2\pi)^{-1} I(\omega_j)$, $j = 1, \ldots, [(n-1)/2]$, as an uncorrelated time series with a trend $f(\omega_j)$ which we wish to estimate. The considerations of Section 1.4 suggest estimating $f(\omega_j)$ by smoothing the series $\{I_n(\omega_j)\}$, with the aid, for example, of the simple moving average filter,

$$(2\pi)^{-1} \sum_{|k| \le m} (2m+1)^{-1} I_n(\omega_{j+k}).$$

More generally we shall consider the class of estimators having the form

$$\hat{f}(\omega_j) = (2\pi)^{-1} \sum_{|k| \le m_n} W_n(k) I_n(\omega_{j+k}), \tag{10.4.1}$$

where $\{m_n\}$ is a sequence of positive integers and $\{W_n(\cdot)\}$ is a sequence of weight functions. For notational simplicity we shall write m for m_n, the dependence on n being understood. In order for this estimate of the spectral density to be consistent (see Theorem 10.4.1 below) we impose the following conditions on m and $\{W_n(\cdot)\}$:

$$m \to \infty \quad \text{and} \quad m/n \to 0 \quad \text{as } n \to \infty, \tag{10.4.2}$$

$$W_n(k) = W_n(-k), \qquad W_n(k) \geq 0, \qquad \text{for all } k, \tag{10.4.3}$$

$$\sum_{|k| \leq m} W_n(k) = 1, \tag{10.4.4}$$

and

$$\sum_{|k| \leq m} W_n^2(k) \to 0 \quad \text{as } n \to \infty. \tag{10.4.5}$$

If $\omega_{j+k} \notin [-\pi, \pi]$, the term $I(\omega_{j+k})$ in (10.4.1) is evaluated by defining I to have period 2π. The same convention will be used to define $f(\omega)$, $\omega \notin [-\pi, \pi]$. We shall refer to the set of weights $\{W_n(k), |k| \leq m\}$ as a filter.

Definition 10.4.1 (Discrete Spectral Average Estimator). The estimator

$$\hat{f}(\omega) = \hat{f}(g(n, \omega)),$$

with $\hat{f}(\omega_j)$ defined by (10.4.1) and m and $\{W_n(\cdot)\}$ satisfying (10.4.2)–(10.4.5), is called a discrete spectral average estimator of $f(\omega)$.

The consistency of discrete spectral average estimators is established in the proof of the following theorem.

Theorem 10.4.1. *Let $\{X_t\}$ be the linear process,*

$$X_t = \sum_{j=-\infty}^{\infty} \psi_j Z_{t-j}, \qquad \{Z_t\} \sim \text{IID}(0, \sigma^2),$$

with $\sum_{j=-\infty}^{\infty} |\psi_j| |j|^{1/2} < \infty$ and $EZ_1^4 < \infty$. If $\hat{f}$ is a discrete spectral average estimator of the spectral density f, then for λ, $\omega \in [0, \pi]$,

(a) $\lim_{n \to \infty} E\hat{f}(\omega) = f(\omega)$

and

(b) $$\lim_{n \to \infty} \left(\sum_{|j| \leq m} W_n^2(j) \right)^{-1} \text{Cov}(\hat{f}(\omega), \hat{f}(\lambda)) = \begin{cases} 2f^2(\omega) & \text{if } \omega = \lambda = 0 \text{ or } \pi, \\ f^2(\omega) & \text{if } 0 < \omega = \lambda < \pi, \\ 0 & \text{if } \omega \neq \lambda. \end{cases}$$

PROOF. (a) From (10.4.1) we have

$$|E\hat{f}(\omega) - f(\omega)| = \left| \sum_{|k| \leq m} W_n(k) \left[(2\pi)^{-1} EI_n(g(n, \omega) + \omega_k) - f(g(n, \omega) + \omega_k) \right. \right.$$
$$\left. \left. + f(g(n, \omega) + \omega_k) - f(\omega) \right] \right|. \tag{10.4.6}$$

The restriction (10.4.2) on m implies that

$$\max_{|k|\le m}|g(n,\omega)+\omega_k-\omega|\to 0 \quad \text{as } n\to\infty.$$

For any given $\varepsilon>0$, this implies by the continuity of f, that

$$\max_{|k|\le m}|f(g(n,\omega)+\omega_k)-f(\omega)|\le \varepsilon/2,$$

for n sufficiently large. Moreover, by Proposition 10.3.1,

$$\max_{|k|\le m}|(2\pi)^{-1}EI_n(g(n,\omega)+\omega_k)-f(g(n,\omega)+\omega_k)|<\varepsilon/2,$$

for n sufficiently large. Noting that $\sum_{|k|\le m}W_n(k)=1$, we see from (10.4.6) that $|E\hat{f}(\omega)-f(\omega)|\le\varepsilon$ for n sufficiently large. Since ε is arbitrary, this implies that $E\hat{f}(\omega)\to f(\omega)$.

(b) From the definition of $\hat{f}$ we have

$$\begin{aligned}&\operatorname{Cov}(\hat{f}(\omega),\hat{f}(\lambda))\\ &\quad=(2\pi)^{-2}\sum_{|j|\le m}\sum_{|k|\le m}W_n(j)W_n(k)\operatorname{Cov}(I_n(g(n,\omega)+\omega_j),I_n(g(n,\lambda)+\omega_k)).\end{aligned}$$

If $\omega\ne\lambda$ and n is sufficiently large, then $g(n,\omega)+\omega_j\ne g(n,\lambda)+\omega_k$ for all $|j|$, $|k|\le m$. Hence, with c_2 as defined in Theorem 10.3.2,

$$\begin{aligned}|\operatorname{Cov}(\hat{f}(\omega),\hat{f}(\lambda))|&=\left|\sum_{|j|\le m}\sum_{|k|\le m}W_n(j)W_n(k)O(n^{-1})\right|\\ &\le c_2n^{-1}\left(\sum_{|j|\le m}W_n(j)\right)^2\\ &\le c_2n^{-1}\left(\sum_{|j|\le m}W_n^2(j)\right)(2m+1).\end{aligned}$$

Since $m/n\to 0$, this proves assertion (b) in the case $\omega\ne\lambda$.

Now suppose that $0<\omega=\lambda<\pi$. Then by Theorem 10.3.2,

$$\begin{aligned}\operatorname{Var}(\hat{f}(\omega))&=(2\pi)^{-2}\sum_{|j|\le m}W_n^2(j)((2\pi)^2f^2(g(n,\omega)+\omega_j)+O(n^{-1/2}))\\ &\quad+(2\pi)^{-2}\sum_{|j|\le m}\sum_{\substack{|k|\le m\\ k\ne j}}W_n(j)W_n(k)O(n^{-1}).\end{aligned}$$

An argument similar to that used in the proof of (a) shows that the first term is equal to

$$\left(\sum_{|j|\le m}W_n^2(j)\right)f^2(\omega)+o\left(\sum_{|j|\le m}W_n^2(j)\right).$$

The second term is bounded by

$$c_2n^{-1}(2\pi)^{-2}\left(\sum_{|j|\le m}W_n(j)\right)^2\le c_2n^{-1}(2\pi)^{-2}\sum_{|j|\le m}W_n^2(j)(2m+1).$$

Consequently

$$\left(\sum_{|j|\leq m} W_n^2(j)\right)^{-1} \operatorname{Var}(\hat{f}(\omega)) \to f^2(\omega).$$

The remaining cases $\omega = \lambda = 0$ or π are handled in a similar fashion. □

Remark 1. The assumption $\sum_{|k|\leq m} W_n^2(k) \to 0$ ensures that $\operatorname{Var}(\hat{f}(\omega)) \to 0$. Since $E\hat{f}(\omega) \to f(\omega)$, this implies that the estimator $\hat{f}(\omega)$ is mean-square consistent for $f(\omega)$. A slight modification of the proof of Theorem 10.4.1 shows in fact that

$$\sup_{-\pi\leq\omega\leq\pi} |E\hat{f}(\omega) - f(\omega)| \to 0$$

and

$$\sup_{-\pi\leq\omega\leq\pi} \operatorname{Var}(\hat{f}(\omega)) \to 0.$$

Hence $\hat{f}$ converges in mean square to f uniformly on $[-\pi, \pi]$, i.e.

$$\sup_{-\pi\leq\omega\leq\pi} E|\hat{f}(\omega) - f(\omega)|^2 = \sup_{-\pi\leq\omega\leq\pi} (\operatorname{Var}(\hat{f}(\omega)) + |E\hat{f}(\omega) - f(\omega)|^2)$$
$$\to 0.$$

Remark 2. Theorem 10.4.1 refers to a zero-mean process $\{X_t\}$. In practice we deal with processes $\{Y_t\}$ having unknown mean μ. The periodogram is then usually computed for the mean-corrected series $\{Y_t - \bar{Y}\}$ where $\bar{Y}$ is the sample mean. The periodograms of $\{Y_t\}$, $\{Y_t - \mu\}$ and $\{Y_t - \bar{Y}\}$ are all identical at the non-zero Fourier frequencies but not at frequency zero. In order to estimate $f(0)$ we therefore ignore the value of the periodogram at frequency 0 and use a slightly modified form of (10.4.1), namely

$$\hat{f}(0) = (2\pi)^{-1}\left[W_n(0)I_n(\omega_1) + 2\sum_{k=1}^{m} W_n(k)I_n(\omega_{k+1})\right]. \tag{10.4.7}$$

Moreover, whenever $I_n(0)$ appears in the moving averages (10.4.1) for $\hat{f}(\omega_j)$, $j = 1, \ldots, [n/2]$, we replace it by $2\pi\hat{f}(0)$ as defined in (10.4.7).

Example 10.4.1. For the simple moving average estimator,

$$W_n(k) = \begin{cases} (2m+1)^{-1} & \text{if } |k| \leq m, \\ 0 & \text{otherwise,} \end{cases}$$

we have $\sum_{|k|\leq m} W_n^2(k) = (2m+1)^{-1}$ so that

$$(2m+1)\operatorname{Var}(\hat{f}(\omega)) \to \begin{cases} 2f^2(\omega) & \text{if } \omega = 0 \text{ or } \pi, \\ f^2(\omega) & \text{if } 0 < \omega < \pi. \end{cases}$$

In choosing a weight function it is necessary to compromise between bias and variance of the spectral estimator. A weight function which assigns roughly equal weights to a broad band of frequencies will produce an estimate of $f(\cdot)$ which, although smooth, may have a large bias, since the estimate of $f(\omega)$ depends on values of I_n at frequencies distant from ω. On the other hand a weight function which assigns most of its weight to a narrow frequency band centered at zero will give an estimator with relatively small bias, but with a large variance. In practice it is advisable to experiment with a range of weight functions and to select the one which appears to strike a satisfactory balance between bias and variance.

EXAMPLE 10.4.2. The periodogram of 160 observations generated from the MA(1) process $X_t = Z_t - .6Z_{t-1}$, $\{Z_t\} \sim \text{WN}(0, 1)$, is displayed in Figure 10.3. Figure 10.4 shows the result of using program SPEC to apply the filter $\{\frac{1}{3}, \frac{1}{3}, \frac{1}{3}\}$ ($W_n(k) = (2m + 1)^{-1}$, $|k| \le m = 1$). As expected with such a small value of m, not much smoothing of the periodogram has occurred. Next we use a more dispersed set of weights, $W_n(0) = W_n(1) = W_n(2) = \frac{3}{21}$, $W_n(3) = \frac{2}{21}$, $W_n(4) = \frac{1}{21}$, producing the smoother spectral estimate shown in Figure 10.5. This particular weight function is obtained by successive application of the filters $\{\frac{1}{3}, \frac{1}{3}, \frac{1}{3}\}$ and $\{\frac{1}{7}, \frac{1}{7}, \frac{1}{7}, \frac{1}{7}, \frac{1}{7}, \frac{1}{7}, \frac{1}{7}\}$ to the periodogram. Thus the estimates in Figure 10.5 (except for the end-values) are obtained by applying the filter $\{\frac{1}{7}, \frac{1}{7}, \frac{1}{7}, \frac{1}{7}, \frac{1}{7}, \frac{1}{7}, \frac{1}{7}\}$ to the estimated spectral density in Figure 10.4. Applying a third filter $\{\frac{1}{11}, \frac{1}{11}, \ldots, \frac{1}{11}, \frac{1}{11}\}$ to the estimate in Figure 10.5 we obtain the still smoother spectral density estimate shown in Figure 10.6. The weight function resulting from successive application of the three filters is shown in the inset of Figure 10.6. Its weights (multiplied by 231) are $\{1, 3, 6, 9, 12, 15, 18, 20, 21, 21, 21, 20, 18, 15, 12, 9, 6, 3, 1\}$. Except for the peak at frequency ω_{75}, the estimate in Figure 10.6 has the same general form as the true spectral density. We shall see in Section 5 that the errors are in fact not large compared with their approximate standard deviations.

EXAMPLE 10.4.3 (The Wölfer Sunspot Numbers). The periodogram for the Wölfer sunspot numbers of Example 1.1.5 is shown in Figure 10.7. Inspecting this graph we notice one main peak at frequency $\omega_{10} = 2\pi(.1)$ (corresponding to a ten-year cycle) and a possible secondary peak at $\omega = \omega_{12}$. In Figure 10.8, the periodogram has been smoothed using the weight function $W_n(0) = W_n(1) = W_n(2) = \frac{3}{21}$, $W_n(3) = \frac{2}{21}$ and $W_n(4) = \frac{1}{21}$, which is obtained by successive application of the two filters $\{\frac{1}{3}, \frac{1}{3}, \frac{1}{3}\}$ and $\{\frac{1}{7}, \frac{1}{7}, \frac{1}{7}, \frac{1}{7}, \frac{1}{7}, \frac{1}{7}, \frac{1}{7}\}$ to the periodogram. In Section 10.6 we shall examine some alternative spectral density estimates for the Wölfer sunspot numbers.

Lag Window Estimators. The spectral density f is often estimated by a function of the form,

$$\hat{f}_L(\omega) = (2\pi)^{-1} \sum_{|h| \le r} w(h/r)\hat{\gamma}(h)e^{-ih\omega}, \tag{10.4.8}$$

where $\hat{\gamma}(\cdot)$ is the sample autocovariance function and $w(x)$ is an even, piecewise

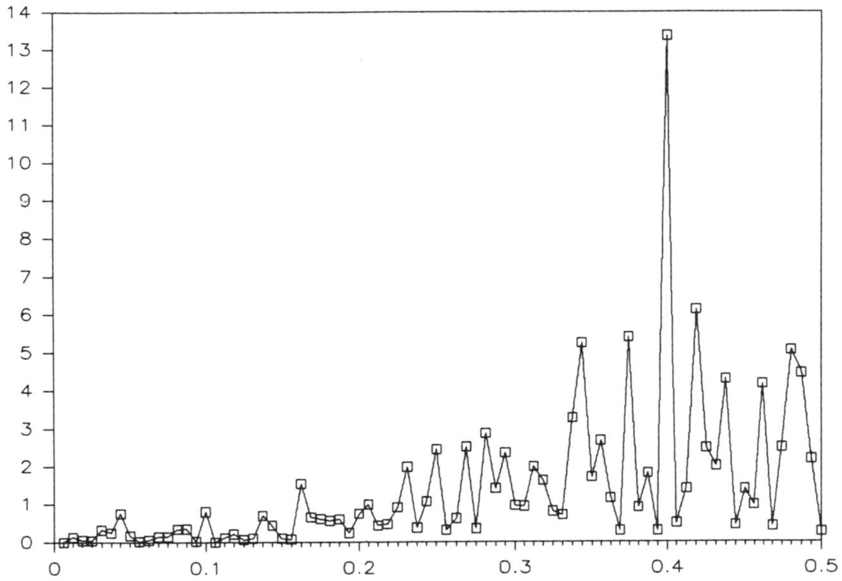

Figure 10.3. The periodogram $I_{160}(2\pi c)$, $0 < c \le 0.5$, of the simulated MA(1) series of Example 10.4.2.

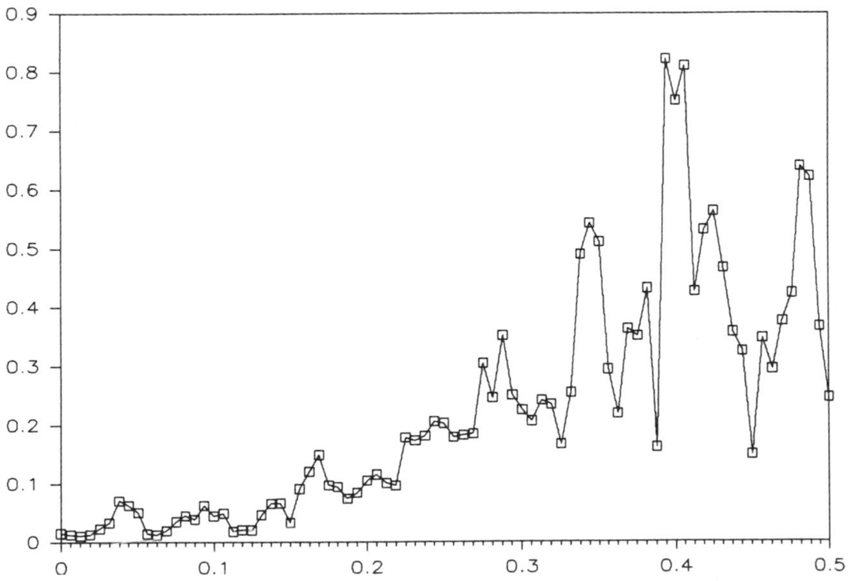

Figure 10.4. The spectral estimate $\hat{f}(2\pi c)$, $0 \le c \le 0.5$, of Example 10.4.2, obtained with the weights $\{\frac{1}{3}, \frac{1}{3}, \frac{1}{3}\}$.

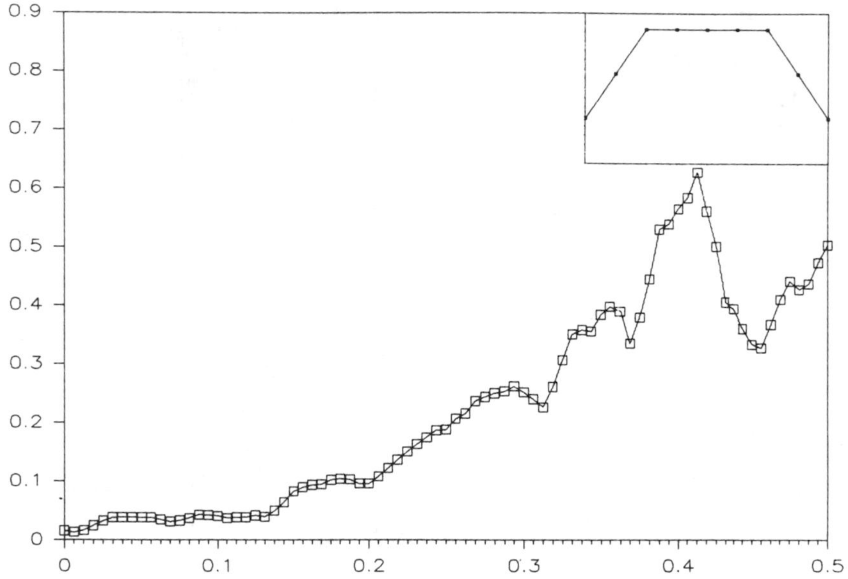

Figure 10.5. The spectral estimate $\hat{f}(2\pi c)$, $0 \leq c \leq 0.5$, of Example 10.4.2, obtained with the inset weight function.

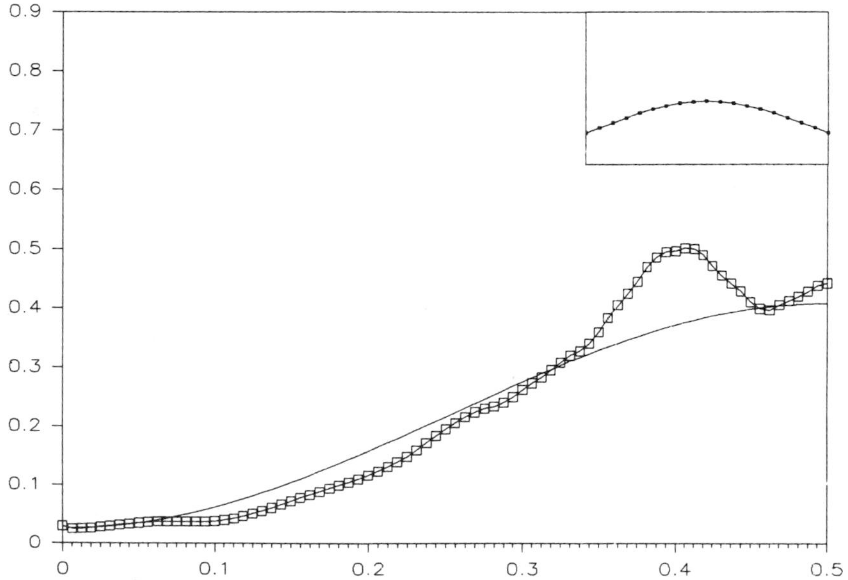

Figure 10.6. The spectral estimate $\hat{f}(2\pi c)$, $0 \leq c \leq 0.5$, of Example 10.4.2, obtained with the inset weight function.

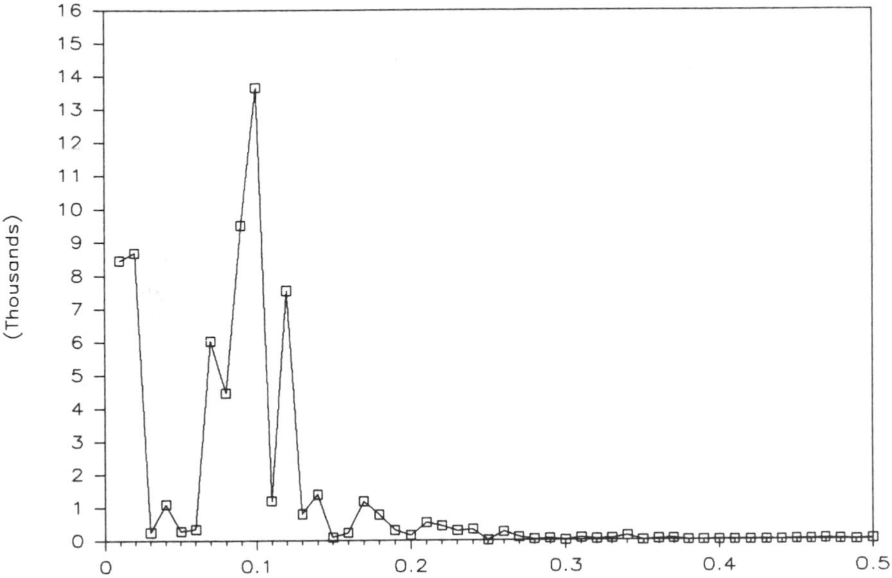

Figure 10.7. The periodogram $I_{100}(2\pi c)$, $0 < c \le 0.5$, of the Wölfer sunspot numbers.

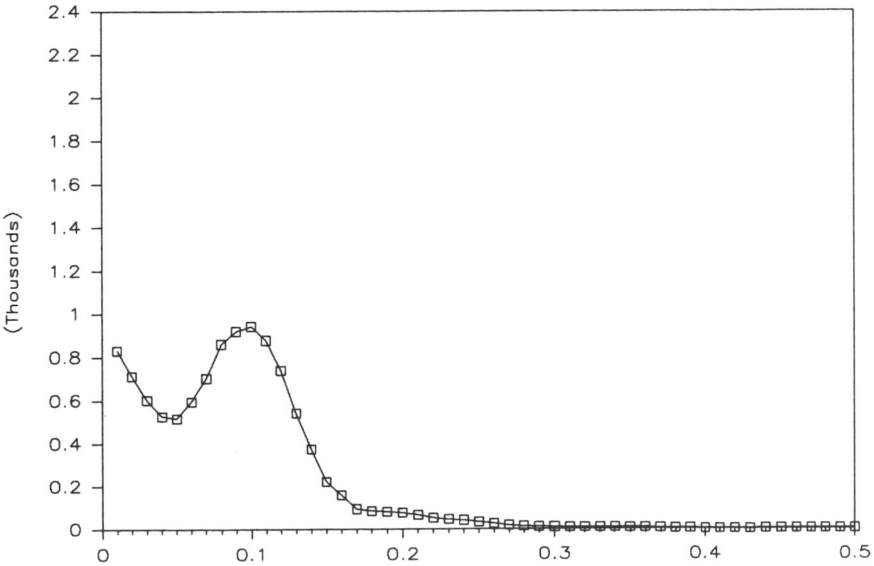

Figure 10.8. The spectral estimate $\hat{f}(2\pi c)$, $0 < c \le 0.5$, of the Wölfer sunspot numbers, obtained with the same weight function as Figure 10.5.

continuous function of x satisfying the conditions,

$$w(0) = 1,$$

$$|w(x)| \le 1, \qquad \text{for all } x,$$

and

$$w(x) = 0, \qquad \text{for } |x| > 1.$$

The function $w(\cdot)$ is called the *lag window*, and the corresponding estimator $\hat{f}_L$ is called the *lag window spectral density estimator*. By setting $w(x) \equiv 1$, $|x| \le 1$, and $r = n$, we obtain $2\pi\hat{f}_L(\omega) = I_n(\omega)$ for all Fourier frequencies $\omega = \omega_j \ne 0$. However if we assume that $r = r_n$ is a function of n such that $r \to \infty$ and $r/n \to 0$ as $n \to \infty$, then $\hat{f}_L$ is a sum of $(2r + 1)$ terms, each with a variance which is $O(n^{-1})$. If $\{r_n\}$ satisfies these conditions and $\{X_t\}$ satisfies the conditions of Theorem 10.4.1, then it can be shown that $\hat{f}_L(\omega)$ is in fact a mean-square consistent estimator of $f(\omega)$.

Although the estimator $\hat{f}_L(\omega)$ and the discrete spectral average estimator $\hat{f}(\omega)$ defined by (10.4.1) appear to be quite different, it is possible to approximate a given lag window estimator by a corresponding average of periodogram ordinates. In order to do this define a *spectral window*,

$$W(\omega) = (2\pi)^{-1} \sum_{|h| \le r} w(h/r)e^{-ih\omega}, \tag{10.4.9}$$

and an extension of the periodogram,

$$\tilde{I}_n(\omega) = \sum_{|h| < n} \hat{\gamma}(h)e^{-ih\omega}.$$

Then $\tilde{I}_n$ coincides with the periodogram I_n at the non-zero Fourier frequencies $2\pi j/n$ and moreover,

$$\hat{\gamma}(h) = (2\pi)^{-1} \int_{-\pi}^{\pi} e^{ih\lambda}\tilde{I}_n(\lambda)\, d\lambda.$$

Substituting this expression into (10.4.8) we get

$$\begin{aligned}
\hat{f}_L(\omega) &= (2\pi)^{-2} \sum_{|h| \le r} w(h/r) \int_{-\pi}^{\pi} e^{-ih(\omega-\lambda)}\tilde{I}_n(\lambda)\, d\lambda \\
&= (2\pi)^{-2} \int_{-\pi}^{\pi} \left(\sum_{|h| \le r} w(h/r)e^{-ih(\omega-\lambda)} \right) \tilde{I}_n(\lambda)\, d\lambda \\
&= (2\pi)^{-1} \int_{-\pi}^{\pi} W(\omega - \lambda)\tilde{I}_n(\lambda)\, d\lambda \\
&= (2\pi)^{-1} \int_{-\pi}^{\pi} W(\lambda)\tilde{I}_n(\omega + \lambda)\, d\lambda.
\end{aligned}$$

Partitioning the interval $[-\pi, \pi]$ at the Fourier frequencies and replacing the last integral by the corresponding Riemann sum, we obtain

$$\hat{f}_L(\omega) \simeq (2\pi)^{-1} \sum_{|j| \le [n/2]} W(\omega_j)\tilde{I}_n(\omega + \omega_j)2\pi/n$$

$$\simeq (2\pi)^{-1} \sum_{|j| \le [n/2]} W(\omega_j)I_n(g(n, \omega) + \omega_j)2\pi/n.$$

Thus we have approximated $\hat{f}_L(\omega)$ by a discrete spectral average with weights

$$W_n(j) = 2\pi W(\omega_j)/n, \qquad |j| \le [n/2]. \tag{10.4.10}$$

(Notice that the approximating spectral average does not necessarily satisfy the constraints (10.4.2)–(10.4.5) imposed earlier.)

From (10.4.10) we have

$$\sum_{|j| \le [n/2]} W_n^2(j) = (2\pi)^2 \sum_{|j| \le [n/2]} W^2(\omega_j)/n^2$$

$$\simeq \frac{2\pi}{n} \int_{-\pi}^{\pi} W^2(\omega)\, d\omega$$

$$= \frac{1}{n} \sum_{|h| \le r} w^2(h/r) \qquad \text{(by (10.4.9))}$$

$$\simeq \frac{r}{n} \int_{-1}^{1} w^2(x)\, dx.$$

Although the approximating spectral average does not satisfy the conditions of Theorem 10.4.1, the conclusion of the theorem suggests that as $n \to \infty$,

$$\frac{n}{r}\operatorname{Var}(\hat{f}_L(\omega)) \to \begin{cases} 2f^2(\omega) \displaystyle\int_{-1}^{1} w^2(x)\, dx & \text{if } \omega = 0 \text{ or } \pi, \\ f^2(\omega) \displaystyle\int_{-1}^{1} w^2(x)\, dx & \text{if } 0 < \omega < \pi. \end{cases} \tag{10.4.11}$$

If $\{X_t\}$ satisfies the conditions of Theorem 10.4.1 and if $\{r_n\}$ satisfies the conditions $r_n \to \infty$ and $r_n/n \to 0$ as $n \to \infty$, then (10.4.11) is in fact true and $E\hat{f}_L(\omega) \to f(\omega)$ for $0 \le \omega \le \pi$. Proofs of these results and further discussion of $\hat{f}_L(\omega)$ can be found in the books of Anderson (1971), Brillinger (1981) and Hannan (1970).

Examples. We conclude this section by listing some commonly used lag windows and the corresponding spectral windows $W(\cdot)$ as defined by (10.4.9).

Example 1 (The Rectangular or Truncated Window). This window has the form

$$w(x) = \begin{cases} 1 & \text{if } |x| \le 1, \\ 0 & \text{otherwise,} \end{cases}$$

and the corresponding spectral window is given by the Dirichlet kernel (see Figure 2.2),

$$W(\omega) = (2\pi)^{-1}\frac{\sin((r+\frac{1}{2})\omega)}{\sin(\omega/2)}. \tag{10.4.12}$$

Observe that $W(\omega)$ is negative for certain values of ω. This may lead to negative estimates of the spectral density at certain frequencies. From (10.4.11) we have, as $n \to \infty$,

$$\mathrm{Var}(\hat{f}_L(\omega)) \sim \frac{2r}{n}f^2(\omega) \quad \text{for } 0 < \omega < \pi.$$

EXAMPLE 2 (The Bartlett or Triangular Window). In this case

$$w(x) = \begin{cases} 1 - |x| & \text{if } |x| \le 1, \\ 0 & \text{if } |x| > 1, \end{cases}$$

and the corresponding spectral window is given by the Fejer kernel (see Figure 2.3),

$$W(\omega) = (2\pi r)^{-1}\frac{\sin^2(r\omega/2)}{\sin^2(\omega/2)}.$$

Since $W(\omega) \ge 0$, this window always gives non-negative spectral density estimates. Moreover, as $n \to \infty$,

$$\mathrm{Var}(\hat{f}_L(\omega)) \sim \frac{r}{n}f^2(\omega)\int_{-1}^{1} w^2(x)\,dx = \frac{2r}{3n}f^2(\omega), \qquad 0 < \omega < \pi.$$

The asymptotic variance is thus smaller than that of the rectangular lag window estimator using the same sequence $\{r_n\}$.

EXAMPLE 3 (The Daniell Window). From (10.4.10) we see that the spectral window,

$$W(\omega) = \begin{cases} r/2\pi, & |\omega| \le \pi/r, \\ 0, & \text{otherwise}, \end{cases}$$

corresponds to the discrete spectral average estimator with weights

$$W_n(j) = (2m+1)^{-1}, \qquad |j| \le m = [n/2r].$$

From (10.4.9) we find that the lag window corresponding to $W(\omega)$ is

$$w(h/r) = \int_{-\pi}^{\pi} W(\omega)e^{ih\omega}\,d\omega = \pi^{-1}(r/h)\sin(\pi h/r),$$

i.e.

$$w(x) = \sin(\pi x)/(\pi x), \qquad -1 \le x \le 1.$$

The corresponding lag window estimator has asymptotic variance

$$\mathrm{Var}(\hat{f}_L(\omega)) \sim rf^2(\omega)/n, \qquad 0 < \omega < \pi.$$

EXAMPLE 4 (The Blackman–Tukey Window). This lag window has the general form

$$w(x) = \begin{cases} 1 - 2a + 2a\cos x, & |x| \le 1, \\ 0, & \text{otherwise}, \end{cases}$$

with corresponding spectral window,

$$W(\omega) = aD_r(\omega - \pi/r) + (1 - 2a)D_r(\omega) + aD_r(\omega + \pi/r),$$

where D_r is the Dirichlet kernel, (10.4.12). The asymptotic variance of the corresponding density estimator is

$$\operatorname{Var}(\hat{f}_L(\omega)) \sim 2r(1 - 4a + 6a^2)f^2(\omega)/n, \qquad 0 < \omega < \pi.$$

The Blackman–Tukey windows with $a = .23$ and $a = .25$ are often referred to as the Tukey–Hamming and Tukey–Hanning windows respectively.

EXAMPLE 5 (The Parzen Window). This lag window is defined to be

$$w(x) = \begin{cases} 1 - 6|x|^2 + 6|x|^3, & |x| < \frac{1}{2}, \\ 2(1 - |x|)^3, & \frac{1}{2} \le |x| \le 1, \\ 0, & \text{otherwise}, \end{cases}$$

with approximate spectral window,

$$W(\omega) = \frac{6}{\pi r^3} \frac{\sin^4(r\omega/4)}{\sin^4(\omega/2)}.$$

The asymptotic variance of the spectral density estimator is

$$\operatorname{Var}(\hat{f}_L(\omega)) \sim .539rf^2(\omega)/n, \qquad 0 < \omega < \pi.$$

Comparison of Lag-Window Estimators. Lag-window estimators may be compared by examining the spectral windows when the values of r for the different estimators are chosen in such a way that the estimators have the same asymptotic variance. Thus to compare the Bartlett and Daniell estimators we plot the spectral windows

$$W_B(\omega) = (2\pi r)^{-1}\sin^2(r\omega/2)/(\sin^2(\omega/2)) \text{ and } W_D'(\omega) = r'/(2\pi), \quad |\omega| \le \pi/r', \tag{10.4.13}$$

where $r' = 2r/3$. Inspection of the graphs (Problem 10.18) reveals that the mass of the window W_B is spread over a broader frequency interval and has secondary peaks or "side-lobes" at some distance from the centre. This means that the Bartlett estimator with the same asymptotic variance as the Daniell estimator is liable (depending on the spectral density being estimated) to exhibit greater bias. For other factors affecting the choice of an appropriate lag window, see Priestley (1981).

The width of the rectangular spectral window which leads to the same asymptotic variance as a given lag-window estimator is sometimes called the bandwidth of the given estimator. For example the Bartlett estimator with parameter r has bandwidth $2\pi/r' = 3\pi/r$.

§10.5 Confidence Intervals for the Spectrum

In this section we provide two approximations to the distribution of the discrete spectral average estimator $\hat{f}(\omega)$ from which confidence intervals for the spectral density $f(\omega)$ can be constructed. Assume that $\{X_t\}$ satisfies the conditions of Theorem 10.4.1 (i.e. $X_t = \sum_{j=-\infty}^{\infty} \psi_j Z_{t-j}, \sum |\psi_j| |j|^{1/2} < \infty$, $\{Z_t\} \sim \text{IID}(0, \sigma^2)$ and $EZ_t^4 < \infty$) and that $\hat{f}$ is the discrete spectral average

$$\hat{f}(\omega_j) = (2\pi)^{-1} \sum_{|k| \le m} W_n(k) I_n(\omega_j + \omega_k). \tag{10.5.1}$$

The χ^2 Approximation. By Theorem 10.3.2, the random variables $I_n(\omega_j + \omega_k)/(\pi f(\omega_j + \omega_k))$, $-j < k < n/2 - j$, are approximately independent and distributed as chi-squared with 2 degrees of freedom. This suggests approximating the distribution of $\hat{f}(\omega_j)$ by the distribution of the corresponding linear combination of independent and identically distributed $\chi^2(2)$ random variables. However, as advocated by Tukey (1949), this distribution may in turn be approximated by the distribution of cY, where c is a constant, $Y \sim \chi^2(\nu)$ and c and ν are found by the method of moments, i.e. by setting the mean and variance of cY equal to the asymptotic mean and variance of $\hat{f}(\omega_j)$. This procedure gives the equations

$$c\nu = f(\omega_j),$$
$$2c^2\nu = \sum_{|k| \le m} W_n^2(k) f^2(\omega_j),$$

from which we find that $c = \sum_{|k| \le m} W_n^2(k) f(\omega_j)/2$ and $\nu = 2/(\sum_{|k| \le m} W_n^2(k))$. The number ν is called the *equivalent degrees of freedom* of the estimator $\hat{f}$. The distribution of $\nu\hat{f}(\omega_j)/f(\omega_j)$ is thus approximated by the chi-squared distribution with ν degrees of freedom, and the interval

$$\left(\frac{\nu\hat{f}(\omega_j)}{\chi^2_{.975}(\nu)}, \frac{\nu\hat{f}(\omega_j)}{\chi^2_{.025}(\nu)}\right), \qquad 0 < \omega_j < \pi, \tag{10.5.2}$$

is an approximate 95% confidence interval for $f(\omega_j)$. By taking logarithms in (10.5.2) we obtain the 95% confidence interval

$$(\ln \hat{f}(\omega_j) + \ln \nu - \ln \chi^2_{.975}(\nu), \ln \hat{f}(\omega_j) + \ln \nu - \ln \chi^2_{.025}(\nu)),$$
$$0 < \omega_j < \pi, \tag{10.5.3}$$

for $\ln f(\omega_j)$. This interval, unlike (10.5.2) has the same width for each $\omega_j \in (0, \pi)$. In Figure 10.9, we have plotted the confidence intervals (10.5.3) for the data

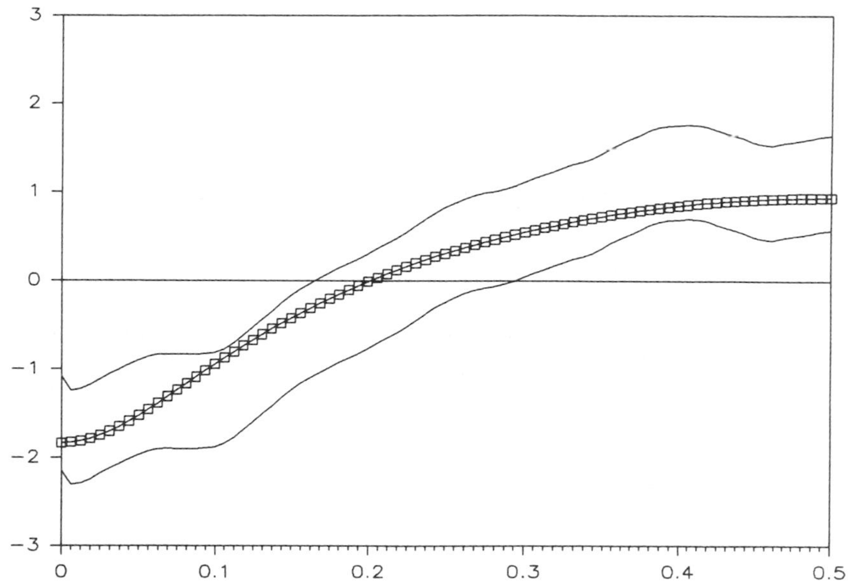

Figure 10.9. 95% confidence intervals for $\ln(2\pi f(2\pi c))$ based on the spectral estimates of Figure 10.6 and a χ^2 approximation. The true function is also shown.

of Example 10.4.2 using the spectral estimate displayed in Figure 10.6. Using the weights specified in Example 10.4.2 we find that $\sum_{|k|\le m} W_n^2(k) = .07052$ and $v = 28.36$ so that (10.5.3) reduces to the interval

$$C_{\omega_j} = (\ln \hat{f}(\omega_j) - .450, \ln \hat{f}(\omega_j) + .617). \tag{10.5.4}$$

Notice that this is a confidence interval for $\ln f(\omega_j)$ only, and the intervals $\{C_{\omega_j}, 0 < \omega_j < \pi\}$ are not to be interpreted as simultaneous 95% confidence intervals for $\{\ln f(\omega_j), 0 < \omega_j < \pi\}$. The probability that C_{ω_j} contains $\ln f(\omega_j)$ for *all* $\omega_j \in (0, \pi)$ is less than .95. However we would expect the intervals C_{ω_j} to include $\ln f(\omega_j)$ for approximately 95% of the frequencies $\omega_j \in (0, \pi)$. As can be seen in Figure 10.9, the true log spectral density lies well within the confidence interval (10.5.4) for all frequencies.

The Normal Approximation. There are two intuitive justifications for making a normal approximation to the distribution of $\hat{f}(\omega_j)$. The first is that if the equivalent number of degrees of freedom v is large (i.e. if $\sum_{|k|\le m} W_n^2(k)$ is small) and if Y is distributed as $\chi^2(v)$, then the distribution of cY can be well approximated by a normal distribution with mean $cv = f(\omega_j)$ and variance $2c^2 v = \sum_{|k|\le m} W_n^2(k) f^2(\omega_j)$, $0 < \omega_j < \pi$. The second is that we may approximate $\hat{f}(\omega_j)$ for n large by a sum of $(2m + 1)$ independent random variables, which by the Lindeberg condition, is $\text{AN}(f(\omega_j), \sum_{|k|\le m} W_n^2(k) f^2(\omega_j))$. Both points of view lead to the approximation $\text{N}(f(\omega_j), \sum_{|k|\le m} W_n^2(k) f^2(\omega_j))$ for the

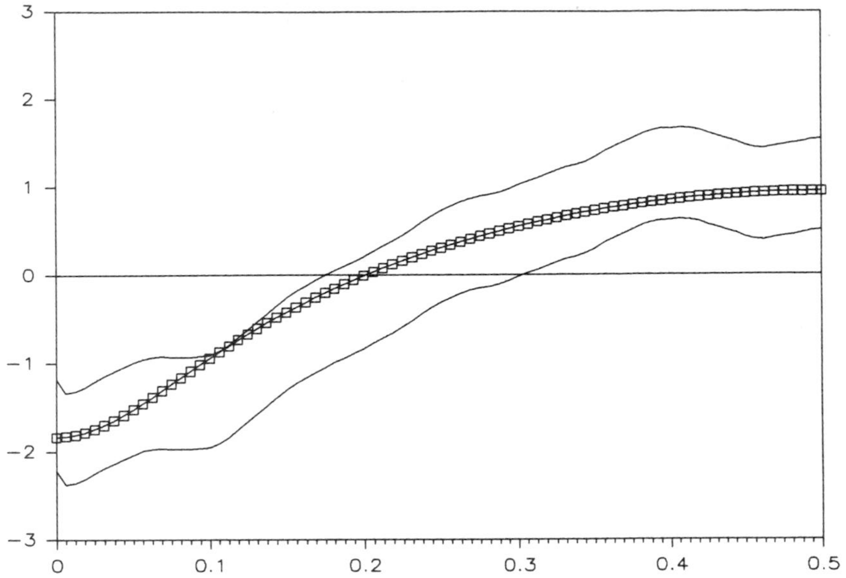

Figure 10.10. 95% confidence intervals for $\ln(2\pi f(2\pi c))$ based on the spectral estimates of Figure 10.6 and a normal approximation. The true function is also shown.

distribution of $\hat{f}(\omega_j)$. Using this approximation we obtain the approximate 95% confidence bounds,

$$\hat{f}(\omega_j) \pm 1.96\left(\sum_{|k|\leq m} W_n^2(k)\right)^{1/2} \hat{f}(\omega_j),$$

for $f(\omega_j)$. Since the width of the confidence interval depends on $\hat{f}(\omega_j)$, it is customary to construct a confidence interval for $\ln f(\omega_j)$. The normal approximation to $\hat{f}(\omega_j)$ implies that $\ln \hat{f}(\omega_j)$ is $\text{AN}(\ln f(\omega_j), \sum_{|k|\leq m} W_n^2(k))$ by Proposition 6.4.1. Approximate 95% confidence bounds for $\ln f(\omega_j)$ are therefore given by

$$\ln \hat{f}(\omega_j) \pm 1.96\left(\sum_{|k|\leq m} W_n^2(k)\right)^{1/2}. \tag{10.5.5}$$

For the spectral estimate shown in Figure 10.6, we have $\sum_{|k|\leq m} W_n^2(k) = .07052$, so that the bounds (10.5.5) become

$$\ln \hat{f}(\omega_j) \pm .520. \tag{10.5.6}$$

These bounds are plotted in Figure 10.10. The width of the intervals (10.5.4) based on the χ^2 approximation is very close to the width of the intervals (10.5.6) based on the normal approximation. However the normal intervals are centered at $\ln \hat{f}(\omega_j)$ and are therefore located below the χ^2 intervals. This can be seen in Figure 10.10 where the spectral density barely touches the upper limit of the confidence interval. For values of $v \geq 20$, there is very little difference between the two approximations.

§10.6 Autoregressive, Maximum Entropy, Moving Average and Maximum Likelihood ARMA Spectral Estimators

The mth order autoregressive estimator $\hat{f}_m(\omega)$ of the spectral density of a stationary time series $\{X_t\}$ is the spectral density of the autoregressive process $\{Y_t\}$ defined by

$$Y_t - \hat{\phi}_{m1} Y_{t-1} - \cdots - \hat{\phi}_{mm} Y_{t-m} = Z_t, \qquad \{Z_t\} \sim \text{WN}(0, \hat{v}_m), \tag{10.6.1}$$

where $\hat{\boldsymbol{\phi}}_m = (\hat{\phi}_{m1}, \ldots, \hat{\phi}_{mm})'$ and $\hat{v}_m$ are the Yule–Walker estimators defined by (8.2.2) and (8.2.3). These estimators can easily be computed recursively using Proposition 8.2.1. Then $\gamma_y(h) = \hat{\gamma}(h)$, $h = 0, \pm 1, \ldots, \pm m$, (see Section 8.1) and

$$\hat{f}_m(\omega) = \frac{\hat{v}_m}{2\pi} |1 - \hat{\phi}_{m1} e^{-i\omega} - \cdots - \hat{\phi}_{mm} e^{-im\omega}|^{-2}. \tag{10.6.2}$$

The choice of m for which the approximating AR(m) process "best" represents the data can be made by minimizing AICC($\hat{\boldsymbol{\phi}}_m$) as defined by (9.3.4). Alternatively the CAT statistic of Parzen (1974) can be minimized. This quantity is defined for $m = 1, 2, \ldots,$ by

$$\text{CAT}(m) = n^{-1} \sum_{j=1}^{m} \tilde{v}_j^{-1} - \tilde{v}_m^{-1},$$

and for $m = 0$ by

$$\text{CAT}(0) = -1 - n^{-1},$$

where

$$\tilde{v}_j := (1 - m/n)^{-1} \hat{v}_j / \hat{v}_0, \qquad j = 1, 2, \ldots.$$

We shall use AICC for choosing m.

The mth order autoregressive estimator $\hat{f}_m(\omega)$ defined by (10.6.2) is the same as the maximum entropy estimator, i.e. the spectral density $\hat{f}$ which maximizes the entropy,

$$E = \int_{-\pi}^{\pi} \ln g(\lambda)\, d\lambda$$

over the class of all densities g which satisfy the constraints,

$$\int_{-\pi}^{\pi} e^{i\lambda h} g(\lambda)\,d\lambda = \hat{\gamma}(h), \qquad h = 0, \pm 1, \ldots, \pm m. \tag{10.6.3}$$

To show this, let $\{W_t\}$ be any zero-mean stationary process with spectral density g satisfying (10.6.3), and let $\tilde{W}_{t+1} = P_{\overline{\text{sp}}\{W_j, -\infty < j \le t\}} W_{t+1}$. Then by Kolmogorov's formula (5.8.1),

$$E(W_{t+1} - \tilde{W}_{t+1})^2 = 2\pi \exp\left\{\frac{1}{2\pi}\int_{-\pi}^{\pi} \ln g(\lambda)\,d\lambda\right\}.$$

Now for any sequence $a_1, \ldots, a_m \in \mathbb{R}$,

$$\begin{aligned} 2\pi \exp\left\{\frac{1}{2\pi}\int_{-\pi}^{\pi} \ln g(\lambda)\,d\lambda\right\} &\le E\left(W_{t+1} - \sum_{j=1}^{m} a_j W_{t+1-j}\right)^2 \\ &= E\left(Y_{t+1} - \sum_{j=1}^{m} a_j Y_{t+1-j}\right)^2, \end{aligned}$$

where $\{Y_t\}$ is the AR(m) process (10.6.1), since $\{Y_t\}$ and $\{W_t\}$ both have autocovariances $\hat{\gamma}(j)$, $0 \le j \le m$. Setting $a_j = \hat{\phi}_{mj}$, $j = 1, \ldots, m$, in the last expression and using Kolmogorov's formula for the process $\{Y_t\}$, we obtain the inequality

$$2\pi \exp\left\{\frac{1}{2\pi}\int_{-\pi}^{\pi} \ln g(\lambda)\,d\lambda\right\} \le 2\pi \exp\left\{\frac{1}{2\pi}\int_{-\pi}^{\pi} \ln \hat{f}_m(\lambda)\,d\lambda\right\},$$

as required. The idea of maximum entropy spectral estimation is due to Burg (1967). Burg's estimates $\hat{\phi}_{m1}, \ldots, \hat{\phi}_{mm}$ in (10.6.1) are however a little different from the Yule–Walker estimates.

The periodogram and the non-parametric window estimators discussed in Section 10.4 are usually less regular in appearance than autoregressive estimators. The non-parametric estimates are valuable for detecting strict periodicities in the data (Section 10.2) and for revealing features of the data which may be smoothed out by autoregressive estimation. The autoregressive procedure however has a much more clearly defined criterion for selecting m than the corresponding criteria to be considered in the selection of a spectral window. In estimating a spectral density it is wise to examine both types of density estimator. Parzen (1978) has also suggested that the cumulative periodogram should be compared with the autoregressive estimate of the spectral distribution function as an aid to autoregressive model selection in the time domain.

In the definition (10.6.2) it is natural to consider replacing the Yule–Walker estimates $\hat{\boldsymbol{\phi}}_m$ and $\hat{v}_m$ by the corresponding maximum likelihood estimates, with m again chosen to minimize the AICC value. In fact there is no need to restrict attention to autoregressive models, although these are convenient since $\hat{\boldsymbol{\phi}}_m$ is asymptotically efficient for an AR(m) process and can be computed very rapidly using Proposition 8.2.1. However there are processes, e.g. a first order moving average with $\theta_1 \simeq 1$, for which autoregressive spectral estimation performs poorly (see Example 10.6.2). To deal with cases of this kind we can use the estimate suggested by Akaike (1974), i.e.

$$\hat{f}(\omega) = \frac{\hat{\sigma}^2}{2\pi} \frac{|1 + \hat{\theta}_1 e^{-i\omega} + \cdots + \hat{\theta}_q e^{-iq\omega}|^2}{|1 - \hat{\phi}_1 e^{-i\omega} - \cdots - \hat{\phi}_p e^{-ip\omega}|^2}, \tag{10.6.4}$$

where $\hat{\boldsymbol{\phi}} = (\hat{\phi}_1, \ldots, \hat{\phi}_p)'$, $\hat{\boldsymbol{\theta}} = (\hat{\theta}_1, \ldots, \hat{\theta}_q)'$ and $\hat{\sigma}^2$ are maximum likelihood estimates of an ARMA(p, q) process fitted to the data, with p and q chosen using the AICC. We shall refer to the function $\hat{f}$ as the maximum likelihood ARMA (or MLARMA) spectral density estimate.

A simpler but less efficient estimator than (10.6.4) which is particularly useful for processes whose MA(∞) representation has rapidly decaying coefficients is the moving average estimator (Brockwell and Davis 1988(a)) given by

$$\hat{g}_m(\omega) = \frac{\hat{v}_m}{2\pi} |1 + \hat{\theta}_{m1} e^{-i\omega} + \cdots + \hat{\theta}_{mm} e^{-im\omega}|^2, \tag{10.6.5}$$

where $\hat{\boldsymbol{\theta}}_m = (\hat{\theta}_{m1}, \ldots, \hat{\theta}_{mm})'$ and $\hat{v}_m$ are the innovation estimates discussed in Section 8.3. Like the autoregressive estimator (10.6.2), $\hat{g}_m(\omega)$ can be calculated very rapidly. The choice of m can again be made by minimizing the AICC value. As is the case for the autoregressive estimator, there are processes for which the moving average estimator performs poorly (e.g. an AR(1) process with $\phi_1 \simeq 1$). The advantage of both estimators over the MLARMA estimator (10.6.4) is the substantial reduction in computation time. Moreover, under specified conditions on the growth of m with n, the asymptotic distributions of the mth order autoregressive and moving average spectral density estimators can be determined for a large class of linear processes (see Berk (1974) and Brockwell and Davis (1988(a)).

Example 10.6.1 (The Wölfer Sunspot Numbers). For the Wölfer sunspot numbers of Example 10.4.3, the minimum AICC model for the mean-corrected data was found to be,

$$X_t - 1.475X_{t-1} + .937X_{t-2} - .218X_{t-3} + .134X_{t-9} = Z_t, \tag{10.6.6}$$

with $\{Z_t\} \sim \text{WN}(0, 197.06)$ and AICC $= 826.25$.

The rescaled periodogram $(2\pi)^{-1} I_{100}(2\pi c_j)$, $c_j = 1/100, 2/100, \ldots, 50/100$, and the MLARMA estimator, $\hat{f}(2\pi c)$, $0 \leq c \leq .50$, i.e. the spectral density of the process (10.6.6), are shown in Figure 10.11.

Figure 10.12 shows the autoregressive estimators $\hat{f}_3(2\pi c)$ (with AICC $= 836.97$) and $\hat{f}_8(2\pi c)$ (with AICC $= 839.54$). The estimator $\hat{f}_3(2\pi c)$ has the smallest AICC value. The estimator $\hat{f}_8(2\pi c)$ which corresponds to the second smallest local minimum and fourth smallest overall AICC value, has a more sharply defined peak (like the periodogram) at frequency ω_{10}.

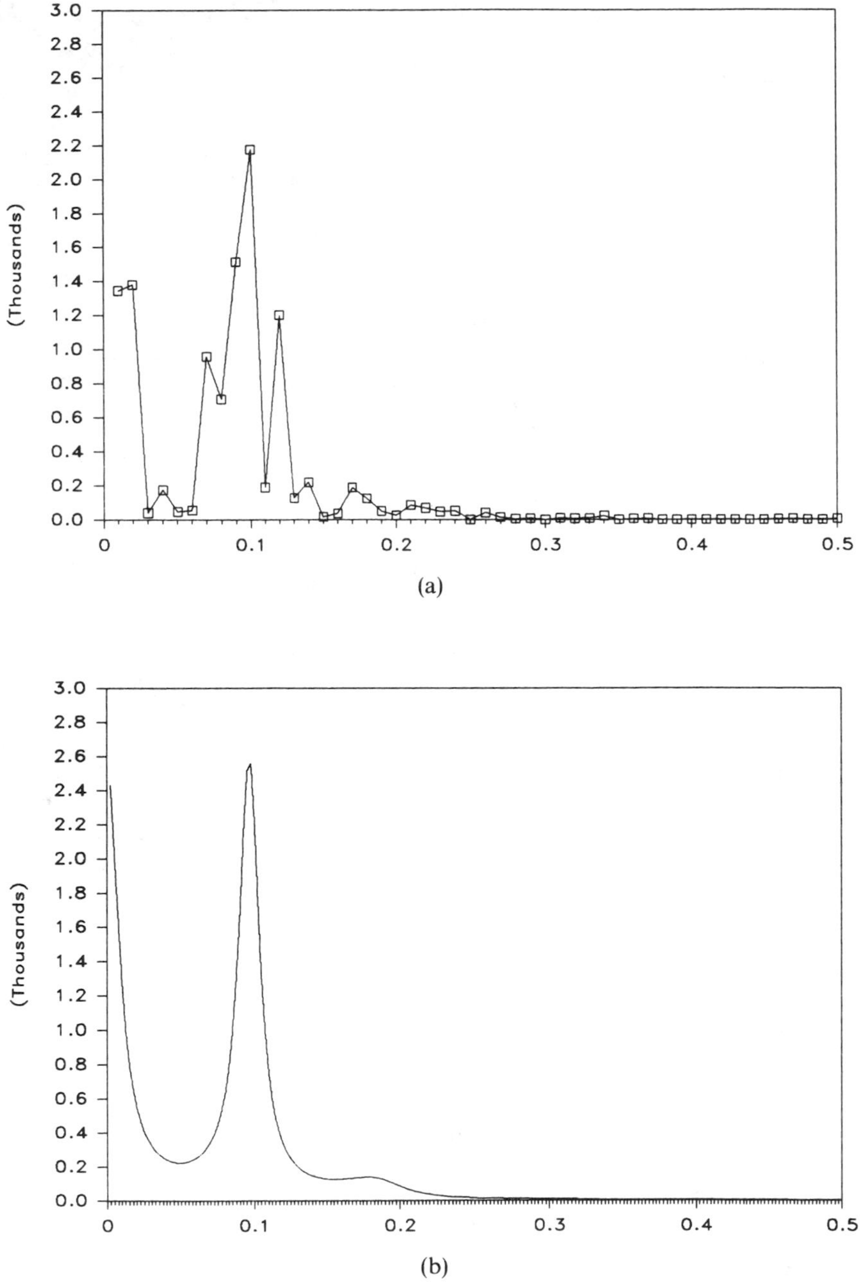

Figure 10.11. (a) The rescaled periodogram $(2\pi)^{-1}I_{100}(2\pi c)$, $0 < c \leq 0.5$, and (b) the maximum likelihood ARMA estimate $\hat{f}(2\pi c)$, for the Wölfer sunspot numbers, Example 10.6.1.

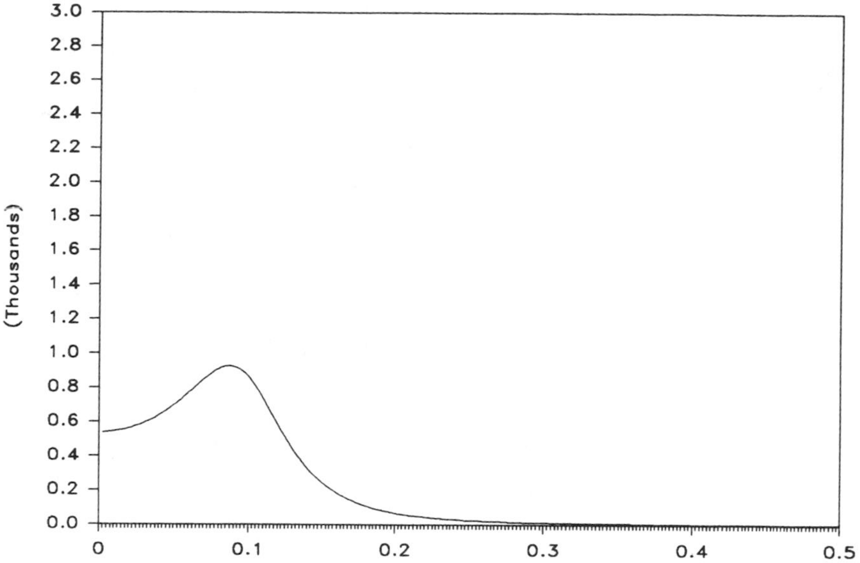

(a)

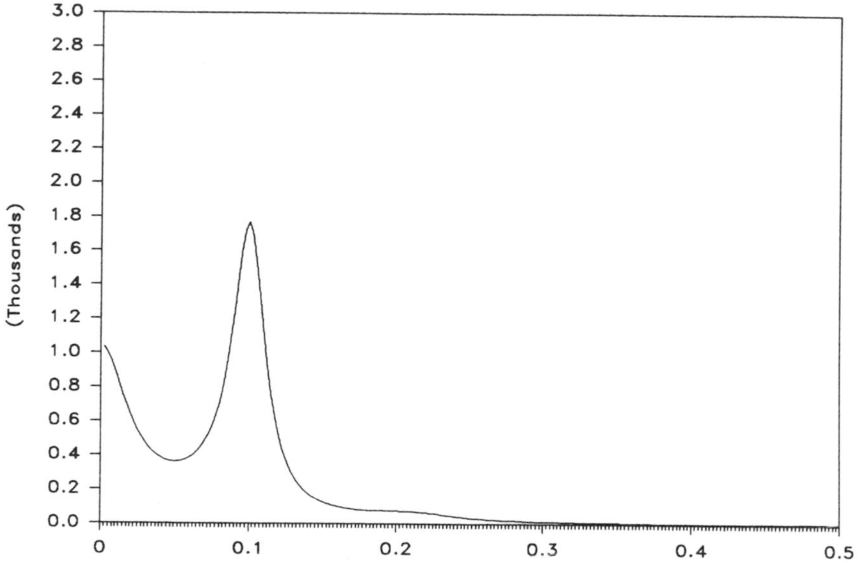

(b)

Figure 10.12. The autoregressive spectral density estimates (a) $\hat{f}_3(2\pi c)$ and (b) $\hat{f}_8(2\pi c)$ for the Wölfer sunspot numbers.

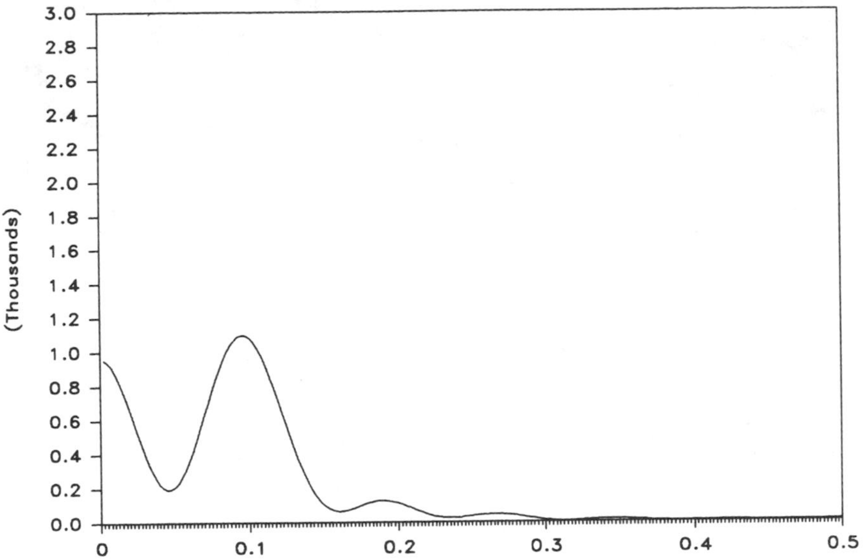

Figure 10.13. The moving average spectral density estimate $\hat{g}_{13}(2\pi c)$ for the Wölfer sunspot numbers.

Observe that there is a close resemblance between $\hat{f}_3(2\pi c)$ and the non-parametric estimate of Figure 10.8.

The moving average estimator with smallest AICC value (848.99) is $\hat{g}_{13}(2\pi c)$ shown in Figure 10.13.

EXAMPLE 10.6.2 (MA(1)). A series of 400 observations was generated using the model

$$X_t = Z_t + Z_{t-1}, \qquad \{Z_t\} \sim \mathrm{WN}(0, 1). \tag{10.6.7}$$

The spectral density of the process,

$$f(2\pi c) = |1 + e^{-i2\pi c}|^2/(2\pi), \qquad 0 \le c \le 0.50,$$

and the rescaled periodogram $(2\pi)^{-1}I_{400}(2\pi c_j)$, $c_j = 1/400, 2/400, \ldots, 200/400$, of the data are shown in Figure 10.14. The data were mean-corrected.

Figure 10.15 shows the autoregressive estimator $\hat{f}_9(2\pi c)$ with (AICC = 1162.66) and the moving average estimator $\hat{g}_6(2\pi c)$ (with AICC = 1152.06). Maximum likelihood estimation gives the minimum AICC ARMA model,

$$X_t - .116 = Z_t + 1.000Z_{t-1}, \qquad \{Z_t\} \sim \mathrm{WN}(0, .980), \tag{10.6.8}$$

with AICC = 1137.72. The MLARMA estimator of the spectral density is therefore $\hat{f}(2\pi c) = .980\ f(2\pi c)$, showing that $\hat{f}$ and f are almost indistinguishable in this example.

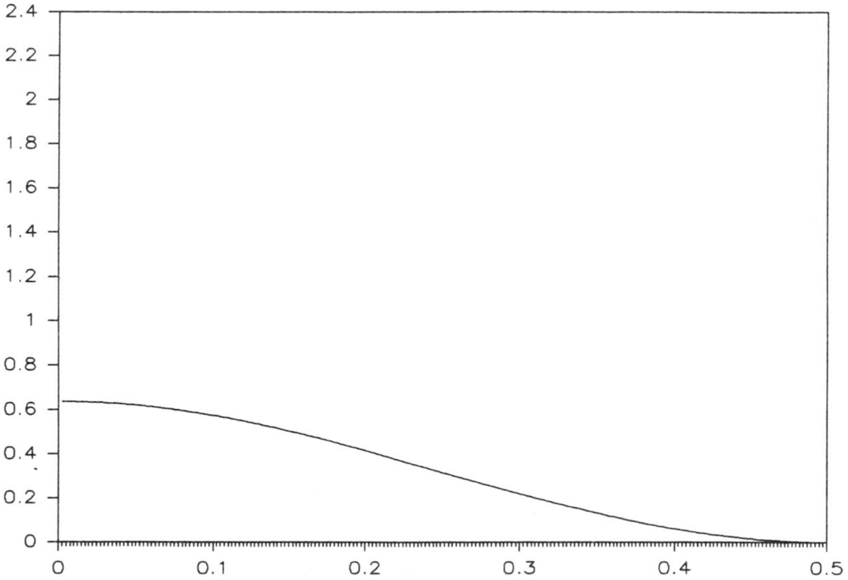

(a)

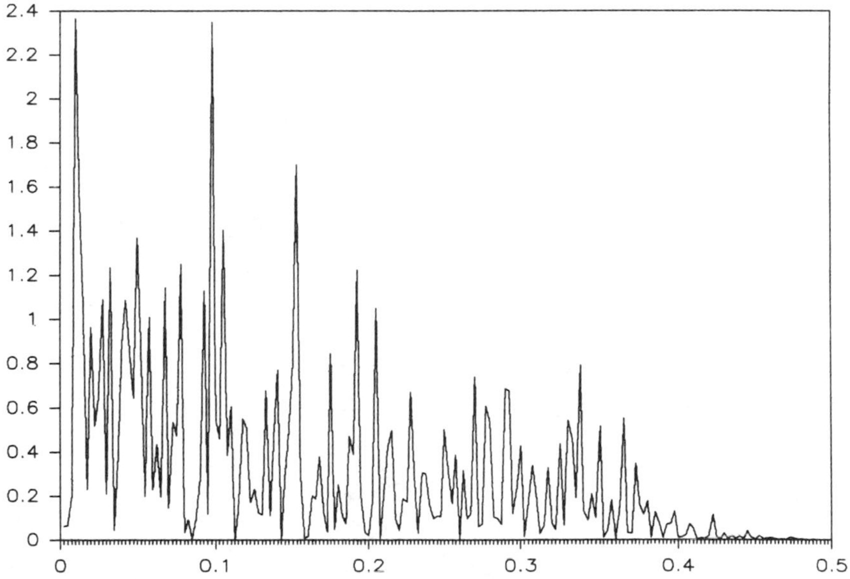

(b)

Figure 10.14. (a) The spectral density $f(2\pi c)$ and (b) the rescaled periodogram of the realization $\{X_1, \ldots, X_{400}\}$ of the process $X_t = Z_t + Z_{t-1}$, $\{Z_t\} \sim \mathrm{WN}(0, 1)$, of Example 10.6.2.

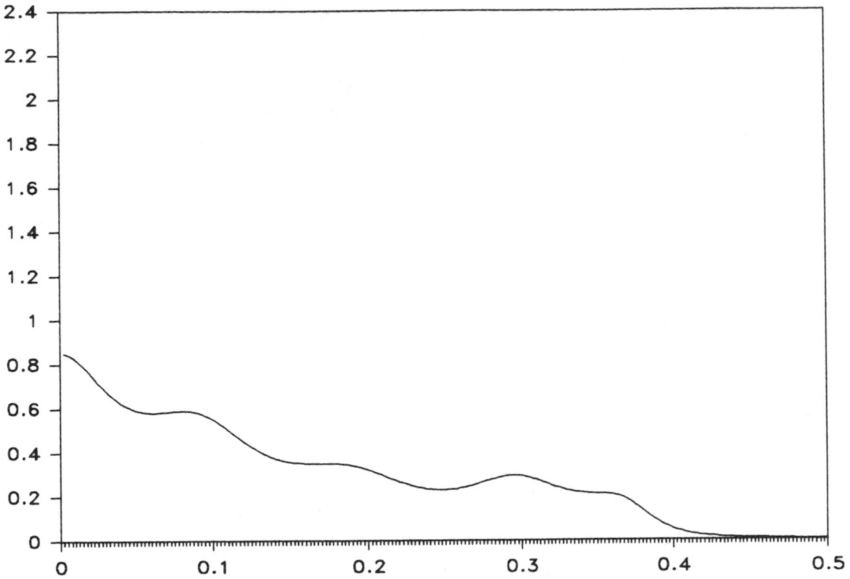

(a)

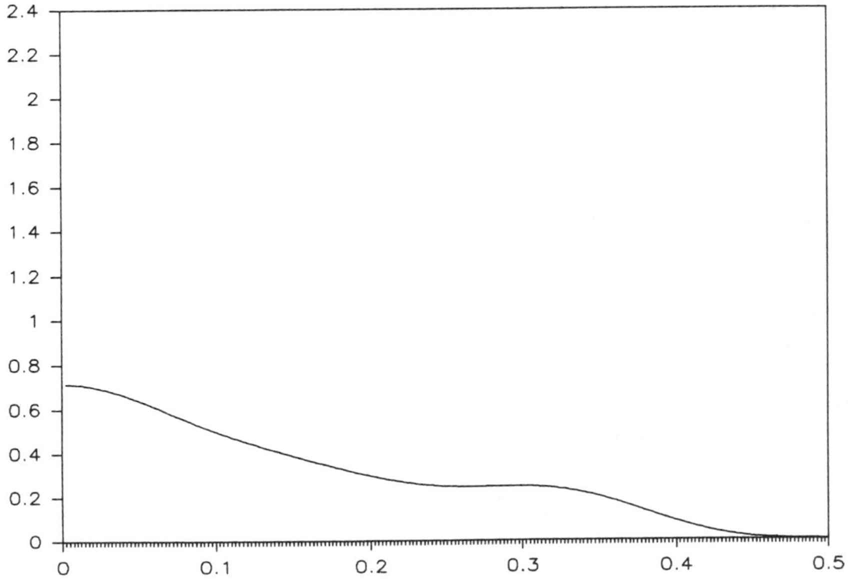

(b)

Figure 10.15. (a) The autoregressive spectral estimate $\hat{f}_9(2\pi c)$ and (b) the moving average estimate $\hat{g}_6(2\pi c)$ for the data of Example 10.6.2.

§10.7 The Fast Fourier Transform (FFT) Algorithm

A major factor in the rapid development of spectral analysis in the past twenty years has been the availability of a very fast technique for computing the discrete Fourier transform (and hence the periodogram) of long series. The algorithm which makes this possible, the FFT algorithm, was developed by Cooley and Tukey (1965) and Gentleman and Sande (1966), although some of the underlying ideas can be traced back to the beginning of this century (Cooley et al., 1967).

We first illustrate the use of the algorithm by examining the computational savings achieved when the number of observations n can be factorized as

$$n = rs. \tag{10.7.1}$$

(The computational speed is increased still more if either r or s can be factorized.) Instead of computing the transform as defined by (10.1.7), i.e.

$$a_j = n^{-1/2} \sum_{t=1}^{n} x_t e^{-i\omega_j t}, \qquad j \in F_n,$$

we shall compute the closely related transform,

$$b_j = \sum_{t=0}^{n-1} x_{t+1} e^{-2\pi ijt/n}, \qquad 0 \le j \le n-1. \tag{10.7.2}$$

Then, by straightforward algebra,

$$a_j = \begin{cases} n^{-1/2} b_j e^{-i\omega_j}, & 0 \le j \le [n/2], \\ n^{-1/2} b_{j+n} e^{-i\omega_j}, & -[(n-1)/2] \le j \le -1. \end{cases} \tag{10.7.3}$$

Under the assumption (10.7.1), each $t \in [0, n-1]$ has a unique representation,

$$t = ru + v, \qquad u \in \{0, \ldots, s-1\}, \quad v \in \{0, \ldots, r-1\}.$$

Hence (10.7.2) can be rewritten as,

$$\begin{aligned} b_j &= \sum_{v=0}^{r-1} \sum_{u=0}^{s-1} x_{ru+v+1} \exp[-2\pi ij(ru+v)/n] \\ &= \sum_{v=0}^{r-1} \exp(-2\pi ijv/n) \sum_{u=0}^{s-1} x_{ru+v+1} \exp(-2\pi iju/s), \end{aligned}$$

i.e.

$$b_j = \sum_{v=0}^{r-1} \exp(-2\pi ijv/n) b_{j,v}, \qquad 0 \le j \le n-1, \tag{10.7.4}$$

where $\{b_{j,v}, 0 \le j \le s-1\}$ is the Fourier transform,

$$b_{j,v} = \sum_{u=0}^{s-1} x_{ru+v+1} \exp(-2\pi iju/s). \tag{10.7.5}$$

If we now define an "operation" to consist of the three steps, the computation of a term $\exp(-2\pi iju/k)$, a complex multiplication and a complex addition, then for each v the calculation of $\{b_{j,v}, 0 \le j \le s-1\}$ requires a total of

$$N_s = s^2$$

operations. Since

$$b_{j+ls,v} = b_{j,v}, \qquad l = 0, 1, 2, \ldots,$$

it suffices to calculate $\{b_{j,v}, 0 \le j \le s-1\}$ in order to determine $\{b_{j,v}, 0 \le j \le n-1\}$. A total of $rN_s = rs^2$ operations is therefore required to determine $\{b_{j,v}, 0 \le v \le r-1, 0 \le j \le n-1\}$. The calculation of $\{b_j, 0 \le j \le n-1\}$ from (10.7.4) then requires another nr operations, giving a total of

$$rN_s + nr = n(r+s) \tag{10.7.6}$$

operations altogether for the computation of $\{b_j, 0 \le j \le n-1\}$. This represents a substantial savings over the n^2 operations required to compute $\{b_j\}$ directly from (10.7.2).

If now s can be factorized as $s = s_1 s_2$, then the number of operations N_s required to compute the Fourier transform (10.7.5) can be reduced by applying the technique of the preceding paragraph, replacing n by s, r by s_1 and s by s_2. From (10.7.6) we see that N_s is then reduced from s^2 to

$$N_s' = s(s_1 + s_2). \tag{10.7.7}$$

Replacing N_s in (10.7.6) by its reduced value N_s', we find that $\{b_j, 0 \le j \le n-1\}$ can now be calculated with

$$rs(s_1 + s_2) + nr = n(r + s_1 + s_2) \tag{10.7.8}$$

operations.

The argument of the preceding paragraph is easily generalized to show that if n has the prime factors $p_1, \ldots, p_k$, then the number of operations can be reduced to $n(p_1 + p_2 + \cdots + p_k)$. In particular if $n = 2^p$, then the number of operations can be reduced to $2n\log_2 n$. The savings in computer time is particularly important for very large n. For the sake of improved computational speed, a small number of dummy observations (equal to the sample mean) is sometimes appended to the data in order to make n highly composite. A small number of observations may also be deleted for the same reason. If n is very large, the resulting periodogram will not be noticeably different from that of the original data, although the Fourier frequencies $\{\omega_j, j \in F_n\}$ will be slightly different. An excellent discussion of the FFT algorithm can be found in the book of Bloomfield (1976).

We conclude this section by showing how the sample autocovariance function of a time series can be calculated by making two applications of the FFT algorithm. Let $\hat{\gamma}(k)$, $|k| < n$, denote the sample autocovariance function of $\{X_1, \ldots, X_n\}$. We first augment the series to $\mathbf{Y} = (Y_1, \ldots, Y_{2n-1})'$, where

$$Y_t = X_t - \bar{X}_n, \qquad t \le n,$$

and

$$Y_t = 0, \qquad t > n.$$

The discrete Fourier transform of **Y** is then

$$a_j = (2n-1)^{-1/2} \sum_{t=1}^{2n-1} Y_t e^{-it\lambda_j}, \qquad \lambda_j = 2\pi j/(2n-1), \quad j \in F_{2n-1}, \qquad (10.7.9)$$

where $F_{2n-1} = \{j \in \mathbb{Z} : -\pi < \lambda_j \le \pi\}$, and the periodogram of **Y** is (by Proposition 10.1.2 and the fact that $\sum_{t=1}^{2n-1} Y_t = 0$)

$$\begin{aligned} I_{2n-1}(\lambda_j) &= \sum_{|k|<2n-1} \hat{\gamma}_Y(k) e^{-ik\lambda_j} \\ &= \frac{n}{2n-1} \sum_{|k|<n} \hat{\gamma}(k) e^{-ik\lambda_j}. \end{aligned}$$

Multiplying each side by $e^{im\lambda_j}$ and summing over $j \in F_{2n-1}$, we get

$$\hat{\gamma}(m) = n^{-1} \sum_{j \in F_{2n-1}} |a_j|^2 e^{im\lambda_j}. \qquad (10.7.10)$$

The autocovariances $\hat{\gamma}(k)$ can thus be computed by taking the two Fourier transforms (10.7.9) and (10.7.10), and using the FFT algorithm in each case. For large n the number of operations required is substantially less than the number of multiplications and additions (of order n^2) required to compute $\hat{\gamma}(k)$, $|k| < n$, from the definition. The fast Fourier transform technique is particularly advantageous for long series, but significant savings can be achieved even for series of length one or two hundred.

§10.8* Derivation of the Asymptotic Behavior of the Maximum Likelihood and Least Squares Estimators of the Coefficients of an ARMA Process

In order to derive the asymptotic properties of the maximum likelihood estimator, it will be convenient to introduce the concept of almost sure convergence.

Definition 10.8.1 (Almost Sure Convergence). A sequence of random variables $\{X_n\}$ is said to converge to the random variable X almost surely or with probability one if

$$P(X_n \text{ converges to } X) = 1.$$

It is implicit that $X, X_1, X_2, \ldots$ are all defined on the same probability space. Almost sure convergence of $\{X_n\}$ to X will be written as $X_n \to X$ a.s.

Remark 1. If $X_n \to X$ a.s. then $X_n \xrightarrow{P} X$. To see this, note that for any $\varepsilon > 0$,

$$
\begin{aligned}
1 &= P(X_n \text{ converges to } X) \\
&\le P\left(\bigcup_{n=1}^{\infty} \bigcap_{k=n}^{\infty} \{|X_k - X| \le \varepsilon\}\right) \\
&= \lim_{n\to\infty} P\left(\bigcap_{k=n}^{\infty} \{|X_k - X| \le \varepsilon\}\right) \\
&\le \liminf_{n\to\infty} P(|X_n - X| \le \varepsilon).
\end{aligned}
$$

The converse is not true, although if $X_n \xrightarrow{P} X$ there exists a subsequence $\{X_{n_j}\}$ of $\{X_n\}$ such that $X_{n_j} \to X$ a.s. (see Billingsley (1986)).

Remark 2. For the two-sided moving average process

$$X_t = \sum_{j=-\infty}^{\infty} \psi_j Z_{t-j}, \qquad \{Z_t\} \sim \text{IID}(0, \sigma^2),$$

with $\sum_j |\psi_j| < \infty$, it can be shown that for $h \in \{0, \pm 1, \dots\}$,

$$\hat{\gamma}(h) = n^{-1} \sum_{t=1}^{n-|h|} (X_t - \bar{X}_n)(X_{t+|h|} - \bar{X}_n) \to \gamma(h) \quad \text{a.s.}$$

and

$$\tilde{\gamma}(h) := n^{-1} \sum_{t=1}^{n-|h|} X_t X_{t+|h|} \to \gamma(h) \quad \text{a.s.} \tag{10.8.1}$$

The proofs of these results are similar to the corresponding proofs of convergence in probability given in Section 7.3 with the strong law of large numbers replacing the weak law.

Strong Consistency of the Estimators

Let $\{X_t\}$ be the causal invertible ARMA(p, q) process,

$$
\begin{aligned}
X_t - \phi_1 X_{t-1} - \cdots - \phi_p X_{t-p} &= Z_t + \theta_1 Z_{t-1} + \cdots + \theta_q Z_{t-q}, \\
&\qquad \{Z_t\} \sim \text{IID}(0, \sigma^2),
\end{aligned}
\tag{10.8.2}
$$

where $\phi(z)$ and $\theta(z)$ have no common zeroes. Let $\boldsymbol{\beta} = (\phi_1, \dots, \phi_p, \theta_1, \dots, \theta_q)'$ and denote by C the parameter set, $C = \{\boldsymbol{\beta} \in \mathbb{R}^{p+q}: \phi(z)\theta(z) \ne 0 \text{ for } |z| \le 1, \phi_p \ne 0, \theta_q \ne 0, \text{ and } \phi(\cdot), \theta(\cdot) \text{ have no common zeroes}\}$.

Remark 3. Notice that $\boldsymbol{\beta}$ can be expressed as a continuous function $\boldsymbol{\beta}(a_1, \dots, a_p, b_1, \dots, b_q)$ of the zeroes $a_1, \dots, a_p$ of $\phi(\cdot)$ and $b_1, \dots, b_q$ of $\theta(\cdot)$. The parameter set C is therefore the image under $\boldsymbol{\beta}$ of the set $\{(a_1, \dots, a_p, b_1, \dots, b_q) : |a_i| > 1, |b_j| > 1 \text{ and } a_i \ne b_j, i = 1, \dots, p, j = 1, \dots, q\}$.

The spectral density $f(\lambda; \boldsymbol{\beta})$ of $\{X_t\}$ can be written in the form,

$$f(\lambda; \boldsymbol{\beta}) = \frac{\sigma^2}{2\pi} g(\lambda; \boldsymbol{\beta}),$$

where

$$g(\lambda; \boldsymbol{\beta}) = \frac{|\theta(e^{-i\lambda})|^2}{|\phi(e^{-i\lambda})|^2}. \tag{10.8.3}$$

Proposition 10.8.1. *Let $\boldsymbol{\beta}_0$ be a fixed vector in C. Then*

$$(2\pi)^{-1} \int_{-\pi}^{\pi} \frac{g(\lambda; \boldsymbol{\beta}_0)}{g(\lambda; \boldsymbol{\beta})} d\lambda > 1$$

for all $\boldsymbol{\beta} \in \bar{C}$ such that $\boldsymbol{\beta} \neq \boldsymbol{\beta}_0$ ($\bar{C}$ denotes the closure of the set C).

Proof. If $\{X_t\}$ is an ARMA(p, q) process with coefficient vector $\boldsymbol{\beta}_0$ and white noise variance σ_0^2, then we can write

$$X_t = \theta_0(B)\phi_0^{-1}(B)Z_t, \qquad \{Z_t\} \sim \text{WN}(0, \sigma_0^2),$$

where $\phi_0(B)$ and $\theta_0(B)$ are the autoregressive and moving average polynomials with coefficients determined by $\boldsymbol{\beta}_0$. Now suppose that $\boldsymbol{\beta} = (\boldsymbol{\phi}', \boldsymbol{\theta}')' \in \bar{C}$, and $\boldsymbol{\beta} \neq \boldsymbol{\beta}_0$. If $|\phi(z)/\theta(z)|$ is unbounded on $|z| \leq 1$ then

$$(2\pi)^{-1} \int_{-\pi}^{\pi} [g(\lambda; \boldsymbol{\beta}_0)/g(\lambda; \boldsymbol{\beta})] \, d\lambda = \infty$$

and the result follows. So suppose $|\phi(z)/\theta(z)|$ is bounded on $|z| \leq 1$ and consider the one-step predictor $-\sum_{j=1}^{\infty} \pi_j X_{t-j}$ of X_t where $\pi(z) = 1 + \sum_{j=1}^{\infty} \pi_j z^j = \phi(z)\theta^{-1}(z)$. Since $\boldsymbol{\beta} \neq \boldsymbol{\beta}_0$, the mean squared error of this predictor is greater than that of the best linear one-step predictor, and hence

$$\sigma_0^2 < E\left(X_t + \sum_{j=1}^{\infty} \pi_j X_{t-j}\right)^2 = E\left(\theta^{-1}(B)\phi(B)X_t\right)^2.$$

But the spectral density of $\theta^{-1}(B)\phi(B)X_t$ is $(\sigma_0^2/2\pi)[g(\lambda; \boldsymbol{\beta}_0)/g(\lambda; \boldsymbol{\beta})]$ and hence

$$\sigma_0^2 < \text{Var}(\theta^{-1}(B)\phi(B)X_t) = \frac{\sigma_0^2}{2\pi} \int_{-\pi}^{\pi} \frac{g(\lambda; \boldsymbol{\beta}_0)}{g(\lambda; \boldsymbol{\beta})} d\lambda,$$

which establishes the proposition. □

The Gaussian likelihood of the vector of observations $\mathbf{X}_n = (X_1, \ldots, X_n)'$ is given by

$$L(\boldsymbol{\beta}, \sigma^2) = (2\pi\sigma^2)^{-n/2} |G_n(\boldsymbol{\beta})|^{-1/2} \exp\left\{-\frac{1}{2\sigma^2} \mathbf{X}_n' G_n^{-1}(\boldsymbol{\beta}) \mathbf{X}_n\right\},$$

where $G_n(\boldsymbol{\beta}) = \sigma^{-2}\Gamma_n(\boldsymbol{\beta})$ and $\Gamma_n(\boldsymbol{\beta})$ is the covariance matrix of $\mathbf{X}_n$. From Section 8.7, the maximum likelihood estimator $\hat{\boldsymbol{\beta}}$ is the value of $\boldsymbol{\beta}$ in C which minimizes

$$l(\boldsymbol{\beta}) = \ln(\mathbf{X}_n' G_n^{-1}(\boldsymbol{\beta})\mathbf{X}_n/n) + n^{-1} \ln \det(G_n(\boldsymbol{\beta})). \tag{10.8.4}$$

The least squares estimator $\tilde{\boldsymbol{\beta}}$ is found by minimizing

$$\tilde{\sigma}_n^2(\boldsymbol{\beta}) = n^{-1} \mathbf{X}_n' G_n^{-1}(\boldsymbol{\beta}) \mathbf{X}_n \tag{10.8.5}$$

with respect to $\boldsymbol{\beta} \in C$. A third estimator $\bar{\boldsymbol{\beta}}$, is found by minimizing

$$\bar{\sigma}_n^2(\beta) = n^{-1} \sum_j I_n(\omega_j)/g(\omega_j; \boldsymbol{\beta}) \tag{10.8.6}$$

with respect to $\boldsymbol{\beta} \in C$, where $I_n(\cdot)$ is the periodogram of $\{X_1, \ldots, X_n\}$ and the sum is taken over all frequencies $\omega_j = 2\pi j/n \in (-\pi, \pi]$. We shall show that the three estimators, $\hat{\boldsymbol{\beta}}$, $\tilde{\boldsymbol{\beta}}$ and $\bar{\boldsymbol{\beta}}$ have the same limit distribution. The argument follows Hannan (1973). See also Whittle (1962), Walker (1964) and Dunsmuir and Hannan (1976).

In the following propositions, assume that $\{X_t\}$ is the ARMA process defined by (10.8.2) with parameter values $\boldsymbol{\beta}_0 \in C$ and $\sigma_0^2 > 0$.

Proposition 10.8.2. *For every* $\boldsymbol{\beta} \in C$,

$$\bar{\sigma}_n^2(\boldsymbol{\beta}) \to \frac{\sigma_0^2}{2\pi} \int_{-\pi}^{\pi} \frac{g(\lambda; \boldsymbol{\beta}_0)}{g(\lambda; \boldsymbol{\beta})} d\lambda \quad \text{a.s.} \tag{10.8.7}$$

Moreover for every $\delta > 0$, *defining* $g_\delta(\lambda; \boldsymbol{\beta}) = (|\theta(e^{-i\lambda})|^2 + \delta)/|\phi(e^{-i\lambda})|^2$,

$$n^{-1} \sum_j \frac{I_n(\omega_j)}{g_\delta(\omega_j; \boldsymbol{\beta})} \to \frac{\sigma_0^2}{2\pi} \int_{-\pi}^{\pi} \frac{g(\lambda; \boldsymbol{\beta}_0)}{g_\delta(\lambda; \boldsymbol{\beta})} d\lambda \tag{10.8.8}$$

uniformly in $\boldsymbol{\beta} \in \bar{C}$ *almost surely.*

Proof. We shall only prove (10.8.8) since the proof of (10.8.7) is similar. Let $q_m(\lambda; \boldsymbol{\beta})$ be the Cesaro mean of the first m Fourier approximations to $g_\delta(\lambda; \boldsymbol{\beta})^{-1}$, given by

$$\begin{aligned} q_m(\lambda; \boldsymbol{\beta}) &= m^{-1} \sum_{j=0}^{m-1} \sum_{|k| \le j} b_k e^{-ik\lambda} \\ &= \sum_{|k|<m} \left(1 - \frac{|k|}{m}\right) b_k e^{-ik\lambda}, \end{aligned}$$

where $b_k = (2\pi)^{-1} \int_{-\pi}^{\pi} e^{ik\lambda}(g_\delta(\lambda; \boldsymbol{\beta}))^{-1} d\lambda$. By the non-negative definiteness of $\{b_k\}$, $q_m(\lambda; \boldsymbol{\beta}) \ge 0$. As a function of $(\lambda, \boldsymbol{\beta})$, $(g_\delta(\lambda; \boldsymbol{\beta}))^{-1}$ is uniformly continuous on the set $[-\pi, \pi] \times \bar{C}$. It therefore follows easily from the proof of Theorem 2.11.1 that $q_m(\lambda; \boldsymbol{\beta})$ converges uniformly to $(g_\delta(\lambda; \boldsymbol{\beta}))^{-1}$ on $[-\pi, \pi] \times \bar{C}$ and in particular that for any $\varepsilon > 0$, there exists an m such that

$$|q_m(\lambda; \boldsymbol{\beta}) - (g_\delta(\lambda; \boldsymbol{\beta}))^{-1}| < \varepsilon$$

for all $(\lambda, \boldsymbol{\beta}) \in [-\pi, \pi] \times \bar{C}$. We can therefore write, for all $\boldsymbol{\beta} \in \bar{C}$,

$$\begin{aligned} &\left| n^{-1} \sum_j \left(\frac{I_n(\omega_j)}{g_\delta(\omega_j; \boldsymbol{\beta})} \right) - n^{-1} \sum_j I_n(\omega_j) q_m(\omega_j; \boldsymbol{\beta}) \right| \\ &\quad = n^{-1} \left| \sum_j I_n(\omega_j)((g_\delta(\omega_j; \boldsymbol{\beta}))^{-1} - q_m(\omega_j; \boldsymbol{\beta})) \right| \qquad (10.8.9) \\ &\quad \le \varepsilon n^{-1} \sum_j I_n(\omega_j) \\ &\quad = \varepsilon \tilde{\gamma}(0) \end{aligned}$$

where the last equality follows from (10.1.9).

Now for $n > m$,

$$\begin{aligned}
n^{-1}\sum_j I_n(\omega_j)q_m(\omega_j;\boldsymbol{\beta})
&= n^{-1}\sum_j \sum_{|h|<n} \tilde{\gamma}(h)e^{-i\omega_j h}\left(\sum_{|k|<m}\left(1-\frac{|k|}{m}\right)b_k e^{i\omega_j k}\right)\\
&= \sum_{|h|<n}\sum_{|k|<m}\tilde{\gamma}(h)\left(1-\frac{|k|}{m}\right)b_k\left(n^{-1}\sum_j e^{-i\omega_j(h-k)}\right)\\
&= \sum_{|k|<m}\tilde{\gamma}(k)\left(1-\frac{|k|}{m}\right)b_k + 2\sum_{k=1}^{m}\tilde{\gamma}(n-k)\left(1-\frac{|k|}{m}\right)b_k. \qquad (10.8.10)
\end{aligned}$$

For k and t fixed, the process $\{X_{t+n-k}, n = 1, 2, \ldots\}$ is strictly stationary and ergodic (see Hannan, 1970) and by a direct application of the ergodic theorem, $n^{-1}X_{t+n-k} \to 0$ a.s. From this it follows that $\tilde{\gamma}(n-k) = n^{-1}\sum_{t=1}^{k} X_t X_{t+n-k} \to 0$ a.s. for each fixed k. The second term in (10.8.10) therefore converges to zero a.s. as $n \to \infty$. By Remark 2, the first term converges to $\sum_{|k|<m}\gamma(k)(1-|k|/m)b_k$ and since b_k is uniformly bounded in $\boldsymbol{\beta}$ and k, we have

$$n^{-1}\sum_j I_n(\omega_j)q_m(\omega_j;\boldsymbol{\beta}) \to \sum_{|k|<m}\gamma(k)\left(1-\frac{|k|}{m}\right)b_k \qquad (10.8.11)$$

uniformly in $\boldsymbol{\beta} \in \bar{C}$ a.s. Moreover

$$\begin{aligned}
&\left|\sum_{|k|<m}\gamma(k)\left(1-\frac{|k|}{m}\right)b_k - \frac{\sigma_0^2}{2\pi}\int_{-\pi}^{\pi}\frac{g(\lambda;\boldsymbol{\beta}_0)}{g_\delta(\lambda;\boldsymbol{\beta})}\,d\lambda\right|\\
&\quad= \left|\int_{-\pi}^{\pi}\sum_{|k|<m}\left(1-\frac{|k|}{m}\right)b_k e^{ik\lambda}f(\lambda;\boldsymbol{\beta}_0)\,d\lambda - \int_{-\pi}^{\pi}(g_\delta(\lambda;\boldsymbol{\beta}))^{-1}f(\lambda;\boldsymbol{\beta}_0)\,d\lambda\right|\\
&\quad\le \int_{-\pi}^{\pi}|q_m(\lambda;\boldsymbol{\beta}) - (g_\delta(\lambda;\boldsymbol{\beta}))^{-1}|f(\lambda;\boldsymbol{\beta}_0)\,d\lambda\\
&\quad\le \varepsilon\gamma(0).
\end{aligned}$$

Since $\tilde{\gamma}(0) \to \gamma(0)$ a.s. we conclude from (10.8.9), (10.8.11) and the above inequality that

$$n^{-1}\sum_j \frac{I_n(\omega_j)}{g_\delta(\omega_j;\boldsymbol{\beta})} \to \frac{\sigma_0^2}{2\pi}\int_{-\pi}^{\pi}\frac{g(\lambda;\boldsymbol{\beta}_0)}{g_\delta(\lambda;\boldsymbol{\beta})}\,d\lambda$$

uniformly in $\boldsymbol{\beta} \in \bar{C}$ a.s. □

Proposition 10.8.3. *There exists an event B with probability one such that for any sequence $\{\boldsymbol{\beta}_n\}$, $\boldsymbol{\beta}_n \in C$ with $\boldsymbol{\beta}_n \to \boldsymbol{\beta}$, we have the following two possibilities:*

(a) *If $\boldsymbol{\beta} \in C$, then for all outcomes in B,*

$$\tilde{\sigma}_n^2(\boldsymbol{\beta}_n) \to \frac{\sigma_0^2}{2\pi}\int_{-\pi}^{\pi}\frac{g(\lambda;\boldsymbol{\beta}_0)}{g(\lambda;\boldsymbol{\beta})}\,d\lambda. \qquad (10.8.12)$$

(b) *If* $\boldsymbol{\beta} \in \partial C$ *(where* ∂C *is the boundary of the set* C*), then for all outcomes in* B,

$$\liminf_{n\to\infty} \tilde{\sigma}_n^2(\boldsymbol{\beta}_n) \geq \frac{\sigma_0^2}{2\pi}\int_{-\pi}^{\pi} \frac{g(\lambda;\boldsymbol{\beta}_0)}{g(\lambda;\boldsymbol{\beta})}\,d\lambda. \tag{10.8.13}$$

PROOF. (a) Since $\boldsymbol{\beta} \in C$, $\inf_\lambda g(\lambda;\boldsymbol{\beta}) > 0$ and $\sup_\lambda g(\lambda;\boldsymbol{\beta}) < \infty$. Consequently for each $\varepsilon \in (0, \inf_\lambda g(\lambda;\boldsymbol{\beta}))$ there exists an N such that

$$\sup_\lambda |g(\lambda;\boldsymbol{\beta}_n) - g(\lambda;\boldsymbol{\beta})| < \varepsilon/2 \quad \text{for } n \geq N. \tag{10.8.14}$$

By Corollary 4.4.2, we can find a polynomial, $a(z) = 1 + a_1 z + \cdots + a_m z^m$, and a positive constant K_m such that $a(z) \neq 0$ for $|z| \leq 1$, and

$$\begin{cases} \sup_\lambda |g(\lambda;\boldsymbol{\beta}) - A_m(\lambda)| < \varepsilon/2, \\ \inf_\lambda A_m(\lambda) > \inf_\lambda g(\lambda;\boldsymbol{\beta})/2 > 0, \end{cases} \tag{10.8.15}$$

where

$$A_m(\lambda) = K_m |a(e^{-i\lambda})|^{-2}.$$

Note that $K_m \to 1$ as $\varepsilon \to 0$. Let H_n be the covariance matrix corresponding to the spectral density $(2\pi)^{-1}A_m(\lambda)$. Then if $\mathbf{y} \in \mathbb{R}^n$ and $\mathbf{y}'\mathbf{y} = 1$,

$$\begin{aligned}
&|\mathbf{y}'G_n(\boldsymbol{\beta}_n)\mathbf{y} - \mathbf{y}'H_n\mathbf{y}| \\
&\quad = \left|(2\pi)^{-1}\int_{-\pi}^{\pi}\left|\sum_{j=1}^n y_j e^{ij\lambda}\right|^2 (g(\lambda;\boldsymbol{\beta}_n) - A_m(\lambda))\,d\lambda\right| \\
&\quad \leq (2\pi)^{-1}\int_{-\pi}^{\pi}\left|\sum_{j=1}^n y_j e^{ij\lambda}\right|^2 (|g(\lambda;\boldsymbol{\beta}_n) - g(\lambda;\boldsymbol{\beta})| + |g(\lambda;\boldsymbol{\beta}) - A_m(\lambda)|)\,d\lambda \\
&\quad \leq \varepsilon(2\pi)^{-1}\int_{-\pi}^{\pi}\left|\sum_{j=1}^n y_j e^{ij\lambda}\right|^2 d\lambda \\
&\quad = \varepsilon \text{ for } n \geq N.
\end{aligned} \tag{10.8.16}$$

Now if $\{Y_t\}$ is an ARMA process with spectral density $(2\pi)^{-1}g(\lambda;\boldsymbol{\beta}_n)$ (and white noise variance 1) then by Proposition 4.5.3 the eigenvalues of $G_n(\boldsymbol{\beta}_n)$ are bounded below by $\inf_\lambda g(\lambda;\boldsymbol{\beta}_n) \geq K > 0$ for some constant K and all n sufficiently large. The same argument also shows that the eigenvalues of H_n are bounded below by $\inf_\lambda A_m(\lambda) > 0$. Thus the eigenvalues of $G_n^{-1}(\boldsymbol{\beta}_n)$ and H_n^{-1} are all less than a constant C_1 (independent of ε) so that

$$\begin{aligned}
|n^{-1}\mathbf{X}_n'(G_n^{-1}(\boldsymbol{\beta}_n) - H_n^{-1})\mathbf{X}_n| &= |n^{-1}\mathbf{X}_n'H_n^{-1}(H_n - G_n(\boldsymbol{\beta}_n))\mathbf{G}_n^{-1}(\boldsymbol{\beta}_n)\mathbf{X}_n| \\
&\leq \varepsilon C_1^2\tilde{\gamma}(0) \quad \text{for } n \geq N.
\end{aligned} \tag{10.8.17}$$

We next consider the asymptotic behavior of $n^{-1}\mathbf{X}_n'H_n^{-1}\mathbf{X}_n$.

Let $\{Y_t\}$ be the AR(m) process with spectral density $(2\pi)^{-1}K_m|a(e^{-i\lambda})|^{-2}$

(and white noise variance K_m). Then by the Gram–Schmidt orthogonalization process, we can choose δ_{jk}, $k = 1, \ldots, j$, $j = 1, \ldots, m$, so that the random variables

$$\begin{aligned}
W_1 &= \delta_{11} Y_1, \\
W_2 &= \delta_{21} Y_1 + \delta_{22} Y_2, \\
&\vdots \\
W_m &= \delta_{m1} Y_1 + \cdots + \delta_{mm} Y_m, \\
W_{m+1} &= a_m Y_1 + \cdots + a_1 Y_m + Y_{m+1}, \\
&\vdots \\
W_n &= a_m Y_{n-m} + \cdots + a_1 Y_{n-1} + Y_n,
\end{aligned}$$

are white noise with mean zero and variance K_m. Then

$$\mathbf{W}_n = T\mathbf{Y}_n,$$

where $\mathbf{W}_n = (W_1, \ldots, W_n)'$, $\mathbf{Y}_n = (Y_1, \ldots, Y_n)'$ and T is the lower trangular matrix

$$T = \begin{bmatrix}
\delta_{11} & & & & & & & \\
\delta_{21} & \delta_{22} & & & & & & \\
\vdots & \vdots & & & & & & \\
\delta_{m1} & \delta_{m2} & \cdots & \delta_{mm} & & & & \\
a_m & a_{m-1} & \cdots & a_1 & 1 & & & \\
 & a_m & \cdots & a_2 & a_1 & 1 & & \\
 & & & & & & & \\
 & & & a_m & \cdots & a_2 & a_1 & 1
\end{bmatrix}. \tag{10.8.18}$$

It follows that $TH_nT' = K_m I$, where I is the $n \times n$ identity matrix, and hence that $H_n^{-1} = (T'T)/(K_m)$. Except for the m^2 components in the upper left corner, and the m^2 components in the bottom right corner, the matrix $H_n^{-1} = [h_{ij}]_{i,j=1}^n$ is the same as the matrix $\tilde{H}_n^{-1} = [\tilde{h}_{ij}]_{i,j=1}^n$ with

$$\tilde{h}_{ij} = \begin{cases} K_m^{-1} \displaystyle\sum_{r=0}^{m-|i-j|} a_r a_{r+|i-j|} & \text{if } |i-j| \le m, \\ 0 & \text{otherwise,} \end{cases} \tag{10.8.19}$$

where $a_0 := 1$. It then follows that

$$\begin{aligned}
&n^{-1}\mathbf{X}_n' H_n^{-1}\mathbf{X}_n - n^{-1}\mathbf{X}_n \tilde{H}_n^{-1}\mathbf{X}_n \\
&\quad = n^{-1} \sum_{i,j=1}^{m} (h_{ij} - \tilde{h}_{ij}) X_i X_j + n^{-1} \sum_{i,j=n-m+1}^{n} (h_{ij} - \tilde{h}_{ij}) X_i X_j \qquad (10.8.20) \\
&\quad \to 0 \text{ a.s.},
\end{aligned}$$

since $n^{-1}X_i \to 0$ a.s. and $n^{-1}X_{n-i} \to 0$ a.s. by an application of the ergodic theorem. It is easy to show that for $n > m$

$$\left| n^{-1}\mathbf{X}_n'\tilde{H}_n^{-1}\mathbf{X}_n - n^{-1}\sum_j \frac{I_n(\omega_j)}{A_m(\omega_j)} \right| = 2\left|\sum_{k=1}^m \tilde{\gamma}(n-k)\tilde{h}_{0k}\right| \le 2C_2 \sum_{k=1}^m |\tilde{\gamma}(n-k)| \tag{10.8.21}$$

and

$$\begin{aligned}\left| n^{-1}\sum_j \frac{I_n(\omega_j)}{A_m(\omega_j)} - n^{-1}\sum_j \frac{I_n(\omega_j)}{g(\omega_j;\boldsymbol{\beta})} \right| &\le C_2^2 n^{-1}\sum_j I_n(\omega_j)|A_m(\omega_j) - g(\omega_j;\boldsymbol{\beta})| \\ &\le C_2^2 \varepsilon\tilde{\gamma}(0),\end{aligned} \tag{10.8.22}$$

where $C_2^{-1} = (\inf_\lambda g(\lambda;\boldsymbol{\beta})/2) > 0$. Combining equations (10.8.17), (10.8.20), (10.8.21) and (10.8.22), we have

$$\begin{aligned}&\left| n^{-1}\mathbf{X}_n' G_n^{-1}(\boldsymbol{\beta}_n)\mathbf{X}_n - n^{-1}\sum_j \frac{I_n(\omega_j)}{g(\omega_j;\boldsymbol{\beta})} \right| \\ &\qquad \le (C_1^2 + C_2^2)\varepsilon\tilde{\gamma}(0) + 2C_2\sum_{k=1}^m |\tilde{\gamma}(n-k)| + |n^{-1}\mathbf{X}_n' H_n^{-1}\mathbf{X}_n - n^{-1}\mathbf{X}_n'\tilde{H}_n^{-1}\mathbf{X}_n|\end{aligned}$$

for all $n \ge N$. Now let $\{\boldsymbol{\beta}_k\}$ be a dense set in C and let S_k denote the probability one event where

$$n^{-1}\sum_j \frac{I_n(\omega_j)}{g(\omega_j;\boldsymbol{\beta}_k)} \to \frac{\sigma_0^2}{2\pi}\int_{-\pi}^{\pi} \frac{g(\lambda;\boldsymbol{\beta}_0)}{g(\lambda;\boldsymbol{\beta}_k)}\,d\lambda.$$

The event $S = \bigcap_{k=1}^\infty S_k$ also has probability one and a routine approximation argument shows that for each outcome in $S \cap \{\tilde{\gamma}(0) \to \gamma(0)\}$,

$$n^{-1}\sum_j \frac{I_n(\omega_j)}{g(\omega_j;\boldsymbol{\beta})} \to \frac{\sigma_0^2}{2\pi}\int_{-\pi}^{\pi} \frac{g(\lambda;\boldsymbol{\beta}_0)}{g(\lambda;\boldsymbol{\beta})}\,d\lambda$$

for each $\boldsymbol{\beta} \in C$. If B_1 denotes the event

$$B_1 = \bigcap_{k=1}^\infty (\{n^{-1}X_k \to 0\} \cap \{n^{-1}X_{n-k} \to 0\}) \cap \{\tilde{\gamma}(0) \to \gamma(0)\} \cap S$$

then $P(B_1) = 1$ and for each outcome in B_1,

$$\limsup_{n\to\infty} \left| n^{-1}\mathbf{X}_n' G_n^{-1}(\boldsymbol{\beta}_n)\mathbf{X}_n - \frac{\sigma_0^2}{2\pi}\int_{-\pi}^{\pi} \frac{g(\lambda;\boldsymbol{\beta}_0)}{g(\lambda;\boldsymbol{\beta})}\,d\lambda \right| \le (C_1^2 + C_2^2)\varepsilon\gamma(0).$$

Since C_1^2 and C_2^2 do not depend on ε and ε is arbitrary, the assertion (10.8.12) follows for each outcome in B_1.

(b) Set $\boldsymbol{\beta}_n = (\boldsymbol{\phi}_n', \boldsymbol{\theta}_n')'$ and $\boldsymbol{\beta} = (\boldsymbol{\phi}', \boldsymbol{\theta}')'$. Since $\boldsymbol{\beta}_n \to \boldsymbol{\beta}$, choose $\boldsymbol{\theta}^\dagger$ such that $\theta^\dagger(z) = 1 + \theta_1^\dagger z + \cdots + \theta_q^\dagger z^q \ne 0$ for $|z| \le 1$ and

$$\sup_\lambda \left| |\theta^\dagger(e^{-i\lambda})|^2 - |\theta(e^{-i\lambda})|^2 \right| < \varepsilon.$$

(If $\theta(z) \ne 0$ for $|z| \le 1$, take $\boldsymbol{\theta}^\dagger = \boldsymbol{\theta}$.) With $\boldsymbol{\beta}_n^\dagger = (\boldsymbol{\phi}_n', \boldsymbol{\theta}^{\dagger\prime})'$, we have

$$g(\lambda;\boldsymbol{\beta}_n) \le \frac{|\theta^\dagger(e^{-i\lambda})|^2 + \varepsilon}{|\phi_n(e^{-i\lambda})|^2} \quad \text{for all sufficiently large } n.$$

By Corollary 4.4.2 there exists a polynomial $b(z) = 1 + b_1 z + \cdots + b_k z^k$ and a positive constant K such that $b(z) \neq 0$ for $|z| \leq 1$ and

$$|\theta^\dagger(e^{-i\lambda})|^2 + \varepsilon \leq \frac{K}{|b(e^{-i\lambda})|^2} \leq |\theta^\dagger(e^{-i\lambda})|^2 + 2\varepsilon \tag{10.8.23}$$

for all λ. Setting $A(\lambda; \boldsymbol{\phi}) = K|1 + a_1 e^{-i\lambda} + \cdots + a_m e^{-im\lambda}|^{-2} = K|b(e^{-i\lambda})|^{-2} \times |\phi(e^{-i\lambda})|^{-2}$ we have

$$g(\lambda; \boldsymbol{\beta}_n) \leq A(\lambda; \boldsymbol{\phi}_n) \quad \text{for all } \lambda \text{ and sufficiently large } n.$$

Define the matrices T, H_n and $\tilde{H}_n^{-1}$ as in the proof of (a) with $A_m(\lambda)$ replaced by $A(\lambda; \boldsymbol{\phi}_n)$. Since the coefficients in the matrix T are bounded in n, we have from (10.8.20) and (10.8.21)

$$n^{-1}\mathbf{X}_n' H_n^{-1} \mathbf{X}_n - n^{-1} \sum_j \frac{I_n(\omega_j)}{A(\omega_j; \boldsymbol{\phi}_n)} \to 0 \quad \text{a.s.} \tag{10.8.24}$$

Since $A^{-1}(\lambda; \boldsymbol{\phi}_n) = K^{-1}|b(e^{-i\lambda})|^2 |\phi(e^{-i\lambda})|^2$ is uniformly bounded, we have, by the argument given in the proof of Proposition 10.8.2 (see (10.8.11)), that

$$n^{-1} \sum_j \frac{I_n(\omega_j)}{A(\omega_j; \boldsymbol{\phi}_n)} \to \frac{\sigma_0^2}{2\pi} \int_{-\pi}^{\pi} \frac{g(\lambda; \boldsymbol{\beta}_0)}{A(\lambda; \boldsymbol{\phi})} d\lambda \quad \text{a.s.} \tag{10.8.25}$$

Also, since $g(\lambda; \boldsymbol{\beta}_n) \leq A(\lambda; \boldsymbol{\phi}_n)$ for all large n, the matrix $H_n - G_n(\boldsymbol{\beta}_n)$ is non-negative definite (see Problem 4.8) and thus the matrix $G_n^{-1}(\boldsymbol{\beta}_n) - H_n^{-1}$ is also non-negative definite (see Rao (1973), p. 70). Thus, by (10.8.24), (10.8.25) and (10.8.23),

$$\begin{aligned} \liminf_{n\to\infty} n^{-1}\mathbf{X}_n' G_n^{-1}(\boldsymbol{\beta}_n)\mathbf{X}_n &\geq \liminf_{n\to\infty} n^{-1}\mathbf{X}_n' H_n^{-1} \mathbf{X}_n \\ &\geq \frac{\sigma_0^2}{2\pi} \int_{-\pi}^{\pi} \frac{g(\lambda; \boldsymbol{\beta}_0)}{A(\lambda; \boldsymbol{\beta})} d\lambda \\ &\geq \frac{\sigma_0^2}{2\pi} \int_{-\pi}^{\pi} \frac{g(\lambda; \boldsymbol{\beta}_0)|\phi(e^{-i\lambda})|^2}{|\theta^\dagger(e^{-i\lambda})|^2 + 2\varepsilon} d\lambda \quad \text{a.s.} \end{aligned}$$

Letting $\boldsymbol{\theta}^\dagger \to \boldsymbol{\theta}$, the lower bound becomes

$$\frac{\sigma_0^2}{2\pi} \int_{-\pi}^{\pi} \frac{g(\lambda; \boldsymbol{\beta}_0)|\phi(e^{-i\lambda})|^2}{|\theta(e^{-i\lambda})|^2 + 2\varepsilon} d\lambda$$

and finally, letting $\varepsilon \to 0$ and applying the monotone convergence theorem, we obtain the desired result, namely

$$\liminf_{n\to\infty} \tilde{\sigma}_n^2(\boldsymbol{\beta}_n) \geq \frac{\sigma_0^2}{2\pi} \int_{-\pi}^{\pi} \frac{g(\lambda; \boldsymbol{\beta}_0)}{g(\lambda; \boldsymbol{\beta})} d\lambda \quad \text{a.s.}$$

As in the proof of (a), we can also find a set B_2 such that $P(B_2) = 1$ and for each outcome in B_2,

$$\liminf_{n\to\infty} \tilde{\sigma}_n^2(\boldsymbol{\beta}_n) \geq \frac{\sigma_0^2}{2\pi} \int_{-\pi}^{\pi} \frac{g(\lambda; \boldsymbol{\beta}_0)}{g(\lambda; \boldsymbol{\beta})} \, d\lambda$$

for any sequence $\boldsymbol{\beta}_n \to \boldsymbol{\beta}$ with $\boldsymbol{\beta} \in \partial C$. □

Proposition 10.8.4. *If* $\boldsymbol{\beta} \in C$, *then* $\ln(\det G_n(\boldsymbol{\beta})) > 0$ *and*

$$n^{-1} \ln(\det G_n(\boldsymbol{\beta})) \to 0$$

as $n \to \infty$.

PROOF. Suppose $\{Y_t\}$ is an ARMA process with spectral density $(2\pi)^{-1} g(\lambda; \boldsymbol{\beta})$. Then $\det G_n(\boldsymbol{\beta}) = r_0 \cdots r_{n-1}$, where $r_t = E(Y_{t+1} - \hat{Y}_{t+1})^2$ (see (8.6.7)). Rewriting the difference equations for $\{Y_t\}$ as $Y_t + \sum_{j=1}^{\infty} \pi_j Y_{t-j} = Z_t$, we see from the definition of $\hat{Y}_{t+1}$ that

$$\begin{aligned}
1 = \operatorname{Var}(Z_{t+1}) &= E\left(Y_{t+1} + \sum_{j=1}^{\infty} \pi_j Y_{t+1-j} \right)^2 \\
&\leq r_t \leq E\left(Y_{t+1} + \sum_{j=1}^{t} \pi_j Y_{t+1-j} \right)^2 \\
&= E\left(Z_{t+1} - \sum_{j=t+1}^{\infty} \pi_j Y_{t+1-j} \right)^2 \\
&= 1 + E\left(\sum_{j=t+1}^{\infty} \pi_j Y_{t+1-j} \right)^2 \\
&\leq 1 + \left(\sum_{j=t+1}^{\infty} |\pi_j| \right)^2 \gamma_Y(0).
\end{aligned}$$

Since $r_t \geq 1$ and $r_0 > 1$, $\det G_n(\boldsymbol{\beta}) = r_0 \cdots r_{n-1} > 1$, and hence $\ln(\det G_n(\boldsymbol{\beta})) > 0$. Moreover, since $(\sum_{j=t+1}^{\infty} |\pi_j|)^2 \gamma_Y(0) \to 0$ as $t \to \infty$, it follows that $r_t \to 1$, so by Cesaro convergence,

$$\begin{aligned}
0 \leq n^{-1} \ln(\det G_n(\boldsymbol{\beta})) &= n^{-1} \sum_{t=0}^{n-1} \ln r_{t-1} \\
&\to 0.
\end{aligned}$$

□

Theorem 10.8.1. *Let* $\hat{\boldsymbol{\beta}}_n$, $\tilde{\boldsymbol{\beta}}_n$ *and* $\bar{\boldsymbol{\beta}}_n$ *be the estimators in C which minimize* $l(\boldsymbol{\beta}) = \ln(\mathbf{X}_n' G_n^{-1}(\boldsymbol{\beta}) \mathbf{X}_n / n) + n^{-1} \ln(\det G_n(\boldsymbol{\beta}))$, $\tilde{\sigma}_n^2(\boldsymbol{\beta}) = n^{-1} \mathbf{X}_n' G_n^{-1}(\boldsymbol{\beta}) \mathbf{X}_n$, *and* $\bar{\sigma}_n^2(\boldsymbol{\beta}) = n^{-1} \sum_j (I_n(\omega_j)/g(\omega_j; \boldsymbol{\beta}))$, *respectively, where* $\{X_t\}$ *is an* ARMA *process with true parameter values* $\boldsymbol{\beta}_0 \in C$ *and* $\sigma_0^2 > 0$. *Then*

(i) $\bar{\boldsymbol{\beta}}_n \to \boldsymbol{\beta}_0$ a.s. *and* $\bar{\sigma}_n^2(\bar{\boldsymbol{\beta}}_n) \to \sigma_0^2$ a.s.,
(ii) $\tilde{\boldsymbol{\beta}}_n \to \boldsymbol{\beta}_0$ a.s. *and* $\tilde{\sigma}_n^2(\tilde{\boldsymbol{\beta}}_n) \to \sigma_0^2$ a.s., and
(iii) $\hat{\boldsymbol{\beta}}_n \to \boldsymbol{\beta}_0$ a.s. *and* $\hat{\sigma}_n^2(\hat{\boldsymbol{\beta}}_n) = n^{-1} \mathbf{X}_n' G_n^{-1}(\hat{\boldsymbol{\beta}}_n) \mathbf{X}_n \to \sigma_0^2$ a.s.

PROOF. Let B be the event given in the statement of Proposition 10.8.3. Then there exists an event $B^* \subset B$ with probability one such that for each outcome in B^*, (10.8.7) holds with $\boldsymbol{\beta} = \boldsymbol{\beta}_0$ and (10.8.8) is valid for all rational $\delta > 0$. We shall therefore prove convergence in (i)–(iii) for each outcome in B^*. So for the remainder of the proof, consider a fixed outcome in B^*.

(i) Suppose $\bar{\boldsymbol{\beta}}_n$ does not converge to $\boldsymbol{\beta}_0$. Then by compactness there exists a subsequence $\{\bar{\boldsymbol{\beta}}_{n_k}\}$ such that $\bar{\boldsymbol{\beta}}_{n_k} \to \boldsymbol{\beta}$ where $\boldsymbol{\beta} \in \bar{C}$ and $\boldsymbol{\beta} \neq \boldsymbol{\beta}_0$. By Proposition 10.8.2, for any rational $\delta > 0$,

$$\liminf_{k\to\infty} \bar{\sigma}^2_{n_k}(\bar{\boldsymbol{\beta}}_{n_k}) \geq \liminf_{k\to\infty} n_k^{-1} \sum_j \frac{I_{n_k}(\omega_j)}{g_\delta(\omega_j; \bar{\boldsymbol{\beta}}_{n_k})}$$
$$= \frac{\sigma_0^2}{2\pi} \int_{-\pi}^{\pi} \frac{g(\lambda; \boldsymbol{\beta}_0)}{g_\delta(\lambda; \boldsymbol{\beta})} d\lambda.$$

However by Proposition (10.8.1),

$$\frac{\sigma_0^2}{2\pi} \int_{-\pi}^{\pi} \frac{g(\lambda; \boldsymbol{\beta}_0)}{g(\lambda; \boldsymbol{\beta})} d\lambda > \sigma_0^2,$$

so by taking δ sufficiently small we have

$$\liminf_{k\to\infty} \bar{\sigma}^2_{n_k}(\bar{\boldsymbol{\beta}}_{n_k}) > \sigma_0^2. \tag{10.8.26}$$

On the other hand, by definition of $\bar{\boldsymbol{\beta}}_n$ and (10.8.7),

$$\limsup_{n\to\infty} \bar{\sigma}^2_n(\bar{\boldsymbol{\beta}}_n) \leq \limsup_{n\to\infty} \bar{\sigma}^2_n(\boldsymbol{\beta}_0)$$
$$= \frac{\sigma_0^2}{2\pi} \int_{-\pi}^{\pi} \frac{g(\lambda; \boldsymbol{\beta}_0)}{g(\lambda; \boldsymbol{\beta}_0)} d\lambda$$
$$= \sigma_0^2$$

which contradicts (10.8.26). Thus we must have $\bar{\boldsymbol{\beta}}_n \to \boldsymbol{\beta}_0$. It now follows quite easily from Proposition 10.8.2 that $\bar{\sigma}^2_n(\bar{\boldsymbol{\beta}}_n) \to \sigma_0^2$.

(ii) As in (i) suppose $\tilde{\boldsymbol{\beta}}_n$ does not converge to $\boldsymbol{\beta}_0$. Then there exists a subsequence $\{\tilde{\boldsymbol{\beta}}_{n_k}\}$ such that $\tilde{\boldsymbol{\beta}}_{n_k} \to \boldsymbol{\beta} \neq \boldsymbol{\beta}_0$ with $\boldsymbol{\beta} \in \bar{C}$. By Propositions 10.8.3 and 10.8.1

$$\liminf_{k\to\infty} \tilde{\sigma}^2_{n_k}(\tilde{\boldsymbol{\beta}}_{n_k}) \geq \frac{\sigma_0^2}{2\pi} \int_{-\pi}^{\pi} \frac{g(\lambda; \boldsymbol{\beta}_0)}{g(\lambda; \boldsymbol{\beta})} d\lambda > \sigma_0^2.$$

But, by Proposition 10.8.3(a) and the definition of $\tilde{\boldsymbol{\beta}}_n$,

$$\limsup_{n\to\infty} \tilde{\sigma}^2_n(\tilde{\boldsymbol{\beta}}_n) \leq \limsup_{n\to\infty} \tilde{\sigma}^2_n(\boldsymbol{\beta}_0) = \sigma_0^2$$

which contradicts the above inequality. Therefore we conclude that $\tilde{\boldsymbol{\beta}}_n \to \boldsymbol{\beta}_0$, and hence, by Proposition 10.8.3(a), that $\tilde{\sigma}^2_n(\tilde{\boldsymbol{\beta}}_n) \to \sigma_0^2$.

(iii) Suppose $\hat{\boldsymbol{\beta}}_{n_k} \to \boldsymbol{\beta} \neq \boldsymbol{\beta}_0$ for some subsequence $\{\hat{\boldsymbol{\beta}}_{n_k}\}$. Then by Propositions 10.8.3 and 10.8.4 and the definition of $\hat{\boldsymbol{\beta}}_n$, we obtain the contradiction

$$\ln(\sigma_0^2) < \liminf_{k\to\infty} \ln(\tilde{\sigma}_{n_k}^2(\hat{\boldsymbol\beta}_{n_k}))$$
$$\leq \liminf_{k\to\infty} l(\hat{\boldsymbol\beta}_{n_k}) \leq \limsup_{n\to\infty} l(\boldsymbol\beta_0)$$
$$= \limsup_{n\to\infty} (n^{-1}\ln(\det G_n(\boldsymbol\beta_0)) + \ln \tilde{\sigma}_n^2(\boldsymbol\beta_0))$$
$$= \ln(\sigma_0^2).$$

Thus $\hat{\boldsymbol\beta}_n \to \boldsymbol\beta_0$ and $\tilde{\sigma}_n^2(\hat{\boldsymbol\beta}_n) \to \sigma_0^2$ by Proposition 10.8.3(a). □

Asymptotic Normality of the Estimators

Theorem 10.8.2. *Under the assumptions of Theorem* 10.8.1,

(i) $\bar{\boldsymbol\beta}_n$ *is* $\mathrm{AN}(\boldsymbol\beta_0, n^{-1}W^{-1}(\boldsymbol\beta_0))$,
(ii) $\tilde{\boldsymbol\beta}_n$ *is* $\mathrm{AN}(\boldsymbol\beta_0, n^{-1}W^{-1}(\boldsymbol\beta_0))$, *and*
(iii) $\hat{\boldsymbol\beta}_n$ *is* $\mathrm{AN}(\boldsymbol\beta_0, n^{-1}W^{-1}(\boldsymbol\beta_0))$,

where

$$W(\boldsymbol\beta_0) = \frac{1}{4\pi}\int_{-\pi}^{\pi}\left[\frac{\partial \ln g(\lambda;\boldsymbol\beta_0)}{\partial\boldsymbol\beta}\right]\left[\frac{\partial \ln g(\lambda;\boldsymbol\beta_0)}{\partial\boldsymbol\beta}\right]' d\lambda.$$

Before proving these results we show the equivalence of the asymptotic covariance matrix $W^{-1}(\boldsymbol\beta_0)$ and the matrix $V(\boldsymbol\beta_0)$ specified in (8.8.3). In order to evaluate the (j,k)-component of $W(\boldsymbol\beta)$ for any given $\boldsymbol\beta \in C$, i.e.

$$W_{jk} = \frac{1}{4\pi}\int_{-\pi}^{\pi}\frac{\partial \ln g(\lambda;\boldsymbol\beta)}{\partial\beta_j}\frac{\partial \ln g(\lambda;\boldsymbol\beta)}{\partial\beta_k}\,d\lambda, \tag{10.8.27}$$

we observe that

$$\ln g(\lambda;\boldsymbol\beta) = \ln\theta(e^{-i\lambda}) + \ln\theta(e^{i\lambda}) - \ln\phi(e^{-i\lambda}) - \ln\phi(e^{i\lambda}),$$

where $\phi(z) = 1 - \phi_1 z - \cdots - \phi_p z^p$ and $\theta(z) = 1 + \theta_1 z + \cdots + \theta_q z^q$. Hence

$$\partial \ln g(\lambda;\boldsymbol\beta)/\partial\phi_j = e^{-ij\lambda}\phi^{-1}(e^{-i\lambda}) + e^{ij\lambda}\phi^{-1}(e^{i\lambda})$$

and

$$\partial \ln g(\lambda;\boldsymbol\beta)/\partial\theta_j = e^{-ij\lambda}\theta^{-1}(e^{-i\lambda}) + e^{ij\lambda}\theta^{-1}(e^{i\lambda}).$$

Substituting in (10.8.27) and noting that for $j, k \geq 1$,

$$\int_{-\pi}^{\pi} e^{i(j+k)\lambda}\phi^{-2}(e^{i\lambda})\,d\lambda = \int_{-\pi}^{\pi} e^{-i(j+k)\lambda}\phi^{-2}(e^{-i\lambda})\,d\lambda = 0,$$

we find that for $j, k \leq p$,

$$W_{jk} = \frac{1}{4\pi}\int_{-\pi}^{\pi}(e^{-i(j-k)\lambda} + e^{i(j-k)\lambda})|\phi(e^{-i\lambda})|^{-2}\,d\lambda = E[U_{t-j+1}U_{t-k+1}],$$

where $\{U_t\}$ is the autoregressive process defined by,

$$\phi(B)U_t = N_t, \qquad \{N_t\} \sim \text{WN}(0, 1). \tag{10.8.28}$$

The same argument shows that for $j, k \geq p + 1$,

$$W_{jk} = \frac{1}{4\pi} \int_{-\pi}^{\pi} (e^{-i(j-k)\lambda} + e^{i(j-k)\lambda}) |\theta(e^{-i\lambda})|^{-2} \, d\lambda = E[V_{t-j+1} V_{t-k+1}],$$

where $\{V_t\}$ is the autoregressive process defined by,

$$\theta(B)V_t = N_t, \qquad \{N_t\} \sim \text{WN}(0, 1). \tag{10.8.29}$$

For $j \leq p$ and $k = p + m \geq p + 1$,

$$\begin{aligned} W_{j,p+m} &= \frac{1}{4\pi} \int_{-\pi}^{\pi} \frac{\partial \ln g(\lambda; \boldsymbol{\beta})}{\partial \phi_j} \frac{\partial \ln g(\lambda; \boldsymbol{\beta})}{\partial \theta_m} \, d\lambda \\ &= \frac{1}{4\pi} \int_{-\pi}^{\pi} [e^{i(m-j)\lambda} \phi^{-1}(e^{-i\lambda}) \theta^{-1}(e^{i\lambda}) + e^{-i(m-j)\lambda} \phi^{-1}(e^{i\lambda}) \theta^{-1}(e^{-i\lambda})] \, d\lambda. \end{aligned}$$

If $\{Z(\lambda), -\pi \leq \lambda \leq \pi\}$ is the orthogonal increment process associated with the process $\{N_t\}$ in (10.8.28) and (10.8.29), we can rewrite $W_{j,p+m}$ as

$$\begin{aligned} W_{j,p+m} &= \frac{1}{2}\left[\left\langle \int_{-\pi}^{\pi} e^{i\lambda(t-j)} \phi^{-1}(e^{-i\lambda}) \, dZ(\lambda), \int_{-\pi}^{\pi} e^{i\lambda(t-m)} \theta^{-1}(e^{-i\lambda}) \, dZ(\lambda) \right\rangle \right. \\ &\quad \left. + \left\langle \int_{-\pi}^{\pi} e^{i\lambda(t-m)} \theta^{-1}(e^{-i\lambda}) \, dZ(\lambda), \int_{-\pi}^{\pi} e^{i\lambda(t-j)} \phi^{-1}(e^{-i\lambda}) \, dZ(\lambda) \right\rangle \right] \\ &= E[U_{t-j+1} V_{t-m+1}], \qquad 1 \leq j \leq p, \ 1 \leq m \leq q, \end{aligned}$$

and by the symmetry of the matrix $W(\boldsymbol{\beta})$,

$$W_{p+m,k} = E[V_{t-m+1} U_{t-k+1}], \qquad 1 \leq m \leq q, \ 1 \leq k \leq p.$$

The expressions for W_{jk} can be written more succinctly in matrix form as

$$W(\boldsymbol{\beta}) = E[\mathbf{Y}_t \mathbf{Y}_t'], \tag{10.8.30}$$

where $\mathbf{Y}_t = (U_t, U_{t-1}, \ldots, U_{t-p+1}, V_t, V_{t-1}, \ldots, V_{t-q+1})'$ and $\{U_t\}$ and $\{V_t\}$ are the autoregressive processes defined by (10.8.28) and (10.8.29) respectively. The expression (10.8.30) is equivalent to (8.8.3). We now return to the proof of Theorem 10.8.2, which is broken up into a series of propositions.

Proposition 10.8.5. *Suppose $I_n(\cdot)$ is the periodogram of $\{X_1, \ldots, X_n\}$ and $I_{n,Z}(\cdot)$ is the periodogram of $\{Z_1, \ldots, Z_n\}$. If $\eta(\cdot)$ is any continuous even function on $[-\pi, \pi]$ with absolutely summable Fourier coefficients $\{a_m, -\infty < m < \infty\}$, then*

$$E\left| n^{-1/2} \sum_j (I_n(\omega_j)\eta(\omega_j) - g(\omega_j; \boldsymbol{\beta}_0) I_{n,Z}(\omega_j)\eta(\omega_j)) \right| \to 0 \tag{10.8.31}$$

as $n \to \infty$.

PROOF. From Theorem 10.3.1, $I_n(\omega_j) - g(\omega_j; \boldsymbol{\beta}_0) I_{n,Z}(\omega_j) = R_n(\omega_j)$ where $R_n(\lambda) = \psi(e^{-i\lambda}) J_Z(\lambda) Y_n(-\lambda) + \psi(e^{i\lambda}) J_Z(-\lambda) Y_n(\lambda) + |Y_n(\lambda)|^2$, $\psi(e^{-i\lambda}) = \sum_{k=0}^{\infty} \psi_k e^{-i\lambda k}$, $J_Z(\lambda) = n^{-1/2} \sum_{t=1}^{n} Z_t e^{-i\lambda t}$, $Y_n(\lambda) = n^{-1/2} \sum_{l=0}^{\infty} \psi_l e^{-i\lambda l} U_{n,l}$ and $U_{n,l} = \sum_{t=1-l}^{n-l} Z_t e^{-i\lambda t} - \sum_{t=1}^{n} Z_t e^{-i\lambda t} = \sum_{r=1}^{l} (Z_{r-l} - Z_{n-l+r}) e^{-i\lambda(r-l)}$. The proof of Theorem 10.3.1 gives $\max_{\omega_k \in [0,\pi]} E|R_n(\omega_k)| = O(n^{-1/2})$, however this result together with the bound

$$E\left| n^{-1/2} \sum_j R_n(\omega_j) \eta(\omega_j) \right| \le n^{1/2} \sup_{\lambda} |\eta(\lambda)| \max_{\omega_k \in [0,\pi]} E|R_n(\omega_k)|,$$

is not good enough to establish (10.8.31). Therefore a more careful analysis is required. Consider

$$n^{-1/2} \sum_j \psi(e^{-i\omega_j}) J_Z(\omega_j) Y_n(-\omega_j) \eta(\omega_j)$$

$$= n^{-3/2} \sum_j \sum_{k=0}^{\infty} \sum_{l=0}^{\infty} \sum_{r=1}^{l} \sum_{m=-\infty}^{\infty} \sum_{t=1}^{n} \psi_k \psi_l a_m Z_t [Z_{r-l} - Z_{n-l+r}] e^{-i\omega_j(k+m-r+t)}.$$

Now for k, l, r and m fixed let $s = r - m - k \bmod(n)$. Then

$$n^{-1} \sum_j e^{-i\omega_j(k+m-r+t)} = n^{-1} \sum_j e^{-i\omega_j(t-s)} = \begin{cases} 0 & \text{if } t \ne s, \\ 1 & \text{if } t = s, \end{cases}$$

which implies that

$$E\left| n^{-1} \sum_{t=1}^{n} \psi_k \psi_l a_m Z_t [Z_{r-l} - Z_{n-l+r}] \sum_j e^{-i\omega_j(k+m-r+t)} \right| \le 2 |\psi_k \psi_l a_m| \sigma_0^2$$

and hence that

$$E\left| n^{-1/2} \sum_j \psi(e^{-i\omega_j}) J_Z(\omega_j) Y_n(-\omega_j) \eta(\omega_j) \right| \le n^{-1/2} 2 \sum_{k=0}^{\infty} \sum_{l=0}^{\infty} \sum_{m=-\infty}^{\infty} |\psi_k \psi_l a_m| \, |l| \sigma_0^2$$

$$\to 0$$

as $n \to \infty$. Since $E|Y_n(\omega_j)|^2 \le 2n^{-1} \sigma_0^2 (\sum_{k=0}^{\infty} |\psi_k| |k|^{1/2})^2$ (see (10.3.13)),

$$E\left| n^{-1/2} \sum_j |Y_n(\omega_j)|^2 \eta(\omega_j) \right| \le 2n^{-1/2} \sigma_0^2 \left(\sum_{k=0}^{\infty} |\psi_k| |k|^{1/2} \right)^2 \sup_{\lambda} |\eta(\lambda)|$$

$$\to 0,$$

whence

$$E\left| n^{-1/2} \sum_j R_n(\omega_j) \eta(\omega_j) \right| \to 0$$

as desired. □

Proposition 10.8.6. *If $\eta(\cdot)$ and its Fourier coefficients $\{a_m\}$ satisfy the conditions of Proposition* 10.8.5 *and if $\sum_{m=1}^{\infty} |a_m| m^{1/2} < \infty$ and $\int_{-\pi}^{\pi} \eta(\lambda) g(\lambda; \boldsymbol{\beta}_0)\, d\lambda = 0$,* *then*

$$n^{-1}\sum_j I_n(\omega_j)\eta(\omega_j) \quad \textit{is } \mathrm{AN}\left(0, n^{-1}\frac{\sigma_0^4}{\pi}\int_{-\pi}^{\pi}\eta^2(\lambda)g^2(\lambda;\boldsymbol{\beta}_0)\,d\lambda\right).$$

PROOF. In view of Proposition 10.8.5, it suffices to show that

$$n^{-1}\sum_j I_{n,Z}(\omega_j)\chi(\omega_j) \quad \text{is } \mathrm{AN}\left(0, n^{-1}\frac{\sigma_0^4}{\pi}\int_{-\pi}^{\pi}\chi^2(\lambda)\,d\lambda\right), \tag{10.8.32}$$

where $\chi(\lambda) = \eta(\lambda)g(\lambda;\boldsymbol{\beta}_0)$. Let $\chi_m(\lambda) = \sum_{|k|\le m} b_k e^{i\lambda k}$, where $b_k = (2\pi)^{-1}\int_{-\pi}^{\pi} e^{-ik\lambda}\chi(\lambda)\,d\lambda$. The assumptions on the Fourier coefficients of $\eta(\lambda)$ together with the geometric decay of $\gamma_X(h)$ imply that $\sum_{k=1}^{\infty}|b_k|k^{1/2} < \infty$, and

$$\chi_m(\lambda) \to \chi(\lambda) = \sum_k b_k e^{ik\lambda} \quad \text{as } m \to \infty,$$

where the convergence is uniform in λ. It also follows from our assumptions that $b_0 = 0$. We next show that for all $\varepsilon > 0$,

$$\lim_{m\to\infty}\limsup_{n\to\infty} P\left(n^{-1/2}\left|\sum_j I_{n,Z}(\omega_j)(\chi(\omega_j) - \chi_m(\omega_j))\right| > \varepsilon\right) = 0. \tag{10.8.33}$$

Observe that

$$\begin{aligned} n^{-1/2}\sum_j I_{n,Z}(\omega_j)(\chi(\omega_j) - \chi_m(\omega_j)) &= n^{-1/2}\sum_j\left(\sum_{|h|<n}\tilde{\gamma}_Z(h)e^{-ih\omega_j}\right)\left(\sum_{|k|>m} b_k e^{ik\omega_j}\right) \\ &= n^{1/2}\sum_{|h|<n}\tilde{\gamma}_Z(h)\left(\sum_{k\in K_h} b_{kn+h}\right), \end{aligned} \tag{10.8.34}$$

where $\tilde{\gamma}_Z(h) = n^{-1}\sum_{t=1}^{n-|h|} Z_t Z_{t+|h|}$ and $K_h = \{k \in Z : |kn + h| > m\}$. For the $h = 0$ term in (10.8.34), we have

$$\begin{aligned} \left|n^{1/2}\tilde{\gamma}_Z(0)\sum_{k\in K_0} b_{kn}\right| &\le 2\tilde{\gamma}_Z(0)n^{1/2}\sum_{k=n}^{\infty}|b_k| \\ &\le 2\tilde{\gamma}_Z(0)\sum_{k=n}^{\infty}|b_k|k^{1/2} \to 0 \quad \text{a.s.} \end{aligned} \tag{10.8.35}$$

by Remark 2 and the summability of $\{|b_k|k^{1/2}\}$. Moreover, since $E\tilde{\gamma}_Z(h) = 0$ for $h \ne 0$ and $E\tilde{\gamma}_Z(h)\tilde{\gamma}_Z(k) = 0$ for $h \ne k$ (see Problem 6.24), we have

$$E\left[2n^{1/2}\sum_{h=1}^{n}\tilde{\gamma}_Z(h)\left(\sum_{k\in K_h} b_{kn+h}\right)\right] = 0$$

and for $n > m$

$$\begin{aligned} \mathrm{Var}\left(2n^{1/2}\sum_{h=1}^{n}\tilde{\gamma}_Z(h)\left(\sum_{k\in K_h} b_{kn+h}\right)\right) &= 4n\sigma_0^4\sum_{h=1}^{n}\frac{n-h}{n^2}\left(\sum_{k\in K_h} b_{kn+h}\right)^2 \\ &\le 4\sigma_0^4\left(\sum_{|k|>m}|b_k|\right)^2. \end{aligned} \tag{10.8.36}$$

Now (10.8.33) follows at once from (10.8.35) and (10.8.36).

To complete the proof it is enough to show, by Proposition 6.3.9 and Problem 6.16, that

$$n^{-1}\sum_j I_{n,Z}(\omega_j)\chi_m(\omega_j) \quad \text{is AN}\left(0, n^{-1}\frac{\sigma_0^4}{\pi}\int_{-\pi}^{\pi}\chi_m^2(\lambda)\,d\lambda\right) \tag{10.8.37}$$

and

$$\int_{-\pi}^{\pi}\chi_m^2(\lambda)\,d\lambda \to \int_{-\pi}^{\pi}\chi^2(\lambda)\,d\lambda \quad \text{as } m\to\infty. \tag{10.8.38}$$

But since $n^{1/2}\tilde{\gamma}_Z(n-k) = n^{-1/2}\sum_{t=1}^{k} Z_t Z_{t+n-k} = o_p(1)$, it follows from Propositions 6.3.3 and 6.3.4 and Problem 6.24 that

$$\begin{aligned}
n^{-1/2}\sum_j I_{n,Z}(\omega_j)\chi_m(\omega_j) &= 2n^{1/2}\sum_{k=1}^{m}\tilde{\gamma}_Z(k)b_k + 2n^{1/2}\sum_{k=1}^{m}\tilde{\gamma}_Z(n-k)b_k \\
&= 2n^{1/2}\sum_{k=1}^{m}\tilde{\gamma}_Z(k)b_k + o_p(1) \\
&\Rightarrow \text{N}\left(0, 4\sigma_0^4\sum_{k=1}^{m} b_k^2\right).
\end{aligned}$$

By Parseval's identity, $4\sigma_0^4\sum_{k=1}^{m} b_k^2 = \sigma_0^4/\pi\int_{-\pi}^{\pi}\chi_m^2(\lambda)\,d\lambda$, which establishes (10.8.37). Finally (10.8.38) follows from the uniform convergence of $\chi_m(\lambda)$ to $\chi(\lambda)$. □

PROOF OF THEOREM 10.8.2. (i) The Taylor-series expansion of $\partial\bar{\sigma}^2(\boldsymbol{\beta}_0)/\partial\boldsymbol{\beta}$ about $\boldsymbol{\beta} = \bar{\boldsymbol{\beta}}_n$ can be written as

$$\begin{aligned}
n^{1/2}\frac{\partial\bar{\sigma}^2(\boldsymbol{\beta}_0)}{\partial\boldsymbol{\beta}} &= n^{1/2}\frac{\partial\bar{\sigma}^2(\bar{\boldsymbol{\beta}}_n)}{\partial\boldsymbol{\beta}} - n^{1/2}\frac{\partial^2\bar{\sigma}^2(\boldsymbol{\beta}_n^{\dagger})}{\partial\boldsymbol{\beta}^2}(\bar{\boldsymbol{\beta}}_n - \boldsymbol{\beta}_0) \\
&= -n^{1/2}\frac{\partial^2\bar{\sigma}^2(\boldsymbol{\beta}_n^{\dagger})}{\partial\boldsymbol{\beta}^2}(\bar{\boldsymbol{\beta}}_n - \boldsymbol{\beta}_0),
\end{aligned}$$

for some $\boldsymbol{\beta}_n^{\dagger}\in C$ satisfying $\|\boldsymbol{\beta}_n^{\dagger} - \bar{\boldsymbol{\beta}}_n\| < \|\bar{\boldsymbol{\beta}}_n - \boldsymbol{\beta}_0\|$ ($\|\cdot\|$ = Euclidean norm). Now

$$\frac{\partial^2\bar{\sigma}^2(\boldsymbol{\beta}_n^{\dagger})}{\partial\boldsymbol{\beta}^2} = n^{-1}\sum_j I_n(\omega_j)\frac{\partial^2 g^{-1}(\omega_j;\boldsymbol{\beta}_n^{\dagger})}{\partial\boldsymbol{\beta}^2}$$

and since $\boldsymbol{\beta}_n^{\dagger}\to\boldsymbol{\beta}_0$ a.s. by Theorem 10.8.1, the proof given for Proposition 10.8.2 can be used to establish the result,

$$\frac{\partial^2\bar{\sigma}^2(\boldsymbol{\beta}_n^{\dagger})}{\partial\boldsymbol{\beta}^2} \to \frac{\sigma_0^2}{2\pi}\int_{-\pi}^{\pi} g(\lambda;\boldsymbol{\beta}_0)\frac{\partial^2 g^{-1}(\lambda;\boldsymbol{\beta}_0)}{\partial\boldsymbol{\beta}^2}\,d\lambda \quad \text{a.s.} \tag{10.8.39}$$

Since $(2\pi)^{-1}g(\lambda;\boldsymbol{\beta})$ is the spectral density of a causal invertible ARMA process with white noise variance equal to one, it follows that $\int_{-\pi}^{\pi}\ln g(\lambda;\boldsymbol{\beta})\,d\lambda = 0$ for all $\boldsymbol{\beta} \in C$, and hence that

$$0 = \frac{\partial^2}{\partial \boldsymbol{\beta}^2} \int_{-\pi}^{\pi} \ln g(\lambda; \boldsymbol{\beta})\, d\lambda$$
$$= \int_{-\pi}^{\pi} g^{-1}(\lambda; \boldsymbol{\beta}) \frac{\partial^2 g(\lambda; \boldsymbol{\beta})}{\partial \boldsymbol{\beta}^2}\, d\lambda - \int_{-\pi}^{\pi} \left[\frac{\partial \ln g(\lambda; \boldsymbol{\beta})}{\partial \boldsymbol{\beta}}\right]\left[\frac{\partial \ln g(\lambda; \boldsymbol{\beta})}{\partial \boldsymbol{\beta}}\right]' d\lambda.$$

Since the last relation holds also with g replaced by g^{-1}, it follows from (10.8.39) that

$$\frac{\partial^2 \bar{\sigma}^2(\boldsymbol{\beta}_n^\dagger)}{\partial \boldsymbol{\beta}^2} \to 2\sigma_0^2 W(\boldsymbol{\beta}_0).$$

Consequently it suffices to show that

$$\frac{\partial \bar{\sigma}^2(\boldsymbol{\beta}_0)}{\partial \boldsymbol{\beta}} \text{ is AN}(\mathbf{0}, n^{-1} 4\sigma_0^4 W(\boldsymbol{\beta}_0)),$$

or equivalently, by the Cramer–Wold device, that

$$\mathbf{c}' \frac{\partial \bar{\sigma}^2(\boldsymbol{\beta}_0)}{\partial \boldsymbol{\beta}} \text{ is AN}(0, n^{-1} 4\sigma_0^4 \mathbf{c}' W(\boldsymbol{\beta}_0)\mathbf{c})$$

for all $\mathbf{c} \in \mathbb{R}^{p+q}$. But

$$\mathbf{c}' \frac{\partial \bar{\sigma}^2(\boldsymbol{\beta}_0)}{\partial \boldsymbol{\beta}} = n^{-1} \sum_j I_n(\omega_j)\eta(\omega_j),$$

where $\eta(\lambda) = \mathbf{c}'\, \partial g^{-1}(\lambda; \boldsymbol{\beta}_0)/\partial \boldsymbol{\beta}$. Now $\eta'(\cdot)$ and $\eta''(\cdot)$ are also continuous functions on $[-\pi, \pi]$, so that by Problem 2.22, the Fourier coefficients of $\eta(\cdot)$ satisfy the assumptions of Proposition 10.8.6 and

$$\int_{-\pi}^{\pi} \eta(\lambda) g(\lambda; \boldsymbol{\beta}_0)\, d\lambda = -\mathbf{c}' \frac{\partial}{\partial \boldsymbol{\beta}} \int_{-\pi}^{\pi} \ln g(\lambda; \boldsymbol{\beta})\, d\lambda \Big|_{\boldsymbol{\beta} = \boldsymbol{\beta}_0} = \mathbf{c}'\mathbf{0} = 0.$$

Hence, invoking Proposition 10.8.6, we have

$$n^{-1} \sum_j I_n(\omega_j)\eta(\omega_j) \text{ is AN}\left(0, n^{-1} \frac{\sigma_0^4}{\pi} \int_{-\pi}^{\pi} \eta^2(\lambda) g^2(\lambda; \boldsymbol{\beta}_0)\, d\lambda\right),$$

and since $(\sigma_0^4/\pi) \int_{-\pi}^{\pi} \eta^2(\lambda) g^2(\lambda; \boldsymbol{\beta}_0)\, d\lambda = 4\sigma_0^4 \mathbf{c}' W(\boldsymbol{\beta}_0)\mathbf{c}'$, the proof of (i) is complete.

(ii) Expanding $\partial \tilde{\sigma}^2(\boldsymbol{\beta}_0)/\partial \boldsymbol{\beta}$ in a Taylor series about the vector $\boldsymbol{\beta} = \tilde{\boldsymbol{\beta}}_n$, we have as in the proof of (i),

$$n^{1/2} \frac{\partial \tilde{\sigma}^2(\boldsymbol{\beta}_0)}{\partial \boldsymbol{\beta}} = -n^{1/2} \frac{\partial^2 \tilde{\sigma}^2(\boldsymbol{\beta}_n^\dagger)}{\partial \boldsymbol{\beta}^2} (\tilde{\boldsymbol{\beta}}_n - \boldsymbol{\beta}_0)$$

for some $\boldsymbol{\beta}_n^\dagger \in C$ with $\boldsymbol{\beta}_n^\dagger \to \boldsymbol{\beta}_0$ a.s. By (i) and Proposition 6.3.3, it suffices to show that

$$\frac{\partial^2 \tilde{\sigma}^2(\boldsymbol{\beta}_n^\dagger)}{\partial \boldsymbol{\beta}^2} - \frac{\partial^2 \bar{\sigma}^2(\boldsymbol{\beta}_n^\dagger)}{\partial \boldsymbol{\beta}^2} \xrightarrow{P} 0 \tag{10.8.40}$$

and

$$n^{1/2}\frac{\partial \tilde{\sigma}^2(\boldsymbol{\beta}_0)}{\partial \beta_k} - n^{1/2}\frac{\partial \bar{\sigma}^2(\boldsymbol{\beta}_0)}{\partial \beta_k} \xrightarrow{P} 0 \quad \text{for } k = 1, \ldots, p + q. \tag{10.8.41}$$

The proofs of (10.8.40) and (10.8.41) follow closely the argument given for the proof of Proposition 10.8.3. We shall only prove (10.8.41).

Since $g(\lambda; \boldsymbol{\beta}_0)$ and $\partial g(\lambda; \boldsymbol{\beta}_0)/\partial \beta_k$ have continuous derivatives of all orders with respect to λ and since $g(\lambda; \boldsymbol{\beta}_0) > 0$ and $|\partial g(\lambda; \boldsymbol{\beta}_0)/\partial \beta_k| > 0$, it follows easily from Problem 2.22 that

$$\begin{cases} \delta(h; \boldsymbol{\beta}_0) := \dfrac{1}{2\pi}\displaystyle\int_{-\pi}^{\pi} e^{ih\lambda} g^{-1}(\lambda; \boldsymbol{\beta}_0)\, d\lambda = O(h^{-4}), \\[2ex] \dfrac{\partial \delta(h; \boldsymbol{\beta}_0)}{\partial \beta_k} = \dfrac{1}{2\pi}\displaystyle\int_{-\pi}^{\pi} e^{ih\lambda} \dfrac{\partial g^{-1}(\lambda; \boldsymbol{\beta}_0)}{\partial \beta_k}\, d\lambda = O(h^{-4}), \end{cases} \tag{10.8.42}$$

as $h \to \infty$. Set

$$q_m(\lambda) = \sum_{|h| \le m} \delta(h; \boldsymbol{\beta}_0) e^{-ih\lambda}.$$

Then

$$\frac{\partial q_m(\lambda)}{\partial \beta_k} = \sum_{|h| \le m} \frac{\partial \delta(h; \boldsymbol{\beta}_0)}{\partial \beta_k} e^{-ih\lambda}.$$

Equations (10.8.42) ensure that if $m = [n^{1/5}]$ (the integer part of $n^{1/5}$), then

$$\begin{cases} \sup\limits_{\lambda} |g^{-1}(\lambda; \boldsymbol{\beta}_0) - q_m(\lambda)| \le \sum\limits_{|h| > m} |\delta(h; \boldsymbol{\beta}_0)| = O(n^{-3/5}), \\[2ex] \sup\limits_{\lambda} \left| \dfrac{\partial g^{-1}(\lambda; \boldsymbol{\beta}_0)}{\partial \beta_k} - \dfrac{\partial q_m(\lambda)}{\partial \beta_k} \right| \le \sum\limits_{|h| > m} \left| \dfrac{\partial \delta(h; \boldsymbol{\beta}_0)}{\partial \beta_k} \right| = O(n^{-3/5}). \end{cases} \tag{10.8.43}$$

As in the proof of Theorem 4.4.3, $q_m(\lambda)$ may be written as

$$q_m(\lambda) = K_m^{-1} |a(e^{-i\lambda})|^2,$$

where $a(z) = 1 + a_1 z + \cdots + a_m z^m \ne 0$ for $|z| \le 1$ and $K_m \to 1$ as $m \to \infty$. Let H_n be the covariance matrix corresponding to the autoregressive spectral density $(2\pi q_m(\lambda))^{-1}$. We shall show that

$$n^{-1/2} \frac{\partial}{\partial \beta_k} \left(\mathbf{X}_n' G_n^{-1}(\boldsymbol{\beta}_0) \mathbf{X}_n - \sum_j I_n(\omega_j) q_m(\omega_j) \right) \xrightarrow{P} 0 \tag{10.8.44}$$

as $n \to \infty$, where $m = [n^{1/5}]$. Once this is accomplished, the result (10.8.41) follows immediately, since by (10.8.43),

$$\left| n^{-1/2} \sum_j I_n(\omega_j) \frac{\partial q_m(\omega_j)}{\partial \beta_k} - n^{-1/2} \sum_j I_n(\omega_j) \frac{\partial g^{-1}(\omega_j; \boldsymbol{\beta}_0)}{\partial \beta_k} \right|$$
$$\leq n^{-1/2} \sum_j I_n(\omega_j) O(n^{-3/5}) = O(n^{-1/10}) \tilde{\gamma}(0) = o_p(1).$$

Throughout the remainder of this proof set $m = [n^{1/5}]$. From the proof of Proposition 10.8.3, the eigenvalues of $G_n^{-1}(\boldsymbol{\beta}_0)$ and H_n^{-1} are uniformly bounded in n. Moreover, the eigenvalues of the matrices $\partial G_n(\boldsymbol{\beta}_0)/\partial \beta_k$ and $\partial H_n(\boldsymbol{\beta}_0)/\partial \beta_k$ are also uniformly bounded in n since $\partial g(\lambda; \boldsymbol{\beta}_0)/\partial \beta_k$ and $\partial q_m^{-1}(\lambda)/\partial \beta_k$ are uniformly bounded (see the proof of Proposition 4.5.3). It is easy to show from (10.8.43) that there exists a positive constant K such that for all $\mathbf{y} \in \mathbb{R}^n$, (cf. (10.8.17)) $\|(G_n^{-1}(\boldsymbol{\beta}_0) - H_n^{-1}(\boldsymbol{\beta}_0))\mathbf{y}\| \leq K n^{-3/5} \|\mathbf{y}\|$, and

$$\left\| \left(\frac{\partial G_n(\boldsymbol{\beta}_0)}{\partial \beta_k} - \frac{\partial H_n(\boldsymbol{\beta}_0)}{\partial \beta_k} \right) \mathbf{y} \right\| \leq K n^{-3/5} \|\mathbf{y}\|.$$

It then follows from a routine calculation that

$$n^{-1/2} \left| \frac{\partial}{\partial \beta_k} (\mathbf{X}_n' G_n^{-1}(\boldsymbol{\beta}_0) \mathbf{X}_n - \mathbf{X}_n' H_n^{-1} \mathbf{X}_n) \right| \leq O(n^{-1/10}) \tilde{\gamma}(0) = o_p(1). \quad (10.8.45)$$

We next compare $n^{-1/2} \partial(\mathbf{X}_n' H_n^{-1} \mathbf{X}_n)/\partial \beta_k$ with $n^{-1/2} \partial(\mathbf{X}_n' \tilde{H}_n^{-1} \mathbf{X}_n)/\partial \beta_k$ where $\tilde{H}_n^{-1}$ is the covariance matrix of an MA(m) process with spectral density $(2\pi)^{-1} q_m(\lambda) = (2\pi K_m)^{-1} |a(e^{-i\lambda})|^2$. Now from the proof of Proposition 10.8.3 (see (10.8.19) and (10.8.20)),

$$n^{-1/2} \frac{\partial}{\partial \beta_k} (\mathbf{X}_n' H_n^{-1} \mathbf{X}_n - \mathbf{X}_n' \tilde{H}_n^{-1} \mathbf{X}_n) = n^{-1/2} \sum_{i,j=1}^{m} \frac{\partial}{\partial \beta_k} (h_{ij} - \tilde{h}_{ij}) X_i X_j$$
$$+ n^{-1/2} \sum_{i,j=n-m+1}^{n} \frac{\partial}{\partial \beta_k} (h_{ij} - \tilde{h}_{ij}) X_i X_j.$$

It follows from (10.8.43) that $\partial(h_{ij} - \tilde{h}_{ij})/\partial \beta_k$ is uniformly bounded in i and j, and since $m = [n^{1/5}]$ the above expression is $o_p(1)$. By the same reasoning,

$$n^{-1/2} \frac{\partial}{\partial \beta_k} \mathbf{X}_n' \tilde{H}_n^{-1} \mathbf{X}_n = n^{-1/2} \frac{\partial}{\partial \beta_k} \sum_j I_n(\omega_j) q_m(\omega_j) + o_p(1)$$

(see (10.8.21)), and with (10.8.45) this establishes (10.8.44) and hence (10.8.41).

(iii) From the Taylor series expansion of $l(\boldsymbol{\beta}_0)$ about $\boldsymbol{\beta} = \hat{\boldsymbol{\beta}}_n$ (cf. (i) and (ii)), it suffices to show that as $n \to \infty$,

$$n^{-1} \frac{\partial^2 \ln(\det G_n(\boldsymbol{\beta}_n^\dagger))}{\partial \boldsymbol{\beta}^2} \xrightarrow{P} 0, \quad (10.8.46)$$

where $\boldsymbol{\beta}_n^\dagger \to \boldsymbol{\beta}_0$ a.s., and

$$n^{-1/2} \frac{\partial \ln(\det G_n(\boldsymbol{\beta}_0))}{\partial \beta_k} \to 0 \quad \text{for } k = 1, \ldots, p + q. \quad (10.8.47)$$

We shall only prove (10.8.47) since the argument for (10.8.46) is similar, but less technical due to the presence of the factor n^{-1}.

As in the proof of Proposition 10.8.4, if $\{Y_t\}$ is an ARMA process with spectral density $(2\pi)^{-1}g(\lambda; \boldsymbol{\beta})$, then $\det G_n(\boldsymbol{\beta}) = r_0(\boldsymbol{\beta}) \cdots r_{n-1}(\boldsymbol{\beta})$, where $r_t(\boldsymbol{\beta}) = E(Y_{t+1} - \hat{Y}_{t+1})^2$. Denote the autocovariance function of $\{Y_t\}$ by $\eta(h; \boldsymbol{\beta})$ and write the difference equations for $\{Y_t\}$ as

$$Y_t + \sum_{j=1}^{\infty} \pi_j(\boldsymbol{\beta}) Y_{t-j} = Z_t, \qquad \{Z_t\} \sim \text{IID}(0, 1). \tag{10.8.48}$$

We have from Corollary 5.1.1,

$$r_t(\boldsymbol{\beta}) = \eta(0; \boldsymbol{\beta}) - \boldsymbol{\eta}_t' G_t^{-1}(\boldsymbol{\beta}) \boldsymbol{\eta}_t,$$

where $\boldsymbol{\eta}_t = (\eta(1; \boldsymbol{\beta}), \ldots, \eta(t; \boldsymbol{\beta}))'$. For notational convenience, we shall often suppress the argument $\boldsymbol{\beta}$ when the dependence on $\boldsymbol{\beta}$ is clear. From (10.8.48), we have

$$\boldsymbol{\eta}_\infty = -G_\infty \boldsymbol{\pi}_\infty,$$

where $\boldsymbol{\eta}_\infty = (\eta(1; \boldsymbol{\beta}), \eta(2; \boldsymbol{\beta}), \ldots)'$, $G_\infty = [\eta(i-j; \boldsymbol{\beta})]_{i,j=1}^{\infty}$ and $\boldsymbol{\pi}_\infty = (\pi(1; \boldsymbol{\beta}), \pi(2; \boldsymbol{\beta}), \ldots)'$. It then follows that

$$\boldsymbol{\pi}_\infty = -G_\infty^{-1} \boldsymbol{\eta}_\infty,$$

and it is easy to show that G_∞^{-1} may be written as

$$G_\infty^{-1} = TT',$$

where

$$T = [\pi_{i-j}(\boldsymbol{\beta})]_{i,j=1}^{\infty},$$

$\pi_0(\boldsymbol{\beta}) = 1$ and $\pi_j(\boldsymbol{\beta}) = 0$ for $j < 0$. We also have from (10.8.48) and the independence of Z_t and $\{Y_{t-1}, Y_{t-2}, \ldots\}$, that

$$\eta(0; \boldsymbol{\beta}) = \boldsymbol{\pi}_\infty' G_\infty \boldsymbol{\pi}_\infty + 1.$$

Consequently, we may write

$$r_t(\boldsymbol{\beta}) = 1 + \boldsymbol{\eta}_\infty' G_\infty^{-1} \boldsymbol{\eta}_\infty - \boldsymbol{\eta}_t' G_t^{-1} \boldsymbol{\eta}_t,$$

and hence

$$\begin{aligned}\frac{\partial r_t(\boldsymbol{\beta}_0)}{\partial \beta_k} &= 2\frac{\partial \boldsymbol{\eta}_\infty'}{\partial \beta_k} G_\infty^{-1} \boldsymbol{\eta}_\infty + \boldsymbol{\eta}_\infty' G_\infty^{-1} \frac{\partial G_\infty}{\partial \beta_k} G_\infty^{-1} \boldsymbol{\eta}_\infty \\ &\quad - 2\frac{\partial \boldsymbol{\eta}_t'}{\partial \beta_k} G_t^{-1} \boldsymbol{\eta}_t + \boldsymbol{\eta}_t' G_t^{-1} \frac{\partial G_t}{\partial \beta_k} G_t^{-1} \boldsymbol{\eta}_t\end{aligned}$$

where all of the terms on the right-hand side are evaluated at $\boldsymbol{\beta} = \boldsymbol{\beta}_0$. Note that if $\boldsymbol{\phi}_t = (\phi_{t1}, \ldots, \phi_{tt})' = G_t^{-1} \boldsymbol{\eta}_t$ is the coefficient vector of the best linear predictor of Y_{t+1} in terms of $(Y_t, \ldots, Y_1)'$, then the above equation reduces to

$$\begin{aligned}\frac{\partial r_t(\boldsymbol{\beta}_0)}{\partial \beta_k} &= -2\left(\frac{\partial \boldsymbol{\eta}_\infty'}{\partial \beta_k}\boldsymbol{\pi}_\infty + \frac{\partial \boldsymbol{\eta}_t'}{\partial \beta_k}\boldsymbol{\phi}_t\right) \\ &\quad - \left(\boldsymbol{\pi}_\infty'\frac{\partial G_\infty}{\partial \beta_k}\boldsymbol{\pi}_\infty - \boldsymbol{\phi}_t'\frac{\partial G_t}{\partial \beta_k}\boldsymbol{\phi}_t\right).\end{aligned} \tag{10.8.49}$$

We next show that the vectors $\boldsymbol{\pi}_t = (\pi_1, \ldots, \pi_t)'$ and $\boldsymbol{\phi}_t$ are not far apart. Observe that

$$Y_{t+1} - \phi_{t1}Y_t - \cdots - \phi_{tt}Y_1 = Y_{t+1} - \hat{Y}_{t+1}$$

and

$$Y_{t+1} + \pi_1 Y_t + \cdots + \pi_t Y_1 = \sum_{j>t} \pi_j Y_{t+1-j} + Z_{t+1}$$

so that the variance of $(\pi_1 + \phi_{t1})Y_t + \cdots + (\pi_t + \phi_{tt})Y_1$ is equal to

$$\begin{aligned}(\boldsymbol{\pi}_t + \boldsymbol{\phi}_t)'G_t(\boldsymbol{\pi}_t + \boldsymbol{\phi}_t) &= \operatorname{Var}\left(\sum_{j>t}\pi_j Y_{t+1-j} + Z_{t+1} - (Y_{t+1} - \hat{Y}_{t+1})\right) \\ &\le 2\operatorname{Var}\left(\sum_{j>t}\pi_j Y_{t+1-j}\right) + 2\operatorname{Var}\left(Z_{t+1} - (Y_{t+1} - \hat{Y}_{t+1})\right) \\ &\le 2\left(\sum_{j>t}|\pi_j|\right)^2 \eta(0, \boldsymbol{\beta}_0) + 2(r_t - 1) \\ &\le 4\left(\sum_{j>t}|\pi_j|\right)^2 \eta(0, \boldsymbol{\beta}_0),\end{aligned}$$

where the last inequality comes from the calculation in the proof of Proposition 10.8.4. Since the eigenvalues of G_t are bounded below by $\inf_\lambda g(\lambda; \boldsymbol{\beta}_0) > L > 0$,

$$\begin{aligned}\sum_{j=1}^{t}(\pi_j + \phi_{tj})^2 &\le L^{-1}(\boldsymbol{\pi}_t + \boldsymbol{\phi}_t)'G_t(\boldsymbol{\pi}_t + \boldsymbol{\phi}_t) \\ &\le 4L^{-1}\left(\sum_{j>t}|\pi_j|\right)^2 \eta(\mathbf{0}, \boldsymbol{\beta}_0) \\ &\le Ks^{-t}\end{aligned} \tag{10.8.50}$$

for some $K > 0$ and $0 < s < 1$. Therefore, from (10.8.49),

$$\begin{aligned}\left|\frac{\partial r_t(\boldsymbol{\beta}_0)}{\partial \beta_k}\right| &\le 2K_1\left(\sum_{j=1}^{t}|\pi_j + \phi_{tj}| + \sum_{j>t}|\pi_j|\right) \\ &\quad + K_1\left(\sum_{i,j=1}^{t}|\pi_i\pi_j - \phi_{ti}\phi_{tj}| + 2\sum_{i>t}|\pi_i|\sum_j|\pi_j|\right)\end{aligned}$$

where $K_1 = (1/2\pi)\int_{-\pi}^{\pi}|\partial g(\lambda; \boldsymbol{\beta}_0)/\partial\beta_k|\,d\lambda$. Since $\sum_{j=1}^{\infty}\pi_j^2 < \infty$ and $\sum_{j=1}^{t}\phi_{tj}^2$ is bounded in t (see (10.8.50)), we have from the Cauchy–Schwarz inequality,

$$\left|\frac{\partial r_t(\boldsymbol{\beta}_0)}{\partial \beta_k}\right| \le K_2 t^{1/2} s^{-t/2} + K_3 \sum_{j>t} |\pi_j| + K_4 t s^{-t/2}$$
$$\le K_5 s_1^{-t}$$

where K_2, K_3, K_4 and K_5 are positive constants and $0 < s_1 < 1$. Since $r_t(\boldsymbol{\beta}_0) \ge 1$, it then follows that

$$\left| n^{-1/2} \frac{\partial \ln(\det G_n(\boldsymbol{\beta}_0)}{\partial \beta_k} \right| \le n^{-1/2} \sum_{t=0}^{n-1} \left|\frac{\partial r_t(\boldsymbol{\beta}_0)}{\partial \beta_k}\right| r_t^{-1}(\boldsymbol{\beta}_0)$$
$$\le n^{-1/2} \sum_{t=0}^{n-1} \left|\frac{\partial r_t(\boldsymbol{\beta}_0)}{\partial \beta_k}\right|$$
$$\le n^{-1/2} K_5 (1 - s_1)^{-1}$$
$$\to 0$$

as $n \to \infty$, which completes the proof of (10.8.47). □

Problems

10.1. The discrete Fourier transform $\{a_j, j \in F_n\}$ of $\{X_1, \ldots, X_n\}$ can be expressed as

$$a_j = J(\omega_j), \qquad \omega_j = 2\pi j/n \in (-\pi, \pi],$$

where

$$J(\lambda) = n^{-1/2} \sum_{t=1}^{n} X_t e^{-it\lambda}, \qquad -\infty < \lambda < \infty.$$

Show that $X_t = (2\pi)^{-1} n^{1/2} \int_{-\pi}^{\pi} J(\lambda) e^{it\lambda}\, d\lambda$. [In this and the following questions we shall refer to $J(\lambda)$, $-\infty < \lambda < \infty$, as the *Fourier transform* of $\{X_1, \ldots, X_n\}$. Note that the periodogram $\{I_n(\omega_j), \omega_j = 2\pi j/n \in (-\pi, \pi]\}$ can be expressed as

$$I_n(\omega_j) = |J(\omega_j)|^2, \qquad j \in F_n.]$$

10.2. Suppose that $z_t = x_t y_t$, $t = 1, \ldots, n$. If J_X, J_Y and J_Z are the Fourier transforms (see Problem 10.1) of $\{x_t\}$, $\{y_t\}$ and $\{z_t\}$ respectively, and if $\xi_j = J_X(2\pi j/n)$, $\eta_j = J_Y(2\pi j/n)$ and $\zeta_j = J_Z(2\pi j/n)$, $j = 0, \pm 1, \ldots$, show that

$$\zeta_j = n^{-1/2} \sum_{k \in F_n} \xi_k \eta_{j-k}.$$

10.3. Suppose that $\{x_t, t = 0, \pm 1, \ldots\}$ has period n and that

$$z_t = \sum_{k=-s}^{s} g_k x_{t-k}, \qquad t = 0, \pm 1, \ldots.$$

If $G(e^{-i\lambda}) = \sum_{j=-s}^{s} g_j e^{-ij\lambda}$, $-\infty < \lambda < \infty$, and J_X, J_Z are the Fourier transforms of $\{x_1, \ldots, x_n\}$ and $\{z_1, \ldots, z_n\}$ respectively, show that

$$J_Z(\omega_j) = G(e^{-i\omega_j}) J_X(\omega_j), \qquad \omega_j = 2\pi j/n \in (-\pi, \pi],$$

and

$$I_Z(\omega_j) = |G(e^{-i\omega j})|^2 I_X(\omega_j), \qquad \omega_j = 2\pi j/n \in (-\pi, \pi].$$

10.4.* Show that the sequence $X_t = e^{i\nu t}$, $t = 1, \ldots, n$, has Fourier transform,

$$J(\lambda) = n^{-1/2} \frac{\sin[n(\lambda - \nu)/2]}{\sin[(\lambda - \nu)/2]} \exp[-i(\lambda - \nu)(n + 1)/2], \qquad -\infty < \lambda < \infty.$$

Use this result to evaluate and plot the periodograms I_8 and I_{10} in the case when $\nu = \pi/4$. [$\pi/4$ is a Fourier frequency for $n = 8$ but not for $n = 10$.] Verify in each case that $\sum_{j \in F_n} I_n(\omega_j) = \sum_{t=1}^n |X_t|^2$.

10.5.* If $J(\cdot)$ is the Fourier transform in Problem 10.4 and $-\pi < \nu < \pi$, determine the frequencies λ such that (a) $|J(\lambda)|^2$ is zero and (b) $|J(\cdot)|^2$ has a local maximum at λ. Let $M = |J(\nu)|^2$ and $M_1 = |J(\lambda_1)|^2$ where λ_1 is the frequency closest to ν (but not equal to ν) at which $|J(\cdot)|^2$ has a local maximum. Show that
(a) $M \to \infty$ as $n \to \infty$,
(b) $M_1/M \to .047190$ and $\lambda_1 \to \nu$ as $n \to \infty$, and
(c) for any *fixed* frequency $\omega \in [-\pi, \pi]$ such that $\omega \neq \nu$, $|J(\omega)|^2 \to 0$ as $n \to \infty$.

10.6. Verify (10.2.11) and show that $\|P_{S(p)}\mathbf{X} - P_{\overline{\text{sp}}\{1\}}\mathbf{X}\|^2$ is independent of $\|\mathbf{X} - P_{S(p)}\mathbf{X}\|^2$.

10.7. The following quarterly sales totals $\{X_t, t = 1, \ldots, 12\}$ were observed over a period of three years: 27, 18, 10, 19, 24, 17, 5, 15, 22, 18, 2, 14. Test at level .05 the hypothesis,

$$X_t = c + Z_t, \qquad \{Z_t\} \sim \text{IID}(0, \sigma^2),$$

where c is constant, against the alternative,

$$X_t = c + S_t + Z_t,$$

where S_t is a deterministic sinusoid with period one year. Repeat the test assuming only that S_t has period one year but is not necessarily sinusoidal.

10.8. Use the computer program SPEC to compute and file the periodogram $I(\omega_j)$, $0 < \omega_j = 2\pi j/n \le \pi$ of the Wölfer sunspot numbers. File also the standardized cumulative periodogram $C(j)$, $1 \le j \le [(n-1)/2]$, defined by (10.2.22). Plot the two periodograms and use the latter to conduct a Kolmogorov–Smirnov test at level .05 of the null hypothesis that $\{X_t\}$ is white noise.

10.9.* Consider the model

$$X_t = \mu + A\cos\omega t + B\sin\omega t + Z_t, \qquad t = 1, 2, \ldots,$$

where $\{Z_t\}$ is iid $\text{N}(0, \sigma^2)$, μ, A, B and σ^2 are unknown parameters and ω is known. If $\bar{X}_n$, $\tilde{A}$ and $\tilde{B}$ are the estimators, $\bar{X}_n = n^{-1}\sum_{t=1}^n X_t$, $\tilde{A} = (2/n)^{1/2}\sum_{t=1}^n (X_t - \bar{X}_n)\cos\omega t$ and $\tilde{B} = (2/n)^{1/2}\sum_{t=1}^n (X_t - \bar{X}_n)\sin\omega t$ of μ, A and B, show that

$$\begin{bmatrix} \bar{X}_n \\ \tilde{A} \\ \tilde{B} \end{bmatrix} \text{ is AN}\left(\begin{bmatrix} \mu \\ A \\ B \end{bmatrix}, \sigma^2 \begin{bmatrix} 2 & 0 & 0 \\ 0 & 2 & 0 \\ 0 & 0 & 1 \end{bmatrix}\right).$$

10.10. Generate 100 observations, $\{X_1, \ldots, X_{100}\}$, of Gaussian white noise. Use the following three procedures to test (at level .05) the null hypothesis, $\{X_t\}$ is Gaussian white noise, against the alternative hypothesis, $\{X_t\}$ is Gaussian white noise with an added deterministic periodic component of unspecified frequency.
(a) Fisher's test.
(b) The Kolmogorov–Smirnov test.
(c) Let ω_j be the frequency at which the periodogram is maximum and apply the test described in Section 10.2(a) using the model $X_t = \mu + A \cos \omega_j t + B \sin \omega_j t + Z_t$. In other words reject the null hypothesis if

$$(100 - 3)I(\omega_j) \bigg/ \left(\sum_{i=1}^{n} X_i^2 - I(0) - 2I(\omega_j) \right) > F_{.95}(2, 97).$$

Is this a reasonable test for hidden periodicities of unspecified frequency?

10.11. Compute the periodogram of the series $\{X_t - X_{t-1}, t = 2, \ldots, 72\}$ where X_t, $t = 1, \ldots, 72$, are the accidental deaths of Example 1.1.6. Use the procedure described in Section 10.2(a) to test for the presence of a deterministic periodic component with frequency $12\pi/71$. (This is the Fourier frequency with period closest to 12.) Apply Fisher's test to the periodogram of the residuals from the fitted model (9.6.6) for $\{X_t\}$.

10.12. For the Lake Huron data of Problem 9.6, estimate the spectral density function using two different discrete spectral average estimators. Construct 95% confidence intervals for the logarithm of the spectral density. Also compute the MLARMA spectral density estimate and compare it with the discrete spectral average estimators.

10.13.* Suppose that $V_1, V_2, \ldots,$ is a sequence of iid exponential random variables with mean one.
(a) Show that $P(\max_{1 \le j \le q} V_j - \ln q \le x) \to e^{-e^{-x}}$ for all x as $q \to \infty$.
(b) Show that $P(\max_{1 \le j \le q} V_j / (q^{-1} \sum_{j=1}^{q} V_j) \le x + \ln q) \to e^{-e^{-x}}$ as $q \to \infty$.
(c) If ξ_q is as defined in (10.2.20) conclude that for large q, $P(\xi_q \ge x) \simeq 1 - \exp\{-qe^{-x}\}$.

10.14. If $\{Z_t\} \sim \text{IID}(0, \sigma^2)$ and $EZ_1^4 < \infty$, establish the inequality, $E(\sum_{j=1}^{m} Z_j)^4 \le mEZ_1^4 + 3m^2\sigma^4$.

10.15. Find approximate values for the mean and variance of the periodogram ordinate $I_{200}(\pi/4)$ of the causal AR(1) process

$$X_t - .5X_{t-1} = Z_t, \qquad \{Z_t\} \sim \text{IID}(0, \sigma^2).$$

Defining $\hat{f}(\omega_j) = (10\pi)^{-1} \sum_{k=-2}^{2} I_{200}(\omega_j + \omega_k)$, $\omega_j = 2\pi j/200$, use the asymptotic distribution of the periodogram ordinates to approximate
(a) the mean and variance of $\hat{f}(\pi/4)$,
(b) the covariance of $\hat{f}(\pi/4)$ and $\hat{f}(26\pi/100)$,
(c) $P(\hat{f}(\pi/4) > 1.1 f(\pi/4))$ where f is the spectral density of $\{X_t\}$,
(d) $P(\max_{1 \le j \le 99} (I_{200}(\omega_j)/f(\omega_j)) > .06 \sum_{j=1}^{99} (J_{200}(\omega_j)/f(\omega_j)))$.

10.16. Show that successive application of two filters $\{a_{-r}, \ldots, a_r\}$ and $\{b_{-s}, \ldots, b_s\}$ to a time series $\{X_t\}$ is equivalent to application of the single filter $\{c_{-r-s}, \ldots, c_{r+s}\}$ where

$$c_k = \sum_{j=-\infty}^{\infty} a_{k-j}b_j = \sum_{j=-\infty}^{\infty} b_{k-j}a_j,$$

and a_j, b_j are defined to be zero for $|j| > r$, s respectively. In Example 10.4.2 show that successive application of the three filters, $3^{-1}\{1, 1, 1\}$, $7^{-1}\{1, \ldots, 1\}$ and $11^{-1}\{1, \ldots, 1\}$ is equivalent to application of the filter $(231)^{-1}\{1, 3, 6, 9, 12, 15, 18, 20, 21, 21, 21, 20, 18, 15, 12, 9, 6, 3, 1\}$.

10.17. If $\sum_{i=1}^{n} X_i = 0$, $I_n(\cdot)$ is the period-2π extension of the periodogram of $\{X_1, \ldots, X_n\}$, and $\hat{f}_D(\omega_j)$, $\omega_j = 2\pi j/n$, is the Daniell estimator,

$$\hat{f}_D(\omega_j) = \frac{1}{2\pi(2m + 1)} \sum_{k=-m}^{m} I_n(\omega_j + \omega_k),$$

show that

$$\hat{f}_D(\omega_j) = \frac{1}{2\pi} \sum_{|k|<n} \lambda_k \hat{\gamma}(k) e^{-ik\omega_j},$$

where $\lambda_k = (2m + 1)^{-1} \sin[(2m + 1)k\pi/n]/[\sin(k\pi/n)]$. Compare this result with the approximate lag window for the Daniell estimator derived in Section 10.4.

10.18. Compare the Bartlett and Daniell spectral density estimators by plotting and examining the spectral windows defined in (10.4.13).

10.19. Derive the equivalent degrees of freedom, asymptotic variance and bandwidth of the Parzen lag-window estimator defined in Section 10.4.

10.20. Simulate 200 observations of the Gaussian AR(2) process,

$$X_t - X_{t-1} + .85X_{t-2} = Z_t, \qquad Z_t \sim \text{WN}(0, 1),$$

and compare the following four spectral density estimators:

(i) the periodogram,
(ii) a discrete spectral average estimator,
(iii) the maximum entropy estimator with m chosen so as to minimize the AICC value,
(iv) the MLARMA spectral density estimator.

Using the discrete spectral average estimator, construct 95% confidence intervals for $\ln f(\lambda)$, $\lambda \in (0, \pi)$, where f is the spectral density of $\{X_t\}$. Does $\ln f(\lambda)$ lie entirely within these bounds? Why does $f(\cdot)$ have such a large peak near $\pi/3$?

10.21.* (a) Let $X_1, \ldots, X_n$ be iid $\text{N}(0, \sigma^2)$ random variables and let $Y_1, \ldots, Y_q$ be the corresponding periodogram ordinates, $Y_j = I_n(\omega_j)$, where $q = [(n - 1)/2]$. Determine the joint density of $Y_1, \ldots, Y_q$ and hence the maximum likelihood estimator of σ^2 based on $Y_1, \ldots, Y_q$.

(b) Derive a pair of equations for the maximum likelihood estimators $\hat{\phi}$ and $\hat{\sigma}^2$ based on the large-sample distribution of the periodogram ordinates $I_n(\lambda_1), \ldots, I_n(\lambda_m)$, $0 < \lambda_1 < \cdots < \lambda_m < \pi$, when $\{X_1, \ldots, X_n\}$ is a sample from the causal AR(1) process, $X_t = \phi X_{t-1} + Z_t$, $\{Z_t\} \sim \text{IID}(0, \sigma^2)$.

10.22.* Show that the partial sum $S_{2n+1}(x)$ of the Fourier series of $I_{[0,\pi]}(x)$ (see (2.8.5)) satisfies

$$S_{2n+1}(x) = \frac{1}{2} + \frac{1}{\pi} \int_0^x \frac{\sin[2(n+1)y]}{\sin y} dy, \qquad x \geq 0.$$

Let x_1 denote the smallest value of x in $(0, \pi]$ at which $S_{2n+1}(\cdot)$ has a local maximum, and let $M_1 = S_{2n+1}(x_1)$. Show that
(a) $\lim_{n\to\infty} x_1 = 0$ and
(b) $\lim_{n\to\infty} M_1 = 1.089367$.
[This persistence as $n \to \infty$ of an "overshoot" of the Fourier series beyond the value of $I_{[0,\pi]}(x)$ on $[0, \pi]$ is called the Gibbs phenomenon.]

CHAPTER 11

Multivariate Time Series

Many time series arising in practice are best considered as components of some vector-valued (multivariate) time series $\{\mathbf{X}_t\}$ whose specification includes not only the serial dependence of each component series $\{X_{ti}\}$ but also the interdependence between different component series $\{X_{ti}\}$ and $\{X_{tj}\}$. From a second order point of view a stationary multivariate time series is determined by its mean vector, $\boldsymbol{\mu} = E\mathbf{X}_t$ and its covariance matrices $\Gamma(h) = E(\mathbf{X}_{t+h}\mathbf{X}_t') - \boldsymbol{\mu}\boldsymbol{\mu}'$, $h = 0, \pm 1, \ldots$. Most of the basic theory of univariate time series extends in a natural way to multivariate series but new problems arise. In this chapter we show how the techniques developed earlier for univariate series are extended to the multivariate case. Estimation of the basic quantities $\boldsymbol{\mu}$ and $\Gamma(\cdot)$ is considered in Section 11.2. In Section 11.3 we introduce multivariate ARMA processes and develop analogues of some of the univariate results in Chapter 3. The prediction of stationary multivariate processes, and in particular of ARMA processes, is treated in Section 11.4 by means of a multivariate generalization of the innovations algorithm used in Chapter 5. This algorithm is then applied in Section 11.5 to simplify the calculation of the Gaussian likelihood of the observations $\{\mathbf{X}_1, \mathbf{X}_2, \ldots, \mathbf{X}_n\}$ of a multivariate ARMA process. Estimation of parameters using maximum likelihood and (for autoregressive models) the Yule–Walker equations is also considered. In Section 11.6 we discuss the cross spectral density of a bivariate stationary process $\{\mathbf{X}_t\}$ and its interpretation in terms of the spectral representation of $\{\mathbf{X}_t\}$. (The spectral representation is discussed in more detail in Section 11.8.) The bivariate periodogram and its asymptotic properties are examined in Section 11.7 and Theorem 11.7.1 gives the asymptotic joint distribution for a linear process of the periodogram matrices at frequencies $\lambda_1, \lambda_2, \ldots, \lambda_m \in (0, \pi)$. Smoothing of the periodogram is used to estimate the cross-spectrum and hence the cross-amplitude spectrum, phase spectrum and squared coherency for which approximate confidence intervals are given. The chapter ends with an introduction to the spectral representation of an m-variate stationary process and multivariate linear filtering.

§11.1 Second Order Properties of Multivariate Time Series

Consider m time series $\{X_{ti}, t = 0, \pm 1, \pm 2, \ldots\}$, $i = 1, \ldots, m$, with $EX_{ti}^2 < \infty$ for all t and all i. If all the finite dimensional joint distributions of the random variables $\{X_{ti}\}$ were multivariate normal, then the distributional properties of $\{X_{ti}\}$ would be completely determined by the means,

$$\mu_{ti} := EX_{ti}, \tag{11.1.1}$$

and covariances,

$$\gamma_{ij}(t + h, t) := E[(X_{t+h,i} - \mu_{t+h,i})(X_{tj} - \mu_{tj})]. \tag{11.1.2}$$

Even when the observations $\{X_{ti}\}$ do not have joint normal distributions, the quantities μ_{ti} and $\gamma_{ij}(t + h, t)$ specify the second-order properties, the covariances providing us with a measure of the dependence, not only between observations in the same series, but also between observations in different series.

It is more convenient when dealing with m interrelated series to use vector notation. Thus we define

$$\mathbf{X}_t := (X_{t1}, \ldots, X_{tm})', \qquad t = 0, \pm 1, \pm 2, \ldots. \tag{11.1.3}$$

The second-order properties of the multivariate time series $\{\mathbf{X}_t\}$ are then specified by the mean vectors,

$$\boldsymbol{\mu}_t := E\mathbf{X}_t = (\mu_{t1}, \ldots, \mu_{tm})', \tag{11.1.4}$$

and covariance matrices,

$$\Gamma(t + h, t) := \mathrm{E}[(\mathbf{X}_{t+h} - \boldsymbol{\mu}_{t+h})(\mathbf{X}_t - \boldsymbol{\mu}_t)'] = [\gamma_{ij}(t + h, t)]_{i,j=1}^m. \tag{11.1.5}$$

Remark. If $\{\mathbf{X}_t\}$ has complex-valued components, then $\Gamma(t + h, t)$ is defined as

$$\Gamma(t + h, t) = E[(\mathbf{X}_{t+h} - \boldsymbol{\mu}_{t+h})(\mathbf{X}_t - \boldsymbol{\mu}_t)^*],$$

where * denotes complex conjugate transpose. However we shall assume except where explicitly stated otherwise that $\mathbf{X}_t$ is real.

As in the univariate case, a particularly important role is played by the class of multivariate stationary time series, defined as follows.

Definition 11.1.1 (Stationary Multivariate Time Series). The series (11.1.3) with means and covariances (11.1.4) and (11.1.5) is said to be stationary if $\boldsymbol{\mu}_t$ and $\Gamma(t+h, t)$, $h = 0, \pm 1, \ldots$, are independent of t.

For a stationary series we shall use the notation,

$$\boldsymbol{\mu} := E\mathbf{X}_t = (\mu_1, \ldots, \mu_m)', \tag{11.1.6}$$

and

$$\Gamma(h) := E[(\mathbf{X}_{t+h} - \boldsymbol{\mu})(\mathbf{X}_t - \boldsymbol{\mu})'] = [\gamma_{ij}(h)]_{i,j=1}^m. \tag{11.1.7}$$

We shall refer to $\boldsymbol{\mu}$ as the mean of the series and to $\Gamma(h)$ as the covariance matrix at lag h. Notice that if $\{\mathbf{X}_t\}$ is stationary with covariance matrix function $\Gamma(\cdot)$, then for each i, $\{X_{ti}\}$ is stationary with covariance function $\gamma_{ii}(\cdot)$. The function $\gamma_{ij}(\cdot)$, $i \neq j$, is called the cross-covariance function of the two series $\{X_{ti}\}$ and $\{X_{tj}\}$. It should be noted that $\gamma_{ij}(\cdot)$ is not in general the same as $\gamma_{ji}(\cdot)$. The correlation matrix function $R(\cdot)$ is defined by

$$R(h) := [\gamma_{ij}(h)/[\gamma_{ii}(0)\gamma_{jj}(0)]^{1/2}]_{i,j=1}^m = [\rho_{ij}(h)]_{i,j=1}^m. \tag{11.1.8}$$

The function $R(\cdot)$ is the covariance matrix function of the normalized series obtained by subtracting $\boldsymbol{\mu}$ from $\mathbf{X}_t$ and then dividing each component by its standard deviation.

The covariance matrix function $\Gamma(\cdot) = [\gamma_{ij}(\cdot)]_{i,j=1}^m$, of a stationary time series $\{\mathbf{X}_t\}$ has the properties,

(i) $\Gamma(h) = \Gamma'(-h)$,
(ii) $|\gamma_{ij}(h)| \leq [\gamma_{ii}(0)\gamma_{jj}(0)]^{1/2}$, $i, j = 1, \ldots, m$,
(iii) $\gamma_{ii}(\cdot)$ is an autocovariance function, $i = 1, \ldots, m$,
(iv) $\sum_{j,k=1}^n \mathbf{a}_j'\Gamma(j-k)\mathbf{a}_k \geq 0$ for all $n \in \{1, 2, \ldots\}$ and $\mathbf{a}_1, \ldots, \mathbf{a}_n \in \mathbb{R}^m$.

The first property follows at once from the definition, the second from the Cauchy–Schwarz inequality, and the third from the observation that $\gamma_{ii}(\cdot)$ is the autocovariance function of the stationary series $\{X_{ti}, t = 0, \pm 1, \ldots\}$. Property (iv) is a statement of the obvious fact that $E(\sum_{j=1}^n \mathbf{a}_j'(\mathbf{X}_j - \boldsymbol{\mu}))^2 \geq 0$. Properties (i), (ii), (iii) and (iv) are shared by the correlation matrix function $R(\cdot) = [\rho_{ij}(\cdot)]_{i,j=1}^m$, which has the additional property,

(v) $\rho_{ii}(0) = 1$.

(A complete characterization of covariance matrix functions of stationary processes is given later in Theorem 11.8.1.)

The correlation $\rho_{ij}(0)$ is the correlation between X_{ti} and X_{tj}, which is generally not equal to 1 if $i \neq j$ (see Example 11.1.1). It is also possible that $|\gamma_{ij}(h)| > |\gamma_{ij}(0)|$ if $i \neq j$ (see Problem 11.1).

EXAMPLE 11.1.1. Consider the bivariate stationary process $\{\mathbf{X}_t\}$ defined by,

$$X_{t1} = Z_t,$$

$$X_{t2} = Z_t + .75Z_{t-10},$$

where $\{Z_t\} \sim \mathrm{WN}(0, 1)$. Elementary calculations yield $\boldsymbol{\mu} = \mathbf{0}$,

$$\Gamma(-10) = \begin{bmatrix} 0 & .75 \\ 0 & .75 \end{bmatrix}, \quad \Gamma(0) = \begin{bmatrix} 1 & 1 \\ 1 & 1.5625 \end{bmatrix}, \quad \Gamma(10) = \begin{bmatrix} 0 & 0 \\ .75 & .75 \end{bmatrix},$$

and $\Gamma(j) = 0$ otherwise. The correlation matrix function is given by

$$R(-10) = \begin{bmatrix} 0 & .6 \\ 0 & .48 \end{bmatrix}, \quad R(0) = \begin{bmatrix} 1 & .8 \\ .8 & 1 \end{bmatrix}, \quad R(10) = \begin{bmatrix} 0 & 0 \\ .6 & .48 \end{bmatrix},$$

and $R(j) = 0$ otherwise.

The simplest multivariate time series is multivariate white noise, defined quite analogously to univariate white noise.

Definition 11.1.2 (Multivariate White Noise). The m-variate series $\{\mathbf{Z}_t, t = 0, \pm 1, \pm 2, \ldots\}$ is said to be white noise with mean $\mathbf{0}$ and covariance matrix $\boldsymbol{\Sigma}$, written

$$\{\mathbf{Z}_t\} \sim \mathrm{WN}(\mathbf{0}, \boldsymbol{\Sigma}), \tag{11.1.9}$$

if and only if $\{\mathbf{Z}_t\}$ is stationary with mean vector $\mathbf{0}$ and covariance matrix function,

$$\Gamma(h) = \begin{cases} \boldsymbol{\Sigma} & \text{if } h = 0. \\ 0, & \text{otherwise.} \end{cases} \tag{11.1.10}$$

We shall also use the notation

$$\{\mathbf{Z}_t\} \sim \mathrm{IID}(\mathbf{0}, \boldsymbol{\Sigma}), \tag{11.1.11}$$

to indicate that the random vectors $\mathbf{Z}_t$, $t = 0, \pm 1, \ldots$, are independently and identically distributed with mean $\mathbf{0}$ and covariance matrix $\boldsymbol{\Sigma}$.

Multivariate white noise $\{\mathbf{Z}_t\}$ is used as a building block from which can be constructed an enormous variety of multivariate time series. The linear processes are those of the form

$$\mathbf{X}_t = \sum_{j=-\infty}^{\infty} C_j \mathbf{Z}_{t-j}, \qquad \{\mathbf{Z}_t\} \sim \mathrm{WN}(\mathbf{0}, \boldsymbol{\Sigma}), \tag{11.1.12}$$

where $\{C_j\}$ is a sequence of matrices whose components are absolutely summable. The linear process $\{\mathbf{X}_t\}$ is stationary (Problem 11.2) with mean $\mathbf{0}$ and covariance matrix function,

$$\Gamma(h) = \sum_{j=-\infty}^{\infty} C_{j+h} \boldsymbol{\Sigma} C_j', \qquad h = 0, \pm 1, \ldots. \tag{11.1.13}$$

We shall reserve the term MA(∞) for a process of the form (11.1.12) with $C_j = 0$, $j < 0$. Thus $\{\mathbf{X}_t\}$ is an MA(∞) process if and only if for some white noise sequence $\{\mathbf{Z}_t\}$,

$$\mathbf{X}_t = \sum_{j=0}^{\infty} C_j \mathbf{Z}_{t-j},$$

where the matrices C_j are again required to have absolutely summable components. Multivariate ARMA processes will be discussed in Section 11.3, where it will be shown in particular that any causal ARMA(p, q) process can be expressed as an MA(∞) process, while any invertible ARMA(p, q) process can be expressed as an AR(∞) process,

$$\sum_{j=0}^{\infty} A_j \mathbf{X}_{t-j} = \mathbf{Z}_t,$$

where the matrices A_j have absolutely summable components.

Provided the covariance matrix function Γ has the property $\sum_{h=-\infty}^{\infty} |\gamma_{ij}(h)| < \infty$, $i, j = 1, \ldots, m$, then Γ has a spectral density matrix function,

$$f(\lambda) = \frac{1}{2\pi} \sum_{h=-\infty}^{\infty} e^{-i\lambda h}\Gamma(h), \qquad -\pi \le \lambda \le \pi, \tag{11.1.14}$$

and Γ can be expressed in terms of f as

$$\Gamma(h) = \int_{-\pi}^{\pi} e^{i\lambda h} f(\lambda)\, d\lambda. \tag{11.1.15}$$

The second order properties of the stationary process $\{\mathbf{X}_t\}$ can therefore be described equivalently in terms of $f(\cdot)$ rather than $\Gamma(\cdot)$. Similarly $\mathbf{X}_t$ has a spectral representation,

$$\mathbf{X}_t = \int_{(-\pi,\pi]} e^{i\lambda t}\, d\mathbf{Z}(\lambda), \tag{11.1.16}$$

where $\{\mathbf{Z}(\lambda), -\pi \le \lambda \le \pi\}$ is a process whose components are orthogonal increment processes satisfying

$$E(dZ_j(\lambda) d\bar{Z}_k(\mu)) = \begin{cases} f_{jk}(\lambda)\, d\lambda & \text{if } \lambda = \mu \\ 0 & \text{if } \lambda \ne \mu. \end{cases} \tag{11.1.17}$$

The spectral representations of $\Gamma(\cdot)$ and $\{\mathbf{X}_t\}$ are discussed in Sections 11.6 and 11.8. They remain valid without absolute summability of $\gamma_{ij}(\cdot)$ provided $f(\lambda)\, d\lambda$ is replaced in (11.1.15) and (11.1.17) by $dF(\lambda)$ (see Section 11.8).

§11.2 Estimation of the Mean and Covariance Function

As in the univariate case, the estimation of the mean vector and cross-correlation function of a stationary multivariate time series plays an important part in describing and modelling the dependence structure between the component time series. Let $\{\mathbf{X}_t = (X_{t1}, \ldots, X_{tm})', -\infty < t < \infty\}$ be an m-dimensional stationary time series with mean vector

$$\boldsymbol{\mu} = (\mu_1, \ldots, \mu_m)' = E\mathbf{X}_t$$

and covariance matrix function

$$\Gamma(h) = E[(\mathbf{X}_{t+h} - \boldsymbol{\mu})(\mathbf{X}_t - \boldsymbol{\mu})'] = [\gamma_{ij}(h)]_{i,j=1}^{m}$$

where $\gamma_{ij}(h) = \mathrm{Cov}(X_{t+h,i}, X_{tj})$. The cross-correlation function between the processes $\{X_{ti}\}$ and $\{X_{tj}\}$ is given by

$$\rho_{ij}(h) = \gamma_{ij}(h)/(\gamma_{ii}(0)\gamma_{jj}(0))^{1/2}, \qquad h = 0, \pm 1, \ldots.$$

Estimation of $\boldsymbol{\mu}$. Based on the observations $\mathbf{X}_1, \ldots, \mathbf{X}_n$, an unbiased estimate of $\boldsymbol{\mu}$ is given by the vector of sample means

$$\bar{\mathbf{X}}_n = \frac{1}{n}\sum_{t=1}^{n} \mathbf{X}_t.$$

Observe that the mean of the jth time series μ_j is estimated by $(1/n)\sum_{t=1}^{n} X_{tj}$. The consistency of the estimator $\bar{\mathbf{X}}_n$ under mild conditions on $\gamma_{ii}(h)$ can be established easily by applying Theorem 7.1.1 to the individual time series $\{X_{ti}\}$, $i = 1, \ldots, m$. This gives the following result.

Proposition 11.2.1. *If $\{\mathbf{X}_t\}$ is a stationary multivariate time series with mean $\boldsymbol{\mu}$ and covariance function $\Gamma(\cdot)$, then as $n \to \infty$*

$$E(\bar{\mathbf{X}}_n - \boldsymbol{\mu})'(\bar{\mathbf{X}}_n - \boldsymbol{\mu}) \to 0 \quad \text{if } \gamma_{ii}(n) \to 0, \quad i = 1, \ldots, m$$

and

$$nE(\bar{\mathbf{X}}_n - \boldsymbol{\mu})'(\bar{\mathbf{X}}_n - \boldsymbol{\mu}) \to \sum_{i=1}^{m} \sum_{h=-\infty}^{\infty} \gamma_{ii}(h) \quad \text{if } \sum_{h=-\infty}^{\infty} |\gamma_{ii}(h)| < \infty, i = 1, \ldots, m.$$

The vector $\bar{\mathbf{X}}_n$ is asymptotically normal under more restrictive assumptions on the process. In particular, if $\{\mathbf{X}_t\}$ is a multivariate moving average process then $\bar{\mathbf{X}}_n$ is asymptotically normal. This result is given in the following proposition.

Proposition 11.2.2. *Let $\{\mathbf{X}_t\}$ be the stationary multivariate time series,*

$$\mathbf{X}_t = \boldsymbol{\mu} + \sum_{k=-\infty}^{\infty} C_k \mathbf{Z}_{t-k}, \qquad \{Z_t\} \sim \mathrm{IID}(\mathbf{0}, \boldsymbol{\Sigma}),$$

where $\{C_k = [C_k(i,j)]_{i,j=1}^{m}\}$ is a sequence of $m \times m$ matrices such that $\sum_{k=-\infty}^{\infty} |C_k(i,j)| < \infty$, $i, j = 1, \ldots, m$. Then

$$\bar{\mathbf{X}}_n \text{ is AN}\left(\boldsymbol{\mu}, \frac{1}{n}\left(\sum_{k=-\infty}^{\infty} C_k\right)\boldsymbol{\Sigma}\left(\sum_{k=-\infty}^{\infty} C_k'\right)\right).$$

PROOF. See Problem 11.3. □

This proposition can be used for constructing confidence regions for $\boldsymbol{\mu}$. For example if the covariance matrix $\Sigma_{\bar{X}} := n^{-1}(\sum_{k=-\infty}^{\infty} C_k)\boldsymbol{\Sigma}(\sum_{k=-\infty}^{\infty} C_k')$ is nonsingular and known, then an asymptotic $(1-\alpha)$ confidence region for $\boldsymbol{\mu}$ is

$$\{\boldsymbol{\mu} \in \mathbb{R}^m : (\boldsymbol{\mu} - \bar{\mathbf{X}}_n)'\Sigma_{\bar{X}}^{-1}(\boldsymbol{\mu} - \bar{\mathbf{X}}_n) \le \chi^2_{1-\alpha}(m)\}. \tag{11.2.1}$$

This region is of little practical use since it is unlikely that $\Sigma_{\bar{X}}$ will be known while $\boldsymbol{\mu}$ is unknown. If we could find a consistent estimate $\hat{\Sigma}_{\bar{X}}$ of $\Sigma_{\bar{X}}$ and replace $\Sigma_{\bar{X}}$ by $\hat{\Sigma}_{\bar{X}}$ in (11.2.1), we would still have an asymptotic $1-\alpha$ confidence region for $\boldsymbol{\mu}$. However, in general, $\Sigma_{\bar{X}}$ is a difficult quantity to estimate. A simpler approach is to construct for each i, individual confidence intervals for μ_i based on $X_{1i}, \ldots, X_{ni}$ which are then combined to form one confidence region for $\boldsymbol{\mu}$. If $f_i(\omega)$ is the spectral density of the ith process, $\{X_{ti}\}$, then by the results of Section 10.4 (see (10.4.11)),

$$2\pi\hat{f}_i(0) := \sum_{|h| \le r} \left(1 - \frac{|h|}{r}\right)\hat{\gamma}_{ii}(h)$$

is a consistent estimator of $2\pi f(0) = \sum_{k=-\infty}^{\infty} \gamma_{ii}(k)$ provided $r = r_n$ is a sequence of numbers satisfying $r_n/n \to 0$ and $r_n \to \infty$. Thus if $\bar{X}_{\cdot i}$ denotes the sample mean of the ith process, and Φ_α is the α-quantile of the standard normal distribution, then by Theorem 7.1.2, the bounds

$$\bar{X}_{\cdot i} \pm \Phi_{1-\alpha/2}(2\pi\hat{f}_i(0)/n)^{1/2}$$

are asymptotic $(1-\alpha)$ confidence bounds for μ_i. Hence

$$\begin{aligned} P(|\mu_i - \bar{X}_{\cdot i}| &\le \Phi_{1-\alpha/2}(2\pi\hat{f}_i(0)/n)^{1/2}, i = 1, \ldots, m) \\ &\ge 1 - \sum_{i=1}^{m} P(|\mu_i - \bar{X}_{\cdot i}| > \Phi_{1-\alpha/2}(2\pi\hat{f}_i(0)/n)^{1/2}) \end{aligned}$$

where the right-hand side converges to $1 - m\alpha$ as $n \to \infty$. Consequently as $n \to \infty$, the region

$$\{\boldsymbol{\mu} : |\mu_i - \bar{X}_{\cdot i}| \le \Phi_{1-(\alpha/(2m))}(2\pi\hat{f}_i(0)/n)^{1/2}, i = 1, \ldots, m\} \tag{11.2.2}$$

has a confidence coefficient of *at least* $1 - \alpha$. For large values of m this confidence region will be substantially larger than an exact $(1-\alpha)$ region. Nevertheless it is easy to construct, and in most applications is of reasonable size provided m is not too large.

Estimation of $\Gamma(h)$. For simplicity we shall assume throughout the remainder of this section that $m = 2$. As in the univariate case, a natural estimate of the covariance matrix $\Gamma(h) = E[(\mathbf{X}_{t+h} - \boldsymbol{\mu})(\mathbf{X}_t - \boldsymbol{\mu})']$ is

$$\hat{\Gamma}(h) = \begin{cases} n^{-1} \sum_{t=1}^{n-h} (\mathbf{X}_{t+h} - \bar{\mathbf{X}}_n)(\mathbf{X}_t - \bar{\mathbf{X}}_n)' & \text{for } 0 \le h \le n-1, \\ n^{-1} \sum_{t=-h+1}^{n} (\mathbf{X}_{t+h} - \bar{\mathbf{X}}_n)(\mathbf{X}_t - \bar{\mathbf{X}}_n)' & \text{for } -n+1 \le h < 0. \end{cases}$$

Writing $\hat{\gamma}_{ij}(h)$ for the (i,j)-component of $\hat{\Gamma}(h)$, $i = 1, 2$, we estimate the cross-correlation function by

$$\hat{\rho}_{ij}(h) = \hat{\gamma}_{ij}(h)(\hat{\gamma}_{ii}(0)\hat{\gamma}_{jj}(0))^{-1/2}.$$

If $i = j$ this reduces to the sample autocorrelation function of the ith series. We first show the weak consistency of the estimator $\hat{\gamma}_{ij}(h)$ (and hence of $\hat{\rho}_{ij}(h)$) for infinite-order moving averages. We then consider the asymptotic distribution of $\hat{\gamma}_{ij}(h)$ and $\hat{\rho}_{ij}(h)$ in some special cases of importance.

Theorem 11.2.1. *Let $\{\mathbf{X}_t\}$ be the bivariate time series*

$$\mathbf{X}_t = \sum_{k=-\infty}^{\infty} C_k \mathbf{Z}_{t-k}, \qquad \{\mathbf{Z}_t = (Z_{t1}, Z_{t2})'\} \sim \text{IID}(\mathbf{0}, \Sigma),$$

where $\{C_k = [C_k(i,j)]_{i,j=1}^2\}$ is a sequence of matrices with $\sum_{k=-\infty}^{\infty} |C_k(i,j)| < \infty$, $i, j = 1, 2$. Then as $n \to \infty$,

$$\hat{\gamma}_{ij}(h) \xrightarrow{P} \gamma_{ij}(h)$$

and

$$\hat{\rho}_{ij}(h) \xrightarrow{P} \rho_{ij}(h)$$

for each fixed $h \geq 0$ and for $i, j = 1, 2$.

PROOF. We shall show that $\hat{\Gamma}(h) \xrightarrow{P} \Gamma(h)$ where convergence in probability of random matrices means convergence in probability of all of the components of the matrix. From the definition of $\hat{\Gamma}(h)$ we have, for $0 \leq h \leq n-1$,

$$\hat{\Gamma}(h) = n^{-1} \sum_{t=1}^{n-h} \mathbf{X}_{t+h}\mathbf{X}_t' - n^{-1}\bar{\mathbf{X}}_n \sum_{t=1}^{n-h} \mathbf{X}_t' - n^{-1} \sum_{t=1}^{n-h} \mathbf{X}_{t+h}\bar{\mathbf{X}}_n' + (n-h)n^{-1}\bar{\mathbf{X}}_n\bar{\mathbf{X}}_n'. \tag{11.2.3}$$

Since $E\mathbf{X} = \mathbf{0}$, we find from Proposition 11.2.1 that $\bar{X}_n = o_p(1)$, $n^{-1}\sum_{t=1}^{n-h} \mathbf{X}_t = o_p(1)$ and $n^{-1}\sum_{t=1}^{n-h} \mathbf{X}_{t+h} = o_p(1)$. Consequently we can write

$$\hat{\Gamma}(h) = \Gamma^*(h) + o_p(1), \tag{11.2.4}$$

where

$$\begin{aligned}\Gamma^*(h) &= n^{-1} \sum_{t=1}^{n} \mathbf{X}_{t+h}\mathbf{X}_t' \\ &= n^{-1} \sum_{t=1}^{n} \sum_{i=-\infty}^{\infty} \sum_{j=-\infty}^{\infty} C_{i+h}\mathbf{Z}_{t-i}\mathbf{Z}_{t-j}'C_j' \\ &= \sum_i \sum_j C_{i+h}\left(n^{-1} \sum_{t=1}^{n} \mathbf{Z}_{t-i}\mathbf{Z}_{t-j}'\right) C_j'.\end{aligned}$$

Observe that for $i \neq j$, the time series $\{Z_{t-i,1}Z_{t-j,2}, t = 0, \pm 1, \ldots\}$ is white noise so that by Theorem 7.1.1, $n^{-1}\sum_{t=1}^{n} Z_{t-i,1}Z_{t-j,2} \xrightarrow{P} 0$. Applying this argument to the other three components of $\mathbf{Z}_{t-i}\mathbf{Z}_{t-j}$, we obtain

$$n^{-1}\sum_{i=1}^{n}\mathbf{Z}_{t-i}\mathbf{Z}_{t-j}\xrightarrow{P}0_{2\times 2},\qquad i\neq j,$$

where $0_{2\times 2}$ denotes the 2×2 zero matrix. Hence for m fixed,

$$G_m^*(h):=\sum_{\substack{i,j=-m\\ i\neq j}}^{m}C_{i+h}\left(n^{-1}\sum_{t=1}^{n}\mathbf{Z}_{t-i}\mathbf{Z}_{t-j}\right)C_j'\xrightarrow{P}0_{2\times 2}.$$

For any matrix A define $|A|$ and EA to be the matrices of absolute values and expected values, respectively, of the elements of A. Then

$$\begin{aligned}
E|G_\infty^*(h)-G_m^*(h)| &= E\left|\sum_{\substack{|i|\text{ or }|j|>m\\ i\neq j}}C_{i+h}n^{-1}\sum_{t=1}^{n}\mathbf{Z}_{t-i}\mathbf{Z}_{t-j}'C_j'\right|\\
&\le \sum_{\substack{|i|\text{ or }|j|>m\\ i\neq j}}|\mathbf{C}_{i+h}|\left(n^{-1}\sum_{t=1}^{n}E|\mathbf{Z}_{t-i}\mathbf{Z}_{t-j}'|\right)|C_j'|\\
&\le \sum_{|i|>m}|C_{i+h}|\left(E|\mathbf{Z}_1\mathbf{Z}_2'|\sum_j|C_j'|\right)\\
&\quad+\left(\sum_i|C_{i+h}|E|\mathbf{Z}_1\mathbf{Z}_2'|\right)\sum_{|j|>m}|C_j'|.
\end{aligned}$$

The latter bound is independent of n and converges to 0 as $m\to\infty$. Hence

$$\lim_{m\to\infty}\limsup_{n\to\infty}E|G_\infty^*(h)-G_m^*(h)|=0,$$

which, by Proposition 6.3.9, implies that

$$G_\infty^*(h)\xrightarrow{P}0_{2\times 2}.$$

Now

$$\begin{aligned}
\Gamma^*(h) &= G_\infty^*(h)+\sum_i C_{i+h}\left(n^{-1}\sum_{t=1}^{n}\mathbf{Z}_{t-i}\mathbf{Z}_{t-i}'\right)C_i'\\
&= G_\infty^*(h)+\sum_i C_{i+h}\left(n^{-1}\sum_{t=1}^{n}\mathbf{Z}_t\mathbf{Z}_t'\right)C_i'+\sum_i C_{i+h}(n^{-1}U_{ni})C_i'
\end{aligned}$$

where $U_{ni}=\sum_{t=1-i}^{n-i}\mathbf{Z}_t\mathbf{Z}_t'-\sum_{t=1}^{n}\mathbf{Z}_t\mathbf{Z}_t'$ is a sum of $2|i|$ random matrices if $|i|<n$ and a sum of $2n$ random matrices if $|i|\ge n$. Hence

$$\begin{aligned}
E\left|\sum_i C_{i+h}(n^{-1}U_{ni})C_i'\right| &\le 2n^{-1}\sum_{|i|\le n}|i||C_{i+h}||\boldsymbol{\Sigma}||C_i'|\\
&\quad+2n^{-1}\sum_{|i|>n}|C_{i+h}||\boldsymbol{\Sigma}||C_i'|
\end{aligned}$$

and by the absolute summability of the components of the matrices $\{C_i\}$, this bound goes to zero as $n \to \infty$. It therefore follows that

$$\Gamma^*(h) = \sum_i C_{i+h}\left(n^{-1}\sum_{t=1}^{n} \mathbf{Z}_t\mathbf{Z}_t'\right)C_i' + o_p(1).$$

By applying the weak law of large numbers to the individual components of $\mathbf{Z}_t\mathbf{Z}_t'$, we find that

$$n^{-1}\sum_{t=1}^{n} \mathbf{Z}_t\mathbf{Z}_t' \xrightarrow{P} \mathbf{\Sigma},$$

and hence

$$\Gamma^*(h) \xrightarrow{P} \sum_i C_{i+h}\mathbf{\Sigma}C_i' = \Gamma(h).$$

Consequently, from (11.2.4),

$$\hat{\Gamma}(h) \xrightarrow{P} \Gamma(h). \tag{11.2.5}$$

The convergence of $\hat{\rho}_{ij}(h)$ to $\rho_{ij}(h)$ follows at once from (11.2.5) and Proposition 6.1.4. □

In general, the derivation of the asymptotic distribution of the sample cross-correlation function is quite complicated even for multivariate moving averages. The methods of Section 7.3 are not immediately adaptable to the multivariate case. An important special case arises when the two component time series are independent moving averages. The asymptotic distribution of $\hat{\rho}_{12}(h)$ for such a process is given in the following theorem.

Theorem 11.2.2. *Suppose that*

$$X_{t1} = \sum_{j=-\infty}^{\infty} \alpha_j Z_{t-j,1}, \qquad \{Z_{t1}\} \sim \text{IID}(0, \sigma_1^2),$$

and

$$X_{t2} = \sum_{j=-\infty}^{\infty} \beta_j Z_{t-j,2}, \qquad \{Z_{t2}\} \sim \text{IID}(0, \sigma_2^2),$$

where the two sequences $\{Z_{t1}\}$ and $\{Z_{t2}\}$ are independent, $\sum_j|\alpha_j| < \infty$ and $\sum_j|\beta_j| < \infty$. Then if $h \geq 0$,

$$\hat{\rho}_{12}(h) \text{ is AN}\left(0, n^{-1}\sum_{j=-\infty}^{\infty} \rho_{11}(j)\rho_{22}(j)\right).$$

If $h, k \geq 0$ and $h \neq k$, then the vector $(\hat{\rho}_{12}(h), \hat{\rho}_{12}(k))'$ is asymptotically normal with mean $\mathbf{0}$, variances as above and covariance

$$n^{-1}\sum_{j=-\infty}^{\infty} \rho_{11}(j)\rho_{22}(j+k-h).$$

PROOF. It follows easily from (11.2.3) and Proposition 11.2.1 that

$$\hat{\gamma}_{12}(h) = \gamma_{12}^*(h) + o_p(n^{-1/2}), \qquad (11.2.6)$$

where

$$\gamma_{12}^*(h) = n^{-1} \sum_{t=1}^{n} X_{t+h,1} X_{t2} = n^{-1} \sum_{t=1}^{n} \sum_i \sum_j \alpha_{i+h} \beta_j Z_{t-i,1} Z_{t-j,2}.$$

Since $E\gamma_{12}^*(h) = 0$, we have

$$n \operatorname{Var}(\gamma_{12}^*(h))$$

$$= n^{-1} \sum_{s=1}^{n} \sum_{t=1}^{n} \sum_i \sum_j \sum_k \sum_l \alpha_{i+h} \beta_j \alpha_{k+h} \beta_l E[Z_{s-i,1} Z_{s-j,2} Z_{t-k,1} Z_{t-l,2}]. \qquad (11.2.7)$$

By the independence assumptions,

$$E[Z_{s-i,1} Z_{s-j,2} Z_{t-k,1} Z_{t-l,2}] = \begin{cases} \sigma_1^2 \sigma_2^2 & \text{if } s - i = t - k \text{ and } s - j = t - l, \\ 0 & \text{otherwise,} \end{cases}$$

so that

$$n \operatorname{Var}(\gamma_{12}^*(h)) = n^{-1} \sum_{s=1}^{n} \sum_{t=1}^{n} \left(\sum_i \alpha_{i+h} \alpha_{t-s+i+h} \right) \sigma_1^2 \left(\sum_j \beta_j \beta_{t-s+j} \right) \sigma_2^2$$

$$= \sum_{|k|<n} \left(1 - \frac{|k|}{n}\right) \gamma_{11}(k) \gamma_{22}(k).$$

Applying the dominated covergence theorem to the last expression, we find that

$$n \operatorname{Var}(\gamma_{12}^*(h)) \to \sum_{j=-\infty}^{\infty} \gamma_{11}(j) \gamma_{22}(j) \quad \text{as } n \to \infty. \qquad (11.2.8)$$

Next we show that $\gamma_{12}^*(h)$ is asymptotically normal. For m fixed, we first consider the $(2m + h)$-dependent, strictly stationary time series, $\{\sum_{|i| \le m} \sum_{|j| \le m} \alpha_i \beta_j Z_{t+h-i,1} Z_{t-j,2}, t = 0, \pm 1, \ldots\}$. By Theorem 6.4.2 and the calculation leading up to (11.2.8),

$$n^{-1} \sum_{t=1}^{n} \sum_{|i| \le m} \sum_{|j| \le m} \alpha_i \beta_j Z_{t+h-i,1} Z_{t-j,2} \text{ is AN}(0, n^{-1} a_m),$$

where

$$a_m = \sum_{|k| \le m} \left(\sum_{|i| \le m} \alpha_i \alpha_{i+|k|} \right) \sigma_1^2 \left(\sum_{|j| \le m} \beta_j \beta_{j+|k|} \right) \sigma_2^2.$$

Now as $m \to \infty$, $a_m \to \sum_j \gamma_{11}(j) \gamma_{22}(j)$. Moreover, the above calculations can be used to show that

$$\lim_{m \to \infty} \limsup_{n \to \infty} nE \left| \gamma_{12}^*(h) - n^{-1} \sum_{t=1}^{n} \sum_{|i| \le m} \sum_{|j| \le m} \alpha_i \beta_j Z_{t+h-i,1} Z_{t-j,2} \right|^2 = 0.$$

This implies, with Proposition 6.3.9, that

$$\gamma_{12}^*(h) \text{ is AN}\left(0, n^{-1} \sum_{k=-\infty}^{\infty} \gamma_{11}(k) \gamma_{22}(k)\right). \qquad (11.2.9)$$

Since $\hat{\gamma}_{11}(0) \xrightarrow{P} \gamma_{11}(0)$ and $\hat{\gamma}_{22}(0) \xrightarrow{P} \gamma_{22}(0)$, we find from (11.2.6), (11.2.9) and Proposition 6.3.8 that

$$\hat{\rho}_{12}(h) = \hat{\gamma}_{12}(h)(\hat{\gamma}_{11}(0)\hat{\gamma}_{22}(0))^{-1/2} \text{ is AN}\left(0, n^{-1} \sum_{j=-\infty}^{\infty} \rho_{11}(j)\rho_{22}(j)\right).$$

Finally, after showing that

$$n \operatorname{Cov}(\gamma_{12}^*(h), \gamma_{12}^*(k)) \to \sum_{j=-\infty}^{\infty} \gamma_{11}(j)\gamma_{22}(j+k-h),$$

the same argument, together with the Cramer–Wold device, can be used to establish the last statement of the theorem. □

This theorem plays an important role in testing for correlation between two processes. If one of the two processes is white noise then $\hat{\rho}_{12}(h)$ is $\text{AN}(0, n^{-1})$ in which case it is straightforward to test the hypothesis that $\rho_{12}(h) = 0$. However, if neither process is white noise, then a value of $|\hat{\rho}_{12}(h)|$ which is large relative to $n^{-1/2}$ does not necessarily indicate that $\rho_{12}(h)$ is different from zero. For example, suppose that $\{X_{t1}\}$ and $\{X_{t2}\}$ are two independent AR(1) processes with $\rho_{11}(h) = \rho_{22}(h) = .8^{|h|}$. Then the asymptotic variance of $\hat{\rho}_{12}(h)$ is $n^{-1}(1 + 2\sum_{k=1}^{\infty}(.64)^k) = 4.556n^{-1}$. It would therefore not be surprising to observe a value of $\hat{\rho}_{12}(h)$ as large as $3n^{-1/2}$ even though $\{X_{t1}\}$ and $\{X_{t2}\}$ are independent. If on the other hand $\rho_{11}(h) = .8^{|h|}$ and $\rho_{22}(h) = (-.8)^{|h|}$, then the asymptotic variance of $\hat{\rho}_{12}(h)$ is $.2195n^{-1}$ and an observed value of $3n^{-1/2}$ for $\hat{\rho}_{12}(h)$ would be very unlikely.

Testing for Independence of Two Stationary Time Series. Since by Theorem 11.2.2 the asymptotic distribution of $\hat{\rho}_{12}(h)$ depends on both $\rho_{11}(\cdot)$ and $\rho_{22}(\cdot)$, any test for independence of the two component series cannot be based solely on estimated values of $\rho_{12}(h)$, $h = 0, \pm 1, \ldots$, without taking into account the nature of the two component series.

This difficulty can be circumvented by "prewhitening" the two series before computing the cross-correlations $\hat{\rho}_{12}(h)$, i.e. by transforming the two series to white noise by application of suitable filters. If $\{X_{t1}\}$ and $\{X_{t2}\}$ are invertible $\text{ARMA}(p, q)$ processes this can be achieved by the transformations,

$$Z_{ti} = \sum_{j=0}^{\infty} \pi_j^{(i)} X_{t-j,i}$$

where $\sum_{j=0}^{\infty} \pi_j^{(i)} z^j = \phi^{(i)}(z)/\theta^{(i)}(z)$, $|z| \le 1$, and $\phi^{(i)}$, $\theta^{(i)}$ are the autoregressive and moving average polynomials of the ith series, $i = 1, 2$.

Since in practice the true model is nearly always unknown and since the data X_{tj}, $t \le 0$, are not available, it is convenient to replace the sequences $\{Z_{ti}\}$, $i = 1, 2$, by the residuals $\{\hat{W}_{ti}, t = 1, \ldots, n\}$ (see (9.4.1)) which, if we assume that the fitted $\text{ARMA}(p, q)$ models are in fact the true models, are white noise sequences for $i = 1, 2$.

To test the hypothesis H_0 that $\{X_{t1}\}$ and $\{X_{t2}\}$ are independent series, we observe that under H_0, the corresponding two prewhitened series $\{Z_{t1}\}$ and $\{Z_{t2}\}$ are also independent. Under H_0, Theorem 11.2.2 implies that the sample autocorrelations $\hat{\rho}_{12}(h)$, $\hat{\rho}_{12}(k)$, $h \neq k$, of $\{Z_{t1}\}$ and $\{Z_{t2}\}$ are asymptotically independent normal with means 0 and variances n^{-1}. An approximate test for independence can therefore be obtained by comparing the values of $|\hat{\rho}_{12}(h)|$ with $1.96n^{-1/2}$, exactly as in Example 7.2.1. If we prewhiten only one of the two original series, say $\{X_{t1}\}$, then under H_0 Theorem 11.2.2 implies that the sample autocorrelations $\tilde{\rho}_{12}(h)$, $\tilde{\rho}_{12}(k)$, $h \neq k$, of $\{Z_{t1}\}$ and $\{X_{t2}\}$ are asymptotically normal with means 0, variances n^{-1} and covariance $n^{-1}\rho_{22}(k-h)$, where $\rho_{22}(\cdot)$ is the autocorrelation function of $\{X_{t2}\}$. Hence for any fixed h, $\tilde{\rho}_{12}(h)$ also falls (under H_0) between the bounds $\pm 1.96n^{-1/2}$ with a probability of approximately .95.

Example 11.2.1. The sample cross-correlation function $\hat{\rho}_{12}(\cdot)$ of a bivariate time series of length $n = 200$ is displayed in Figure 11.1. Without knowing the correlation function of each process, it is impossible to decide if the two processes are uncorrelated with one another. Note that several of the values of $\hat{\rho}_{12}(h)$ lie outside the bounds $\pm 1.96n^{-1/2} = \pm.139$. Based on the sample autocorrelation function and partial autocorrelation function of the first process, we modelled $\{X_{t1}\}$ as an AR(1) process. The sample cross-correlation function $\tilde{\rho}_{12}(\cdot)$ between the residuals $(\hat{W}_{t1}, t = 1, \ldots, 200\}$ for this model and $\{X_{t2}, t = 1, \ldots, 200\}$ is given in Figure 11.2. All except one of the values $\tilde{\rho}_{12}(h)$ lie between the bounds $\pm.139$, suggesting by Theorem 11.2.2, that the time

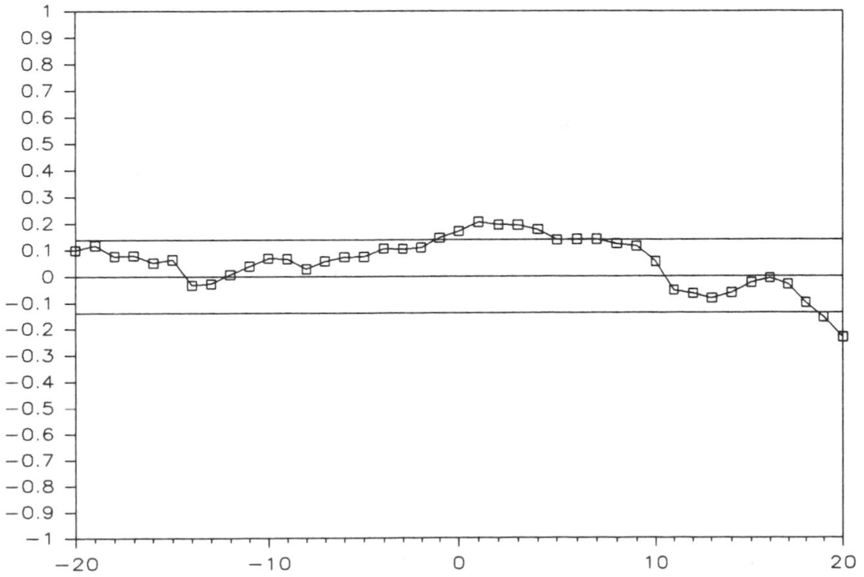

Figure 11.1. The sample cross-correlation function $\hat{\rho}_{12}(h)$ between $\{X_{t1}\}$ and $\{X_{t2}\}$, Example 11.2.1, showing the bounds $\pm 1.96n^{-1/2}$.

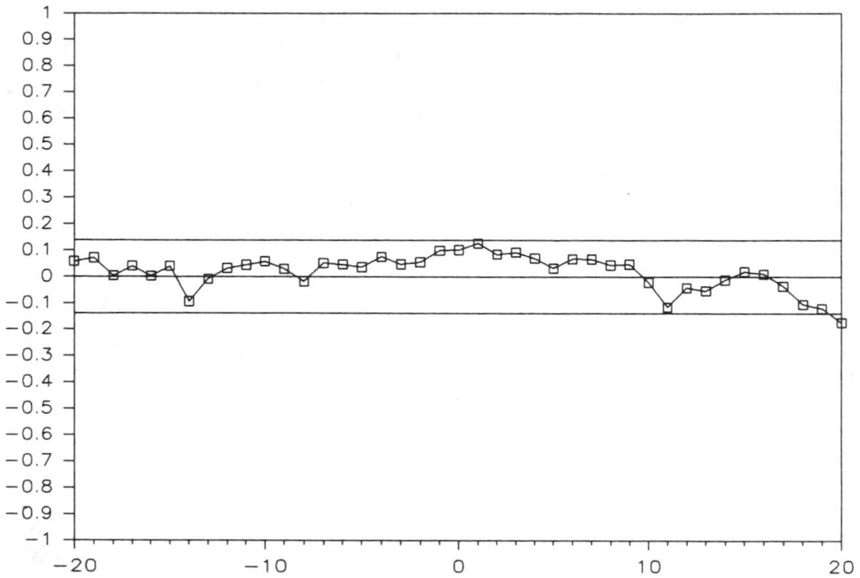

Figure 11.2. The sample cross-correlation function between the residuals $\{W_{t1}\}$ and $\{X_{t2}\}$, Example 11.2.1, showing the bounds $\pm 1.96n^{-1/2}$.

series $\{\hat{W}_{t1}\}$ (and hence $\{X_{t1}\}$) is uncorrelated with the series $\{X_{t2}\}$. The data for this example were in fact generated from two independent AR(1) processes and the cross-correlations were computed using the program TRANS.

EXAMPLE 11.2.2 (Sales with a Leading Indicator). In this example we consider the sales data $\{Y_{t2}, t = 1, \ldots, 150\}$ with leading indicator $\{Y_{t1}, t = 1, \ldots, 150\}$ given by Box and Jenkins (1976), p. 537. The autocorrelation functions of $\{Y_{t1}\}$ and $\{Y_{t2}\}$ suggest that both series are non-stationary. Application of the operator $(1 - B)$ yields the two differenced series $\{D_{t1}\}$ and $\{D_{t2}\}$ whose properties are compatible with those of low order ARMA processes. Using the program PEST, it is found that the models

$$D_{t1} - .0228 = Z_{t1} - .474Z_{t-1,1}, \qquad \{Z_{t1}\} \sim \text{WN}(0, .0779), \tag{11.2.10}$$

and

$$D_{t2} - .838D_{t-1,2} - .0676 = Z_{t2} - .610Z_{t-1,2}, \qquad \{Z_{t2}\} \sim \text{WN}(0, 1.754), \tag{11.2.11}$$

provide a good fit to the series $\{D_{t1}\}$ and $\{D_{t2}\}$, yielding the "whitened" series of residuals $\{\hat{W}_{t1}\}$ and $\{\hat{W}_{t2}\}$ with sample variances .0779 and 1.754 respectively.

The sample cross-correlation function of $\{D_{t1}\}$ and $\{D_{t2}\}$ is shown in Figure 11.3. Without taking into account the autocorrelation structures of $\{D_{t1}\}$ and $\{D_{t2}\}$ it is not possible to draw any conclusions from this function.

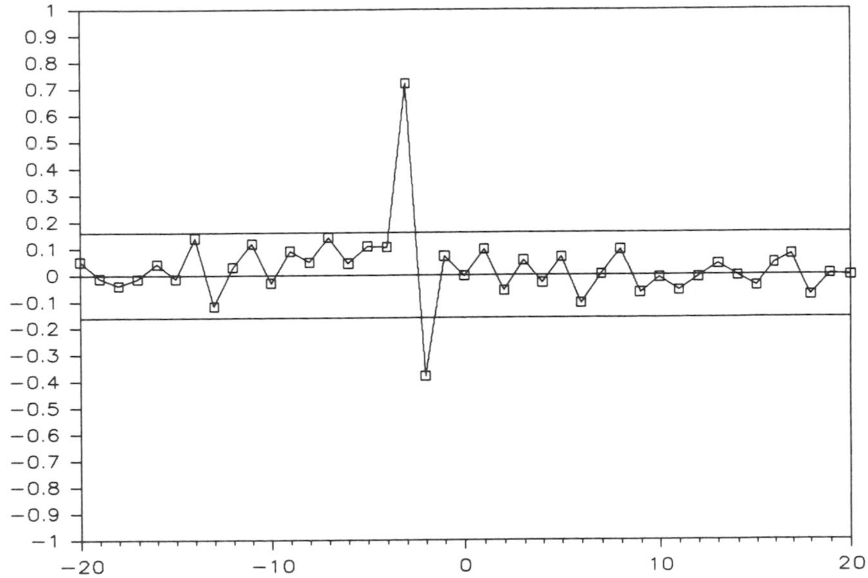

Figure 11.3. The sample cross-correlation function between $\{D_{t1}\}$ and $\{D_{t2}\}$, Example 11.2.2.

Examination of the sample cross-correlation function of the whitened series $\{\hat{W}_{t1}\}$ and $\{\hat{W}_{t2}\}$ is however much more informative. From Figure 11.4 it is apparent that there is one large sample cross-correlation (between $\hat{W}_{t1}$ and $\hat{W}_{t+3,2}$) and that the others are all between $\pm 1.96n^{-1/2}$. Under the assumption that $\{\hat{W}_{t1}\}$ and $\{\hat{W}_{t2}\}$ are jointly Gaussian, Bartlett's formula (see Corollary 11.2.1 below) indicates the compatibility of the cross-correlations with a model for which

$$\rho_{12}(-3) \neq 0$$

and

$$\rho_{12}(h) = 0, \qquad h \neq -3.$$

The value $\hat{\rho}_{12}(-3) = .969$ suggests the model,

$$\hat{W}_{t2} = 4.74\hat{W}_{t-3,1} + N_t, \tag{11.2.12}$$

where the stationary noise $\{N_t\}$ has small variance compared with $\{\hat{W}_{t2}\}$ and is uncorrelated with $\{\hat{W}_{t1}\}$. The coefficient 4.74 is the square root of the ratio of sample variances of $\{\hat{W}_{t2}\}$ and $\{\hat{W}_{t1}\}$. A study of the sample values of $\{\hat{W}_{t2} - 4.74\,\hat{W}_{t-3,1}\}$ suggests the model for $\{N_t\}$,

$$(1 + .345B)N_t = U_t, \qquad \{U_t\} \sim \text{WN}(0, .0782). \tag{11.2.13}$$

Finally, replacing $\{Z_{t1}\}$ and $\{Z_{t2}\}$ in (11.2.10) and (11.2.11) by $\{\hat{W}_{t1}\}$ and $\{\hat{W}_{t2}\}$ and using (11.2.12) and (11.2.13), we obtain a model relating $\{D_{t1}\}$, $\{D_{t2}\}$ and $\{U_t\}$, namely,

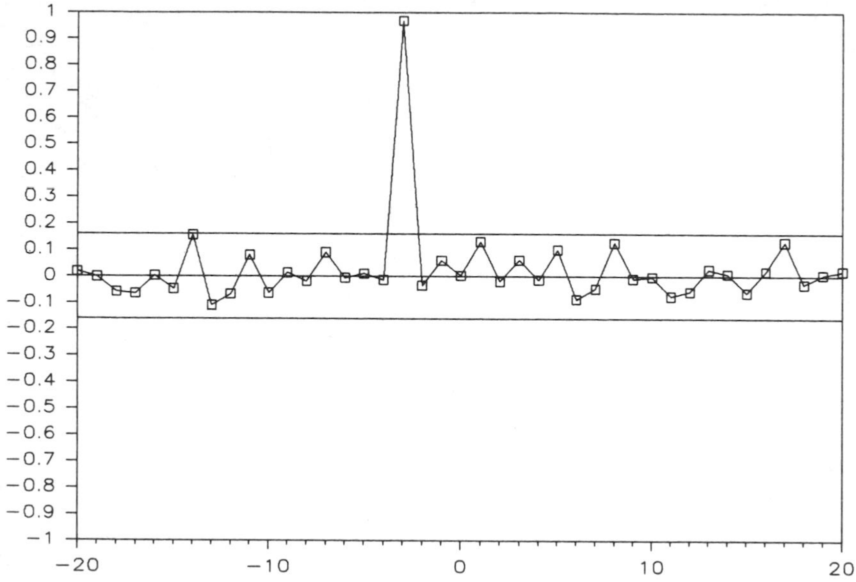

Figure 11.4. The sample cross-correlation function between the whitened series $\{W_{t1}\}$ and $\{W_{t2}\}$, Example 11.2.2.

$$D_{t2} + .0765 = (1 - .610B)(1 - .838B)^{-1}[4.74(1 - .474B)^{-1}D_{t-3,1} + (1 + .345B)^{-1}U_t].$$

This model should be compared with the one derived later (Section 13.1) by the more systematic technique of transfer function modelling.

Theorem 11.2.3 (Bartlett's Formula). *If $\{\mathbf{X}_t\}$ is a bivariate Gaussian process (i.e. if all of the finite dimensional distributions of $\{(X_{t1}, X_{t2})', t = 0, \pm 1, \ldots\}$ are multivariate normal) and if the autocovariances satisfy*

$$\sum_{h=-\infty}^{\infty} |\gamma_{ij}(h)| < \infty, \qquad i, j = 1, 2,$$

then

$$\begin{aligned}\lim_{n\to\infty} n\,\mathrm{Cov}(\hat\rho_{12}(h), \hat\rho_{12}(k)) = \sum_{j=-\infty}^{\infty} [&\rho_{11}(j)\rho_{22}(j+k-h) + \rho_{12}(j+k)\rho_{21}(j-h) \\ &- \rho_{12}(h)\{\rho_{11}(j)\rho_{21}(j+k) + \rho_{22}(j)\rho_{21}(j-k)\} \\ &- \rho_{12}(j)\{\rho_{11}(j)\rho_{21}(j+h) + \rho_{22}(j)\rho_{21}(j-k)\} \\ &+ \rho_{12}(h)\rho_{12}(k)\{\tfrac{1}{2}\rho_{11}^2(j) + \rho_{12}^2(j) + \tfrac{1}{2}\rho_{22}^2(j)\}].\end{aligned}$$

[*See Bartlett* (1955).]

Corollary 11.2.1. *If* $\{\mathbf{X}_t\}$ *satisfies the conditions of Theorem* 11.2.3, *if either* $\{X_{t1}\}$ *or* $\{X_{t2}\}$ *is white noise, and if*

$$\rho_{12}(h) = 0, \qquad h \notin [a, b],$$

then

$$\lim_{n\to\infty} n \operatorname{Var}(\hat{\rho}_{12}(h)) = 1, \qquad h \notin [a, b].$$

Proof. The limit is evaluated by direct application of Theorem 11.2.3. □

§11.3 Multivariate ARMA Processes

As in the univariate case, we can define an extremely useful class of multivariate stationary processes $\{\mathbf{X}_t\}$, by requiring that $\{\mathbf{X}_t\}$ should satisfy a set of linear difference equations with constant coefficients.

Definition 11.3.1 (Multivariate ARMA(p, q) Process). $\{\mathbf{X}_t, t = 0, \pm 1, \ldots\}$ is an m-variate ARMA(p, q) process if $\{\mathbf{X}_t\}$ is a stationary solution of the difference equations,

$$\mathbf{X}_t - \Phi_1 \mathbf{X}_{t-1} - \cdots - \Phi_p \mathbf{X}_{t-p} = \mathbf{Z}_t + \Theta_1 \mathbf{Z}_{t-1} + \cdots + \Theta_q \mathbf{Z}_{t-q}, \quad (11.3.1)$$

where $\Phi_1, \ldots, \Phi_p, \Theta_1, \ldots, \Theta_q$ are real $m \times m$ matrices and $\{\mathbf{Z}_t\} \sim \mathrm{WN}(\mathbf{0}, \Sigma)$.

The equations (11.3.1) can be written in the more compact form

$$\Phi(B)\mathbf{X}_t = \Theta(B)\mathbf{Z}_t, \qquad \{\mathbf{Z}_t\} \sim \mathrm{WN}(\mathbf{0}, \Sigma), \quad (11.3.2)$$

where $\Phi(z) := I - \Phi_1 z - \cdots - \Phi_p z^p$ and $\Theta(z) := I + \Theta_1 z + \cdots + \Theta_q z^q$ are matrix-valued polynomials, I is the $m \times m$ identity matrix and B as usual denotes the backward shift operator. (Each component of the matrices $\Phi(z)$, $\Theta(z)$ is a polynomial with real coefficients and degree less than or equal to p, q respectively.)

Example 11.3.1 (Multivariate AR(1) Process). This process satisfies

$$\mathbf{X}_t = \Phi \mathbf{X}_{t-1} + \mathbf{Z}_t, \qquad \{\mathbf{Z}_t\} \sim \mathrm{WN}(\mathbf{0}, \Sigma). \quad (11.3.3)$$

By exactly the same argument used in Example 3.2.1, we can express $\mathbf{X}_t$ as

$$\mathbf{X}_t = \sum_{j=0}^{\infty} \Phi^j \mathbf{Z}_{t-j}, \quad (11.3.4)$$

provided all the eigenvalues of Φ are less than 1 in absolute value, i.e. provided

$$\det(I - z\Phi) \neq 0 \quad \text{for all } z \in \mathbb{C} \text{ such that } |z| \leq 1. \quad (11.3.5)$$

If this condition is satisfied then the series (11.3.4) converges (componentwise) both in mean square and absolutely with probability 1. Moreover it is the unique stationary solution of (11.3.3). The condition (11.3.5) is the multivariate analogue of the condition $|\phi| < 1$, required for the existence of the causal representation (11.3.4) in the univariate case.

Causality and invertibility of a general ARMA(p, q) model are defined precisely as in Definitions 3.1.3 and 3.1.4 respectively, the only difference being that the coefficients ψ_j, π_j in the representations $X_t = \sum_{j=0}^{\infty} \psi_j Z_{t-j}$ and $Z_t = \sum_{j=0}^{\infty} \pi_j X_{t-j}$, are replaced by matrices Ψ_j and Π_j whose components are required to be absolutely summable. The following two theorems provide us with criteria for causality and invertibility analogous to those of Theorems 3.1.1 and 3.1.2.

Theorem 11.3.1 (Causality Criterion). *If*

$$\det \Phi(z) \neq 0 \quad \textit{for all } z \in \mathbb{C} \textit{ such that } |z| \leq 1, \tag{11.3.6}$$

then (11.3.2) *has exactly one stationary solution,*

$$\mathbf{X}_t = \sum_{j=0}^{\infty} \Psi_j \mathbf{Z}_{t-j}, \tag{11.3.7}$$

where the matrices Ψ_j *are determined uniquely by*

$$\Psi(z) := \sum_{j=0}^{\infty} \Psi_j z^j = \Phi^{-1}(z)\Theta(z), \qquad |z| \leq 1. \tag{11.3.8}$$

Proof. The condition (11.3.6) implies that there exists $\varepsilon > 0$ such that $\Phi^{-1}(z)$ exists for $|z| < 1 + \varepsilon$. Since each of the m^2 elements of $\Phi^{-1}(z)$ is a rational function of z with no singularities in $\{|z| < 1 + \varepsilon\}$, $\Phi^{-1}(z)$ has the power series expansion,

$$\Phi^{-1}(z) = \sum_{j=0}^{\infty} A_j z^j = A(z), \qquad |z| < 1 + \varepsilon.$$

Consequently $A_j(1 + \varepsilon/2)^j \to 0$ (componentwise) as $j \to \infty$, so there exists $K \in (0, \infty)$ independent of j, such that all the components of A_j are bounded in absolute value by $K(1 + \varepsilon/2)^{-j}$, $j = 0, 1, 2, \ldots$. In particular this implies absolute summability of the components of the matrices A_j. Moreover we have

$$A(z)\Phi(z) = I \quad \text{for } |z| \leq 1,$$

where I is the $(m \times m)$ identity matrix. By Proposition 3.1.1, if $\{\mathbf{X}_t\}$ is a stationary solution of (11.3.2) we can apply the operator $A(B)$ to each side of this equation to obtain

$$\mathbf{X}_t = A(B)\Theta(B)\mathbf{Z}_t.$$

Thus we have the desired representation,

$$\mathbf{X}_t = \sum_{j=0}^{\infty} \Psi_j \mathbf{Z}_{t-j},$$

where the sequence $\{\Psi_j\}$ is determined by (11.3.8).

Conversely if $\mathbf{X}_t = \sum_{j=0}^{\infty} \Psi_j \mathbf{Z}_{t-j}$ with $\{\Psi_j\}$ defined by (11.3.8) then

$$\Phi(B)\mathbf{X}_t = \Phi(B)\Psi(B)\mathbf{Z}_t = \Theta(B)\mathbf{Z}_t,$$

showing that $\{\Psi(B)\mathbf{Z}_t\}$ is a stationary solution of (11.3.6).

Combining the results of the two preceding paragraphs we conclude that if $\det \Phi(z) \neq 0$ for $|z| \leq 1$, then the unique stationary solution of (11.3.2) is the causal solution (11.3.7). □

Since the analogous criterion for invertibility is established in the same way (see also the proof of Theorem 3.1.2), we shall simply state the result and leave the proof as an exercise.

Theorem 11.3.2 (Invertibility Criterion). *If*

$$\det \Theta(z) \neq 0 \quad \textit{for all } z \in \mathbb{C} \textit{ such that } |z| \leq 1, \tag{11.3.9}$$

and $\{\mathbf{X}_t\}$ *is a stationary solution of* (11.3.2) *then*

$$\mathbf{Z}_t = \sum_{j=0}^{\infty} \Pi_j \mathbf{X}_{t-j}, \tag{11.3.10}$$

where the matrices Π_j *are determined uniquely by*

$$\Pi(z) := \sum_{j=0}^{\infty} \Pi_j z^j = \Theta^{-1}(z)\Phi(z), \qquad |z| \leq 1. \tag{11.3.11}$$

PROOF. Problem 11.4. □

Remark. The matrices Ψ_j and Π_j of Theorems 11.3.1 and 11.3.2 can easily be found recursively from the equations

$$\Psi_0 = I = \Pi_0,$$

$$\Psi_j = \sum_{i=1}^{j} \Phi_i \Psi_{j-i} + \Theta_j, \qquad j = 1, 2, \ldots, \tag{11.3.12}$$

and

$$\Pi_j = -\sum_{i=1}^{j} \Theta_i \Pi_{j-i} - \Phi_j, \qquad j = 1, 2, \ldots, \tag{11.3.13}$$

where $\Theta_j = 0$, $j > q$, and $\Phi_i = 0$, $i > p$. These equations are established by comparing coefficients of z^j in the power series identities (11.3.8) and (11.3.11) after multiplying through by $\Phi(z)$ and $\Theta(z)$ respectively.

EXAMPLE 11.3.2. For the multivariate ARMA(1, 1) process with $\Phi_1 = \begin{bmatrix} .5 & .5 \\ 0 & .5 \end{bmatrix}$

and $\Theta_1 = \Phi_1'$, an elementary calculation using (11.3.8) gives

$$\Psi(z) = (1 - .5z)^{-2} \begin{bmatrix} 1 & .5z(1 + .5z) \\ .5z(1 - .5z) & 1 - .25z^2 \end{bmatrix}, \qquad |z| \le 1. \quad (11.3.14)$$

The coefficient matrices Ψ_j in the representation (11.3.7) are found, either by expanding each component of (11.3.14) or by using the recursion relation (11.3.12), to be $\Psi_0 = I$ and

$$\Psi_j = 2^{-j} \begin{bmatrix} j + 1 & 2j - 1 \\ 1 & 2 \end{bmatrix}, \qquad j = 1, 2, \ldots.$$

It is a simple matter to carry out the same calculations for any multivariate ARMA process satisfying (11.3.6), although the algebra becomes tedious for larger values of m. The calculation of the matrices Π_j for an invertible ARMA process is of course quite analogous. For numerical calculation of the coefficient matrices, the simplest method is to use the recursions (11.3.12) and (11.3.13).

The Covariance Matrix Function of a Causal ARMA Process

From the representation (11.3.7) we can express the covariance matrix $\Gamma(h) = E(\mathbf{X}_{t+h}\mathbf{X}_t')$ of the causal process (11.3.1) as

$$\Gamma(h) = \sum_{k=0}^{\infty} \Psi_{h+k} \Sigma \Psi_k', \qquad h = 0, \pm 1, \pm 2, \ldots,$$

where the matrices Ψ_j are found from (11.3.8) or (11.3.12) and $\Psi_j := 0$ for $j < 0$. It is not difficult to show (Problem 11.5) that there exists $\varepsilon \in (0, 1)$ and a constant K such that the components $\gamma_{ij}(h)$ of $\Gamma(h)$ satisfy $|\gamma_{ij}(h)| < K\varepsilon^{|h|}$ for all i, j and h.

The covariance matrices $\Gamma(h)$, $h = 0, \pm 1, \pm 2, \ldots$, can be determined by solving the Yule–Walker equations,

$$\Gamma(j) - \sum_{r=1}^{p} \Phi_r \Gamma(j - r) = \sum_{j \le r \le q} \Theta_r \Sigma \Psi_{r-j}', \qquad j = 0, 1, 2, \ldots, \quad (11.3.15)$$

obtained by post-multiplying (11.3.1) by $\mathbf{X}_{t-j}'$ and taking expectations. The first $(p + 1)$ of the equations (11.3.15) can be solved for the components of $\Gamma(0), \ldots, \Gamma(p)$ using the fact that $\Gamma(-h) = \Gamma'(h)$. The remaining equations then give $\Gamma(p + 1), \Gamma(p + 2), \ldots$ recursively.

The covariance matrix generating function is defined (cf. (3.5.1)) as

$$G(z) = \sum_{h=-\infty}^{\infty} \Gamma(h) z^h, \quad (11.3.16)$$

which can be expressed (Problem 11.7) as

$$G(z) = \Psi(z) \Sigma \Psi'(z^{-1}) = \Phi^{-1}(z) \Theta(z) \Sigma \Theta'(z^{-1}) \Phi'^{-1}(z^{-1}). \quad (11.3.17)$$

§11.4 Best Linear Predictors of Second Order Random Vectors

Let $\{\mathbf{X}_t = (X_{t1}, \ldots, X_{tm})', t = 0, \pm 1, \pm 2, \ldots\}$ be an m-variate time series with mean $E\mathbf{X}_t = \mathbf{0}$ and covariance function given by the $m \times m$ matrix,

$$K(i, j) = E(\mathbf{X}_i \mathbf{X}_j').$$

If $\mathbf{Y} = (Y_1, \ldots, Y_m)'$ is a random vector with finite second moment, we define

$$P(\mathbf{Y} | \mathbf{X}_1, \ldots, \mathbf{X}_n) = (P_{S_n} Y_1, \ldots, P_{S_n} Y_m)', \qquad (11.4.1)$$

where $S_n = \overline{\mathrm{sp}}\{X_{tj}, t = 1, \ldots, n;\ j = 1, \ldots, m\}$. If $\mathbf{U} = (U_1, \ldots, U_m)'$ is a random vector, we shall say that $\mathbf{U} \in S_n$ if $U_i \in S_n, i = 1, \ldots, m$. It then follows from the projection theorem that the vector $P(\mathbf{Y} | \mathbf{X}_1, \ldots, \mathbf{X}_n)$ is characterized by the two properties:

$$P(\mathbf{Y} | \mathbf{X}_1, \ldots, \mathbf{X}_n) \in S_n \qquad (11.4.2)$$

and

$$\mathbf{Y} - P(\mathbf{Y} | \mathbf{X}_1, \ldots, \mathbf{X}_n) \perp \mathbf{X}_{n+1-i}, \qquad i = 1, \ldots, n, \qquad (11.4.3)$$

where we say that two m-dimensional random vectors $\mathbf{X}$ and $\mathbf{Y}$ are orthogonal (written $\mathbf{X} \perp \mathbf{Y}$) if $E(\mathbf{X}\mathbf{Y}') = 0_{m \times m}$.

The best linear predictor of $\mathbf{X}_{n+1}$ based on the observations $\mathbf{X}_1, \ldots, \mathbf{X}_n$ is obtained on replacing $\mathbf{Y}$ by $\mathbf{X}_{n+1}$ in (11.4.1), i.e.

$$\hat{\mathbf{X}}_{n+1} = \begin{cases} \mathbf{0}, & \text{if } n = 0, \\ P(\mathbf{X}_{n+1} | \mathbf{X}_1, \ldots, \mathbf{X}_n), & \text{if } n \geq 1. \end{cases}$$

Since $\hat{\mathbf{X}}_{n+1} \in S_n$, there exist $m \times m$ matrices $\Phi_{n1}, \ldots, \Phi_{nn}$ such that

$$\hat{\mathbf{X}}_{n+1} = \Phi_{n1}\mathbf{X}_n + \cdots + \Phi_{nn}\mathbf{X}_1, \qquad n = 1, 2, \ldots. \qquad (11.4.4)$$

Moreover, from (11.4.3), we have $\mathbf{X}_{n+1} - \hat{\mathbf{X}}_{n+1} \perp \mathbf{X}_{n+1-i}$, $i = 1, \ldots, n$, or equivalently,

$$E(\hat{\mathbf{X}}_{n+1}\mathbf{X}_{n+1-i}') = E(\mathbf{X}_{n+1}\mathbf{X}_{n+1-i}'), \qquad i = 1, \ldots, n. \qquad (11.4.5)$$

When $\hat{\mathbf{X}}_{n+1}$ is replaced by the expression in (11.4.4), these prediction equations become

$$\sum_{j=1}^{n} \Phi_{nj} K(n + 1 - j, n + 1 - i) = K(n + 1, n + 1 - i), \qquad i = 1, \ldots, n.$$

In the case when $\{\mathbf{X}_t\}$ is stationary with $K(i, j) = \Gamma(i - j)$, the prediction equations simplify to the m-dimensional analogues of (5.1.5), i.e.

$$\sum_{j=1}^{n} \Phi_{nj}\Gamma(i - j) = \Gamma(i), \qquad i = 1, \ldots, n. \qquad (11.4.6)$$

The coefficients $\{\Phi_{nj}\}$ may be computed recursively using the multivariate Durbin–Levinson algorithm given by Whittle (1963). Unlike the univariate algorithm, however, the multivariate version requires the simultaneous solution of two sets of equations, one arising in the calculation of the forward predictor, $P(\mathbf{X}_{n+1}|\mathbf{X}_1,\ldots,\mathbf{X}_n)$, and the other in the calculation of the backward predictor, $P(\mathbf{X}_0|\mathbf{X}_1,\ldots,\mathbf{X}_n)$. Let $\tilde{\Phi}_{n1},\ldots,\tilde{\Phi}_{nn}$ be $m \times m$ coefficient matrices satisfying

$$P(\mathbf{X}_0|\mathbf{X}_1,\ldots,\mathbf{X}_n) = \tilde{\Phi}_{n1}\mathbf{X}_1 + \cdots + \tilde{\Phi}_{nn}\mathbf{X}_n, \qquad n = 1, 2, \ldots. \tag{11.4.7}$$

Then from (11.4.3),

$$\sum_{j=1}^{n} \tilde{\Phi}_{nj}\Gamma(j-i) = \Gamma(-i), \qquad i = 1, \ldots, n. \tag{11.4.8}$$

The two prediction error covariance matrices will be denoted by

$$V_n = E(\mathbf{X}_{n+1} - \hat{\mathbf{X}}_{n+1})(\mathbf{X}_{n+1} - \hat{\mathbf{X}}_{n+1})',$$

$$\tilde{V}_n = E(\mathbf{X}_0 - P(\mathbf{X}_0|\mathbf{X}_1,\ldots,\mathbf{X}_n))(\mathbf{X}_0 - P(\mathbf{X}_0|\mathbf{X}_1,\ldots,\mathbf{X}_n))'.$$

Observe from (11.4.5) that for $n \geq 1$,

$$\begin{aligned} V_n &= E[(\mathbf{X}_{n+1} - \hat{\mathbf{X}}_{n+1})\mathbf{X}_{n+1}'] \\ &= \Gamma(0) - \Phi_{n1}\Gamma(-1) - \cdots - \Phi_{nn}\Gamma(-n) \end{aligned} \tag{11.4.9}$$

and similarly that

$$\tilde{V}_n = \Gamma(0) - \tilde{\Phi}_{n1}\Gamma(1) - \cdots - \tilde{\Phi}_{nn}\Gamma(n). \tag{11.4.10}$$

We also need to introduce the matrices

$$\begin{aligned} \Delta_n &= E[(\mathbf{X}_{n+1} - \hat{\mathbf{X}}_{n+1})\mathbf{X}_0'] \\ &= \Gamma(n+1) - \Phi_{n1}\Gamma(n) - \cdots - \Phi_{nn}\Gamma(1), \end{aligned} \tag{11.4.11}$$

and

$$\begin{aligned} \tilde{\Delta}_n &= E[(\mathbf{X}_0 - P(\mathbf{X}_0|\mathbf{X}_1,\ldots,\mathbf{X}_n))\mathbf{X}_{n+1}'] \\ &= \Gamma(-n-1) - \tilde{\Phi}_{n1}\Gamma(-n) - \cdots - \tilde{\Phi}_{nn}\Gamma(-1). \end{aligned} \tag{11.4.12}$$

Proposition 11.4.1. (The Multivariate Durbin–Levinson Algorithm). *Let $\{\mathbf{X}_t\}$ be a stationary m-dimensional time series with $E\mathbf{X}_t = \mathbf{0}$ and autocovariance function $\Gamma(h) = E(\mathbf{X}_{t+h}\mathbf{X}_t')$. If the covariance matrix of the nm components of $\mathbf{X}_1,\ldots,\mathbf{X}_n$ is nonsingular for every $n \geq 1$, then the coefficients $\{\Phi_{nj}\}$, $\{\tilde{\Phi}_{nj}\}$ in* (11.4.4) *and* (11.4.7) *satisfy, for $n \geq 1$,*

$$\begin{aligned} \Phi_{nn} &= \Delta_{n-1}\tilde{V}_{n-1}^{-1}, \\ \tilde{\Phi}_{nn} &= \tilde{\Delta}_{n-1}V_{n-1}^{-1}, \\ \Phi_{nk} &= \Phi_{n-1,k} - \Phi_{nn}\tilde{\Phi}_{n-1,n-k}, \qquad k = 1,\ldots,n-1, \\ \tilde{\Phi}_{nk} &= \tilde{\Phi}_{n-1,k} - \tilde{\Phi}_{nn}\Phi_{n-1,n-k}, \qquad k = 1,\ldots,n-1, \end{aligned} \tag{11.4.13}$$

where V_n, $\tilde{V}_n$, Δ_n, $\tilde{\Delta}_n$ *are given by* (11.4.9)–(11.4.12) *with*

$$V_0 = \tilde{V}_0 = \Gamma(0)$$

and

$$\Delta_0 = \tilde{\Delta}_0' = \Gamma(1).$$

PROOF. The proof of this result parallels the argument given in the univariate case, Proposition 5.2.1. For $n = 1$, the result follows immediately from (11.4.6) and (11.4.8) so we shall assume that $n > 1$. The multivariate version of (5.2.6) is

$$\hat{\mathbf{X}}_{n+1} = P(\mathbf{X}_{n+1}|\mathbf{X}_2, \dots, \mathbf{X}_n) + A\mathbf{U} \tag{11.4.14}$$

where $\mathbf{U} = \mathbf{X}_1 - P(\mathbf{X}_1|\mathbf{X}_2, \dots, \mathbf{X}_n)$ and A is an $m \times m$ matrix chosen to satisfy the orthogonality condition

$$\mathbf{X}_{n+1} - A\mathbf{U} \perp \mathbf{U}$$

i.e.,

$$E(\mathbf{X}_{n+1}\mathbf{U}') = AE(\mathbf{U}\mathbf{U}'). \tag{11.4.15}$$

By stationarity,

$$P(\mathbf{X}_{n+1}|\mathbf{X}_2, \dots, \mathbf{X}_n) = \Phi_{n-1,1}X_n + \cdots + \Phi_{n-1,n-1}\mathbf{X}_2, \tag{11.4.16}$$

$$\mathbf{U} = \mathbf{X}_1 - \tilde{\Phi}_{n-1,1}\mathbf{X}_2 - \cdots - \tilde{\Phi}_{n-1,n-1}\mathbf{X}_n, \tag{11.4.17}$$

and

$$E(\mathbf{U}\mathbf{U}') = \tilde{V}_{n-1}. \tag{11.4.18}$$

It now follows from (11.4.3), (11.4.11), (11.4.15) and (11.4.18) that

$$\begin{aligned}
A &= E(\mathbf{X}_{n+1}\mathbf{U}')\tilde{V}_{n-1}^{-1} \\
&= E[(\mathbf{X}_{n+1} - P(\mathbf{X}_{n+1}|\mathbf{X}_2, \dots, \mathbf{X}_n))\mathbf{U}']\tilde{V}_{n-1}^{-1} \\
&= E[(\mathbf{X}_{n+1} - P(\mathbf{X}_{n+1}|\mathbf{X}_2, \dots, \mathbf{X}_n))\mathbf{X}_1']\tilde{V}_{n-1}^{-1} \\
&= [\Gamma(n) - \Phi_{n-1,1}\Gamma(n-1) - \cdots - \Phi_{n-1,n-1}\Gamma(1)]\tilde{V}_{n-1}^{-1} \\
&= \Delta_{n-1}\tilde{V}_{n-1}^{-1}.
\end{aligned} \tag{11.4.19}$$

Combining equations (11.4.14), (11.4.16) and (11.4.17), we have

$$\hat{\mathbf{X}}_{n+1} = A\mathbf{X}_1 + \sum_{j=1}^{n-1} (\Phi_{n-1,j} - A\tilde{\Phi}_{n-1,n-j})\mathbf{X}_{n+1-j}$$

which, together with (11.4.19), proves one half of the recursions (11.4.13). A symmetric argument establishes the other half and completes the proof. □

Remark 1. In the univariate case, $\Gamma(h) = \Gamma(-h)$, so that the two equations (11.4.6) and (11.4.8) are identical. This implies that $\Phi_{nj} = \tilde{\Phi}_{nj}$ for all j and n. The equations (11.4.13) then reduce to the univariate recursions (5.2.3) and (5.2.4).

Remark 2. If for a fixed $p \geq 1$, the covariance matrix of $(\mathbf{X}'_{p+1}, \ldots, \mathbf{X}'_1)'$ is nonsingular, then the matrix polynomial $\Phi(z) = I - \Phi_{p1}z - \cdots - \Phi_{pp}z^p$ is causal in the sense that $\det \Phi(z) \neq 0$ for all $z \in \mathbb{C}$ such that $|z| \leq 1$ (cf. Problem 8.3). To prove this, let $\{\boldsymbol{\eta}_t\}$ be the stationary mp-variate time series

$$\boldsymbol{\eta}_t = \begin{bmatrix} \mathbf{X}_{t+p-1} \\ \vdots \\ \mathbf{X}_{t-1} \\ \mathbf{X}_t \end{bmatrix}.$$

Applying Proposition 11.4.1 to this process with $n = 1$, we obtain

$$\boldsymbol{\eta}_2 = \boldsymbol{\eta}_2 - \hat{\boldsymbol{\eta}}_2 + \hat{\boldsymbol{\eta}}_2$$

where

$$\hat{\boldsymbol{\eta}}_2 = P(\boldsymbol{\eta}_2 | \boldsymbol{\eta}_1) = M\boldsymbol{\eta}_1$$

with $M = E(\boldsymbol{\eta}_2\boldsymbol{\eta}'_1)[E(\boldsymbol{\eta}_1\boldsymbol{\eta}'_1)]^{-1}$ and

$$\boldsymbol{\eta}_2 - \hat{\boldsymbol{\eta}}_2 \perp \boldsymbol{\eta}_1. \tag{11.4.20}$$

It is easily seen, from the composition of the vectors $\boldsymbol{\eta}_2$ and $\boldsymbol{\eta}_1$ and stationarity, that the matrix M has the form

$$M = \begin{bmatrix} \Phi_{p1} & \Phi_{p2} & \cdots & \Phi_{p,p-1} & \Phi_{pp} \\ I & 0 & \cdots & 0 & 0 \\ \vdots & \vdots & & \vdots & \vdots \\ 0 & 0 & \cdots & 0 & 0 \\ 0 & 0 & \cdots & I & 0 \end{bmatrix}, \tag{11.4.21}$$

and since $\det(zI - M) = z^{mp}\det(\Phi(z^{-1}))$ (see Problem 11.8), it suffices to show that the eigenvalues of M all have modulus less than one. Let $\Gamma = E(\boldsymbol{\eta}_1\boldsymbol{\eta}'_1)$, which is positive definite by assumption, and observe that from the orthogonality relation (11.4.20),

$$E(\boldsymbol{\eta}_2 - \hat{\boldsymbol{\eta}}_2)(\boldsymbol{\eta}_2 - \hat{\boldsymbol{\eta}}_2)' = \Gamma - M\Gamma M'.$$

If λ is an eigenvalue of M with corresponding left eigenvector $\mathbf{a}$, i.e. $\mathbf{a}^*M = \lambda\mathbf{a}^*$ where $\mathbf{a}^*$ denotes the complex-conjugate transpose of $\mathbf{a}$, then

$$\begin{aligned} E|\mathbf{a}^*(\boldsymbol{\eta}_2 - \hat{\boldsymbol{\eta}}_2)|^2 &= \mathbf{a}^*\Gamma\mathbf{a} - \mathbf{a}^*M\Gamma M'\mathbf{a} \\ &= \mathbf{a}^*\Gamma\mathbf{a} - |\lambda|^2\mathbf{a}^*\Gamma\mathbf{a} \\ &= \mathbf{a}^*\Gamma\mathbf{a}(1 - |\lambda|^2). \end{aligned}$$

Since Γ is positive definite, we must have $|\lambda| \leq 1$. The case $|\lambda| = 1$ is precluded since this would imply that

$$\mathbf{a}^*(\boldsymbol{\eta}_2 - \hat{\boldsymbol{\eta}}_2) = 0,$$

which in turn implies that the covariance matrix of $(\mathbf{X}'_{p+1}, \ldots, \mathbf{X}'_1)'$ is singular, a contradiction. Thus we conclude that $\det \Phi(z) \neq 0$ for all $|z| \leq 1$.

We next extend the innovations algorithm for computing the best one-step predictor to a general m-variate time series with mean zero. From the definition of S_n, it is clear that

$$S_n = \overline{\text{sp}}\{X_{tj} - \hat{X}_{tj}, j = 1, \ldots, m;\ t = 1, \ldots, n\},$$

so that we may write

$$\hat{\mathbf{X}}_{n+1} = \sum_{j=1}^{n} \Theta_{nj}(\mathbf{X}_{n+1-j} - \hat{\mathbf{X}}_{n+1-j}),$$

where $\{\Theta_{nj}, j = 1, \ldots, n\}$ is a sequence of $m \times m$ matrices which can be found recursively using the following algorithm. The recursions are identical to those given in the univariate case (Proposition 5.2.2) and, in contrast to the Durbin–Levinson recursions, involve only one set of predictor coefficients.

Proposition 11.4.2 (The Multivariate Innovations Algorithm). *Let $\{X_t\}$ be an m-dimensional time series with mean $E\mathbf{X}_t = \mathbf{0}$ for all t and with covariance function $K(i, j) = E(\mathbf{X}_i\mathbf{X}'_j)$. If the covariance matrix of the nm components of $\mathbf{X}_1, \ldots, \mathbf{X}_n$ is nonsingular for every $n \geq 1$, then the one-step predictors $\hat{\mathbf{X}}_{n+1}$, $n \geq 0$, and their prediction error covariance matrices V_n, $n \geq 1$, are given by*

$$\hat{\mathbf{X}}_{n+1} = \begin{cases} \mathbf{0}, & \text{if } n = 0, \\ \displaystyle\sum_{j=1}^{n} \Theta_{nj}(\mathbf{X}_{n+1-j} - \hat{\mathbf{X}}_{n+1-j}) & \text{if } n \geq 1, \end{cases} \tag{11.4.22}$$

and

$$\begin{cases} V_0 = K(1, 1) \\ \Theta_{n,n-k} = \left(K(n+1, k+1) - \displaystyle\sum_{j=0}^{k-1} \Theta_{n,n-j} V_j \Theta'_{k,k-j}\right) V_k^{-1}, \\ \qquad\qquad\qquad\qquad k = 0, \ldots, n-1, \\ V_n = K(n+1, n+1) - \displaystyle\sum_{j=0}^{n-1} \Theta_{n,n-j} V_j \Theta'_{n,n-j}. \end{cases} \tag{11.4.23}$$

(The recursions are solved in the order V_0; Θ_{11}, V_1; Θ_{22}, Θ_{21}, V_2; Θ_{33}, Θ_{32}, Θ_{31}, V_3; ....)

Proof. For $i < j$, $\mathbf{X}_i - \hat{\mathbf{X}}_i \in S_{j-1}$ and since each component of $\mathbf{X}_j - \hat{\mathbf{X}}_j$ is orthogonal to S_{j-1} by the prediction equations, we have

$$(\mathbf{X}_i - \hat{\mathbf{X}}_i) \perp (\mathbf{X}_j - \hat{\mathbf{X}}_j) \quad \text{if } i \neq j. \tag{11.4.24}$$

Post multiplying both sides of (11.4.22) by $(\mathbf{X}_{k+1} - \hat{\mathbf{X}}_{k+1})'$, $0 \leq k \leq n$, and taking expectations, we find from (11.4.24) that

$$E\hat{\mathbf{X}}_{n+1}(\mathbf{X}_{k+1} - \hat{\mathbf{X}}_{k+1})' = \Theta_{n,n-k}V_k.$$

Since $(\mathbf{X}_{n+1} - \hat{\mathbf{X}}_{n+1}) \perp (\mathbf{X}_{k+1} - \hat{\mathbf{X}}_{k+1})$ (see (11.4.3)), we have

$$E\mathbf{X}_{n+1}(\mathbf{X}_{k+1} - \hat{\mathbf{X}}_{k+1})' = E\hat{\mathbf{X}}_{n+1}(\mathbf{X}_{k+1} - \hat{\mathbf{X}}_{k+1})' = \Theta_{n,n-k}V_k. \quad (11.4.25)$$

Replacing $\hat{\mathbf{X}}_{k+1}$ in (11.4.25) by its representation given in (11.4.22), we obtain

$$\Theta_{n,n-k}V_k = K(n+1, k+1) - \sum_{j=0}^{k-1} E\mathbf{X}_{n+1}(\mathbf{X}_{j+1} - \hat{\mathbf{X}}_{j+1})'\Theta'_{k,k-j},$$

which, by (11.4.25), implies that

$$\Theta_{n,n-k}V_k = K(n+1, k+1) - \sum_{j=0}^{k-1} \Theta_{n,n-j}V_j\Theta'_{k,k-j}.$$

Since the covariance matrix of $\mathbf{X}_1, \ldots, \mathbf{X}_n$ is nonsingular by assumption, V_k is nonsingular and hence

$$\Theta_{n,n-k} = \left(K(n+1, k+1) - \sum_{j=0}^{k-1} \Theta_{n,n-j}V_j\Theta'_{k,k-j}\right)V_k^{-1}.$$

Finally we have

$$\mathbf{X}_{n+1} = \mathbf{X}_{n+1} - \hat{\mathbf{X}}_{n+1} + \sum_{j=0}^{n-1} \Theta_{n,n-j}(\mathbf{X}_{j+1} - \hat{\mathbf{X}}_{j+1}),$$

which, by the orthogonality of the set $\{\mathbf{X}_j - \hat{\mathbf{X}}_j, j = 1, \ldots, n+1\}$, implies that

$$K(n+1, n+1) = V_n + \sum_{j=0}^{n-1} \Theta_{n,n-j}V_j\Theta'_{n,n-j}$$

as desired. □

Recursive Prediction of an ARMA(p, q) Process

Let $\{\mathbf{X}_t\}$ be an m-dimensional causal ARMA(p, q) process

$$\Phi(B)\mathbf{X}_t = \Theta(B)\mathbf{Z}_t, \qquad \{\mathbf{Z}_t\} \sim \mathrm{WN}(\mathbf{0}, \Sigma),$$

where $\Phi(B) = I - \Phi_1 B - \cdots - \Phi_p B^p$, $\Theta(B) = I + \Theta_1 B + \cdots + \Theta_q B^q$, $\det \Sigma \neq 0$ and I is the $m \times m$ identity matrix. As in Section 5.3, there is a substantial savings in computation if the innovations algorithm is applied to the transformed process

$$\begin{cases} \mathbf{W}_t = \mathbf{X}_t, & t = 1, \ldots, \max(p, q), \\ \mathbf{W}_t = \Phi(B)\mathbf{X}_t, & t > \max(p, q), \end{cases} \quad (11.4.26)$$

rather than to $\{\mathbf{X}_t\}$ itself. If the covariance function of the $\{\mathbf{X}_t\}$ process is denoted by $\Gamma(\cdot)$, then the covariance function $K(i, j) = E(\mathbf{W}_i\mathbf{W}_j')$ is found to be

$$K(i,j) = \begin{cases} \Gamma(i-j) & \text{if } 1 \le i \le j \le l, \\ \Gamma(i-j) - \sum_{r=1}^{p} \Phi_r \Gamma(i+r-j) & \text{if } 1 \le i \le l < j \le 2l, \\ \sum_{r=0}^{q} \Theta_r \Sigma \Theta'_{r+j-i} & \text{if } l < i \le j \le i+q, \\ 0 & \text{if } l < i \text{ and } i+q < j, \\ K'(i,j) & \text{if } j < i, \end{cases} \tag{11.4.27}$$

where $l = \max(p, q)$ and by convention $\mathbf{\Theta}_j = 0_{m \times m}$ for $j > q$. The advantage of working with this process is that the covariance matrix is zero when $|i - j| > q$, $i, j > l$. The argument leading up to equations (5.3.9) carries over practically verbatim in the multivariate setting to give

$$\hat{\mathbf{X}}_{n+1} = \begin{cases} \sum_{j=1}^{n-1} \Theta_{nj}(\mathbf{X}_{n+1-j} - \hat{\mathbf{X}}_{n+1-j}) & \text{if } 1 \le n \le l, \\ \Phi_1 \mathbf{X}_n + \cdots + \Phi_p \mathbf{X}_{n+1-p} + \sum_{j=1}^{q} \Theta_{nj}(\mathbf{X}_{n+1-j} - \hat{\mathbf{X}}_{n+1-j}) & \text{if } n > l, \end{cases} \tag{11.4.28}$$

and

$$E(\mathbf{X}_{n+1} - \hat{\mathbf{X}}_{n+1})(\mathbf{X}_{n+1} - \hat{\mathbf{X}}_{n+1})' = V_n,$$

where Θ_{nj}, $j = 1, \ldots, n$ and V_n are found from (11.4.23) with $K(i, j)$ as in (11.4.27).

Remark 3. In the one dimensional case, the coefficients $\theta_{nj}, j = 1, \ldots, q$ do not depend on the white noise variance σ^2 (see Remark 1 of Section 5.3). However, in the multivariate case, the coefficients Θ_{nj} of $\mathbf{X}_{n+1-j} - \hat{\mathbf{X}}_{n+1-j}$ will typically depend on Σ.

Remark 4. In the case when $\{\mathbf{X}_t\}$ is also invertible, $\mathbf{X}_{n+1} - \hat{\mathbf{X}}_{n+1}$ is an approximation to $\mathbf{Z}_{n+1}$ for n large in the sense that

$$E(\mathbf{X}_{n+1} - \hat{\mathbf{X}}_{n+1} - \mathbf{Z}_{n+1})(\mathbf{X}_{n+1} - \hat{\mathbf{X}}_{n+1} - \mathbf{Z}_{n+1})' \to 0 \quad \text{as } n \to \infty.$$

It follows (see Problem 11.12) that as $n \to \infty$,

$$\Theta_{nj} \to \Theta_j, \qquad j = 1, \ldots, q,$$

and

$$V_n \to \Sigma.$$

Example 11.4.1 (Prediction of an ARMA(1, 1)). Let $\mathbf{X}_t$ be the ARMA(1, 1) process

$$\mathbf{X}_t - \Phi \mathbf{X}_{t-1} = \mathbf{Z}_t + \Theta \mathbf{Z}_{t-1}, \qquad \{\mathbf{Z}_t\} \sim \mathrm{WN}(0, \Sigma) \tag{11.4.29}$$

with $\det(I - \Phi z) \ne 0$ for $|z| \le 1$. From (11.4.28), we see that

$$\hat{\mathbf{X}}_{n+1} = \Phi \mathbf{X}_n + \Theta_{n1}(\mathbf{X}_n - \hat{\mathbf{X}}_n), \qquad n \ge 1. \tag{11.4.30}$$

The covariance function for the process $\{\mathbf{W}_t\}$ defined by (11.4.26) is given by

$$K(i,j) = \begin{cases} \Gamma(0), & i, j = 1, \\ \Sigma\Theta', & 1 \le i, j = i + 1, \\ \Sigma + \Theta\Sigma\Theta', & 1 < i = j, \\ 0, & 1 \le i, j > i + 1, \\ K'(i,j), & j < i. \end{cases}$$

As in Example 5.3.3, the recursions in (11.4.23) simplify to

$$\begin{cases} V_0 = \Gamma(0), \\ \Theta_{n1} = \Theta\Sigma V_{n-1}^{-1}, \\ V_n = \Sigma + \Theta\Sigma\Theta' - \Theta_{n1} V_{n-1} \Theta_{n1}'. \end{cases} \tag{11.4.31}$$

In order to start this recursion, it is necessary first to compute $\Gamma(0)$. From (11.3.15) we obtain the two matrix equations

$$\Gamma(0) - \Phi\Gamma'(1) = \Sigma + \Theta\Sigma(\Phi' + \Theta'),$$

$$\Gamma(1) - \Phi\Gamma(0) = \Theta\Sigma.$$

Substituting $\Gamma(1) = \Phi\Gamma(0) + \Theta\Sigma$ into the first expression, we obtain the single matrix equation,

$$\Gamma(0) - \Phi\Gamma(0)\Phi' = \Phi\Sigma\Theta' + \Theta\Sigma\Phi' + \Sigma + \Theta\Sigma\Theta', \tag{11.4.32}$$

which is equivalent to a set of linear equations which can be solved for the components of $\Gamma(0)$.

Ten observations $\mathbf{X}_1, \ldots, \mathbf{X}_{10}$ were generated from the two-dimensional ARMA(1, 1) process

$$\begin{bmatrix} X_{t1} \\ X_{t2} \end{bmatrix} - \begin{bmatrix} .7 & 0 \\ 0 & .6 \end{bmatrix} \begin{bmatrix} X_{t-1,1} \\ X_{t-1,2} \end{bmatrix} = \begin{bmatrix} Z_{t1} \\ Z_{t2} \end{bmatrix} + \begin{bmatrix} .5 & .6 \\ -.7 & .8 \end{bmatrix} \begin{bmatrix} Z_{t-1,1} \\ Z_{t-1,2} \end{bmatrix} \tag{11.4.33}$$

where $\{\mathbf{Z}_t\}$ is a sequence of iid $N(\left[\begin{smallmatrix} 0 \\ 0 \end{smallmatrix}\right], \left[\begin{smallmatrix} 1 & .71 \\ .71 & 2 \end{smallmatrix}\right])$ random vectors. The values of $\hat{\mathbf{X}}_{n+1}$, V_n and Θ_{n1} for $n = 0, 1, \ldots, 10$, computed from equations (11.4.30)–(11.4.32), are displayed in Table 11.1. Notice that the matrices V_n and Θ_{n1} are converging rapidly to the matrices Σ and Θ, respectively.

Once $\hat{\mathbf{X}}_1, \ldots, \hat{\mathbf{X}}_n$ are found from equations (11.4.28), it is a simple matter to compute the h-step predictors of the process. As in Section 5.3 (see equations (5.3.15)), the h-step predictors $P_{S_n}\mathbf{X}_{n+h}$, $h = 1, 2, \ldots$, satisfy

$$P_{S_n}\mathbf{X}_{n+h} = \begin{cases} \sum_{j=h}^{n+h-1} \Theta_{n+h-1,j}(\mathbf{X}_{n+h-j} - \hat{\mathbf{X}}_{n+h-j}), & 1 \le h \le l - n \\ \sum_{i=1}^{p} \Phi_i P_{S_n}\mathbf{X}_{n+h-i} + \sum_{h \le j \le q} \Theta_{n+h-1,j}(\mathbf{X}_{n+h-j} - \hat{\mathbf{X}}_{n+h-j}), & \\ & h > l - n \end{cases} \tag{11.4.34}$$

Table 11.1. Calculation of $\hat{\mathbf{X}}_n$ for Data from the ARMA(1, 1) Process of Example 11.4.1

n	$\mathbf{X}_{n+1}$	V_n	Θ_{n1}	$\hat{\mathbf{X}}_{n+1}$
0	$\begin{bmatrix} -1.875 \\ 1.693 \end{bmatrix}$	$\begin{bmatrix} 7.240 & 3.701 \\ 3.701 & 6.716 \end{bmatrix}$		$\begin{bmatrix} 0 \\ 0 \end{bmatrix}$
1	$\begin{bmatrix} -2.518 \\ -.030 \end{bmatrix}$	$\begin{bmatrix} 2.035 & 1.060 \\ 1.060 & 2.688 \end{bmatrix}$	$\begin{bmatrix} .013 & .224 \\ -.142 & .243 \end{bmatrix}$	$\begin{bmatrix} -.958 \\ 1.693 \end{bmatrix}$
2	$\begin{bmatrix} -3.002 \\ -1.057 \end{bmatrix}$	$\begin{bmatrix} 1.436 & .777 \\ .777 & 2.323 \end{bmatrix}$	$\begin{bmatrix} .193 & .502 \\ -.351 & .549 \end{bmatrix}$	$\begin{bmatrix} -2.930 \\ -.417 \end{bmatrix}$
3	$\begin{bmatrix} -2.454 \\ -1.038 \end{bmatrix}$	$\begin{bmatrix} 1.215 & .740 \\ .740 & 2.238 \end{bmatrix}$	$\begin{bmatrix} .345 & .554 \\ -.426 & .617 \end{bmatrix}$	$\begin{bmatrix} -2.481 \\ -1.000 \end{bmatrix}$
4	$\begin{bmatrix} -1.119 \\ -1.086 \end{bmatrix}$	$\begin{bmatrix} 1.141 & .750 \\ .750 & 2.177 \end{bmatrix}$	$\begin{bmatrix} .424 & .555 \\ -.512 & .662 \end{bmatrix}$	$\begin{bmatrix} -1.728 \\ -.662 \end{bmatrix}$
5	$\begin{bmatrix} -.720 \\ -.455 \end{bmatrix}$	$\begin{bmatrix} 1.113 & .744 \\ .744 & 2.119 \end{bmatrix}$	$\begin{bmatrix} .442 & .562 \\ -.580 & .707 \end{bmatrix}$	$\begin{bmatrix} -.073 \\ -1.304 \end{bmatrix}$
6	$\begin{bmatrix} -2.738 \\ .962 \end{bmatrix}$	$\begin{bmatrix} 1.085 & .728 \\ .728 & 2.084 \end{bmatrix}$	$\begin{bmatrix} .446 & .577 \\ -.610 & .735 \end{bmatrix}$	$\begin{bmatrix} .001 \\ .331 \end{bmatrix}$
7	$\begin{bmatrix} -2.565 \\ 1.992 \end{bmatrix}$	$\begin{bmatrix} 1.059 & .721 \\ .721 & 2.069 \end{bmatrix}$	$\begin{bmatrix} .461 & .585 \\ -.623 & .747 \end{bmatrix}$	$\begin{bmatrix} -2.809 \\ 2.754 \end{bmatrix}$
8	$\begin{bmatrix} -4.603 \\ 2.434 \end{bmatrix}$	$\begin{bmatrix} 1.045 & .722 \\ .722 & 2.057 \end{bmatrix}$	$\begin{bmatrix} .475 & .586 \\ -.639 & .756 \end{bmatrix}$	$\begin{bmatrix} -2.126 \\ .463 \end{bmatrix}$
9	$\begin{bmatrix} -2.689 \\ 2.118 \end{bmatrix}$	$\begin{bmatrix} 1.038 & .721 \\ .721 & 2.042 \end{bmatrix}$	$\begin{bmatrix} .480 & .587 \\ -.657 & .767 \end{bmatrix}$	$\begin{bmatrix} -3.254 \\ 4.598 \end{bmatrix}$
10		$\begin{bmatrix} 1.030 & .717 \\ .717 & 2.032 \end{bmatrix}$	$\begin{bmatrix} .481 & .591 \\ -.666 & .775 \end{bmatrix}$	$\begin{bmatrix} -3.077 \\ -1.029 \end{bmatrix}$

where for fixed n, the predictors $P_{S_n}\mathbf{X}_{n+1}, P_{S_n}\mathbf{X}_{n+2}, P_{S_n}\mathbf{X}_{n+3}, \ldots$ are determined recursively from (11.4.34). Of course in most applications $n > l = \max(p, q)$, in which case the second of the two relations in (11.4.34) applies. For the ARMA(1, 1) process of Example 11.4.1 we have for $h \geq 1$,

$$\begin{aligned} P_{S_n}\mathbf{X}_{n+h} &= \Phi P_{S_n}\mathbf{X}_{n+h-1} \\ &= \cdots \\ &= \Phi^{h-1} P_{S_n}\mathbf{X}_{n+1} \\ &= \begin{bmatrix} (.7)^{h-1}\hat{X}_{n+1,1} \\ (.6)^{h-1}\hat{X}_{n+1,2} \end{bmatrix}. \end{aligned}$$

More generally, let us fix n and define $\mathbf{g}(h) := P_{S_n}\mathbf{X}_{n+h}$. Then $\mathbf{g}(h)$ satisfies the multivariate homogeneous difference equation,

$$\mathbf{g}(h) - \Phi_1\mathbf{g}(h-1) - \cdots - \Phi_p\mathbf{g}(h-p) = \mathbf{0}, \qquad \text{for } h > q, \quad (11.4.35)$$

with initial conditions,

$$\mathbf{g}(q-i) = P_{S_n}\mathbf{X}_{n+q-i}, \qquad i = 0, \ldots, p-1.$$

By appealing to the theory of multivariate homogeneous difference equations, it is often possible to find a convenient representation for $\mathbf{g}(h)$ and hence $P_{S_n}\mathbf{X}_{n+h}$ by solving (11.4.35).

§11.5 Estimation for Multivariate ARMA Processes

If $\{\mathbf{X}_t\}$ is a causal m-variate ARMA(p, q) process,

$$\mathbf{X}_t - \Phi_1\mathbf{X}_{t-1} - \cdots - \Phi_p\mathbf{X}_{t-p} = \mathbf{Z}_t + \Theta_1\mathbf{Z}_{t-1} + \cdots + \Theta_q\mathbf{Z}_{t-q}, \quad (11.5.1)$$

where $\{\mathbf{Z}_t\} \sim \mathrm{WN}(\mathbf{0}, \Sigma)$, then the Gaussian likelihood of $\{\mathbf{X}_1, \ldots, \mathbf{X}_n\}$ can be determined with the aid of the multivariate innovations algorithm and the technique used in Section 8.7 for the univariate case.

For an arbitrary m-variate Gaussian process $\{\mathbf{X}_t\}$ with mean $\mathbf{0}$ and covariance matrices

$$K(i,j) = E(\mathbf{X}_i\mathbf{X}_j'),$$

we can determine the exact likelihood of $\{\mathbf{X}_1, \ldots, \mathbf{X}_n\}$ as in Section 8.6. Let $\mathbf{X}$ denote the nm-component column vector of observations, $\mathbf{X} := (\mathbf{X}_1', \ldots, \mathbf{X}_n')'$ and let $\hat{\mathbf{X}} := (\hat{\mathbf{X}}_1', \ldots, \hat{\mathbf{X}}_n')'$ where $\hat{\mathbf{X}}_1, \ldots, \hat{\mathbf{X}}_n$ are the one-step predictors defined in Section 11.4. Assume that $\Gamma_n := E(\mathbf{X}\mathbf{X}')$ is non-singular for every n and let Θ_{jk} and V_j be the coefficient and covariance matrices defined in Proposition 11.4.2, with $\Theta_{i0} = I$ and $\Theta_{ij} = 0, j < 0, i = 0, 1, 2, \ldots$. Then, introducing the $(nm \times nm)$ matrices,

$$C = [\Theta_{i,i-j}]_{i,j=0}^{n-1} \quad (11.5.2)$$

and

$$D = \mathrm{diag}\{V_0, \ldots, V_{n-1}\}, \quad (11.5.3)$$

we find by precisely the same steps as in Section 8.6 that the likelihood of $\{\mathbf{X}_1, \ldots, \mathbf{X}_n\}$ is

$$L(\Gamma_n) = (2\pi)^{-nm/2}\left(\prod_{j=1}^{n} \det V_{j-1}\right)^{-1/2} \exp\left\{-\frac{1}{2}\sum_{j=1}^{n} (\mathbf{X}_j - \hat{\mathbf{X}}_j)' V_{j-1}^{-1}(\mathbf{X}_j - \hat{\mathbf{X}}_j)\right\}, \quad (11.5.4)$$

where the one-step predictors $\hat{\mathbf{X}}_j$ and the corresponding error covariance matrices $V_{j-1}, j = 1, \ldots, n$, are found from Proposition 11.4.2. Notice that the calculation of $L(\Gamma_n)$ involves operations on vectors and square matrices of dimension m only.

To compute the Gaussian likelihood of $\{\mathbf{X}_1, \ldots, \mathbf{X}_n\}$ for the ARMA process (11.5.1) we proceed as in Section 8.7. First we introduce the process $\{\mathbf{W}_t\}$

defined by (11.4.26) with covariance matrices $K(i, j) = E(\mathbf{W}_i \mathbf{W}_j')$ given by (11.4.27). Applying the multivariate innovations algorithm to the transformed process $\{\mathbf{W}_t\}$ gives the coefficients Θ_{jk} and error covariance matrices V_j in the representation of (11.4.28) of $\hat{\mathbf{X}}_{j+1}$. Since $\mathbf{X}_j - \hat{\mathbf{X}}_j = \mathbf{W}_j - \hat{\mathbf{W}}_j$, $j = 1, 2, \ldots$, it follows from (11.5.4) that the Gaussian likelihood $L(\boldsymbol{\Phi}, \boldsymbol{\Theta}, \boldsymbol{\Sigma})$ of $\{\mathbf{X}_1, \ldots, \mathbf{X}_n\}$ can be written as

$$L(\boldsymbol{\Phi}, \boldsymbol{\Theta}, \boldsymbol{\Sigma}) = (2\pi)^{-nm/2}\left(\prod_{j=1}^{n} \det V_{j-1}\right)^{-1/2} \times \exp\left\{-\frac{1}{2}\sum_{j=1}^{n} (\mathbf{X}_j - \hat{\mathbf{X}}_j)' V_{j-1}^{-1}(\mathbf{X}_j - \hat{\mathbf{X}}_j)\right\}, \tag{11.5.5}$$

where $\hat{\mathbf{X}}_j$ is found from (11.4.28) and Θ_{jk}, V_j are found by applying Proposition 11.4.2 to the covariance matrix (11.4.27).

In view of Remark 3 of Section 11.4, it is not possible to compute maximum likelihood estimators of $\boldsymbol{\Phi}$ and $\boldsymbol{\Theta}$ independently of $\boldsymbol{\Sigma}$ as in the univariate case. Maximization of the likelihood must be performed with respect to all the parameters of $\boldsymbol{\Phi}$, $\boldsymbol{\Theta}$ and $\boldsymbol{\Sigma}$ simultaneously. The potentially large number of parameters involved makes the determination of maximum likelihood estimators much more difficult from a numerical point of view than the corresponding univariate problem. However the maximization can be performed with the aid of efficient non-linear optimization algorithms.

A fundamental difficulty in the estimation of parameters for mixed ARMA models arises from the question of identifiability. The spectral density matrix of the process (11.5.1) is

$$f(\omega) = \frac{1}{2\pi} \Phi^{-1}(e^{-i\omega})\Theta(e^{-i\omega})\Sigma\Theta'(e^{i\omega})\Phi'^{-1}(e^{i\omega}).$$

The covariance matrix function, or equivalently the spectral density matrix function $f(\cdot)$, of a causal invertible ARMA process does not uniquely determine Σ, $\Phi(\cdot)$ and $\Theta(\cdot)$ unless further conditions are imposed (see Dunsmuir and Hannan (1976)). Non-identifiability of a model results in a likelihood surface which does not have a unique maximum. The identifiability problem arises only when $p > 0$ and $q > 0$. For a causal autoregressive or invertible moving average process, the coefficient matrices and the white noise covariance matrix Σ are uniquely determined by the second order properties of the process.

It is particularly important in the maximum likelihood estimation of multivariate ARMA parameters, to have good initial estimates of the parameters since the likelihood function may have many local maxima which are much smaller than the global maximum. Jones (1984) recommends initial fitting of univariate models to each component of the series to give an initial approximation with uncorrelated components.

Order selection for multivariate ARMA models can be made by minimizing a multivariate analogue of (9.3.4), namely

$$\text{AICC} = -2\ln L(\Phi_1, \ldots, \Phi_p, \Theta_1, \ldots, \Theta_q, \Sigma) + 2(k+1)nm/(nm-k-2),$$

where $k = (p+q)m^2$.

Spectral methods of estimation for multivariate ARMA parameters are also frequently used. A discussion of these (as well as some time domain methods) is given in Anderson (1980).

Estimation for Autoregressive Processes Using the Durbin–Levinson Algorithm

There is a simple alternative estimation procedure, based on the multivariate Durbin–Levison algorithm, for fitting autoregressions of increasing order. This is analogous to the preliminary estimation procedure for autoregressions in the univariate case discussed in Section 8.2. Suppose we have observations $\mathbf{x}_1, \ldots, \mathbf{x}_n$ of a zero-mean stationary m-variate time series and let $\hat{\Gamma}(0), \ldots, \hat{\Gamma}(n-1)$ be the sample covariance function estimates. Then the fitted AR(p) process ($p < n$) is

$$\mathbf{X}_t = \hat{\Phi}_{p1}\mathbf{X}_{t-1} + \cdots + \hat{\Phi}_{pp}\mathbf{X}_{t-p} + \mathbf{Z}_t, \qquad \{\mathbf{Z}_t\} \sim \text{WN}(\mathbf{0}, \hat{V}_p),$$

where the coefficients $\hat{\Phi}_{p1}, \ldots, \hat{\Phi}_{pp}$ and $\hat{V}_p$ are computed recursively from Proposition 11.4.1 with $\Gamma(h)$ replaced by $\hat{\Gamma}(h)$, $h = 0, \ldots, n-1$. The order p of the autoregression may be chosen to minimize

$$\text{AICC} = -2\ln L(\hat{\Phi}_{p1}, \ldots, \hat{\Phi}_{pp}, \hat{V}_p) + 2(pm^2+1)nm/(nm-pm^2-2).$$

Example 11.5.1 (Sales with a Leading Indicator). In this example we fit an autoregressive model to the bivariate time series of Example 11.2.2. Let

$$X_{t1} = (1-B)Y_{t1} - .0228, \qquad t = 1, \ldots, 149,$$

and

$$X_{t2} = (1-B)Y_{t2} - .420 \qquad t = 1, \ldots, 149,$$

where $\{Y_{t1}\}$ and $\{Y_{t2}\}$, $t = 0, \ldots, 149$, are the leading indicator and sales data respectively. The order of the minimum AICC autoregressive model for $(X_{t1}, X_{t2})'$, computed using the program ARVEC, is $p = 5$ with parameter estimates given by

$$\hat{\Phi}_{51} = \begin{bmatrix} -.517 & .024 \\ -.019 & -.051 \end{bmatrix}, \quad \hat{\Phi}_{52} = \begin{bmatrix} -.192 & -.018 \\ .047 & .250 \end{bmatrix}, \quad \hat{\Phi}_{53} = \begin{bmatrix} -.073 & .010 \\ 4.678 & .207 \end{bmatrix},$$

$$\hat{\Phi}_{54} = \begin{bmatrix} -.032 & -.009 \\ 3.664 & .004 \end{bmatrix}, \quad \hat{\Phi}_{55} = \begin{bmatrix} .022 & .011 \\ 1.300 & .029 \end{bmatrix}, \quad \hat{V}_5 = \begin{bmatrix} .076 & -.003 \\ -.003 & .095 \end{bmatrix},$$

and AICC = 114.94. Since the upper right component of each of the coefficient estimates is near 0, we may model the $\{X_{t1}\}$ process separately from $\{X_{t2}\}$. The MA(1) model

$$X_{t1} = (1 - .474B)U_t, \qquad \{U_t\} \sim \text{WN}(0, .0779) \tag{11.5.6}$$

provides an adequate fit to the series $\{X_{t1}\}$. Inspecting the bottom row of the coefficient matrices, $\hat{\Phi}_{51}, \ldots, \hat{\Phi}_{55}$, and deleting those elements which are near 0, we arrive at the approximate relation between $\{X_{t1}\}$ and $\{X_{t2}\}$ given by

$$\begin{aligned} X_{t2} = {} & .250X_{t-2,2} + .207X_{t-3,2} + 4.678X_{t-3,1} + 3.664X_{t-4,1} \\ & + 1.300X_{t-5,1} + W_t \end{aligned}$$

or, equivalently,

$$X_{t2} = \frac{4.678B^3(1 + .783B + .278B^2)}{(1 - .250B^2 - .207B^3)} X_{t1} + (1 - .250B^2 - .207B^3)^{-1} W_t \tag{11.5.7}$$

where $\{W_t\} \sim \text{WN}(0, .095)$. Moreover, since the estimated noise covariance matrix is essentially diagonal, it follows that the two sequences $\{X_{t1}\}$ and $\{W_t\}$ are uncorrelated. This reduced model (11.5.6) and (11.5.7) is an example of a transfer function model which expresses the output series $\{X_{t2}\}$ as the output of a linear filter with input $\{X_{t1}\}$ plus added noise. The model (11.5.6) and (11.5.7) is similar to the model found later in Section 13.1 (see (13.1.23)) using transfer function techniques.

Assuming that the fitted AR(5) model is the true model for $\{\mathbf{X}_t := (X_{t1}, X_{t2})'\}$, the one- and two-step ahead predictors of $\mathbf{X}_{150}$ and $\mathbf{X}_{151}$ are

$$\begin{aligned} P_{S_{149}}\mathbf{X}_{150} &= \hat{\Phi}_{51}\mathbf{X}_{149} + \cdots + \hat{\Phi}_{55}\mathbf{X}_{145} \\ &= \begin{bmatrix} .163 \\ -.217 \end{bmatrix} \end{aligned}$$

and

$$\begin{aligned} P_{S_{149}}\mathbf{X}_{151} &= \hat{\Phi}_{51}\hat{\mathbf{X}}_{150} + \hat{\Phi}_{52}X_{149} + \cdots + \hat{\Phi}_{55}\mathbf{X}_{146} \\ &= \begin{bmatrix} -.027 \\ .816 \end{bmatrix}, \end{aligned}$$

with error covariance matrices

$$E[(\mathbf{X}_{150} - P_{S_{149}}\mathbf{X}_{150})(\mathbf{X}_{150} - P_{S_{149}}\mathbf{X}_{150})'] = \hat{V}_5 = \begin{bmatrix} .076 & -.003 \\ -.003 & .095 \end{bmatrix},$$

$$\begin{aligned} E[(\mathbf{X}_{151} - P_{S_{149}}\mathbf{X}_{151})(\mathbf{X}_{151} - P_{S_{149}}\mathbf{X}_{151})'] &= \hat{\Phi}_{51}\hat{V}_5\hat{\Phi}'_{51} + \hat{V}_5 \\ &= \begin{bmatrix} .0964 & -.0024 \\ -.0024 & .0953 \end{bmatrix}. \end{aligned}$$

Forecasting future values of the original data $\mathbf{Y}_t = (Y_{t1}, Y_{t2})'$ is analogous to the forecasting of univariate ARIMA models discussed in Section 9.5. Let $P_{149}(\cdot)$ denote the operator $P(\cdot|\mathbf{1}, \mathbf{Y}_0, \dots, \mathbf{Y}_{149})$ where $\mathbf{1} = (1, 1)'$ and assume, as in the univariate case, that $\mathbf{Y}_0 \perp \mathbf{X}_1, \dots, \mathbf{X}_{149}$. Then, defining S_n as in (11.4.1), we find (see Problem 11.9) that

$$P_{149}\mathbf{Y}_{150} = \begin{bmatrix} .0228 \\ .420 \end{bmatrix} + P_{S_{149}}\mathbf{X}_{150} + \mathbf{Y}_{149}$$

$$= \begin{bmatrix} .0228 \\ .420 \end{bmatrix} + \begin{bmatrix} .163 \\ -.217 \end{bmatrix} + \begin{bmatrix} 13.4 \\ 262.7 \end{bmatrix} = \begin{bmatrix} 13.59 \\ 262.90 \end{bmatrix}$$

and

$$P_{149}\mathbf{Y}_{151} = \begin{bmatrix} .0228 \\ .420 \end{bmatrix} + P_{S_{149}}\mathbf{X}_{151} + P_{149}\mathbf{Y}_{150}$$

$$= \begin{bmatrix} .0228 \\ .420 \end{bmatrix} + \begin{bmatrix} -.027 \\ .816 \end{bmatrix} + \begin{bmatrix} 13.59 \\ 262.90 \end{bmatrix} = \begin{bmatrix} 13.59 \\ 264.14 \end{bmatrix},$$

with error covariance matrices

$$E[(\mathbf{Y}_{150} - P_{149}\mathbf{Y}_{150})(\mathbf{Y}_{150} - P_{149}\mathbf{Y}_{151})'] = \hat{V}_5 = \begin{bmatrix} .076 & -.003 \\ -.003 & .095 \end{bmatrix}$$

and

$$E[(\mathbf{Y}_{151} - P_{149}\mathbf{Y}_{151})(\mathbf{Y}_{151} - P_{149}\mathbf{Y}_{151})'] = (I + \hat{\Phi}_{51})\hat{V}_5(I + \hat{\Phi}_{51})' + \hat{V}_5$$

$$= \begin{bmatrix} .094 & -.003 \\ -.003 & .181 \end{bmatrix}.$$

These predicted values, computed using the program ARVEC, are in close agreement with those obtained from the transfer function model of Section 13.1 (see (13.1.27) and (13.1.28). Although the two models produce roughly the same prediction mean squared errors for the leading indicator data, the AR model gives substantially larger values for the sales data (see (13.1.29) and (13.1.30)).

§11.6 The Cross Spectrum

Recall from Chapter 4 that if $\{X_t\}$ is a stationary time series with absolutely summable autocovariance function $\gamma(\cdot)$, then $\{X_t\}$ has a spectral density (Corollary 4.3.2) given by

$$f(\lambda) = \frac{1}{2\pi} \sum_{h=-\infty}^{\infty} e^{-ih\lambda}\gamma(h), \qquad -\pi \le \lambda \le \pi, \tag{11.6.1}$$

and the autocovariance function can then be expressed as

$$\gamma(h) = \int_{-\pi}^{\pi} e^{ih\lambda} f(\lambda)\,d\lambda. \tag{11.6.2}$$

By Theorem 4.8.2, the process $\{X_t\}$ has a corresponding spectral representation,

$$X_t = \int_{(-\pi,\pi]} e^{it\lambda}\,dZ(\lambda), \tag{11.6.3}$$

where $\{Z(\lambda), -\pi \le \lambda \le \pi\}$ is an orthogonal increment process satisfying

$$\int_{\lambda_1}^{\lambda_2} f(\lambda)\,d\lambda = E|Z(\lambda_2) - Z(\lambda_1)|^2, \qquad -\pi \le \lambda_1 \le \lambda_2 \le \pi,$$

the latter expression representing the contribution to the variance of $\{X_t\}$ from harmonic components with frequencies in the interval $(\lambda_1, \lambda_2]$.

In this section we shall consider analogous representations for a bivariate stationary time series, $\mathbf{X}_t = (X_{t1}, X_{t2})'$, with mean zero and covariances $\gamma_{ij}(h) = E(X_{t+h,i}X_{tj})$ satisfying

$$\sum_{h=-\infty}^{\infty} |\gamma_{ij}(h)| < \infty, \qquad i, j = 1, 2. \tag{11.6.4}$$

Although we shall confine our discussion to bivariate time series, the ideas can easily be extended to higher dimensions and to series whose covariances are not absolutely summable (see Section 11.8).

Definition 11.6.1 (The Cross Spectrum). If $\{\mathbf{X}_t\}$ is a stationary bivariate time series with mean $\mathbf{0}$ and covariance matrix function $\Gamma(\cdot)$ satisfying (11.6.4), then the function

$$f_{12}(\lambda) = \frac{1}{2\pi} \sum_{h=-\infty}^{\infty} e^{-ih\lambda} \gamma_{12}(h), \qquad \lambda \in [-\pi, \pi],$$

is called the cross spectrum or cross spectral density of $\{X_{t1}\}$ and $\{X_{t2}\}$. The matrix

$$f(\lambda) = \frac{1}{2\pi} \sum_{h=-\infty}^{\infty} e^{-ih\lambda} \Gamma(h) = \begin{bmatrix} f_{11}(h) & f_{12}(h) \\ f_{21}(h) & f_{22}(h) \end{bmatrix}$$

is called the spectral density matrix or spectrum of $\{\mathbf{X}_t\}$.

The spectral representations of $\gamma_{ij}(h)$ and $\Gamma(h)$ follow at once from this definition. Thus

$$\gamma_{ij}(h) = \int_{-\pi}^{\pi} e^{ih\lambda} f_{ij}(\lambda)\,d\lambda, \qquad i, j = 1, 2, \tag{11.6.5}$$

and

$$\Gamma(h) = \int_{-\pi}^{\pi} e^{ih\lambda} f(\lambda)\,d\lambda.$$

The function $f_{ii}(\cdot)$ is the spectral density of the univariate series $\{X_{ti}\}$ as defined in Chapter 4, and is therefore real-valued and symmetric about zero. However since $\gamma_{ij}(\cdot)$, $i \neq j$, is not in general symmetric about zero, the cross spectrum $f_{ij}(\cdot)$ is typically complex-valued.

If $\{Z_i(\lambda), -\pi \leq \lambda \leq \pi\}$ is the orthogonal increment process in the spectral representation of the univariate series $\{X_{ti}\}$, then we know from Chapter 4 that

$$X_{ti} = \int_{(-\pi,\pi]} e^{it\lambda}\, dZ_i(\lambda) \tag{11.6.6}$$

and

$$f_{ii}(\lambda)\, d\lambda = E|dZ_i(\lambda)|^2, \tag{11.6.7}$$

the latter being an abbreviation for $\int_{\lambda_1}^{\lambda_2} f_{ii}(\lambda)\, d\lambda = E|Z_i(\lambda_2) - Z_i(\lambda_1)|^2$, $-\pi \leq \lambda_1 \leq \lambda_2 \leq \pi$. The cross spectrum $f_{ij}(\lambda)$ has a similar interpretation, namely

$$f_{ij}(\lambda)\, d\lambda = E(dZ_i(\lambda)\overline{dZ_j(\lambda)}), \tag{11.6.8}$$

which is shorthand for $\int_{\lambda_1}^{\lambda_2} f_{ij}(\lambda)\, d\lambda = E[(Z_i(\lambda_2) - Z_i(\lambda_1))(\overline{Z_j(\lambda_2)} - \overline{Z_j(\lambda_1)})]$, $-\pi \leq \lambda_1 \leq \lambda_2 \leq \pi$. As shown in Section 11.8, the processes $\{Z_1(\lambda)\}$ and $\{Z_2(\lambda)\}$ have the additional property,

$$E(dZ_i(\lambda)\overline{dZ_j(\mu)}) = 0 \quad \text{for } \lambda \neq \mu \text{ and } i, j = 1, 2.$$

The relation (4.7.5) for univariate processes extends in the bivariate case to

$$E\left[\int_{(-\pi,\pi]} g(\lambda)\, dZ_i(\lambda) \overline{\int_{(-\pi,\pi]} h(\lambda)\, dZ_j(\lambda)}\right] = \int_{-\pi}^{\pi} g(\lambda)\overline{h(\lambda)} f_{ij}(\lambda)\, d\lambda, \quad i, j = 1, 2, \tag{11.6.9}$$

for all functions g and h which are square integrable with respect to f_{ii} and f_{jj} respectively (see Remark 1 of Section 11.8). From (11.6.8) we see that

$$f_{21}(\lambda) = \overline{f_{12}(\lambda)}.$$

This implies that the matrices $f(\lambda)$ are Hermitian, i.e. that $f(\lambda) = f^*(\lambda)$ where $*$ denotes complex conjugate transpose. Moreover if $\mathbf{a} = (a_1, a_2)' \in \mathbb{C}^2$ then $\mathbf{a}^* f(\lambda)\mathbf{a}$ is the spectral density of $\{\mathbf{a}^*\mathbf{X}_t\}$. Consequently $\mathbf{a}^* f(\lambda)\mathbf{a} \geq 0$ for all $\mathbf{a} \in \mathbb{C}^2$, i.e. the matrix $f(\lambda)$ is non-negative definite.

The correlation between $dZ_1(\lambda)$ and $dZ_2(\lambda)$ is called the *coherency* or *coherence*, $\mathscr{K}_{12}(\lambda)$, at frequency λ. From (11.6.7) and (11.6.8) we have

$$\mathscr{K}_{12}(\lambda) = f_{12}(\lambda)/[f_{11}(\lambda)f_{22}(\lambda)]^{1/2}. \tag{11.6.10}$$

By the Cauchy–Schwarz inequality, the *squared coherency function* $|\mathscr{K}_{12}(\lambda)|^2$ satisfies the inequalities,

$$0 \leq |\mathscr{K}_{12}(\lambda)|^2 \leq 1, \qquad -\pi \leq \lambda \leq \pi,$$

and a value near one indicates a strong linear relationship between $dZ_1(\lambda)$ and $dZ_2(\lambda)$.

Since $f_{12}(\lambda)$ is complex-valued, it can be expressed as

$$f_{12}(\lambda) = c_{12}(\lambda) - iq_{12}(\lambda),$$

where

$$c_{12}(\lambda) = \mathrm{Re}\{f_{12}(\lambda)\}$$

and

$$q_{12}(\lambda) = -\mathrm{Im}\{f_{12}(\lambda)\}.$$

The function $c_{12}(\lambda)$ is called the *cospectrum* of $\{X_{t1}\}$ and $\{X_{t2}\}$, and $q_{12}(\lambda)$ is called the *quadrature spectrum.*

Alternatively $f_{12}(\lambda)$ can be expressed in polar coordinates as

$$f_{12}(\lambda) = \alpha_{12}(\lambda)\exp[i\phi_{12}(\lambda)],$$

where

$$\alpha_{12}(\lambda) = (c_{12}^2(\lambda) + q_{12}^2(\lambda))^{1/2}$$

is called the *amplitude spectrum* and

$$\phi_{12}(\lambda) = \arg(c_{12}(\lambda) - iq_{12}(\lambda)) \in (-\pi, \pi],$$

the *phase spectrum* of $\{X_{t1}\}$ and $\{X_{t2}\}$. The coherency is related to the phase and amplitude spectra by

$$\mathscr{K}_{12}(\lambda) = \alpha_{12}(\lambda)[f_{11}(\lambda)f_{22}(\lambda)]^{-1/2}\exp[i\phi_{12}(\lambda)] = |\mathscr{K}_{12}(\lambda)|\exp[i\phi_{12}(\lambda)].$$

EXAMPLE 11.6.1. Let $\{\mathbf{X}_t\}$ be the process defined in Example 11.1.1, i.e.

$$X_{t1} = Z_t,$$

$$X_{t2} = Z_t + .75Z_{t-10},$$

where $\{Z_t\} \sim \mathrm{WN}(0, 1)$. Then

$$f(\lambda) = \frac{1}{2\pi}[\Gamma(-10)e^{10i\lambda} + \Gamma(0) + \Gamma(10)e^{-10i\lambda}]$$

and

$$f_{12}(\lambda) = \frac{1}{2\pi}[1 + .75\cos(10\lambda) + .75i\sin(10\lambda)]$$

$$= \alpha_{12}(\lambda)\exp[i\phi_{12}(\lambda)],$$

where the amplitude spectrum $\alpha_{12}(\lambda)$ is

$$\alpha_{12}(\lambda) = \frac{1}{2\pi}[1.5625 + 1.5\cos(10\lambda)]^{1/2},$$

and

$$\tan\phi_{12}(\lambda) = .75\sin(10\lambda)/[1 + .75\cos(10\lambda)].$$

Since $f_{11}(\lambda) = (2\pi)^{-1}$ and $f_{22}(\lambda) = (2\pi)^{-1}(1.5625 + 1.5\cos(10\lambda))$, the squared coherency is

$$|\mathscr{K}_{12}(\lambda)|^2 = \alpha_{12}^2(\lambda)/[f_{11}(\lambda)f_{22}(\lambda)] = 1, \qquad -\pi \le \lambda \le \pi.$$

Remark 1. The last result is a special case of the more general result that $|\mathscr{K}_{12}(\lambda)|^2 = 1$, $-\pi \le \lambda \le \pi$, whenever $\{X_{t1}\}$ and $\{X_{t2}\}$ are related by a time-invariant linear filter. Thus if

$$X_{t2} = \sum_{j=-\infty}^{\infty} \psi_j X_{t-j,1}$$

where $\sum_j |\psi_j| < \infty$, then by Theorem 4.10.1,

$$X_{t2} = \int_{(-\pi,\pi]} \left(\sum_j \psi_j e^{-ij\lambda}\right) e^{it\lambda}\, dZ_1(\lambda).$$

Hence $dZ_2(\lambda) = \sum_j \psi_j e^{-ij\lambda}\, dZ_1(\lambda)$, $-\pi \le \lambda \le \pi$. Since $dZ_2(\lambda)$ and $dZ_1(\lambda)$ are linearly related for all λ, the squared absolute correlation between $dZ_1(\lambda)$ and $dZ_2(\lambda)$, i.e. $|\mathscr{K}_{12}(\lambda)|^2$, is 1 for all λ. This result can also be obtained by observing that

$$\begin{aligned} E(X_{t+h,2}\bar{X}_{t1}) &= E\left[\int_{(-\pi,\pi]} \sum_j \psi_j e^{i(t+h-j)\lambda}\, dZ_1(\lambda) \overline{\int_{(-\pi,\pi]} e^{it\lambda}\, dZ_1(\lambda)}\right] \\ &= \int_{-\pi}^{\pi} \left(\sum_j \psi_j e^{-ij\lambda}\right) e^{ih\lambda} f_{11}(\lambda)\, d\lambda, \end{aligned}$$

whence

$$f_{21}(\lambda) = \sum_j \psi_j e^{-ij\lambda} f_{11}(\lambda).$$

Substituting in (11.6.10) and using the fact that $f_{22}(\lambda) = |\sum_j \psi_j e^{-ij\lambda}|^2 f_{11}(\lambda)$, we obtain the same result, i.e. $|\mathscr{K}_{12}(\lambda)|^2 = 1$, $-\pi \le \lambda \le \pi$.

Remark 2. If $\{X_{t1}\}$ and $\{X_{t2}\}$ have squared coherency $|\mathscr{K}_{12}(\lambda)|^2$ and if linear filters are applied to each process giving

$$Y_{t1} = \sum_{j=-\infty}^{\infty} \alpha_j X_{t-j,1}$$

and

$$Y_{t2} = \sum_{j=-\infty}^{\infty} \beta_j X_{t-j,2}$$

where $\sum_j |\alpha_j| < \infty$ and $\sum_j |\beta_j| < \infty$, then $\{Y_{t1}\}$ and $\{Y_{t2}\}$ have the same squared coherency $|\mathscr{K}_{12}(\lambda)|^2$. This can be seen by considering the spectral representations $X_{tk} = \int_{(-\pi,\pi]} e^{it\nu}\, dZ_k(\nu)$, $Y_{tk} = \int_{(-\pi,\pi]} e^{it\nu}\, dZ_{Y_k}(\nu)$, and observing, from Theorem 4.10.1, that

$$dZ_{Y_1}(\nu) = \sum_j \alpha_j e^{-ij\nu}\, dZ_1(\nu)$$

and

$$dZ_{Y_2}(\nu) = \sum_j \beta_j e^{-ij\nu}\, dZ_2(\nu).$$

From these linear relations it follows at once that the correlation between $dZ_{Y_1}(\nu)$ and $dZ_{Y_2}(\nu)$ is the same as that between $dZ_1(\nu)$ and $dZ_2(\nu)$.

Remark 3. Let $\{\mathbf{X}_t\}$ be a bivariate stationary series and consider the problem of finding a time-invariant linear filter $\Psi = \{\psi_j\}$ which minimizes $E|X_{t2} - \sum_{j=-\infty}^{\infty} \psi_j X_{t-j,1}|^2$. If Ψ is any time-invariant linear filter with transfer function

$$\psi(e^{-i\nu}) = \sum_{j=-\infty}^{\infty} \psi_j e^{-ij\nu}, \qquad \int_{-\pi}^{\pi} |\psi(e^{-i\nu})|^2 f_{11}(\nu)\, d\nu < \infty,$$

then using (11.6.6) and (11.6.9) we can write

$$\begin{aligned} E\left|X_{t2} - \sum_{j=-\infty}^{\infty} \psi_j X_{t-j,1}\right|^2 &= E\left|\int_{-\pi}^{\pi} e^{it\nu}\, dZ_2(\nu) - \int_{-\pi}^{\pi} \psi(e^{-i\nu}) e^{it\nu}\, dZ_1(\nu)\right|^2 \\ &= \int_{-\pi}^{\pi} [f_{22}(\nu) - \psi(e^{-i\nu}) f_{12}(\nu) - \overline{\psi(e^{-i\nu}) f_{12}(\nu)} \\ &\quad + |\psi(e^{-i\nu})|^2 f_{11}(\nu)]\, d\nu \\ &= \int_{-\pi}^{\pi} E|dZ_2(\nu) - \psi(e^{-i\nu})\, dZ_1(\nu)|^2. \end{aligned}$$

It is easy to check (Problem 11.13) that the integrand is minimized for each ν if

$$\psi(e^{-i\nu}) = \frac{E(dZ_2(\nu)\overline{dZ_1(\nu)})}{E|dZ_1(\nu)|^2} = f_{21}(\nu)/f_{11}(\nu), \tag{11.6.11}$$

and the spectral density of $\sum_j \psi_j X_{t-j,1}$ is then $f_{2|1}(\nu) = |f_{21}(\nu)|^2/f_{11}(\nu)$. The density $f_{2|1}$ is thus the spectral density of the linearly filtered version of $\{X_{t1}\}$ which is the best mean square approximation to $\{X_{t2}\}$. We also observe that

$$|\mathscr{K}_{12}(\lambda)|^2 = \frac{f_{2|1}(\nu)}{f_{22}(\nu)}, \tag{11.6.12}$$

so that $|\mathscr{K}_{12}(\lambda)|^2$ can be interpreted as the proportion of the variance of $\{X_{t2}\}$ at frequency ν which can be attributed to a linear relationship between $\{X_{t2}\}$ and $\{X_{t1}\}$.

Remark 4. If $\{X_{t1}\}$ and $\{X_{t2}\}$ are uncorrelated, then by Definition 11.6.1, $f_{12}(\nu) = 0$, $-\pi \le \nu \le \pi$, from which it follows that $|\mathscr{K}_{12}(\lambda)|^2 = 0$, $-\pi \le \nu \le \pi$.

EXAMPLE 11.6.2. Consider the bivariate series defined by

$$\begin{aligned} X_{t1} &= Z_{t1}, \\ X_{t2} &= \phi X_{t-d,1} + Z_{t2}, \end{aligned}$$

where $\phi > 0$ and $\{Z_t\} \sim \mathrm{WN}(\mathbf{0}, \sigma^2 I)$. The cross covariance between $\{X_{t1}\}$ and $\{X_{t2}\}$ is

$$\gamma_{12}(h) = \begin{cases} \phi\sigma^2 & \text{if } h = -d, \\ 0 & \text{otherwise,} \end{cases}$$

and the cross spectrum is therefore

$$f_{12}(\lambda) = (2\pi)^{-1}\phi\sigma^2 e^{id\lambda}.$$

The amplitude and phase spectra are clearly

$$\alpha_{12}(\lambda) = (2\pi)^{-1}\phi\sigma^2$$

and

$$\phi_{12}(\lambda) = (d\lambda + \pi)\mathrm{mod}(2\pi) - \pi.$$

(The constraint $-\pi < \phi_{12}(\lambda) \le \pi$ means that the graph of $\phi_{12}(\lambda)$, $-\pi < \lambda \le \pi$, instead of being a straight line through the origin with slope d, consists of $2r + 1$ parallel lines, where r is the largest integer less than $(d + 1)/2$. Each line has slope d and one of them passes through the origin.) Since $f_{11}(\lambda) = \sigma^2/(2\pi)$ and $f_{22}(\lambda) = \sigma^2(1 + \phi^2)/2\pi$, the squared coherency is

$$|\mathscr{K}_{12}(\lambda)|^2 = |f_{12}(\lambda)|^2/[f_{11}(\lambda)f_{22}(\lambda)] = \phi^2/(1 + \phi^2), \qquad -\pi \le \lambda \le \pi.$$

Remark 5. In the preceding example the series $\{X_{t2}\}$ is a lagged multiple of $\{X_{t1}\}$ with added uncorrelated noise. The lag is precisely the slope of the phase spectrum ϕ_{12}. In general of course the phase spectrum will not be piecewise linear with constant slope, however $\phi_{12}(\lambda)$ can still be regarded as a measure of the phase lag of $\{X_{t2}\}$ behind $\{X_{t1}\}$ at frequency λ in the sense that

$$f_{12}(\lambda)\,d\lambda = \alpha_{12}(\lambda)e^{i\phi_{12}(\lambda)}\,d\lambda = E[|dZ_1(\lambda)|\,|dZ_2(\lambda)|\,e^{i(\Theta_1(\lambda) - \Theta_2(\lambda))}],$$

where $\Theta_i(\lambda) = \arg(dZ_i(\lambda))$, $i = 1, 2$. We say that X_{t2} lags d time units behind X_{t1} at frequency λ if $\exp(it\lambda)\,dZ_2(\lambda) = \exp(i(t - d)\lambda)\,dZ_1(\lambda)$. We can then write

$$f_{12}(\lambda)\,d\lambda = \mathrm{Cov}(dZ_1(\lambda), \exp(-id\lambda)\,dZ_1(\lambda)) = \exp(id\lambda)f_{11}(\lambda)\,d\lambda.$$

Hence $\phi_{12}(\lambda) = \arg(f_{12}(\lambda)) = (d\lambda + \pi)\mathrm{mod}(2\pi)$ and $\phi'_{12}(\lambda) = d$. In view of its interpretation as a time lag, $\phi'_{12}(\lambda)$ is known as the *group delay* at frequency λ.

EXAMPLE 11.6.3 (An Econometrics Model). The mean corrected price and supply of a commodity at time t are sometimes represented by X_{t1} and X_{t2} respectively, where

$$\begin{cases} X_{t1} = -\phi_1 X_{t2} + U_t, & 0 < \phi_1 < 1, \\ X_{t2} = \phi_2 X_{t-1,1} + V_t, & 0 < \phi_2 < 1, \end{cases} \tag{11.6.13}$$

where $\{U_t\} \sim \mathrm{WN}(0, \sigma_U^2)$, $\{V_t\} \sim \mathrm{WN}(0, \sigma_V^2)$ and $\{U_t\}$, $\{V_t\}$ are uncorrelated. We now replace each term in these equations by its spectral representation. Noting that the resulting equations are valid for all t, we obtain the following equations for the orthogonal increment processes Z_1, Z_2, Z_U and Z_V in the spectral representations of $\{X_{t1}\}$, $\{X_{t2}\}$, $\{U_t\}$ and $\{V_t\}$:

$$dZ_1(\lambda) = -\phi_1\, dZ_2(\lambda) + dZ_U(\lambda),$$

and

$$dZ_2(\lambda) = \phi_2 e^{-i\lambda}\, dZ_1(\lambda) + dZ_V(\lambda).$$

Solving for $dZ_1(\lambda)$ and $dZ_2(\lambda)$, we obtain

$$dZ_1(\lambda) = (1 + \phi_1\phi_2 e^{-i\lambda})^{-1}[-\phi_1\, dZ_V(\lambda) + dZ_U(\lambda)]$$

and

$$dZ_2(\lambda) = (1 + \phi_1\phi_2 e^{-i\lambda})^{-1}[dZ_V(\lambda) + \phi_2 e^{-i\lambda}\, dZ_U(\lambda)].$$

From (11.6.8) and (11.6.9) it follows that

$$f_{11}(\lambda) = |1 + \phi_1\phi_2 e^{-i\lambda}|^{-2}(\sigma_U^2 + \phi_1^2\sigma_V^2)/(2\pi),$$

$$f_{22}(\lambda) = |1 + \phi_1\phi_2 e^{-i\lambda}|^{-2}(\sigma_V^2 + \phi_2^2\sigma_U^2)/(2\pi),$$

and

$$f_{12}(\lambda) = |1 + \phi_1\phi_2 e^{-i\lambda}|^{-2}(\phi_2\sigma_U^2 \cos\lambda - \phi_1\sigma_V^2 + i\phi_2\sigma_U^2 \sin\lambda)/(2\pi).$$

The squared coherency is therefore, by (11.6.10),

$$|\mathscr{K}_{12}(\lambda)|^2 = \frac{\phi_2^2\sigma_U^4 + \phi_1^2\sigma_V^4 - 2\phi_1\phi_2\sigma_U^2\sigma_V^2\cos\lambda}{\phi_2^2\sigma_U^4 + \phi_1^2\sigma_V^4 + (1 + \phi_1^2\phi_2^2)\sigma_U^2\sigma_V^2},$$

and

$$\tan\phi_{12}(\lambda) = \phi_2\sigma_U^2 \sin\lambda/(\phi_2\sigma_U^2\cos\lambda - \phi_1\sigma_V^2).$$

Notice that the squared coherency is largest at high frequencies. This suggests that the linear relationship between price and supply is strongest at high frequencies. Notice also that for λ close to π,

$$\phi_{12}'(\lambda) \sim \frac{(\phi_2\sigma_U^2\cos\lambda - \phi_1\sigma_V^2)\phi_2\sigma_U^2\cos\lambda}{(\phi_2\sigma_U^2\cos\lambda - \phi_1\sigma_V^2)^2}$$

$$\sim \frac{\phi_1\phi_2\sigma_U^2\sigma_V^2 + \phi_2^2\sigma_U^4}{(\phi_1\sigma_V^2 + \phi_2\sigma_U^2)^2} > 0,$$

indicating that price leads supply at high frequencies as might be expected. In the special case $\phi_1 = 0$, we recover the model of Example 11.6.2 with $d = 1$, for which $\phi_{12}(\lambda) = (\lambda + \pi)\text{mod}(2\pi) - \pi$ and $\phi_{12}'(\lambda) = 1$.

Example 11.6.4 (Linear Filtering with Added Uncorrelated Noise). Suppose that $\Psi = \{\psi_j, j = 0, \pm 1, \ldots\}$ is an absolutely summable time-invariant linear filter and that $\{X_{t1}\}$ is a zero-mean stationary process with spectral density $f_{11}(\lambda)$. Let $\{N_t\}$ be a zero-mean stationary process uncorrelated with $\{X_{t1}\}$ and with spectral density $f_N(\lambda)$. We then define the filtered process with added noise,

$$X_{t2} = \sum_{j=-\infty}^{\infty} \psi_j X_{t-j,1} + N_t. \tag{11.6.14}$$

Since $\{X_{t1}\}$ and $\{N_t\}$ are uncorrelated, the spectral density of $\{X_{t2}\}$ is

$$f_{22}(\lambda) = |\psi(e^{-i\lambda})|^2 f_{11}(\lambda) + f_N(\lambda), \tag{11.6.15}$$

where $\psi(e^{-i\lambda}) = \sum_{j=-\infty}^{\infty} \psi_j e^{-ij\lambda}$. Corresponding to (11.6.14) we can also write

$$dZ_2(\lambda) = \psi(e^{-i\lambda})\,dZ_1(\lambda) + dZ_N(\lambda), \tag{11.6.16}$$

where Z_2, Z_1 and Z_N are the orthogonal increment processes in the spectral representations of $\{X_{t2}\}$, $\{X_{t1}\}$ and $\{N_t\}$. From (11.6.16),

$$E(dZ_2(\lambda)\overline{dZ_1(\lambda)}) = \psi(e^{-i\lambda}) f_{11}(\lambda)\,d\lambda$$

and hence

$$f_{21}(\lambda) = \psi(e^{-i\lambda}) f_{11}(\lambda).$$

The amplitude spectrum is

$$\alpha_{21}(\lambda) = |\psi(e^{-i\lambda})| f_{11}(\lambda),$$

and since f_{11} is real-valued, the phase spectrum coincides with the phase gain of the filter, i.e.

$$\phi_{21}(\lambda) = \arg(\psi(e^{-i\lambda})).$$

In the case of a simple delay filter with lag d, i.e. $\psi_j = 0$, $j \neq d$, $\phi_{21}(\lambda) = \arg(e^{-id\lambda}) = (-d\lambda + \pi)\text{mod}(2\pi) - \pi$, indicating that $\{X_{t1}\}$ leads $\{X_{t2}\}$ by d as expected.

The transfer function $\psi(e^{-i\cdot})$ of the filter, and hence the weights $\{\psi_j\}$, can be found from the relation

$$\psi(e^{-i\lambda}) = f_{21}(\lambda)/f_{11}(\lambda) = \alpha_{21}(\lambda)\exp[i\phi_{21}(\lambda)]/f_{11}(\lambda), \tag{11.6.17}$$

quite independently of the noise sequence $\{N_t\}$. From (11.6.15) and (11.6.17) we also have the relation,

$$\begin{aligned} f_{22}(\lambda) &= |f_{21}(\lambda)|^2/f_{11}(\lambda) + f_N(\lambda) \\ &= |\mathscr{K}_{21}(\lambda)|^2 f_{22}(\lambda) + f_N(\lambda), \end{aligned}$$

where $|\mathscr{K}_{21}(\lambda)|^2$ is the squared coherency between $\{X_{t2}\}$ and $\{X_{t1}\}$. Hence

$$f_N(\lambda) = (1 - |\mathscr{K}_{21}(\lambda)|^2) f_{22}(\lambda), \tag{11.6.18}$$

and by integrating both sides, we obtain

$$\sigma_N^2 := \text{Var}(N_t) = \int_{-\pi}^{\pi} (1 - |\mathscr{K}_{21}(\lambda)|^2) f_{22}(\lambda)\,d\lambda.$$

In the next section we discuss the estimation of $f_{11}(\lambda)$, $f_{22}(\lambda)$ and $f_{12}(\lambda)$ from n pairs of observations, $(X_{t1}, X_{t2})'$, $t = 1, \ldots, n$. For the model (11.6.14), these estimates can then be used in the equations (11.6.17) and (11.6.18) to estimate the transfer function of the filter and the spectral density of the noise sequence $\{N_t\}$.

§11.7 Estimating the Cross Spectrum

Let $\{\mathbf{X}_t\}$ be a stationary bivariate time series with $E\mathbf{X}_t = \boldsymbol{\mu}$ and $E(\mathbf{X}_{t+h}\mathbf{X}_t') - \boldsymbol{\mu}\boldsymbol{\mu}' = \Gamma(h)$, where the covariance matrices $\Gamma(h)$ have absolutely summable components. The spectral density matrix function of $\{\mathbf{X}_t\}$ is defined by

$$f(\lambda) = \begin{bmatrix} f_{11}(\lambda) & f_{12}(\lambda) \\ f_{21}(\lambda) & f_{22}(\lambda) \end{bmatrix} = (2\pi)^{-1} \sum_{h=-\infty}^{\infty} \Gamma(h) e^{-ih\lambda}, \qquad -\pi \le \lambda \le \pi.$$

In this section we shall consider estimation of $f(\lambda)$ by smoothing the multivariate periodogram of $\{\mathbf{X}_1, \ldots, \mathbf{X}_n\}$. First we derive bivariate analogues of the asymptotic results of Sections 10.3 and 10.4. We then discuss inference for the squared coherency, the amplitude spectrum and the phase spectrum which were defined in Section 11.6.

The discrete Fourier transform of $\{\mathbf{X}_1, \ldots, \mathbf{X}_n\}$ is defined by

$$\mathbf{J}(\omega_j) = n^{-1/2} \sum_{t=1}^{n} \mathbf{X}_t e^{-it\omega_j},$$

where $\omega_j = 2\pi j/n$, $-[(n-1)/2] \le j \le [n/2]$, are the n Fourier frequencies introduced in Section 10.1. The periodogram of $\{\mathbf{X}_1, \ldots, \mathbf{X}_n\}$ is defined at each of these frequencies ω_j to be the 2×2 matrix,

$$I_n(\omega_j) = \mathbf{J}(\omega_j)\mathbf{J}^*(\omega_j),$$

where * denotes complex conjugate transpose. As in Section 10.3 the definition is extended to all frequencies $\omega \in [-\pi, \pi]$ by setting

$$I_n(\omega) = \begin{cases} I_n(g(n,\omega)) & \text{if } \omega \ge 0, \\ I_n^*(g(n,-\omega)) & \text{if } \omega < 0, \end{cases} \tag{11.7.1}$$

where $g(n,\omega)$, $0 \le \omega \le \pi$, is the multiple of $2\pi/n$ closest to ω (the smaller one if there are two). We shall suppress the subscript n and write $I_{ij}(\omega)$, $i, j = 1, \ldots, n$, for the components of $I_n(\omega)$. Observe that $I_{ii}(\omega)$ is the periodogram of the univariate observations $\{X_{1i}, \ldots, X_{ni}\}$. The function $I_{12}(\omega)$ is called the cross periodogram. At the Fourier frequency ω_k it has the value,

$$I_{12}(\omega_k) = n^{-1}\left(\sum_{t=1}^{n} X_{t1} e^{-it\omega_k}\right)\left(\sum_{t=1}^{n} X_{t2} e^{it\omega_k}\right).$$

Asymptotic Properties of the Periodogram

Since the next two propositions are straightforward extensions of Propositions 10.1.2 and 10.3.1, the proofs are left to the reader.

Proposition 11.7.1. *If ω_j is any non-zero Fourier frequency and $\bar{\mathbf{X}}_n = n^{-1}\sum_{t=1}^{n} \mathbf{X}_t$, then*

$$I_n(\omega_j) = \sum_{|k|<n} \hat{\Gamma}(k) e^{-ik\omega_j}$$

where $\hat{\Gamma}(k) = n^{-1}\sum_{t=1}^{n-k}(\mathbf{X}_{t+k} - \bar{\mathbf{X}}_n)(\mathbf{X}_t - \bar{\mathbf{X}}_n)', k \geq 0,$ *and* $\hat{\Gamma}(k) = \hat{\Gamma}'(-k), k < 0.$ *The periodogram at frequency zero is*

$$I_n(0) = n\bar{\mathbf{X}}_n\bar{\mathbf{X}}_n'.$$

Proposition 11.7.2. *If* $\{\mathbf{X}_t\}$ *is a stationary bivariate time series with mean* $\boldsymbol{\mu}$ *and covariance matrices* $\Gamma(h)$ *having absolutely summable components, then*

$$\text{(i)}\ \ EI_n(0) - n\boldsymbol{\mu}\boldsymbol{\mu}' \to 2\pi f(0)$$

and

$$\text{(ii)}\ \ EI_n(\omega) \to 2\pi f(\omega), \qquad \text{if } \omega \neq 0$$

where $f(\cdot)$ *is the spectral matrix function of* $\{\mathbf{X}_t\}$.

We now turn to the asymptotic distribution and asymptotic covariances of the periodogram values of a linear process. In order to describe the asymptotic distribution it is convenient first to define the complex multivariate normal distribution.

Definition 11.7.1 (The Complex Multivariate Normal Distribution). If $\Sigma = \Sigma_1 + i\Sigma_2$ is a complex-valued $m \times m$ matrix such that $\Sigma = \Sigma^*$ and $\mathbf{a}^*\Sigma\mathbf{a} \geq 0$ for all $\mathbf{a} \in \mathbb{C}^m$, then we say that $\mathbf{Y} = \mathbf{Y}_1 + i\mathbf{Y}_2$ is a complex-valued multivariate normal random vector with mean $\boldsymbol{\mu} = \boldsymbol{\mu}_1 + i\boldsymbol{\mu}_2$ and covariance matrix Σ if

$$\begin{bmatrix}\mathbf{Y}_1\\ \mathbf{Y}_2\end{bmatrix} \sim \mathrm{N}\left(\begin{bmatrix}\boldsymbol{\mu}_1\\ \boldsymbol{\mu}_2\end{bmatrix}, \frac{1}{2}\begin{bmatrix}\Sigma_1 & -\Sigma_2\\ \Sigma_2 & \Sigma_1\end{bmatrix}\right). \tag{11.7.2}$$

We then write $\mathbf{Y} \sim \mathrm{N_c}(\boldsymbol{\mu}, \Sigma)$. If $\mathbf{Y}^{(n)} = \mathbf{Y}_1^{(n)} + i\mathbf{Y}_2^{(n)}, n = 1, 2, \ldots,$ we say that $\mathbf{Y}^{(n)}$ is $\mathrm{AN_c}(\boldsymbol{\mu}^{(n)}, \Sigma^{(n)})$ if $\begin{bmatrix}\mathbf{Y}_1^{(n)}\\ \mathbf{Y}_2^{(n)}\end{bmatrix}$ is $\mathrm{AN}\left(\begin{bmatrix}\boldsymbol{\mu}_1^{(n)}\\ \boldsymbol{\mu}_2^{(n)}\end{bmatrix}, \frac{1}{2}\begin{bmatrix}\Sigma_1^{(n)} & -\Sigma_2^{(n)}\\ \Sigma_2^{(n)} & \Sigma_1^{(n)}\end{bmatrix}\right)$, where each $\Sigma^{(n)} = \Sigma_1^{(n)} + i\Sigma_2^{(n)}$ satisfies the conditions imposed on Σ. These guarantee (Problem 11.16) that the matrix in (11.7.2) is a real covariance matrix.

Proposition 11.7.3. *Suppose that* $\{\mathbf{Z}_t\} \sim \mathrm{IID}(\mathbf{0}, \boldsymbol{\Sigma})$ *where* $\boldsymbol{\Sigma}$ *is non-singular, and let* $I_n(\omega), -\pi \leq \omega \leq \pi$, *denote the periodogram of* $\{\mathbf{Z}_1, \ldots, \mathbf{Z}_n\}$ *as defined by* (11.7.1).

(i) *If* $0 < \lambda_1 < \cdots < \lambda_m < \pi$ *then the matrices* $I_n(\lambda_1), \ldots, I_n(\lambda_m)$ *converge jointly in distribution as* $n \to \infty$ *to independent random matrices, each distributed as* $\mathbf{Y}_k\mathbf{Y}_k^*$ *where* $\mathbf{Y}_k \sim \mathrm{N_c}(\mathbf{0}, \boldsymbol{\Sigma})$.

(ii) *If* $EZ_{ti}^4 < \infty, i = 1, 2,$ *and* ω_j, ω_k *are Fourier frequencies in* $[0, \pi]$, *then*

$$\mathrm{Cov}[I_{pq}(\omega_j), I_{rs}(\omega_k)] = \begin{cases}\sigma_{pr}\sigma_{sq} + \kappa_{pqrs}/n, & 0 < \omega_j = \omega_k < \pi,\\ (\sigma_{pr}\sigma_{sq} + \sigma_{ps}\sigma_{qr}) + \kappa_{pqrs}/n, & \omega_j = \omega_k = 0 \text{ or } \pi,\\ \kappa_{pqrs}/n, & \omega_j \neq \omega_k,\end{cases}$$

where $I_{pq}(\cdot)$ *is the* (p,q)*-element of* $I_n(\cdot)$, σ_{pq} *is the* (p,q)*-element of* $\boldsymbol{\Sigma}$, *and* κ_{pqrs} *is the fourth cumulant between* Z_{tp}, Z_{tq}, Z_{tr} *and* Z_{ts}. *(See Hannan* (1970), *p.* 23.)

Proof. (i) For an arbitrary frequency $\lambda \in (0, \pi)$ define

$$\mathbf{J}(\lambda) = n^{-1/2} \sum_{t=1}^{n} \mathbf{Z}_t e^{-itg(n,\lambda)}.$$

We first show that $\mathbf{J}(\lambda)$ is $\mathrm{AN}_c(\mathbf{0}, \boldsymbol{\Sigma})$ (see Definition 11.7.1). We can rewrite $\mathbf{J}(\lambda)$ as

$$\mathbf{J}(\lambda) = n^{-1/2} \sum_{t=1}^{n} [\mathbf{Z}_t \cos(tg(n,\lambda)) - i\mathbf{Z}_t \sin(tg(n,\lambda))].$$

Now the four-dimensional random vector,

$$\mathbf{U}_n := n^{-1/2} \sum_{t=1}^{n} \begin{bmatrix} \mathbf{Z}_t \cos(tg(n,\lambda)) \\ \mathbf{Z}_t \sin(tg(n,\lambda)) \end{bmatrix},$$

is a sum of independent random vectors and for $g(n,\lambda) \in (0, \pi)$ we can write (see Problem 11.17)

$$E(\mathbf{U}_n \mathbf{U}_n') = \frac{1}{2} \begin{bmatrix} \boldsymbol{\Sigma} & 0 \\ 0 & \boldsymbol{\Sigma} \end{bmatrix}. \tag{11.7.3}$$

Applying the Cramer–Wold device and the Lindeberg condition as in the proof of Proposition 10.3.2, we find that

$$\mathbf{U}_n \text{ is AN}\left(\mathbf{0}, \frac{1}{2} \begin{bmatrix} \boldsymbol{\Sigma} & 0 \\ 0 & \boldsymbol{\Sigma} \end{bmatrix}\right).$$

This is equivalent, by Definition 11.7.1, to the statement that

$$\mathbf{J}(\lambda) \text{ is } \mathrm{AN}_c(\mathbf{0}, \boldsymbol{\Sigma}).$$

(Note that a complex normal random vector with real covariance matrix Σ has uncorrelated real and imaginary parts each with covariance matrix $\Sigma/2$.) It then follows by Proposition 6.3.4 that $I_n(\lambda) \Rightarrow \mathbf{YY}^*$ where $\mathbf{Y} \sim N(\mathbf{0}, \boldsymbol{\Sigma})$.

For $\omega \neq \lambda$, a computation analogous to the one giving (11.7.3) yields

$$E[\mathbf{J}(\lambda)\mathbf{J}^*(\omega)] = 0$$

for all n sufficiently large. Since $\mathbf{J}(\lambda)$ and $\mathbf{J}(\omega)$ are asymptotically joint normal, it follows that they are asymptotically independent. Extending this argument to the distinct frequencies $0 < \lambda_1 < \cdots < \lambda_m < \pi$, we find that $J(\lambda_1), \ldots, J(\lambda_m)$ and hence $I_n(\lambda_1), \ldots, I_n(\lambda_m)$ are asymptotically independent.

(ii) The proof is essentially the same as that of Proposition 10.3.2 and is therefore omitted. (See also Hannan (1970), p. 249.) □

As in Section 10.3, a corresponding result (Theorem 11.7.1) holds also for linear processes. Before stating it we shall relate the periodogram of a linear process to the periodogram of the underlying white noise sequence.

Proposition 11.7.4. *Let $\{\mathbf{X}_t\}$ be the linear process,*

$$\mathbf{X}_t = \sum_{k=-\infty}^{\infty} C_k \mathbf{Z}_{t-k}, \qquad \{\mathbf{Z}_t\} \sim \mathrm{IID}(0, \mathbf{\Sigma}), \tag{11.7.4}$$

where $\mathbf{\Sigma}$ *is non-singular and the components of the matrices* C_k *satisfy* $\sum_{k=-\infty}^{\infty} |C_k(i,j)|\,|k|^{1/2} < \infty, i,j = 1, 2$. *Let* $I_{n,X}(\cdot)$ *and* $I_{n,Z}(\cdot)$ *be the periodograms of* $\{\mathbf{X}_1, \ldots, \mathbf{X}_n\}$ *and* $\{\mathbf{Z}_1, \ldots, \mathbf{Z}_n\}$ *respectively. If* $EZ_{ti}^4 < \infty$, $i = 1, 2$, *and* $C(e^{-i\omega}) := \sum_{k=-\infty}^{\infty} C_k e^{-ik\omega}$, *then for each Fourier frequency* $\omega_k \in [0, \pi]$,

$$I_{n,X}(\omega_k) = C(e^{-i\omega_k}) I_{n,Z}(\omega_k) C'(e^{i\omega_k}) + R_n(\omega_k),$$

where the components of $R_n(\omega_k)$ *satisfy*

$$\max_{\omega_k \in [0,\pi]} E|R_{n,ij}(\omega_k)|^2 = O(n^{-1}), \qquad i,j = 1, 2.$$

PROOF. The argument follows that in the proof of Theorem 10.3.1. (See also Hannan (1970), p. 248.) □

Theorem 11.7.1. *Let* $\{\mathbf{X}_t\}$ *be the linear process defined by* (11.7.4) *with periodogram* $I_n(\lambda) = [I_{ij}(\lambda)]_{i,j=1}^2$, $-\pi \le \lambda \le \pi$.

(i) *If* $0 < \lambda_1 < \cdots < \lambda_m < \pi$ *then the matrices* $I_n(\lambda_1), \ldots, I_n(\lambda_m)$ *converge jointly in distribution as* $n \to \infty$ *to independent random matrices, the* k^{th} *of which is distributed as* $\mathbf{W}_k \mathbf{W}_k^*$ *where* $\mathbf{W}_k \sim N_c(\mathbf{0}, 2\pi f(\lambda_k))$ *and* f *is the spectral density matrix of* $\{\mathbf{X}_t\}$.

(ii) *If* $\omega_j = 2\pi j/n \in [0, \pi]$ *and* $\omega_k = 2\pi k/n \in [0, \pi]$, *then*

$$\mathrm{Cov}(I_{pq}(\omega_j), I_{rs}(\omega_k)) = \begin{cases} (2\pi)^2[f_{pr}(\omega_j)f_{sq}(\omega_j) + f_{ps}(\omega_j)f_{qr}(\omega_j)] + O(n^{-1/2}) & \text{if } \omega_j = \omega_k = 0 \text{ or } \pi, \\ (2\pi)^2 f_{pr}(\omega_j)f_{sq}(\omega_j) + O(n^{-1/2}) & \text{if } 0 < \omega_j = \omega_k < \pi, \\ O(n^{-1}) & \text{if } \omega_j \ne \omega_k, \end{cases}$$

where the terms $O(n^{-1/2})$ and $O(n^{-1})$ *can be bounded uniformly in* j *and* k *by* $c_1 n^{-1/2}$ *and* $c_2 n^{-1}$ *respectively for some positive constants* c_1 *and* c_2.

PROOF. The proof is left to the reader. (See the proof of Theorem 10.3.2 and Hannan (1970), pp. 224 and 249.) □

Smoothing the Periodogram

As in Section 10.4, a consistent estimator of the spectral matrix of the linear process (11.7.4) can be obtained by smoothing the periodogram. Let $\{m_n\}$ and $\{W_n(\cdot)\}$ be sequences of integers and (scalar) weight functions respectively, satisfying conditions (10.4.2)–(10.4.5). We define the discrete spectral average estimator $\hat{f}$ by

$$\hat{f}(\omega) := (2\pi)^{-1} \sum_{|k| \le m_n} W_n(k) I_n(g(n, \omega) + \omega_k), \qquad 0 \le \omega \le \pi. \tag{11.7.5}$$

In order to evaluate $\hat{f}(\omega)$, $0 \le \omega \le \pi$, we define I_n to have period 2π and replace $I_n(0)$ whenever it appears in (11.7.5) by

$$\hat{f}(0) := (2\pi)^{-1} \operatorname{Re}\left\{W_n(0)I_n(\omega_1) + 2\sum_{k=1}^{m} W_n(k)I_n(\omega_{k+1})\right\}.$$

We have applied the same weight function to all four components of $I_n(\omega)$ in order to facilitate the statement and derivation of the properties of $\hat{f}(\omega)$. It is frequently advantageous however to choose a different weight-function sequence for each component of $I_n(\cdot)$ since the components may have quite diverse characteristics. For a discussion of choosing weight functions to match the characteristics of $I_n(\cdot)$ see Chapter 9 of Priestley (1981).

The following theorem asserts the consistency of the estimator $\hat{f}(\omega)$. It is a simple consequence of Theorem 11.7.1.

Theorem 11.7.2. *If* $\{\mathbf{X}_t\}$ *is the linear process defined by* (11.7.4) *and* $\hat{f}(\omega) = [\hat{f}_{ij}(\omega)]_{i,j=1}^2$ *is the discrete spectral average estimator defined by* (11.7.5), *then for* $\lambda, \omega \in [0, \pi]$,

(a) $\lim_{n\to\infty} E\hat{f}(\omega) = f(\omega)$

and

(b) $$\lim_{n\to\infty} \frac{\operatorname{Cov}(\hat{f}_{pq}(\omega), \hat{f}_{rs}(\lambda))}{\sum_{|k|\le m} W_n^2(k)} = \begin{cases} f_{pr}(\omega)f_{sq}(\omega) + f_{ps}(\omega)f_{qr}(\omega) & \text{if } \omega = \lambda = 0 \text{ or } \pi, \\ f_{pr}(\omega)f_{sq}(\omega) & \text{if } 0 < \omega = \lambda < \pi, \\ 0 & \text{if } \omega \ne \lambda. \end{cases}$$

(*Recall that if X and Y are complex-valued,* $\operatorname{Cov}(X, Y) = E(X\bar{Y}) - (EX)(\overline{EY})$.)

The cospectrum $c_{12}(\omega) = [f_{12}(\omega) + f_{21}(\omega)]/2$ and the quadrature spectrum $q_{12}(\omega) = i[f_{12}(\omega) - f_{21}(\omega)]/2$ will be estimated by

$$\hat{c}_{12}(\omega) = [\hat{f}_{12}(\omega) + \hat{f}_{21}(\omega)]/2$$

and

$$\hat{q}_{12}(\omega) = i[\hat{f}_{12}(\omega) - \hat{f}_{21}(\omega)]/2$$

respectively. By Theorem 11.7.2(b) we find, under the conditions specified, that the real-valued random vector $(\sum_{|k|\le m} W_n^2(k))^{-1}(\hat{f}_{11}(\omega), \hat{f}_{22}(\omega), \hat{c}_{12}(\omega), \hat{q}_{12}(\omega))'$, $0 < \omega < \pi$, has asymptotic covariance matrix,

$$V = \begin{bmatrix} f_{11}^2 & |f_{12}|^2 & f_{11}c_{12} & f_{11}q_{12} \\ |f_{12}|^2 & f_{22}^2 & f_{22}c_{12} & f_{22}q_{12} \\ f_{11}c_{12} & f_{22}c_{12} & \frac{1}{2}(f_{11}f_{22} + c_{12}^2 - q_{12}^2) & c_{12}q_{12} \\ f_{11}q_{12} & f_{22}q_{12} & c_{12}q_{12} & \frac{1}{2}(f_{11}f_{22} + q_{12}^2 - c_{12}^2) \end{bmatrix}, \tag{11.7.6}$$

where the argument ω has been suppressed. Moreover we can express $(\hat{f}_{11}(\omega), \hat{f}_{22}(\omega), \hat{c}_{12}(\omega), \hat{q}_{12}(\omega))'$ as the sum of $(2m + 1)$ random vectors,

$$\begin{bmatrix} \hat{f}_{11}(\omega) \\ \hat{f}_{22}(\omega) \\ \hat{c}_{12}(\omega) \\ \hat{q}_{12}(\omega) \end{bmatrix} = \sum_{|k| \le m} W_n(k) \begin{bmatrix} I_{11}(g(n,\omega) + \omega_k) \\ I_{22}(g(n,\omega) + \omega_k) \\ \mathrm{Re}\{I_{12}(g(n,\omega) + \omega_k)\} \\ -Im\{I_{12}(g(n,\omega) + \omega_k)\} \end{bmatrix},$$

where the summands, by Theorem 11.7.1, are asymptotically independent. This suggests that

$$(\hat{f}_{11}(\omega), \hat{f}_{22}(\omega), \hat{c}_{12}(\omega), \hat{q}_{12}(\omega))' \quad \text{is AN}((f_{11}(\omega), f_{22}(\omega), c_{12}(\omega), q_{12}(\omega))', a_n^2 V) \tag{11.7.7}$$

where $a_n^2 = \sum_{|k| \le m} W_n^2(k)$ and V is defined by (11.7.6). We shall base our statistical inference for the spectrum on the asymptotic distribution (11.7.7). For a proof of (11.7.7) in the case when $\hat{f}(\omega)$ is a lag window spectral estimate, see Hannan (1970), p. 289.

Estimation of the Cross-Amplitude Spectrum

To estimate $\alpha_{12}(\omega) = |f_{12}(\omega)| = |c_{12}(\omega) - iq_{12}(\omega)|$ we shall use

$$\hat{\alpha}_{12}(\omega) := (\hat{c}_{12}^2(\omega) + \hat{q}_{12}^2(\omega))^{1/2} = h(\hat{c}_{12}(\omega), \hat{q}_{12}(\omega)).$$

By (11.7.7) and Proposition 6.4.3 applied to $h(x, y) = (x^2 + y^2)^{1/2}$, we find that if $\alpha_{12}(\omega) > 0$, then

$$\hat{\alpha}_{12}(\omega) \sim \text{AN}(\alpha_{12}(\omega), a_n^2 \sigma_\alpha^2(\omega)),$$

where

$$\sigma_\alpha^2(\omega) = \left(\frac{\partial h}{\partial x}\right)^2 v_{33} + \left(\frac{\partial h}{\partial y}\right)^2 v_{44} + 2\left(\frac{\partial h}{\partial x}\right)\left(\frac{\partial h}{\partial y}\right) v_{34},$$

v_{ij} is the (i, j)-element of the matrix defined by (11.7.6), and the derivatives of h are evaluated at $(c_{12}(\omega), q_{12}(\omega))$. Calculating the derivatives and simplifying, we find that if the squared coherency, $|\mathscr{K}_{12}(\omega)|^2$, is strictly positive then

$$\hat{\alpha}_{12}(\omega) \quad \text{is AN}(\alpha_{12}(\omega),\ a_n^2 \alpha_{12}^2(\omega)(|\mathscr{K}_{12}(\omega)|^{-2} + 1)/2). \tag{11.7.8}$$

Observe that for small values of $|\mathscr{K}_{12}(\omega)|^2$, the asymptotic variance of $\hat{\alpha}_{12}(\omega)$ is large

Estimation of the Phase Spectrum

The phase spectrum $\phi_{12}(\omega) = \arg f_{12}(\omega)$ will be estimated by

$$\hat{\phi}_{12}(\omega) := \arg(\hat{c}_{12}(\omega) - i\hat{q}_{12}(\omega)) \in (-\pi, \pi].$$

If $|\mathscr{K}_{12}(\omega)|^2 > 0$, then by (11.7.7) and Proposition 6.4.3,

$$\hat{\phi}_{12}(\omega) \quad \text{is AN}(\phi_{12}(\omega),\ a_n^2\alpha_{12}^2(\omega)(|\mathscr{K}_{12}(\omega)|^{-2} - 1)/2). \tag{11.7.9}$$

The asymptotic variance of $\hat{\phi}_{12}(\omega)$, like that of $\hat{\alpha}_{12}(\omega)$, is large if $|\mathscr{K}_{12}(\omega)|^2$ is small

In the case when $\mathscr{K}_{12}(\omega) = 0$, both $c_{12}(\omega)$ and $q_{12}(\omega)$ are zero, so from (11.7.7) and (11.7.6)

$$\begin{bmatrix} \hat{c}_{12}(\omega) \\ \hat{q}_{12}(\omega) \end{bmatrix} \quad \text{is AN}\left(\begin{bmatrix} 0 \\ 0 \end{bmatrix}, \tfrac{1}{2}a_n^2 \begin{bmatrix} f_{11}f_{22} & 0 \\ 0 & f_{11}f_{22} \end{bmatrix}\right).$$

As $\hat{\phi}_{12}(\omega) = \arg(\hat{c}_{12}(\omega) - i\hat{q}_{12}(\omega)) = \arg[(a_n^2 f_{11}f_{22}/2)^{-1/2}(\hat{c}_{12}(\omega) - i\hat{q}_{12}(\omega))]$, we conclude from Proposition 6.3.4 that

$$\hat{\phi}_{12}(\omega) \Rightarrow \arg(U_1 + iU_2),$$

where U_1 and U_2 are independent standard normal random variables. Since U_1/U_2 has a Cauchy distribution, it is a routine exercise in distribution theory to show that $\arg(U_1 + iU_2)$ is uniformly distributed on $(-\pi, \pi)$. Hence if n is large and $\mathscr{K}_{12}(\omega) = 0$, $\hat{\phi}_{12}(\omega)$ is approximately uniformly distributed on $(-\pi, \pi)$.

From (11.7.9) we obtain the approximate 95% confidence bounds for $\phi_{12}(\omega)$,

$$\hat{\phi}_{12}(\omega) \pm 1.96 a_n \hat{\alpha}_{12}(\omega)(|\hat{\mathscr{K}}_{12}(\omega)|^{-2} - 1)^{1/2}/2^{1/2},$$

where $|\hat{\mathscr{K}}_{12}(\omega)|^2$ is the estimated squared coherency,

$$|\hat{\mathscr{K}}_{12}(\omega)|^2 = \hat{\alpha}_{12}^2(\omega)/[\hat{f}_{11}(\omega)\hat{f}_{22}(\omega)],$$

and it is assumed that $|\hat{\mathscr{K}}_{12}(\omega)|^2 > 0$.

Hannan (1970), p. 257, discusses an alternative method for constructing a confidence region for $\phi_{12}(\omega)$ in the case when $W_n(k) = (2m + 1)^{-1}$ for $|k| \le m$ and $W(k) = 0$ for $|k| > m$. He shows that if the distribution of the periodogram is replaced by the asymptotic distributions of Theorem 11.7.1, then the event E has probability $(1 - \alpha)$, where

$$E = \left\{|\sin(\hat{\phi}_{12}(\omega) - \phi_{12}(\omega))| \le \left[\frac{1 - |\hat{\mathscr{K}}_{12}(\omega)|^2}{4m|\hat{\mathscr{K}}_{12}(\omega)|^2}\right]^{1/2} t_{1-\alpha/2}(4m)\right\}$$

and $t_{1-\alpha/2}(4m)$ is the $(1 - \alpha/2)$-quantile of the t-distribution with $4m$ degrees of freedom. For given values of $\hat{\phi}_{12}(\omega)$ and $|\hat{\mathscr{K}}_{12}(\omega)|$, the set of $\phi_{12}(\omega)$ values satisfying the inequality which defines E is therefore a $100(1 - \alpha)\%$ confidence region for $\phi_{12}(\omega)$. If the right-hand side of the inequality is greater than or equal to 1 (as will be the case if $|\hat{\mathscr{K}}_{12}(\omega)|^2$ is sufficiently small), then we obtain the uninformative confidence interval $(-\pi, \pi]$ for $\phi_{12}(\omega)$. On the other hand if the right-hand side is less than one, let us denote its arcsin (in $[0, \pi/2)$) by ϕ^*. Our confidence region then consists of values $\phi_{12}(\omega)$ such that

$$|\sin(\hat{\phi}_{12}(\omega) - \phi_{12}(\omega))| \le \sin \phi^*,$$

i.e. such that

$$\hat{\phi}_{12}(\omega) - \phi^* \le \phi_{12}(\omega) \le \hat{\phi}_{12}(\omega) + \phi^*, \tag{11.7.10}$$

or

$$\hat{\phi}_{12}(\omega) + \pi - \phi^* \le \phi_{12}(\omega) \le \hat{\phi}_{12}(\omega) + \pi + \phi^*.$$

The confidence region can thus be represented as a union of two subintervals of the unit circle whose centers are diametrically opposed (at $\hat{\phi}_{12}(\omega)$ and $\hat{\phi}_{12}(\omega) + \pi$) and whose arc lengths are $2\phi^*$. If $|\hat{\mathscr{K}}_{12}(\omega)|^2$ is close to one, then we normally choose the interval centered at $\hat{\phi}_{12}(\omega)$, since the other interval corresponds to a sign change in both $c_{12}(\omega)$ and $q_{12}(\omega)$ which is unlikely if $|\mathscr{K}_{12}(\omega)|^2$ is close to one.

Estimation of the Absolute Coherency

The squared coherency $|\mathscr{K}_{12}(\omega)|^2$ is estimated by $|\hat{\mathscr{K}}_{12}(\omega)|^2$ where

$$\begin{aligned} |\hat{\mathscr{K}}_{12}(\omega)| &= [\hat{c}_{12}^2(\omega) + \hat{q}_{12}^2(\omega)]^{1/2}/[\hat{f}_{11}(\omega)\hat{f}_{22}(\omega)]^{1/2} \\ &= h(\hat{f}_{11}(\omega), \hat{f}_{22}(\omega), \hat{c}_{12}(\omega), \hat{q}_{12}(\omega)). \end{aligned}$$

If $|\hat{\mathscr{K}}_{12}(\omega)| > 0$, then by (11.7.7) and Proposition 6.4.3,

$$|\hat{\mathscr{K}}_{12}(\omega)| \quad \text{is AN}(|\mathscr{K}_{12}(\omega)|,\ a_n^2(1 - |\mathscr{K}_{12}(\omega)|^2)^2/2), \tag{11.7.11}$$

giving the approximate 95% confidence bounds,

$$|\hat{\mathscr{K}}_{12}(\omega)| \pm 1.96 a_n (1 - |\hat{\mathscr{K}}_{12}(\omega)|^2)/\sqrt{2},$$

for $|\mathscr{K}_{12}(\omega)|$.

Since $d[\tanh^{-1}(x)]/dx = d\left[\frac{1}{2}\ln\left(\frac{1+x}{1-x}\right)\right]/dx = (1 - x^2)^{-1}$, it follows from Proposition 6.4.3 that

$$\tanh^{-1}(|\hat{\mathscr{K}}_{12}(\omega)|) \quad \text{is AN}(\tanh^{-1}(|\mathscr{K}_{12}(\omega)|), a_n^2/2). \tag{11.7.12}$$

From (11.7.12) we obtain the constant-width large-sample $100(1 - \alpha)\%$ confidence interval,

$$(\tanh^{-1}(|\hat{\mathscr{K}}_{12}(\omega)|) - \Phi_{1-\alpha/2} a_n/\sqrt{2}, \tanh^{-1}(|\hat{\mathscr{K}}_{12}(\omega)|) + \Phi_{1-\alpha/2} a_n/\sqrt{2}),$$

for $\tanh^{-1}(|\mathscr{K}_{12}(\omega)|)$. The corresponding $100(1 - \alpha)\%$ confidence region for $|\mathscr{K}_{12}(\omega)|$ is the intersection with $[0, 1]$ of the interval

$$\begin{aligned} (\tanh[&\tanh^{-1}(|\hat{\mathscr{K}}_{12}(\omega)|) - \Phi_{1-\alpha/2} a_n/\sqrt{2}], \\ &\tanh[\tanh^{-1}(|\hat{\mathscr{K}}_{12}(\omega)|) + \Phi_{1-\alpha/2} a_n/\sqrt{2}]), \end{aligned} \tag{11.7.13}$$

assuming still that $|\mathscr{K}_{12}(\omega)| > 0$.

If the weight function $W_n(k)$ in (11.7.5) has the form $W_n(k) = (2m + 1)^{-1}$ for $|k| \le m$ and $W_n(k) = 0$, $|k| > m$, then the hypothesis $|\mathscr{K}_{12}(\omega)| = 0$ can be tested against the alternative hypothesis $|\mathscr{K}_{12}(\omega)| > 0$ using the statistic,

$$Y = 2m|\hat{\mathscr{K}}_{12}(\omega)|^2/[1 - |\hat{\mathscr{K}}_{12}(\omega)|^2].$$

Under the approximating asymptotic distribution of Theorem 11.7.1, it can be shown that $|\hat{\mathscr{K}}_{12}(\omega)|^2$ is distributed as the squarc of a multiple correlation coefficient, so that $Y \sim F(2, 4m)$ under the hypothesis that $|\mathscr{K}_{12}(\omega)| = 0$. (See Hannan (1970), p. 254.) We therefore reject the hypothesis $|\mathscr{K}_{12}(\omega)| = 0$ if

$$Y > F_{1-\alpha}(2, 4m) \tag{11.7.14}$$

where $F_{1-\alpha}(2, 4m)$ is the $(1 - \alpha)$-quantile of the F distribution with 2 and $4m$ degress of freedom. The power of this test has been tabulated for numerous values of $|\mathscr{K}_{12}(\omega)| > 0$ by Amos and Koopmans (1963).

Example 11.7.1 (Sales with a Leading Indicator). Estimates of the spectral density for the two differenced series $\{D_{t1}\}$ and $\{D_{t2}\}$ in Example 11.2.2 are shown in Figures 11.5 and 11.6. Both estimates were obtained by smoothing the respective periodograms with the same weight function $W_n(k) = \frac{1}{13}$, $|k| \leq 6$. From the graphs, it is clear that the power is concentrated at high frequencies for the leading indicator series and at low frequencies for the sales series.

The estimated absolute coherency, $|\hat{\mathscr{K}}_{12}(\omega)|$ is shown in Figure 11.7 with

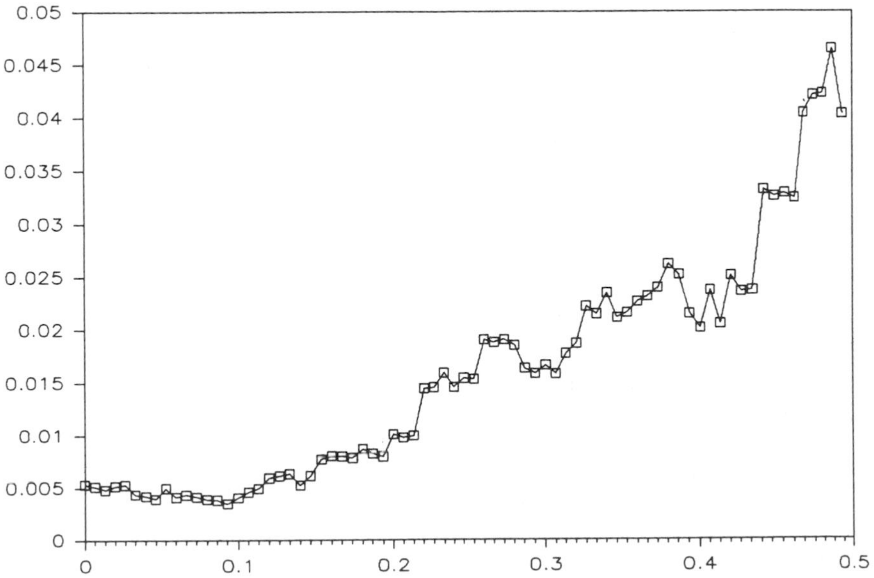

Figure 11.5. The spectral density estimate $\hat{f}_1(2\pi c)$, $0 \leq c \leq 0.5$, for the differenced leading indicator series of Example 11.7.1.

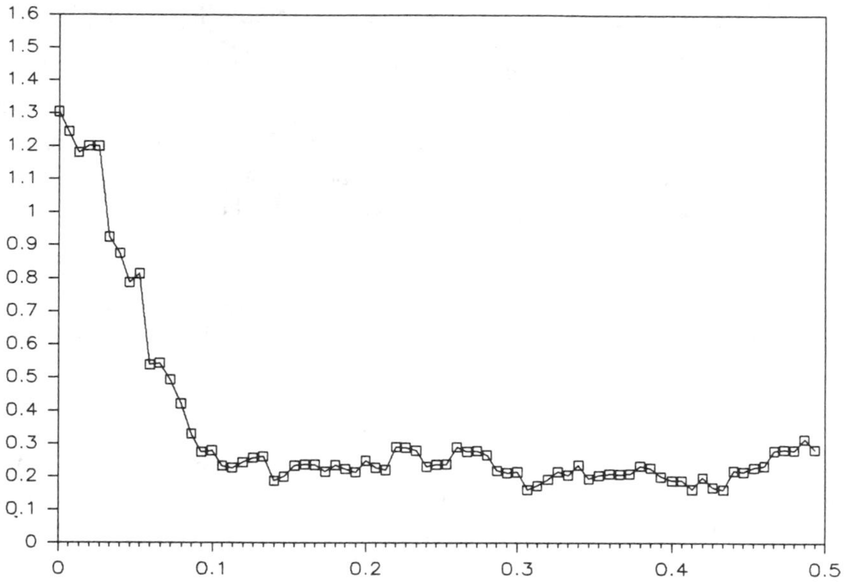

Figure 11.6. The spectral density estimate $\hat{f}_2(2\pi c)$, $0 \le c \le 0.5$, for the differenced sales data of Example 11.7.1.

corresponding 95% confidence intervals computed from 11.7.13. The confidence intervals for $|\mathscr{K}_{12}(\omega)|$ are bounded away from zero for all ω, suggesting that the coherency is positive at all frequencies. To test the hypothesis H_0 : $|\mathscr{K}_{12}(\omega)| = 0$ at level $\alpha = .05$, we use the rejection region (11.7.14). Since $m = 6$, we reject H_0 if

$$\frac{12|\hat{\mathscr{K}}_{12}(\omega)|^2}{1 - |\hat{\mathscr{K}}_{12}(\omega)|^2} > F_{.95}(2, 24) = 3.40,$$

i.e. if $|\hat{\mathscr{K}}_{12}(\omega)| > .470$. Applying this test to $|\hat{\mathscr{K}}_{12}(\omega)|$, we find that the hypothesis $|\mathscr{K}_{12}(\omega)| = 0$ is rejected for all $\omega \in (0, \pi)$. In fact the same conclusions hold even at level $\alpha = .005$. We therefore conclude that the two series are correlated at each frequency. The estimated phase spectrum $\hat{\phi}_{12}(\omega)$ is shown with the 95% confidence intervals from (11.7.10) in Figure 11.8. The confidence intervals for $\phi_{12}(\omega)$ are quite narrow at each ω owing to the large values of $|\hat{\mathscr{K}}_{12}(\omega)|$. Observe that the graph of $\hat{\phi}_{12}(\omega)$ is piecewise linear with slope 4.1 at low frequencies and slope 2.7 at the other frequencies. This is evidence, supported by the earlier analysis of the cross correlation function in Example 11.2.2, that $\{D_{t1}\}$ leads $\{D_{t2}\}$ by approximately 3 time units. A transfer function model for these two series which incorporates a delay of 3 time units is discussed in Example 13.1.1. The results shown in Figures 11.5–11.8 were obtained using the program SPEC.

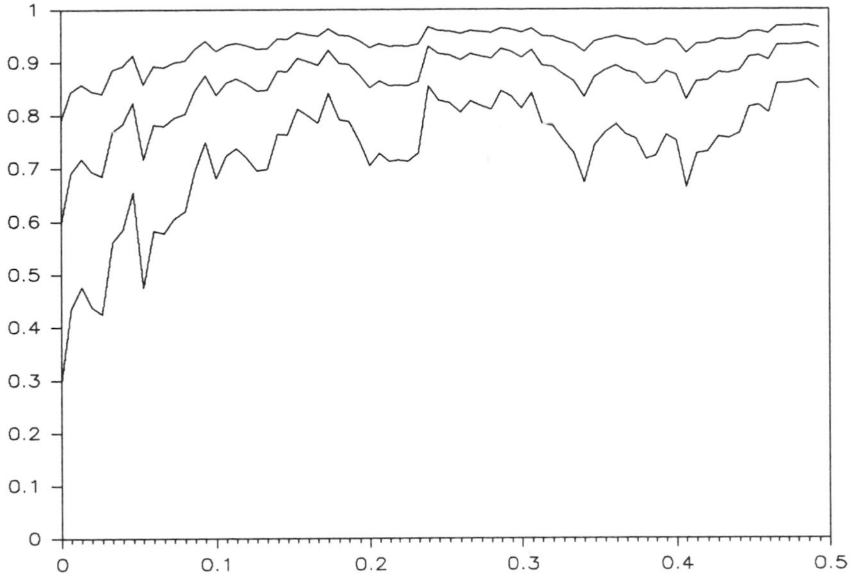

Figure 11.7. The estimated absolute coherency $|\hat{K}_{12}(2\pi c)|$ for the differenced leading indicator and sales series of Example 11.7.1, showing 95% confidence limits.

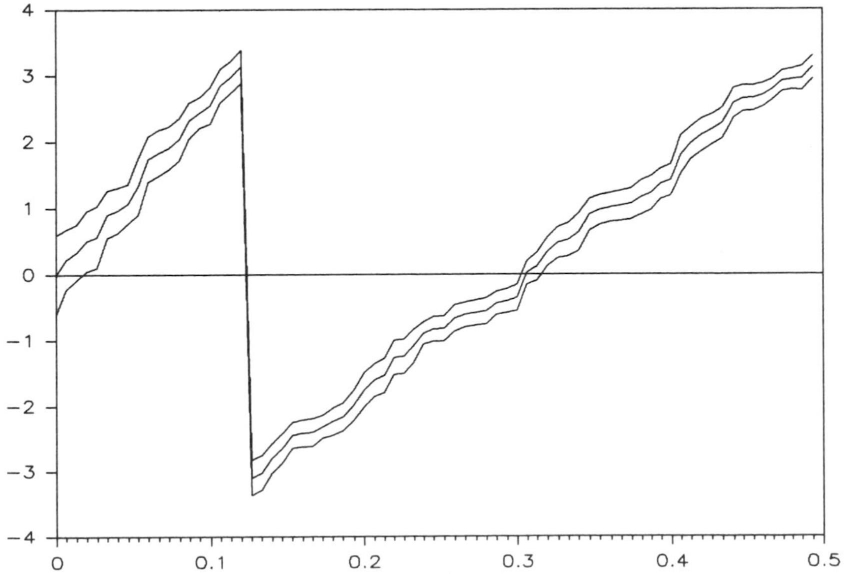

Figure 11.8. The estimated phase spectrum, $\hat{\phi}_{12}(2\pi c)$, for the differenced leading indicator and sales series, showing 95% confidence limits.

§11.8* The Spectral Representation of a Multivariate Stationary Time Series

In this section we state the multivariate versions of the spectral representation Theorems 4.3.1 and 4.8.2. For detailed proofs see Gihman and Skorohod (1974) or Hannan (1970). All processes are assumed to be defined on the probability space $(\Omega, \mathscr{F}, P)$.

Theorem 11.8.1. *$\Gamma(\cdot)$ is the covariance matrix function of an m-variate stationary process $\{\mathbf{X}_t, t = 0, \pm 1, \ldots\}$ if and only if*

$$\Gamma(h) = \int_{(-\pi,\pi]} e^{ihv}\, dF(v), \qquad h = 0, \pm 1, \ldots,$$

where $F(\cdot)$ is an $m \times m$ matrix distribution function on $[-\pi, \pi]$. (We shall use this term to mean that $F(-\pi) = 0$, $F(\cdot)$ is right-continuous and $(F(\mu) - F(\lambda))$ is non-negative definite for all $\lambda \le \mu$, i.e. $\infty > \mathbf{a}^*(F(\mu) - F(\lambda))\mathbf{a} \ge 0$ for all $\mathbf{a} \in \mathbb{C}^m$, where $\mathbf{a}^*$ denotes the complex conjugate transpose of $\mathbf{a}$.) F is called the spectral distribution matrix *of $\{\mathbf{X}_t\}$ or of $\Gamma(\cdot)$. Each component $F_{jk}(\cdot)$ of $F(\cdot)$ is a complex-valued distribution function and $\int_{(-\pi,\pi]} e^{ihv}\, dF(v)$ is the matrix whose (j, k)-component is $\int_{(-\pi,\pi]} e^{ihv}\, dF_{jk}(v)$.*

PROOF. See Gihman and Skorohod (1974), p. 217. □

In order to state the spectral representation of $\{\mathbf{X}_t\}$, we need the concept of a (right-continuous) vector-valued orthogonal increment process $\{\mathbf{Z}(\lambda), -\pi \le \lambda \le \pi\}$. For this we use Definition 4.6.1, replacing $\langle X, Y \rangle$ by $E\mathbf{X}\mathbf{Y}^*$ and $\|\mathbf{X}\|^2$ by $E\mathbf{X}\mathbf{X}^*$. Specifically, we shall say that $\{\mathbf{Z}(\lambda), -\pi \le \lambda \le \pi\}$ is a vector-valued orthogonal increment process if

(i) the components of the matrix $E(\mathbf{Z}(\lambda)\mathbf{Z}^*(\lambda))$ are finite, $-\pi \le \lambda \le \pi$,
(ii) $E\mathbf{Z}(\lambda) = \mathbf{0}$, $-\pi \le \lambda \le \pi$,
(iii) $E(\mathbf{Z}(\lambda_4) - \mathbf{Z}(\lambda_3))(\mathbf{Z}(\lambda_2) - \mathbf{Z}(\lambda_1))^* = 0$ if $(\lambda_1, \lambda_2] \cap (\lambda_3, \lambda_4] = \phi$, and
(iv) $E(\mathbf{Z}(\lambda + \delta) - \mathbf{Z}(\lambda))(\mathbf{Z}(\lambda + \delta) - \mathbf{Z}(\lambda))^* \to 0$ as $\delta \downarrow 0$.

Corresponding to any process $\{\mathbf{Z}(\lambda), -\pi \le \lambda \le \pi\}$ satisfying these four properties, there is a unique matrix distribution function G on $[-\pi, \pi]$ such that

$$G(\mu) - G(\lambda) = E[(\mathbf{Z}(\mu) - \mathbf{Z}(\lambda))(\mathbf{Z}(\mu) - \mathbf{Z}(\lambda))^*], \qquad \lambda \le \mu. \quad (11.8.1)$$

In shorthand notation the relation between the matrix distribution function G and $\{\mathbf{Z}(\lambda), -\pi \le \lambda \le \pi\}$ can be expressed as

$$E(d\mathbf{Z}(\lambda)\, d\mathbf{Z}^*(\mu)) = \delta_{\lambda,\mu}\, dG(\lambda) = \begin{cases} dG(\lambda) & \text{if } \mu = \lambda, \\ 0 & \text{otherwise.} \end{cases}$$

Standard Brownian motion $\{\mathbf{B}(\lambda), -\pi \le \lambda \le \pi\}$ with values in $\mathbb{R}^m$ and

$\mathbf{B}(-\pi) = \mathbf{0}$ is an orthogonal increment process with $G(\lambda) = (\lambda + \pi)I$ where I is the $(m \times m)$ identity matrix. The fact that $G(\lambda)$ is diagonal in this particular case reflects the orthogonality of $B_i(\lambda), B_j(\lambda), i \neq j$, for m-dimensional Brownian motion. It is not generally the case that $G(\lambda)$ is diagonal; in fact from (11.8.1) the (i, j)-element of $dG(\lambda)$ is the covariance, $E(dZ_i(\lambda)\overline{dZ_j(\lambda)})$.

The stochastic integral $I(f)$ with respect to $\{\mathbf{Z}(\lambda)\}$ is defined for functions f which are square integrable with respect to the distribution function $G_0 := \sum_{i=1}^{m} G_{ii}$ as follows. For functions of the form

$$f(\lambda) = \sum_{i=0}^{n} f_i I_{(\lambda_i, \lambda_{i+1}]}(\lambda), \qquad -\pi = \lambda_0 < \lambda_1 < \cdots < \lambda_{n+1} = \pi, \quad (11.8.2)$$

we define

$$I(f) := \sum_{i=0}^{n} f_i[\mathbf{Z}(\lambda_{i+1}) - \mathbf{Z}(\lambda_i)]. \quad (11.8.3)$$

This mapping is then extended to a Hilbert space isomorphism I of $L^2(G_0)$ into $L^2(\mathbf{Z})$, where $L^2(\mathbf{Z})$ is the closure in $L^2(\Omega, \mathscr{F}, P)$ of the set of all linear combinations of the form (11.8.3) with arbitrary complex coefficients f_i. The inner product in $L^2(\mathbf{Z})$ is defined by

$$\langle \mathbf{Y}_1, \mathbf{Y}_2 \rangle := E(\mathbf{Y}_2^* \mathbf{Y}_1). \quad (11.8.4)$$

Definition 11.8.1. If $\{\mathbf{Z}(\lambda), -\pi \leq \lambda \leq \pi\}$ is an m-variate orthogonal increment process with $E(d\mathbf{Z}(\lambda)\, d\mathbf{Z}^*(\mu)) = \delta_{\lambda,\mu}\, dG(\lambda)$ and $G_0 = \sum_{i=1}^{m} G_{ii}$, then for any $f \in L^2(G_0)$ we define the stochastic integral $\int_{(-\pi,\pi]} f(\nu)\, d\mathbf{Z}(\nu)$ to be the random vector $I(f) \in L^2(\mathbf{Z})$ with I defined as above.

The stochastic integral has properties analogous to (4.7.4)–(4.7.7), namely

$$E(I(f)) = \mathbf{0},$$

$$I(a_1 f + a_2 g) = a_1 I(f) + a_2 I(g), \qquad a_1, a_2 \in \mathbb{C},$$

$$\langle I(f), I(g) \rangle = \int_{(-\pi,\pi]} f(\nu)\bar{g}(\nu)\, dG_0(\nu),$$

and

$$\langle I(f_n), I(g_n) \rangle \to \int_{(-\pi,\pi]} f(\nu)\bar{g}(\nu)\, dG_0(\nu) \quad \text{if} \quad f_n \xrightarrow{L^2(G_0)} f \quad \text{and} \quad g_n \xrightarrow{L^2(G_0)} g,$$

with the additional property,

$$E(I(f)I(g)^*) = \int_{(-\pi,\pi]} f(\nu)\bar{g}(\nu)\, dG(\nu).$$

Now suppose that $\{\mathbf{X}_t\}$ is a zero-mean m-variate stationary process with spectral distribution matrix $F(\cdot)$ as in Theorem 11.8.1. Let $\mathscr{H}$ be the set of random vectors of the form

$$\mathbf{Y} = \sum_{j=1}^{n} a_j \mathbf{X}_{t_j}, \qquad t_j \in \mathbb{Z}, n \geq 1, \tag{11.8.5}$$

and let $\bar{\mathscr{H}}$ denote the closure in $L^2(\Omega, \mathscr{F}, P)$ of $\mathscr{H}$. The inner product in $\bar{\mathscr{H}}$ is defined by (11.8.4). Define $\mathscr{K}$ to be the (not necessarily closed) subspace $\overline{\text{sp}}\{e^{it\cdot}, t \in \mathbb{Z}\}$ of $L^2(F_0)$ where $F_0 = \sum_{i=1}^m F_{ii}$. If $\bar{\mathscr{K}}$ denotes the closure of $\mathscr{K}$ in $L^2(F_0)$ then, as in Section 4.8, $\bar{\mathscr{K}} = L^2(F_0)$. The mapping T defined by

$$T\left(\sum_{j=1}^{n} a_j \mathbf{X}_{t_j}\right) = \sum_{j=1}^{n} a_j e^{it_j\cdot}, \tag{11.8.6}$$

can be extended as in Section 4.8 to a Hilbert space isomorphism of $\bar{\mathscr{H}}$ onto $L^2(F_0)$, which by Theorem 11.8.1 has the property that

$$E[(T^{-1}f)(T^{-1}g)^*] = \int_{(-\pi,\pi]} f(v)\bar{g}(v)\,dF(v), \qquad f, g \in L^2(F_0). \tag{11.8.7}$$

Consequently the process $\{\mathbf{Z}(\lambda), -\pi \leq \lambda \leq \pi\}$ defined by

$$\mathbf{Z}(\lambda) = T^{-1} I_{(-\pi,\lambda]}(\cdot), \qquad -\pi \leq \lambda \leq \pi, \tag{11.8.8}$$

is an orthogonal increment process, and the matrix distribution function associated with $\{\mathbf{Z}(\lambda)\}$ is precisely the spectral distribution matrix F of $\{\mathbf{X}_t\}$ appearing in Theorem 11.8.1. The spectral representation of $\{\mathbf{X}_t\}$ is then established by first showing that

$$T^{-1}f = \int_{(-\pi,\pi]} f(v)\,d\mathbf{Z}(v), \qquad f \in L^2(F_0), \tag{11.8.9}$$

then setting $f(v) = e^{itv}$ and using (11.8.6).

Theorem 11.8.2 (The Spectral Representation Theorem). *If $\{\mathbf{X}_t\}$ is a stationary sequence with mean zero and spectral distribution matrix $F(\cdot)$, then there exists a right-continuous orthogonal increment process $\{\mathbf{Z}(\lambda), -\pi \leq \lambda \leq \pi\}$ such that*

(i) $E[(\mathbf{Z}(\lambda) - \mathbf{Z}(-\pi))(\mathbf{Z}(\lambda) - \mathbf{Z}(-\pi))^*] = F(\lambda), \qquad -\pi \leq \lambda \leq \pi,$

and

(ii) $\mathbf{X}_t = \int_{(-\pi,\pi]} e^{itv}\,d\mathbf{Z}(v)$ *with probability* 1.

PROOF. The steps are the same as those in the proof of Theorem 4.8.2, the process $\{\mathbf{Z}(\lambda)\}$ being defined by (11.8.8). □

The following corollary is established by the same argument which we used to prove Corollary 4.8.1.

Corollary 11.8.1. *If $\{\mathbf{X}_t\}$ is a zero-mean stationary sequence then there exists a right-continuous orthogonal increment process $\{\mathbf{Z}(\lambda), -\pi \leq \lambda \leq \pi\}$ such that*

$\mathbf{Z}(-\pi) = \mathbf{0}$ *and*

$$\mathbf{X}_t = \int_{(-\pi,\pi]} e^{it\lambda}\, d\mathbf{Z}(v) \quad \textit{with probability } 1.$$

If $\{\mathbf{Y}(\lambda)\}$ *and* $\{\mathbf{Z}(\lambda)\}$ *are two such processes, then*

$$P(\mathbf{Y}(\lambda) = \mathbf{Z}(\lambda) = 1 \ \textit{for each } \lambda \in [-\pi, \pi].$$

Remark 1. Equations (11.8.7) and (11.8.9) imply that for any functions f, $g \in L^2(F_0)$,

$$E\left[\int_{(-\pi,\pi]} f(\lambda)\, dZ_i(\lambda) \overline{\int_{(-\pi,\pi]} g(\lambda)\, dZ_j(\lambda)}\right] = \int_{(-\pi,\pi]} f(\lambda)\overline{g(\lambda)}\, dF_{ij}(\lambda).$$

It can be shown (Problem 11.22) that the same relation then holds for all $f \in L^2(F_{ii})$ and $g \in L^2(F_{jj})$.

Remark 2. As in the univariate case we also have the important result that $\mathbf{Y} \in \overline{\mathscr{H}}$ (see (11.8.5)) if and only if there exists a function $g \in L^2(F_0)$ such that

$$\mathbf{Y} = \int_{(-\pi,\pi]} g(v)\, d\mathbf{Z}(v) \qquad \text{with probability } 1.$$

Remark 3. In many important cases of interest (in particular if $\{\mathbf{X}_t\}$ is an ARMA process) the spectral distribution matrix $F(\cdot)$ has the form,

$$F(\omega) = \int_{-\pi}^{\omega} f(v)\, dv, \qquad -\pi \le \omega \le \pi.$$

Then $f(\cdot)$ is called the *spectral density matrix* of the process. In the case when $\sum_{h=-\infty}^{\infty} |\gamma_{ij}(h)| < \infty$ for all $i, j \in \{1, \dots, m\}$ we have the simple relations (11.1.14) and (11.1.15) connecting $\Gamma(\cdot)$ and $f(\cdot)$.

Time Invariant Linear Filters

The spectral representation of a stationary m-variate time series is particularly useful when dealing with time-invariant linear filters. These are defined for m-variate series just as in Section 4.10, the only difference being that the coefficients H_j of the filter $H = \{H_j, j = 0, \pm 1, \dots\}$ are now $(l \times m)$ matrices instead of scalars. In particular an absolutely summable TLF has the property that the elements of the matrices are absolutely summable and a causal TLF has the property that $H_j = 0$ for $j < 0$.

If $\{\mathbf{Y}_t\}$ is obtained from $\{\mathbf{X}_t\}$ by application of the absolutely summable $(l \times m)$ TLF $H = \{H_j\}$, then

$$\mathbf{Y}_t = \sum_{j=-\infty}^{\infty} H_j \mathbf{X}_{t-j}. \tag{11.8.10}$$

The following theorem expresses the spectral representation of $\{\mathbf{Y}_t\}$ in terms of that of $\{\mathbf{X}_t\}$.

Theorem 11.8.3. *If $H = \{H_j\}$ is an absolutely summable ($l \times m$) TLF and $\{\mathbf{X}_t\}$ is any zero-mean m-variate stationary process with spectral representation,*

$$\mathbf{X}_t = \int_{(-\pi,\pi]} e^{itv}\, d\mathbf{Z}(v),$$

and spectral distribution matrix F, then the l-variate process (11.8.10) *is stationary with spectral representation*

$$\mathbf{Y}_t = \int_{(-\pi,\pi]} e^{itv} H(e^{-iv})\, d\mathbf{Z}(v),$$

and spectral distribution matrix $F_{\mathbf{Y}}$ satisfying

$$dF_{\mathbf{Y}}(v) = H(e^{-iv})\, dF(v) H'(e^{iv}),$$

where

$$H(e^{iv}) = \sum_{j=-\infty}^{\infty} H_j e^{ijv}.$$

Proof. The proof of the representation of $\mathbf{Y}_t$ is the same as that of Theorem 4.10.1. Since $\mathbf{Y}_t$ is a stochastic integral with respect to the orthogonal increment process $\{\mathbf{W}(v)\}$ with $d\mathbf{W}(v) = H(e^{-iv})\, d\mathbf{Z}(v)$, it follows at once that $E\mathbf{Y}_t = \mathbf{0}$ and that $\mathbf{Y}_t$ is stationary with

$$dF_{\mathbf{Y}}(v) = E(d\mathbf{W}(v)\, d\mathbf{W}^*(v)) = H(e^{-iv})\, dF(v) H'(e^{iv})$$

and

$$E(\mathbf{Y}_{t+h}\mathbf{Y}_t^*) = \int_{(-\pi,\pi]} e^{ihv}\, dF_{\mathbf{Y}}(v). \qquad \square$$

Remark 4. The spectral representation decomposes $\mathbf{X}_t$ into a sum of sinusoids

$$e^{itv}\, d\mathbf{Z}(v), \qquad -\pi \le v \le \pi.$$

The effect of the TLF H is to produce corresponding components

$$e^{itv} H(e^{-iv})\, d\mathbf{Z}(v), \qquad -\pi \le v \le \pi,$$

which combine to form the filtered process $\{\mathbf{Y}_t\}$. The function $H(e^{-iv})$, $-\pi \le v \le \pi$, is called the *matrix transfer function* of the filter $H = \{H_j\}$.

Example 11.8.1. The causal ARMA(p, q) process,

$$\Phi(B)\mathbf{X}_t = \Theta(B)\mathbf{Z}_t, \qquad \{\mathbf{Z}_t\} \sim \mathrm{WN}(0, \Sigma),$$

can be written (by Theorem 11.3.1) in the form

$$\mathbf{X}_t = \sum_{j=0}^{\infty} \Psi_j \mathbf{Z}_{t-j},$$

where $\sum_{j=0}^{\infty} \Psi_j z^j = \Phi^{-1}(z)\Theta(z)$, $|z| \le 1$. Hence $\{\mathbf{X}_t\}$ is obtained from $\{\mathbf{Z}_t\}$ by application of the causal TLF $\{\Psi_j, j = 0, 1, 2, \dots\}$, with matrix transfer function,

$$\Psi(e^{-iv}) = \Phi^{-1}(e^{-iv})\Theta(e^{-iv}), \qquad -\pi \le v \le \pi.$$

By Theorem 11.8.3 the spectral distribution matrix of $\mathbf{X}$ therefore has density matrix,

$$f_{\mathbf{X}}(v) = \frac{1}{2\pi}\Phi^{-1}(e^{-iv})\Theta(e^{-iv})\Sigma\Theta'(e^{iv})\Phi'^{-1}(e^{iv}), \qquad -\pi \le v \le \pi.$$

EXAMPLE 11.8.2. The spectral representation of any linear combination of components of $\mathbf{X}_t$ is easily found from Theorem 11.8.3. Thus if

$$Y_t = \mathbf{a}^*\mathbf{X}_t \quad \text{where } \mathbf{a} \in \mathbb{C}^m,$$

then

$$Y_t = \int_{(-\pi,\pi]} e^{itv}\, dZ_Y(v),$$

where

$$Z_Y(v) = \mathbf{a}^*\mathbf{Z}(v),$$

and

$$dF_Y(v) = E(dZ_Y(v)\, dZ_Y^*(v)) = \mathbf{a}^*\, dF(v)\mathbf{a}.$$

The same argument is easily extended to the case when Y_t is a linear combination of components of $\mathbf{X}_t, \mathbf{X}_{t-1}, \dots$.

Problems

11.1. Let $\{Y_t\}$ be a stationary process and define the bivariate process, $X_{t1} = Y_t$, $X_{t2} = Y_{t-d}$ where $d \ne 0$. Show that $\{(X_{t1}, X_{t2})'\}$ is stationary and express its cross-correlation function in terms of the autocorrelation function of $\{Y_t\}$. If $\rho_Y(h) \to 0$ as $h \to \infty$ show that there exists a lag k such that $\rho_{12}(k) > \rho_{12}(0)$.

11.2. Show that the linear process defined in (11.1.12) is stationary with mean $\mathbf{0}$ and covariance matrix function given by (11.1.13).

11.3.* Prove Proposition 11.2.2.

11.4. Prove Theorem 11.3.2.

11.5. If $\{\mathbf{X}_t\}$ is a causal ARMA process, show that there exists $\varepsilon \in (0, 1)$ and a constant K such that $|\gamma_{ij}(h)| \le K\varepsilon^{|h|}$ for all i, j and h.

11.6. Determine the covariance matrix function of the ARMA(1, 1) process defined in (11.4.33).

11.7. If $G(z) = \sum_{h=-\infty}^{\infty} \Gamma(h)z^h$ is the covariance matrix generating function of an ARMA process, show that $G(z) = \Phi^{-1}(z)\Theta(z)\Sigma\Theta'(z^{-1})\Phi'^{-1}(z^{-1})$.

11.8. For the matrix M in (11.4.21), show that $\det(zI - M) = z^{mp}\det(\Phi(z^{-1}))$ where $\Phi(z) = I - \Phi_{p1}z - \cdots - \Phi_{pp}z^p$.

11.9. (a) Let $\{\mathbf{X}_t\}$ be a causal multivariate AR(p) process satisfying the recursions

$$\mathbf{X}_t = \Phi_1\mathbf{X}_{t-1} + \cdots + \Phi_p\mathbf{X}_{t-p} + \mathbf{Z}_t, \qquad \{Z_t\} \sim \mathrm{WN}(\mathbf{0}, \Sigma).$$

For $n > p$ write down recursion relations for the predictors, $P_{S_n}\mathbf{X}_{n+h}$, $h \geq 0$, and find explicit expressions for the error covariance matrices in terms of the AR coefficients and Σ when $h = 1$, 2 and 3.

(b) Suppose now that $\{\mathbf{Y}_t\}$ is the multivariate ARIMA(p, 1, 0) process satisfying $\nabla\mathbf{Y}_t = \mathbf{X}_t$, where $\{\mathbf{X}_t\}$ is the AR process in (a). Assuming that $\mathbf{Y}_0 \perp \mathbf{X}_t$, $t \geq 1$, show that

$$P(\mathbf{Y}_{n+h}|\mathbf{Y}_0, \mathbf{Y}_1, \ldots, \mathbf{Y}_n) = \mathbf{Y}_n + \sum_{j=1}^{h} P_{S_n}\mathbf{X}_{n+j}$$

and derive the error covariance matrices when $h = 1$, 2 and 3. Compare these results with those obtained in Example 11.5.1.

11.10. Use the program ARVEC to analyze the bivariate time series, X_{t1}, X_{t2}, $t = 1, \ldots, 200$ (Series J and K respectively in the Appendix). Use the minimum AICC model to predict $(X_{t,1}, X_{t,2})$, $t = 201, 202, 203$ and estimate the error covariance matrices of the predictors.

11.11. Derive methods for simulating multivariate Gaussian processes and multivariate Gaussian ARMA processes analogous to the univariate methods specified in Problems 8.16 and 8.17.

11.12. Let $\{\mathbf{X}_t\}$ be the invertible MA(q) process

$$\mathbf{X}_t = \mathbf{Z}_t + \Theta_1\mathbf{Z}_{t-1} + \cdots + \Theta_q\mathbf{Z}_{t-q}, \qquad \{\mathbf{Z}_t\} \sim \mathrm{WN}(\mathbf{0}, \Sigma),$$

where Σ is non-singular. Show that as $n \to \infty$,
(a) $E(\mathbf{X}_{n+1} - \hat{\mathbf{X}}_{n+1} - \mathbf{Z}_{n+1})(\mathbf{X}_{n+1} - \hat{\mathbf{X}}_{n+1} - \mathbf{Z}_{n+1})' \to 0$,
(b) $V_n \to \Sigma$, and
(c) $\Theta_{nj} \to \Theta_j$, $j = 1, \ldots, q$.
(For (c), note that $\Theta_j = E(\mathbf{X}_{n+1}\mathbf{Z}'_{n+1-j})\Sigma^{-1}$ and $\Theta_{nj} = E(X_{n+1}(X_{n+1-j} - \hat{\mathbf{X}}_{n+1-j})')V_{n-j}^{-1}$.)

11.13. If X and Y are complex-valued random variables, show that $E|Y - aX|^2$ is minimum when $a = E(Y\bar{X})/E|X|^2$.

11.14. Show that the bivariate time series $(X_{t1}, X_{t2})'$ defined in (11.6.14) is stationary.

11.15. If A and its complex conjugate $\bar{A}$ are uncorrelated complex-valued random variables such that $EA = 0$ and $E|A|^2 = \sigma^2$, find the mean and covariance matrix of the real and imaginary parts of A. If $X_t = \sum_{j=1}^{n}(A_je^{i\lambda_jt} + \bar{A}_je^{-i\lambda_jt})$, $0 < \lambda_1 < \cdots < \lambda_n < \pi$, where $\{A_j, \bar{A}_j, j = 1, \ldots, n\}$ are uncorrelated, $EA_j = 0$ and $E|A_j|^2 = \sigma_j^2/2$, $j = 1, \ldots, n$, express X_t as $\sum_{j=1}^{n}[B_j\cos(\lambda_jt) + C_j\sin(\lambda_jt)]$ and find the mean and variance of B_j and C_j.

11.16. If $\mathbf{Y}$ is a complex-valued random vector with covariance matrix $\Sigma := E(\mathbf{Y} - \boldsymbol{\mu})(\mathbf{Y} - \boldsymbol{\mu})^* = \Sigma_1 + i\Sigma_2$, verify that the matrix

$$\frac{1}{2}\begin{bmatrix} \Sigma_1 & -\Sigma_2 \\ \Sigma_2 & \Sigma_1 \end{bmatrix}$$

is the covariance matrix of a real-valued random vector.

11.17. Let

$$\mathbf{U}_n = n^{-1/2} \sum_{t=1}^{n} \begin{bmatrix} \mathbf{Z}_t \cos(t\omega_j) \\ \mathbf{Z}_t \sin(t\omega_j) \end{bmatrix},$$

where $\{\mathbf{Z}_t\}$ is bivariate white noise with mean $\mathbf{0}$ and covariance matrix $\boldsymbol{\Sigma}$, and $\omega_j = 2\pi j/n \in (0, \pi)$. Show that $E\mathbf{U}_n\mathbf{U}_n' = \frac{1}{2}\left[\begin{smallmatrix} \boldsymbol{\Sigma} & 0 \\ 0 & \boldsymbol{\Sigma} \end{smallmatrix}\right]$.

11.18. If U_1 and U_2 are independent standard normal random variables, show that U_2/U_1 has a Cauchy distribution and that $\arg(U_1 + iU_2)$ is uniformly distributed on $(-\pi, \pi)$.

11.19. Verify the calculation of the asymptotic variances in equations (11.7.8), (11.7.9) and (11.7.11).

11.20. Let $\{X_{t1}, t = 1, \ldots, 63\}$ and $\{X_{t2}, t = 1, \ldots, 63\}$ denote the differenced series $\{\nabla \ln Y_{t1}\}$, $\{\nabla \ln Y_{t2}\}$ where $\{Y_{t1}\}$ and $\{Y_{t2}\}$ are the annual mink and muskrat trappings (Appendix A, series H and I respectively).

(a) Compute the sample cross correlation function of $\{X_{t1}\}$ and $\{X_{t2}\}$ for lags between -30 and $+30$ using the program TRANS
(b) Test for independence of the two series.

11.21. With $\{X_{t1}\}$ and $\{X_{t2}\}$ as in Problem 11.20, estimate the absolute coherency, $|K_{12}(\lambda)|$ and phase spectrum $\phi_{12}(\lambda)$, $0 \le \lambda \le \pi$, using SPEC. What do these functions tell you about the relation between the two series? Compute approximate 95% confidence intervals for $|K_{12}(\lambda)|$ and $\phi_{12}(\lambda)$.

11.22.* Prove Remark 1 of Section 11.8.

11.23.* Let $\{\mathbf{X}_t\}$ be a bivariate stationary process with mean $\mathbf{0}$ and a continuous spectral distribution matrix F. Use Problem 4.25 and Theorem 11.8.2 to show that $\{\mathbf{X}_t\}$ has the spectral representation

$$X_{tj} = 2\int_{(0,\pi]} \cos(vt)\, dU_j(v) + 2\int_{(0,\pi]} \sin(vt)\, dV_j(v), \qquad j = 1, 2,$$

where $\{\mathbf{U}(\lambda) = (U_1(\lambda), U_2(\lambda))'\}$ and $\{\mathbf{V}(\lambda) = (V_1(\lambda), V_2(\lambda))'\}$ are bivariate orthogonal increment processes on $[0, \pi]$ with

$$E(d\mathbf{U}(\lambda)\, d\mathbf{U}'(\mu)) = 2^{-1}\delta_{\lambda,\mu} \operatorname{Re}\{dF(\lambda)\},$$

$$E(d\mathbf{V}(\lambda)\, d\mathbf{V}'(\mu)) = 2^{-1}\delta_{\lambda,\mu} \operatorname{Re}\{dF(\lambda)\},$$

and

$$E(d\mathbf{U}(\lambda)\, d\mathbf{V}'(\mu)) = 2^{-1}\delta_{\lambda,\mu} \operatorname{Im}\{dF(\lambda)\}.$$

If $\{\mathbf{X}_t\}$ has spectral density matrix $f(\lambda)$, then

$$c_{12}(\lambda) = 2^{-1}\operatorname{Cov}(dU_1(\lambda), dU_2(\lambda)) = 2^{-1}\operatorname{Cov}(dV_1(\lambda), dV_2(\lambda))$$

and

$$q_{12}(\lambda) = 2^{-1}\operatorname{Cov}(dV_1(\lambda), dU_2(\lambda)) = -2^{-1}\operatorname{Cov}(dV_2(\lambda), dU_1(\lambda)),$$

where $c_{12}(\lambda)$ is the cospectrum and $q_{12}(\lambda)$ is the quadrature spectrum. Thus $c_{12}(\lambda)$ and $q_{12}(\lambda)$ can be interpreted as the covariance between the "in-phase" and "out of phase" components of the two processes $\{X_{t1}\}$ and $\{X_{t2}\}$ at frequency λ.

CHAPTER 12

State-Space Models and the Kalman Recursions

In recent years, state-space representations and the associated Kalman recursions have had a profound impact on time series analysis and many related areas. The techniques were originally developed in connection with the control of linear systems (for accounts of this subject, see the books of Davis and Vinter (1985) and Hannan and Deistler (1988)). The general form of the state-space model needed for the applications in this chapter is defined in Section 12.1, where some illustrative examples are also given. The Kalman recursions are developed in Section 12.2 and applied in Section 12.3 to the analysis of ARMA and ARIMA processes with missing values. In Section 12.4 we examine the fundamental concepts of controllability and observability and their relevance to the determination of the minimal dimension of a state-space representation. Section 12.5 deals with recursive Bayesian state estimation, which can be used (at least in principle) to compute conditional expectations for a large class of not necessarily Gaussian processes. Further applications of the Bayesian approach can be found in the papers of Sorenson and Alspach (1971), Kitagawa (1987) and Grunwald, Raftery and Guttorp (1989).

§12.1 State-Space Models

In this section we shall illustrate some of the many time-series models which can be represented in linear *state-space* form. By this we mean that the series $\{\mathbf{Y}_t, t = 1, 2, \dots\}$ satisfies an equation of the form

$$\mathbf{Y}_t = G_t\mathbf{X}_t + \mathbf{W}_t, \qquad t = 1, 2, \dots, \tag{12.1.1}$$

where

$$\mathbf{X}_{t+1} = F_t \mathbf{X}_t + \mathbf{V}_t, \qquad t = 1, 2, \dots. \tag{12.1.2}$$

The equation (12.1.2) can be interpreted as describing the evolution of the *state* $\mathbf{X}_t$ of a system at time t (a $v \times 1$ vector) in terms of a known sequence of $v \times v$ matrices $F_1, F_2, \dots$ and the sequence of random vectors $\mathbf{X}_1, \mathbf{V}_1, \mathbf{V}_2, \dots$. Equation (12.1.1) then defines a sequence of *observations*, $\mathbf{Y}_t$, which are obtained by applying a linear transformation to $\mathbf{X}_t$ and adding a random noise vector, $\mathbf{W}_t$, $t = 1, 2, \dots$. (The equation (12.1.2) is generalized in control theory to include an additional term $H_t \mathbf{u}_t$ on the right, representing the effect of applying a **control** $\mathbf{u}_t$ at time t for the purpose of influencing $\mathbf{X}_{t+1}$.)

Assumptions. Before proceeding further, we list the assumptions to be used in the analysis of the *state equation* (12.1.2) and the *observation equation* (12.1.1):

(a) $F_1, F_2, \dots$ is a sequence of specified $v \times v$ matrices.
(b) $G_1, G_2, \dots$ is a sequence of specified $w \times v$ matrices.
(c) $\{\mathbf{X}_1, (\mathbf{V}_t', \mathbf{W}_t')', t = 1, 2, \dots\}$ is an orthogonal sequence of random vectors with finite second moments. (The random vectors $\mathbf{X}$ and $\mathbf{Y}$ are said to be *orthogonal*, written $\mathbf{X} \perp \mathbf{Y}$, if the matrix $E(\mathbf{X}\mathbf{Y}')$ is zero.)
(d) $E\mathbf{V}_t = \mathbf{0}$ and $E\mathbf{W}_t = \mathbf{0}$ for all t.
(e) $E(\mathbf{V}_t\mathbf{V}_t') = Q_t$, $E(\mathbf{W}_t\mathbf{W}_t') = R_t$, $E(\mathbf{V}_t\mathbf{W}_t') = S_t$, where $\{Q_t\}$, $\{R_t\}$ and $\{S_t\}$ are specified sequences of $v \times v$, $w \times w$ and $v \times w$ matrices respectively.

Remark 1. In many important special cases (and in all the examples of this section) the matrices F_t, G_t, Q_t, R_t and S_t will be independent of t, in which case we shall suppress the subscripts.

Remark 2. It follows from the observation equation (12.1.1) and the state equation (12.1.2) that $\mathbf{X}_t$ and $\mathbf{Y}_t$ have the functional forms, for $t = 2, 3, \dots,$

$$\mathbf{X}_t = f_t(\mathbf{X}_1, \mathbf{V}_1, \dots, \mathbf{V}_{t-1}) \tag{12.1.3}$$

and

$$\mathbf{Y}_t = g_t(\mathbf{X}_1, \mathbf{V}_1, \dots, \mathbf{V}_{t-1}, \mathbf{W}_t). \tag{12.1.4}$$

Remark 3. From Remark 2 and Assumption (c) it is clear that we have the orthogonality relations,

$$\mathbf{V}_t \perp \mathbf{X}_s, \qquad \mathbf{V}_t \perp \mathbf{Y}_s, \qquad 1 \le s \le t,$$

and

$$\mathbf{W}_t \perp \mathbf{X}_s, \quad 1 \le s \le t, \qquad \mathbf{W}_t \perp \mathbf{Y}_s, \quad 1 \le s < t.$$

As already indicated, it is possible to formulate a great variety of time-series (and other) models in state-space form. It is clear also from the definition that neither $\{\mathbf{X}_t\}$ nor $\{\mathbf{Y}_t\}$ is necessarily stationary. The beauty of a state-space representation, when one can be found, lies in the simple structure of the state equation (12.1.2) which permits relatively simple analysis of the process $\{\mathbf{X}_t\}$. The behaviour of $\{\mathbf{Y}_t\}$ is then easy to determine from that of $\{\mathbf{X}_t\}$ using the observation equation (12.1.1). If the sequence $\{\mathbf{X}_1, \mathbf{V}_1, \mathbf{V}_2, \ldots\}$ is independent, then $\{\mathbf{X}_t\}$ has the Markov property, i.e. the distribution of $\mathbf{X}_{t+1}$ given $\mathbf{X}_t, \ldots, \mathbf{X}_1$ is the same as the distribution of $\mathbf{X}_{t+1}$ given $\mathbf{X}_t$. This is a property possessed by many physical systems provided we include sufficiently many components in the specification of the state $\mathbf{X}_t$ (for example, we may choose the state-vector in such a way that $\mathbf{X}_t$ includes components of $\mathbf{X}_{t-1}$ for each t).

To illustrate the versatility of state-space models, we now consider some examples. More can be found in subsequent sections and in the books of Aoki (1987) and Hannan and Deistler (1988). The paper of Harvey (1984) shows how state-space models provide a unifying framework for a variety of statistical forecasting techniques.

Example 12.1.1 (A Randomly Varying Trend With Added Noise). If β is constant, $\{V_t\} \sim \mathrm{WN}(0, \sigma^2)$ and Z_1 is a random variable uncorrelated with $\{V_t, t = 1, 2, \ldots\}$, then the process $\{Z_t, t = 1, 2, \ldots\}$ defined by

$$Z_{t+1} = Z_t + \beta + V_t = Z_1 + \beta t + V_1 + \cdots + V_t, \qquad t = 1, 2, \ldots, \tag{12.1.5}$$

has approximately linear sample-paths if σ is small (perfectly linear if $\sigma = 0$). The sequence $\{V_t\}$ introduces random variation into the slope of the sample-paths. To construct a state-space representation for $\{Z_t\}$ we introduce the vector

$$\mathbf{X}_t = (Z_t, \beta)'.$$

Then (12.1.5) can be written in the equivalent form,

$$\mathbf{X}_{t+1} = \begin{bmatrix} 1 & 1 \\ 0 & 1 \end{bmatrix} \mathbf{X}_t + \mathbf{V}_t, \qquad t = 1, 2, \ldots, \tag{12.1.6}$$

where $\mathbf{V}_t = (V_t, 0)'$. The process $\{Z_t\}$ is then determined by the observation equation, $Z_t = [1 \quad 0]\mathbf{X}_t$. A further random noise component can be added to Z_t, giving rise to the sequence

$$Y_t = [1 \quad 0]\mathbf{X}_t + W_t, \qquad t = 1, 2, \ldots, \tag{12.1.7}$$

where $\{W_t\} \sim \mathrm{WN}(0, v^2\}$. If $\{\mathbf{X}_1, V_1, W_1, V_2, W_2, \ldots\}$ is an orthogonal sequence, the equations (12.1.6) and (12.1.7) constitute a state-space representation of the process $\{Y_t\}$, which is a model for data with randomly varying trend and added noise. For this model we have

$$v = 2,\ w = 1,\ F = \begin{bmatrix} 1 & 1 \\ 0 & 1 \end{bmatrix},\ G = [1 \quad 0],\ Q = \begin{bmatrix} \sigma^2 & 0 \\ 0 & 0 \end{bmatrix},\ R = v^2 \text{ and } S = 0.$$

EXAMPLE 12.1.2 (A Seasonal Series with Noise). The classical decomposition (1.4.12) considered earlier in Chapter 1 expressed the time series $\{X_t\}$ as a sum of trend, seasonal and noise components. The seasonal component (with period d) was a sequence $\{s_t\}$ with the properties $s_{t+d} = s_t$ and $\sum_{t=1}^{d} s_t = 0$. Such a sequence can be generated, for *any* values of $s_1, s_0, \ldots, s_{-d+3}$, by means of the recursions,

$$s_{t+1} = -s_t - \cdots - s_{t-d+2}, \qquad t = 1, 2, \ldots, \tag{12.1.8}$$

A somewhat more general seasonal component $\{Y_t\}$, allowing for random deviations from strict periodicity, is obtained by adding a term V_t to the right side of (12.1.18), where $\{V_t\}$ is white noise with mean zero. This leads to the recursion relations,

$$Y_{t+1} = -Y_t - \cdots - Y_{t-d+2} + V_t, \qquad t = 1, 2, \ldots. \tag{12.1.9}$$

To find a state-space representation for $\{Y_t\}$ we introduce the $(d-1)$-dimensional state vector,

$$\mathbf{X}_t = (Y_t, Y_{t-1}, \ldots, Y_{t-d+2})'.$$

The series $\{Y_t\}$ is then given by the observation equation,

$$Y_t = [1 \quad 0 \quad 0 \quad \cdots \quad 0]\mathbf{X}_t, \qquad t = 1, 2, \ldots,$$

where $\{\mathbf{X}_t\}$ satisfies the state equation,

$$\mathbf{X}_{t+1} = F\mathbf{X}_t + \mathbf{V}_t, \qquad t = 1, 2, \ldots,$$

with $\mathbf{V}_t = (V_t, 0, \ldots, 0)'$ and

$$F = \begin{bmatrix} -1 & -1 & \cdots & -1 & -1 \\ 1 & 0 & \cdots & 0 & 0 \\ 0 & 1 & \cdots & 0 & 0 \\ \vdots & \vdots & \ddots & \vdots & \vdots \\ 0 & 0 & \cdots & 1 & 0 \end{bmatrix}.$$

EXAMPLE 12.1.3 (A Randomly Varying Trend with Seasonal and Noise Components). Such a series can be constructed by adding the two series in Examples 12.1.1 and 12.1.2. (Addition of series with state-space representations is in fact always possible by means of the following construction. See Problem 12.2.) We introduce the state-vector

$$\mathbf{X}_t = \begin{bmatrix} \mathbf{X}_t^1 \\ \mathbf{X}_t^2 \end{bmatrix},$$

where $\mathbf{X}_t^1$ and $\mathbf{X}_t^2$ are the state vectors in Examples 12.1.1 and 12.1.2 respectively. We then have the following representation for $\{Y_t\}$, the sum of the two series whose state-space representations were given in Examples

12.1.1 and 12.1.2. The state equation is

$$\mathbf{X}_{t+1} = \begin{bmatrix} F_1 & 0 \\ 0 & F_2 \end{bmatrix} \mathbf{X}_t + \begin{bmatrix} \mathbf{V}_t^1 \\ \mathbf{V}_t^2 \end{bmatrix}, \tag{12.1.10}$$

where F_1, F_2 are the coefficient matrices and $\{\mathbf{V}_t^1\}$, $\{\mathbf{V}_t^2\}$ are the noise vectors in the state equations of Examples 12.1.1 and 12.1.2 respectively. The observation equation is

$$Y_t = [1 \quad 0 \quad 1 \quad 0 \quad \cdots \quad 0]\mathbf{X}_t + W_t, \tag{12.1.11}$$

where $\{W_t\}$ is the noise sequence in (12.1.7). If the sequence of random vectors, $\{\mathbf{X}_1, \mathbf{V}_1^1, \mathbf{V}_1^2, W_1, \mathbf{V}_2^1, \mathbf{V}_2^2, W_2, \ldots\}$, is orthogonal, the equations (12.1.10) and (12.1.11) constitute a state-space representation for $\{Y_t\}$ satisfying assumptions (a)–(e).

We shall be concerned particularly in this chapter with the use of state-space representations and the Kalman recursions in the analysis of ARMA processes. In order to deal with such processes we shall need to consider state and observation equations which are defined for all $t \in \{0, \pm 1, \ldots\}$.

Stationary State-Space Models Defined for $t \in \{0, \pm 1, \ldots\}$

Consider the observation and state equations,

$$\mathbf{Y}_t = G\mathbf{X}_t + \mathbf{W}_t, \qquad t = 0, \pm 1, \ldots, \tag{12.1.12}$$

$$\mathbf{X}_{t+1} = F\mathbf{X}_t + \mathbf{V}_t, \qquad t = 0, \pm 1, \ldots, \tag{12.1.13}$$

where F and G are $v \times v$ and $w \times v$ matrices respectively, $\{\mathbf{V}_t\} \sim \mathrm{WN}(\mathbf{0}, Q)$, $\{\mathbf{W}_t\} \sim \mathrm{WN}(\mathbf{0}, R)$, $E(\mathbf{V}_t\mathbf{W}_t') = S$ for all t and $\mathbf{V}_s \perp \mathbf{W}_t$ for all $s \neq t$.

The state equation (12.1.13) is said to be *stable* (or *causal*) if the matrix F has all its eigenvalues in the interior of the unit circle, or equivalently if $\det(I - Fz) \neq 0$ for all $z \in \mathbb{C}$ such that $|z| \leq 1$. The matrix F is then also said to be stable.

In the stable case the equations (12.1.13) have the unique stationary solution (Problem 12.3) given by

$$\mathbf{X}_t = \sum_{j=0}^{\infty} F^j \mathbf{V}_{t-j-1}. \tag{12.1.14}$$

The corresponding sequence of observations,

$$\mathbf{Y}_t = \mathbf{W}_t + \sum_{j=0}^{\infty} GF^j \mathbf{V}_{t-j-1},$$

is also stationary.

EXAMPLE 12.1.4 (State-Space Representation of a Causal AR(p) Process). Consider the AR(p) process defined by

$$X_{t+1} = \phi_1 X_t + \phi_2 X_{t-1} + \cdots + \phi_p X_{t-p+1} + Z_{t+1}, \qquad t = 0, \pm 1, \ldots, \tag{12.1.15}$$

where $\{Z_t\} \sim \text{WN}(0, \sigma^2)$, and $\phi(z) := 1 - \phi_1 z - \cdots - \phi_p z^p$ is non-zero for $|z| \le 1$. To express $\{X_t\}$ in state-space form we simply introduce the state vectors,

$$\mathbf{X}_t = \begin{bmatrix} X_{t-p+1} \\ X_{t-p+2} \\ \vdots \\ X_t \end{bmatrix}, \qquad t = 0, \pm 1, \ldots. \tag{12.1.16}$$

If at time t we observe $Y_t = X_t$, then from (12.1.15) and (12.1.16) we obtain the observation equation,

$$Y_t = [0 \quad 0 \quad 0 \quad \cdots \quad 1]\mathbf{X}_t, \qquad t = 0, \pm 1, \ldots, \tag{12.1.17}$$

and state equation,

$$\mathbf{X}_{t+1} = \begin{bmatrix} 0 & 1 & 0 & \cdots & 0 \\ 0 & 0 & 1 & \cdots & 0 \\ \vdots & \vdots & \vdots & \ddots & \vdots \\ 0 & 0 & 0 & \cdots & 1 \\ \phi_p & \phi_{p-1} & \phi_{p-2} & \cdots & \phi_1 \end{bmatrix} \mathbf{X}_t + \begin{bmatrix} 0 \\ 0 \\ \vdots \\ 0 \\ 1 \end{bmatrix} Z_{t+1}, \qquad t = 0, \pm 1, \ldots. \tag{12.1.18}$$

Remark 4. In Example 12.1.4 the causality condition, $\phi(z) \ne 0$ for $|z| \le 1$, is equivalent to the condition that the state equation (12.1.18) is stable, since the eigenvalues of the coefficient matrix F in (12.1.18) are simply the reciprocals of the zeroes of $\phi(z)$ (Problem 12.4). The unique stationary solution of (12.1.18) determines a stationary solution of the AR(p) equation (12.1.15), which therefore coincides with the unique stationary solution specified in Remark 2 of Section 3.1.

Remark 5. If equations (12.1.17) and (12.1.18) are postulated to hold only for $t = 1, 2, \ldots$, and if $\mathbf{X}_1$ is a random vector such that $\{\mathbf{X}_1, Z_1, Z_2, \ldots\}$ is an orthogonal sequence, then we have a state-space representation for $\{Y_t\}$ of the type defined earlier by (12.1.1) and (12.1.2). The resulting process $\{Y_t\}$ is well-defined, regardless of whether or not the state equation is stable, but it will not in general be stationary. It will be stationary if the state equation is stable and if $\mathbf{X}_1$ is defined by (12.1.16) with $X_t = \sum_{j=0}^{\infty} \psi_j Z_{t-j}$, $t = 1, 0, \ldots, 2 - p$, and $\psi(z) = 1/\phi(z)$, $|z| \le 1$.

EXAMPLE 12.1.5 (State-Space Representation of a Causal ARMA(p, q) Process). State-space representations are not unique. We shall give two representations for an ARMA(p, q) process. The first follows easily from Example 12.1.4 and the second (Example 12.1.6 below) has a state-space with the smallest possible dimension (more will be said on this topic in Section 12.4). Consider the causal ARMA(p, q) process defined by

$$\phi(B)Y_t = \theta(B)Z_t, \qquad t = 0, \pm 1, \ldots, \tag{12.1.19}$$

where $\{Z_t\} \sim \text{WN}(0, \sigma^2)$ and $\phi(z) \neq 0$ for $|z| \leq 1$. Let

$$r = \max(p, q+1), \qquad \phi_j = 0 \quad \text{for } j > p, \quad \theta_j = 0 \quad \text{for } j > q \text{ and } \theta_0 = 1.$$

Then it is clear from (12.1.19) that we can write

$$Y_t = [\theta_{r-1} \quad \theta_{r-2} \quad \cdots \quad \theta_0]\mathbf{X}_t, \tag{12.1.20}$$

where

$$\mathbf{X}_t = \begin{bmatrix} X_{t-r+1} \\ X_{t-r+2} \\ \vdots \\ X_t \end{bmatrix} \tag{12.1.21}$$

and

$$\phi(B)X_t = Z_t, \qquad t = 0, \pm 1, \ldots. \tag{12.1.22}$$

But from Example 12.1.4 we can write

$$\mathbf{X}_{t+1} = \begin{bmatrix} 0 & 1 & 0 & \cdots & 0 \\ 0 & 0 & 1 & \cdots & 0 \\ \vdots & \vdots & \vdots & \ddots & \vdots \\ 0 & 0 & 0 & \cdots & 1 \\ \phi_r & \phi_{r-1} & \phi_{r-2} & \cdots & \phi_1 \end{bmatrix} \mathbf{X}_t + \begin{bmatrix} 0 \\ 0 \\ \vdots \\ 0 \\ 1 \end{bmatrix} Z_{t+1}, \qquad t = 0, \pm 1, \ldots. \tag{12.1.23}$$

Equations (12.1.20) and (12.1.23) are the required observation and state equations. The causality assumption implies that (12.1.23) has a unique stationary solution which determines a stationary sequence $\{Y_t\}$ through the observation equation (12.1.20). It is easy to check that this sequence satisfies the ARMA equations (12.1.19) and therefore coincides with their unique stationary solution.

EXAMPLE 12.1.6 (The Canonical Observable Representation of a Causal ARMA(p, q) Process). Consider the ARMA(p, q) process $\{Y_t\}$ defined by (12.1.19). We shall now establish a lower dimensional state-space representation than the one derived in Example 12.1.5. Let

$$m = \max(p, q) \quad \text{and} \quad \phi_j = 0 \quad \text{for } j > p.$$

Then

$$Y_t = [1 \quad 0 \quad 0 \quad \cdots \quad 0]\mathbf{X}_t + Z_t, \qquad t = 0, \pm 1, \ldots, \tag{12.1.24}$$

where $\{\mathbf{X}_t\}$ is the unique stationary solution of

$$\mathbf{X}_{t+1} = \begin{bmatrix} 0 & 1 & 0 & \cdots & 0 \\ 0 & 0 & 1 & \cdots & 0 \\ \vdots & \vdots & \vdots & \ddots & \vdots \\ 0 & 0 & 0 & \cdots & 1 \\ \phi_m & \phi_{m-1} & \phi_{m-2} & \cdots & \phi_1 \end{bmatrix} \mathbf{X}_t + \begin{bmatrix} \psi_1 \\ \psi_2 \\ \vdots \\ \psi_{m-1} \\ \psi_m \end{bmatrix} Z_t, \qquad t = 0, \pm 1, \ldots, \tag{12.1.25}$$

and $\psi_1, \ldots, \psi_m$ are the coefficients of $z, z^2, \ldots, z^m$ in the power series expansion of $\theta(z)/\phi(z)$, $|z| \le 1$. (If $m = 1$ the coefficients of $\mathbf{X}_t$ in (12.1.24) and (12.1.25) are 1 and ϕ_1 respectively.)

Proof. The result will be proved by showing that if $\{\mathbf{X}_t\}$ is the unique stationary solution of (12.1.25) and if $\{Y_t\}$ is defined by (12.1.24), then $\{Y_t\}$ is the unique stationary solution of (12.1.19). Let F and G denote the matrix coefficients of $\mathbf{X}_t$ in (12.1.25) and (12.1.24) respectively, and let $H = (\psi_1, \psi_2, \ldots, \psi_m)'$. Then

$$GF^{i-1}H = \psi_i, \qquad i = 1, \ldots, m, \tag{12.1.26}$$

and, since $\det(zI - F) = z^m - \phi_1 z^{m-1} - \cdots - \phi_m$ (see Problem 12.4), the Cayley–Hamilton Theorem implies that

$$F^m - \phi_1 F^{m-1} - \cdots - \phi_m I = 0. \tag{12.1.27}$$

From (12.1.24) and (12.1.25) we have

$$\begin{aligned} Y_t &= G\mathbf{X}_t + Z_t, \\ Y_{t+1} &= GF\mathbf{X}_t + GHZ_t + Z_{t+1}, \\ &\vdots \\ Y_{t+m} &= GF^m\mathbf{X}_t + GF^{m-1}HZ_t + \cdots + Z_{t+m}. \end{aligned}$$

These equations, together with (12.1.26) and (12.1.27), imply that

$$Y_{t+m} - \phi_1 Y_{t+m-1} - \cdots - \phi_m Y_t$$
$$= [-\phi_m \cdots -\phi_1 \quad 1] \begin{bmatrix} 1 & 0 & 0 & \cdots & 0 \\ \psi_1 & 1 & 0 & \cdots & 0 \\ \psi_2 & \psi_1 & 1 & \cdots & 0 \\ \vdots & \vdots & \vdots & \ddots & \vdots \\ \psi_m & \psi_{m-1} & \psi_{m-2} & \cdots & 1 \end{bmatrix} \begin{bmatrix} Z_t \\ Z_{t+1} \\ Z_{t+2} \\ \vdots \\ Z_{t+m} \end{bmatrix}.$$

Since $\psi_1, \ldots, \psi_m$ are the coefficients of $z, z^2, \ldots, z^m$ in the power series expansion of $\theta(z)/\phi(z)$, i.e. $\psi_j = \phi_1\psi_{j-1} + \phi_2\psi_{j-2} + \cdots + \phi_j + \theta_j$ as in (3.3.3), we conclude that $\{Y_t\}$ satisfies the ARMA equations,

$$Y_{t+m} - \phi_1 Y_{t+m-1} - \cdots - \phi_m Y_t = [\theta_m \quad \theta_{m-1} \quad \cdots \quad \theta_1 \quad 1] \begin{bmatrix} Z_t \\ Z_{t+1} \\ Z_{t+2} \\ \vdots \\ Z_{t+m} \end{bmatrix}. \tag{12.1.28}$$

Thus the stationary process $\{Y_t\}$ defined by (12.1.24) and (12.1.25) satisfies $\phi(B)Y_t = \theta(B)Z_t$, $t = 0, \pm 1, \ldots$, and therefore coincides with the unique stationary solution of these equations.

EXAMPLE 12.1.7 (State-Space Representation of an ARIMA(p, d, q) Process). If $\{Y_t\}$ is an ARIMA(p, d, q) process with $\{\nabla^d Y_t\}$ satisfying (12.1.19), then by the preceding example $\{\nabla^d Y_t\}$ has the representation,

$$\nabla^d Y_t = G\mathbf{X}_t + Z_t, \qquad t = 0, \pm 1, \ldots, \tag{12.1.29}$$

where $\{\mathbf{X}_t\}$ is the unique stationary solution of the state equation,

$$\mathbf{X}_{t+1} = F\mathbf{X}_t + HZ_t,$$

and G, F and H are the coefficient matrices in (12.1.24) and (12.1.25). Let A and B be the $d \times 1$ and $d \times d$ matrices defined by $A = B = 1$ if $d = 1$ and

$$A = \begin{bmatrix} 0 \\ 0 \\ \vdots \\ 0 \\ 1 \end{bmatrix}, \qquad B = \begin{bmatrix} 0 & 1 & 0 & \cdots & 0 \\ 0 & 0 & 1 & \cdots & 0 \\ \vdots & \vdots & \vdots & & \vdots \\ 0 & 0 & 0 & \cdots & 1 \\ (-1)^{d+1}\binom{d}{d} & (-1)^{d}\binom{d}{d-1} & (-1)^{d-1}\binom{d}{d-2} & \cdots & d \end{bmatrix}$$

if $d > 1$. Then since

$$Y_t = \nabla^d Y_t - \sum_{j=1}^{d} \binom{d}{j} (-1)^j Y_{t-j}, \tag{12.1.30}$$

the vector

$$\mathbf{Y}_{t-1} := (Y_{t-d}, \ldots, Y_{t-1})'$$

satisfies the equation,

$$\mathbf{Y}_t = A\nabla^d Y_t + B\mathbf{Y}_{t-1} = AG\mathbf{X}_t + B\mathbf{Y}_{t-1} + AZ_t.$$

Defining a new state vector $\mathbf{T}_t$ by stacking $\mathbf{X}_t$ and $\mathbf{Y}_{t-1}$, we therefore obtain the state equation,

$$\mathbf{T}_{t+1} := \begin{bmatrix} \mathbf{X}_{t+1} \\ \mathbf{Y}_t \end{bmatrix} = \begin{bmatrix} F & 0 \\ AG & B \end{bmatrix} \mathbf{T}_t + \begin{bmatrix} H \\ A \end{bmatrix} Z_t, \qquad t = 1, 2, \ldots, \tag{12.1.31}$$

and the observation equation, from (12.1.29) and (12.1.30),

$$Y_t = \begin{bmatrix} G & (-1)^{d+1}\binom{d}{d} & (-1)^{d}\binom{d}{d-1} & (-1)^{d-1}\binom{d}{d-2} & \cdots & d \end{bmatrix} \begin{bmatrix} \mathbf{X}_t \\ \mathbf{Y}_{t-1} \end{bmatrix} + Z_t, \qquad t = 1, 2, \ldots, \quad (12.1.32)$$

with initial condition,

$$\mathbf{T}_1 = \begin{bmatrix} \mathbf{X}_1 \\ \mathbf{Y}_0 \end{bmatrix} = \begin{bmatrix} \sum_{j=0}^{\infty} F^j H Z_{-j} \\ \mathbf{Y}_0 \end{bmatrix},$$

and the orthogonality conditions,

$$\mathbf{Y}_0 \perp Z_t, \qquad t = 0, \pm 1, \ldots, \quad (12.1.33)$$

where $\mathbf{Y}_0 = (Y_{1-d}, Y_{2-d}, \ldots, Y_0)'$. The conditions (12.1.33), which are satisfied in particular if $\mathbf{Y}_0$ is considered to be non-random and equal to the vector of *observed* values $(y_{1-d}, y_{2-d}, \ldots, y_0)'$, are imposed to ensure that the assumptions (a)–(e) are satisfied. They also imply that $\mathbf{X}_1 \perp \mathbf{Y}_0$ and $\mathbf{Y}_0 \perp \nabla^d Y_t$, $t \geq 1$, as required earlier in Section 9.5 for prediction of ARIMA processes.

State-space models for more general ARIMA processes (e.g. $\{Y_t\}$ such that $\{\nabla\nabla_{12} Y_t\}$ is an ARMA(p, q) process) can be constructed in the same way. (See Problem 12.9.)

For the ARIMA(1, 1, 1) process defined by

$$(1 - \phi B)(1 - B)Y_t = (1 + \theta B)Z_t, \qquad \{Z_t\} \sim \text{WN}(0, \sigma^2),$$

the vectors $\mathbf{X}_t$ and $\mathbf{Y}_{t-1}$ reduce to $\mathbf{X}_t = X_t$ and $\mathbf{Y}_{t-1} = Y_{t-1}$. The state-space representation (12.1.31) and (12.1.32) becomes (Problem 12.7)

$$Y_t = \begin{bmatrix} 1 & 1 \end{bmatrix} \begin{bmatrix} X_t \\ Y_{t-1} \end{bmatrix} + Z_t, \quad (12.1.34)$$

where

$$\begin{bmatrix} X_{t+1} \\ Y_t \end{bmatrix} = \begin{bmatrix} \phi & 0 \\ 1 & 1 \end{bmatrix} \begin{bmatrix} X_t \\ Y_{t-1} \end{bmatrix} + \begin{bmatrix} \phi + \theta \\ 1 \end{bmatrix} Z_t, \qquad t = 1, 2, \ldots, \quad (12.1.35)$$

and

$$\begin{bmatrix} X_1 \\ Y_0 \end{bmatrix} = \begin{bmatrix} (\phi + \theta)\sum_{j=0}^{\infty} \phi^j Z_{-j} \\ Y_0 \end{bmatrix}. \quad (12.1.36)$$

EXAMPLE 12.1.8 (An ARMA Process with Observational Error). The canonical observable representation of Example 12.1.6 is not always the most convenient representation to employ. For example, if instead of observing the ARMA process $\{Y_t\}$, we observe

$$U_t = Y_t + N_t,$$

where $\{N_t\}$ is a white-noise sequence uncorrelated with $\{Y_t\}$, then a state-space representation for $\{U_t\}$ is immediately obtained by retaining the state equation (12.1.23) of Example 12.1.5 and replacing the observation equation (12.1.20) by

$$U_t = [\theta_{r-1} \quad \theta_{r-2} \quad \cdots \quad \theta_0]\mathbf{X}_t + N_t.$$

(The state-space model in Example 12.1.6 can also be adapted to allow for added observational noise.)

State-space representations have many virtues, their generality, their ease of analysis, and the availability of the Kalman recursions which make least squares prediction and estimation a routine matter. Applications of the latter will be discussed later in this chapter. We conclude this section with a simple application of the state-space representation of Example 12.1.6 to the determination of the autocovariance function of a causal ARMA process.

The Autocovariance Function of a Causal ARMA Process

If $\{Y_t\}$ is the causal ARMA process defined by (12.1.19), then we know from Example 12.1.6 and (12.1.14) that

$$Y_t = G\mathbf{X}_t + Z_t,$$

where

$$G = [1 \quad 0 \quad 0 \quad \cdots \quad 0]$$

and

$$\mathbf{X}_t = HZ_{t-1} + FHZ_{t-2} + F^2HZ_{t-3} + \cdots,$$

with the square matrix F and the column vector H as in (12.1.25). It follows at once from these equations that

$$Y_t = Z_t + \sum_{j=1}^{\infty} GF^{j-1}HZ_{t-j}. \tag{12.1.37}$$

Hence

$$\gamma_Y(k) = \begin{cases} \sigma^2[1 + \sum_{j=1}^{\infty} GF^{j-1}HH'F'^{j-1}G'] & \text{if } k = 0, \\ \sigma^2[GF^{|k|-1}H + \sum_{j=1}^{\infty} GF^{j-1}HH'F'^{|k|+j-1}G'] & \text{if } k \neq 0. \end{cases}$$

The coefficients ψ_j in the representation $Y_t = \sum_{j=0}^{\infty} \psi_j Z_{t-j}$ can be read from (12.1.37) as

$$\psi_j = \begin{cases} 1 & \text{if } j = 0, \\ GF^{j-1}H & \text{if } j \geq 1, \end{cases}$$

which shows in particular that ψ_j converges to zero geometrically as $j \to \infty$.

This argument, unlike those used in Chapter 3, does not require any knowledge of the general theory of difference equations.

§12.2 The Kalman Recursions

In this section we shall consider three fundamental problems associated with the state-space model defined by (12.1.1) and (12.1.2) under the assumptions (a)–(e) of Section 12.1. These are all concerned with finding best (in the sense of minimum mean-square error) linear estimates of the state-vector $\mathbf{X}_t$ in terms of the observations $\mathbf{Y}_1, \mathbf{Y}_2, \ldots,$ and a random vector $\mathbf{Y}_0$ satisfying the conditions

$$\mathbf{Y}_0 \perp \mathbf{V}_t \quad \text{and} \quad \mathbf{Y}_0 \perp \mathbf{W}_t \quad \text{for all } t. \tag{12.2.1}$$

The vector $\mathbf{Y}_0$ will depend on the type of estimates required. In many (but not all) applications, $\mathbf{Y}_0$ will be the degenerate random vector $\mathbf{Y}_0 = \mathbf{1} = (1, 1, \ldots, 1)'$. Estimation of $\mathbf{X}_t$ in terms of

(a) $\mathbf{Y}_0, \ldots, \mathbf{Y}_{t-1}$ defines the *prediction problem*,
(b) $\mathbf{Y}_0, \ldots, \mathbf{Y}_t$ defines the *filtering problem*, and
(c) $\mathbf{Y}_0, \ldots, \mathbf{Y}_n$ defines the *smoothing problem* (in which it is assumed that $n > t$).

Each of these problems can be solved recursively using an appropriate set of Kalman recursions which will be established in this section. Before we can do so however we need to clarify the meaning of best linear estimate in this context.

Definition 12.2.1. The best one-step linear predictor, $\hat{\mathbf{X}}_t$, of $\mathbf{X}_t = (X_{t1}, \ldots, X_{tv})'$ is the random vector whose ith component, $i = 1, \ldots, v$, is the best linear predictor of X_{ti} in terms of all the components of the t vectors, $\mathbf{Y}_0$, $\mathbf{Y}_1, \ldots, \mathbf{Y}_{t-1}$. More generally the best estimator $\mathbf{X}_{t|k}$ of $\mathbf{X}_t$ is the random vector whose ith component, $i = 1, \ldots, v$, is the best linear estimator of X_{ti} in terms of all the components of $\mathbf{Y}_0, \mathbf{Y}_1, \ldots, \mathbf{Y}_k$. The latter notation covers all three problems (a), (b) and (c) with $k = t - 1$, t and n respectively. In particular $\hat{\mathbf{X}}_t = \mathbf{X}_{t|t-1}$. The corresponding error covariance matrices are defined to be

$$\Omega_{t|k} = E[(\mathbf{X}_t - \mathbf{X}_{t|k})(\mathbf{X}_t - \mathbf{X}_{t|k})'].$$

The Projection $P(\mathbf{X}|\mathbf{Y}_0, \ldots, \mathbf{Y}_t)$ of a Second-Order Random Vector $\mathbf{X}$

In order to find $\hat{\mathbf{X}}_t$ (and more generally $\mathbf{X}_{t|k}$) we introduce (cf. Section 11.4) the projections $P(\mathbf{X}|\mathbf{Y}_0, \ldots, \mathbf{Y}_t)$, where $\mathbf{X}, \mathbf{Y}_0, \ldots, \mathbf{Y}_t$ are jointly distributed random vectors with finite second moments. If $\mathbf{X}$ is a v-component random vector with finite second moments we shall say that $\mathbf{X} \in L_2^v$.

Definition 12.2.2. If $\mathbf{X} \in L_2^v$, and $\mathbf{Y}_0, \mathbf{Y}_1, \mathbf{Y}_2, \ldots$ have finite second moments, then we define $P(\mathbf{X}|\mathbf{Y}_0, \ldots, \mathbf{Y}_t)$ to be the random v-vector whose ith component is the projection $P(X_i|S)$ of the ith component of $\mathbf{X}$ onto the span, S, of all of the components of $\mathbf{Y}_0, \ldots, \mathbf{Y}_t$. We shall abbreviate the notation by writing

$$P_t(\mathbf{X}) := P(\mathbf{X}|\mathbf{Y}_0, \ldots, \mathbf{Y}_t), \qquad t = 0, 1, 2, \ldots,$$

throughout this chapter. The operator P_t is defined on $\bigcup_{v=1}^{\infty} L_2^v$.

Remark 1. By the definition of $P(X_i|S)$, $P_t(\mathbf{X})$ is the unique random vector with components in S such that

$$[\mathbf{X} - P_t(\mathbf{X})] \perp \mathbf{Y}_s, \qquad s = 0, \ldots, t.$$

(See (11.4.2) and (11.4.3).)

Remark 2. For any fixed v, $P_t(\cdot)$ is a projection operator on the Hilbert space L_2^v with inner product $\langle \mathbf{X}, \mathbf{Y} \rangle = \sum_{i=1}^{v} E(X_i Y_i)$ (see Problem 12.10). Orthogonality of $\mathbf{X}$ and $\mathbf{Y}$ with respect to this inner product however is not equivalent to the definition $E(\mathbf{X}\mathbf{Y}') = 0$. We shall continue to use the latter.

Remark 3. If all the components of $\mathbf{X}$, $\mathbf{Y}_1, \ldots, \mathbf{Y}_t$ are jointly normally distributed and $\mathbf{Y}_0 = \mathbf{1}$, then

$$P_t(\mathbf{X}) = E(\mathbf{X}|\mathbf{Y}_1, \ldots, \mathbf{Y}_t), \qquad t \geq 1.$$

Remark 4. P_t is linear in the sense that if A is any $k \times v$ matrix and $\mathbf{X}, \mathbf{V} \in L_2^v$ then

$$P_t(A\mathbf{X}) = AP_t(\mathbf{X})$$

and

$$P_t(\mathbf{X} + \mathbf{V}) = P_t(\mathbf{X}) + P_t(\mathbf{V}).$$

Remark 5. If $\mathbf{Y} \in L_2^w$ and $\mathbf{X} \in L_2^v$, then

$$P(\mathbf{X}|\mathbf{Y}) = M\mathbf{Y},$$

where M is the $v \times w$ matrix, $M = E(\mathbf{X}\mathbf{Y}')[E(\mathbf{Y}\mathbf{Y}')]^{-1}$ and $[E(\mathbf{Y}\mathbf{Y}')]^{-1}$ is any generalized inverse of $E(\mathbf{Y}\mathbf{Y}')$. (A generalized inverse of a matrix S is a matrix S^{-1} such that $SS^{-1}S = S$. Every matrix has at least one. See Problem 12.11).

Proposition 12.2.1. *If* $\{\mathbf{X}_t\}$ *and* $\{\mathbf{Y}_t\}$ *are defined as in* (12.1.1) *and* (12.1.2), *then for* $t, s \geq 1$,

$$\mathbf{X}_{t|s} = P_s(\mathbf{X}_t), \tag{12.2.2}$$

and in particular,

$$\hat{\mathbf{X}}_t = P_{t-1}(\mathbf{X}_t), \tag{12.2.3}$$

where $\hat{\mathbf{X}}_t$ *and* $\mathbf{X}_{t|s}$ *are as in Definition* 12.2.1 *and* $\mathbf{Y}_0$ *satisfies* (12.2.1).

Proof. The result is an immediate consequence of Definitions 12.2.1 and 12.2.2. □

We turn next to the derivation of the Kalman recursions for the one-step predictors of $\mathbf{X}_t$ in the state-space model defined by (12.1.1) and (12.1.2).

Proposition 12.2.2 (Kalman Prediction). *Suppose that*

$$\mathbf{X}_{t+1} = F_t\mathbf{X}_t + \mathbf{V}_t, \qquad t = 1, 2, \ldots, \tag{12.2.4}$$

and

$$\mathbf{Y}_t = G_t\mathbf{X}_t + \mathbf{W}_t, \qquad t = 1, 2, \ldots, \tag{12.2.5}$$

where

$$E\mathbf{U}_t = E\begin{bmatrix}\mathbf{V}_t\\ \mathbf{W}_t\end{bmatrix} = \mathbf{0}, \qquad E(\mathbf{U}_t\mathbf{U}_t') = \begin{bmatrix}Q_t & S_t\\ S_t' & R_t\end{bmatrix},$$

$\mathbf{X}_1, \mathbf{U}_1, \mathbf{U}_2, \ldots,$ *are uncorrelated, and* $\mathbf{Y}_0$ *satisfies* (12.2.1). *Then the one-step predictors,*

$$\hat{\mathbf{X}}_t = P_{t-1}\mathbf{X}_t,$$

and the error covariance matrices,

$$\Omega_t = E[(\mathbf{X}_t - \hat{\mathbf{X}}_t)(\mathbf{X}_t - \hat{\mathbf{X}}_t)'],$$

are uniquely determined by the initial conditions,

$$\hat{\mathbf{X}}_1 = P(\mathbf{X}_1|\mathbf{Y}_0), \qquad \Pi_1 = E(\mathbf{X}_1\mathbf{X}_1'), \qquad \Psi_1 = E(\hat{\mathbf{X}}_1\hat{\mathbf{X}}_1'), \qquad \Omega_1 = \Pi_1 - \Psi_1,$$

and the recursions, for $t = 1, 2, \ldots,$

$$\begin{cases}\Delta_t &= G_t\Omega_t G_t' + R_t,\\ \Theta_t &= F_t\Omega_t G_t' + S_t,\\ \Pi_{t+1} &= F_t\Pi_t F_t' + Q_t,\\ \Psi_{t+1} &= F_t\Psi_t F_t' + \Theta_t\Delta_t^{-1}\Theta_t',\\ \Omega_{t+1} &= \Pi_{t+1} - \Psi_{t+1},\end{cases} \tag{12.2.6}$$

$$\hat{\mathbf{X}}_{t+1} = F_t\hat{\mathbf{X}}_t + \Theta_t\Delta_t^{-1}(\mathbf{Y}_t - G_t\hat{\mathbf{X}}_t), \tag{12.2.7}$$

where Δ_t^{-1} *is any generalized inverse of* Δ_t.

Proof. We shall make use of the *innovations*, $\mathbf{I}_t$, defined by $\mathbf{I}_0 = \mathbf{Y}_0$ and

$$\mathbf{I}_t = \mathbf{Y}_t - P_{t-1}\mathbf{Y}_t = \mathbf{Y}_t - G_t\hat{\mathbf{X}}_t = G_t(\mathbf{X}_t - \hat{\mathbf{X}}_t) + \mathbf{W}_t, \qquad t = 1, 2, \ldots.$$

The sequence $\{\mathbf{I}_t\}$ is orthogonal by Remark 1. Using Remarks 4 and 5 and the relation,

$$P_t(\cdot) = P_{t-1}(\cdot) + P(\cdot|\mathbf{I}_t), \tag{12.2.8}$$

(see Problem 12.12), we find that

$$\begin{aligned}\hat{\mathbf{X}}_{t+1} &= P_{t-1}\mathbf{X}_{t+1} + P(\mathbf{X}_{t+1}|\mathbf{I}_t)\\ &= P_{t-1}(F_t\mathbf{X}_t + \mathbf{V}_t) + \Theta_t\Delta_t^{-1}\mathbf{I}_t\\ &= F_t\hat{\mathbf{X}}_t + \Theta_t\Delta_t^{-1}\mathbf{I}_t,\end{aligned} \tag{12.2.9}$$

where

$$\begin{aligned}\Delta_t &= E(\mathbf{I}_t\mathbf{I}_t') = G_t\Omega_t G_t' + R_t,\\ \Theta_t &= E(\mathbf{X}_{t+1}\mathbf{I}_t') = E[(F_t\mathbf{X}_t + \mathbf{V}_t)([\mathbf{X}_t - \hat{\mathbf{X}}_t]'G_t' + \mathbf{W}_t')]\\ &= F_t\Omega_t G_t' + S_t.\end{aligned}$$

To evaluate Δ_t, Θ_t and Ω_t recursively, we observe that

$$\Omega_{t+1} = E(\mathbf{X}_{t+1}\mathbf{X}_{t+1}') - E(\hat{\mathbf{X}}_{t+1}\hat{\mathbf{X}}_{t+1}') = \Pi_{t+1} - \Psi_{t+1},$$

where, from (12.2.4) and (12.2.9),

$$\Pi_{t+1} = F_t\Pi_t F_t' + Q_t.$$

and

$$\Psi_{t+1} = F_t\Psi_t F_t' + \Theta_t\Delta_t^{-1}\Theta_t'. \qquad \square$$

Remark 6. The initial state predictor $\hat{\mathbf{X}}_1$ is found using Remark 5. In the important special case when $\mathbf{Y}_0 = \mathbf{1}$, it reduces to $E\mathbf{X}_1$.

h-Step Prediction of $\{\mathbf{Y}_t\}$ Using the Kalman Recursions

The results of Proposition 12.2.2 lead to a very simple algorithm for the recursive calculation of the best linear mean-square predictors, $P_t\mathbf{Y}_{t+h}$, $h = 1, 2, \ldots$. From (12.2.9), (12.2.4), (12.2.5), (12.2.7) and Remark 2 in Section 12.1, we find that

$$P_t\mathbf{X}_{t+1} = F_t P_{t-1}\mathbf{X}_t + \Theta_t\Delta_t^{-1}(\mathbf{Y}_t - P_{t-1}\mathbf{Y}_t), \tag{12.2.10}$$

$$\begin{aligned}P_t\mathbf{X}_{t+h} &= F_{t+h-1}P_t\mathbf{X}_{t+h-1}\\ &\ \ \vdots\\ &= F_{t+h-1}F_{t+h-2}\cdots F_{t+1}P_t\mathbf{X}_{t+1}, \qquad h = 2, 3, \ldots,\end{aligned} \tag{12.2.11}$$

and

$$P_t\mathbf{Y}_{t+h} = G_{t+h}P_t\mathbf{X}_{t+h}, \qquad h = 1, 2, \ldots. \tag{12.2.12}$$

From the relation,

$$\mathbf{X}_{t+h} - P_t\mathbf{X}_{t+h} = F_{t+h-1}(\mathbf{X}_{t+h-1} - P_t\mathbf{X}_{t+h-1}) + \mathbf{V}_{t+h-1}, \qquad h = 2, 3, \ldots,$$

we find that $\Omega_t^{(h)} := E[(\mathbf{X}_{t+h} - P_t\mathbf{X}_{t+h})(\mathbf{X}_{t+h} - P_t\mathbf{X}_{t+h})']$ satisfies the recursions,

$$\Omega_t^{(h)} = F_{t+h-1}\Omega_t^{(h-1)}F'_{t+h-1} + Q_{t+h-1}, \qquad h = 2, 3, \ldots, \tag{12.2.13}$$

with $\Omega_t^{(1)} = \Omega_{t+1}$. Then from (12.2.5) and (12.2.12) it follows that $\Sigma_t^{(h)} := E[(\mathbf{Y}_{t+h} - P_t\mathbf{Y}_{t+h})(\mathbf{Y}_{t+h} - P_t\mathbf{Y}_{t+h})']$ is given by

$$\Sigma_t^{(h)} = G_{t+h}\Omega_t^{(h)}G'_{t+h} + R_{t+h}, \qquad h = 1, 2, \ldots. \tag{12.2.14}$$

Proposition 12.2.3 (Kalman Filtering). *Under the conditions of Proposition* 12.2.2, *and with the same notation, the estimates* $\mathbf{X}_{t|t} = P_t\mathbf{X}_t$ *and the error covariance matrices* $\Omega_{t|t} = E[(\mathbf{X}_t - \mathbf{X}_{t|t})(\mathbf{X}_t - \mathbf{X}_{t|t})']$ *are determined by the relations,*

$$P_t\mathbf{X}_t = P_{t-1}\mathbf{X}_t + \Omega_t G_t'\Delta_t^{-1}(\mathbf{Y}_t - G_t\hat{\mathbf{X}}_t), \tag{12.2.15}$$

and

$$\Omega_{t|t} = \Omega_t - \Omega_t G_t'\Delta_t^{-1}G_t\Omega_t'. \tag{12.2.16}$$

Proof. From (12.2.8) it follows that

$$P_t\mathbf{X}_t = P_{t-1}\mathbf{X}_t + M\mathbf{I}_t,$$

where

$$\begin{aligned} M &= E(\mathbf{X}_t\mathbf{I}_t')[E(\mathbf{I}_t\mathbf{I}_t')]^{-1} \\ &= E[\mathbf{X}_t(G_t(\mathbf{X}_t - \hat{\mathbf{X}}_t) + W_t)']\Delta_t^{-1} \\ &= \Omega_t G_t'\Delta_t^{-1}. \end{aligned} \tag{12.2.17}$$

To establish (12.2.16) we write

$$\mathbf{X}_t - P_{t-1}\mathbf{X}_t = \mathbf{X}_t - P_t\mathbf{X}_t + P_t\mathbf{X}_t - P_{t-1}\mathbf{X}_t = \mathbf{X}_t - P_t\mathbf{X}_t + M\mathbf{I}_t.$$

Using (12.2.17) and the orthogonality of $\mathbf{X}_t - P_t\mathbf{X}_t$ and $M\mathbf{I}_t$, we find from the last equation that

$$\Omega_t = \Omega_{t|t} + \Omega_t G_t'\Delta_t^{-1}G_t\Omega_t',$$

as required. □

Proposition 12.2.4 (Kalman Fixed Point Smoothing). *Under the conditions of Proposition* 12.2.2, *and with the same notation, the estimates* $\mathbf{X}_{t|n} = P_n\mathbf{X}_t$, *and the error covariance matrices* $\Omega_{t|n} = E[(\mathbf{X}_t - \mathbf{X}_{t|n})(\mathbf{X}_t - \mathbf{X}_{t|n})']$ *are determined for fixed t by the following recursions, which can be solved successively for* $n = t, t+1, \ldots$:

$$P_n\mathbf{X}_t = P_{n-1}\mathbf{X}_t + \Omega_{t,n}G_n'\Delta_n^{-1}(\mathbf{Y}_n - G_n\hat{\mathbf{X}}_n), \tag{12.2.18}$$

$$\Omega_{t,n+1} = \Omega_{t,n}[F_n - \Theta_n\Delta_n^{-1}G_n]', \tag{12.2.19}$$

$$\Omega_{t|n} = \Omega_{t|n-1} - \Omega_{t,n}G_n'\Delta_n^{-1}G_n\Omega_{t,n}', \tag{12.2.20}$$

with initial conditions, $P_{t-1}\mathbf{X}_t = \hat{\mathbf{X}}_t$ *and* $\Omega_{t,t} = \Omega_{t|t-1} = \Omega_t$ (*found from Proposition* 12.2.2).

PROOF. Using (12.2.8) we can write $P_n\mathbf{X}_t = P_{n-1}\mathbf{X}_t + C\mathbf{I}_n$, where $\mathbf{I}_n = G_n(\mathbf{X}_n - \hat{\mathbf{X}}_n) + \mathbf{W}_n$. By Remark 5 above,

$$C = E[\mathbf{X}_t(G_n(\mathbf{X}_n - \hat{\mathbf{X}}_n) + \mathbf{W}_n)'][E(\mathbf{I}_n\mathbf{I}_n')]^{-1} = \Omega_{t,n}G_n'\Delta_n^{-1}, \qquad (12.2.21)$$

where $\Omega_{t,n} := E[(\mathbf{X}_t - \hat{\mathbf{X}}_t)(\mathbf{X}_n - \hat{\mathbf{X}}_n)']$. It follows now from (12.2.4), (12.2.10), the orthogonality of $\mathbf{V}_n$ and $\mathbf{W}_n$ with $\mathbf{X}_t - \hat{\mathbf{X}}_t$, and the definition of $\Omega_{t,n}$ that

$$\Omega_{t,n+1} = E[(\mathbf{X}_t - \hat{\mathbf{X}}_t)(\mathbf{X}_n - \hat{\mathbf{X}}_n)'(F_n - \Theta_n\Delta_n^{-1}G_n)'] = \Omega_{t,n}[F_n - \Theta_n\Delta_n^{-1}G_n]'$$

thus establishing (12.2.19). To establish (12.2.20) we write

$$\mathbf{X}_t - P_n\mathbf{X}_t = \mathbf{X}_t - P_{n-1}\mathbf{X}_t - C\mathbf{I}_n.$$

Using (12.2.21) and the orthogonality of $\mathbf{X}_t - P_n\mathbf{X}_t$ and $\mathbf{I}_n$, the last equation then gives

$$\Omega_{t|n} = \Omega_{t|n-1} - \Omega_{t,n}G_n'\Delta_n^{-1}G_n\Omega_{t,n}', \qquad n = t, t+1, \ldots,$$

as required. □

EXAMPLE 12.2.1 (A Non-Stationary State-Space Model). Consider the univariate non-stationary model defined by

$$X_{t+1} = 2X_t + V_t, \qquad t = 1, 2, \ldots,$$

and

$$Y_t = X_t + W_t, \qquad t = 1, 2, \ldots,$$

where

$$X_1 = 1, \quad Y_0 = 1, \quad \text{and} \quad \mathbf{U}_t = \begin{bmatrix} V_t \\ W_t \end{bmatrix} \sim \text{WN}\left(\mathbf{0}, \begin{bmatrix} 1 & 0 \\ 0 & 1 \end{bmatrix}\right).$$

We seek state estimates in terms of $1, Y_1, Y_2, \ldots$, and therefore choose $Y_0 = 1$. In the notation of Proposition 12.2.2 we have $\Pi_1 = \Psi_1 = 1$, $\Omega_1 = 0$, and the recursions,

$$\begin{cases} \Delta_t &= \Omega_t + 1, \\ \Theta_t &= 2\Omega_t, \\ \Pi_{t+1} &= 4\Pi_t + 1 = \frac{1}{3}(4^{t+1} - 1), \\ \Psi_{t+1} &= 4\Psi_t + \Theta_t^2\Delta_t^{-1} = 4\Psi_t + \dfrac{4\Omega_t^2}{1 + \Omega_t}, \\ \Omega_{t+1} &= \Pi_{t+1} - \Psi_{t+1} = \frac{1}{3}(4^{t+1} - 1) - \Psi_{t+1}. \end{cases} \qquad (12.2.22)$$

Setting $\Psi_{t+1} = -\Omega_{t+1} + \frac{1}{3}(4^{t+1} - 1)$, we find from the fourth equation that

$$-\Omega_{t+1} + \tfrac{1}{3}(4^{t+1} - 1) = 4(-\Omega_t + \tfrac{1}{3}(4^t - 1)) + \frac{4\Omega_t^2}{1 + \Omega_t}.$$

This yields the recursion

$$\Omega_{t+1} = \frac{1 + 5\Omega_t}{1 + \Omega_t},$$

from which it can be shown that

$$\Omega_t = \frac{4 + 2\sqrt{5} - (\sqrt{5} - 1)c^{2-t}}{2 + (\sqrt{5} + 3)c^{2-t}}, \qquad \text{where } c = \tfrac{1}{2}(7 + 3\sqrt{5}). \quad (12.2.23)$$

We can now write the solution of (12.2.22) as

$$\begin{cases} \Delta_t = \Omega_t + 1, \\ \Theta_t = 2\Omega_t, \\ \Pi_t = \tfrac{1}{3}(4^t - 1), \\ \Psi_t = \tfrac{1}{3}(4^t - 1) - \Omega_t, \end{cases} \qquad (12.2.24)$$

with Ω_t as in (12.2.23).

The equations for the estimators and mean squared errors as derived in the preceding propositions can be made quite explicit for this example. Thus from Proposition 12.2.2 we find that the one-step predictor of X_{t+1} satisfies the recursions,

$$\hat{X}_{t+1} = 2\hat{X}_t + \frac{2\Omega_t}{1 + \Omega_t}(Y_t - \hat{X}_t), \qquad \text{with } \hat{X}_1 = 1,$$

and with mean squared error Ω_{t+1} given by (12.2.23). Similarly, from (12.2.12) and (12.2.14), the one-step predictor of Y_{t+1} and its mean squared error are given by,

$$P_t Y_{t+1} = \hat{X}_{t+1}$$

and

$$\Sigma_t^{(1)} = \Omega_{t+1} + 1.$$

The filtered state estimate for X_{t+1} and its mean squared error are found from Proposition 12.2.3 to be

$$P_{t+1}X_{t+1} = \hat{X}_{t+1} + \frac{\Omega_{t+1}}{1 + \Omega_{t+1}}(Y_{t+1} - \hat{X}_{t+1})$$

and

$$\Omega_{t+1|t+1} = \frac{\Omega_{t+1}}{1 + \Omega_{t+1}}.$$

Finally the smoothed estimate of X_{t+1} based on $Y_0, Y_1, \ldots, Y_{t+2}$ is found, using Proposition 12.2.4 and some simple algebra, to be

$$P_{t+2}X_{t+1} = \hat{X}_{t+1} + \frac{\Omega_{t+1}}{1 + \Omega_{t+1}}\left((Y_{t+1} - \hat{X}_{t+1}) + \frac{2}{1 + \Omega_{t+1}}(Y_{t+2} - \hat{X}_{t+2})\right),$$

with mean squared error,

$$\Omega_{t+1|t+2} = \frac{\Omega_{t+1}}{1+\Omega_{t+1}} - \frac{4\Omega_{t+1}^2}{(1+\Omega_{t+1})^2(1+\Omega_{t+2})}.$$

It is clear from (12.2.23) that the mean squared error, Ω_t, of the one-step predictor of the state, X_t, converges as $t \to \infty$. In fact we have, as $t \to \infty$,

$$\Omega_{t|t-1} = \Omega_t \to 2 + \sqrt{5}, \qquad \Omega_{t|t} \to \frac{\sqrt{5}+1}{4} \quad \text{and} \quad \Omega_{t|t+1} \to \frac{\sqrt{5}-1}{4}.$$

These results demonstrate the improvement in estimation of the state X_t as we go from one-step prediction to filtering to smoothing based on the observed data $Y_0, \ldots, Y_{t+1}$.

Remark 7. For more complex state-space models it is not feasible to derive explicit algebraic expressions for the coefficients and mean squared errors as in Example 12.2.1. Numerical solution of the Kalman recursions is however relatively straightforward.

EXAMPLE 12.2.2 (Prediction of an ARIMA(p, 1, q) Process). In Example 12.1.7, we derived the following state-space model for the ARIMA(p, 1, q) process $\{Y_t\}$:

$$Y_t = [G \quad 1]\begin{bmatrix} \mathbf{X}_t \\ Y_{t-1} \end{bmatrix} + Z_t, \qquad t = 1, 2, \ldots, \tag{12.2.25}$$

where

$$\begin{bmatrix} \mathbf{X}_{t+1} \\ Y_t \end{bmatrix} = \begin{bmatrix} F & 0 \\ G & 1 \end{bmatrix}\begin{bmatrix} \mathbf{X}_t \\ Y_{t-1} \end{bmatrix} + \begin{bmatrix} H \\ 1 \end{bmatrix} Z_t, \qquad t = 1, 2, \ldots,$$

$$\mathbf{X}_1 = \sum_{j=0}^{\infty} F^j H Z_{t-j}, \tag{12.2.26}$$

$$Y_0 \perp Z_t, \qquad t = 0, \pm 1, \ldots,$$

and the matrices, F, G and H are as specified in Example 12.1.7. Note that Y_0 (and the corresponding innovation $I_0 = Y_0$) here refers to the first observation of the ARIMA series and *not* to the constant value 1. The operator $P_t(\cdot)$, as usual, denotes projection onto $\overline{\text{sp}}\{Y_0, \ldots, Y_t\}$. Letting $\mathbf{T}_t$ denote the state vector $(\mathbf{X}_t', Y_{t-1})'$ at time t, the initial conditions for the recursions (12.2.6) and (12.2.7) are therefore

$$\hat{\mathbf{T}}_1 = P_0 \mathbf{T}_1 = \begin{bmatrix} 0 \\ Y_0 \end{bmatrix}, \qquad \Pi_1 = E(\mathbf{T}_1 \mathbf{T}_1') = \begin{bmatrix} E(\mathbf{X}_1 \mathbf{X}_1') & 0 \\ 0 & EY_0^2 \end{bmatrix},$$

$$\Psi_1 = E(\hat{\mathbf{T}}_1 \hat{\mathbf{T}}_1') = \begin{bmatrix} 0 & 0 \\ 0 & EY_0^2 \end{bmatrix}, \qquad \Omega_1 = \Pi_1 - \Psi_1.$$

The recursions (12.2.6) and (12.2.7) for the one-step predictors $\hat{\mathbf{T}}_t$ and the error covariance matrices $\Omega_t = E[(\mathbf{T}_t - \hat{\mathbf{T}}_t)(\mathbf{T}_t - \hat{\mathbf{T}}_t)']$ can now be solved. The h-step predictors and mean squared errors for the ARIMA process $\{Y_t\}$ are then found from (12.2.11)–(12.2.14).

Remark 8. It is worth noting in the preceding example, since $X_t = Y_t - Y_{t-1}$ is orthogonal to Y_0, $t \geq 1$, that

$$P_t \mathbf{X}_{t+1} = P(\mathbf{X}_{t+1} | X_1, \ldots, X_t) = \mathbf{X}^*_{t+1},$$

where $\mathbf{X}^*_{t+1}$ is the best linear predictor of $\mathbf{X}_{t+1}$ based on $X_1, \ldots, X_t$. Consequently, the one-step predictors of the state-vectors $\mathbf{T}_t = (\mathbf{X}'_t, Y_{t-1})'$ are

$$\begin{bmatrix} P_{t-1}\mathbf{X}_t \\ Y_{t-1} \end{bmatrix} = \begin{bmatrix} \mathbf{X}^*_t \\ Y_{t-1} \end{bmatrix}, \qquad t = 1, 2, \ldots,$$

with error covariance matrices,

$$\Omega_t = \begin{bmatrix} \Omega^*_t & 0 \\ 0 & 0 \end{bmatrix},$$

where $\mathbf{X}^*_t$ and Ω^*_t are computed by applying the recursions (12.2.6) and (12.2.7) to the state-space model for the ARMA process $\{X_t\}$. Applying (12.2.11) and (12.2.12) to the model (12.2.25), (12.2.26) we see in particular that

$$P_t Y_{t+1} = G P_t \mathbf{X}_{t+1} + Y_t = G\mathbf{X}^*_{t+1} + Y_t$$

and

$$E(Y_{t+1} - P_t Y_{t+1})^2 = G\Omega^*_{t+1}G' + \sigma^2.$$

Remark 9. In view of the matrix manipulations associated with state-space representations, the forecasting of ARIMA models by the method described in Section 9.5 is simpler and more direct than the method described above. However, if there are missing observations in the data set, the state-space representation is much more convenient for prediction, parameter estimation and the estimation of missing values. These problems are treated in the next section.

§12.3 State-Space Models with Missing Observations

State-space representations and the associated Kalman recursions are ideally suited to the precise analysis of data with missing values, as was pointed out by Jones (1980) in the context of maximum likelihood estimation for ARMA processes. In this section we shall deal with two missing-value problems for state-space models. The first is the evaluation of the (Gaussian) likelihood based on $\{\mathbf{Y}_{i_1}, \ldots, \mathbf{Y}_{i_r}\}$ where $i_1, i_2, \ldots, i_r$ are positive integers such that $1 \leq i_1 < i_2 < \cdots < i_r \leq n$. (This allows for observation of the process $\{\mathbf{Y}_t\}$ at irregular intervals, or equivalently for the possibility that $(n - r)$ observations are missing from the sequence $\{\mathbf{Y}_1, \ldots, \mathbf{Y}_n\}$.) The solution of this problem will enable us, in particular, to carry out maximum likelihood estimation for ARMA and ARIMA processes with missing values. The second problem to be considered is the minimum mean squared error estimation of the missing values themselves.

The Gaussian Likelihood of $\{\mathbf{Y}_{i_1}, \ldots, \mathbf{Y}_{i_r}\}$, $1 \le i_1 < i_2 < \cdots < i_r \le n$

Consider the state-space model defined by equations (12.1.1) and (12.1.2) and suppose that the model is completely parameterized by the components of the vector $\boldsymbol{\theta}$. If there are no missing observations, i.e. if $r = n$ and $i_j = j$, $j = 1, \ldots, n$, then the likelihood of the observations $\{\mathbf{Y}_1, \ldots, \mathbf{Y}_n\}$ is easily found as in (11.5.4) to be

$$L(\boldsymbol{\theta}; \mathbf{Y}_1, \ldots, \mathbf{Y}_n) = (2\pi)^{-nw/2}\left(\prod_{j=1}^{n} \det \Sigma_j\right)^{-1/2} \exp\left\{-\frac{1}{2}\sum_{j=1}^{n} (\mathbf{Y}_j - \hat{\mathbf{Y}}_j)'\Sigma_j^{-1}(\mathbf{Y}_j - \hat{\mathbf{Y}}_j)\right\},$$

where $\hat{\mathbf{Y}}_j = P_{j-1}\mathbf{Y}_j$ and $\Sigma_j = \Sigma_j^{(1)}, j \ge 1$, are the one-step predictors and error covariance matrices found from (12.2.12) and (12.2.14) with $\mathbf{Y}_0 := \mathbf{1}$.

To deal with the more general case of possibly irregularly spaced observations $\{\mathbf{Y}_{i_1}, \ldots, \mathbf{Y}_{i_r}\}$, we introduce a new series $\{\mathbf{Y}_t^*\}$, related to the process $\{\mathbf{X}_t\}$ by the modified observation equations, $\mathbf{Y}_0^* = \mathbf{Y}_0$ and

$$\mathbf{Y}_t^* = G_t^*\mathbf{X}_t + \mathbf{W}_t^*, \qquad t = 1, 2, \ldots, \tag{12.3.1}$$

where

$$G_t^* = \begin{cases} G_t & \text{if } t \in \{i_1, \ldots, i_r\}, \\ 0 & \text{otherwise,} \end{cases} \qquad \mathbf{W}_t^* = \begin{cases} \mathbf{W}_t & \text{if } t \in \{i_1, \ldots, i_r\}, \\ \mathbf{N}_t & \text{otherwise,} \end{cases} \tag{12.3.2}$$

and $\{\mathbf{N}_t\}$ is iid with

$$\mathbf{N}_t \sim \mathrm{N}(\mathbf{0}, I_{w \times w}), \quad \mathbf{N}_s \perp \mathbf{X}_1, \quad \mathbf{N}_s \perp \begin{bmatrix} \mathbf{V}_t \\ \mathbf{W}_t \end{bmatrix}, \quad s, t = 0, \pm 1, \ldots. \tag{12.3.3}$$

Equations (12.3.1) and (12.1.2) constitute a state-space representation for the new series $\{\mathbf{Y}_t^*\}$, which coincides with $\{\mathbf{Y}_t\}$ at each $t \in \{i_1, i_2, \ldots, i_r\}$, and at other times takes random values which are independent of $\{\mathbf{Y}_t\}$ with a distribution independent of $\boldsymbol{\theta}$.

Let $L_1(\boldsymbol{\theta}; \mathbf{y}_{i_1}, \ldots, \mathbf{y}_{i_r})$ be the Gaussian likelihood based on the observed values $\mathbf{y}_{i_1}, \ldots, \mathbf{y}_{i_r}$ of $\mathbf{Y}_{i_1}, \ldots, \mathbf{Y}_{i_r}$ under the model defined by (12.1.1) and (12.1.2). Corresponding to these observed values, we define a new sequence,

$\mathbf{y}_1^*, \ldots, \mathbf{y}_n^*$, by

$$\mathbf{y}_t^* = \begin{cases} \mathbf{y}_t & \text{if } t \in \{i_1, \ldots, i_r\}, \\ \mathbf{0} & \text{otherwise.} \end{cases} \tag{12.3.4}$$

Then it is clear from the preceding paragraph that

$$L_1(\boldsymbol{\theta}; \mathbf{y}_{i_1}, \ldots, \mathbf{y}_{i_r}) = (2\pi)^{(n-r)w/2} L_2(\boldsymbol{\theta}; \mathbf{y}_1^*, \ldots, \mathbf{y}_n^*), \tag{12.3.5}$$

where L_2 denotes the Gaussian likelihood under the model defined by (12.3.1) and (12.1.2).

In view of (12.3.5) we can now compute the required likelihood L_1 of the realized values $\{\mathbf{y}_t, t = i_1, \ldots, i_r\}$ as follows:

(i) Define the sequence $\{\mathbf{y}_t^*, t = 1, \ldots, n\}$ as in (12.3.4).

(ii) Find the one-step predictors $\hat{\mathbf{Y}}_t^*$ of $\mathbf{Y}_t^*$, and their error covariance matrices Σ_t^*, using Proposition 12.2.2 and the equations (12.2.12) and (12.2.14) applied to the state-space representation, (12.3.1) and (12.1.2) of $\{\mathbf{Y}_t^*\}$. Denote the realized values of the predictors, based on the observation sequence $\{\mathbf{y}_t^*\}$, by $\{\hat{\mathbf{y}}_t^*\}$.

(iii) The required Gaussian likelihood of the irregularly spaced observations, $\{\mathbf{y}_{i_1}, \ldots, \mathbf{y}_{i_r}\}$, is then, by (12.3.5),

$$L_1(\boldsymbol{\theta}; \mathbf{y}_{i_1}, \ldots, \mathbf{y}_{i_r}) = (2\pi)^{-rw/2}\left(\prod_{j=1}^{n} \det \Sigma_j^*\right)^{-1/2} \exp\left\{-\frac{1}{2}\sum_{j=1}^{n} (\mathbf{y}_j^* - \hat{\mathbf{y}}_j^*)'\Sigma_j^{*-1}(\mathbf{y}_j^* - \hat{\mathbf{y}}_j^*)\right\}. \tag{12.3.6}$$

EXAMPLE 12.3.1 (An AR(1) Series with One Missing Observation). Let $\{Y_t\}$ be the causal AR(1) process defined by

$$Y_t - \phi Y_{t-1} = Z_t, \qquad \{Z_t\} \sim \mathrm{WN}(0, \sigma^2).$$

To find the Gaussian likelihood of the observations y_1, y_3, y_4 and y_5 of Y_1, Y_3, Y_4 and Y_5 we follow the steps outlined above.

(i) Set $y_i^* = y_i$, $i = 1, 3, 4, 5$ and $y_2^* = 0$.

(ii) We start with the state-space model for $\{Y_t\}$ from Example 12.1.4, i.e. $Y_t = X_t$, $X_{t+1} = \phi X_t + Z_{t+1}$. The corresponding model for $\{Y_t^*\}$ is then, from (12.3.1),

$$X_{t+1} = F_t X_t + V_t, \qquad t = 1, 2, \ldots,$$

$$Y_t^* = G_t^* X_t + W_t^*, \qquad t = 1, 2, \ldots,$$

where

$$F_t = \phi, \quad G_t^* = \begin{cases} 1 & \text{if } t \neq 2, \\ 0 & \text{if } t = 2, \end{cases} \qquad V_t = Z_{t+1}, \qquad W_t^* = \begin{cases} 0 & \text{if } t \neq 2, \\ N_t & \text{if } t = 2, \end{cases}$$

$$Q_t = \sigma^2, \qquad R_t^* = \begin{cases} 0 & \text{if } t \neq 2, \\ 1 & \text{if } t = 2, \end{cases} \qquad S_t^* = 0,$$

and $X_1 = \sum_{j=0}^{\infty} \phi^j Z_{1-j}$ (see Remark 5 of Section 12.1). Starting from the initial conditions,

$$\hat{X}_1 = 0, \quad \Pi_1 = \sigma^2/(1-\phi^2), \quad \Psi_1 = 0, \quad \Omega_1 = \sigma^2/(1-\phi^2),$$

and applying the recursions (12.2.6), we find (Problem 12.16) that

$$\Theta_t \Delta_t^{-1} = \begin{cases} \phi & \text{if } t = 1, 3, 4, 5, \\ 0 & \text{if } t = 2, \end{cases} \qquad \Omega_t = \begin{cases} \sigma^2/(1-\phi^2) & \text{if } t = 1, \\ \sigma^2(1+\phi^2) & \text{if } t = 3, \\ \sigma^2 & \text{if } t = 2, 4, 5, \end{cases}$$

and

$$\hat{X}_1 = 0, \quad \hat{X}_2 = \phi Y_1, \quad \hat{X}_3 = \phi^2 Y_1, \quad \hat{X}_4 = \phi Y_3, \quad \hat{X}_5 = \phi Y_4.$$

From (12.2.12) and (12.2.14) with $h = 1$, we find that

$$\hat{Y}_1^* = 0, \quad \hat{Y}_2^* = 0, \quad \hat{Y}_3^* = \phi^2 Y_1, \quad \hat{Y}_4^* = \phi Y_3, \quad \hat{Y}_5^* = \phi Y_4,$$

with corresponding mean squared errors,

$$\Sigma_1^* = \sigma^2/(1-\phi^2), \quad \Sigma_2^* = 1, \quad \Sigma_3^* = \sigma^2(1+\phi^2), \quad \Sigma_4^* = \sigma^2, \quad \Sigma_5^* = \sigma^2.$$

(iii) From the preceding calculations we can now write the likelihood of the original data as

$$L_1(\phi, \sigma^2; y_1, y_3, y_4, y_5) = \sigma^{-4}(2\pi)^{-2}[(1-\phi^2)/(1+\phi^2)]^{1/2}$$
$$\cdot \exp\left\{ -\frac{1}{2\sigma^2} \left\{ y_1^2(1-\phi^2) + \frac{(y_3 - \phi^2 y_1)^2}{1+\phi^2} + (y_4 - \phi y_3)^2 + (y_5 - \phi y_4)^2 \right\}\right\}.$$

EXAMPLE 12.3.2 (An ARIMA(1, 1, 1) Series with One Missing Observation). Suppose we have observations y_0, y_1, y_3, y_4 and y_5 of the ARIMA(1, 1, 1) process defined in Example 12.1.7. The Gaussian likelihood of the observations y_1, y_3, y_4 and y_5 conditional on $Y_0 = y_0$ can be computed by a slight modification of the method described above. Instead of setting $\mathbf{Y}_0 = \mathbf{1}$ as in the calculation of unconditional likelihoods, we take as $\mathbf{Y}_0$ the one-component vector consisting of the first observation Y_0 of the process. The calculation of the conditional likelihood is then performed as follows:

(i) Set $y_i^* = y_i$, $i = 0, 1, 3, 4, 5$ and $y_2^* = 0$.

(ii) The one-step predictors $\hat{\mathbf{Y}}_t^*$, $t = 1, 2, \ldots$, and their mean squared errors are evaluated from Proposition 12.2.2, equations (12.2.12) and (12.2.14) and the state-space model for $\{Y_t^*\}$ derived from Example 12.1.7, i.e.

$$\mathbf{X}_{t+1} = F_t \mathbf{X}_t + \mathbf{V}_t, \qquad t = 1, 2, \ldots,$$
$$Y_t^* = G_t^* \mathbf{X}_t + W_t^*, \qquad t = 1, 2, \ldots,$$

where

$$F_t = \begin{bmatrix} \phi & 0 \\ 1 & 1 \end{bmatrix}, \qquad G_t^* = \begin{cases} [1 \quad 1] & \text{if } t \neq 2, \\ [0 \quad 0] & \text{if } t = 2, \end{cases}$$

$$\mathbf{V}_t = \begin{bmatrix} \phi + \theta \\ 1 \end{bmatrix} Z_t, \qquad W_t^* = \begin{cases} Z_t & \text{if } t \neq 2, \\ N_t & \text{if } t = 2, \end{cases}$$

$$Q_t = \sigma^2 \begin{bmatrix} (\phi + \theta)^2 & \phi + \theta \\ \phi + \theta & 1 \end{bmatrix}, \qquad R_t^* = \begin{cases} \sigma^2 & \text{if } t \neq 2, \\ 1 & \text{if } t = 2, \end{cases}$$

$$S_t^* = \begin{cases} \sigma^2 \begin{bmatrix} \phi + \theta \\ 1 \end{bmatrix} & \text{if } t \neq 2, \\ \mathbf{0} & \text{if } t = 2, \end{cases} \quad \text{and} \quad \mathbf{X}_1 = \begin{bmatrix} (\phi + \theta) \sum_{j=0}^{\infty} \phi^j Z_{-j} \\ y_0 \end{bmatrix}.$$

Starting from the initial conditions,

$$\hat{\mathbf{X}}_1 = \begin{bmatrix} 0 \\ y_0 \end{bmatrix}, \qquad \Pi_1 = \begin{bmatrix} \dfrac{\sigma^2(\phi + \theta)^2}{1 - \phi^2} & 0 \\ 0 & y_0^2 \end{bmatrix},$$

$$\Psi_1 = \begin{bmatrix} 0 & 0 \\ 0 & y_0^2 \end{bmatrix}, \qquad \Omega_1 = \begin{bmatrix} \dfrac{\sigma^2(\phi + \theta)^2}{1 - \phi^2} & 0 \\ 0 & 0 \end{bmatrix},$$

and using the recursions (12.2.6), (12.2.12) and (12.2.14), we obtain the predicted values, $\hat{y}_1^*, \ldots, \hat{y}_5^*$, and mean squared errors, $\Sigma_1^*, \ldots, \Sigma_5^*$ ($\hat{y}_t^*$ is the best linear predictor of Y_t^* in terms of $Y_0^*, Y_1^*, \ldots, Y_{t-1}^*$, evaluated at $y_0, y_1^*, \ldots, y_{t-1}^*$).

(iii) The required likelihood of the observations y_1, y_3, y_4, y_5, conditional on $Y_0 = y_0$, is found by direct substitution into (12.3.6). We shall not attempt to write it algebraically.

Remark 1. If we are given observations $y_{1-d}, y_{2-d}, \ldots, y_0, y_{i_1}, y_{i_2}, \ldots, y_{i_r}$ of an ARIMA(p, d, q) process at times $1 - d, 2 - d, \ldots, 0, i_1, \ldots, i_r$, where $1 \le i_1 < i_2 < \cdots < i_r \le n$, we can use the representation (12.1.31) and (12.1.32) with the same argument as in Example 12.3.2 to find the Gaussian likelihood of $y_{i_1}, \ldots, y_{i_r}$ conditional on $Y_{1-d} = y_{1-d}, Y_{2-d} = y_{2-d}, \ldots, Y_0 = y_0$. (Missing values among the first d observations $y_{1-d}, y_{2-d}, \ldots, y_0$ can be handled by treating them as unknown parameters for likelihood maximization.) A similar analysis can be carried out using more general differencing operators of the form $(1 - B)^d(1 - B^s)^D$ (see Problem 12.9). The dimension of the state vector constructed in this way is $\max(p + d + sD, q)$. Different approaches to maximum likelihood estimation for ARIMA processes with missing values can be found in Ansley and Kohn (1985) and Bell and Hillmer (1990).

Remark 2. Observation vectors from which some but not all components are missing can be handled using arguments similar to those used above. For details see Brockwell, Davis and Salehi (1990).

Estimation of Missing Observations for State-Space Models

Given that we observe only $\mathbf{Y}_{i_1}, \mathbf{Y}_{i_2}, \ldots, \mathbf{Y}_{i_r}$, $1 \le i_1 < i_2 < \cdots < i_r \le n$, where $\{\mathbf{Y}_t\}$ has the state-space representation (12.1.1) and (12.1.2), we now consider the problem of finding the minimum mean square error estimators $P(\mathbf{Y}_t | \mathbf{Y}_0, \mathbf{Y}_{i_1}, \ldots, \mathbf{Y}_{i_r})$ of $\mathbf{Y}_t$, $1 \le t \le n$ where $\mathbf{Y}_0 = \mathbf{1}$. To handle this problem we again use the modified process $\{\mathbf{Y}_t^*\}$ defined by (12.3.1) and (12.1.2) with $\mathbf{Y}_0^* = \mathbf{1}$. Since $\mathbf{Y}_s^* = \mathbf{Y}_s$ for $s \in \{i_1, \ldots, i_r\}$ and $\mathbf{Y}_s^* \perp \mathbf{X}_t, \mathbf{Y}_0$ for $1 \le t \le n$ and $s \notin \{0, i_1, \ldots, i_r\}$, we immediately obtain the minimum mean squared error state estimators,

$$P(\mathbf{X}_t | \mathbf{Y}_0, \mathbf{Y}_{i_1}, \ldots, \mathbf{Y}_{i_r}) = P(\mathbf{X}_t | \mathbf{Y}_0^*, \mathbf{Y}_1^*, \ldots, \mathbf{Y}_n^*), \qquad 1 \le t \le n. \quad (12.3.7)$$

The right-hand side can be evaluated by direct application of the Kalman fixed point smoothing algorithm (Proposition 12.2.4) to the state-space model (12.3.1) and (12.1.2). For computational purposes the observed values of $\mathbf{Y}_t^*$, $t \notin \{0, i_1, \ldots, i_r\}$ are quite immaterial. They may, for example, all be set equal to zero, giving the sequence of *observations* of $\mathbf{Y}_t^*$ defined in (12.3.4).

In order to evaluate $P(\mathbf{Y}_t | \mathbf{Y}_0, \mathbf{Y}_{i_1}, \ldots, \mathbf{Y}_{i_r})$, $1 \le t \le n$, we use (12.3.7) and the relation,

$$\mathbf{Y}_t = G_t \mathbf{X}_t + \mathbf{W}_t. \quad (12.3.8)$$

Under the assumption that

$$E(\mathbf{V}_t \mathbf{W}_t') = S_t = 0, \qquad t = 1, \ldots, n, \quad (12.3.9)$$

we find from (12.3.8) that

$$P(\mathbf{Y}_t | \mathbf{Y}_0, \mathbf{Y}_{i_1}, \ldots, \mathbf{Y}_{i_r}) = G_t P(\mathbf{X}_t | \mathbf{Y}_0^*, \mathbf{Y}_1^*, \ldots, \mathbf{Y}_n^*). \quad (12.3.10)$$

It is essential, in estimating missing observations of $\mathbf{Y}_t$ with (12.3.10), to use a state-space representation for $\{\mathbf{Y}_t\}$ which satisfies (12.3.9). The ARMA state-space representation in Example 12.1.5 satisfies this condition, but the one in Example 12.1.6 does not.

EXAMPLE 12.3.3 (An AR(1) Series with One Missing Observation). Consider the problem of estimating the missing value Y_2 in Example 12.3.1 in terms of $Y_0 = 1$, Y_1, Y_3, Y_4 and Y_5. We start from the state-space model, $X_{t+1} = \phi X_t + Z_{t+1}$, $Y_t = X_t$, for $\{Y_t\}$, which satisfies the required condition (12.3.9). The corresponding model for $\{Y_t^*\}$ is the one used in Example 12.3.1. Applying Proposition 12.2.4 to the latter model, we find that

$$P_1 X_2 = \phi Y_1, \qquad P_2 X_2 = \phi Y_1, \qquad P_3 X_2 = \frac{\phi(Y_1 + Y_3)}{(1 + \phi^2)},$$

$$P_4 X_2 = P_3 X_2, \qquad P_5 X_2 = P_3 X_2,$$

$$\Omega_{2,2} = \sigma^2, \qquad \Omega_{2,3} = \phi\sigma^2, \qquad \Omega_{2,t} = 0, \quad t \ge 4,$$

and

$$\Omega_{2|1} = \sigma^2, \qquad \Omega_{2|2} = \sigma^2, \qquad \Omega_{2|t} = \frac{\sigma^2}{(1 + \phi^2)}, \quad t \geq 3,$$

where $P_t(\cdot)$ here denotes $P(\cdot | Y_0^*, \ldots, Y_t^*)$ and $\Omega_{t,n}$, $\Omega_{t|n}$ are defined correspondingly. Since the condition (12.3.9) is satisfied, we deduce from (12.3.10) that the minimum mean squared error estimator of the missing value Y_2 is

$$P_5 Y_2 = P_5 X_2 = \frac{\phi(Y_1 + Y_3)}{(1 + \phi^2)},$$

with mean squared error,

$$\Omega_{2|5} = \frac{\sigma^2}{(1 + \phi^2)}.$$

EXAMPLE 12.3.4 (Estimating Missing Observations of an ARIMA(p, d, q) Process). Suppose we are given observations $Y_{1-d}, Y_{2-d}, \ldots, Y_0, Y_{i_1}, \ldots, Y_{i_r}$ ($1 \leq i_1 < i_2 \cdots < i_r \leq n$) of an ARIMA($p, d, q$) process. We wish to find the best linear estimates of the missing values Y_t, $t \notin \{i_1, \ldots, i_r\}$, in terms of Y_t, $t \in \{i_1, \ldots, i_r\}$ and the components of $\mathbf{Y}_0 := (Y_{1-d}, Y_{2-d}, \ldots, Y_0)'$. This can be done exactly as described above provided we start with a state-space representation of the ARIMA series $\{Y_t\}$ which satisfies (12.3.9) and we apply the Kalman recursions to the state-space model for $\{Y_t^*\}$. Although the representation in Example 12.1.7 does not satisfy (12.3.9), it is quite easy to contruct another which does by starting from the model in Example 12.1.5 for $\{\nabla^d Y_t\}$ and following the same steps as in Example 12.1.7. This gives (Problem 12.8),

$$Y_t = [G \quad (-1)^{d+1}\tbinom{d}{d} \quad (-1)^d \tbinom{d}{d-1} \quad (-1)^{d-1}\tbinom{d}{d-2} \quad \cdots \quad d] \begin{bmatrix} \mathbf{X}_t \\ \mathbf{Y}_{t-1} \end{bmatrix}, \tag{12.3.11}$$

where

$$\begin{bmatrix} \mathbf{X}_{t+1} \\ \mathbf{Y}_t \end{bmatrix} = \begin{bmatrix} F & 0 \\ AG & B \end{bmatrix} \begin{bmatrix} \mathbf{X}_t \\ \mathbf{Y}_{t-1} \end{bmatrix} + \begin{bmatrix} H \\ \mathbf{0} \end{bmatrix} Z_{t+1}, \tag{12.3.12}$$

for $t = 1, 2, \ldots$. The matrices G, F and H are the coefficients in (12.1.20) and (12.1.23) and A and B are defined as in Example 12.1.7. We assume, as in Example 12.1.7, that

$$\mathbf{Y}_0 \perp Z_t, \qquad t = 0, \pm 1, \ldots. \tag{12.3.13}$$

This model clearly satisfies (12.3.9). Missing observations can therefore be estimated by introducing the corresponding model for $\{Y_t^*\}$ and using (12.3.7) and (12.3.10).

§12.4 Controllability and Observability

In this section we introduce the concepts of controllability and observability, which provide a useful criterion (Proposition 12.4.6) for determining whether or not the state vector in a given state-space model for $\{\mathbf{Y}_t\}$ has the smallest possible dimension. Consider the model (with $\mathbf{X}_t$ stationary),

$$\begin{aligned} \mathbf{X}_{t+1} &= F\mathbf{X}_t + \mathbf{V}_t, \qquad t = 0, \pm 1, \ldots, \\ \mathbf{Y}_t &= G\mathbf{X}_t + \mathbf{W}_t, \qquad t = 0, \pm 1, \ldots, \end{aligned} \tag{12.4.1}$$

where $\left\{\begin{bmatrix} \mathbf{V}_t \\ \mathbf{W}_t \end{bmatrix}\right\} \sim \mathrm{WN}\left(\mathbf{0}, \begin{bmatrix} Q & S \\ S' & R \end{bmatrix}\right)$ and F satisfies the stability condition

$$\det(I - Fz) \neq 0 \quad \text{for } |z| \leq 1. \tag{12.4.2}$$

From Section 12.1, $\mathbf{X}_t$ and $\mathbf{Y}_t$ have the representations

$$\mathbf{X}_t = \sum_{j=1}^{\infty} F^{j-1}\mathbf{V}_{t-j},$$

$$\mathbf{Y}_t = \sum_{j=1}^{\infty} GF^{j-1}\mathbf{V}_{t-j} + \mathbf{W}_t, \tag{12.4.3}$$

and $(\mathbf{X}_t', \mathbf{Y}_t')'$ is a stationary multivariate time series (by a simple generalization of Problem 11.14).

To discuss controllability and observability, we introduce the subclass of stationary state-space models for $\{\mathbf{Y}_t\}$ defined by the equations

$$\begin{aligned} \mathbf{X}_{t+1} &= F\mathbf{X}_t + H\mathbf{Z}_t, \qquad t = 0, \pm 1, \ldots, \\ \mathbf{Y}_t &= G\mathbf{X}_t + \mathbf{Z}_t, \qquad t = 0, \pm 1, \ldots, \end{aligned} \tag{12.4.4}$$

where $\{\mathbf{Z}_t\} \sim \mathrm{WN}(\mathbf{0}, \Sigma)$ with dimension v, H is a $v \times w$ matrix and F is stable. If the noise vector, $\mathbf{Z}_t$, is the same as the innovation of $\mathbf{Y}_t$, i.e. if

$$\mathbf{Z}_t = \mathbf{Y}_t - P(\mathbf{Y}_t | \mathbf{Y}_j, -\infty < j < t) \tag{12.4.5}$$

then we refer to the model (12.4.4) as an *innovations representation*. Obviously if $\{\mathbf{Y}_t\}$ has an innovations representation then it has a representation of the form (12.4.1) with stable F. The converse of this statement is established below as Proposition 12.4.1.

Remark 1. Even with the restriction that the white noise sequence $\{\mathbf{Z}_t\}$ satisfies (12.4.5), the matrices F, G and H in the innovations representation (12.4.4) are not uniquely determined (see Example 12.4.2). However, if Σ is non-singular then the sequence of matrices $\{GF^{j-1}H, \ j = 1, 2, \ldots\}$, is necessarily the same for all innovations representation of $\{\mathbf{Y}_t\}$ since from

$$\mathbf{Y}_t = \sum_{j=1}^{\infty} GF^{j-1}H\mathbf{Z}_{t-j} + \mathbf{Z}_t,$$

it follows that $GF^{j-1}H = E(\mathbf{Y}_t\mathbf{Z}'_{t-j})\Sigma^{-1}$ with $\mathbf{Z}_t$ given by (12.4.5).

Proposition 12.4.1. *Under assumption* (12.4.2), *the state-space model* (12.4.1) *has an innovations representation, i.e. a representation of the form* (12.4.4) *with noise vector* $\mathbf{Z}_t$ *defined by* (12.4.5).

PROOF. For the state vectors defined in (12.4.1), set

$$\mathbf{X}(t|s) = P(\mathbf{X}_t|\mathbf{Y}_j, -\infty < j \le s).$$

Then, with $\mathbf{Z}_t$ defined by (12.4.5), $\mathbf{Z}_t \perp \mathbf{Y}_j$, $j < t$, so that by Problem 12.12,

$$P(\cdot|\mathbf{Y}_j, j \le t) = P(\cdot|\mathbf{Y}_j, j < t) + P(\cdot|\mathbf{Z}_t).$$

Hence by Remark 5 of Section 12.2 and the orthogonality of $\mathbf{V}_t$ and $\mathbf{Y}_j, j < t$,

$$\begin{aligned}\mathbf{X}(t+1|t) &= P(\mathbf{X}_{t+1}|\mathbf{Y}_j, j < t) + H\mathbf{Z}_t \\ &= P(F\mathbf{X}_t + \mathbf{V}_t|\mathbf{Y}_j, j < t) + H\mathbf{Z}_t \\ &= F\mathbf{X}(t|t-1) + H\mathbf{Z}_t, \end{aligned} \tag{12.4.6}$$

where

$$H = E(\mathbf{X}_{t+1}\mathbf{Z}'_t)[E(\mathbf{Z}_t\mathbf{Z}'_t)]^{-1},$$

and $[E(\mathbf{Z}_t\mathbf{Z}'_t)]^{-1}$ is any generalized inverse of $E(\mathbf{Z}_t\mathbf{Z}'_t)$. Since $(\mathbf{X}'_t, \mathbf{Y}'_t)'$ is stationary, H can be chosen to be independent of t. Finally, since $\mathbf{W}_t \perp \mathbf{Y}_j$, $j < t$,

$$\begin{aligned}P(\mathbf{Y}_t|\mathbf{Y}_j, j < t) &= P(G\mathbf{X}_t + \mathbf{W}_t|\mathbf{Y}_j, j < t) \\ &= G\mathbf{X}(t|t-1).\end{aligned}$$

Together with (12.4.6), this gives the innovations representation,

$$\begin{aligned}\mathbf{X}(t+1|t) &= F\mathbf{X}(t|t-1) + H\mathbf{Z}_t, \\ \mathbf{Y}_t &= G\mathbf{X}(t|t-1) + \mathbf{Z}_t,\end{aligned} \tag{12.4.7}$$

as required. □

EXAMPLE 12.4.1. The canonical representation (12.1.24), (12.1.25) of the causal ARMA process $\{Y_t\}$ satisfying

$$Y_t = \phi_1 Y_{t-1} + \cdots + \phi_p Y_{t-p} + Z_t + \theta_1 Z_{t-1} + \cdots + \theta_q Z_{t-q}, \qquad \{Z_t\} \sim \mathrm{WN}(0, \sigma^2), \tag{12.4.8}$$

has the form (12.4.4). It is also an innovations representation if (12.4.8) is invertible (Problem 12.19). Assuming that $Z_t \in \overline{\mathrm{sp}}\{Y_s, -\infty < s \le t\}$ and defining

$$Y(t|s) = P(Y_t|Y_j, -\infty < j \le s),$$

we now show how the canonical representation arises naturally if we seek a model of the form (12.4.7) with

$$\mathbf{X}(t|t-1) = (Y(t|t-1), \ldots, Y(t-1+m|t-1))'$$

and $m = \max(p, q)$. Since we are also assuming the causality of (12.4.8), we have $\overline{\text{sp}}\{Y_j, j \le s\} = \overline{\text{sp}}\{Z_j, j \le s\}$, so that from (5.5.4)

$$\begin{aligned} Y(t+j|t) &= \sum_{k=j}^{\infty} \psi_k Z_{t+j-k} \\ &= Y(t+j|t-1) + \psi_j Z_t, \qquad j = 0, \ldots, m = \max(p, q). \end{aligned} \tag{12.4.9}$$

Replacing t by $t + m$ in (12.4.8) and projecting both sides onto $\overline{\text{sp}}\{Y_j, j \le t\}$, we obtain from (12.4.9) and the identity, $\psi_m = \sum_{k=1}^{m} \phi_k \psi_{m-k} + \theta_m$ (see (3.3.3)), the relation

$$\begin{aligned} Y(t+m|t) &= \sum_{k=1}^{m} \phi_k Y(t+m-k|t) + \theta_m Z_t \\ &= \sum_{k=1}^{m} \phi_k Y(t+m-k|t-1) + \left(\sum_{k=1}^{m} \phi_k \psi_{m-k} + \theta_m\right) Z_t \\ &= \sum_{k=1}^{m} \phi_k Y(t+m-k|t-1) + \psi_m Z_t. \end{aligned} \tag{12.4.10}$$

Now the state-vector defined by

$$\mathbf{X}(t|t-1) = (Y(t|t-1), \ldots, Y(t-1+m|t-1))'$$

satisfies the state equation

$$\mathbf{X}(t+1|t) = F\mathbf{X}(t-1|t) + HZ_t,$$

where F and H are the matrices in Example 12.1.6 (use (12.4.9) for the first $r-1$ rows and (12.4.10) for the last row). Together with the observation equation,

$$\begin{aligned} Y_t &= Y(t|t-1) + Y_t - Y(t|t-1) \\ &= [1 \quad 0 \quad \ldots \quad 0]\mathbf{X}(t|t-1) + Z_t, \end{aligned}$$

this yields the canonical state-space model of Example 12.1.6.

Definition 12.4.1 (Controllability). The state-space model (12.4.4) is said to be controllable if for any two vectors $\mathbf{x}_a$ and $\mathbf{x}_b$, there exists an integer k and noise inputs, $\mathbf{Z}_1, \ldots, \mathbf{Z}_k$ such that $\mathbf{X}_k = \mathbf{x}_b$ when $\mathbf{X}_0 = \mathbf{x}_a$.

In other words, the state-space model is controllable, if by judicious choice of the noise inputs, $\mathbf{Z}_1, \mathbf{Z}_2, \ldots$, the state vector $\mathbf{X}_t$ can be made to pass from $\mathbf{x}_a$ to $\mathbf{x}_b$. In such a case, we have

$$\begin{aligned} \mathbf{X}_0 &= \mathbf{x}_a, \\ \mathbf{X}_1 &= F\mathbf{x}_a + H\mathbf{Z}_1, \\ &\vdots \\ \mathbf{X}_k &= \mathbf{x}_b = F^k\mathbf{x}_a + H\mathbf{Z}_k + FH\mathbf{Z}_{k-1} + \cdots + F^{k-1}H\mathbf{Z}_1, \end{aligned} \tag{12.4.11}$$

and hence

$$\begin{aligned}\mathbf{x}_b - F^k\mathbf{x}_a &= [H \quad FH \quad \cdots \quad F^{k-1}H][\mathbf{Z}'_k \quad \mathbf{Z}'_{k-1} \quad \cdots \quad \mathbf{Z}'_1]' \\ &= C_k[\mathbf{Z}'_k \quad \cdots \quad \mathbf{Z}'_1]',\end{aligned}$$

where

$$C_k = [H \quad FH \quad \cdots \quad F^{k-1}H]. \tag{12.4.12}$$

From these equations, we see that controllability is in fact a property of the two matrices F and H. We therefore say that the pair (F, H) is *controllable* if and only if the model (12.4.4) is controllable.

Proposition 12.4.2. *The state-space model* (12.4.4), *or, equivalently, the pair* (F, H) *is controllable if and only if* C_v *has rank* v *(where* v *is the dimension of* $\mathbf{X}_t$*).*

PROOF. The matrix C_v is called the *controllability matrix*. If C_v has rank v then the state can be made to pass from $\mathbf{x}_a$ to $\mathbf{x}_b$ in v time steps by choosing

$$[\mathbf{Z}'_v \quad \cdots \quad \mathbf{Z}'_1]' = C'_v(C_vC'_v)^{-1}(\mathbf{x}_b - F^v\mathbf{x}_a).$$

Recall from Remark 2 of Section 2.5 that $C_vC'_v$ is non-singular if C_v has full rank.

To establish the converse, suppose that (F, H) is controllable. If $\lambda(z) = \det(F - zI)$ is the characteristic polynomial of F, then by the Cayley–Hamilton Theorem, $\lambda(F) = 0$, so that there exist constants $\beta_1, \ldots, \beta_v$ such that

$$F^v = \beta_v F^{v-1} + \cdots + \beta_2 F + \beta_1 I. \tag{12.4.13}$$

More generally, for any $k \geq 1$, there exist constants $\beta_1, \ldots, \beta_v$ (which depend on k) such that

$$F^k = \beta_v F^{v-1} + \cdots + \beta_2 F + \beta_1 I. \tag{12.4.14}$$

This is immediate for $k \leq v$. For $k > v$ it follows from (12.4.13) by induction. Now if C_v does not have full rank, there exists a non-zero v-vector $\mathbf{y}$ such that

$$\mathbf{y}'C_v = \mathbf{y}'[H \quad FH \quad \cdots \quad F^{v-1}H] = \mathbf{0}',$$

which, in conjunction with (12.4.14), implies that

$$\mathbf{y}'F^jH = \mathbf{0}', \quad \text{for } j = 0, 1, \ldots,$$

Choosing $\mathbf{x}_a = \mathbf{0}$ and $\mathbf{x}_b = \mathbf{y}$, we have from (12.4.11) and the preceding equation that

$$\mathbf{y}'\mathbf{y} = \mathbf{y}'(F^k\mathbf{x}_a + H\mathbf{Z}_k + FH\mathbf{Z}_{k-1} + \cdots + F^{k-1}H\mathbf{Z}_1) = 0,$$

which contradicts the fact that $\mathbf{y} \neq \mathbf{0}$. Thus C_v must have full rank. □

Remark 2. From the proof of this proposition, we also see that

$$\begin{aligned}\operatorname{rank}(C_k) &\leq \operatorname{rank}(C_v) \quad \text{for } k \leq v,\\ \operatorname{rank}(C_k) &= \operatorname{rank}(C_v) \quad \text{for } k > v.\end{aligned}$$

For $k \le v$ this is obvious and, for $k > v$, it follows from (12.4.14) since the columns of $F^{v+j}H$, $j \ge 0$, are in the linear span of the columns of C_v.

EXAMPLE 12.4.2. Suppose $v = 2$ and $w = 1$ with

$$F = \begin{bmatrix} .5 & 1 \\ 0 & .5 \end{bmatrix} \quad \text{and} \quad H = \begin{bmatrix} 2 \\ 0 \end{bmatrix}.$$

Then

$$C_2 = \begin{bmatrix} 2 & 1 \\ 0 & 0 \end{bmatrix}$$

has rank one so that (F, H) is not controllable. In this example,

$$F^{j-1}H = \begin{bmatrix} .5^{j-2} \\ 0 \end{bmatrix} \quad \text{for } j \ge 1,$$

so that by replacing $\mathbf{V}_t$ and $\mathbf{W}_t$ in (12.4.3) by HZ_t and Z_t, respectively, we have

$$\mathbf{X}_t = \sum_{j=1}^{\infty} \begin{bmatrix} .5^{j-2} \\ 0 \end{bmatrix} Z_{t-j},$$

$$Y_t = \sum_{j=1}^{\infty} G \begin{bmatrix} .5^{j-2} \\ 0 \end{bmatrix} Z_{t-j} + Z_t.$$

Since the second component in $\mathbf{X}_t$ plays no role in these equations, we can eliminate it from the state-vector through the transformation $\tilde{\mathbf{X}}_t = [1 \;\; 0]\mathbf{X}_t = X_{t1}$. Using these new state variables, the state-space system is now controllable with state-space equations given by

$$\tilde{\mathbf{X}}_{t+1} = .5\tilde{\mathbf{X}}_t + 2Z_t,$$

$$Y_t = G \begin{bmatrix} 1 \\ 0 \end{bmatrix} \tilde{\mathbf{X}}_t + Z_t.$$

This example is a special case of a more general result, Proposition 12.4.3, which says that any non-controllable state-space model may be transformed into a controllable system whose state-vector has dimension equal to $\operatorname{rank}(C_v)$.

EXAMPLE 12.4.3. Let F and H denote the coefficient matrices in the state equation (12.1.25) of the canonical observable representation of an ARMA(p, q) process. Here $v = m = \max(p, q)$ and since

$$F^j H = [\psi_{j+1} \quad \cdots \quad \psi_{j+v}]', \qquad j = 0, 1, \ldots,$$

we have

$$C_v = [\psi_{i+j-1}]_{i,j=1}^{v} = \begin{bmatrix} \psi_1 & \psi_2 & \cdots & \psi_v \\ \psi_2 & \psi_3 & \cdots & \psi_{v+1} \\ \vdots & \vdots & \vdots & \vdots \\ \psi_v & \psi_{v+1} & \cdots & \psi_{2v-1} \end{bmatrix},$$

where ψ_j are the coefficients in the power series

$$\psi(z) = 1 + \psi_1 z + \psi_2 z^2 + \cdots = \frac{\theta(z)}{\phi(z)}. \tag{12.4.15}$$

If C_v is singular, then there exists a non-zero vector, $\mathbf{a} = (a_{v-1}, \ldots, a_0)'$ such that

$$C_v \mathbf{a} = \mathbf{0}. \tag{12.4.16}$$

and hence

$$a_0 \psi_k + a_1 \psi_{k-1} + \cdots + a_{v-1} \psi_{k-v+1} = 0, \qquad k = v, v+1, \ldots, 2v-1. \tag{12.4.17}$$

Multiplying the left side of (12.4.16) by the vector $(\phi_v, \ldots, \phi_1)$ and using (3.3.4) with $j > v$, we find that (12.4.17) also holds with $k = 2v$. Repeating this same argument with C_v replaced by the matrix $[\psi_{i+j}]_{i,j=1}^{v}$ (which satisfies equation (12.4.16) by what we have just shown), we see that (12.4.17) holds with $k = 2v + 1$. Continuing in this fashion, we conclude that (12.4.17) is valid for all $k \geq v$ which implies that $a(z)\psi(z)$ is a polynomial of degree at most $v - 1$, viz.

$$a(z)\psi(z) = a(z)\frac{\theta(z)}{\phi(z)} = b_0 + b_1 z + \cdots + b_{v-1} z^{v-1} = b(z),$$

where $a(z) = a_0 + a_1 z + \cdots + a_{v-1} z^{v-1}$. In particular, $\phi(z)$ must divide $a(z)$. This implies that $p \leq v - 1$ and, since $v = \max(p, q)$, that $v = q > p$. But since $\phi(z)$ divides $a(z)$, $a(z)\psi(z) = b(z)$ is a polynomial of degree at least $q > v - 1 \geq \deg(b(z))$, a contradiction. Therefore C_v must have full rank.

Proposition 12.4.3. *If the state-space model* (12.4.4) *is not controllable and* $k = \operatorname{rank}(C_v)$, *then there exists a stationary sequence of k-dimensional state-vectors* $\{\tilde{\mathbf{X}}_t\}$ *and matrices* $\tilde{F}$, $\tilde{H}$, *and* $\tilde{G}$ *such that* $\tilde{F}$ *is stable,* $(\tilde{F}, \tilde{H})$ *is controllable and*

$$\begin{aligned} \tilde{\mathbf{X}}_{t+1} &= \tilde{F}\tilde{\mathbf{X}}_t + \tilde{H}\mathbf{Z}_t, \\ \mathbf{Y}_t &= \tilde{G}\tilde{\mathbf{X}}_t + \mathbf{Z}_t. \end{aligned} \tag{12.4.18}$$

PROOF. For any matrix M, let $\mathscr{R}(M)$ denote the range or column space of M. By assumption $\operatorname{rank}(C_v) = k < v$ so that there exist v linearly independent vectors, $\mathbf{v}_1, \ldots, \mathbf{v}_v$, which can be indexed so that $\mathscr{R}(C_v) = \overline{\operatorname{sp}}\{\mathbf{v}_1, \ldots, \mathbf{v}_k\}$. Let T denote the non-singular matrix

$$T = [\mathbf{v}_1 \quad \mathbf{v}_2 \quad \cdots \quad \mathbf{v}_v].$$

Observe that

$$\mathscr{R}(F[\mathbf{v}_1 \quad \cdots \quad \mathbf{v}_k]) = \mathscr{R}(FC_v) \subseteq \mathscr{R}(C_{v+1}) = \mathscr{R}(C_v) = \overline{\operatorname{sp}}\{\mathbf{v}_1, \ldots, \mathbf{v}_k\}$$

where the second equality follows from Remark 2. Now set

$$\hat{F} = T^{-1}FT \quad \text{and} \quad \hat{H} = T^{-1}H,$$

so that

$$T\hat{F} = FT \quad \text{and} \quad T\hat{H} = H. \tag{12.4.19}$$

By partitioning $\hat{F}$ as $\begin{bmatrix} \hat{F}_{11} & \hat{F}_{12} \\ \hat{F}_{21} & \hat{F}_{22} \end{bmatrix}$ and considering only the first k columns of the equation in (12.4.19), we obtain

$$[\mathbf{v}_1 \quad \cdots \quad \mathbf{v}_v]\begin{bmatrix} \hat{F}_{11} \\ \hat{F}_{21} \end{bmatrix} = F[\mathbf{v}_1 \quad \cdots \quad \mathbf{v}_k].$$

Since the columns of the product on the right belong to $\overline{\text{sp}}\{\mathbf{v}_1, \ldots, \mathbf{v}_k\}$ and since $\mathbf{v}_1, \ldots, \mathbf{v}_v$ are linearly independent, it follows that $\hat{F}_{21} = 0$. Similarly, by writing $\hat{H} = \begin{bmatrix} \hat{H}_1 \\ \hat{H}_2 \end{bmatrix}$ with $\hat{H}_1$ a $k \times w$ matrix and noting that $\mathscr{R}(H) \subseteq \overline{\text{sp}}\{\mathbf{v}_1, \ldots, \mathbf{v}_k\}$, we conclude that $\hat{H}_2 = 0$.

The matrices appearing in the statement of the proposition are now defined to be

$$\tilde{F} = \hat{F}_{11}, \quad \tilde{H} = \hat{H}_1, \quad \text{and} \quad \tilde{G} = GT\begin{bmatrix} I_{k \times k} \\ 0 \end{bmatrix},$$

where $I_{k \times k}$ is the k-dimensional identity matrix. To verify that $\tilde{F}$, $\tilde{G}$ and $\tilde{H}$ have the required properties, observe that

$$\hat{F}^{j-1}\hat{H} = \begin{bmatrix} \tilde{F}^{j-1}\tilde{H} \\ 0 \end{bmatrix}$$

and

$$\begin{aligned} &\text{rank}[\tilde{H} \quad \tilde{F}\tilde{H} \quad \cdots \quad \tilde{F}^{v-1}\tilde{H}] \\ &\quad = \text{rank}[\hat{H} \quad \hat{F}\hat{H} \quad \cdots \quad \hat{F}^{v-1}\hat{H}] \\ &\quad = \text{rank}[T^{-1}\hat{H} \quad (T^{-1}\hat{F}T)(T^{-1}\hat{H}) \quad \cdots \quad (T^{-1}\hat{F}^{v-1}T)(T^{-1}\hat{H})] \\ &\quad = \text{rank}[H \quad FH \quad \cdots \quad F^{v-1}H] \\ &\quad = \text{rank}(C_v) = k. \end{aligned}$$

By Remark 2 the pair $(\tilde{F}, \tilde{H})$ is therefore controllable. In addition, $\tilde{F}$ satisfies the stability condition (12.4.2) since its eigenvalues form a subset of the eigenvalues of $\hat{F}$ which in turn are equal to the eigenvalues of F. Now let $\mathbf{X}_t$ be the unique stationary solution of the state equation

$$\tilde{\mathbf{X}}_{t+1} = \tilde{F}\tilde{\mathbf{X}}_t + \tilde{H}\mathbf{Z}_t.$$

Then $\mathbf{Y}_t$ satisfies the observation equation

$$\mathbf{Y}_t = \tilde{G}\mathbf{X}_t + \mathbf{Z}_t,$$

since we know from (12.4.4) that $\mathbf{Y}_t = \mathbf{Z}_t + \sum_{j=1}^{\infty} GF^{j-1}H\mathbf{Z}_{t-j}$, and since

$$\begin{aligned}\tilde{G}\tilde{F}^{j-1}\tilde{H} &= GT\begin{bmatrix} I_{k\times k} \\ 0 \end{bmatrix}\hat{F}^{j-1}\hat{H} \\ &= G(T\hat{F}^{j-1}T^{-1})(T\hat{H}) \\ &= GF^{j-1}H, \qquad j = 1, 2, \ldots. \end{aligned}$$ □

Definition 12.4.2 (Observability). The state-space system (12.4.4) is observable if the state $\mathbf{X}_0$ can be completely determined from the observations $\mathbf{Y}_0$, $\mathbf{Y}_1, \ldots$ when $\mathbf{Z}_0 = \mathbf{Z}_1 = \cdots = \mathbf{0}$.

For a system to be observable, $\mathbf{X}_0$ must be uniquely determined by the sequence of values

$$G\mathbf{X}_0,\ GF\mathbf{X}_0,\ GF^2\mathbf{X}_0, \ldots.$$

Thus observability is a property of the two matrices F and G and we shall say that the pair (F, G) is *observable* if and only if the system (12.4.4) is observable. If the $v \times kw$ matrix

$$O_k' := [G' \ \ F'G' \ \ \cdots \ \ F'^{k-1}G']$$

has rank v for some k, then we can express $\mathbf{X}_0$ as

$$\begin{aligned}\mathbf{X}_0 &= (O_k'O_k)^{-1}O_k'\begin{bmatrix} G\mathbf{X}_0 \\ GF\mathbf{X}_0 \\ \vdots \\ GF^{k-1}\mathbf{X}_0 \end{bmatrix} \\ &= (O_k'O_k)^{-1}O_k'(O_k\mathbf{X}_0),\end{aligned}$$

showing that (F, G) is observable in this case.

Proposition 12.4.4. *The pair of matrices (F, G) is observable if and only if O_v has rank v. In particular, (F, G) is observable if and only if (F', G') is controllable.*

The matrix O_v is referred to as the *observability matrix*.

Proof. The discussion leading up to the statement of the proposition shows that the condition $\text{rank}(O_v) = v$ is sufficient for observability. To establish the necessity suppose that (F, G) is observable and O_v is not of full rank. Then there exists a non-zero vector $\mathbf{y}$ such that $O_v\mathbf{y} = \mathbf{0}$. This implies that

$$GF^{j-1}\mathbf{y} = \mathbf{0}$$

for $j = 1, \ldots, v$, and hence for all $j \geq 1$ (by (12.4.14)). It is also true that

$$GF^{j-1}\mathbf{0} = \mathbf{0}$$

showing that the sequence $GF^{j-1}\mathbf{X}_0, j = 1, 2, \ldots,$ is the same for $\mathbf{X}_0 = \mathbf{y}$ as for $\mathbf{X}_0 = \mathbf{0}$. This contradicts the assumed observability of (F, G), and hence rank(O_v) must be v. The last statement of the proposition is an immediate consequence of Proposition 12.4.2 and the observation that $O_v' = C_v$ where C_v is the controllability matrix corresponding to (F', G'). □

EXAMPLE 12.4.3 (*cont.*). The canonical observable state-space model for an ARMA process given in Example 12.4.6 is observable. In this case $v = m = \max(p, q)$ and GF^{j-1} is the row-vector,

$$GF^{j-1} = [\delta_{jk}]_{k=1}^{v}, \qquad j = 1, \ldots, v.$$

from which it follows that the observability matrix O_v is the v-dimensional identity matrix.

If (F, G) is not observable, then we can proceed as in Proposition 12.4.3 to construct two matrices $\tilde{F}$ and $\tilde{G}$ such that $\tilde{F}$ has dimension $k = \text{rank}(O_v)$ and $(\tilde{F}, \tilde{G})$ is observable. We state this result without proof in the following proposition.

Proposition 12.4.5. *If the state-space model* (12.4.4) *is not observable and* $k = \text{rank}(O_v)$, *then there exists a stationary sequence of k-dimensional state vectors* $\{\hat{\mathbf{X}}_t\}$ *and matrices* $\hat{F}$, $\hat{H}$ *and* $\hat{G}$ *such that* $\hat{F}$ *is stable,* $(\hat{F}, \hat{G})$ *is observable and*

$$\hat{\mathbf{X}}_{t+1} = \hat{F}\hat{\mathbf{X}}_t + \hat{H}\mathbf{Z}_t,$$

$$\mathbf{Y}_t = \hat{G}\hat{\mathbf{X}}_t + \mathbf{Z}_t. \tag{12.4.20}$$

The state-space model defined by (12.4.4) and (12.4.5) is said to be *minimal* or of *minimum dimension* if the coefficient matrix F has dimension less than or equal to that of the corresponding matrix in any other state-space model for $\{\mathbf{Y}_t\}$. A minimal state-space model is necessarily controllable and observable; otherwise, by Propositions 12.4.3 and 12.4.4, the state equation can be reduced in dimension. Conversely, controllable and observable innovations models with non-singular innovations covariance are minimal, as shown below in Proposition 12.4.6. This result provides a useful means of checking for minimality, and a simple procedure (successive application of Propositions 12.4.3 and 12.4.5) for constructing minimal state-space models. It implies in particular that the canonical observable model for a causal invertible ARMA process given in Example 12.4.1 is minimal.

Proposition 12.4.6. *The innovations model defined by equations* (12.4.4) *and* (12.4.5), *with* $\mathfrak{T}$ *non-singular, is minimal if and only if it is controllable and observable.*

PROOF. The necessity of the conditions has already been established. To show sufficiency, consider two controllable and observable state-space models satisfying (12.4.4) and (12.4.5), with coefficient matrices (F, G, H) and

$(\tilde{F}, \tilde{G}, \tilde{H})$ and with state dimensions v and $\tilde{v}$ respectively. It suffices to show that $v = \tilde{v}$. Suppose that $\tilde{v} < v$. From Remark 1 it follows that

$$GF^{j-1}H = \tilde{G}\tilde{F}^{j-1}\tilde{H}, \qquad j = 1, 2, \ldots,$$

and hence, multiplying the observability and controllability matrices for each model, we obtain

$$O_v C_v = \begin{bmatrix} GH & GFH & \cdots & GF^{v-1}H \\ GFH & GF^2H & \cdots & GF^vH \\ \vdots & \vdots & \cdots & \vdots \\ GF^{v-1}H & GF^vH & \cdots & GF^{2v-1}H \end{bmatrix} = \tilde{O}_v \tilde{C}_v \qquad (12.4.21)$$

Since O_v and C_v have rank v, $\mathscr{R}(C_v) = \mathbb{R}^v$, $\mathscr{R}(O_v C_v) = \mathscr{R}(O_v)$ and hence $\operatorname{rank}(O_v C_v) = v$. On the other hand by (12.4.21), $\mathscr{R}(O_v C_v) \subseteq \mathscr{R}(\tilde{O}_v)$, and since $\operatorname{rank}(\tilde{O}_v) = \tilde{v}$ (Remark 2), we obtain the contradiction $\tilde{v} \geq \operatorname{rank}(O_v C_v) = v$. Thus $\tilde{v} = v$ as was to be shown. □

§12.5 Recursive Bayesian State Estimation

As in Section 12.1, we consider a sequence of v-dimensional state-vectors $\{\mathbf{X}_t, t \geq 1\}$ and a sequence of w-dimensional observation vectors $\{\mathbf{Y}_t, t \geq 1\}$. It will be convenient to write $\mathbf{Y}^{(t)}$ for the wt-dimensional column vector $\mathbf{Y}^{(t)} = (\mathbf{Y}'_1, \mathbf{Y}'_2, \ldots, \mathbf{Y}'_t)'$.

In place of the observation and state equations (12.1.1) and (12.1.2), we now assume the existence, for each t, of specified conditional probability densities of $\mathbf{Y}_t$ given $(\mathbf{X}_t, \mathbf{Y}^{(t-1)})$ and of $\mathbf{X}_{t+1}$ given $(\mathbf{X}_t, \mathbf{Y}^{(t)})$. We also assume that these densities are independent of $\mathbf{Y}^{(t-1)}$ and $\mathbf{Y}^{(t)}$, respectively. Thus the observation and state equations are replaced by the collection of conditional densities,

$$p_t^{(o)}(\mathbf{y}_t | \mathbf{x}_t) := p_{\mathbf{Y}_t|\mathbf{X}_t}(\mathbf{y}_t | \mathbf{x}_t), \qquad t = 1, 2, \ldots, \qquad (12.5.1)$$

and

$$p_t^{(s)}(\mathbf{x}_{t+1} | \mathbf{x}_t) := p_{\mathbf{X}_{t+1}|X_t}(\mathbf{x}_{t+1} | \mathbf{x}_t), \qquad t = 1, 2, \ldots, \qquad (12.5.2)$$

with respect to measures v and μ, respectively. We shall also assume that the initial state $\mathbf{X}_1$ has a probability density p_1 with respect to μ. If all the conditions of this paragraph are satisfied, we shall say that the densities $\{p_1, p_t^{(o)}, p_t^{(s)}, t = 1, 2, \ldots\}$ define a *Bayesian state-space model* for $\{\mathbf{Y}_t\}$.

In order to solve the *filtering* and *prediction* problems in this setting, we shall determine the conditional densities $p_t^{(f)}(\mathbf{x}_t | \mathbf{y}^{(t)})$ of $\mathbf{X}_t$ given $\mathbf{Y}^{(t)}$, and $p_t^{(p)}(\mathbf{x}_t | \mathbf{y}^{(t-1)})$ of $\mathbf{X}_t$ given $\mathbf{Y}^{(t-1)}$, respectively. The minimum mean squared error estimates of $\mathbf{X}_t$ based on $\mathbf{Y}^{(t)}$ and $\mathbf{Y}^{(t-1)}$ can then be computed as the conditional expectations, $E(\mathbf{X}_t | \mathbf{Y}^{(t)})$ and $E(\mathbf{X}_t | \mathbf{Y}^{(t-1)})$.

The required conditional densities $p_t^{(f)}$ and $p_t^{(p)}$ can be determined from $\{p_1, p_t^{(o)}, p_t^{(s)}, t = 1, 2, \ldots\}$ and the following recursions, the first of which is obtained by a direct application of Bayes's Theorem using the assumption that the distribution of $\mathbf{Y}_t$ given $(\mathbf{X}_t, \mathbf{Y}^{(t-1)})$ does not depend on $\mathbf{Y}^{(t-1)}$:

$$p_t^{(f)}(\mathbf{x}_t|\mathbf{y}^{(t)}) = p_t^{(o)}(\mathbf{y}_t|\mathbf{x}_t)p_t^{(p)}(\mathbf{x}_t|\mathbf{y}^{(t-1)})/c_t(\mathbf{y}^{(t)}), \tag{12.5.3}$$

$$p_{t+1}^{(p)}(\mathbf{x}_{t+1}|\mathbf{y}^{(t)}) = \int p_t^{(f)}(\mathbf{x}_t|\mathbf{y}^{(t)})p_t^{(s)}(\mathbf{x}_{t+1}|\mathbf{x}_t)\, d\mu(\mathbf{x}_t), \tag{12.5.4}$$

where

$$c_t(\mathbf{y}^{(t)}) = p_{\mathbf{Y}_t|\mathbf{Y}^{(t-1)}}(\mathbf{y}_t|\mathbf{y}^{(t-1)}).$$

The initial condition needed to solve these recursions is

$$p_1^{(p)}(\mathbf{x}_1|\mathbf{y}^{(0)}) = p_1(\mathbf{x}_1). \tag{12.5.5}$$

The constant $c_t(\mathbf{y}^{(t)})$ appearing in (12.5.3) is just a scale factor, determined by the condition $\int p_t^{(f)}(\mathbf{x}_t|\mathbf{y}^{(t)})\, d\mu(\mathbf{x}_t) = 1$.

EXAMPLES 12.5.1. Consider the linear state-space model defined by (12.1.1) and (12.1.2) and suppose, in addition to the assumptions made earlier, that $\{\mathbf{X}_1, \mathbf{W}_1, \mathbf{V}_1, \mathbf{W}_2, \mathbf{V}_2, \ldots\}$ is a sequence of independent Gaussian random vectors, each having a non-degenerate multivariate normal density.

Using the same notation as in Section 12.1, we can reformulate this system as a Bayesian state-space model, characterized by the Gaussian probability densities,

$$p_1(\mathbf{x}_1) = n(\mathbf{x}_1; E\mathbf{X}_1, \Omega_1), \tag{12.5.6}$$

$$p_t^{(o)}(\mathbf{y}_t|\mathbf{X}_t) = n(\mathbf{y}_t; G_t\mathbf{X}_t, R_t), \tag{12.5.7}$$

$$p_t^{(s)}(\mathbf{x}_{t+1}|\mathbf{X}_t) = n(\mathbf{x}_{t+1}; F_t\mathbf{X}_t, Q_t), \tag{12.5.8}$$

where $n(\mathbf{x}; \boldsymbol{\mu}, \Sigma)$ denotes the multivariate normal density with mean $\boldsymbol{\mu}$ and covariance matrix Σ. Note that in this formulation, we assume that $S_t = E(\mathbf{V}_t\mathbf{W}_t') = 0$ in order to satisfy the requirement that the distribution of $\mathbf{X}_{t+1}$ given $(\mathbf{X}_t, \mathbf{Y}^{(t)})$ does not depend on $\mathbf{Y}^{(t)}$.

To solve the filtering and prediction problems in the Bayesian framework, we first observe that the conditional densities, $p_t^{(f)}$ and $p_t^{(p)}$ are both normal. We shall write them (using notation analogous to that of Section 12.2) as

$$p_t^{(f)}(\mathbf{x}_t|\mathbf{Y}^{(t)}) = n(\mathbf{x}_t; \mathbf{X}_{t|t}, \Omega_{t|t}) \tag{12.5.9}$$

and

$$p_{t+1}^{(p)}(\mathbf{x}_{t+1}|\mathbf{Y}^{(t)}) = n(\mathbf{x}_{t+1}; \hat{\mathbf{X}}_{t+1}, \Omega_{t+1}). \tag{12.5.10}$$

From (12.5.4), (12.5.8), (12.5.9) and (12.5.10), we find that

$$\hat{\mathbf{X}}_{t+1} = F_t\mathbf{X}_{t|t}$$

and

$$\Omega_{t+1} = F_t \Omega_{t|t} F_t' + Q_t.$$

Substituting the corresponding density $n(\mathbf{x}_t; \hat{\mathbf{X}}_t, \Omega_t)$ for $p_t^{(p)}(\mathbf{x}_t | \mathbf{Y}^{(t-1)})$ in (12.5.3) and (12.5.7), we find that

$$\begin{aligned}\Omega_{t|t}^{-1} &= G_t' R_t^{-1} G_t + \Omega_t^{-1} \\ &= G_t' R_t^{-1} G_t + (F_{t-1} \Omega_{t-1|t-1} F_{t-1}' + Q_{t-1})^{-1}\end{aligned}$$

and

$$\mathbf{X}_{t|t} = \hat{\mathbf{X}}_t + \Omega_{t|t} G_t' R_t^{-1} (\mathbf{Y}_t - G_t \hat{\mathbf{X}}_t).$$

From (12.5.3) with $p_1^{(p)}(\mathbf{x}_1 | \mathbf{y}^{(0)}) = n(\mathbf{x}_1; E\mathbf{X}_1, \Omega_1)$ we obtain the initial conditions,

$$\mathbf{X}_{1|1} = E\mathbf{X}_1 + \Omega_{1|1} G_1' R_1^{-1} (\mathbf{Y}_1 - E\mathbf{X}_1)$$

and

$$\Omega_{1|1}^{-1} = G_1' R_1^{-1} G_1 + \Omega_1^{-1}.$$

Remark 1. Under the assumptions made in Example 12.5.1, the recursions of Propositions 12.2.2 and 12.2.3 give the same results for $\hat{\mathbf{X}}_t$ and $\mathbf{X}_{t|t}$, since for Gaussian systems best linear mean-square estimation is equivalent to best mean-square estimation. Note that the recursions of Example 12.5.1 require stronger assumptions on the covariance matrices (including existence of R_t^{-1}) than the recursions of Section 12.2, which require only the assumptions (a)–(e) of Section 12.1.

EXAMPLE 12.5.2. Application of the results of Example 12.5.1 to Example 12.2.1 (with $X_1 = 1$ and the additional assumption that the sequence $\{W_1, V_1, W_2, V_2, \ldots\}$ is Gaussian) immediately furnishes the recursions,

$$\begin{aligned}\Omega_{t|t}^{-1} &= 1 + (4\Omega_{t-1|t-1} + 1)^{-1}, \\ X_{t|t} &= 2X_{t-1|t-1} + \Omega_{t|t}(Y_t - 2X_{t-1|t-1}), \\ \hat{X}_{t+1} &= 2X_{t|t}, \\ \Omega_{t+1} &= 4\Omega_{t|t} + 1,\end{aligned}$$

with initial conditions $X_{1|1} = 1$ and $\Omega_{1|1} = 0$. It is easy to check that these recursions are equivalent to those derived earlier in Example 12.2.1.

EXAMPLE 12.5.3 (A Non-Gaussian Example). In general the solution of the recursions (12.5.3) and (12.5.4) presents substantial computational problems. Numerical methods for dealing with non-Gaussian models are discussed by Sorenson and Alspach (1971) and Kitagawa (1987). Here we shall illustrate the recursions (12.5.3) and (12.5.4) in a very simple special case. Consider the state equation,

$$X_t = aX_{t-1}, \tag{12.5.11}$$

with observation density (relative to counting measure on the non-negative integers),

$$p_t^{(o)}(y_t|x_t) = \frac{(\pi x_t)^{y_t} e^{-\pi x_t}}{y_t!}, \qquad y_t = 0, 1, \dots, \tag{12.5.12}$$

and initial density (with respect to Lebesgue measure),

$$p_1(x) = \frac{\lambda^\alpha x^{\alpha-1} e^{-\lambda x}}{\Gamma(\alpha)}, \qquad x \geq 0. \tag{12.5.13}$$

(This is a simplified model for the evolution of the number X_t of individuals at time t infected with a rare disease in which X_t is treated as a continuous rather than an integer-valued random variable. The observation Y_t represents the number of infected individuals observed in a random sample consisting of a small fraction π of the population at time t.) Although there is no density $p_t^{(s)}(x_t|x_{t-1})$ with respect to Lebesgue measure corresponding to the state equation (12.5.11), it is clear that the recursion (12.5.4) is replaced in this case by the relation,

$$p_t^{(p)}(x_t|\mathbf{y}^{(t-1)}) = a^{-1} p_{t-1}^{(f)}(a^{-1}x_t|\mathbf{y}^{(t-1)}), \tag{12.5.14}$$

while the recursion (12.5.3) is exactly as before. The filtering and prediction densities $p_t^{(f)}$ and $p_t^{(p)}$ are both with respect to Lebesgue measure. Solving for $p_1^{(f)}$ from (12.5.13) and the initial condition (12.5.5) and then successively substituting in the recursions (12.5.14) and (12.5.3), we easily find that

$$p_t^{(f)}(x_t|\mathbf{y}^{(t)}) = \lambda_t^{\alpha_t} x_t^{\alpha_t - 1} e^{-\alpha_t x_t}/\Gamma(\alpha_t), \qquad t \geq 0, \tag{12.5.15}$$

and

$$p_{t+1}^{(p)}(x_{t+1}|\mathbf{y}^{(t)}) = \left(\frac{\lambda_t}{a}\right)^{\alpha_t} x_{t+1}^{\alpha_t - 1} e^{-\lambda_t x_{t+1}/a}/\Gamma(\alpha_t), \qquad t \geq 0,$$

where $\alpha_t = \alpha + y_1 + \cdots + y_t$ and $\lambda_t = \lambda a^{1-t} + \pi(1 - a^{-t})/(1 - a^{-1})$. In particular the minimum mean squared error estimate of x_t based on $\mathbf{y}^{(t)}$ is the conditional expectation α_t/λ_t. The minimum mean squared error is the variance of the distribution defined by (12.5.15), i.e. α_t/λ_t^2.

Problems

12.1. For the state-space model (12.1.1) and (12.1.2), show that

$$\mathbf{X}_{t+1} = F_t F_{t-1} \cdots F_1 \mathbf{X}_1 + \mathbf{V}_t + F_t \mathbf{V}_{t-1} + \cdots + F_t F_{t-1} \cdots F_2 \mathbf{V}_1$$

and

$$\begin{aligned}\mathbf{Y}_t = {} & G_t(F_{t-1} F_{t-2} \cdots F_1 \mathbf{X}_1) + G_t \mathbf{V}_{t-1} + G_t F_{t-1} \mathbf{V}_{t-2} + \cdots \\ & + G_t F_{t-1} F_{t-2} \cdots F_2 \mathbf{V}_1 + \mathbf{W}_t.\end{aligned}$$

These expressions define f_t and g_t in (12.1.3) and (12.1.4). Specialize to the case when $F_t = F$ and $G_t = G$ for all t.

12.2. Consider the two state-space models

$$\begin{cases} \mathbf{X}_{t+1,1} = F_1\mathbf{X}_{t1} + \mathbf{V}_{t1}, \\ \mathbf{Y}_{t,1} = G_1\mathbf{X}_{t1} + \mathbf{W}_{t1}, \end{cases}$$

and

$$\begin{cases} \mathbf{X}_{t+1,2} = F_2\mathbf{X}_{t2} + \mathbf{V}_{t2}, \\ \mathbf{Y}_{t2} = G_2\mathbf{X}_{t2} + \mathbf{W}_{t2}, \end{cases}$$

where $\{(\mathbf{V}'_{t1}, \mathbf{W}'_{t1}, \mathbf{V}'_{t2}, \mathbf{W}'_{t2})'\}$ is white noise. Derive a state-space representation for $\{(\mathbf{Y}'_{t1}, \mathbf{Y}'_{t2})'\}$.

12.3. Show that the unique stationary solutions to equations (12.1.13) and (12.1.12) are given by the infinite series

$$\mathbf{X}_t = \sum_{j=0}^{\infty} F^j\mathbf{V}_{t-j-1}$$

and

$$\mathbf{Y}_t = \mathbf{W}_t + \sum_{j=0}^{\infty} GF^j\mathbf{V}_{t-j-1},$$

which converge in mean square provided $\det(I - Fz) \neq 0$ for $|z| \leq 1$. Conclude that $\{(\mathbf{X}'_t, \mathbf{Y}'_t)'\}$ is a multivariate stationary process. (Hint: Show that there exists an $\varepsilon > 0$ such that $(I - Fz)^{-1}$ has the power series representation, $\sum_{j=0}^{\infty} F^j z^j$, in the region $|z| < 1 + \varepsilon$.)

12.4. Let F be the coefficient of $\mathbf{X}_t$ in the state equation (12.1.18) for the causal AR(p) process

$$X_t - \phi_1 X_{t-1} - \cdots - \phi_p X_{t-p} = Z_t, \qquad \{Z_t\} \sim \text{WN}(0, \sigma^2).$$

Establish the stability of (12.1.18) by showing that the eigenvalues of F are equal to the reciprocals of the zeros of the autoregressive polynomial $\phi(z)$. In particular, show that

$$\det(zI - F) = z^p\phi(z^{-1}).$$

12.5. Let $\{\mathbf{X}_t\}$ be the unique stationary solution of the state equation (12.1.23) and suppose that $\{Y_t\}$ is defined by (12.1.20). Show that $\{Y_t\}$ must be the unique stationary solution of the ARMA equations (12.1.19).

12.6. Let $\{Y_t\}$ be the MA(1) process

$$Y_t = Z_t + \theta Z_{t-1}, \qquad \{Z_t\} \sim \text{WN}(0, \sigma^2).$$

(a) Show that $\{Y_t\}$ has the state-space representation

$$Y_t = [1 \quad 0]\mathbf{X}_t,$$

where $\{\mathbf{X}_t\}$ is the unique stationary solution of

$$\mathbf{X}_{t+1} = \begin{bmatrix} 0 & 1 \\ 0 & 0 \end{bmatrix}\mathbf{X}_t + \begin{bmatrix} 1 \\ \theta \end{bmatrix} Z_{t+1}.$$

In particular, show that the state-vector $\mathbf{X}_t$ may be written as

$$\mathbf{X}_t = \begin{bmatrix} 1 & \theta \\ \theta & 0 \end{bmatrix} \begin{bmatrix} Z_t \\ Z_{t-1} \end{bmatrix}.$$

(b) Display the state-space model for $\{Y_t\}$ obtained from Example 12.1.6.

12.7. Verify equations (12.1.34)–(12.1.36) for an ARIMA(1, 1, 1) process.

12.8. Let $\{Y_t\}$ be an ARIMA(p, d, q) process. By using the state-space model in Example 12.1.5 show that $\{Y_t\}$ has the representation

$$Y_t = G\mathbf{X}_t$$

with

$$\mathbf{X}_{t+1} = F\mathbf{X}_t + HZ_{t+1}$$

for $t = 1, 2, \ldots$ and suitably chosen matrices F, G and H. Write down the explicit form of the observation and state equations for an ARIMA(1, 1, 1) process and compare with equations (12.1.34)–(12.1.36).

12.9. Following the technique of Example 12.1.7, write down a state-space model for $\{Y_t\}$ where $\{\nabla\nabla_{12} Y_t\}$ is an ARMA(p, q) process.

12.10. Show that the set L_2^v of random v-vectors with all components in $L^2(\Omega, \mathscr{F}, P)$ is a Hilbert space if we define the inner product to be $\langle \mathbf{X}, \mathbf{Y} \rangle = \sum_{i=1}^{v} E(X_i Y_i)$ for all $\mathbf{X}, \mathbf{Y} \in L_2^v$. If $\mathbf{X}, \mathbf{Y}_0, \ldots, \mathbf{Y}_r \in L_2^v$ show that $P(\mathbf{X} | \mathbf{Y}_0, \ldots, \mathbf{Y}_r)$ as in Definition 12.2.2 is the projection of $\mathbf{X}$ (in this Hilbert space) onto S^v, the closed linear subspace of L_2^v consisting of all vectors of the form $C_0 \mathbf{Y}_0 + \cdots + C_r \mathbf{Y}_r$, where $C_0, \ldots, C_r$ are constant matrices.

12.11. Prove Remark 5 of Section 12.2. Note also that if the linear equation,

$$S\mathbf{x} = \mathbf{b},$$

has a solution, then $\mathbf{x} = S^{-1}\mathbf{b}$ is a solution for any generalized inverse S^{-1} of S. (If $S\mathbf{y} = \mathbf{b}$ for some vector $\mathbf{y}$ then $S(S^{-1}\mathbf{b}) = SS^{-1}S\mathbf{y} = S\mathbf{y} = \mathbf{b}$.)

12.12. Let $\mathscr{M}_1$ and $\mathscr{M}_2$ be two closed subspaces of a Hilbert space $\mathscr{H}$ and suppose that $\mathscr{M}_1 \perp \mathscr{M}_2$ (i.e. $x \perp y$ for all $x \in \mathscr{M}_1$ and $y \in \mathscr{M}_2$). Show that

$$\mathscr{P}_{\mathscr{M}_1 \oplus \mathscr{M}_2} = P_{\mathscr{M}_1} + P_{\mathscr{M}_2},$$

where $\mathscr{M}_1 \oplus \mathscr{M}_2$ is the closed subspace $\{x + y : x \in \mathscr{M}_1, y \in \mathscr{M}_2\}$. Note that (12.2.8) follows immediately from this identity.

12.13. The mass of a body grows according to the rule

$$X_{t+1} = aX_t + V_t, \qquad a > 1,$$

where X_1 is known to be 10 exactly and $\{V_t\} \sim \mathrm{WN}(0, 1)$. At time t we observe

$$Y_t = X_t + W_t,$$

where $\{W_t\} \sim \text{WN}(0, 1)$ and $\{W_t\}$ is uncorrelated with $\{V_t\}$. If P_t denotes projection (in $L^2(\Omega, \mathscr{F}, P)$) onto $\overline{\text{sp}}\{1, Y_1, \ldots, Y_t\}$, $t \geq 1$, and P_0 denotes projection onto $\overline{\text{sp}}\{1\}$,

(a) express σ_{t+1}^2 in terms of σ_t^2, where

$$\sigma_t^2 := E(X_t - P_{t-1}X_t)^2, \qquad t = 1, 2, \ldots,$$

(b) express $P_t X_{t+1}$ in terms of σ_t^2, Y_t and $P_{t-1}X_t$,

(c) evaluate $P_2 X_3$ and its mean squared error if $Y_2 = 12$, and $a = 1.5$,

(d) assuming that $\lim_{t\to\infty} \sigma_t^2$ exists, determine its value.

12.14. Use the representation found in Problem 12.6(a) to derive a recursive scheme for computing the best linear one-step predictors $\hat{Y}_t$ based on $Y_1, \ldots, Y_{t-1}$ and their mean squared errors.

12.15. Consider the state-space model defined by (12.2.4) and (12.2.5) with $F_t = F$ and $G_t = G$ for all t and let $k > h \geq 1$. Show that

$$E(\mathbf{X}_{t+k} - P_t\mathbf{X}_{t+k})(\mathbf{X}_{t+h} - P_t\mathbf{X}_{t+h})' = F^{k-h}\Omega_t^{(h)}$$

and

$$E(\mathbf{Y}_{t+k} - P\mathbf{Y}_{t+k})(\mathbf{Y}_{t+h} - P_t\mathbf{Y}_{t+h})' = GF^{k-h}\Omega_t^{(h)}G' + GF^{k-h}S_{t+h}.$$

12.16. Verify the calculation of $\Theta_t\Delta_t^{-1}$ and Ω_t in Example 12.3.1.

12.17. Verify the calculation of $P_5 X_2$ and its mean squared error in Example 12.3.3.

12.18. Let $y_1 = -.210$, $y_2 = .968$, $y_4 = .618$ and $y_5 = -.880$ be observed values of the MA(1) process

$$Y_t = Z_t + .5Z_{t-1}, \qquad \{Z_t\} \sim \text{WN}(0, 1).$$

(a) Compute $P(Y_6 | Y_1, Y_2, Y_4, Y_5)$ and its mean squared error.

(b) Compute $P(Y_7 | Y_1, Y_2, Y_4, Y_5)$ and its mean squared error.

(c) Compute $P(Y_3 | Y_1, Y_2, Y_4, Y_5)$ and its mean squared error.

(d) Substitute the value found in (c) for the missing observation y_3 and evaluate $P(Y_6 | Y_1, Y_2, Y_3, Y_4, Y_5)$ using the enlarged data set.

(e) Explain in terms of projection operators why the results of (a) and (d) are the same.

12.19. Show that the state-space representation (12.1.24), (12.1.25) of a causal invertible ARMA(p, q) process is also an innovations representation.

12.20. Consider the non-invertible MA(1) process,

$$Y_t = Z_t + 2Z_{t-1}, \qquad \{Z_t\} \sim \text{WN}(0, 1).$$

Find an innovations representation of $\{Y_t\}$ (i.e. a state-space model of the form (12.4.4) which satisfies (12.4.5)).

12.21. Let $\{V_t\}$ be a sequence of independent exponential random variables with $EV_t = t^{-1}$ and suppose that $\{X_t, t \geq 1\}$ and $\{Y_t, t \geq 1\}$ are the state and observation random variables, respectively, of the state-space system,

$$X_1 = V_1,$$

$$X_t = X_{t-1} + V_t, \qquad t = 2, 3, \ldots,$$

where the distribution of the observation Y_t, conditional on the random variables X_1, Y_2, $1 \le s < t$, is Poisson with mean X_t.

(a) Determine the densities $\{p_1, p_t^{(o)}, p_t^{(s)}, t \ge 1\}$, in the Bayesian state-space model for $\{Y_t\}$.

(b) Show, using (12.5.3)–(12.5.5), that

$$p_1^{(f)}(x_1 | y_1) = \frac{2^{1+y_1}}{\Gamma(1 + y_1)} x_1^{y_1} e^{-2x_1}, \qquad x_1 \ge 0,$$

and

$$p_2^{(p)}(x_2 | y_1) = \frac{2^{2+y_1}}{\Gamma(2 + y_1)} x_2^{1+y_1} e^{-2x_2}, \qquad x_2 > 0.$$

(c) Show that

$$p_t^{(f)}(x_t | \mathbf{y}^{(t)}) = \frac{(t + 1)^{t+\alpha_t}}{\Gamma(t + \alpha_t)} x_t^{t-1+\alpha_t} e^{-(t+1)x_t}, \qquad x_t > 0,$$

and

$$p_{t+1}^{(p)}(x_{t+1} | \mathbf{y}^{(t)}) = \frac{(t + 1)^{t+1+\alpha_t}}{\Gamma(t + 1 + \alpha_t)} x_{t+1}^{t+\alpha_t} e^{-(t+1)x_{t+1}}, \qquad x_{t+1} > 0,$$

where $\alpha_t = y_1 + \cdots + y_t$.

(d) Conclude from (c) that the minimum mean squared error estimates of X_t and X_{t+1} based on $Y_1, \ldots, Y_t$, are

$$X_{t|t} = \frac{t + Y_1 + \cdots + Y_t}{t + 1}$$

and

$$\hat{X}_{t+1} = \frac{t + 1 + Y_1 + \cdots + Y_t}{t + 1},$$

respectively.

CHAPTER 13

Further Topics

In this final chapter we touch on a variety of topics of special interest. In Section 13.1 we consider transfer function models, designed to exploit, for predictive purposes, the relationship between two time series when one leads the other. Section 13.2 deals with long-memory models, characterized by very slow convergence to zero of the autocorrelations $\rho(h)$ as $h \to \infty$. Such models are suggested by numerous observed series in hydrology and economics. In Section 13.3 we examine linear time-series models with infinite variance and in Section 13.4 we briefly consider non-linear models and their applications.

§13.1 Transfer Function Modelling

In this section we consider the problem of estimating the transfer function of a linear filter when the output includes added uncorrelated noise. Suppose that $\{X_{t1}\}$ and $\{X_{t2}\}$ are, respectively, the input and output of the transfer function model

$$X_{t2} = \sum_{j=0}^{\infty} t_j X_{t-j,1} + N_t, \tag{13.1.1}$$

where $T = \{t_j, j = 0, 1, \dots\}$ is a causal time-invariant linear filter and $\{N_t\}$ is a zero-mean stationary process, uncorrelated with the input process $\{X_{t1}\}$. Suppose also that $\{X_{t1}\}$ is a zero-mean stationary time series. Then the bivariate process $\{(X_{t1}, X_{t2})'\}$ is also stationary. From the analysis of Example 11.6.4, the transfer function $T(e^{-i\lambda}) = \sum_{j=0}^{\infty} t_j e^{-ij\lambda}$, $-\pi < \lambda \leq \pi$, can be expressed in terms of the spectrum of $\{(X_{t1}, X_{t2})'\}$ (see 11.6.17)) as

$$T(e^{-i\lambda}) = f_{21}(\lambda)/f_{11}(\lambda). \tag{13.1.2}$$

The analogous time-domain equation which relates the weights $\{t_j\}$ to the cross covariances is

$$\gamma_{21}(k) = \sum_{j=0}^{\infty} t_j \gamma_{11}(k-j). \tag{13.1.3}$$

This equation is obtained by multiplying each side of (13.1.1) by $X_{t-k,1}$ and then taking expectations.

The equations (13.1.2) and (13.1.3) simplify a great deal if the input process happens to be white noise. For example, if $\{X_{t1}\} \sim \mathrm{WN}(0, \sigma_1^2)$, then we can immediately identify t_k from (13.1.3) as

$$t_k = \gamma_{21}(k)/\sigma_1^2. \tag{13.1.4}$$

This observation suggests that "pre-whitening" of the input process might simplify the identification of an appropriate transfer-function model and at the same time provide simple preliminary estimates of the coefficients t_k.

If $\{X_{t1}\}$ can be represented as an invertible ARMA(p, q) process,

$$\phi(B)X_{t1} = \theta(B)Z_t, \qquad \{Z_t\} \sim \mathrm{WN}(0, \sigma_z^2), \tag{13.1.5}$$

then application of the filter $\pi(B) = \phi(B)\theta^{-1}(B)$ to $\{X_{t1}\}$ will produce the whitened series $\{Z_t\}$. Now applying the operator $\pi(B)$ to each side of (13.1.1) and letting $Y_t = \pi(B)X_{t2}$, we obtain the relation,

$$Y_t = \sum_{j=0}^{\infty} t_j Z_{t-j} + N_t',$$

where

$$N_t' = \pi(B)N_t,$$

and $\{N_t'\}$ is a zero-mean stationary process, uncorrelated with $\{Z_t\}$. The same arguments which gave (13.1.2) and (13.1.4) therefore yield, when applied to $(Z_t, Y_t)'$,

$$T(e^{-i\lambda}) = 2\pi\sigma_z^{-2} f_{YZ}(\lambda) = \sigma_Z^{-2} \sum_{h=-\infty}^{\infty} \gamma_{YZ}(h) e^{-ih\lambda}$$

and

$$t_j = \rho_{YZ}(j)\sigma_Y/\sigma_Z,$$

where $\rho_{YZ}(\cdot)$ is the cross-correlation function of $\{Y_t\}$ and $\{Z_t\}$, $f_{YZ}(\cdot)$ is the cross spectrum, $\sigma_Z^2 = \mathrm{Var}(Z_t)$ and $\sigma_Y^2 = \mathrm{Var}(Y_t)$.

Given the observations $\{(X_{t1}, X_{t2})', t = 1, \ldots, n\}$, the results of the previous paragraph suggest the following procedure for estimating $\{t_j\}$ and analyzing the noise $\{N_t\}$ in the model (13.1.1):

(1) Fit an ARMA model to $\{X_{t1}\}$ and file the residuals $\{\hat{Z}_1, \dots, \hat{Z}_n\}$. Let $\hat{\boldsymbol{\phi}}$ and $\hat{\boldsymbol{\theta}}$ denote the maximum likelihood estimates of the autoregressive and moving average parameters and let $\hat{\boldsymbol{\sigma}}_Z^2$ be the maximum likelihood estimate of the variance of $\{Z_t\}$.

(2) Apply the operator $\hat{\pi}(B) = \hat{\phi}(B)\hat{\theta}^{-1}(B)$ to $\{X_{t2}\}$, giving the series $\{\hat{Y}_1, \dots, \hat{Y}_n\}$. The values $\hat{Y}_t$ can be computed as the residuals obtained by running the computer program PEST with initial coefficients $\hat{\boldsymbol{\phi}}$, $\hat{\boldsymbol{\theta}}$ and using Option 8 with 0 iterations. Let $\hat{\sigma}_Y^2$ denote the sample variance of $\hat{Y}_t$.

(3) Compute the sample cross-correlation function $\hat{\rho}_{YZ}(h)$ between $\{\hat{Y}_t\}$ and $\{\hat{Z}_t\}$. Comparison of $\hat{\rho}_{YZ}(h)$ with the bounds $\pm 1.96n^{-1/2}$ gives a preliminary indication of the lags h at which $\rho_{YZ}(h)$ is significantly different from zero. A more refined check can be carried out by using Bartlett's formula (Theorem 11.2.3) for the asymptotic variance of $\hat{\rho}_{YZ}(h)$. Under the assumptions that $\{\hat{Z}_t\} \sim \text{WN}(0, \hat{\sigma}_Z^2)$ and $\{(\hat{Y}_t, \hat{Z}_t)'\}$ is a stationary Gaussian process,

$$n \operatorname{Var}(\hat{\rho}_{YZ}(h)) \sim 1 - \rho_{YZ}^2(h)\left[1.5 - \sum_{k=-\infty}^{\infty} (\rho_{YZ}^2(k) + \rho_{YY}^2(k)/2)\right]$$

$$+ \sum_{k=-\infty}^{\infty} [\rho_{YZ}(h+k)\rho_{YZ}(h-k) - 2\rho_{YZ}(h)\rho_{YZ}(-k)\rho_{YY}^2(h+k)].$$

In order to check the hypothesis H_0 that $\rho_{YZ}(h) = 0$, $h \notin [a, b]$, where a and b are integers, we note from Corollary 11.2.1 that under H_0,

$$\operatorname{Var}(\hat{\rho}_{YZ}(h)) \sim n^{-1} \quad \text{for } h \notin [a, b].$$

We can therefore check the hypothesis H_0 by comparing $\hat{\rho}_{YZ}$, $h \notin [a, b]$ with the bounds $\pm 1.96n^{-1/2}$. Observe that $\rho_{YZ}(h)$ should be zero for $h < 0$ if the model (13.1.1) is valid.

(4) Preliminary estimates of t_h for lags h at which $\rho_{YZ}(h)$ is found to be significantly different from zero are

$$\hat{t}_h = \hat{\rho}_{YZ}(h)\hat{\sigma}_Y/\hat{\sigma}_Z.$$

For other values of h the preliminary estimates are $\hat{t}_h = 0$. Let $m \geq 0$ be the largest value of j such that $\hat{t}_j$ is non-zero and let $b \geq 0$ be the smallest such value. Then b is known as the *delay parameter* of the filter $\{\hat{t}_j\}$. If m is very large and if the coefficients $\{\hat{t}_j\}$ are approximately related by difference equations of the form

$$\hat{t}_j - v_1\hat{t}_{j-1} - \cdots - v_p\hat{t}_{j-p} = 0, \qquad j \geq b + p,$$

then $\hat{T}(B) = \sum_{j=b}^{m} \hat{t}_j B^j$ can be represented approximately, using fewer parameters, as

$$\hat{T}(B) = w_0(1 - v_1 B - \cdots - v_p B_p)^{-1} B^b.$$

In particular, if $\hat{t}_j = 0, j < b$, and $\hat{t}_j = w_0 v_1^{j-b}, j \geq b$, then

$$\hat{T}(B) = w_0(1 - v_1 B)^{-1} B^b. \tag{13.1.6}$$

Box and Jenkins (1970) recommend choosing $\hat{T}(B)$ to be a ratio of two polynomials, however the degrees of the polynomials are often difficult to estimate from $\{\hat{t}_j\}$. The primary objective at this stage is to find a parametric function which provides an adequate approximation to $\hat{T}(B)$ without introducing too large a number of parameters. If $\hat{T}(B)$ is represented as $\hat{T}(B) = B^b w(B) v^{-1}(B) = B^b(w_0 + w_1 B + \cdots + w_q B^q)(1 - v_1 B - \cdots - v_p B^p)^{-1}$ with $v(z) \neq 0$ for $|z| \leq 1$, then we define $m = \max(q + b, p)$.

(5) The noise sequence $\{N_t, t = m + 1, \ldots, n\}$ is estimated by

$$\hat{N}_t = X_{t2} - \hat{T}(B) X_{t1}.$$

(We set $\hat{N}_t = 0$, $t \leq m$, in order to compute $\hat{N}_t$, $t > m = \max(b + q, p)$.)

(6) Preliminary identification of a suitable model for the noise sequence is carried out by fitting a causal invertible ARMA model

$$\phi^{(N)}(B)\hat{N}_t = \theta^{(N)}(B) W_t, \qquad \{W_t\} \sim \mathrm{WN}(0, \sigma_W^2),$$

to the estimated noise $\hat{N}_{m+1}, \ldots, \hat{N}_n$.

(7) Selection of the parameters b, p and q and the orders p_2 and q_2 of $\phi^{(N)}(\cdot)$ and $\theta^{(N)}(\cdot)$ gives the preliminary model,

$$\phi^{(N)}(B) v(B) X_{t2} = B^b \phi^{(N)}(B) w(B) X_{t1} + \theta^{(N)}(B) v(B) W_t,$$

where $\hat{T}(B) = B^b w(B) v^{-1}(B)$ as in step (4). For this model we can compute $\hat{W}_t(\mathbf{w}, \mathbf{v}, \boldsymbol{\phi}^{(N)}, \boldsymbol{\theta}^{(N)})$, $t > m^* = \max(p_2 + p, b + p_2 + q)$, by setting $\hat{W}_t = 0$ for $t \leq m^*$. The parameters $\mathbf{w}$, $\mathbf{v}$, $\boldsymbol{\phi}^{(N)}$ and $\boldsymbol{\theta}^{(N)}$ can then be estimated by minimizing

$$\sum_{t=m^*+1}^{n} \hat{W}_t^2(\mathbf{w}, \mathbf{v}, \boldsymbol{\phi}^{(N)}, \boldsymbol{\theta}^{(N)}) \tag{13.1.7}$$

subject to the constraints that $\phi^{(N)}(z)$, $\theta^{(N)}(z)$ and $v(z)$ are all non-zero for $|z| \leq 1$. The preliminary estimates from steps (4) and (6) can be used as initial values in the minimization and the minimization may be carried out using the program TRANS. Alternatively, the parameters can be estimated by maximum likelihood, as discussed in Section 12.3, using a state-space representation for the transfer function model (see (13.1.19) and (13.1.20)).

(8) From the least squares estimators of the parameters of $T(B)$, a new estimated noise sequence can be computed as in step (5) and checked for compatibility with the ARMA model for $\{N_t\}$ fitted by the least squares procedure. If the new estimated noise sequence suggests different orders for $\phi^{(N)}(\cdot)$ and $\theta^{(N)}(\cdot)$, the least squares procedure in step (7) can be repeated using the new orders.

(9) To test for goodness of fit, the residuals from the ARMA fitting in steps (1) and (6) should both be checked as described in Section 9.4. The sample cross-correlations of the two residual series $\{\hat{Z}_t, t > m^*\}$ and

$\{\hat{W}_t, t > m^*\}$ should also be compared with the bounds $\pm 1.96/\sqrt{n}$ in order to check the hypothesis that the sequences $\{N_t\}$ and $\{Z_t\}$ are uncorrelated.

EXAMPLE 13.1.1 (Sales with a Leading Indicator). In this example we fit a transfer function model to the bivariate time series of Example 11.2.2. Let

$$X_{t1} = (1 - B)Y_{t1} - .0228, \qquad t = 1, \dots, 149,$$

and

$$X_{t2} = (1 - B)Y_{t2} - .420, \qquad t = 1, \dots, 149,$$

where $\{Y_{t1}\}$ and $\{Y_{t2}\}$, $t = 0, \dots, 149$, are the leading indicator and sales data respectively. It was found in Example 11.2.2 that $\{X_{t1}\}$ can be modelled as the zero mean ARMA process,

$$X_{t1} = (1 - .474B)Z_t, \qquad \{Z_t\} \sim \text{WN}(0, .0779).$$

We can therefore whiten the series by application of the filter $\hat{\pi}(B) = (1 - .474B)^{-1}$. Applying $\hat{\pi}(B)$ to both $\{X_{t1}\}$ and $\{X_{t2}\}$ we obtain

$$\hat{Z}_t = (1 - .474B)^{-1}X_{t1}, \qquad \hat{\sigma}_Z^2 = .0779,$$

and

$$\hat{Y}_t = (1 - .474B)^{-1}X_{t2}, \qquad \hat{\sigma}_Y^2 = 4.0217.$$

These calculations and the filing of the series $\{\hat{Z}_t\}$ and $\{\hat{Y}_t\}$ were carried out using the program PEST as described in step (2). The sample correlation function $\hat{\rho}_{YZ}(h)$ of $\{\hat{Y}_t\}$ and $\{\hat{Z}_t\}$, computed using the program TRANS, is shown in Figure 13.1. Comparison of $\hat{\rho}_{YZ}(h)$ with the bounds $\pm 1.96(149)^{-1/2} = \pm .161$ suggests that $\rho_{YZ}(h) = 0$ for $h < 3$. Since $\hat{t}_j = \hat{\rho}_{YZ}(j)\hat{\sigma}_Y/\hat{\sigma}_Z$ is decreasing approximately geometrically for $j \geq 3$, we take $T(B)$ to have the form (13.1.6), i.e.

$$T(B) = w_0(1 - v_1 B)^{-1}B^3.$$

Preliminary estimates of w_0 and v_1 are given by $\hat{w}_0 = \hat{t}_3 = 4.86$ and $\hat{v}_1 = \hat{t}_4/\hat{t}_3 = .698$. The estimated noise sequence is obtained from the equation

$$\hat{N}_t = X_{t2} - 4.86B^3(1 - .698B)^{-1}X_{t1}, \qquad t = 4, 5, \dots, 149.$$

Examination of this sequence using the program PEST leads to the MA(1) model,

$$\hat{N}_t = (1 - .307B)W_t, \qquad \{W_t\} \sim \text{WN}(0, .0644).$$

Substituting these preliminary noise and transfer function models into equation (13.1.1) then gives

$$X_{t2} = 4.86B^3(1 - .698B)^{-1}X_{t1} + (1 - .307B)W_t, \qquad \{W_t\} \sim \text{WN}(0, .0644).$$

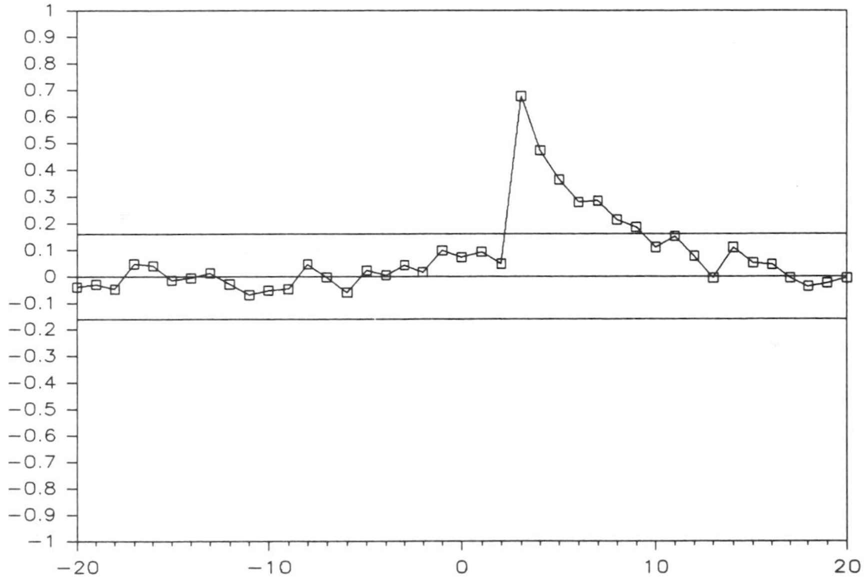

Figure 13.1. The sample cross-correlation function $\hat{\rho}_{YZ}(h)$, $-20 \le h \le 20$, of Example 13.1.1.

Now minimizing the sum of squares (13.1.7) with respect to the parameters $(w_0, v_1, \theta_1^{(N)})$ using the program TRANS, we obtain the least squares model

$$X_{t2} = 4.717B^3(1 - .724B)^{-1}X_{t1} + (1 - .582B)W_t, \ \{W_t\} \sim \text{WN}(0, .0486), \tag{13.1.8}$$

where

$$X_{t1} = (1 - .474B)Z_t, \qquad \{Z_t\} \sim \text{WN}(0, .0779).$$

Notice the reduced white noise variance of $\{W_t\}$ in the least squares model as compared with the preliminary model.

The sample autocorrelation and partial autocorrelation functions for the series

$$\hat{N}_t = X_{t2} - 4.717B^3(1 - .724B)^{-1}X_{t1}$$

are shown in Figure 13.2. These graphs strongly indicate that the MA(1) model is appropriate for the noise process. Moreover the residuals $\hat{W}_t$ obtained from the least squares model (13.1.8) pass the diagnostic tests for white noise as described in Section 9.4, and the sample cross-correlations between the residuals $\hat{W}_t$ and $\hat{Z}_t$, $t = 4, \ldots, 149$, are found to lie between the bounds $\pm 1.96/\sqrt{144}$ for all lags between ± 20.

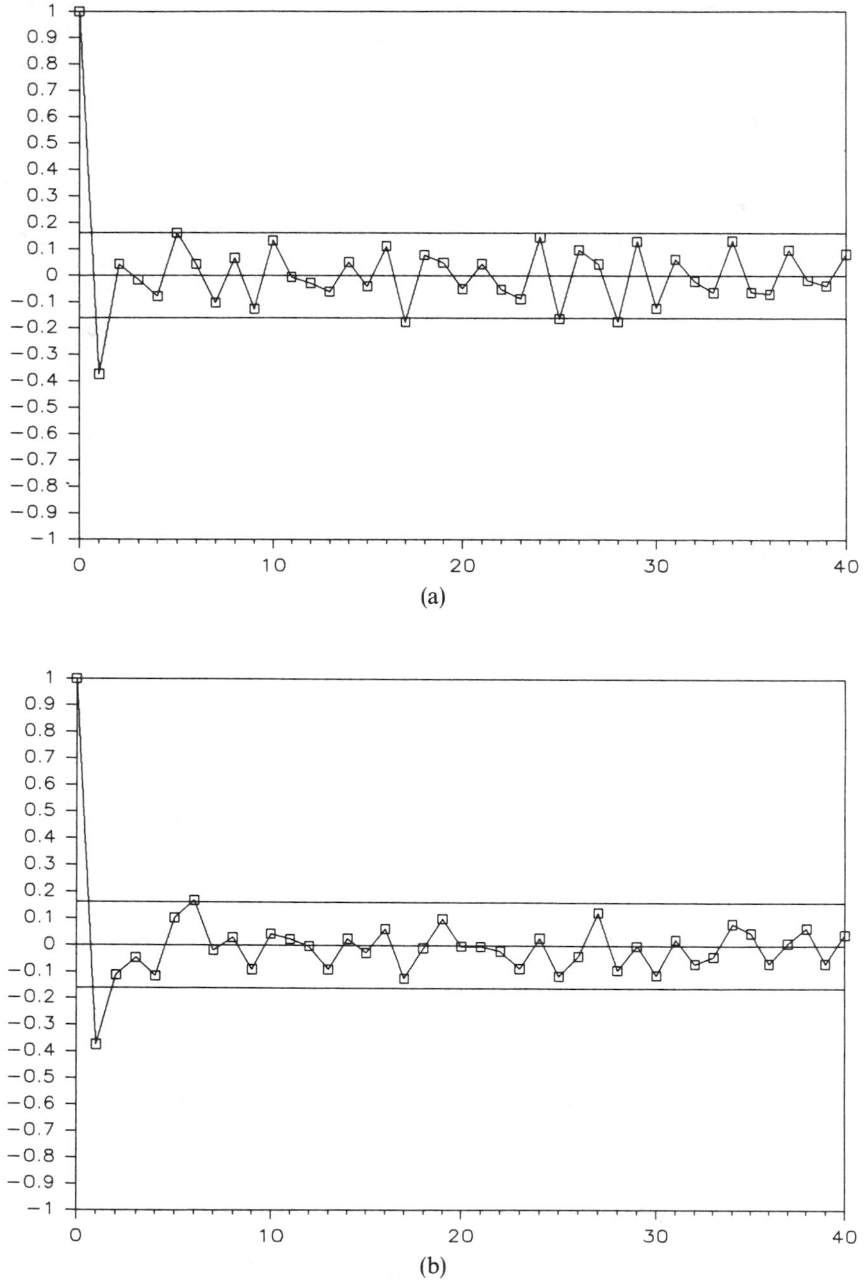

(a)

(b)

Figure 13.2. The sample ACF (a) and PACF (b) of the estimated noise sequence $\hat{N}_t = X_{t2} - 4.717B^3(1 - .724B)^{-1}X_{t1}$ of Example 13.1.1.

A State-Space Representation of the Series $\{(X_{t1}, X_{t2})'\}$†

A major goal of transfer function modelling is to provide more accurate prediction of $X_{n+h,2}$ than can be obtained by modelling $\{X_{t2}\}$ as a univariate series and projecting $X_{n+h,2}$ onto $\overline{\text{sp}}\{X_{t2}, 1 \le t \le n\}$. Instead, we predict $X_{n+h,2}$ using

$$P(X_{n+h,2} | X_{t1}, X_{t2}, 1 \le t \le n). \tag{13.1.9}$$

To facilitate the computation of this predictor, we shall now derive a state-space representation of the input–output series $\{(X_{t1}, X_{t2})'\}$ which is equivalent to the transfer function model. We shall then apply the Kalman recursions of Section 12.2. (The state-space representation can also be used to compute the Gaussian likelihood of $\{(X_{t1}, X_{t2})', t = 1, \ldots, n\}$ and hence to find maximum likelihood estimates of the model parameters. Model selection and missing values can also be handled with the aid of the state-space representation; for details, see Brockwell, Davis and Salehi (1990).)

The transfer function model described above in steps (1)–(8) can be written as

$$X_{t2} = B^b w(B) v^{-1}(B) X_{t1} + N_t \tag{13.1.10}$$

where $\{X_{t1}\}$ and $\{N_t\}$ are the causal and invertible ARMA processes

$$\phi(B) X_{t1} = \theta(B) Z_t, \qquad \{Z_t\} \sim \text{WN}(0, \sigma_Z^2), \tag{13.1.11}$$

$$\phi^{(N)}(B) N_t = \theta^{(N)}(B) W_t, \qquad \{W_t\} \sim \text{WN}(0, \sigma_W^2), \tag{13.1.12}$$

$\{Z_t\}$ is uncorrelated with $\{W_t\}$, and $v(z) \ne 0$ for $|z| \le 1$. By Example 12.1.6, $\{X_{t1}\}$ and $\{N_t\}$ have the state-space representations

$$\begin{aligned} \mathbf{x}_{t+1,1} &= F_1 \mathbf{x}_{t1} + H_1 Z_t, \\ X_{t1} &= G_1 \mathbf{x}_{t1} + Z_t, \end{aligned} \tag{13.1.13}$$

and

$$\begin{aligned} \mathbf{n}_{t+1} &= F^{(N)} \mathbf{n}_t + H^{(N)} W_t, \\ N_t &= G^{(N)} \mathbf{n}_t + W_t, \end{aligned} \tag{13.1.14}$$

where (F_1, G_1, H_1) and $(F^{(N)}, G^{(N)}, H^{(N)})$ are defined in terms of the autoregressive and moving average polynomials for $\{X_t\}$ and $\{N_t\}$, respectively, as in Example 12.1.6. In the same manner, define the triple (F_2, G_2, H_2) in terms of the "autoregressive" and "moving average" polynomials $v(z)$ and $z^b w(z)$ in (13.1.10). From (13.1.10), it is easy to see that $\{X_{t2}\}$ has the representation (see Problem 13.2),

$$X_{t2} = G_2 \mathbf{x}_{t2} + \delta X_{t1} + N_t, \tag{13.1.15}$$

where $\delta = w_0$ if $b = 0$ and 0 otherwise, and $\{\mathbf{x}_{t2}\}$ is the unique stationary solution of

$$\mathbf{x}_{t+1,2} = F_2 \mathbf{x}_{t2} + H_2 X_{t1}. \tag{13.1.16}$$

† Pages 513–517 may be omitted without loss of continuity.

Substituting from (13.1.13) and (13.1.14) into (13.1.15) and (13.1.16), we obtain

$$X_{t2} = G_2\mathbf{x}_{t2} + \delta G_1\mathbf{x}_{t1} + \delta Z_t + G^{(N)}\mathbf{n}_t + W_t, \tag{13.1.17}$$

$$\mathbf{x}_{t+1,2} = F_2\mathbf{x}_{t2} + H_2G_1\mathbf{x}_{t1} + H_2Z_t. \tag{13.1.18}$$

By combining (13.1.13), (13.1.14), (13.1.17) and (13.1.18), the required state-space representation for the process, $\{(X_{t1}, X_{t2})', t = 1, 2, \ldots\}$, can now be written down as

$$\begin{bmatrix} X_{t1} \\ X_{t2} \end{bmatrix} = \begin{bmatrix} G_1 & 0 & 0 \\ \delta G_1 & G_2 & G^{(N)} \end{bmatrix} \boldsymbol{\eta}_t + \begin{bmatrix} 1 & 0 \\ \delta & 1 \end{bmatrix} \begin{bmatrix} Z_t \\ W_t \end{bmatrix}, \tag{13.1.19}$$

where $\{\boldsymbol{\eta}_t = (\mathbf{x}'_{t1}, \mathbf{x}'_{t2}, \mathbf{n}'_t)'\}$ is the unique stationary solution of the state equation,

$$\boldsymbol{\eta}_{t+1} = \begin{bmatrix} F_1 & 0 & 0 \\ H_2G_1 & F_2 & 0 \\ 0 & 0 & F^{(N)} \end{bmatrix} \boldsymbol{\eta}_t + \begin{bmatrix} H_1 & 0 \\ H_2 & 0 \\ 0 & H^{(N)} \end{bmatrix} \begin{bmatrix} Z_t \\ W_t \end{bmatrix}. \tag{13.1.20}$$

EXAMPLE 13.1.1 (*cont.*). The state-space model for the differenced and mean-corrected leading indicator-sales data (with $b = 3$, $w(B) = w_0$, $v(B) = 1 - v_1B$, $\phi(B) = 1$, $\theta(B) = 1 + \theta_1 B$, $\phi^{(N)}(B) = 1$ and $\theta^{(N)}(B) = 1 + \theta_1^{(N)}B$) is

$$\begin{bmatrix} X_{t1} \\ X_{t2} \end{bmatrix} = \begin{bmatrix} 1 & 0 & 0 & 0 & 0 \\ 0 & 1 & 0 & 0 & 1 \end{bmatrix} \boldsymbol{\eta}_t + \begin{bmatrix} Z_t \\ W_t \end{bmatrix}, \tag{13.1.21}$$

where $\{\boldsymbol{\eta}_t = (\mathbf{x}'_{t1}, \mathbf{x}'_{t2}, \mathbf{n}'_t)'\}$ is the unique stationary solution of the state equation,

$$\boldsymbol{\eta}_{t+1} = \begin{bmatrix} 0 & 0 & 0 & 0 & 0 \\ 0 & 0 & 1 & 0 & 0 \\ 0 & 0 & 0 & 1 & 0 \\ w_0 & 0 & 0 & v_1 & 0 \\ 0 & 0 & 0 & 0 & 0 \end{bmatrix} \boldsymbol{\eta}_t + \begin{bmatrix} \theta_1 & 0 \\ 0 & 0 \\ 0 & 0 \\ w_0 & 0 \\ 0 & \theta_1^{(N)} \end{bmatrix} \begin{bmatrix} Z_t \\ W_t \end{bmatrix}. \tag{13.1.22}$$

We can estimate the model parameters in (13.1.21) and (13.1.22) by maximizing the Gaussian likelihood (11.5.4) of $\{(X_{t1}, X_{t2})', t = 1, \ldots, 149\}$, using (12.2.12) and (12.2.14) to determine the one-step predictors and their error covariance matrices. This leads to the fitted model,

$$\begin{aligned} X_{t1} &= (1 - .476B)Z_t, \qquad \{Z_t\} \sim \mathrm{WN}(0, .0768), \\ X_{t2} &= 4.701B^3(1 - .726B)^{-1}X_{t1} + N_t, \\ N_t &= (1 - .621B)W_t, \qquad \{W_t\} \sim \mathrm{WN}(0, .0457), \end{aligned} \tag{13.1.23}$$

which differs only slightly from the least squares model (13.1.8).

It is possible to construct a state-space model for the *original* leading indicator and sales data, $\{\mathbf{Y}_t := (Y_{t1}, Y_{t2})', t = 0, 1, \ldots, 149\}$, at the expense of increasing the dimension of the state-vector by two. The analysis is similar to that given in Example 12.1.7 for ARIMA processes. Thus we rewrite the model (13.1.21)–(13.1.22) as

$$\mathbf{X}_t = G\boldsymbol{\eta}_t + \begin{bmatrix} Z_t \\ W_t \end{bmatrix}, \tag{13.1.24}$$

$$\boldsymbol{\eta}_{t+1} = F\boldsymbol{\eta}_t + \mathbf{V}_t.$$

Observing that

$$\begin{aligned}
\mathbf{Y}_t - t\begin{bmatrix} .0228 \\ .420 \end{bmatrix} &= \begin{bmatrix} Y_{t1} \\ Y_{t2} \end{bmatrix} - t\begin{bmatrix} .0228 \\ .420 \end{bmatrix} \\
&= \begin{bmatrix} \nabla Y_{t1} \\ \nabla Y_{t2} \end{bmatrix} - \begin{bmatrix} .0228 \\ .420 \end{bmatrix} + \begin{bmatrix} Y_{t-1,1} \\ Y_{t-1,2} \end{bmatrix} - (t-1)\begin{bmatrix} .0228 \\ .420 \end{bmatrix} \\
&= \mathbf{X}_t + \mathbf{Y}_{t-1} - (t-1)\begin{bmatrix} .0228 \\ .420 \end{bmatrix} \\
&= G\boldsymbol{\eta}_t + \begin{bmatrix} Z_t \\ W_t \end{bmatrix} + \mathbf{Y}_{t-1} - (t-1)\begin{bmatrix} .0228 \\ .420 \end{bmatrix},
\end{aligned}$$

we introduce the state vectors, $\mathbf{T}_{t+1} = (\boldsymbol{\eta}'_{t+1}, \mathbf{Y}'_t - t\boldsymbol{\mu}')'$, where $\boldsymbol{\mu}' = (.0228, .420)$. It then follows from the preceding equation and (13.1.24) that $\{\mathbf{Y}_t - t\boldsymbol{\mu}\}$ has the state-space representation, for $t = 1, 2, \ldots,$

$$\mathbf{Y}_t - t\boldsymbol{\mu} = [G \quad I_{2\times 2}]\mathbf{T}_t + \begin{bmatrix} Z_t \\ W_t \end{bmatrix}, \tag{13.1.25}$$

with state equation,

$$\mathbf{T}_{t+1} = \begin{bmatrix} F & 0 \\ G & I_{2\times 2} \end{bmatrix}\mathbf{T}_t + \begin{bmatrix} \mathbf{V}_t \\ Z_t \\ W_t \end{bmatrix}, \tag{13.1.26}$$

initial condition,

$$\mathbf{T}_1 = \begin{bmatrix} \boldsymbol{\eta}_1 \\ \mathbf{Y}_0 \end{bmatrix} = \begin{bmatrix} \sum_{j=0}^{\infty} F^j \mathbf{V}_{-j} \\ \mathbf{Y}_0 \end{bmatrix},$$

and orthogonality conditions,

$$\mathbf{Y}_0 \perp \begin{bmatrix} Z_t \\ W_t \end{bmatrix}, \qquad t = 0, \pm 1, \ldots.$$

To forecast future sales, we apply the Kalman recursions, (12.2.10)–(12.2.14) to this state-space model to evaluate $P_n(\mathbf{Y}_{n+h} - (n+h)\boldsymbol{\mu})$, where $P_n(\cdot)$ denotes projection onto

$$\overline{\text{sp}}\{\mathbf{Y}_0, \mathbf{Y}_1 - \boldsymbol{\mu}, \ldots, \mathbf{Y}_n - n\boldsymbol{\mu}\} = \overline{\text{sp}}\{\mathbf{Y}_0, \mathbf{X}_1, \ldots, \mathbf{X}_n\}.$$

Then the required predictor $P(\mathbf{Y}_{n+h}|\mathbf{1}, \mathbf{Y}_0, \ldots, \mathbf{Y}_n)$ is given by

$$P(\mathbf{Y}_{n+h}|\mathbf{1}, \mathbf{Y}_0, \ldots, \mathbf{Y}_n) = (n+h)\boldsymbol{\mu} + P_n(\mathbf{Y}_{n+h} - (n+h)\boldsymbol{\mu}).$$

As in the case of an ARIMA process (see Remark 8, Section 12.2), this predictor can be computed more directly as follows. Since $\mathbf{Y}_0$ is orthogonal to $\mathbf{X}_1, \mathbf{X}_2, \ldots$ and $\boldsymbol{\eta}_1, \boldsymbol{\eta}_2, \ldots$, we can write

$$P_0\boldsymbol{\eta}_1 = E\boldsymbol{\eta}_1 = \mathbf{0}$$

and

$$P_{t-1}\boldsymbol{\eta}_t = P(\boldsymbol{\eta}_t | \mathbf{X}_1, \ldots, \mathbf{X}_{t-1}) = \hat{\boldsymbol{\eta}}_t \quad \text{for } t \geq 2.$$

Similarly, $P_t\boldsymbol{\eta}_{t+h} = P(\boldsymbol{\eta}_{t+h}|\mathbf{X}_1, \ldots, \mathbf{X}_t)$ and $P_t\mathbf{X}_{t+h} = GP_t\boldsymbol{\eta}_{t+h}$ for all $t \geq 1$ and $h \geq 1$. Both $\hat{\boldsymbol{\eta}}_t$ and its error covariance matrix,

$$\Omega_{\eta,t} = E(\boldsymbol{\eta}_t - \hat{\boldsymbol{\eta}}_t)(\boldsymbol{\eta}_t - \hat{\boldsymbol{\eta}}_t)',$$

can be computed recursively by applying Proposition 12.2.2 to the model (13.1.24) with initial conditions

$$\hat{\boldsymbol{\eta}}_1 = \mathbf{0},$$

$$\boldsymbol{\Psi}_{\eta,1} = 0,$$

and

$$\Pi_{\eta,1} = E(\boldsymbol{\eta}_1\boldsymbol{\eta}_1') = \sum_{j=0}^{\infty} F^j Q_1 F'^j$$

$$= \begin{bmatrix} 0.0174 & 0.0000 & 0.0000 & -0.1719 & 0.0000 \\ 0.0000 & 1.9215 & 0.5872 & 0.4263 & 0.0000 \\ 0.0000 & 0.5872 & 1.9215 & 0.5872 & 0.0000 \\ -0.1719 & 0.4263 & 0.5872 & 1.9215 & 0.0000 \\ 0.0000 & 0.0000 & 0.0000 & 0.0000 & 0.0176 \end{bmatrix},$$

where Q_1 is the covariance matrix of V_1 (see Problem 13.4). Consequently, the one-step predictors for the state-vectors, $\mathbf{T}_t = (\boldsymbol{\eta}_t', \mathbf{Y}_{t-1}' - (t-1)'\boldsymbol{\mu}')'$, in (13.1.26) are

$$P_{t-1}\mathbf{T}_t = \begin{bmatrix} \hat{\boldsymbol{\eta}}_t \\ \mathbf{Y}_{t-1} - (t-1)\boldsymbol{\mu} \end{bmatrix},$$

with error covariance matrices given by

$$\Omega_t = \begin{bmatrix} \Omega_{\eta,t} & 0 \\ 0 & 0 \end{bmatrix},$$

for $t \geq 1$. It follows from (12.2.12), (13.1.24) and (13.1.25) that

$$P_{149}(\mathbf{Y}_{150} - 150\boldsymbol{\mu}) = [G \quad I_{2\times 2}]\begin{bmatrix} \hat{\boldsymbol{\eta}}_{150} \\ \mathbf{Y}_{149} - 149\boldsymbol{\mu} \end{bmatrix}$$

whence

$$P(\mathbf{Y}_{150}|\mathbf{1}, \mathbf{Y}_0, \ldots, \mathbf{Y}_{149}) = \boldsymbol{\mu} + \hat{\mathbf{X}}_{150} + \mathbf{Y}_{149}$$

$$= \begin{bmatrix} .0228 \\ .420 \end{bmatrix} + \begin{bmatrix} .138 \\ -.232 \end{bmatrix} + \begin{bmatrix} 13.4 \\ 262.7 \end{bmatrix} = \begin{bmatrix} 13.56 \\ 262.89 \end{bmatrix}. \tag{13.1.27}$$

Similarly,

$$P(\mathbf{Y}_{151}|\mathbf{1}, \mathbf{Y}_0, \ldots, \mathbf{Y}_{149}) = \begin{bmatrix} .0228 \\ .420 \end{bmatrix} + P_{149}\mathbf{X}_{151} + P(\mathbf{Y}_{150}|\mathbf{1}, \mathbf{Y}_0, \ldots, \mathbf{Y}_{149})$$

$$= \begin{bmatrix} .0228 \\ .420 \end{bmatrix} + \begin{bmatrix} 0 \\ .921 \end{bmatrix} + \begin{bmatrix} 13.56 \\ 262.89 \end{bmatrix} = \begin{bmatrix} 13.58 \\ 264.23 \end{bmatrix}. \tag{13.1.28}$$

The corresponding one- and two-step error covariance matrices, computed from (12.2.13) and (12.2.14), are found to be

$$\Sigma_{149}^{(1)} = [G \ I_{2\times 2}] \begin{bmatrix} \Omega_{\eta,149}^{(1)} & 0 \\ 0 & 0 \end{bmatrix} [G \quad I_{2\times 2}]' + \begin{bmatrix} .0768 & 0 \\ 0 & .0457 \end{bmatrix}$$

$$= E(\mathbf{X}_{150} - \hat{\mathbf{X}}_{150})(\mathbf{X}_{150} - \hat{\mathbf{X}}_{150})'$$

$$= \begin{bmatrix} .0768 & 0 \\ 0 & .0457 \end{bmatrix}, \tag{13.1.29}$$

and

$$\Sigma_{149}^{(2)} = [G \quad I_{2\times 2}]\Omega_{149}^{(2)}[G \quad I_{2\times 2}]' + \begin{bmatrix} .0768 & 0 \\ 0 & .0457 \end{bmatrix}$$

$$= \begin{bmatrix} .0979 & 0 \\ 0 & .0523 \end{bmatrix}, \tag{13.1.30}$$

where

$$\Omega_{149}^{(2)} = \begin{bmatrix} F & 0 \\ G & I_{2\times 2} \end{bmatrix} \begin{bmatrix} \Omega_{\eta,149}^{(1)} & 0 \\ 0 & 0 \end{bmatrix} \begin{bmatrix} F' & G' \\ 0 & I_{2\times 2} \end{bmatrix} + \tilde{Q}_1.$$

and $\tilde{Q}_1$ is the covariance matrix of $(\mathbf{V}_1', Z_1, W_1)'$.

Prediction Based on the Infinite Past

For the transfer function model described by (13.1.10)–(13.1.12), prediction of $X_{n+h,2}$ based on the infinite past $\{(X_{t1}, X_{t2})', -\infty < t \le n\}$, is substantially simpler than that based on $\{(X_{t1}, X_{t2})', 1 \le t \le n\}$. The infinite-past predictor, moreover, gives a good approximation to the finite-past predictor provided n is sufficiently large.

The transfer function model (13.1.10)–(13.1.12) can be rewritten as

$$X_{t2} = T(B)X_{t1} + \beta(B)W_t, \tag{13.1.31}$$

$$X_{t1} = \theta(B)\phi^{-1}(B)Z_t, \tag{13.1.32}$$

where $\beta(B) = \theta^{(N)}(B)/\phi^{(N)}(B)$. Eliminating X_{t1} gives

$$X_{t2} = \sum_{j=0}^{\infty} \alpha_j Z_{t-j} + \sum_{j=0}^{\infty} \beta_j W_{t-j}, \tag{13.1.33}$$

where $\alpha(B) = T(B)\theta(B)/\phi(B)$. Our objective is to compute

$$\tilde{P}_n X_{n+h,2} := P_{\overline{sp}\{X_{t1}, X_{t2}, -\infty < t \le n\}} X_{n+h,2}.$$

Since $\{X_{t1}\}$ and $\{N_t\}$ are assumed to be causal invertible ARMA processes, it follows that $\overline{sp}\{(X_{t1}, X_{t2})', -\infty < t \le n\} = \overline{sp}\{(Z_t, W_t)', -\infty < t \le n\}$. Using the fact that $\{Z_t\}$ and $\{W_t\}$ are uncorrelated, we see at once from (13.1.33) that

$$\tilde{P}_n X_{n+h,2} = \sum_{j=h}^{\infty} \alpha_j Z_{n+h-j} + \sum_{j=h}^{\infty} \beta_j W_{n+h-j}. \tag{13.1.34}$$

Setting $t = n + h$ in (13.1.33) and subtracting (13.1.34) gives the mean squared error,

$$E(X_{n+h,2} - \tilde{P}_n X_{n+h,2})^2 = \sigma_Z^2 \sum_{j=0}^{h-1} \alpha_j^2 + \sigma_W^2 \sum_{j=0}^{h-1} \beta_j^2. \tag{13.1.35}$$

To compute the predictors $\tilde{P}_n X_{n+h,2}$ we proceed as follows. Rewrite (13.1.31) as

$$A(B)X_{t2} = B^b U(B) X_{t1} + V(B) W_t, \tag{13.1.36}$$

where A, U and V are polynomials of the form,

$$A(B) = 1 - A_1 B - \cdots - A_a B^a,$$

$$U(B) = U_o + U_1 B + \cdots + U_u B^u,$$

and

$$V(B) = 1 + V_1 B + \cdots + V_u B^v.$$

Applying the operator $\tilde{P}_n$ to equation (13.1.36) with $t = n + h$, we obtain

$$\tilde{P}_n X_{n+h,2} = \sum_{j=1}^{a} A_j \tilde{P}_n X_{n+h-j,2} + \sum_{j=0}^{u} U_j \tilde{P}_n X_{n+h-b-j,1} + \sum_{j=h}^{v} V_j W_{n+h-j}, \tag{13.1.37}$$

where the last sum is zero if $h > v$.

Since $\{X_{t1}\}$ is uncorrelated with $\{W_t\}$, $\tilde{P}_n X_{j1} = P_{\overline{sp}\{X_{t1}, -\infty < t \le n\}} X_{j1}$, and the second term in (13.1.37) is therefore obtained by predicting the univariate series $\{X_{t1}\}$ as described in Chapter 5 using the model (13.1.32). In keeping with our assumption that n is large, we can replace $\tilde{P}_n X_j$ by the finite past predictor obtained from the program PEST.

The values W_j, $j \le n$, are replaced by their estimated values $\hat{W}_j$ from the least squares estimation in step (7) of the modelling procedure.

Equations (13.1.37) can now be solved recursively for the predictors $\tilde{P}_n X_{n+1,2}, \tilde{P}_n X_{n+2,2}, \tilde{P}_n X_{n+3,2}, \ldots$.

EXAMPLE 13.1.2 (Sales with a Leading Indicator). Applying the preceding results to the series $\{(X_{t1}, X_{t2})', 1 \le t \le 149\}$ of Example 13.1.1, we find from (13.1.23) and (13.1.37) that

$$\tilde{P}_{149} X_{150,2} = .726 X_{149,2} + 4.701 X_{147,1} - 1.347 W_{149} + .451 W_{148}.$$

Replacing $X_{149,2}$ and $X_{147,1}$ by the observed values .08 and $-.093$ and W_{149} and W_{148} by the sample estimates $-.0706$ and .1449, we obtain the predicted value

$$\tilde{P}_{149}X_{150,2} = -.219.$$

Similarly, setting $X_{148,1} = .237$, we find that

$$\begin{aligned}\tilde{P}_{149}X_{151,2} &= .726\tilde{P}_{149}X_{150,2} + 4.701X_{148,1} + .451W_{149}\\ &= .923.\end{aligned}$$

In terms of the original sales data $\{Y_{t2}\}$ we have $Y_{149,2} = 262.7$ and

$$Y_{t2} = Y_{t-1,2} + X_{t2} + .420.$$

Hence the predictors of actual sales are

$$\begin{cases} P^*_{149}Y_{150,2} = 262.7 - .219 + .420 = 262.90, \\ P^*_{149}Y_{151,2} = 262.90 + .923 + .420 = 264.24, \end{cases} \tag{13.1.38}$$

where P^*_{149} means projection onto $\overline{\text{sp}}\{1, Y_{149,2}, (X_{s,1}, X_{s,2})', \ -\infty < s \le 149\}$. These values are in close agreement with (13.1.27) and (13.1.28) obtained earlier by projecting onto $\overline{\text{sp}}\{1, (Y_{t1}, Y_{t2})', \ 0 \le t \le 149\}$. Since our model for the sales data is

$$(1 - B)Y_{t2} = .420 + 4.701B^3(1 - .476B)(1 - .726B)^{-1}Z_t + (1 - .621B)W_t,$$

an argument precisely analogous to the one giving (13.1.35) yields the mean squared errors,

$$E(Y_{149+h,2} - P^*_{149}Y_{149+h,2})^2 = \sigma_Z^2 \sum_{j=0}^{h-1} \alpha_j^{*2} + \sigma_W^2 \sum_{j=0}^{h-1} \beta_j^{*2},$$

where

$$\sum_{j=0}^{\infty} \alpha_j^* z^j = 4.701z^3(1 - .476z)(1 - .726z)^{-1}(1 - z)^{-1}$$

and

$$\sum_{j=0}^{\infty} \beta_j^* z^j = (1 - .621z)(1 - z)^{-1}.$$

For $h = 1$ and 2 we obtain

$$\begin{cases} E(Y_{150,2} - P^*_{149}Y_{150,2})^2 = .0457, \\ E(Y_{151,2} - P^*_{149}Y_{151,2})^2 = .0523, \end{cases} \tag{13.1.39}$$

in agreement with the finite-past mean squared errors in (13.1.29) and (13.1.30).

It is interesting to examine the improvement obtained by using the transfer function model rather than fitting a univariate model to the sales data alone.

If we adopt the latter course we find the model,

$$X_{t2} - .249X_{t-1,2} - .199X_{t-2,2} = U_t,$$

where $\{U_t\} \sim \text{WN}(0, 1.794)$ and $X_{t2} = Y_{t2} - Y_{t-1,2} - .420$. The corresponding predictors of $Y_{150,2}$ and $Y_{151,2}$ are easily found from the program PEST to be 263.14 and 263.58 with mean squared errors 1.794 and 4.593 respectively. These mean squared errors are dramatically worse than those obtained using the transfer function model.

§13.2 Long Memory Processes

An ARMA process $\{X_t\}$ is often referred to as a short memory process since the covariance (or dependence) between X_1 and X_{1+k} decreases rapidly as $k \to \infty$. In fact we know from Chapter 3 that the autocorrelation function is geometrically bounded, i.e.

$$|\rho(k)| \le Cr^k, \qquad k = 1, 2, \ldots,$$

where $C > 0$ and $0 < r < 1$. A long memory process is a stationary process for which

$$\rho(k) \sim Ck^{2d-1} \quad \text{as } k \to \infty, \tag{13.2.1}$$

where $C \neq 0$ and $d < .5$. [Some authors make a distinction between "intermediate memory" processes for which $d < 0$ and hence $\sum_{k=-\infty}^{\infty} |\rho(k)| < \infty$, and "long memory" processes for which $0 < d < .5$ and $\sum_{k=-\infty}^{\infty} |\rho(k)| = \infty$.]

There is evidence that long memory processes occur quite frequently in fields as diverse as hydrology and economics (see Hurst (1951), Lawrance and Kottegoda (1977), Hipel and McLeod (1978), and Granger (1980)). In this section we extend the class of ARMA processes as in Hosking (1981) and Granger and Joyeux (1980) to include processes whose autocorrelation functions have the asymptotic behaviour (13.2.1). While a long memory process can always be approximated by an ARMA(p, q) process (see Sections 4.4 and 8.1), the orders p and q required to achieve a reasonably good approximation may be so large as to make parameter estimation extremely difficult.

For any real number $d > -1$, we define the difference operator $\nabla^d = (1 - B)^d$ by means of the binomial expansion,

$$\nabla^d = (1 - B)^d = \sum_{j=0}^{\infty} \pi_j B^j,$$

where

$$\pi_j = \frac{\Gamma(j - d)}{\Gamma(j + 1)\Gamma(-d)} = \prod_{0<k\le j} \frac{k - 1 - d}{k}, \qquad j = 0, 1, 2, \ldots, \tag{13.2.2}$$

and $\Gamma(\cdot)$ is the gamma function,

$$\Gamma(x) := \begin{cases} \displaystyle\int_0^\infty t^{x-1} e^{-t}\, dt, & x > 0, \\ \infty, & x = 0, \\ x^{-1}\Gamma(1 + x), & x < 0. \end{cases}$$

Definition 13.2.1 (The ARIMA (0,d,0) Process). The process $\{X_t, t=0, \pm 1, \ldots\}$ is said to be an ARIMA (0, d, 0) process with $d \in (-.5, .5)$ if $\{X_t\}$ is a stationary solution with zero mean of the difference equations,

$$\nabla^d X_t = Z_t, \qquad \text{where } \{Z_t\} \sim \text{WN}(0, \sigma^2). \tag{13.2.3}$$

The process $\{X_t\}$ is often called *fractionally integrated noise.*

Remark 1. Throughout this section convergence of sequences of random variables means convergence in mean square.

Remark 2. Implicit in Definition 13.2.1 is the requirement that the series $\nabla^d X_t = \sum_{j=0}^{\infty} \pi_j X_{t-j}$ with $\{\pi_j\}$ as in (13.2.2), should be mean square convergent. This implies, by Theorem 4.10.1, that if X_t has the spectral representation $X_t = \int_{(-\pi,\pi]} e^{it\lambda}\, dZ_X(\lambda)$ then

$$\nabla^d X_t = \int_{(-\pi, \pi]} e^{it\lambda}(1 - e^{-i\lambda})^d\, dZ_X(\lambda). \tag{13.2.4}$$

Remark 3. In view of the representation (13.2.3) of $\{Z_t\}$ we say that $\{X_t\}$ is invertible, even though the coefficients $\{\pi_j\}$ may not be absolutely summable as in the corresponding representation of $\{Z_t\}$ for an invertible ARMA process. We shall say that $\{X_t\}$ is causal if X_t can be expressed as

$$X_t = \sum_{j=0}^{\infty} \psi_j Z_{t-j}$$

where $\sum_{j=0}^{\infty} \psi_j^2 < \infty$. The existence of a stationary causal solution of (13.2.3) and the covariance properties of the solution are established in the following theorem.

Theorem 13.2.1. *If* $d \in (-.5, .5)$ *then there is a unique purely nondeterministic stationary solution* $\{X_t\}$ *of* (13.2.3) *given by*

$$X_t = \sum_{j=0}^{\infty} \psi_j Z_{t-j} = \nabla^{-d} Z_t, \tag{13.2.5}$$

where

$$\psi_j = \frac{\Gamma(j+d)}{\Gamma(j+1)\Gamma(d)} = \prod_{0<k\le j} \frac{k-1+d}{k}, \qquad j = 0, 1, 2, \ldots. \tag{13.2.6}$$

Denoting by $f(\cdot)$, $\gamma(\cdot)$, $\rho(\cdot)$ and $\alpha(\cdot)$ the spectral density, autocovariance function, autocorrelation function and partial autocorrelation function respectively of $\{X_t\}$, we have

$$f(\lambda) = |1 - e^{-i\lambda}|^{-2d}\sigma^2/(2\pi) = |2\sin(\lambda/2)|^{-2d}\sigma^2/(2\pi), \qquad -\pi \le \lambda \le \pi, \tag{13.2.7}$$

$$\gamma(0) = \sigma^2\Gamma(1-2d)/\Gamma^2(1-d), \tag{13.2.8}$$

$$\rho(h) = \frac{\Gamma(h+d)\Gamma(1-d)}{\Gamma(h-d+1)\Gamma(d)} = \prod_{0<k\le h} \frac{k-1+d}{k-d}, \qquad h = 1, 2, \ldots, \tag{13.2.9}$$

and

$$\alpha(h) = d/(h-d), \qquad h = 1, 2, \ldots. \tag{13.2.10}$$

Remark 4. Applying Stirling's formula, $\Gamma(x) \sim \sqrt{2\pi}e^{-x+1}(x-1)^{x-1/2}$ as $x \to \infty$, to (13.2.2), (13.2.6) and (13.2.9), we obtain

$$\pi_j \sim j^{-d-1}/\Gamma(-d) \qquad \text{as } j \to \infty, \tag{13.2.11}$$

$$\psi_j \sim j^{d-1}/\Gamma(d) \qquad \text{as } j \to \infty, \tag{13.2.12}$$

and

$$\rho(h) \sim h^{2d-1}\Gamma(1-d)/\Gamma(d) \qquad \text{as } h \to \infty. \tag{13.2.13}$$

Fractionally integrated noise with $d \neq 0$ is thus a long memory process in the sense of Definition 13.2.1.

Remark 5. Since $\sin\lambda \sim \lambda$ as $\lambda \to 0$, we see from (13.2.7) that

$$f(\lambda) \sim \lambda^{-2d}\sigma^2/(2\pi) \quad \text{as } \lambda \to 0, \tag{13.2.14}$$

showing that $f(0)$ is finite if and only if $d \le 0$. The asymptotic behaviour (13.2.14) of $f(\lambda)$ as $\lambda \to 0$ suggests an alternative frequency-domain definition of long memory process which could be used instead of (13.2.1).

Proof of Theorem 13.2.1. We shall give the proof only for $0 < d < .5$ since the proof for $-.5 < d < 0$ is quite similar and the case $d = 0$ is trivial.

From (13.2.12) it follows that $\sum_{j=0}^{\infty} \psi_j^2 < \infty$ so that

$$\sum_{j=0}^{n} \psi_j e^{-ij\cdot} \to (1 - e^{-i\cdot})^{-d} \quad \text{as } n \to \infty, \tag{13.2.15}$$

where convergence is in $L^2(d\lambda)$ and $d\lambda$ denotes Lebesgue measure. By Theorem 4.10.1,

$$(1 - B)^{-d} Z_t := \sum_{j=0}^{\infty} \psi_j Z_{t-j}$$

is a well-defined stationary process and if $\{Z_t\}$ has the spectral representation $Z_t = \int_{(-\pi,\pi]} e^{i\lambda t} \, dW(\lambda)$, then

$$(1 - B)^{-d} Z_t = \int_{(-\pi,\pi]} e^{i\lambda t}(1 - e^{-i\lambda})^{-d} \, dW(\lambda).$$

Since $\sum_{j=0}^{\infty} |\pi_j| < \infty$ (by (13.2.11)), we can apply the operator $(1 - B)^d = \sum_{j=0}^{\infty} \pi_j B^j$ to $(1 - B)^{-d} Z_t$ (see Remark 1 in Section 4.10), giving

$$(1 - B)^d (1 - B)^{-d} Z_t = \int_{(-\pi,\pi]} e^{i\lambda t} \, dW(\lambda) = Z_t.$$

Hence $\{X_t\}$ as defined by (13.2.5) satisfies (13.2.3).

To establish uniqueness, let $\{Y_t\}$ be any purely nondeterministic stationary solution of (13.2.3). If $\{Y_t\}$ has the spectral representation,

$$Y_t = \int_{(-\pi,\pi]} e^{it\lambda} \, dZ_Y(\lambda),$$

then by (13.2.4), the process $\{(1 - B)^d Y_t\}$ has spectral representation,

$$(1 - B)^d Y_t = \int_{(-\pi,\pi]} e^{it\lambda}(1 - e^{-i\lambda})^d \, dZ_Y(\lambda)$$

and spectral density $\sigma^2/(2\pi)$. By (13.2.15), Theorem 4.10.1 and the continuity of F_Y at 0,

$$\begin{aligned}
(1 - B)^{-d} Z_t &= (1 - B)^{-d}(1 - B)^d Y_t \\
&= \lim_{n \to \infty} \left(\sum_{j=0}^{n} \psi_j B^j \right) (1 - B)^d Y_t \\
&= \int_{(-\pi,\pi]} (1 - e^{-i\lambda})^{-d}(1 - e^{-i\lambda})^d e^{it\lambda} \, dZ_Y(\lambda) \\
&= Y_t.
\end{aligned}$$

Hence $Y_t = (1 - B)^{-d} Z_t = X_t$.

By (4.10.8) the spectral density of $\{X_t\}$ is

$$f(\lambda) = |1 - e^{-i\lambda}|^{-2d} \sigma^2/(2\pi) = |2 \sin(\lambda/2)|^{-2d} \sigma^2/(2\pi).$$

The autocovariances are

$$\begin{aligned}
\gamma(h) &= \int_{-\pi}^{\pi} e^{ih\lambda} f(\lambda) \, d\lambda \\
&= \frac{\sigma^2}{\pi} \int_0^{\pi} \cos(h\lambda)(2 \sin(\lambda/2))^{-2d} \, d\lambda \\
&= \frac{(-1)^h \Gamma(1 - 2d)}{\Gamma(h - d + 1)\Gamma(1 - h - d)} \sigma^2, \qquad h = 0, 1, 2, \ldots .
\end{aligned}$$

where the last expression is derived with the aid of the identity,

$$\int_0^\pi \cos(hx)\sin^{\nu-1}(x)\,dx = \frac{\pi\cos(h\pi/2)\Gamma(\nu+1)2^{1-\nu}}{\nu\Gamma((\nu+h+1)/2)\Gamma((\nu-h+1)/2)}$$

(see Gradshteyn and Ryzhik (1965), p. 372). The autocorrelations (13.2.9) can be written down at once from the expression for $\gamma(h)$. To determine the partial autocorrelation function we write the best linear predictor of X_{n+1} in terms of $X_n, \ldots, X_1$ as

$$\hat{X}_{n+1} = \phi_{n1}X_n + \cdots + \phi_{nn}X_1$$

and compute the coefficients ϕ_{nj} from the Durbin–Levinson algorithm (Proposition 5.2.1). An induction argument gives, for $n = 1, 2, 3, \ldots,$

$$\phi_{nj} = -\binom{n}{j}\frac{\Gamma(j-d)\Gamma(n-d-j+1)}{\Gamma(-d)\Gamma(n-d+1)}, \qquad j = 1, \ldots, n,$$

whence

$$\alpha(h) = \phi_{hh} = -\frac{\Gamma(h-d)\Gamma(1-d)}{\Gamma(-d)\Gamma(h-d+1)} = \frac{d}{h-d}. \qquad \square$$

Fractionally integrated noise processes themselves are of limited value in modelling long memory data since the two parameters d and σ^2 allow only a very limited class of possible autocovariance functions. However they can be used as building blocks to generate a much more general class of long memory processes whose covariances at small lags are capable of assuming a great variety of different forms. These processes were introduced independently by Granger and Joyeux (1980) and Hosking (1981).

Definition 13.2.2 (The ARIMA(p, d, q) Process with $d \in (-.5, .5)$). $\{X_t, t = 0, \pm 1, \ldots\}$ is said to be an ARIMA(p, d, q) process with $d \in (-.5, .5)$ or a fractionally integrated ARMA(p, q) process if $\{X_t\}$ is stationary and satisfies the difference equations,

$$\phi(B)\nabla^d X_t = \theta(B)Z_t \qquad (13.2.16)$$

where $\{Z_t\}$ is white noise and ϕ, θ are polynomials of degrees p, q respectively.

Clearly $\{X_t\}$ is an ARIMA(p, d, q) process with $d \in (-.5, .5)$ if and only if $\nabla^d X_t$ is an ARMA(p, q) process. If $\theta(z) \neq 0$ for $|z| \leq 1$ then the sequence $Y_t = \phi(B)\theta^{-1}(B)X_t$ satisfies

$$\nabla^d Y_t = Z_t$$

and

$$\phi(B)X_t = \theta(B)Y_t,$$

so that $\{X_t\}$ can be regarded as an ARMA(p, q) process driven by fractionally integrated noise.

Theorem 13.2.2 (Existence and Uniqueness of a Stationary Solution of (13.2.16)). *Suppose that* $d \in (-5, .5)$ *and that* $\phi(\cdot)$ *and* $\theta(\cdot)$ *have no common zeroes.*

(a) *If* $\phi(z) \neq 0$ *for* $|z| = 1$ *then there is a unique purely nondeterministic stationary solution of* (13.2.16) *given by*

$$X_t = \sum_{j=-\infty}^{\infty} \psi_j \nabla^{-d} Z_{t-j}$$

where $\psi(z) = \sum_{j=-\infty}^{\infty} \psi_j z^j = \theta(z)/\phi(z)$.

(b) *The solution* $\{X_t\}$ *is causal if and only if* $\phi(z) \neq 0$ *for* $|z| \leq 1$.

(c) *The solution* $\{X_t\}$ *is invertible if and only if* $\theta(z) \neq 0$ for $|z| \leq 1$.

(d) *If the solution* $\{X_t\}$ *is causal and invertible then its autocorrelation function* $\rho(\cdot)$ *and spectral density* $f(\cdot)$ *satisfy, for* $d \neq 0$,

$$\rho(h) \sim Ch^{2d-1} \quad \text{as } h \to \infty, \tag{13.2.17}$$

where $C \neq 0$, *and*

$$f(\lambda) = \frac{\sigma^2}{2\pi} \frac{|\theta(e^{-i\lambda})|^2}{|\phi(e^{-i\lambda})|^2} |1 - e^{-i\lambda}|^{-2d} \sim \frac{\sigma^2}{2\pi} [\theta(1)/\phi(1)]^2 \lambda^{-2d} \tag{13.2.18}$$

as $\lambda \to 0$.

PROOF. We omit the proofs of (a), (b) and (c) since they are similar to the arguments given for Theorems 3.1.1–3.1.3 with Theorem 4.10.1 replacing Proposition 3.1.2.

If $\{X_t\}$ is causal then $\phi(z) \neq 0$ for $|z| \leq 1$ and we can write

$$X_t = \psi(B) Y_t = \sum_{j=0}^{\infty} \psi_j Y_{t-j},$$

where

$$Y_t = \nabla^{-d} Z_t$$

is fractionally integrated noise. If $\gamma_Y(\cdot)$ is the autocovariance function of $\{Y_t\}$, then by Proposition 3.1.1 (with $\psi_j := 0, j < 0$),

$$\begin{aligned} \operatorname{Cov}(X_{t+h}, X_t) &= \sum_j \sum_k \psi_j \psi_k \gamma_Y(h - j + k) \\ &= \sum_k \sum_j \psi_j \psi_{j+k} \gamma_Y(h - k), \end{aligned}$$

i.e.

$$\gamma_X(h) = \sum_k \tilde{\gamma}(k) \gamma_Y(h - k), \tag{13.2.19}$$

where $\tilde{\gamma}(k) = \sum_j \psi_j \psi_{j+k}$ is the autocovariance function of an ARMA(p, q) process with $\sigma^2 = 1$. If follows that $|\tilde{\gamma}(k)| < Cr^k$, $k = 0, 1, 2, \ldots$, for some $C > 0$

and $r \in (0, 1)$ and hence that

$$h^{1-2d} \sum_{|k|>\sqrt{h}} |\tilde{\gamma}(k)| \to 0 \quad \text{as } h \to \infty. \tag{13.2.20}$$

From (13.2.19) we have

$$h^{1-2d}\gamma_X(h) = h^{1-2d} \sum_{|k|>\sqrt{h}} \tilde{\gamma}(k)\gamma_Y(h-k) + \sum_{|k|\le\sqrt{h}} \tilde{\gamma}(k)h^{1-2d}\gamma_Y(h-k). \tag{13.2.21}$$

The first term on the right converges to zero as $h \to \infty$ by (13.2.20). By (13.2.13) there is a constant $C \neq 0$ (since we are assuming that $d \neq 0$) such that

$$\gamma_Y(h-k) \sim C(h-k)^{2d-1} \sim Ch^{2d-1}$$

uniformly on the set $|k| \le \sqrt{h}$. Hence

$$\sum_{|k|\le\sqrt{h}} \tilde{\gamma}(k)h^{1-2d}\gamma_Y(h-k) \to C\sum_k \tilde{\gamma}(k) \text{ as } h \to \infty.$$

Now letting $h \to \infty$ in (13.2.21) gives the result (13.2.17).

Finally from (4.4.3) and (13.2.7) the spectral density of $\{X_t\}$ is

$$\begin{aligned} f(\lambda) &= |\theta(e^{-i\lambda})|^2 |\phi(e^{-i\lambda})|^{-2} f_Y(\lambda) \\ &= \frac{\sigma^2}{2\pi}|\theta(e^{-i\lambda})|^2 |\phi(e^{-i\lambda})|^{-2} |1 - e^{-i\lambda}|^{-2d} \\ &\sim \frac{\sigma^2}{2\pi}[\theta(1)/\phi(1)]^2 \lambda^{-2d} \quad \text{as } \lambda \to 0. \end{aligned}$$

Remark 6. A formula involving only the gamma and hypergeometric functions is given in Sowell (1990) for computing the autocovariance function of an ARIMA(p, d, q) process when the autoregressive polynomial $\phi(z)$ has distinct zeroes.

Remark 7 (The ARIMA(p, d, q) Process with $d \le -.5$). This is a stationary process $\{X_t\}$ satisfying

$$\phi(B)X_t = \theta(B)\nabla^{-d}Z_t, \qquad \{Z_t\} \sim \text{WN}(0, \sigma^2). \tag{13.2.22}$$

It is not difficult to show that (13.2.22) has a unique stationary solution.

$$X_t = \phi^{-1}(B)\theta(B)\nabla^{-d}Z_t.$$

The solution however is not invertible. Notice that if $\{X_t\}$ is an ARIMA(p, d, q) process with $d < .5$ then $\{(1 - B)X_t\}$ is an ARIMA($p, d - 1, q$) process. In particular if $0 < d < .5$, the effect of applying the operator $(1 - B)$ is to transform the long memory process into an intermediate memory process (with zero spectral density at frequency zero).

Parameter Estimation for ARIMA(p, d, q) Processes with $d \in (-.5, .5)$

Estimation of The Mean. Let $\{X_t\}$ be the causal invertible ARIMA(p, d, q) process defined by

$$\phi(B)\nabla^d(X_t - \mu) = \theta(B)Z_t, \qquad \{Z_t\} \sim \text{WN}(0, \sigma^2), \qquad d \in (-.5, .5). \quad (13.2.23)$$

A natural estimator of the mean $EX_t = \mu$ is the sample mean,

$$\bar{X}_n = n^{-1}(X_1 + \cdots + X_n).$$

Since the autocorrelation function $\rho(\cdot)$ of $\{X_t\}$ satisfies $\rho(h) \to 0$ as $h \to \infty$, we conclude from Theorem 7.1.1 that

$$E(\bar{X}_n - \mu)^2 \to 0 \quad \text{as } n \to \infty,$$

and that

$$nE(\bar{X}_n - \mu)^2 \to \begin{cases} 0 & \text{if } -.5 < d < 0, \\ \infty & \text{if } 0 < d < .5. \end{cases}$$

Using (13.2.13) we can derive (Problem 13.6) the more refined result,

$$n^{1-2d}E(\bar{X}_n - \mu)^2 \to C \quad \text{for } d \in (-.5, .5),$$

where C is a positive constant. For long memory processes the sample mean may not be asymptotically normal (see Taqqu (1975)).

Estimation of the Autocorrelation Function, $\rho(\cdot)$. The function $\rho(\cdot)$ is usually estimated by means of the sample autocorrelation function $\hat{\rho}(\cdot)$. In the case $-.5 < d < 0$, $\{X_t\}$ has the moving average representation

$$X_t = \mu + \sum_{j=0}^{\infty} \psi_j Z_{t-j},$$

with $\sum_{j=0}^{\infty} |\psi_j| < \infty$. If in addition $\{Z_t\} \sim \text{IID}(0, \sigma^2)$ and $EZ_1^4 < \infty$ then $n^{1/2}(\hat{\rho}(h) - \rho(h))$ is asymptotically normal with mean zero and variance given by Bartlett's formula, (7.2.5). If $0 < d < .5$ the situation is much more complicated; partial results for the case when $\{Z_t\}$ is Gaussian can be found in Fox and Taqqu (1986).

Estimation of d, $\boldsymbol{\phi}$ and $\boldsymbol{\theta}$

(a) *Maximum likelihood.* The Gaussian likelihood of $\mathbf{X} = (X_1, \ldots, X_n)'$ for the process (13.2.23) with $\mu = 0$ can be expressed (cf. (8.7.4)) as

$$L(\boldsymbol{\beta}, \sigma^2) = (2\pi\sigma^2)^{-n/2}(r_0, \ldots, r_{n-1})^{-1/2} \exp\left\{-\frac{1}{2\sigma^2}\sum_{j=1}^{n}(X_j - \hat{X}_j)^2/r_{j-1}\right\},$$

where $\boldsymbol{\beta} = (d, \phi_1, \ldots, \phi_p, \theta_1, \ldots, \theta_q)'$, $\hat{X}_j, j = 1, \ldots, n$, are the one-step predictors and $r_{j-1} = \sigma^{-2}E(X_j - \hat{X}_j)^2, j = 1, \ldots, n$. The maximum likelihood estimators $\hat{\boldsymbol{\beta}}$ and $\hat{\sigma}^2$ can be found by maximizing $L(\boldsymbol{\beta}, \sigma^2)$ with respect to $\boldsymbol{\beta}$ and σ^2. By the same arguments used in Section 8.7 we find that

$$\hat{\sigma}^2 = n^{-1}S(\hat{\boldsymbol{\beta}}),$$

where

$$S(\hat{\boldsymbol{\beta}}) = \sum_{j=1}^{n} (X_j - \hat{X}_j)^2/r_{j-1},$$

and $\hat{\boldsymbol{\beta}}$ is the value of $\boldsymbol{\beta}$ which minimizes

$$l(\boldsymbol{\beta}) = \ln(S(\boldsymbol{\beta})/n) + n^{-1}\sum_{j=1}^{n} \ln r_{j-1}. \tag{13.2.24}$$

For $\{Z_t\}$ Gaussian, it has been shown by Yajima (1985) in the case $p = q = 0$, $d > 0$, and argued by Li and McLeod (1986) in the case $d > 0$, that

$$\hat{\boldsymbol{\beta}} \text{ is AN}(\boldsymbol{\beta}, n^{-1}W^{-1}(\boldsymbol{\beta})) \tag{13.2.25}$$

where $W(\boldsymbol{\beta})$ is the $(p + q + 1) \times (p + q + 1)$ matrix whose (j, k) element is

$$W_{jk}(\boldsymbol{\beta}) = \frac{1}{4\pi}\int_{-\pi}^{\pi} \frac{\partial \ln g(\lambda; \boldsymbol{\beta})}{\partial \beta_j} \frac{\partial \ln g(\lambda; \boldsymbol{\beta})}{\partial \beta_k}\, d\lambda,$$

and $\sigma^2 g(\cdot; \boldsymbol{\beta})/(2\pi)$ is the spectral density of the process. The asymptotic behaviour of $\hat{\boldsymbol{\beta}}$ is unknown in the case $d < 0$. Direct calculation of $l(\boldsymbol{\beta})$ from (13.2.24) is slow, especially for large n, partly on account of the difficulty involved in computing the autocovariance function of the process (13.2.23), and partly because the device used in Section 8.7 to express $\hat{X}_j$ in terms of only q innovations and p observations cannot be applied when $d \neq 0$.

It is therefore convenient to consider the approximation to $l(\boldsymbol{\beta})$,

$$l_a(\boldsymbol{\beta}) = \ln \frac{1}{n}\sum_j \frac{I_n(\omega_j)}{g(\omega_j; \boldsymbol{\beta})},$$

where $I_n(\cdot)$ is the periodogram of the series $\{X_1, \ldots, X_n\}$ and the sum is over all non-zero Fourier frequencies $\omega_j = 2\pi j/n \in (-\pi, \pi]$. Hannan (1973) and Fox and Taqqu (1986) show that the estimator $\tilde{\boldsymbol{\beta}}$ which minimizes $l_a(\boldsymbol{\beta})$ is consistent and, if $d > 0$, that $\tilde{\boldsymbol{\beta}}$ has the same limit distribution as in (13.2.25). The white noise variance is estimated by

$$\tilde{\sigma}^2 = \frac{1}{n}\sum_j \frac{I_n(\omega_j)}{g(\omega_j; \tilde{\boldsymbol{\beta}})}.$$

The approximation $l_a(\boldsymbol{\beta})$ to $l(\boldsymbol{\beta})$ does not account for the determinant term $n^{-1}\sum_{j=1}^{n} \ln r_{j-1} = n^{-1} \ln \det(\sigma^{-2}\Gamma_n)$ where $\Gamma_n = E(\mathbf{X}\mathbf{X}')$. Although

$$n^{-1}\sum_{j=1}^{n} \ln r_{j-1} \to 0 \quad \text{as} \quad n \to \infty,$$

this expression may have a non-negligible effect on the minimization of $l(\boldsymbol{\beta})$ even for series of several hundred observations. A convenient approximation to the determinant term can be found from Proposition 4.5.2, namely

$$n^{-1}\ln\left(\prod_j g(\omega_j;\boldsymbol{\beta})\right) = n^{-1}\sum_j \ln g(\omega_j;\boldsymbol{\beta}).$$

Adding this term to $l_a(\boldsymbol{\beta})$, we arrive at a second approximation to l given by

$$l_b(\boldsymbol{\beta}) := l_a(\boldsymbol{\beta}) + n^{-1}\sum_j \ln g(\omega_j;\boldsymbol{\beta}). \tag{13.2.26}$$

Estimation based on minimizing $l_b(\cdot)$ has been studied by Rice (1979) in a more general setting. For ARIMA(p, d, q) processes with $d \in (-.5, .5)$, empirical studies show that the estimates which minimize l_b tend to have less bias than those which minimize l_a.

(b) *A regression method.* The second method is based on the form of the spectral density

$$f(\lambda) = |1 - e^{-i\lambda}|^{-2d} f_U(\lambda), \tag{13.2.27}$$

where

$$f_U(\lambda) = \frac{\sigma^2}{2\pi}\frac{|\theta(e^{-i\lambda})|^2}{|\phi(e^{-i\lambda})|^2} \tag{13.2.28}$$

is the spectral density of the ARMA(p, q) process,

$$U_t = \nabla^d X_t. \tag{13.2.29}$$

Taking logarithms in (13.2.27) gives

$$\ln f(\lambda) = \ln f_U(0) - d\ln|1 - e^{-i\lambda}|^2 + \ln[f_U(\lambda)/f_U(0)]. \tag{13.2.30}$$

Replacing λ in (13.2.30) by the Fourier frequency $\omega_j = 2\pi j/n \in (0, \pi)$ and adding $\ln I_n(\omega_j)$ to both sides, we obtain

$$\begin{aligned}\ln I_n(\omega_j) = \ln f_U(0) - d\ln|1 - e^{-i\omega_j}|^2 + \ln(I_n(\omega_j)/f(\omega_j))\\ + \ln(f_U(\omega_j)/f_U(0)).\end{aligned} \tag{13.2.31}$$

Now if ω_j is near zero, say $\omega_j \le \omega_m$ where ω_m is small, then the last term is negligible compared with the others on the right-hand side, so we can write (13.2.31) as the simple linear regression equation,

$$Y_j = a + bx_j + \varepsilon_j, \qquad j = 1, \ldots, m, \tag{13.2.32}$$

where $Y_j = \ln I_n(\omega_j)$, $x_j = \ln|1 - e^{-i\omega_j}|^2$, $\varepsilon_j = \ln(I_n(\omega_j)/f(\omega_j))$, $a = \ln f_U(0)$ and $b = -d$. This suggests estimating d by least-squares regression of $Y_1, \ldots, Y_m$ on $x_1, \ldots, x_m$. When this regression is carried out, we find that the least-squares estimator $\hat{d}$ of d is given by

$$\hat{d} = -\sum_{i=1}^{m} (x_i - \bar{x})(Y_i - \bar{Y}) \Big/ \sum_{i=1}^{m} (x_i - \bar{x})^2. \tag{13.2.33}$$

Geweke and Porter-Hudak (1983) argue that when $-.5 < d < 0$ there exists a sequence m such that $(\ln n)^2/m \to 0$ as $n \to \infty$ and

$$\hat{d} \text{ is AN}\left(d, \pi^2 \Big/ \left[6 \sum_{i=1}^{m} (x_i - \bar{x})^2\right]\right) \quad \text{as } n \to \infty. \tag{13.2.34}$$

Notice that $\pi^2/6$ is the variance of the asymptotic distribution of $\ln(I(\lambda)/f(\lambda))$ for any fixed $\lambda \in (0, \pi)$.

Having estimated d, we must now estimate the ARMA parameters $\boldsymbol{\phi}$ and $\boldsymbol{\theta}$. Since $X_t = \nabla^{-d} U_t$ where $\{U_t\}$ is an ARMA(p, q) process, we find from (10.3.12) (replacing Z by U) that

$$J_X(\lambda) = (1 - e^{-i\lambda})^{-d} J_U(\lambda) + Y_n(\lambda) \tag{13.2.35}$$

where $J_X(\cdot)$ and $J_U(\cdot)$ are the discrete Fourier transforms of $\{X_1, \ldots, X_n\}$ and $\{U_1, \ldots, U_n\}$ respectively. Ignoring the error term $Y_n(\lambda)$ (which converges in probability to zero as $n \to \infty$) and replacing d by $\hat{d}$, we obtain the approximate relation

$$J_U(\lambda) = (1 - e^{-i\lambda})^{\hat{d}} J_X(\lambda). \tag{13.2.36}$$

If now we apply the inverse Fourier transform to each side of (13.2.36) we obtain the estimates of U_t,

$$\tilde{U}_t = n^{-1/2} \sum_j e^{i\omega_j t} (1 - e^{-i\omega_j})^{\hat{d}} J_X(\omega_j), \qquad t = 1, \ldots, n, \tag{13.2.37}$$

where the sum is taken over all Fourier frequencies $\omega_j \in (-\pi, \pi]$ (omitting the zero-frequency term if $d < 0$). Estimates of p, q, $\boldsymbol{\phi}$ and $\boldsymbol{\theta}$ are then obtained by applying the techniques of Chapter 9 to the series $\{\tilde{U}_t\}$.

The virtue of the regression method is that it permits estimation of d without knowledge of p and q. The values $\{\tilde{U}_t\}$ then permit tentative identification of p and q using the methods already developed for ARMA processes. Final estimates of the parameters are obtained by application of the approximate likelihood method described in (a).

Example 13.2.1. We now fit a fractionally integrated ARMA model to the data $\{X_t, t = 1, \ldots, 200\}$ shown in Figure 13.3. The sample autocorrelation function (Figure 13.4) suggests that the series is long-memory or perhaps even non-stationary. Proceeding under the assumption that the series is stationary, we shall fit an ARIMA(p, d, q) model with $d \in (-.5, .5)$. The first step is to estimate d using (13.2.33). Table 13.1 shows the values of the regression estimate $\hat{d}$ for values of m up to 40. The simulations of Geweke and Porter-Hudak (1983) suggest the choice $m = n^{.5}$ or 14 in this case. In fact from the table we see that the variation in $\hat{d}$ is rather small over the range $13 \le m \le 35$. It appears however that the term $\ln(f_U(\omega_j)/f_U(0))$ in (13.2.30) is not negligible

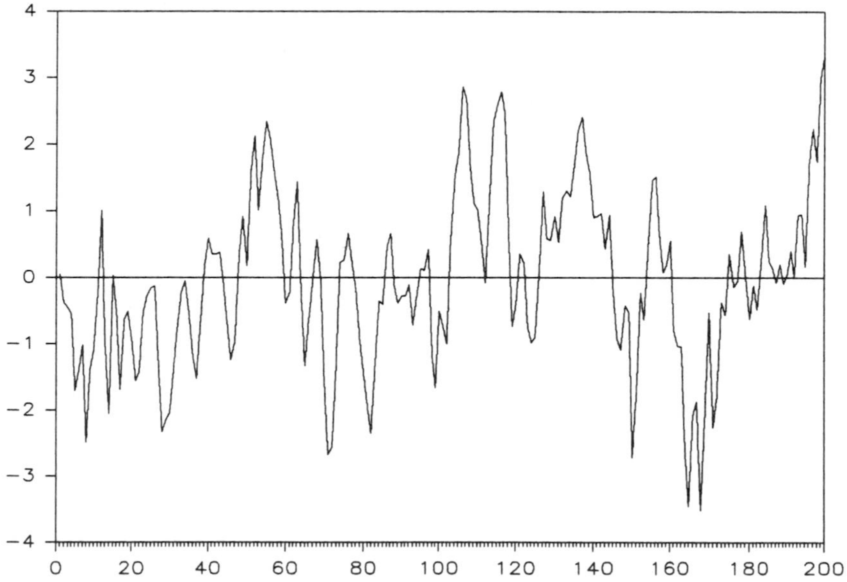

Figure 13.3. The data $\{X_t, t = 1, \ldots, 200\}$ of Example 13.1.

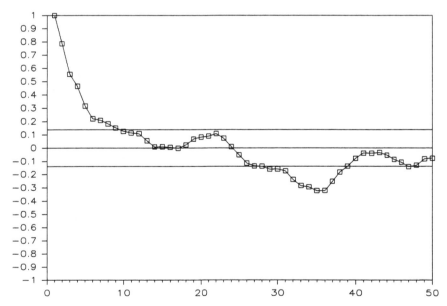

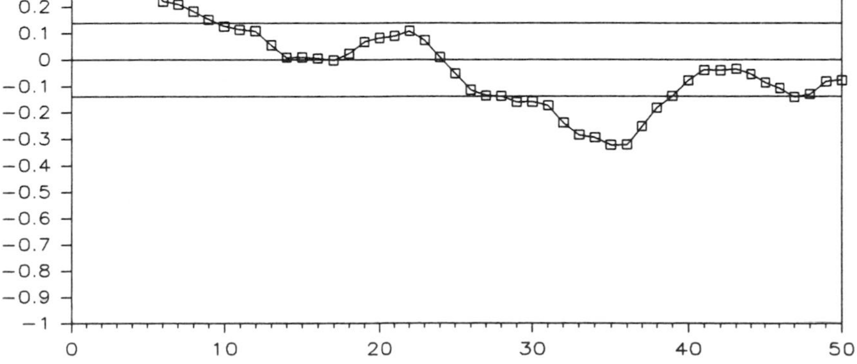

Figure 13.4. The sample autocorrelation function of the data shown in Figure 13.3.

Table 13.1. Values of the Regression Estimator $\hat{d}$ in Example 13.2.1

m	13	14	15	16	17	18	19	20	25	30	35	40	45
$\hat{d}$	.342	.371	.299	.356	.411	.421	.370	.334	.370	.409	.424	.521	.562

for $j \geq 40$. We take as our estimate the value of $\hat{d}$ when m is 14, i.e. $\hat{d} = .371$, with estimated variance $\pi^2/[6(\sum_{i=1}^{14}(x_i - \bar{x})^2] = .0531$. An approximate 95% confidence interval for d is therefore given by $(-.081, .500)$. (Although the asymptotic distribution (13.2.34) is discussed by Geweke and Porter-Hudak only for $d < 0$, their simulation results support the validity of this distribution even in the case when $0 < d < .5$.)

Estimated values of $U_t = \nabla^d(X_t + .0434)$ are next found from (13.2.37). The sample autocorrelation function of the estimates $(\tilde{U}_t\}$ (Figure (13.5) strongly suggests an MA(1) process, and maximum likelihood estimation of the parameters gives the preliminary model,

$$\nabla^{.375}(X_t + .0434) = Z_t + .816Z_{t-1}, \qquad \{Z_t\} \sim \text{WN}(0, .489). \qquad (13.2.38)$$

Finally we reestimate the parameters of the ARIMA$(0, d, 1)$ model by minimizing the function $l_b(d, \theta)$ defined in (13.2.26). The resulting model is

$$\nabla^{.396}(X_t + .0434) = Z_t + .812Z_{t-1}, \qquad \{Z_t\} \sim \text{WN}(0, .514), \qquad (13.2.39)$$

which is very similar both to the preliminary model (13.2.38) and to the model

$$\nabla^{.4}X_t = Z_t + .8Z_t, \qquad \{Z_t\} \sim \text{WN}(0, .483), \qquad (13.2.40)$$

from which the series $\{X_t\}$ was generated.

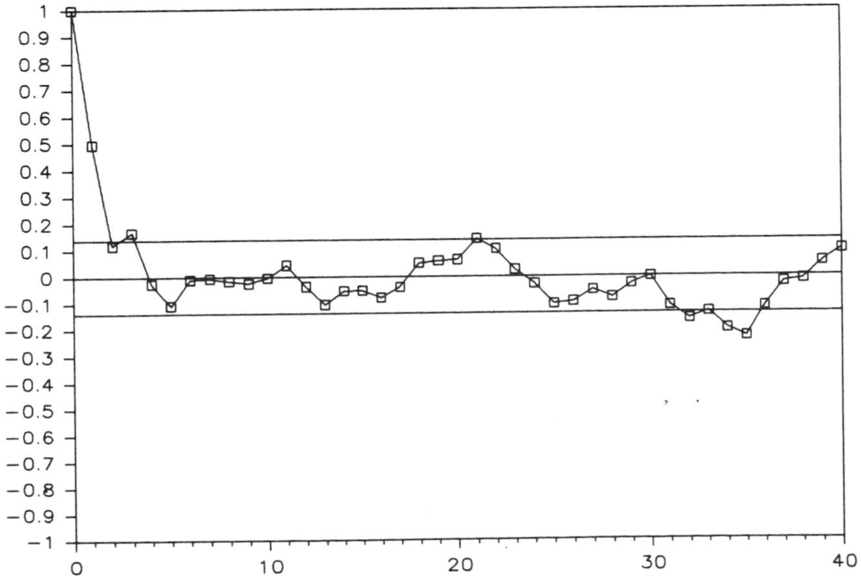

Figure 13.5. The sample autocorrelation function of the estimates $\tilde{U}_t$ of $\nabla^{.371}(X_t + .0433)$, Example 13.2.1.

Prediction of an ARIMA(p, d, q) Process, $d \in (-.5, .5)$

Let $\{X_t\}$ be the causal invertible ARIMA(p, d, q) process,

$$\phi(B)\nabla^d X_t = \theta(B)Z_t, \qquad \{Z_t\} \sim \text{WN}(0, \sigma^2), \qquad d \in (-.5, .5) \tag{13.2.41}$$

The innovations algorithm can be applied to the covariance function of $\{X_t\}$ to compute the best linear predictor of X_{n+h} in terms of $X_1, \ldots, X_n$. For large n however it is much simpler to consider the approximation

$$\tilde{X}_{n+h} := P_{\overline{\text{sp}}\{X_j, -\infty < j \le n\}} X_{n+h}.$$

Since we are assuming causality and invertibility we can write

$$X_t = \sum_{j=0}^{\infty} \psi_j Z_{t-j}$$

and

$$Z_t = \sum_{j=0}^{\infty} \pi_j X_{t-j},$$

where $\sum_{j=0}^{\infty} \psi_j z^j = \theta(z)\phi^{-1}(z)(1 - z)^{-d}$ and $\sum_{j=0}^{\infty} \pi_j z^j = \phi(z)\theta^{-1}(z)(1 - z)^d$, $|z| < 1$. Theorem 5.5.1 can be extended to include the process (13.2.41), giving

$$\tilde{X}_{n+h} = -\sum_{j=1}^{\infty} \pi_j \tilde{X}_{n+h-j} = \sum_{j=h}^{\infty} \psi_j Z_{n+h-j} \tag{13.2.42}$$

and

$$\tilde{\sigma}_n^2(h) = E(X_{n+h} - \tilde{X}_{n+h})^2 = \sigma^2 \sum_{j=0}^{h-1} \psi_j^2. \tag{13.2.43}$$

Predicted values of $X_{201}, \ldots, X_{230}$ were computed for the data of Example (13.2.1) using the fitted model (13.2.39). The predictors $\tilde{X}_{200+h}$ were found using a truncated and mean-corrected version of (13.2.42), namely

$$\tilde{X}_{200+h} + .0433 = -\sum_{j=1}^{199+h} \pi_j(\tilde{X}_{200+h-j} + .0433),$$

from which $\tilde{X}_{201}, \tilde{X}_{202}, \ldots,$ can be found recursively. The predictors are shown with the corresponding observed values $X_{201}, \ldots, X_{230}$ in Figure 13.6. Also shown are the predictors based on the ARMA (3, 3) model,

$$\begin{aligned}(1 - .132B + .037B^2 - .407B^3)(X_t + .0433)& \\ = Z_t + 1.061Z_{t-1} + .491Z_{t-2} + .218Z_{t-3},& \qquad \{Z_t\} \sim \text{WN}(0, .440),\end{aligned}$$

which was fitted to the data $\{X_t, t = 1, \ldots, 200\}$ using the methods of Section 9.2. The predictors based on the latter model converge much more rapidly to the mean value, $-.0433$, than those based on the long memory model.

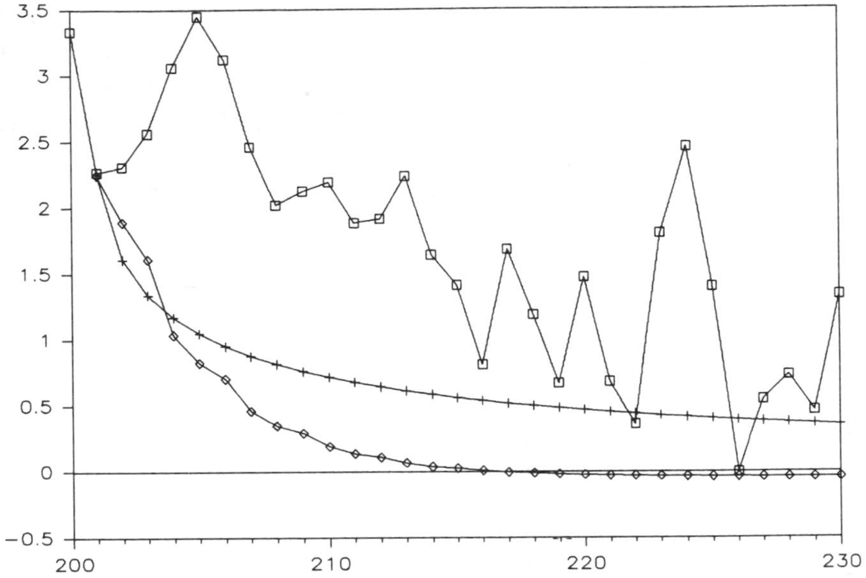

Figure 13.6. The data $X_{201}, \ldots, X_{230}$ of Example 13.2.1 showing the predictors based on the ARMA model (lower) and the long memory model (upper).

The average squared errors of the 30 predictors are 1.43 for the long memory model and 2.35 for the ARMA model. Although the ARMA model appears to fit the data very well, with an *estimated* one-step mean square prediction variance .440 as compared with .514 for the long memory model, the value .440 is a low estimate of the true value (.483) for the process (13.2.40) from which the series was generated. Both predictors will, as the lead time $h \to \infty$, have bias $-.0433$ and asymptotic variance 2.173, the variance of the generating process. It is interesting to compare the rates of approach of the two predictor variances to their asymptotic values. This is done in Table 13.2 which shows the ratio $\tilde{\sigma}_h^2/\tilde{\sigma}_n^2(\infty)$ for both models as computed from (13.2.43). It is apparent that the ARMA predictors are not appreciably better than the mean for lead times of 10 or more, while the value of the long memory predictor persists for much greater lead times.

Table 13.2. $\tilde{\sigma}_n^2(h)/\tilde{\sigma}_n^2(\infty)$ for the Long Memory and ARMA Models for the Data of Example 13.2.1

h	1	2	3	4	5	10	20	30	40	5
Long memory model	.250	.606	.685	.728	.757	.830	.887	.914	.932	.9
ARMA model	.261	.632	.730	.844	.922	.993	1.000	1.000	1.000	1.0

§13.3 Linear Processes with Infinite Variance

There has recently been a great deal of interest in modelling time series using ARMA processes with infinite variance. Examples where such models appear to be appropriate have been found by Stuck and Kleiner (1974), who considered telephone signals, and Fama (1965), who modelled stock market prices. Any time series which exhibits sharp spikes or occasional bursts of outlying observations suggests the possible use of an infinite variance model. In this section we shall restrict attention to processes generated by application of a linear filter to an iid sequence, $\{Z_t, t = 0, \pm 1, \dots\}$, of random variables whose distribution F has Pareto-like tails, i.e.

$$\begin{cases} x^\alpha(1 - F(x)) = x^\alpha P(Z_t > x) \to pC, & \text{as } x \to \infty, \\ x^\alpha F(-x) = x^\alpha P(Z_t \le -x) \to qC, & \text{as } x \to \infty, \end{cases} \tag{13.3.1}$$

where $0 < \alpha < 2$, $0 \le p = 1 - q \le 1$, and C is a finite positive constant which we shall call the *dispersion*, $\text{disp}(Z_t)$, of the random variable Z_t. From (13.3.1) we can write

$$x^\alpha(1 - F(x) + F(-x)) = x^\alpha P(|Z_t| > x) \to C \quad \text{as } x \to \infty. \tag{13.3.2}$$

A straightforward calculation (Problem (13.7)) shows that

$$\begin{cases} E|Z_t|^\delta = \infty & \text{if } \delta \ge \alpha \\ E|Z_t|^\delta < \infty & \text{if } \delta < \alpha. \end{cases} \tag{13.3.3}$$

Hence $\text{Var}(Z_t) = \infty$ for $0 < \alpha < 2$ and $E|Z_t| < \infty$ only if $1 < \alpha < 2$. An important class of distributions satisfying (13.3.1) consists of the non-normal stable distributions.

Definition 13.3.1 (Stable Distributions). A random variable Z is said to be stable, or to have a stable distribution, if for every positive integer n there exist constants, $a_n > 0$ and b_n, such that the sum $Z_1 + \cdots + Z_n$ has the same distribution as $a_n Z + b_n$ for all iid random variables $Z_1, \dots, Z_n$, with the same distribution as Z.

Properties of a Stable Random Variable, Z

Some of the important properties of Z are listed below. For an extensive discussion of stable random variables see Feller (1971), pp. 568–583, but note the error in sign in equation (3.18).

1. The characteristic function, $\phi(u) = E\exp(iuZ)$, is given by

$$\phi(u) = \begin{cases} \exp\{iu\beta - d|u|^\alpha(1 - i\theta\,\text{sgn}(u)\tan(\pi\alpha/2))\} & \text{if } \alpha \ne 1, \\ \exp\{iu\beta - d|u|(1 + i\theta(2/\pi)\text{sgn}(u)\ln|u|)\} & \text{if } \alpha = 1, \end{cases} \tag{13.3.4}$$

where $\operatorname{sgn}(u)$ is $u/|u|$ if $u \neq 0$, and zero otherwise. The parameters $\alpha \in (0, 2]$, $\beta \in \mathbb{R}$, $d^{1/\alpha} \in [0, \infty)$ and $\theta \in [-1, 1]$ are known as the exponent, location, scale and symmetry parameters respectively.

2. If $\alpha = 2$ then $Z \sim N(\beta, 2d)$.
3. If $\theta = 0$ then the distribution of Z is symmetric about β. The symmetric stable distributions (i.e. those which are symmetric about 0) have characteristic functions of the form
$$\phi(u) = e^{-d|u|^\alpha}. \tag{13.3.5}$$
4. If $\alpha = 1$ and $\theta = 0$ then Z has the Cauchy distribution with probability density $f(z) = (d/\pi)[d^2 + (z - \beta)^2]^{-1}$, $z \in \mathbb{R}$.
5. The symmetric stable distributions satisfy the property of Definition 13.3.1 with $a_n = n^{1/\alpha}$ and $b_n = 0$, since if $Z, Z_1, \ldots, Z_n$ all have the characteristic function (13.3.5) and $Z_1, \ldots, Z_n$ are independent, then
$$E\exp[iu(Z_1 + \cdots + Z_n)] = e^{-nd|u|^\alpha} = E\exp[iuZn^{1/\alpha}].$$
6. If F is the distribution function of Z and $\alpha \in (0, 2)$, then (13.3.1) is satisfied with $p = (1 + \theta)/2$ and
$$C = \begin{cases} d/(\Gamma(1 - \alpha)\cos(\pi\alpha/2)) & \text{if } \alpha \neq 1, \\ 2d/\pi & \text{if } \alpha = 1. \end{cases} \tag{13.3.6}$$

In the following proposition, we provide sufficient conditions under which the sum $\sum_{j=-\infty}^{\infty} \psi_j Z_{t-j}$ exists when $\{Z_t\}$ is an iid sequence satisfying (13.3.1).

Proposition 13.3.1. *Let $\{Z_t\}$ be an iid sequence of random variables satisfying (13.3.1). If $\{\psi_j\}$ is a sequence of constants such that*
$$\sum_{j=-\infty}^{\infty} |\psi_j|^\delta < \infty \quad \text{for some } \delta \in (0, \alpha) \cap [0, 1], \tag{13.3.7}$$
then the infinite series,
$$\sum_{j=-\infty}^{\infty} \psi_j Z_{t-j},$$
converges absolutely with probability one.

Proof. First consider the case $1 < \alpha < 2$. Then by (13.3.3), $E|Z_1| < \infty$ and hence
$$E\left(\sum_{j=-\infty}^{\infty} |\psi_j Z_{t-j}|\right) = \sum_{j=-\infty}^{\infty} |\psi_j| E|Z_{t-j}|$$
$$= \sum_{j=-\infty}^{\infty} |\psi_j| E|Z_1| < \infty.$$

Thus $\sum_{j=-\infty}^{\infty} |\psi_j Z_{t-j}|$ is finite with probability one.

Now suppose $0 < \alpha < 1$. Since $0 < \delta < 1$, we can apply the triangle inequality $|x + y|^\delta \leq |x|^\delta + |y|^\delta$ to the infinite sum $\sum_{j=-\infty}^{\infty} \psi_j Z_{t-j}$. Making use of (13.3.3) we then find that

$$E\left(\sum_{j=-\infty}^{\infty}|\psi_j Z_{t-j}|\right)^{\delta} \le E\left(\sum_{j=-\infty}^{\infty}|\psi_j|^{\delta}|Z_{t-j}|^{\delta}\right)$$

$$= \sum_{j=-\infty}^{\infty}|\psi_j|^{\delta}E|Z_1|^{\delta}$$

$$< \infty.$$

Hence $\sum_{j=-\infty}^{\infty}|\psi_j Z_{t-j}| < \infty$ with probability one. □

Remark 1. The distribution of the infinite sum $\sum_{j=-\infty}^{\infty}\psi_j Z_{t-j}$ satisfies (13.3.2). Specifically,

$$x^{\alpha}P\left(\left|\sum_{j=-\infty}^{\infty}\psi_j Z_{t-j}\right| > x\right) \to \left(\sum_{j=-\infty}^{\infty}|\psi_j|^{\alpha}\right)C$$

(see Cline, 1983).

Remark 2. If Z_1 has a symmetric stable distribution with characteristic function $e^{-d|t|^{\alpha}}$ (and dispersion C given by (13.3.6)), then $\sum_{j=-\infty}^{\infty}\psi_j Z_{t-j}$ also has a symmetric stable distribution with dispersion $\tilde{C} = C\sum_{j=-\infty}^{\infty}|\psi_j|^{\alpha}$.

Remark 3. The process defined by

$$X_t = \sum_{j=-\infty}^{\infty}\psi_j Z_{t-j}, \tag{13.3.8}$$

where $\{\psi_j\}$ and $\{Z_t\}$ satisfy the assumptions of Proposition 13.3.1, exists with probability one and is strictly stationary, i.e. the joint distribution of $(X_1, \ldots, X_k)'$ is the same as that of $(X_{1+h}, \ldots, X_{k+h})'$ for all integers h and positive integers k (see Problem 13.8). In particular if the coefficients ψ_j are chosen so that $\psi_j = 0$ for $j < 0$ and

$$\sum_{j=0}^{\infty}\psi_j z^j = \theta(z)/\phi(z), \qquad |z| \le 1, \tag{13.3.9}$$

where $\theta(z) = 1 + \theta_1 z + \cdots + \theta_q z^q$ and $\phi(z) = 1 - \phi_1 z - \cdots - \phi_p z^p \ne 0$ for $|z| \le 1$, then it is easy to show that $\{X_t\}$ as defined by (13.3.8) satisfies the ARMA equations $\phi(B)X_t = \theta(B)Z_t$. We record this result as a proposition.

Proposition 13.3.2. *Let $\{Z_t\}$ be an iid sequence of random variables with distribution function F satisfying* (13.3.1). *Then if $\theta(\cdot)$ and $\phi(\cdot)$ are polynomials such that $\phi(z) \ne 0$ for $|z| \le 1$, the difference equations*

$$\phi(B)X_t = \theta(B)Z_t, \tag{13.3.10}$$

have the unique strictly stationary solution,

$$X_t = \sum_{j=0}^{\infty} \psi_j Z_{t-j}, \tag{13.3.11}$$

where the coefficients $\{\psi_j\}$ *are determined by the relation* (13.3.9). *If in addition* $\phi(z)$ *and* $\theta(z)$ *have no common zeroes, then the process* (13.3.11) *is invertible if and only if* $\theta(z) \neq 0$ *for* $|z| \leq 1$.

PROOF. The series (13.3.11) converges absolutely with probability one by Proposition 13.3.1. The fact that it is the unique strictly stationary solution of (13.3.10) is established by an argument similar to that used in the proof of Theorem 3.1.1. Invertibility is established by arguments similar to those in the proof of Theorem 3.1.2. See Problem 13.9. □

Although the process $\{X_t\}$ defined by (13.3.8) is strictly stationary it is not second-order stationary since by Remark 1 and (13.3.3), $E|X_t|^2 = \infty$. Nevertheless we can still define, for such a process, an analogue of the autocorrelation function, namely

$$\rho(h) := \frac{\sum_j \psi_j \psi_{j+h}}{\sum_j \psi_j^2}, \qquad h = 1, 2, \ldots. \tag{13.3.12}$$

We use the same notation as for the autocorrelation function of a second-order stationary process since if $\{Z_t\}$ is replaced in (13.3.8) by a finite variance white noise sequence, then (13.3.12) coincides with the autocorrelation function of $\{X_t\}$. Our point of view in this section however is that $\rho(h)$ is simply a function of the coefficients $\{\psi_j\}$ in the representation (13.3.8), or as a function of the coefficients $\{\phi_j\}$ and $\{\theta_j\}$ if $\{X_t\}$ is an ARMA process defined as in (13.3.10).

We can estimate $\rho(h)$ using the sample autocorrelation function,

$$\tilde{\rho}(h) = \sum_{t=1}^{n-h} X_t X_{t+h} \bigg/ \sum_{t=1}^{n} X_t^2, \qquad h = 1, 2, \ldots,$$

but it is by no means clear that $\tilde{\rho}(h)$ is even a consistent estimator of $\rho(h)$. However, from the following theorem of Davis and Resnick (1986), we find that $\tilde{\rho}(h)$ is not only consistent but has other good properties as an estimator of $\rho(h)$.

Theorem 13.3.1. *Let* $\{Z_t\}$ *be an iid symmetric sequence of random variables satisfying* (13.3.1) *and let* $\{X_t\}$ *be the strictly stationary process,*

$$X_t = \sum_{j=-\infty}^{\infty} \psi_j Z_{t-j},$$

where

$$\sum_{j=-\infty}^{\infty} |j| |\psi_j|^{\delta} < \infty \quad \textit{for some } \delta \in (0, \alpha) \cap [0, 1].$$

Then for any positive integer h,

$$(n/ln(n))^{1/\alpha}(\tilde{\rho}(1) - \rho(1), \dots, \tilde{\rho}(h) - \rho(h))' \Rightarrow (Y_1, \dots, Y_h)', \tag{13.3.13}$$

where

$$Y_k = \sum_{j=1}^{\infty} (\rho(k+j) + \rho(k-j) - 2\rho(j)\rho(k))S_j/S_0, \qquad k = 1, \dots, h,$$

and $S_0, S_1, \dots,$ are independent stable random variables; S_0 is positive stable with characteristic function,

$$\begin{aligned} &E\exp(iuS_0) \\ &\quad = \exp\{-C\Gamma(1-\alpha/2)\cos(\pi\alpha/4)|u|^{\alpha/2}(1 - i\,\mathrm{sgn}(u)\tan(\pi\alpha/4))\} \end{aligned} \tag{13.3.14}$$

and $S_1, S_2, \dots,$ are iid with characteristic function,

$$E\exp(iuS_1) = \begin{cases} \exp\{-C^2\Gamma(1-\alpha)\cos(\pi\alpha/2)|u|^{\alpha}\} & \text{if } \alpha \neq 1, \\ \exp\{-C^2\pi|u|/2\} & \text{if } \alpha = 1. \end{cases} \tag{13.3.15}$$

If $\alpha > 1$ then (13.3.13) is also true when $\tilde{\rho}(\theta)$ is replaced by its mean-corrected version, $\hat{\rho}(h) = \sum_{t=1}^{n-h}(X_t - \bar{X})(X_{t+h} - \bar{X})/\sum_{t=1}^{n}(X_t - \bar{X})^2$, where $\bar{X} = n^{-1}(X_1 + \cdots + X_n)$.

It follows at once from this theorem that $\tilde{\rho}(h) \xrightarrow{P} \rho(h)$, and more specifically that

$$\tilde{\rho}(h) - \rho(h) = O_p([n/ln(n)]^{-1/\alpha}) = o_p(n^{-1/\beta})$$

for all $\beta > \alpha$. This rate of convergence to zero compares favourably with the slower rate, $O_p(n^{-1/2})$, for the difference $\tilde{\rho}(h) - \rho(h)$ in the finite variance case.

The form of the asymptotic distribution of $\tilde{\rho}(h)$ can be somewhat simplified. In order to do this, note that Y_h has the same distribution as

$$\left(\sum_{j=1}^{\infty} |\rho(h+j) + \rho(h-j) - 2\rho(j)\rho(h)|^{\alpha}\right)^{1/\alpha} U/V, \tag{13.3.16}$$

where V (≥ 0) and U are independent random variables with characteristic functions given by (13.3.14) and (13.3.15) respectively with $C = 1$. Percentiles of the distribution of U/V can be found either by simulation of independent copies of U/V or by numerical integration of the joint density of (U, V) over an appropriate region. Except when $\alpha = 1$, the joint density of U and V cannot be written down in closed form. In the case $\alpha = 1$, U is a Cauchy random variable with probability density $f_U(u) = \frac{1}{2}[\pi^2/4 + u^2]^{-1}$ (see Property 4 of stable random variables), and V is a non-negative stable random variable with density (see Feller (1971)), $f_V(v) = \frac{1}{2}v^{-3/2}e^{-\pi/(4v)}$, $v \geq 0$. The distribution function of U/V is therefore given by

$$\begin{aligned} P(U/V \leq x) &= \int_0^{\infty} P(U \leq xy) f_V(y)\,dy \\ &= \int_0^{\infty} 2^{-1/2}(\pi w)^{-3/2}[\arctan(xw) + (\pi/2)]\exp(-1/(2w))\,dw. \end{aligned} \tag{13.3.17}$$

Notice also that U/V has the same distribution as the product of a standard Cauchy random variable (with probability density $\pi^{-1}(1 + x^2)^{-1}$) and an independent random variable distributed as $\chi^2(1)$.

EXAMPLE 13.3.1 (An Infinite Variance Moving Average Process). Let $\{X_t\}$ be the MA(q) process,

$$X_t = Z_t + \theta_1 Z_{t-1} + \cdots + \theta_q Z_{t-q},$$

where the sequence $\{Z_t\}$ satisfies the assumptions of Theorem 13.3.1. Since $\rho(h) = 0$ for $|h| > q$, the theorem implies in this case that

$$(n/ln(n))^{1/\alpha}(\tilde{\rho}(h) - \rho(h)) \Rightarrow \left(1 + 2\sum_{j=1}^{q} |\rho(j)|^{\alpha}\right)^{1/\alpha} U/V, \qquad h > q,$$

where the right-hand side reduces to U/V if $q = 0$.

Two hundred simulated values of the MA(1) process

$$X_t = Z_t + .4Z_t, \tag{13.3.18}$$

with $\{Z_t\}$ an iid standard Cauchy sequence (i.e. $Ee^{iuZ_1} = e^{-|u|}$), are shown in Figure 13.7. The corresponding function $\tilde{\rho}(\theta)$ is shown in Figure 13.8. Except for the value at lag 7, the graph of $\tilde{\rho}(h)$ does suggest that the data is a realization of an MA(1) process. Furthermore the moment estimator, $\tilde{\theta}$, of θ is .394, agreeing well with the true value $\theta = .40$. ($\tilde{\theta}$ is the root in $[-1, 1]$ of $\tilde{\rho}(1) = \theta/(1 + \theta^2)$. If there is no such root, we define $\tilde{\theta} = \text{sgn}(\tilde{\rho}(1))$ as in Section 8.5.)

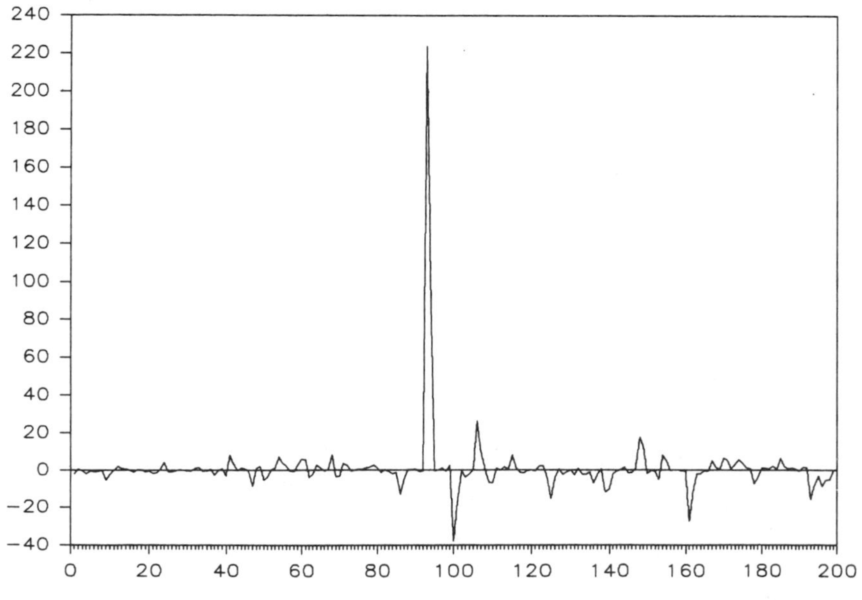

Figure 13.7. Two hundred simulated values of the MA(1) process, $X_t = Z_t + .4Z_{t-1}$, where $\{Z_t\}$ is an iid standard Cauchy sequence.

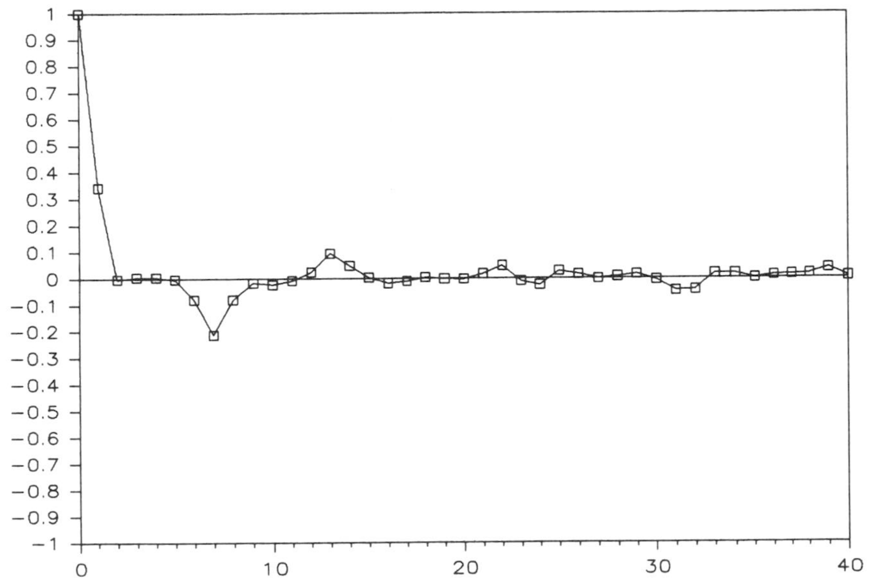

Figure 13.8. The function $\tilde{\rho}(h)$ for the simulated Cauchy MA(1) series of Example 13.3.1.

The .975 quantile of U/V for the process (13.3.18) is found numerically from (13.3.17) to have the value 12.4. By Theorem 13.3.1, approximately 95% confidence bounds for $\rho(1)$ are therefore given by

$$\tilde{\rho}(1) \pm 12.4(|1 - 2\tilde{\rho}^2(1)| + |\tilde{\rho}(1)|)(\ln(n)/n) = .341 \pm .364.$$

These are not particularly informative bounds when $n = 200$, but the difference between them decreases rapidly as n increases. In simulation studies it has been found moreover that $\tilde{\rho}(h)$ gives good estimates of $\rho(h)$ even when $n = 200$. Ten thousand samples of $\{X_1, \ldots, X_{200}\}$ for the process (13.3.18) gave 10,000 values of $\tilde{\rho}(1)$, from which the sample mean and variance were found to be .341 and .0024 respectively. For a finite-variance MA(1) process, Bartlett's formula gives the value, $v = (1 - 3\rho^2(1) + 4\rho^4(1))/n$, for the asymptotic variance of $\hat{\rho}(1)$. Setting $n = 200$ and $\rho(1) = .4/(1 + .4^2) = .345$, we find that $v = .00350$. Thus the sample variance of $\tilde{\rho}(1)$ for 200 observations of the Cauchy process (13.3.18) compares favourably with the asymptotic approximation to the variance of $\hat{\rho}(1)$ for 200 observations of the corresponding finite-variance process. Analogous remarks apply to the moment estimator, $\tilde{\theta}$, of the coefficient of the MA(1) process. From our 10,000 realizations of $\{X_1, \ldots, X_{200}\}$, the sample mean and variance of $\tilde{\theta}$ were found to be .401 and .00701 respectively. The variance of the moment estimator, $\hat{\theta}$, for a finite-variance MA(1) process is $n^{-1}(1 + \theta^2 + 4\theta^4 + \theta^6 + \theta^8)/(1 - \theta^2)^2$ (see Section 8.5). When $n = 200$ and $\theta = .4$ this has the value .00898, which is somewhat larger than the observed sample variance, .00701, of $\hat{\theta}$ for the Cauchy process.

EXAMPLE 13.3.2 (An Infinite Variance AR(1) Process). Figure 13.9 shows 200 simulated values $\{X_1, \ldots, X_{200}\}$ of the AR(1) process,

$$X_t = .7X_{t-1} + Z_t$$

where $\{Z_t\}$ is again an iid Cauchy sequence with $Ee^{iuZ_1} = e^{-|u|}$. Each observed spike in the graph corresponds to a large value of Z_t. Starting from each spike, the absolute value of X_t decays geometrically and then fluctuates near zero until the next large value of Z_t gives rise to a new spike. The graph of $\tilde{\rho}(h)$ resembles a geometrically decreasing function as would be expected from a finite-variance AR(1) process (Figure 13.10). The "Yule–Walker" estimate of ϕ is $\hat{\rho}(1) = .697$, which is remarkably close to the true value, $\phi = .7$. From 10,000 simulations of the sequence $\{X_1, \ldots, X_{200}\}$, the sample mean of $\tilde{\rho}(1)$ was found to be .692 and the sample variance was .0025. For an AR(1) process with finite variance, the asymptotic variance of $\hat{\rho}(1)$ is $(1 - \phi^2)/n$ (see Example 7.2.3). When $n = 200$ and $\phi = .7$, this is equal to .00255, almost the same as the observed sample variance in the simulation experiment. The performance of the estimator $\tilde{\rho}(1)$ of ϕ in this case is thus very close, from the point of view of sample variance, to that of the Yule–Walker estimator in the finite variance case.

Linear Prediction of ARMA *Processes with Infinite Variance.* Let $\{X_t\}$ be the strictly stationary ARMA process defined by (13.3.7) with $\phi(z)\theta(z) \neq 0$ for all $z \in \mathbb{C}$ such that $|z| \leq 1$. Suppose also that the iid sequence $\{Z_t\}$ satisfies

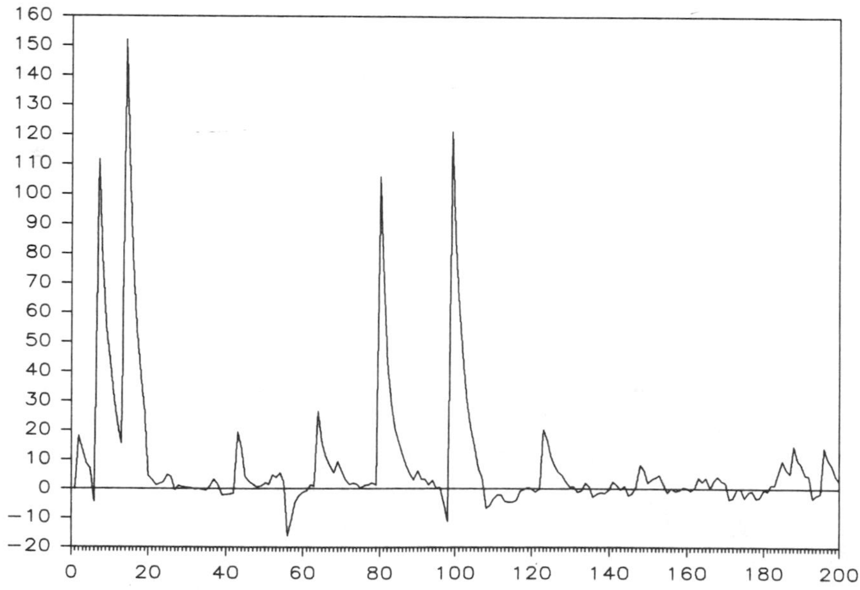

Figure 13.9. Two hundred simulated values of the AR(1) process, $X_t = .7X_{t-1} + Z_t$, where $\{Z_t\}$ is an iid standard Cauchy sequence.

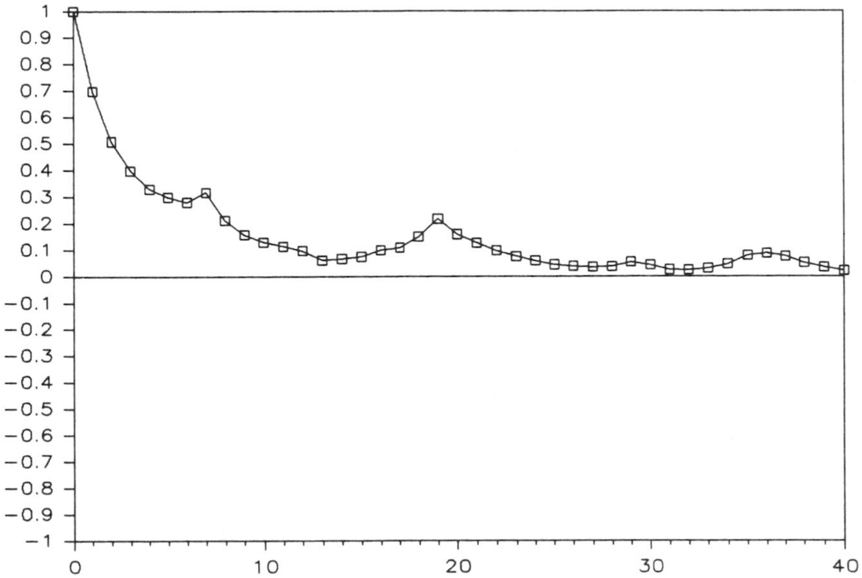

Figure 13.10. The function $\tilde{\rho}(h)$ for the simulated Cauchy AR(1) series of Example 12.5.2.

(13.3.1). In assessing the performance of the linear predictor,

$$\hat{X}_{n+1} = a_{n1}X_n + a_{n2}X_{n-1} + \cdots + a_{nn}X_1, \qquad (13.3.19)$$

we cannot consider $E(X_{n+1} - \hat{X}_{n+1})^2$ as we did for second order processes since this expectation is infinite. Other criteria for choosing a "best" predictor which have been suggested include minimization of the expected absolute error (when $\alpha > 1$), and the use of a pseudo-spectral technique (Cambanis and Soltani (1982)). Here we shall consider just one criterion, namely minimization of the error dispersion (see (13.3.1)). Using (13.3.11) we can rewrite $\hat{X}_{n+1}$ in the form

$$\hat{X}_{n+1} = \sum_{j=0}^{\infty} (a_{n1}\psi_j + a_{n2}\psi_{j-1} + \cdots + a_{nn}\psi_{j-n+1})Z_{n-j}, \qquad (13.3.20)$$

and using (13.3.11) again we obtain

$$\begin{aligned} X_{n+1} - \hat{X}_{n+1} \\ = Z_{n+1} + \sum_{j=0}^{\infty} (\psi_{j+1} - a_{n1}\psi_j - a_{n2}\psi_{j-1} - \cdots - a_{nn}\psi_{j-n+1})Z_{n-j}. \end{aligned} \qquad (13.3.21)$$

Since $\{Z_t\}$ is assumed to have dispersion C, it follows from Remark 2 that

$$\operatorname{disp}(X_{n+1} - \hat{X}_{n+1}) = C\left(1 + \sum_{j=0}^{\infty} |\psi_{j+1} - a_{n1}\psi_j - \cdots - a_{nn}\psi_{j-n+1}|^\alpha\right). \qquad (13.3.22)$$

In the special case when Z_t has the symmetric stable distribution with exponent $\alpha \in (0, 2)$ and scale parameter $d^{1/\alpha}$ (i.e. $Ee^{i\theta Z_t} = \exp(-d|\theta|^\alpha)$), the dispersion of Z_t (see Property 6) is $C = d/[\Gamma(1 - \alpha)\cos(\pi\alpha/2)]$, $\alpha \neq 1$, and $C = 2d/\pi$, $\alpha = 1$. The prediction error is also symmetric stable with dispersion given by (13.3.22). Minimization of (13.3.22) is therefore equivalent to minimization of the scale parameter of the error distribution and hence to minimization of $P(|X_{n+1} - \hat{X}_{n+1}| > \varepsilon)$ for every $\varepsilon > 0$. The minimum dispersion criterion is useful also in regression problems (Blattberg and Sargent (1971)) and Kalman filtering problems (Stuck (1978)) associated with stable sequences. For general sequences $\{Z_t\}$ satisfying (13.3.1) the minimum dispersion criterion minimizes the tail probabilities of the distribution of the prediction error.

The minimization of (13.3.22) for $\alpha \in (0, 2)$ is rather more complicated than in the case $\alpha = 2$ and the best predictor is not in general unique. For a general discussion of this problem (and the related problem of finding h-step predictors) see Cline and Brockwell (1985). Here we shall simply state the results for an MA(1) process and, when Z_t has a Cauchy distribution, compare the minimum dispersion predictor $\hat{X}_{n+1} = \sum_{j=1}^n a_{nj} X_{n+1-j}$ with the predictor $X_n^* = \sum_{j=1}^n \phi_{nj} X_{n+1-j}$ obtained by assuming that $\{Z_t\}$ has finite variance.

Proposition 13.3.3. *If $X_t = Z_t + \theta Z_{t-1}$ where $\{Z_t\}$ is an iid sequence with distribution function satisfying* (13.3.1), *then the minimum dispersion linear predictor $\hat{X}_{n+1}$ of X_{n+1} based on $X_1, \ldots, X_n$ is*

$$\hat{X}_{n+1} = -\sum_{j=1}^{n} (-\theta)^j X_{n+1-j} \quad \text{if } \alpha \le 1,$$

$$\hat{X}_{n+1} = -\sum_{j=1}^{n} (-\theta)^j \frac{1 - \eta^{n+1-j}}{1 - \eta^{n+1}} X_{n+1-j} \quad \text{if } \alpha > 1,$$

where $\eta = |\theta|^{\alpha/(\alpha-1)}$. The error dispersion of $\hat{X}_{n+1}$ is

$$C[1 + |\theta|^{(n+1)\alpha}] \quad \text{if } \alpha \le 1,$$

$$C\left[1 + |\theta|^{(n+1)\alpha}\left(\frac{1 - \eta}{1 - \eta^n}\right)^{\alpha-1}\right] \quad \text{if } \alpha > 1.$$

The minimum dispersion h-step predictor, $h \ge 1$, is zero with error dispersion $C[1 + |\theta|^\alpha]$.

Proof. See Cline and Brockwell (1985). □

Example 13.3.3 (Linear Prediction of a Cauchy MA(1) Process). Suppose that

$$X_t = Z_t + \theta Z_{t-1}, \qquad |\theta| < 1, \tag{13.3.23}$$

where $\{Z_t\}$ is an iid standard Cauchy sequence, i.e. $Ee^{iuZ_t} = e^{-|u|}$. Then condition (13.3.1) is satisfied with $p = q = \frac{1}{2}$ and $C = 2/\pi$. By Proposition

13.3.3, the minimum dispersion one-step predictor is

$$\hat{X}_{n+1} = -\sum_{j=1}^{n} (-\theta)^j X_{n+1-j}, \tag{13.3.24}$$

with corresponding error dispersion,

$$\operatorname{disp}(X_{n+1} - \hat{X}_{n+1}) = \frac{2}{\pi}(1 + |\theta|^{n+1}). \tag{13.3.25}$$

If now we imagine $\{Z_t\}$ in (13.3.23) to have finite variance and compute the best linear mean square predictor X_{n+1}^*, we find from Problem 3.10 that

$$(1 - \theta^{2n+2})X_{n+1}^* = -\sum_{j=1}^{n} [(-\theta)^j - (-\theta)^{2n+2-j}]X_{n+1-j}, \tag{13.3.26}$$

and hence that

$$\begin{aligned}(1 - \theta^{2n+2})&(X_{n+1} - X_{n+1}^*)\\ &= (1 - \theta^{2n+2})Z_{n+1} - (-\theta)^{n+1}(1 - \theta^2)\sum_{j=0}^{n} (-\theta)^j Z_j.\end{aligned} \tag{13.3.27}$$

From (13.3.27) we can easily compute the error dispersion when the mean-square linear predictor X_{n+1}^* is applied to the Cauchy process (13.3.23). We find that

$$\operatorname{disp}(X_{n+1} - X_{n+1}^*) = \frac{2}{\pi}\left(1 + |\theta|^{n+1}\frac{1 + |\theta|}{1 + |\theta|^{n+1}}\right), \tag{13.3.28}$$

which is clearly greater than the dispersion of $(X_{n+1} - \hat{X}_{n+1})$ in (13.3.25).

The minimum dispersion linear predictor of X_{n+1} based on $\{X_j, -\infty < j \le n\}$ turns out to be the same (for a causal invertible ARMA process) as the best linear mean square predictor computed on the assumption that $\{Z_t\}$ has finite variance. The dispersion of the one-step prediction error is just the dispersion of $\{Z_t\}$ ($2/\pi$ in Example 13.3.3).

Although we have only considered *linear* prediction in this section, we should not forget the potential for improved prediction of infinite variance (and finite variance) processes using predictors which are non-linear in the observations. In the next section we give a brief introduction to non-linear time-series models, with particular reference to one of the families of non-linear models ("threshold models") which have been found useful in practice.

§13.4 Threshold Models

Linear processes of the form

$$X_t = \sum_{j=0}^{\infty} \psi_j Z_{t-j}, \qquad \{Z_t\} \sim \text{IID}(0, \sigma^2), \tag{13.4.1}$$

where $Z_t \in \mathcal{M}_t = \overline{\text{sp}}\{X_s, -\infty < s \le t\}$, play an important role in time series analysis since for such processes the best mean square predictor, $E(X_{t+h}|X_s, -\infty < s \le t)$ and the best linear predictor, $P_{\mathcal{M}_t}X_{t+h}$, are identical. (In fact for the linear process (13.4.1) with $\{Z_t\} \sim \text{WN}(0, \sigma^2)$, the two predictors are identical if and only if $\{Z_t\}$ is a martingale difference sequence relative to $\{X_t\}$, i.e. if and only if $E(Z_{t+1}|X_s, -\infty < s \le t) = 0$ for all t (see Problem 13.11).) The Wold decomposition (Section 5.7) ensures that any purely non-deterministic stationary process can be expressed in the form (13.4.1) with $\{Z_t\} \sim \text{WN}(0, \sigma^2)$, but the process $\{Z_t\}$ is generally not an iid sequence and the best mean square predictor of X_{t+h} may be quite different from the best linear predictor. However, in the case when $\{X_t\}$ is a purely non-deterministic Gaussian stationary process, the sequence $\{Z_t\}$ in the Wold decomposition is Gaussian and therefore iid. Every stationary purely non-deterministic Gaussian process can therefore be generated by applying a *causal* linear filter to an iid Gaussian sequence. We shall therefore refer to such processes as Gaussian linear processes. They have the desirable property (like the more general linear process (13.4.1)) that $P_{\mathcal{M}_t}X_{t+h} = E(X_{t+h}|X_s, -\infty < s \le t)$.

Many of the time series encountered in practice exhibit characteristics not shown by linear Gaussian processes and so in order to obtain good models and predictors for such series it is necessary to relax either the Gaussian or the linear assumption. In the previous section we examined a class of non-Gaussian (infinite variance) linear processes. In this section we shall provide a glimpse of the rapidly expanding area of non-linear time series modelling and illustrate this with a threshold model proposed by Tong (1983) for the lynx data (Series G, Appendix A).

Properties of Gaussian linear processes which are sometimes found to be violated by observed time series are the following. A Gaussian linear process $\{X_t\}$ is reversible in the sense that $(X_{t_1}, \dots, X_{t_n})'$ has the same distribution as $(X_{t_n}, \dots, X_{t_1})'$. (Except in a few special cases, linear, and hence ARMA processes, are *reversible* if and only if they are Gaussian (Weiss (1975), Breidt and Davis (1990)).) Deviations from this property are suggested by sample-paths which rise to their maxima and fall away at different rates (see, for example, the Wölfer sunspot numbers, Figure 1.5, and the logarithms to base 10 of the lynx data, Figure 13.11). Gaussian linear processes do not exhibit sudden bursts of outlying values as are sometimes observed in practice. Such behaviour can however be shown by non-linear processes (and by processes with infinite variance). Other characteristics suggesting deviation from a Gaussian linear model are discussed by Tong (1983).

If we restrict attention to second order properties of a time series, it will clearly not be possible to decide on the appropriateness or otherwise of a Gaussian linear model. To resolve this question we consider moments of order greater than two.

Let $\{X_t\}$ be a process which, for some $k \ge 3$, satisfies $\sup_t E|X_t|^k < \infty$ and

$$E(X_{t_0}X_{t_1}\cdots X_{t_j}) = E(X_{t_0+h}X_{t_1+h}\cdots X_{t_j+h}),$$

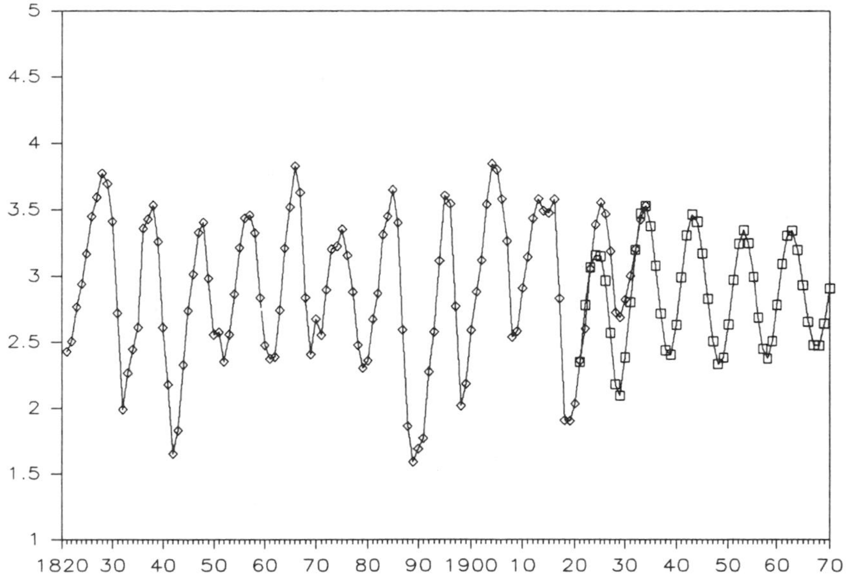

Figure 13.11. The logarithms to base 10 of the Canadian lynx series (1821–1934), showing 50 predicted values based on the observations up to 1920 and the autoregressive model (13.4.3).

for all $t_0, t_1, \ldots, t_j, h \in \{0, \pm 1, \ldots\}$ and for all $j \in \{0, 1, \ldots, k-1\}$. The kth order cumulant $C_k(r_1, \ldots, r_{k-1})$ of $\{X_t\}$ is then defined as the joint cumulant of the random variables, $X_t, X_{t+r_1}, \ldots, X_{t+r_{k-1}}$, i.e. as the coefficient of $i^k z_1 z_2 \cdots z_k$ in the Taylor expansion about $(0, \ldots, 0)$ of

$$\chi(z_1, \ldots, z_k) := \ln E[\exp(iz_1 X_t + iz_2 X_{t+r_1} + \cdots + iz_k X_{t+r_{k-1}})].$$

In particular, the third order cumulant function C_3 of $\{X_t\}$ coincides with the third order central moment function, i.e.

$$C_3(r, s) = E[(X_t - \mu)(X_{t+r} - \mu)(X_{t+s} - \mu)], \qquad r, s \in \{0, \pm 1, \ldots\}, \tag{13.4.2}$$

where $\mu = EX_t$. If $\sum_r \sum_s |C_3(r, s)| < \infty$, we define the third order polyspectral density (or bispectral density) of $\{X_t\}$ to be the Fourier transform,

$$f_3(\omega_1, \omega_2) = \frac{1}{(2\pi)^2} \sum_{r=-\infty}^{\infty} \sum_{s=-\infty}^{\infty} C_3(r, s) e^{-ir\omega_1 - is\omega_2}, \qquad -\pi \le \omega_1, \omega_2 \le \pi,$$

in which case

$$C_3(r, s) = \int_{-\pi}^{\pi} \int_{-\pi}^{\pi} e^{ir\omega_1 + is\omega_2} f_3(\omega_1, \omega_2)\, d\omega_1\, d\omega_2.$$

[More generally, if the kth order cumulants $C_k(r_1, \ldots, r_{k-1})$, of $\{X_t\}$ are absolutely summable, we define the kth order polyspectral density as the

Fourier transform of C_k. For details see Rosenblatt (1985) and Priestley (1988).]

If $\{X_t\}$ is a Gaussian linear process, it follows from Problem 13.12 that the cumulant function C_3 of $\{X_t\}$ is identically zero. (The same is also true of all the cumulant functions C_k with $k > 3$.) Consequently $f_3(\omega_1, \omega_2) = 0$ for all $\omega_1, \omega_2 \in [-\pi, \pi]$. Appropriateness of a Gaussian linear model for a given data set can therefore be checked by using the data to test the null hypothesis, $f_3 = 0$. For details of such a test, see Subba-Rao and Gabr (1984).

If $\{X_t\}$ is a linear process of the form (13.4.1) with $E|Z_t|^3 < \infty$, $EZ_t^3 = \eta$ and $\sum_{j=0}^{\infty} |\psi_j| < \infty$, it can be shown from (13.4.2) (see Problem 13.12) that the third order cumulant function of $\{X_t\}$ is given by

$$C_3(r, s) = \eta \sum_{i=-\infty}^{\infty} \psi_i \psi_{i+r} \psi_{i+s}, \tag{13.4.3}$$

(with $\psi_j = 0$ for $j < 0$) and hence that $\{X_t\}$ has bispectral density,

$$f_3(\omega_1, \omega_2) = \frac{\eta}{4\pi^2} \psi(e^{i(\omega_1+\omega_2)})\psi(e^{-i\omega_1})\psi(e^{-i\omega_2}), \tag{13.4.4}$$

where $\psi(z) := \sum_{j=0}^{\infty} \psi_j z^j$. By Theorem 4.4.1, the spectral density of $\{X_t\}$ is

$$f(\omega) = \frac{\sigma^2}{2\pi} |\psi(e^{-i\omega})|^2.$$

Hence

$$\phi(\omega_1, \omega_2) := \frac{|f_3(\omega_1, \omega_2)|^2}{f(\omega_1)f(\omega_2)f(\omega_1 + \omega_2)} = \frac{\eta^2}{2\pi\sigma^6}.$$

Appropriateness of the linear process (13.4.1) for modelling a given data set can therefore be checked by using the data to test for constancy of $\phi(\omega_1, \omega_2)$ (see Subba-Rao and Gabr (1984)).

If it is decided that a linear Gaussian model is not appropriate, there is a choice of several families of non-linear processes which have been found useful for modelling purposes. These include bilinear models, autoregressive models with random coefficients and threshold models. Excellent accounts of these are available in the books of Subba-Rao and Gabr (1984), Nicholls and Quinn (1982) and Tong (1990) respectively.

Threshold models can be regarded as piecewise linear models in which the linear relationship varies with the values of the process. For example if $R^{(i)}$, $i = 1, \ldots, k$, is a partition of $\mathbb{R}^p$, and $\{Z_t\} \sim \text{IID}(0, 1)$, then the k difference equations,

$$X_t = \sigma^{(i)} Z_t + \sum_{j=1}^{p} \phi_j^{(i)} X_{t-j}, \qquad (X_{t-1}, \ldots, X_{t-p})' \in R^{(i)}, \qquad i = 1, \ldots, k, \tag{13.4.5}$$

defined a threshold AR(p) model. Model identification and parameter estimation for threshold models can be carried out in a manner similar to that for linear models using maximum likelihood and the AIC criterion. Details can be found in the book of Tong (1990). It is sometimes useful to express threshold models in state-space form (cf. Section 12.1). For example, the model (13.4.5) can be re-expressed as

$$X_t = [0 \quad 0 \quad 0 \quad \cdots \quad 1]\mathbf{S}_t,$$

where $\mathbf{S}_t := (X_{t-p+1}, X_{t-p+2}, \ldots, X_t)'$ satisfies the state equation,

$$\mathbf{S}_{t+1} = F_t\mathbf{S}_t + H_t Z_{t+1}.$$

This state representation differs from those in Chapter 12 in that the matrices F_t and H_t now depend on $\mathbf{S}_t$. Thus

$$F_t = \begin{bmatrix} 0 & 1 & 0 & \cdots & 0 \\ 0 & 0 & 1 & \cdots & 0 \\ \vdots & \vdots & \vdots & & \vdots \\ 0 & 0 & 0 & \cdots & 1 \\ \phi_p^{(i)} & \phi_{p-1}^{(i)} & \phi_{p-2}^{(i)} & \cdots & \phi_1^{(i)} \end{bmatrix}, \qquad H_t = \begin{bmatrix} 0 \\ 0 \\ \vdots \\ 0 \\ \sigma^{(i)} \end{bmatrix}, \qquad \text{if } \mathbf{S}_i \in R^{(i)}.$$

As an illustration of the use of threshold models, Tong identifies the following model for the logarithm to base 10 of the lynx data (1821–1920):

$$X_t = \begin{cases} .802 + 1.068X_{t-1} - .207X_{t-2} + .171X_{t-3} - .453X_{t-4} \\ \qquad + .224X_{t-5} - .033X_{t-6} + .174Z_t, \qquad X_{t-2} \le 3.05, \\ 2.296 + 1.425X_{t-1} - 1.080X_{t-2} - .091X_{t-3} + .237Z_t, \qquad X_{t-2} > 3.05, \end{cases} \tag{13.4.6}$$

where $\{Z_t\} \sim \text{IID}(0, 1)$. The model (13.4.6) is said to have a delay parameter $b = 2$ since the form of the equation specifying X_t is dependent on the value of X_{t-2}. It is easy to compute the best mean square predictor $E(X_{n+h}|X_t, t \le n)$ for the model (13.4.6) if $h = 1$ or $h = 2$ but not if $h > 2$ (see Problem 13.13). More generally, if the delay parameter is b, computation of the best predictor is easy if $h < b$ but is extremely complicated otherwise. A natural approximation procedure for computing the best predictors given $X_1, \ldots, X_n$ is to set $Z_t = 0$, $t \ge n + 1$, in the recursions defining the process and then to solve recursively for $X_{n+1}, X_{n+2}, \ldots$. Tong (1983), p. 187, refers to the predictors obtained in this way as the values of the *eventual forecast function*. For the logarithms of the lynx data and the model (13.4.6) the eventual forecast function exhibits a stable limit cycle of period 9 years with values (2.61, 2.67, 2.82, 3.02, 3.25, 3.41, 3.37, 3.13, 2.80). An alternative technique suggested by Tong for computing h-step predictors when $h > b$ and the data is nearly cyclic with period T is to fit a new model with delay parameter $kT + b$ where k is a positive integer. Under the new model, prediction can then be carried out for values of h up to $kT + b$. The most satisfactory general procedure for forecasting threshold models is to simulate future values of the process using the fitted model and the observed data $\{X_1, \ldots, X_n\}$. From N simulated values of X_{n+h} we can construct a histogram which estimates the conditional density of X_{n+h} given the data. This procedure is implemented in the software package STAR of H. Tong (see Tong (1990)).

It is interesting to compare the non-linear model (13.4.6) for the logarithms of the lynx data with the minimum AICC autoregressive model found for the same series using the program PEST, namely

$$\begin{aligned} X_t = 1.123 &+ 1.084X_{t-1} - .477X_{t-2} + .265X_{t-3} - .218X_{t-4} \\ &+ .180X_{t-9} - .224X_{t-12} + Z_t, \qquad \{Z_t\} \sim \mathrm{WN}(0, .0396) \end{aligned} \tag{13.4.7}$$

The best linear mean-square h-step predictors, $h = 1, 2, \ldots, 50$, for the years 1921–1970 were found from (13.4.7). They are shown with the observed values of the series (1821–1934) in Figure 13.11. As can be seen from the graph, the h-step predictors execute slowly damped oscillations about the mean (2.880) of the first 100 observations. As $h \to \infty$ the predictors converge to 2.880.

Figures 13.12 and 13.13 show respectively 150 simulated values of the processes (13.4.7) and (13.4.6). Both simulated series exhibit the approximate 9 year cycles of the data itself.

In Table 13.3 we show the last 14 observed values of X_t with the corresponding one-step predictors $\hat{X}_t$ based on (13.4.7) and the predictors $\tilde{X}_t = E(X_t | X_s, s < t)$ based on (13.4.6).

The relative performance of the predictors can be assessed by computing $s = (S/14)^{1/2}$ where S is the sum of squares of the prediction errors for

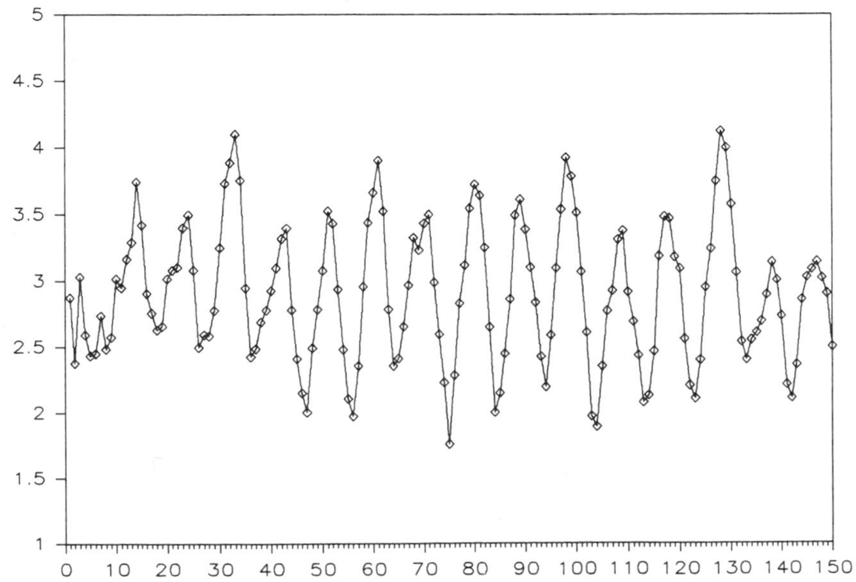

Figure 13.12. One hundred and fifty simulated values of the autoregressive model (13.4.7) for the transformed lynx series.

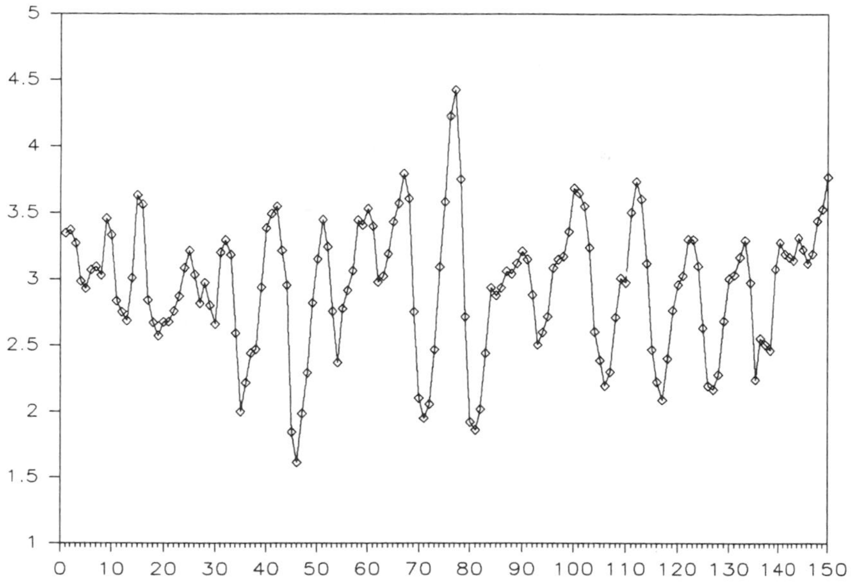

Figure 13.13. One hundred and fifty simulated values of the threshold model (13.4.6) for the transformed lynx series.

Table 13.3. The Transformed Lynx Data $\{X_t, t = 101, \dots, 114\}$ with the One-Step Predictors, $\hat{X}_t$, Based on the Autoregressive Model (13.4.7), and $\tilde{X}_t$, Based on the Threshold Model (13.4.6)

t	X_t	$\hat{X}_t$	$\tilde{X}_t$
101	2.360	2.349	2.311
102	2.601	2.793	2.877
103	3.054	2.865	2.911
104	3.386	3.231	3.370
105	3.553	3.354	3.588
106	3.468	3.329	3.426
107	3.187	2.984	3.094
108	2.724	2.668	2.771
109	2.686	2.432	2.422
110	2.821	2.822	2.764
111	3.000	2.969	2.940
112	3.201	3.242	3.246
113	3.424	3.406	3.370
114	3.531	3.545	3.447

$t = 101, \ldots, 114$. From the table we find that the value of s for the autoregressive predictor $\hat{X}_t$ is .138, while for the threshold predictor $\tilde{X}_t$, the values of s is reduced to .120.

A bilinear model for the log lynx data can be found on p. 204 of Subba-Rao and Gabr (1984) and an AR(2) model with random coefficients on p.143 of Nicholls and Quinn (1982). The values of s for predictors based on these models are .115 and .116 respectively. These values indicate the improvements attainable in this example by consideration of non-linear models.

Problems

13.1. Suppose that $\{X_{t1}\}$ and $\{X_{t2}\}$ are related by the transfer function model,

$$X_{t2} = \frac{2.5B}{1 - .7B} X_{t1} + W_t,$$

$$(1 - .8B)X_{t1} = Z_t,$$

where $\{Z_t\} \sim \text{WN}(0, 1)$, $\{W_t\} \sim \text{WN}(0, 2)$, and $\{Z_t\}$ and $\{W_t\}$ are uncorrelated.

(a) Write down a state-space model for $\{(X_{t1}, X_{t2})'\}$.

(b) If $W_{100} = 1.3$, $X_{100,2} = 2.4$, $X_{100,1} = 3.15$, find the best linear predictors of $X_{101,2}$ and $X_{102,2}$ based on $\{X_{t1}, X_{t2}, t \leq 100\}$.

(c) Find the mean squared errors of the predictors in (b).

13.2. Show that the output, $\{X_{t2}\}$, of a transfer function model satisfies the equation (13.1.15).

13.3. Consider the transfer function model

$$X_{t2} = \frac{B^2}{1 - .5B} X_{t1} + W_t,$$

$$(1 + .5B)X_{t1} = Z_t,$$

where $\{Z_t\}$ and $\{W_t\}$ are uncorrelated WN(0, 1) sequences. Let

$$P_t = P_{\overline{\text{sp}}\{X_{s1}, X_{s2}, -\infty < s \leq t\}},$$

$$Q_t = P_{\overline{\text{sp}}\{X_{s1}, -\infty < s \leq t\}},$$

$$R_t = P_{\overline{\text{sp}}\{X_{s2}, -\infty < s \leq t\}}.$$

(a) Show that $P_n X_{n+1,1} = Q_n X_{n+1,1}$ and hence evaluate $P_n X_{n+1,1}$.

(b) Express $P_n X_{n+1,2}$ and $P_n X_{n+2,2}$ in terms of $X_{j1}, X_{j2}, j \leq n$ and W_n.

(c) Evaluate $E(X_{n+1,2} - P_n X_{n+1,2})^2$, $E(X_{n+2,2} - P_n X_{n+2,2})^2$.

(d) Show that the univariate process $\{X_{t2}\}$ has the autocovariance function of an ARMA process and specify the process. (Hint: consider $(1 - .25B^2)X_{t2}$.)

(e) Use (d) to compute $E(X_{n+1,2} - R_n X_{n+1,2})^2$.

13.4. Verify the calculations of $\Pi_{n,1}$ in Example 13.1.1 (cont.).

13.5. Find a transfer function model relating the input and output series X_{t1}, X_{t2}, $t = 1, \ldots, 200$ (Series J and K respectively in the Appendix). Use the model to predict $X_{201,2}$, $X_{202,2}$ and $X_{203,2}$. Compare the predictors and their mean squared errors with the corresponding predictors and mean squared errors obtained by modelling $\{X_{t2}\}$ as a univariate ARMA process and with the results of Problem 11.10.

13.6. If $\{X_t\}$ is the causal invertible ARIMA(p, d, q) process defined by (13.2.23) and $\bar{X}_n = n^{-1}(X_1 + \cdots + X_n)$, show that

$$n^{1-2d}E(\bar{X}_n - \mu)^2 \to C,$$

where C is a positive constant.

13.7. Verify the properties (13.3.3) for a random variable Z_1 whose distribution satisfies (13.3.1).

13.8. Show that the linear process (13.3.8) is strictly stationary if $\{\psi_j\}$ and $\{Z_t\}$ satisfy the conditions of Proposition 13.3.1.

13.9. Prove Proposition 13.3.2. (Note that Proposition 13.3.1 remains valid for strictly stationary sequences $\{Z_t\}$ satisfying (13.3.1), and use arguments similar to those used in the proofs of Theorems 3.1.1 and 3.1.2.)

13.10. Modify Example 13.3.3 by supposing that $\{Z_t\}$ is iid with $Ee^{iuZ_t} = e^{-|u|^\alpha}$, $u \in \mathbb{R}$, $0 < \alpha < 2$. Use Proposition 13.3.3 to show that for each fixed sample size n and coefficient $\theta \in (-1, 1)$, the ratio of the error dispersion of X^*_{n+1} (see (13.3.28)) to that of $\hat{X}_{n+1}$, converge as $\alpha \to 0$ to $1 + n/2$.

13.11. Let $\{X_t\}$ be the process,

$$X_t = \sum_{j=0}^{\infty} \psi_j Z_{t-j}, \quad \{Z_t\} \sim \mathrm{WN}(0, \sigma^2),$$

where $\psi_0 \neq 0$, $\sum_{j=0}^{\infty} \psi_j^2 < \infty$ and $Z_t \in \mathcal{M}_t = \overline{\mathrm{sp}}\{X_s, -\infty < s \le t\}$. Show that

$$E(X_{t+1} | X_s, -\infty < s \le t) = P_{\mathcal{M}_t} X_{t+1}$$

if and only if $\{Z_t\}$ is a martingale difference sequence, i.e. if and only if $E(Z_{t+1} | X_s, -\infty < s \le t) = 0$ for all t.

13.12. If $\{X_t\}$ is the linear process (13.4.1) with $\{Z_t\} \sim \mathrm{IID}(0, \sigma^2)$ and $\eta = EZ_t^3$, show that the third order cumulant function of $\{X_t\}$ is given by

$$C_3(r, s) = \eta \sum_{i=-\infty}^{\infty} \psi_i \psi_{i+r} \psi_{i+s}.$$

Use this result to establish equation (13.4.4). Conclude that if $\{X_t\}$ is a Gaussian linear process, then $C_3(r, s) \equiv 0$ and $f_3(\omega_1, \omega_2) \equiv 0$.

13.13. Evaluate the best mean square predictors $E(X_{t+h} | X_s, -\infty < s \le t)$, $h = 1, 2$ for the threshold model (13.4.6).

APPENDIX

Data Sets

All of the following data sets are listed by columns.

Series A. Level of Lake Huron in Feet (Reduced by 570), 1875–1972

10.38	11.17	9.35	8.19	8.05	6.84	9.38	7.13	9.89
11.86	10.53	8.82	8.67	7.79	6.85	9.10	9.10	9.96
10.97	10.01	9.32	9.55	6.75	6.90	7.95	8.25	
10.80	9.91	9.01	8.92	6.75	7.79	8.12	7.91	
9.79	9.14	9.00	8.09	7.82	8.18	9.75	6.89	
10.39	9.16	9.90	9.37	8.64	7.51	10.85	5.96	
10.42	9.55	9.83	10.13	10.58	7.23	10.41	6.80	
10.82	9.67	9.72	10.14	9.48	8.42	9.96	7.68	
11.40	8.44	9.89	9.51	7.38	9.61	9.61	8.38	
11.32	8.24	10.01	9.24	6.90	9.05	8.76	8.52	
11.44	9.10	9.37	8.66	6.94	9.26	8.18	9.74	
11.68	9.09	8.69	8.86	6.24	9.22	7.21	9.31	

Series B. Dow Jones Utilities Index, Aug. 28–Dec. 18, 1972

110.94	109.60	109.53	112.06	114.65	119.70	124.11	123.18
110.69	109.31	109.89	111.96	115.06	119.28	124.14	122.67
110.43	109.31	110.56	111.68	115.86	119.66	123.37	122.73
110.56	109.25	110.56	111.36	116.40	120.14	123.02	122.86
110.75	109.02	110.72	111.42	116.44	120.97	122.86	122.67
110.84	108.54	111.23	112.00	116.88	121.13	123.02	122.09
110.46	108.77	111.48	112.22	118.07	121.55	123.11	122.00
110.56	109.02	111.58	112.70	118.51	121.96	123.05	121.23
110.46	109.44	111.90	113.15	119.28	122.26	123.05	
110.05	109.38	112.19	114.36	119.79	123.79	122.83	

Series C. Private Housing Units Started, U.S.A. (Monthly). [Makridakis 922]

1361	1614	1361	1086	1380	1368	1828	2252
1278	1639	1433	1119	1520	1358	1741	2382
1443	1763	1423	1046	1466	1507	1910	2481
1524	1779	1438	843	1554	1381	1986	2485
1483	1622	1478	961	1408	1229	2049	2421
1404	1491	1488	990	1405	1327	2026	2366
1450	1603	1529	1067	1512	1085	2083	2481
1517	1820	1432	1123	1495	1305	2158	2289
1324	1517	1482	1056	1556	1319	2041	2365
1533	1448	1452	1091	1569	1264	2128	2084
1622	1467	1460	1304	1630	1290	2182	2266
1564	1550	1656	1248	1548	1385	2295	2067
1244	1562	1370	1364	1769	1517	2494	2123
1456	1569	1378	1407	1705	1399	2390	2051
1534	1455	1394	1421	1561	1534	2334	1874
1689	1524	1352	1491	1524	1580	2249	1677
1641	1486	1265	1538	1583	1647	2221	1724
1588	1484	1194	1308	1528	1893	2254	1526

Series D. Industrial Production, Austria (Quarterly). [Makridakis 337]

54.1	59.6	66.5	72.3	76.0	93.1	105.5	122.6
59.5	63.6	72.0	78.1	85.1	103.5	115.4	131.9
56.5	60.4	67.8	72.4	80.5	96.4	108.0	120.5
63.9	66.3	75.6	82.6	89.1	107.2	129.9	130.7
57.8	60.6	69.2	72.9	84.8	101.7	112.4	115.7
62.0	66.8	74.1	79.5	94.2	109.5	123.6	119.7
58.5	63.2	70.7	72.6	89.5	101.3	114.9	109.7
65.0	71.0	77.8	82.8	99.3	112.6	131.0	125.1

Series E. Industrial Production, Spain (Monthly). [Makridakis 868]

128	108	156	132	101	152	161	123
134	142	160	139	141	153	160	175
133	146	161	139	122	152	167	175
141	149	149	137	145	153	178	176
134	141	118	144	148	113	167	174
142	156	147	146	137	151	176	
143	151	158	149	137	159	173	
136	160	146	142	155	165	164	

Series F. General Index of Industrial Production (Monthly). [Makridakis 904]

96	110	135	147	160	184	175	205
97	111	138	115	161	179	182	208
99	114	132	150	167	181	182	216
100	113	106	152	167	179	185	216
102	109	134	158	164	187	191	210
106	91	132	154	165	185	191	169
102	116	136	155	173	183	188	217
80	118	132	159	179	177	139	213
104	123	133	160	181	176	189	220
104	121	140	163	182	130	190	217
107	119	142	167	175	176	199	
102	125	146	160	131	175	193	
100	131	148	162	183	181	195	
107	132	149	126	181	176	200	

Series G. Annual Canadian Lynx Trappings, 1821–1934

269	2285	377	6721	469	3495	1388	2935
321	2685	225	4254	736	587	2713	1537
585	3409	360	687	2042	105	3800	529
871	1824	731	255	2811	153	3091	485
1475	409	1638	473	4431	387	2985	662
2821	151	2725	358	2511	758	3790	1000
3928	45	2871	784	389	1307	674	1590
5943	68	2119	1594	73	3465	81	2657
4950	213	684	1676	39	6991	80	3396
2577	546	299	2251	49	6313	108	
523	1033	236	1426	59	3794	229	
98	2129	245	756	188	1836	399	
184	2536	552	299	377	345	1132	
279	957	1623	201	1292	382	2432	
409	361	3311	229	4031	808	3574	

Series H. Annual Mink Trappings, 1848–1911

37123	61581	61727	39266	35072	83023	70229	54673
34712	61951	60334	44740	36160	40748	76365	55996
29619	76231	51404	60429	45600	35396	70407	60053
21151	63264	58451	72273	47508	29479	41839	39169
24859	44730	73575	79214	52290	42264	45978	21534
25152	31094	74343	79060	110824	58171	47813	17857
42375	49452	27708	84244	76503	50815	57620	21788
50839	43961	31985	62590	64303	51285	66549	33008

Series I. Annual Muskrat Trappings, 1848–1911

224347	258806	509769	703789	478078	344878	813159	924439
179075	302267	418370	767896	829034	223614	551716	1056253
175472	313502	320824	671982	1029296	322160	568934	695070
194682	254246	412164	523802	1069183	574742	701487	407472
292530	177291	618081	583319	1083067	806103	767741	172418
493952	206020	414173	437121	817003	934646	928199	302195
512291	335385	232251	486030	347050	648687	1650214	749142
345626	357060	443999	499727	380132	674811	1488287	963597

Series J. Simulated Input Series

6.94	5.98	3.31	3.69	4.81	5.06	4.70	0.87
3.48	5.07	5.56	7.40	5.46	4.93	6.29	9.26
5.92	6.43	5.06	3.18	5.01	6.00	1.98	1.29
5.11	3.64	4.28	6.92	3.05	5.59	6.71	7.59
7.00	6.84	3.89	3.40	6.48	6.24	5.39	3.90
2.23	1.49	7.47	7.65	6.35	3.79	5.73	4.37
5.78	7.39	3.60	3.42	3.86	4.51	5.55	6.00
5.26	2.36	8.88	5.91	5.01	6.00	5.88	3.56
5.19	5.24	1.26	3.91	6.49	4.76	4.27	5.92
4.83	4.63	7.78	6.99	3.43	5.55	4.34	4.38
4.48	6.87	2.76	2.15	7.70	4.74	4.26	6.24
5.98	2.81	9.23	7.17	2.13	5.61	4.98	6.57
4.45	6.27	2.95	3.94	8.47	6.52	3.38	3.05
5.73	4.98	6.38	3.87	2.41	5.95	6.14	6.43
6.38	3.45	4.35	7.61	5.93	4.38	5.45	4.03
2.47	4.90	5.33	3.73	4.69	5.95	5.92	5.86
7.43	4.11	5.00	5.08	5.29	5.80	2.62	4.82
2.73	5.40	4.87	6.61	6.44	4.87	7.68	2.90
5.90	6.38	6.74	1.94	5.14	6.47	3.37	5.65
5.19	3.94	3.37	6.17	5.93	4.01	3.90	5.14
5.08	4.96	6.56	5.73	3.95	5.57	7.01	3.17
4.28	6.35	3.45	4.14	5.87	6.30	1.24	6.73
3.64	2.75	4.54	5.76	4.64	3.34	9.03	4.43
6.01	6.03	4.95	3.56	7.25	6.38	1.08	3.99
3.08	4.76	4.34	5.44	5.00	4.73	7.86	4.46

Series K. Simulated Output Series

15.21	12.79	12.36	10.70	18.88	13.78	8.69	21.28
21.13	14.12	15.92	11.71	14.20	19.22	16.46	10.00
14.14	7.77	14.00	10.65	16.58	14.90	14.26	18.47
18.22	11.03	7.05	7.61	17.32	13.06	11.23	4.75
8.74	13.36	10.19	17.73	16.85	10.99	12.90	23.36
17.33	17.36	10.78	9.71	18.32	17.73	5.82	6.40
17.10	12.82	9.21	18.58	16.11	21.11	15.13	17.65
22.20	22.16	7.36	15.04	19.17	22.43	18.50	13.67
4.44	8.83	17.01	24.44	21.18	16.87	17.66	14.15
11.62	22.66	12.14	14.33	16.17	14.56	17.37	14.16
17.50	10.65	24.86	17.18	12.02	17.60	17.78	10.63
18.59	10.72	12.34	12.21	21.46	14.45	13.96	17.78
14.77	6.77	23.49	21.90	17.35	14.89	13.89	11.99
14.63	15.73	12.76	12.28	22.76	18.00	11.98	13.65
17.16	7.26	26.97	20.73	14.03	24.21	13.02	17.64
13.44	12.82	14.95	11.76	25.30	22.95	10.64	11.00
15.63	12.38	22.94	8.82	13.46	19.59	18.09	17.07
20.99	7.26	15.88	16.89	17.17	14.67	19.92	15.07
11.98	7.90	20.30	12.58	12.98	21.08	24.02	16.24
20.13	12.11	19.13	13.63	20.27	24.71	13.62	14.92
11.90	14.56	17.26	19.38	22.75	20.18	24.69	11.62
13.94	17.11	19.99	9.80	20.75	22.18	13.59	16.75
14.06	16.15	15.76	17.85	23.21	18.08	10.97	16.54
19.71	17.12	19.82	19.01	17.67	17.29	16.25	10.82
15.73	19.97	11.09	18.91	17.24	16.30	4.93	18.45

Bibliography

Ahlfors, L.V. (1953), *Complex Analysis*, McGraw-Hill, New York.

Akaike, H. (1969), Fitting autoregressive models for prediction, *Annals of the Institute of Statistical Mathematics*, Tokyo, *21*, 243–247.

Akaike, H. (1973), Information theory and an extension of the maximum likelihood principle, *2nd International Symposium on Information Theory*, B.N. Petrov and F. Csaki (eds.), Akademiai Kiado, Budapest, 267–281.

Akaike, H. (1978), Time series analysis and control through parametric models, *Applied Time Series Analysis*, D.F. Findley (ed.), Academic Press, New York.

Amos, D.E. and Koopmans, L.H. (1963), Tables of the distribution of coherences for stationary bivariate Gaussian processes, Sandia Corporation Monograph SCR-483.

Anderson, O.D. (1976), *Time Series Analysis and Forecasting. The Box–Jenkins Approach*, Butterworths, London.

Anderson, T.W. (1971), *The Statistical Analysis of Time Series*, John Wiley, New York.

Anderson, T.W. (1980), Maximum likelihood estimation for vector autoregressive moving average models, *Directions in Time Series*, D.R. Brillinger and G.C. Tiao (eds.), Institute of Mathematical Statistics, 80–111.

Ansley, C.F. (1979), An algorithm for the exact likelihood of a mixed autoregressive-moving average process, *Biometrika*, *66*, 59–65.

Ansley, C.F. and Kohn, R. (1985), On the estimation of ARIMA models with missing values, *Time Series Analysis of Irregularly Observed Data*, E. Parzen (ed.), Springer Lecture Notes in Statistics, 25, 9–37.

Aoki, M. (1987), *State Space Modelling of Time Series*, Springer-Verlag, Berlin.

Ash, R.B. (1972), *Real Analysis and Probability*, Academic Press, New York.

Ash, R.B. and Gardner, M.F. (1975), *Topics in Stochastic Processes*, Academic Press, New York.

Bartlett, M.S. (1955), *An Introduction to Stochastic Processes*, Cambridge University Press.

Bell, W. and Hillmer, S. (1990), Initializing the Kalman filter for non-stationary time series models, Research Report, U.S. Bureau of the Census.

Berk, K.N. (1974), Consistent autoregressive spectral estimates, *Ann. Statist.*, *2*, 489–502.

Billingsley, P. (1986), *Probability and Measure*, 2nd ed., Wiley-Interscience, New York.
Birkhoff, G. and Mac Lane, S. (1965), *A Survey of Modern Algebra*, MacMillan, New York.
Blattberg, R. and Sargent, T. (1971), Regression with non-Gaussian stable disturbances: Some sampling results, *Econometrica*, *39*, 501–510.
Bloomfield, P. (1976), *Fourier Analysis of Time Series: An Introduction*, John Wiley, New York.
Box, G.E.P. and Jenkins, G.M. (1970), *Time Series Analysis: Forecasting and Control*, Holden-Day, San Francisco.
Box, G.E.P. and Pierce, D.A. (1970), Distribution of residual autocorrelations in autoregressive-integrated moving average time series models. *J. Amer. Statist. Assoc.* *65*, 1509–1526.
Breidt, F.J. and Davis, R.A. (1991). Time-reversibility, identifiability and independence of innovations for stationary time series. *J. Time Series Analysis*, *13*, 377–390.
Breiman, L. (1968), *Probability*, Addison-Wesley, Reading, Massachusetts.
Brillinger, D.R. (1965), An introduction to polyspectra, *Annals of Math. Statist.*, *36*, 1351–1374.
Brillinger, D.R. (1981), *Time Series Analysis: Data Analysis and Theory*, Holt, Rinehart & Winston, New York.
Brillinger, D.R. and Rosenblatt, M. (1967), Asymptotic theory of estimates of kth order spectra, *Spectral Analysis of Time Series*, B. Harris (ed.), John Wiley, New York, 153–188.
Brillinger, D.R. and Rosenblatt, M. (1967), Computation and interpretation of kth order spectra, *Spectral Analysis of Time Series*, B. Harris (ed.), John Wiley, New York, 189–232.
Brockwell, P.J. and Davis, R.A. (1988a), Applications of innovation representations in time series analysis, *Probability and Statistics, Essays in Honor of Franklin A. Graybill*, J.N. Srivastava (ed.), Elsevier, Amsterdam, 61–84.
Brockwell, P.J. and Davis, R.A. (1988b), Simple consistent estimation of the coefficients of a linear filter, *Stoch. Processes and Their Applications*, *22*, 47–59.
Brockwell, P.J., Davis, R.A. and Salehi, H. (1990), A state-space approach to transfer function modelling, in *Inference from Stochastic Processes*, I.V. Basawa and N.U. Prabhu (eds).
Burg, J.P. (1967), Maximum entropy spectral analysis, 37th Annual International S.E.G. Meeting, Oklahoma City, Oklahoma.
Cambanis, S. and Soltani, A.R. (1982), Prediction of stable processes: Spectral and moving average representations, *Z. Wahrscheinlichkeitstheorie verw. Geb.*, *66*, 593–612.
Chatfield, C. (1984), *The Analysis of Time Series: An Introduction*, 3rd ed., Chapman and Hall, London.
Churchill, R.V. (1969), *Fourier Series and Boundary Value Problems*, McGraw-Hill, New York.
Cline, D.B.H. (1983), Estimation and linear prediction for regression, autoregression and ARMA with infinite variance data, Ph.D. Dissertation, Statistics Department, Colorado State University.
Cline, D.B.H. and Brockwell, P.J. (1985), Linear prediction of ARMA processes with infinite variance, *Stoch. Processes and Their Applications*, *19*, 281–296.
Cooley, J.W. and Tukey, J.W. (1965), An algorithm for the machine calculation of complex Fourier series, *Math. Comp.*, *19*, 297–301.
Cooley, J.W., Lewis, P.A.W. and Welch, P.D. (1967), Historical notes on the fast Fourier transform, *IEEE Trans. Electroacoustics*, *AU-15*, 76–79.

Davies, N., Triggs, C.M. and Newbold, P. (1977), Significance levels of the Box–Pierce portmanteau statistic in finite samples, *Biometrika*, *64*, 517–522.
Davis, H.F. (1963), *Fourier Series and Orthogonal Functions*, Allyn & Bacon, Boston.
Davis, M.H.A. and Vinter, R.B. (1985), *Stochastic Modelling and Control*, Chapman and Hall, London.
Davis, R.A. and Resnick, S.I. (1986), Limit theory for the sample covariance and correlation functions of moving averages, *Ann. Statist.*, *14*, 533–558.
Doob, J.L. (1953), *Stochastic Processes*, John Wiley, New York.
Dunsmuir, W. and Hannan, E.J. (1976), Vector linear time series models, *Adv. Appl. Prob.*, *8*, 339–364.
Duong, Q.P. (1984), On the choice of the order of autoregressive models: a ranking and selection approach, *J. Time Series Anal.*, *5*, 145–157.
Fama, E. (1965), Behavior of stock market prices, *J. Bus. U. Chicago*, *38*, 34–105.
Feller, W. (1971), *An Introduction to Probability Theory and Its Applications*, Vol. 2, 2nd ed., John Wiley, New York.
Fox, R. and Taqqu, M.S. (1986), Large sample properties of parameter estimates for strongly dependent stationary Gaussian time series, *Ann. Statist.*, *14*, 517–532.
Fuller, W.A. (1976), *Introduction to Statistical Time Series*, John Wiley, New York.
Gardner, G., Harvey, A.C. and Phillips, G.D.A. (1980), An algorithm for exact maximum likelihood estimation of autoregressive-moving average models by means of Kalman filtering, *Applied Statistics*, *29*, 311–322.
Gentleman, W.M. and Sande, G. (1966), Fast Fourier transforms for fun and profit, AFIPS, Proc. 1966 Fall Joint Computer Conference, Spartan, Washington, *28*, 563–578.
Geweke, J. and Porter-Hudak, S. (1983), The estimation and application of long-memory time series models, *J. Time Series Analysis*, *4*, 221–238.
Gihman, I.I. and Skorohod, A.V. (1974), *The Theory of Stochastic Processes I*, translated by S. Kotz, Springer-Verlag, Berlin.
de Gooijer, J.G., Abraham, B., Gould, A. and Robinson, L. (1985), Methods of determining the order of an autoregressive-moving average process: A survey, *Int. Statist, Review*, *53*, 301–329.
Gradshteyn, I.S. and Ryzhik, I.M. (1965), *Tables of Integrals, Series and Products*, Academic Press, New York.
Granger, C.W. (1980), Long memory relationships and the aggregation of dynamic models, *Econometrics*, *14*, 227–238.
Granger, C.W.J. and Andersen, A.P. (1978), Non-linear time series modelling, *Applied Time Series Analysis*, D.F. Findley (ed.), Academic Press, New York.
Granger, C.W. and Joyeux, R. (1980), An introduction to long-memory time series models and fractional differencing, *J. Time Series Analysis*, *1*, 15–29.
Gray, H.L., Kelley, G.D. and McIntire, D.D. (1978), A new approach to ARMA modelling, *Comm. Statist.*, *B 7*, 1–77.
Graybill, F.A. (1983), *Matrices with Applications in Statistics*, Wadsworth, Belmont, California.
Grenander, U. and Rosenblatt, M. (1957), *Statistical Analysis of Stationary Time Series*, John Wiley, New York.
Grunwald, G.K., Raftery, A.E. and Guttorp, P. (1989), Time series of continuous proportions, Statistics Department Report, 6, The University of Melbourne.
Hannan, E.J. (1970), *Multiple Time Series*, John Wiley, New York.
Hannan, E.J. (1973), The asymptotic theory of linear time series models, *J. Appl. Prob.*, *10*, 130–145.
Hannan, E.J. (1980). The estimation of the order of an ARMA process. *Ann. Statist.* *8*, 1071–1081.

Hannan, E.J. and Deistler, M. (1988), *The Statistical Theory of Linear Systems*, John Wiley, New York.

Harvey, A.C. (1984), A unified view of statistical forecasting procedures, *J. Forecasting*, *3*, 245–275.

Hosking, J.R.M. (1981), Fractional differencing, *Biometrika*, *68*, 165–176.

Hurst, H. (1951), Long-term storage capacity of reservoirs, *Trans. Amer. Soc. Civil Engrs.*, *116*, 778–808.

Hurvich, C.M. and Tsai, C.L. (1989). Regression and time series model selection in small samples. *Biometrika* *76*, 297–307.

Jagers, P. (1975), *Branching Processes with Biological Applications*, Wiley-Interscience, London.

Jones, R.H. (1975), Fitting autoregressions, *J. Amer. Statist. Assoc.*, *70*, 590–592.

Jones, R.H. (1980), Maximum likelihood fitting of ARMA models to time series with missing observations, *Technometrics*, *22*, 389–395.

Jones, R.H. (1985), Fitting multivariate models to unequally spaced data, *Time Series Analysis of Irregularly Observed Data*, E. Parzen (ed.), Springer Lecture Notes in Statistics, *25*, 158–188.

Kailath, T. (1968), An innovations approach to least squares estimation—Part I: Linear filtering in additive white noise, *IEEE Transactions on Automatic Control*, *AC-13*, 646–654.

Kailath, T. (1970), The innovations approach to detection and estimation theory, *Proceedings IEEE*, *58*, 680–695.

Kalman, R.E. (1960), A new approach to linear filtering and prediction problems, *Trans. ASME, J. Basic Eng.*, *83D*, 35–45.

Kendall, M.G. and Stuart, A. (1976), *The Advanced Theory of Statistics*, Vol. 3, Griffin, London.

Kitagawa, G. (1987), Non-Gaussian state-space modelling of non-stationary time series, *J.A.S.A*, *82* (with discussion), 1032–1063.

Koopmans, L.H. (1974), *The Spectral Analysis of Time Series*, Academic Press, New York.

Lamperti, J. (1966), *Probability*, Benjamin, New York.

Lawrance, A.J. and Kottegoda, N.T. (1977), Stochastic modelling of riverflow time series, *J. Roy. Statist. Soc. Ser. A*, *140*, 1–47.

Lehmann, E.L. (1983), *Theory of Point Estimation*, John Wiley, New York.

Li, W.K. and McLeod, A.I. (1986), Fractional time series modelling, *Biometrika*, *73*, 217–221.

Lii, K.S. and Rosenblatt, M. (1982), Deconvolution and estimation of transfer function phase and coefficients for non-Gaussian linear processes, *Ann. Statist.*, *10*, 1195–1208.

Ljung, G.M. and Box, G.E.P. (1978), On a measure of lack of fit in time series models, *Biometrika*, *65*, 297–303.

McLeod, A.I. and Hipel, K.W. (1978), Preservation of the rescaled adjusted range, I. A reassessment of the Hurst phenomenon, *Water Resources Res.*, *14*, 491–508.

McLeod, A.I. and Li, W.K. (1983), Diagnostic checking ARMA time series models using squared-residual autocorrelations, *J. Time Series Analysis*, *4*, 269–273.

Mage, D.T. (1982), An objective graphical method for testing normal distributional assumptions using probability plots, *American Statistician*, *36*, 116–120.

Makridakis, S., Andersen, A., Carbone, R., Fildes, R., Hibon, M., Lewandowski, R., Newton, J., Parzen, E. and Winkler, R. (1984), *The Forecasting Accuracy of Major Time Series Methods*, John Wiley, New York.

Melard, G. (1984), A fast algorithm for the exact likelihood of moving average models, *Applied Statistics*, *33*, 104–114.

Mood, A.M., Graybill, F.A. and Boes, D.C. (1974), *Introduction to the Theory of Statistics*, McGraw-Hill, New York.

Nicholls, D.F. and Quinn, B.G. (1982), *Random Coefficient Autoregressive Models: An Introduction*, Springer Lecture Notes in Statistics, *11*.

Parzen, E. (1974), Some recent advances in time series modelling, *IEEE Transactions on Automatic Control*, *AC-19*, 723–730.

Parzen, E. (1978), Time series modeling, spectral analysis and forecasting, *Directions in Time Series*, D.R. Brillinger and G.C. Tiao (eds.), Institute of Mathematical Statistics, 80–111.

Priestley, M.B. (1981), *Spectral Analysis and Time Series*, Vols. 1 and 2, Academic Press, New York.

Priestley, M.B. (1988), *Non-linear and Non-stationary Time Series Analysis*, Academic Press, London.

Rao, C.R. (1973), *Linear Statistical Inference and Its Applications*, 2nd ed., John Wiley, New York.

Rice, J. (1979). On the estimation of the parameters of a power spectrum, *J. Multivariate Analysis*, *9*, 378–392.

Rissanen, J. (1973), A fast algorithm for optimum linear predictors, *IEEE Transactions on Automatic Control*, *AC-18*, 555.

Rissanen, J. and Barbosa, L. (1969), Properties of infinite covariance matrices and stability of optimum predictors, *Information Sci.*, *1*, 221–236.

Rosenblatt, M. (1985), *Stationary Sequences and Random Fields*, Birkhäuser, Boston.

Schweppe, F.C. (1965), Evaluation of likelihood functions for Gaussian signals, *IEEE Transactions on Information Theory*, *IT-11*, 61–70.

Seeley, R.T. (1970), *Calculus of Several Variables*, Scott Foresman, Glenview, Illinois.

Serfling, R.J. (1980), *Approximation Theorems of Mathematical Statistics*, John Wiley, New York.

Shapiro, S.S. and Francia, R.S. (1972), An approximate analysis of variance test for normality, *J. Amer. Statist. Assoc.*, *67*, 215–216.

Shibata, R. (1976), Selection of the order of an autoregressive model by Akaike's information criterion, *Biometrika*, *63*, 117–126.

Shibata, R. (1980). Asymptotically efficient selection of the order of the model for estimating parameters of a linear process. *Ann. Statist.* *8*, 147–164.

Simmons, G.F. (1963), *Introduction to Topology and Modern Analysis*, McGraw-Hill, New York.

Sorenson, H.W. and Alspach, D.L. (1971), Recursive Bayesian estimation using Gaussian sums, *Automatica*, *7*, 465–479.

Sowell, F.B. (1990). Maximum likelihood estimation of stationary univariate fractionally integrated time series models, *J. of Econometrics*, *53*, 165–188.

Stuck, B.W. (1978), Minimum error dispersion linear filtering of scalar symmetric stable processes, *IEEE Transactions on Automatic Control*, *AC-23*, 507–509.

Stuck, B.W. and Kleiner, B. (1974), A statistical analysis of telephone noise, *The Bell System Technical Journal*, *53*, 1263–1320.

Subba Rao, T. and Gabr, M.M. (1984), *An Introduction to Bispectral Analysis and Bilinear Time Series Models*, Springer Lecture Notes in Statistics, *24*.

Taqqu, M.S. (1975), Weak convergence to fractional Brownian motion and to the Rosenblatt process, *Z. Wahrscheinlichkeitstheorie verw. Geb.*, *31*, 287–302.

Tong, H. (1983), *Threshold Models in Non-linear Time Series Analysis*, Springer Lecture Notes in Statistics, *21*.

Tong, H. (1990), *Non-linear Time Series: A Dynamical Systems Approach*, Oxford University Press, Oxford.

Tukey, J. (1949), The sampling theory of power spectrum estimates, Proc. Symp. on Applications of Autocorrelation Analysis to Physical Problems, NAVEXOS-P-735, Office of Naval Research, Washington, 47–67.

Walker, A.M. (1964), Asymptotic properties of least squares estimates of the parameters of the spectrum of a stationary non-deterministic time series, *J. Aust. Math. Soc.*, *4*, 363–384.

Wampler, S. (1988), Missing values in time series analysis, Statistics Department, Colorado State University.

Weiss, G. (1975), Time-reversibility of linear stochastic processes, *J. Appl. Prob.*, *12*, 831–836.

Whittle, P. (1962), Gaussian estimation in stationary time series, *Bull. Int. Statist. Inst.*, *39*, 105–129.

Whittle, P. (1963), On the fitting of multivariate autoregressions and the approximate canonical factorization of a special density matrix, *Biometrika*, *40*, 129–134.

Whittle, P. (1983), *Prediction and Regulation by Linear Least-Square Methods*, 2nd ed., University of Minnesota, Minneapolis.

Wilson, G.T. (1969), Factorization of the generating function of a pure moving average process, *SIAM J. Num. Analysis*, *6*, 1–7.

Yajima, Y. (1985), On estimation of long-memory time series models, *The Australian J. Statistics*, *27*, 303–320.

Index

Springer Series in Statistics

(continued from p. ii)

Pratt/Gibbons: Concepts of Nonparametric Theory.
Read/Cressie: Goodness-of-Fit Statistics for Discrete Multivariate Data.
Reinsel: Elements of Multivariate Time Series Analysis.
Reiss: A Course on Point Processes.
Reiss: Approximate Distributions of Order Statistics: With Applications to Nonparametric Statistics.
Rieder: Robust Asymptotic Statistics.
Rosenbaum: Observational Studies.
Ross: Nonlinear Estimation.
Sachs: Applied Statistics: A Handbook of Techniques, 2nd edition.
Särndal/Swensson/Wretman: Model Assisted Survey Sampling.
Schervish: Theory of Statistics.
Seneta: Non-Negative Matrices and Markov Chains, 2nd edition.
Shao/Tu: The Jackknife and Bootstrap.
Siegmund: Sequential Analysis: Tests and Confidence Intervals.
Tanner: Tools for Statistical Inference: Methods for the Exploration of Posterior Distributions and Likelihood Functions, 2nd edition.
Tong: The Multivariate Normal Distribution.
Vapnik: Estimation of Dependences Based on Empirical Data.
Weerahandi: Exact Statistical Methods for Data Analysis.
West/Harrison: Bayesian Forecasting and Dynamic Models.
Wolter: Introduction to Variance Estimation.
Yaglom: Correlation Theory of Stationary and Related Random Functions I: Basic Results.
Yaglom: Correlation Theory of Stationary and Related Random Functions II: Supplementary Notes and References.